Vegetable Production and Practices

Vegetable Production and Practices

Gregory E. Welbaum, Ph.D.

Visiting Professor
Department of Plant Sciences
Seed Biotechnology Center
University of California, Davis
Davis, California
USA

and

Professor, Department of Horticulture
Virginia Polytechnic Institute and State University
Blacksburg, Virginia
USA

www.cabi.org

CABI is a trading name of CAB International

CABI
Nosworthy Way
Wallingford
Oxfordshire OX10 8DE
UK

CABI
38 Chauncy Street
Suite 1002
Boston, MA 02111
USA

Tel: +44 (0)1491 832111
Fax: +44 (0)1491 833508
E-mail: info@cabi.org
Website: www.cabi.org

Tel: +1 800 552 3083 (toll free)
Tel: +1 617 395 4051
E-mail: cabi-nao@cabi.org

© Gregory E. Welbaum 2015. All rights reserved. No part of this publication may be reproduced in any form or by any means, electronically, mechanically, by photocopying, recording or otherwise, without the prior permission of the copyright owners.

A catalogue record for this book is available from the British Library, London, UK.

Library of Congress Cataloging-in-Publication Data

Welbaum, G.E. (Gregory E.)
  Vegetable production and practices / Gregory E. Welbaum, Ph.D., Visiting Professor, Department of Plant Sciences, Seed Biotechnology Center, University of California, Davis, Davis, California, USA and Professor, Department of Horticulture, Virginia Polytechnic Institute and State University Blacksburg, Virginia, USA.
     pages cm
  Includes bibliographical references and index.
  ISBN 978-1-78064-534-6 (hardback : alk. paper)--ISBN 978-1-84593-802-4 (pbk. : alk. paper)
1. Vegetables. I. Title.

SB320.9W45 2014
635--dc23

2014021237

Paperback ISBN-13: 978 1 84593 802 4
Hardback ISBN-13: 978 1 78064 534 6

Commissioning editors: Sarah Hulbert and Charlotte Hammond
Editorial assistant: Alexandra Lainsbury
Production editor: Claire Sissen

Typeset by SPi, Pondicherry, India.
Printed and bound by CPI Group (UK) Ltd, Croydon, CR0 4YY.

Contents

Preface		vii
Acknowledgements		ix
Chapter 1	Vegetable History, Nomenclature, and Classification	1
Chapter 2	Tillage and Cropping Systems	16
Chapter 3	Vegetable Seeds and Crop Establishment	27
Chapter 4	Fertilization and Mineral Nutrition Requirements for Growing Vegetables	47
Chapter 5	Irrigation of Vegetable Crops	66
Chapter 6	Mulches	80
Chapter 7	Protected Culture	90
Chapter 8	Organic and Sustainable Vegetable Production	107
Chapter 9	Vegetable Safety	127
Chapter 10	Family Cucurbitaceae	136
	Cucumber	136
	Netted and Mixed Melons	146
	Watermelon	156
	Pumpkins and Squash	165
Chapter 11	Family Solanaceae	176
	Potato	176
	Peppers	187
	Eggplant	201
	Tomato	204
Chapter 12	Family Asteraceae	222
	Lettuce, Endive, and Chicory	222
	Globe Artichoke and Cardoon	239
Chapter 13	Family Poaceae	248
	Sweet Corn, Popcorn, and Ornamental Corn	248
Chapter 14	Family Amaryllidaceae, Subfamily Allioideae	267
	Onions and Garlic	267
	Leek, Shallots, and Chives	286
Chapter 15	Family Convolvulaceae	289
	Sweetpotato	289
Chapter 16	Family Brassicaceae	304
	Genus *Brassica*	304

Asian Brassicas		335
Root Crops		341
Other Crops		345

Chapter 17 Family Amaranthaceae, Subfamily Chenopodiaceae 349
    Beets and Chard 349
    Spinach 358
    New Zealand Spinach and Orach 366

Chapter 18 Family Asparagaceae 369
    Asparagus 369

Chapter 19 Family Polygonaceae 382
    Rhubarb 382

Chapter 20 Family Fabaceae 390
    Beans and Peas 390

Chapter 21 Family Apiaceae 415
    Carrot 415
    Celery 426
    Minor Crops in the Family Apiaceae 435

Chapter 22 Family Agaricaceae 441
    Mushrooms 441

Index 453

Online supplementary material

Please visit http://www.cabi.org/openresources/45346 to view
the pictures in colour and to access supplementary material such as videos.

Preface

The topic of vegetable crop production is vital to human health and should be of interest to everyone. The study of vegetable crops is a fundamental discipline because it deals with the needs of mankind rather than our wants. Each one of us eats vegetables, lots of them, to stay healthy. According to the World Health Organization, the average adult has an annual per capita vegetable consumption estimated to be 102 kg (240 lb). There is growing awareness that the way we grow vegetables has a profound effect not only on our health but also our environment. As the average person in developed countries becomes more and more detached from vegetable production, misconceptions grow about the safety and sustainability of production practices. One of the goals of this book is to provide a scientifically based discussion of some of the current and traditional issues surrounding vegetable production, handling and consumption. Hopefully, this book will educate students about how vegetables are actually produced as well as offer ideas and stimulate thinking to improve future practices.

I developed an interest in vegetable crops at an early age. Having grown up on a diverse family farm in west-central Ohio, I became interested in vegetables because of their greater profit potential and amazing diversity compared to the corn, soybeans, wheat and hay we were growing. I remember planting my own vegetable garden at age 8. I started growing vegetables commercially on our farm and selling them from a roadside vegetable market in western Miami County, Ohio during the 1970s. I attended the Ohio State University primarily to learn more about vegetable crop production from Dr. Kenneth Alban, Dale Kretchman and others. Wanting to know even more, I ventured across the country to UC Davis and earned an M.S. in Vegetable Crops in 1979. I returned in the 1980s to get a Ph.D. in Plant Physiology studying the water relations of cantaloupe seed development for my research. While at Davis, I was a teaching assistant (TA) for Systematic Olericulture, World Vegetable Crops, and Vegetable Seed Production and Physiology and was fortunate to be mentored by several notable scientists: M. Allen Stevens, Kent J. Bradford, Vince E. Rubatzky, Mas Yamaguchi, Oscar A. Lorenz, Jim Harrington, Ron Voss and many others. The book *World Vegetables: Principles, Production and Nutritive Values* is based on the syllabus used by Mas Yamaguchi to teach the World Vegetable Crops class that I once TAed. This book updates and references portions of Vince and Mas's outstanding text.

This book is based on the class Horticulture 4764 at Virginia Tech that I have taught since 1992. I developed an online version of this course that has been taught as an asynchronous web-based course since 1999. In the book, I have tried to share my own personal knowledge of vegetables through my experiences as a grower, farm owner, international agriculturist and scientist whenever possible. The book is intended to have global appeal for use outside North America with pictures and discussions of various aspects of vegetable production from many different states and countries based on my international experiences in Canada, China, Taiwan, the Netherlands, Israel, India, Germany and Greece. Both US and metric units are provided throughout. In some instances, units have been rounded to give a practical rather than an exact conversion. For example, for a measurement of stake height or bed width of 1.8 m, the exact conversion is 5 ft 11 in. However, a stake or bed would be sold or constructed to a measurement of 6 ft for practical purposes.

I have wanted to covert my teaching website to a book for several years and a sabbatical in the Department of Plant and Environmental Sciences at UC Davis allowed me finally to do so.

The book consists of 22 chapters. The first nine chapters cover general background information that applies to all vegetable crops. I have included chapters about particularly timely topics such as vegetable safety, where the causes of biological contamination are discussed. The chapter on seeds and stand establishment includes a section on the genetic improvement of vegetables to dispel the common notion that genetic modification is a new phenomenon. The chapter entitled Organic and Sustainable Vegetable Production provides a brief history of the organic movement and how it may evolve in the future. This book is not intended to advocate for either conventional or organic production. I have described plant requirements for growth and development and described some of the more serious crop pests. Examples are given of how both methods could be used for production to let the reader decide which approach may be best suited. The chapter on protected culture covers the many ways vegetables can be grown out of season or in optimized environments. The chapter on Fertilization and Mineral Nutrition Requirements for Growing Vegetables takes a somewhat unique approach of using basic plant physiology to educate readers about how plants use minerals to grow and develop. The tillage and cropping systems chapter describes intensive monoculture systems but provides alternatives that may work better in certain situations.

Chapters 10–22 cover many of the important vegetable crops of the world in detail. In addition to tomato, potato and lettuce, a discussion of several less traditional vegetables like rhubarb, horseradish and mushrooms is provided to broaden knowledge and add perspective. To add context to crop production, chapters on specific vegetables contain a brief history of the crop, basic botanical information, human nutritional information and descriptions of diverse types and cultivars. Also included is fundamental information on germination, establishment, fertilization, crop management, crop pests, the status of transgenic crops, harvesting and recommendations for postharvest handling. This allows the book to be used as a reference by home gardeners, food industry professionals and vegetable producers. Creating this book was frustrating because time and space do not allow discussion of all topics and pictures that I had hoped to include.

Acknowledgements

Many people helped prepare and review this book. I would like to thank Maura Wood and Ashley Wills for their help compiling material. I wish to thank the following people who provided chapter reviews: David Schmale III, Josh Freeman, Richard Veilluex, Anthony Bratsch, Carl Cantaluppi, Dale Marshall and Bernard Zandstra. I thank Carl Estes for his help in creating figures on vegetable nutrition. I especially appreciate the editing and formatting efforts of Pris Sears who polished my crude drafts into a formatted manuscript. I would like to thank Kent J. Bradford for providing support and allowing me to finish this book while on sabbatical at UC Davis. I apologize to my family I-mo, Whitney and Stephanie for spending so much time away from them to finish this book. I thank my mother, Ruth Welbaum, for encouraging me to pursue a career as a scientist and my brother, Bob Welbaum, for helping to proof read certain chapters.

1 Vegetable History, Nomenclature, and Classification

Introduction

All societies and ethnic groups eat vegetables because they are essential for maintaining human health. In simple terms, modern vegetable science deals with growing herbaceous plants for human consumption to meet basic nutritional needs. As the world's population grows, the demand for vegetables will continue to grow as well. Vegetable science, sometimes called olericulture, is one of the most dynamic and important fields of the agricultural sciences. The importance of vegetables has never been greater.

What is a Vegetable?

Most definitions of a vegetable are not botanically based. Vegetable definitions are rather arbitrary by nature and commonly based on usage rather than plant morphology. For example, one widely used definition of a vegetable is: a herbaceous plant or portion of a plant that is eaten whole or in part, raw or cooked, generally with an entree or in a salad but not as a dessert. Of course there are exceptions to this definition. Rhubarb, watermelon and cantaloupes are all considered vegetables but commonly used as desserts. Mushrooms are fungi and not plants but are generally considered to be vegetables, and their production is described in a later chapter.

Since "vegetable" is not a botanical term, some vegetables botanically speaking are also fruits. In a botanical sense, a fruit describes a ripened ovary containing seeds together with adjacent parts that are eaten at maturity. For example, tomato, pepper, bean and cantaloupe botanically speaking meet the definition of fruit, but because of the way they are traditionally used and produced they are considered to be vegetables. Therefore, since there are essentially two classification systems, some commodities may be classified as a vegetable based on their usage while at the same time they are botanically fruits.

Vegetables are a horticultural food crop. Other horticultural food crops include small fruits and tree fruits, which are usually grown as perennials. Vegetable crops may be either annuals or perennials. From a production standpoint, a vegetable crop may be defined as a high-value crop that is intensively managed and requires special care after harvest. Agriculturalists often segregate agronomic or "field" crops into a separate category as crops that are extensively grown and less intensively managed in comparison to vegetable and the other horticultural crops. Wheat, cotton, soybean, sugarcane and rice are all examples of agronomic crops. Many agronomic crops are grains that are planted and destructively mechanically harvested at full physiological maturity at the end of their life cycles. In contrast, many vegetable crops are harvested in an immature state, while still fragile, so great care must be taken to preserve their quality from the field to the consumer.

There are some exceptions to these general production definitions. Some agronomic crops such as tobacco are intensively managed and are valuable in a monetary sense but are considered agronomic for historical reasons. Irish potato is considered to be a vegetable by some but an agronomic crop to others depending on the region, type and scale of production. Corn and soybeans can be either agronomic or vegetable crops depending on the cultivars grown, their stage of maturity at harvest and their end use. While these definitions are not perfect, they do have value in allowing us to group different crops and types of production to understand better the more unique aspects of vegetable production, management and handling compared to other crops. So in summary, "vegetable" is a term based on the usage of herbaceous plants or portions of plants that are eaten whole or in part, raw or cooked, generally with an entree or in a salad but not as a dessert, that are intensively managed and may require special care after harvest to maintain quality.

The Evolution of Vegetable Production

Over the years, world vegetable production has increased. For example, there was over a four-fold increase in world vegetable production from 1970 until 2009 (FAO, 2011). The long-term increase has largely been the result of a series of technological advances. The first were labor-saving technologies such as the moldboard plow, and power equipment such as trucks, tractors and harvesting equipment. Subsurface drainage systems were developed to improve productivity of chronically water-logged soils. During the first half of the 20th century, low-cost commercial fertilizers were mass-produced, leading to dramatic improvements in fertility management and productivity. In the 1930s, new plant-breeding techniques led to development of more productive cultivars. One of the major genetic advancements during this period was the development of F-1 hybrid cultivars, which increased the productivity of some vegetables by 30% or more. Many view the period after World War II as the chemical era of agriculture as many synthetic pesticides, e.g. fungicides, insecticides and herbicides, became readily available. Chemicals were intended to more easily control crop pests to enable a large-scale, single-crop production system called monoculture where only a single crop is grown in a field and all other vegetation is excluded. The monoculture system and the related technologies like chemical pest control and concentrated synthetic-fertilizer usage were designed in part to reduce the amount of labor needed in agriculture so that one person could effectively manage more crop-production acreage and increase productivity.

During the 1970s and 1980s, concerns about human health issues led to interest in and the development of more sustainable and low-input approaches to vegetable production. Also in the past 50 years, conservation tillage practices have been developed to reduce soil erosion, decreasing the number of passes made over a field to decrease soil compaction. Plastic mulches were developed to modify soil temperatures, control weeds, reduce leaching and use less water to produce vegetables. Trickle-irrigation systems were developed to precisely apply water and nutrients to vegetable crops throughout the season. Raised-bed production systems improved drainage, encouraged better root development and reduced disease. During this same time frame, computer technology also impacted both production and management issues, increasing production efficiency.

Despite these advances, there seems to be declining interest in vegetable science as a discipline in the academic circles of many developed countries. Government sponsorship of research on applied aspects of vegetable science in many developed countries has declined in recent years as funding has been diverted to more basic research in plant science. There is a perception that research on vegetable-crop production is no longer a high priority in countries with developed agriculture industries because most of the production-related questions have already been answered or are being addressed by the private sector. Some believe that production has become "cookbook", so advanced training is no longer necessary.

Most agriculturalists would agree that research has played an important role in advancing vegetable science to its current state. Innovative new research will be necessary in the future to meet the needs of a growing global population for safe, nutritious and sustainably produced vegetables. The need for primary information about vegetable production for the developing world remains great. Both now and in the future, research and education programs will be needed to address food safety and security issues.

In many developed countries, only a very small percentage of people are involved in food production. This has led to misconceptions and uncertainty about where vegetables come from, the effect production has on the environment, and whether or not vegetables are safe to eat. Consumers need to be educated about their food supply to intelligently discuss timely topics such as the risks and benefits of transgenic crops. Accurate scientific information needs to be readily available about this and other key issues. Vegetable science is truly fundamental to our very existence because it is a discipline that deals with the basic needs of the human race rather than its wants.

The following is a partial list of challenges that face commercial vegetable growers in many areas of the world; novel research approaches will be required to solve them. Understanding and controlling the causes of biological contamination on vegetable crops is a major challenge. Developing highly productive and efficient sustainable vegetable production systems is another. Improving the quality and nutritional value of commercial vegetables is an ongoing challenge. There is an opportunity to fully utilize the Global Positioning System (GPS) and its related technologies to improve vegetable-production efficiency. Decreasing the environmental footprint caused by intensive commercial vegetable production while using less water and energy will only increase in importance in the future.

In developed countries of the Americas and Europe, the commercial vegetable industry is dominated by large corporations with multiple production areas that can continuously produce a wide range of crops to supply supermarket chains throughout the year (Cook, 2001). Grocery chains prefer dealing with single suppliers for convenience, economies of size and uniformity of product. Vegetable industries in North America and Europe have evolved so that vegetables are grown in regions best suited for a particular crop for shipment, often to distant markets in population centers. The shift from local production to optimum production away from population centers has been occurring for decades (Cook, 2001). In the USA, Canada and some European countries, studies show that the average vegetable purchased in a supermarket traveled over 2,400 km (1,500) miles from the field before it was sold (Carlsson-Kanyama, 1997; Pirog and Benjamin, 2003). Shipping vegetables from distant markets is very energy intensive and the current realities of energy costs bring into the question the sustainability of growing vegetables for distant markets. Some experts are proposing a return to more locally grown produce by smaller farmers to improve vegetable quality, stimulate local economies, reduce energy usage and improve sustainability. However, the convenience and diversity of supermarket shopping appeals to many consumers, so commercial groceries are not likely to disappear anytime soon. These and other timely issues remind us that the vegetable industry is not static and needs to evolve to meet challenges over time just like other industries.

Much of this volume is devoted to describing commercial vegetable production practices. However, it is important to consider some of the key peripheral influences that affect production. Vegetable growers do not function in a vacuum and increasingly must adhere to local laws on water quality, waste handling, chemical usage and noise abatement. When discussing individual practices of vegetable production, we sometimes lose sight of these other influences. It is easy to forget that most vegetable-production operations are businesses that need to be profitable to survive. Production practices must make business sense in order to be adopted. We will not consider the financial aspects of vegetable production in detail because they are beyond the scope of this book, but the reader should remember that profitability plays a major role in the production practices that are selected by farmers.

Vegetable Diversity

The diversity of vegetables consumed by humans varies widely with both geography and culture. It is difficult to estimate the diversity of vegetables grown around the world, but the number of unique vegetables is likely in the thousands. Vegetables are increasingly traded internationally. It is common to buy fresh vegetables in northern European countries that were grown in southern Europe or Israel. In North America it is common to buy vegetables grown in Holland, Guatemala, Chile or New Zealand. Vegetable growers in the USA ship fresh broccoli and melons to Southeast Asia. Improved transportation systems make the long-distance movement of vegetables possible across regions and among countries. Efficient transportation enables the long-distance transport of vegetables for sale in areas where they are out of season or cannot be grown due to unfavorable climate. Improved postharvest handling of vegetables allows maintaining marketable quality for longer periods through improved genetics and postharvest handling conditions. For example, slow-ripening durable cultivars of tomato have been developed to better withstand long-distance transport. The diversity of vegetables available to consumers has also been increased through new introductions and cultivar development. Cultivars of colored lettuce with diverse leaf shapes add color and appeal to salads, an array of colored bell peppers are visually attractive, purple asparagus, yellow-fleshed watermelon, green cauliflower and purple broccoli are other examples of diverse types that appeal to consumers. The production of new vegetables often occurs following their introduction.

Domestication and History of Vegetables

It is interesting to briefly discuss the history to understand the origins of the vegetables we eat today. Scientific evidence shows that the vegetables of today did not suddenly appear on earth, but have evolved over a very long period of time thanks to natural selection, human selection, and more recently plant breeding and other forms of genetic improvement. Ethnobotany is the scientific study of the relationships that exist between people and plants. From the studies of anthropologists and enthobotanists, we know that ancient humans were not farmers.

Early humans existed on earth over 2 million years ago and gathered all kinds of plants from the wild to supplement food obtained from hunting and fishing. The ample food supply allowed humans to establish residence rather than pursue a nomadic existence. Having a stable food supply and a place to live allowed people to think about things other than their personal survival, such as growing crops rather than gathering them. About this point, humans began the long process of food-crop domestication. Domestication describes a selection process conducted by humans to produce plants that have more desirable traits. Several conditions are believed to have been necessary, or at the very least helpful, in leading humans to domesticate plants:

1. Fire was used to clear land of vegetation in some areas. The "slash and burn" method is still used in some areas to clear land.
2. A temperate or subtropical climate with distinct wet and dry seasons helped facilitate seed-crop production. The Middle East was especially suited to seed-crop production because the dry summer climate was conducive to the evolution of large-seeded annual plants, and the variety of elevations in the region led to a great variety of species (Balfour-Paul, 1996).
3. Location was an important consideration. Since river valleys were subject to periodic flooding, grasslands were difficult to till with primitive tools and rainforests were difficult to clear, early humans most likely used open woodlands to attempt cultivation.
4. Areas with a diversity of plants and abundant animal supplies for food were prime targets for crop domestication.

Conscious cultivation and trait selection of plants may have occurred in what is today Syria as early as c. 11,050 BC, but this appears to have been a localized phenomenon rather than a definitive step towards domestication (Hillman et al., 2001). The earliest human attempts at plant domestication likely occurred in southwestern Asia and the Middle East about 10,000 years ago (Zohary and Hopf, 1988). By 10,000 BC, the bottle gourd, *Lagenaria siceraria*, appears to have been domesticated as a container and not a vegetable before the invention of ceramic technology. Interestingly, the domesticated bottle gourd reached the Americas from Asia as early as 8000 BC, most likely due to the migration of peoples from Asia who had domesticated the crop 2,000 years earlier (Erickson et al., 2005). By comparison, cereal crops were likely first domesticated around 9000 BC in the Fertile Crescent between the Tigris and Euphrates rivers region in the Middle East (Balfour-Paul, 1996). Starchy seeds were gathered as food during the dry season. The domestication process began when starchy seeds that were dropped or spilled near dwellings during the dry season germinated and grew rapidly due to the lack of competition during the rainy season. These adventitious plants benefited from the increased fertility near dwellings and protection from animal predation. Vegetative plants like sweetpotato were likely collected in the tropics, semi-tropical regions of the Americas, Southeast Asia and Africa. When digging wild roots and tubers with sharp sticks, some of the roots undoubtedly remained in the ground to regrow. Some of the roots and tubers may have dropped or been discarded near encampments, only to regrow in areas with adequate rainfall.

Harvesting unintentional plantings evolved into the practices of intentionally planting seeds or vegetative plant parts into soil. The domestication of vegetables resulted from years of human selection of plants with more desirable traits. The first domesticated vegetable crops were generally annuals with large seeds, or fruits including legumes such as peas, and not herbaceous vegetable crops (Zohary and Hopf, 1988).

As domestication took place, humans began to shift from a hunter-gatherer society to a settled agricultural society. The establishment of agricultural societies led to the first city states and eventually the rise of civilization itself. So crop domestication was a key development in human evolution whose importance should not be discounted. Continued domestication was gradual, a process of trial and error that occurred intermittently. Outside the Middle East/West Asia, very different species were domesticated. In the Americas cultivation of squash, corn, potato and beans was adapted. In East Asia millet, rice and soy were the most important early crops (Zohary and Hopf, 1988).

Genetic Change and Germplasm Preservation

Domestication renders plants dependent on humans for their continued existence. Many domesticated plants show little or no resemblance to their natural ancestors (Smartt and Simmonds, 1995). The divergence between wild and cultivated plants occurred because of different mechanisms that ultimately resulted in genetic changes. For example, corn ears are now dozens of times the size of those of wild teosinte, which is believed to be the closest non-domesticated relative of corn. Modern cantaloupe cultivars have much larger fruit than their wild relatives and have been selected for crack resistance to the point that the seeds inside often die before being released from the fruit, so the plants can no longer survive in the wild (Welbaum, 1993). Seed stalks have difficulty breaking through the heads of crisphead lettuce unless slashed with a knife to allow flowering to occur.

The diversion between crops and native plants occurred because large populations of wild plants that were first selected for domestication were genetically very diverse. Saving seeds from wild plants selected for desirable traits obtained through natural hybridization and genetic recombination resulted in crops possessing traits that were desirable to humans but different from the average plant in a wild population. Another cause for the divergence between wild and cultivated plants was genetic mutation. Natural gene mutations occurred infrequently and most of the new characteristics that resulted from mutation were deleterious, leading to their elimination. Occasionally, mutations produced desirable qualities, like increased size, for example, that benefited humans. Through natural mutations, hybridizations and chromosome duplications, plants slowly changed over thousands of years. Natural genetic changes, combined with breeding crop plants in more recent times, gradually changed some plants so they no longer resembled their wild progenitors. Plant breeding is the art and science of changing the genetics of plants for the benefit of mankind using many different techniques. These range from simply selecting plants with desirable characteristics for propagation to more complex techniques (Poehlman and Sleeper, 1995). It is sometimes forgotten that plant breeding has been practiced for thousands of years, since near the beginning of human civilization.

The domestication of food crops by humans has caused our food supply to become increasingly dependent on a limited number of cultivars for many vegetables. Preserving genetic resources has become a higher priority for many groups in recent years. International collaboration on germplasm preservation issues has increased as well. Many countries maintain germplasm collections, such as the United States National Center for Genetic Resources Preservation (NCGRP) in Fort Collins, Colorado, the Kew Royal Botanical Garden's Millennium Seed Bank located at Wakehurst Place in Sussex, England, and the N.I. Vavilov All-Russian Scientific Research Institute of Plant Industry in St. Petersburg, Russia, where the world's first seed bank and one of the world's largest collections of plant genetic material is maintained. The Svalbard Global Seed Vault is a secure international seed bank located on the Norwegian island of Spitsbergen that preserves a wide variety of plant seeds, including vegetables, in an underground cavern. The seeds are duplicate samples, or "spare" copies, of seeds held in gene banks worldwide. The seed vault and other germplasm collections provide insurance against the loss of seeds in other gene banks, as well as a refuge for seeds in the case of large-scale regional or global crises. These germplasm collections are also available for genetic improvement of crop plants and provide sources of novel genes that are currently not needed but may be useful in the future.

The news media often describes modern transgenic crops that have been engineered using various recombinant DNA technologies as "genetically modified". Unfortunately, this terminology leads us to believe that vegetables and other crops were not changed genetically until the 1980s when recombinant DNA technologies for crop improvement were created. The above discussion illustrates that genetic modification of plants is a natural continuous process caused by mutation, hybridization, chromosome duplication, human selection, and plant breeding that has produced the vegetable crops grown throughout the world today. We will discuss the genetic improvement of vegetables in greater detail in a later section.

Centers of Origin

The Russian botanist Nikolai Vavilov described Centers of Origin as locations where the original center for the domestication of certain crop plants likely occurred (Ladizinsky, 1998). Vavilov believed that plants were not domesticated somewhere in the world at random, but there were specific regions where the domestication started (Fig. 1.1). It is believed that the center of origin is also the center of genetic diversity.

Vavilov centers are regions where a high diversity of crop-related plants can be found. These plants, called crop wild relatives, represent the progenitors of domesticated crop plants and are studied by ethnobotanists and geneticists for use in the genetic improvement of modern cultivars or for preservation in germplasm collections. Almost all the vegetables grown today were domesticated from Centers of Crop Origin (Table 1.1).

Domesticated plant species often differ from their wild relatives in predictable ways. These differences are often called the "domestication syndrome" and include the following changes that were necessary for the domestication of wild plants to be successful:

1. Gigantism: the size of plants, reproductive organs and seeds increased through natural genetic selection. This also includes changes in biomass allocation (more in fruits, roots or stems, depending on human selection). Gigantism is the reason why wild plants and berries are smaller than their domesticated relatives.

2. Seeds: during domestication, seeds generally became larger, decreased in number, lost the ability to shatter (fall off the plant) prematurely, and lost dormancy. Most modern vegetable seeds lack dormancy compared to wild plants.

3. Maturity: domesticated plants mature earlier and have more concentrated seed production.

4. Defenses: domesticated plants have reduced physical (e.g. thorns) and chemical defenses.

Many vegetables familiar to us today were unknown to Europeans before Columbus discovered the New World. Indigenous peoples of the Western Hemisphere domesticated many of our most important world vegetables long before the Europeans were aware of their existence. When vegetables were taken from their original habitat in the New World and introduced to Europe, some were initially very successful because there were fewer diseases and insects and lack of competition. An example of this would be the "Irish" potato, whose center of origin is South America. The Irish potato grew well initially in Ireland after its introduction because the climate was favorable and there were no diseases. The potato became a staple crop that the Irish people came to be overly reliant upon. Eventually diseases caused potato crop failures in Ireland, resulting in famine and emigration that altered the history of the entire country.

Vegetable Classification Systems

A systematic method for grouping different plants is important for identifying and cataloguing the large volume of information about the many diverse plants that exist on earth. A proper and efficient classification system can result in more efficient use of information. Of the hundreds of thousands of

Fig. 1.1. Primary centers of origin of cultivated plants (Ladizinsky, 1998). Center descriptions are in Table 1.1.

Table 1.1. The following crops have been identified as Centers of Crop Origin in both the New and Old World as shown in Fig. 1.1 (Schery, 1954). The numbers for each center refer to regions in Fig. 1.1.

New World
1. South Mexican and Central American Center – southern sections of south Mexico and Central America.
Corn (*Zea mays*)
Common bean (*Phaseolus vulgaris*)
Lima bean (*Phaseolus lunatus*)
Malabar gourd (*Cucurbita ficifolia*)
Winter pumpkin (*Cucurbita moschata*)
Chayote (*Sechium edule*)
Sweetpotato (*Ipomoea batatas*)
Arrowroot (*Maranta arundinacea*)
Pepper (*Capsicum annuum*)
2. Northern South American Center – Peru, Ecuador, Bolivia
Andean potato (*Solanum andigenum*)
Potato, common (*Solanum tuberosum*) (24 chromosomes)
Starchy corn (*Zea mays*)
Lima bean (*Phaseolus lunatus*) secondary center
Common bean (*Phaseolus vulgaris*) secondary center
Edible canna (*Canna edulis*)
Pepino (*Solanum muricatum*)
Tomato (*Solanum lycopersicum*)
Ground cherry (*Physalis peruviana*)
Pumpkin (*Cucurbita maxima*)
Pepper (*Capsicum annuum*)
2A. Chiloe Center – islands near coast of Chile
Potato, common (*Solanum tuberosum*) (48 chromosomes)
2B. Brazilian-Paraguayan Center
Cassava (*Manihot esculenta*)
3. Mediterranean Center – includes the borders of the Mediterranean Sea
Pea (*Pisum sativum*)
Garden beet (*Beta vulgaris*)
Cabbage (*Brassica oleracea*)
Turnip (*Brassica rapa*)
Lettuce (*Lactuca sativa*)
Celery (*Apium graveolens*)
Chicory (*Cichorium intybus*)
Asparagus (*Asparagus officinalis*)
Parsnip (*Pastinaca sativa*)
Rhubarb (*Rheum officinale*)
4. Near-Eastern Center – Asia Minor – Transcaucasia, Iran and Turkey
Lentil (*Lens esculenta*)
Lupine (*Lupinus albus*)
5. Ethiopian (Abyssinian) Center – Ethiopia and Somali Republic
Cowpea (*Vigna unguiculata*)
Garden cress (*Lepidium sativum*)
Okra (*Hibiscus esculentus*)
6. Central Asiatic Center – northwest India (Punjab and Kashmir), Afghanistan
Pea (*Pisum sativum*)
Horse bean (*Vicia faba*)
Mung bean (*Phaseolus radiata*)
Mustard (*Brassica juncea*)
Onion (*Allium cepa*)
Garlic (*Allium sativum*)

Continued

Table 1.1. Continued.

Spinach (*Spinacia oleracea*)
Carrot (*Daucus carota*)
7. Northeast India and Myanmar (Burma) Center
Mung bean (*Phaseolus aureus*)
Cowpea (*Vigna sinensis*)
Eggplant (*Solanum melongena*)
Taro (*Colocasia esculenta*)
Cucumber (*Cucumis sativus*)
Yam (*Dioscorea alata*)
7A. Indo-Malayan Center (Indo-China and Malay Archipelago)
Banana (*Musa paradisiaca*) (starchy fruit used as vegetable)
Breadfruit (*Artocarpus communis*)
8. Chinese Center – mountains of central and western China and adjacent lowlands.
Soybean (*Glycine max*)
Chinese yam (*Dioscorea batatas*)
Radish (*Raphanus sativus*)
Chinese cabbage (*Brassica campestris*) (Chinensis and Pekinensis groups)
Onion (*Allium chinense* and *A. fistulosum*)
Cucumber (*Cucumis satvus*)

plants known in the world, only several hundred are used as vegetables. However, in order to manage information about plants, some system of classification, preferably having universal applicability, is beneficial. Many different classification systems have been developed, but the value of any system depends on its usefulness. To be useful, a classification system needs to be simple to use, accessible to all, and stable over time. The following section introduces different classification systems that are useful in the study of vegetable crops.

Fresh versus processed

One of the most common classifications of vegetables is whether they are consumed fresh or processed. Since vegetables are perishable, they are often canned, frozen or dried for long-term preservation. Of course, vegetables can also be eaten fresh or lightly processed in a prewashed ready-to-eat form that is more convenient for consumers. The lightly or minimally processed value-added category has resonated with consumers in many countries because there is a demand for the flavor and nutritional benefits of fresh vegetables without the labor and extra time needed for preparing raw vegetables for eating.

Fresh
A – raw: no washing or packaging
B – washed and wrapped, but not ready to eat
C – lightly (or minimally) processed, cut, washed and ready to eat with no further preparation

Processed
A – canned
B – frozen
C – dried

Thermo-classification of vegetables

Classification by climate was probably one of the earliest attempts to group plants and it is still widely used today. Through long-term observation, plants were grouped by optimum growing temperatures as either warm- or cool-season plants. Cool-season vegetables show a preference during most of their development for mean growing temperatures between 10°C and 18°C (50–64°F). Some have frost (surface ice formation) and freeze tolerance and for most, the edible tissues are leaves, stems or roots. Frost is defined as ice crystals that form on the surface of vegetables, while freezing is ice formation inside plant tissues. Familiar examples of cool-season vegetables are cabbage, lettuce, spinach, potato, and carrot.

In contrast, warm-season crops exhibit a preference for mean temperatures from 18°C to 30°C (64–86°F) during most of their growth and development. Warm-season vegetables are often of tropical origin and are perennial plants, but are grown in temperate regions as annuals. Warm-season crops are intolerant of frost and many are chilling sensitive. Chilling injury is primarily a disorder of crops of tropical and subtropical origin and is not the same as freezing injury. Chilling injury generally occurs when vegetables are exposed to 10–13°C (50–55°F). Chilling-sensitive crops have a short storage life at low temperatures and deteriorate more rapidly when returned to warm temperatures because of damage to cell membranes. Botanically, the edible tissues of many warm-season vegetables are reproductive organs. Examples include tomato, melons, and beans. Temperature classifications based on growing temperature or seasonal periods provide us with useful generalities, but upon closer examination there are overlaps and exceptions. Thermo-classification is useful to classify plants grown in temperature zones because in the tropics, the distinction between warm- and cool-season vegetables is less clear (Rubatzky and Yamaguchi, 1997).

Cool-season crops
Group A: prefer average monthly temperatures of 15.6–18.3°C (60–65°F). Intolerant of 21.1–23.9°C (70–75°F) and slightly tolerant of freezing; e.g. spinach, cabbage, parsnip, broccoli, radish, beet, turnip, rutabaga, cauliflower
Group B: prefer 15.6–18.3°C (60–65°F). Intolerant of 21.1–23.9°C (70–75°F). Damaged near maturity by freezing; e.g. lettuce, celery, artichoke, endive, carrot, chard
Group C: adapted to 12.8–23.9°C (55–75°F). Tolerant of frost; e.g. asparagus (mature plants), garlic, kale, Brussels sprouts

Warm-season crops
Group D: adapted to 18.3–29.4°C (65–85°F). Intolerant of frost. Chilling sensitive and damaged by exposure to or storage at temperatures below 13°C (55°F); e.g. sweet corn, pepper, snap bean, squash, pumpkin, lima bean, cucumber, tomato, muskmelon
Group E: long-season crops; prefer temperatures above 21.1°C (70°F). Chilling sensitive and damaged by exposure to or storage at temperatures below 13°C (55°F); e.g. watermelon, sweetpotato, eggplant, okra

The following are some other examples of useful classification systems that are frequently used to group vegetables. Additional classification systems also exist.

Classification of vegetables based on use, botany or a combination of both

Potherbs or greens
Spinach, kale, New Zealand spinach, mustard, chard, collards, dandelion

Salad crops
Celery, chicory, lettuce, watercress, endive

Cole crops (all are members of *Brassica oleracea* except Chinese cabbage)
Cabbage, Brussels sprouts, cauliflower, kohlrabi, sprouting broccoli, Chinese cabbage

Root crops (refers to crops which have a fleshy taproot)
Beet, turnip, carrot, rutabaga, parsnip, radish, salsify, celeriac

Bulb crops (*Allium* spp.).
Onion, garlic, leek, shallot, Welsh onion

Pulses or legumes
Peas, beans (including dry-seeded or agronomic forms)

Cucurbits (all members of the Cucurbitaceae)
Cucumber, pumpkin, muskmelon, squash, watermelon, several oriental crops

Solanaceous fruits (members of the Solanaceae)
Tomato, eggplant, pepper, husk tomato

Classification by edible part

Root
Enlarged taproot; e.g. beet, rutabaga, carrot, turnip, radish, parsnip, salsify, celeriac
Enlarged lateral root; e.g. sweetpotato, cassava

Stem
Above ground – not starchy; e.g. asparagus and kohlrabi
Below ground – starchy; e.g. white or Irish potato, Jerusalem artichoke, yam, taro, Andean tubers (oca, ullucu, anu)

Leaf
Onion group – leaf bases eaten (except chives); e.g. onion, garlic, leek, chive, shallot
Broad-leaved plants
Salad use; e.g. lettuce, chicory, celery (petiole only), endive, cabbage
Cooked (may include tender stems in some); e.g. spinach, kale, chicory, chard, vegetable amaranth, Chinese cabbage, mustard, dandelion, cardoon (petiole only), rhubarb (petiole only)

Immature flowers
Cauliflower, broccoli raab, broccoli, globe artichoke

Fruit
Immature; e.g. pea, chayote, okra, snap bean, summer squash, sweet corn, lima bean, cucumber, eggplant, broad bean
Mature; e.g. gourd family (cucurbits: pumpkin, winter squash, muskmelon, watermelon), tomato family (tomato, pepper, husk tomato)

Classification of vegetables according to salt tolerance

Salt concentration upper limits based on saturation soil extracts.

High salt tolerance: 7,700 ppm
Garden beets, kale, asparagus, spinach

Medium salt tolerance: 6,400 ppm
Tomato, broccoli, cabbage, peppers, cauliflower, lettuce, sweet corn, white potato, carrot, onion, peas, squash, cucumber, cantaloupe

Low salt tolerance: 2,600 ppm
Radish, green bean

All other vegetables: 1,900 ppm

Classification of vegetables according to tolerance to soil acidity

Slightly tolerant: pH 6.8–6.0
Asparagus, celery, muskmelon, beet, spinach, New Zealand spinach, broccoli, Chinese cabbage, okra, cabbage, leek, onion, cauliflower, lettuce, spinach

Moderately tolerant: pH 6.8–5.5
Bean, horseradish, pumpkin, Brussels sprouts, kohlrabi, radish, carrot, parsley, squash, cucumber, pea, tomato, eggplant, pepper, turnip, garlic

Very tolerant: pH 6.8–5.0
Chicory, rhubarb, endive, sweetpotato, potato, watermelon

Classification by root depth into soil

Shallow: 80 cm (36 in)
Cabbage, lettuce, onion, potato, spinach, sweet corn

Medium: 80–160 cm (36–72 in)
Beans, beets, carrot, cucumber, eggplant, summer squash, peas

Deep: 160 cm (as deep as 72 in)
Globe artichoke, asparagus, sweetpotato, tomato

Botanical classification system

All of the classification systems mentioned thus far overlap and although they are generally useful, they are inadequate for precise identification. The botanical classification system is the most precise, universal, and useful. Botanical classification is largely based on the variability among plants with regard to flower type, morphology, and sexual compatibility. All plants are considered as one community.

The famous botanist L.H. Bailey lists four subcommunities of plants. These are: (i) algae and fungi; (ii) mosses and liverworts; (iii) ferns; and (iv) seed plants. Vegetables are primarily seed-bearing plants, but by general agreement mushrooms, a fungus, are included (Bailey, 1949). The subgroupings most useful to the discussion of vegetables are family, genus, species, and cultivars (Table 1.2). The scientific names in this system are in Latin, which adds stability since this language will not change. This classification system, also known as the Latin binomial, was published as *Species Plantarum* in 1753 by Linnaeus and tends to be the most exact and accepted internationally.

Although the botanical classification system has stood the test of time relatively well, modern genetic tools have shown certain classifications based on plant morphology to be incorrect. For example, the Latin binomial for tomato has been changed from the long-standing *Lycopersicon esculentum* to *Solanum lycopersicon* based on molecular phylogenetic analyses (Bohs, 2005). More changes in the botanical system of vegetables can be anticipated in the years ahead as additional molecular phylogenetic studies are conducted. In some cases, molecular analyses show that genetic differences are not as significant as the phenotype or appearance of a vegetable may suggest. An example of this occurs with the classification of seemingly different vegetables into separate botanic varieties (also called subspecies). This confusion in nomenclature is due to the fact that some molecular genetic studies have shown that major morphological differences may be caused by single or few genes and therefore do not warrant classification into separate botanical varieties or as subspecies. An example of this would be cauliflower (family: Brassicaceae; genus: *Brassica*; species: *oleracea*; botanical variety: botrytis). Molecular genetic studies have shown that a single gene controls the development of undifferentiated flower primordial (curd) and if

Table 1.2. Botanical classification of vegetables. This table includes only monocots and dicots. Some vegetables like mushrooms are fungi and do not fit into either category (Rubatzky and Yamaguchi, 1997; GRIN, 2010; ePIC, 2011; USDA, 2011).

I. Monocotyledoneae

Amaryllidaceae, Subfamily Allioideae (formerly Alliaceae) – Amaryllis Family
Allium ampeloprasum var. ampeloprasum   Elephant garlic
Allium ampeloprasum var. kurrat   Kurrat
Allium ampeloprasum var. porrum   Leek
Allium tricoccum   Wild leek, ramp
Allium cepa (Aggregatum group)   Shallot
Allium cepa (Cepa group)   Common onion
Allium cepa (Proliferum group)   Egyptian onion
Allium chinense   Chinese onion, Chinese scallion, Oriental onion
Allium fistulosum   Welsh onion, Japanese bunching onion
Allium sativum   Garlic
Allium schoenoprasum   Chive
Allium tuberosum Chinese chives

Dioscoreaceae – Yam Family
Dioscorea alata   Water yam, winged yam, purple yam
Dioscorea bulbifera   Air potato, varahi, kaachil
Dioscorea cayenensis   Yellow yam
Dioscorea esculenta   Lesser yam
Dioscorea opposita   Chinese yam
Dioscorea rotundata   White yam
Dioscorea trifida   Cush-cush yam, Indian yam, napi

Poaceae (old name Gramineae) – Grass Family
Zea mays var. indentata   Dent corn, field corn
Zea mays var. indurata   Flint corn, ornamental corn
Zea mays var. everta   Popcorn
Zea mays var. saccharata   Sweet corn

Asparagaceae (formerly Liliaceae)
Asparagus officinalis   Asparagus

II. Dicotyledoneae
Most vegetables fall into this category.

Amaranthaceae (subfamily Chenopodiaceae, formerly family) – Goosefoot Family
Atriplex hortensis   Orach
Beta vulgaris var. cicla   Chard
Beta vulgaris var. crassa   Beet
Spinacia oleracea   Spinach

Asteraceae (Compositae outdated name) – Sunflower Family
Artemisia dracunculus   Tarragon
Cichorium endivia   Endive, escarole
Cichorium intybus   Chicory, radicchio
Cynara cardunculus   Cardoon
Cynara scolymus   Globe artichoke
Helianthus tuberosus   Jerusalem artichoke
Lactuca sativa   Lettuce
Taraxacum officinalis   Dandelion
Tragopogon porrifolius   Salsify, oyster plant

Convolvulaceae – Morning-glory Family
Ipomoea aquatica   Water spinach
Ipomoea batatas   Sweetpotato

Continued

Table 1.2. Continued.

Brassicaceae (Cruciferae outdated name) – Mustard Family
Armoracia rusticana   Horseradish
Sinapis alba   White mustard
Brassica juncea   Leaf mustard
Brassica napus (Napobrassica group)   Rutabaga
Brassica napus (Pabularia group)   Siberian kale
Brassica nigra   Black mustard
Brassica oleracea (Acephala group)   Kale, collard
Brassica oleracea (Alboglabra group)   Chinese kale
Brassica oleracea (Botrytis group)   Cauliflower
Brassica oleracea (Capitata group)   Cabbage
Brassica oleracea (Gemmifera group)   Brussels sprouts
Brassica oleracea (Gongylodes group)   Kohlrabi
Brassica oleracea (Italica group)   Broccoli
Brassica oleracea (Costata group)   Tronchuda cabbage
Brassica rapa (Chinensis group)   Chinese cabbage (nonheading), pak-choi
Brassica rapa (Pekinensis group)   Chinese cabbage (heading), pe-tsai
Brassica rapa (Perviridis group)   Spinach mustard
Brassica rapa (Rapifera group)   Turnip
Brassica rapa (Ruvo group)   Broccoli raab, rapini
Lepidium sativum   Garden cress
Crambe maritime   Sea kale
Nasturtium officinale   Watercress
Raphanus sativus   Radish

Cucurbitaceae – Gourd Family
Citrullus lanatus   Watermelon
Cucumis melo (Inodorus group)   Honeydew melon, casaba, crenshaw, persian
Cucumis melo (Cantaloupensis group)   Muskmelon, cantaloupe
Cucumis sativus   Cucumber
Cucurbita maxima   Winter squash, pumpkin, banana squash, buttercup squash, hubbard squash
Cucurbita argyrosperma   Green cushaw, Japanese pie pumpkin, silver-seed gourd
Cucurbita moschata   Butternut squash, calabaza, cheese pumpkin, golden cushaw
Cucurbica pepo   Pumpkin, acorn squash, summer squash, marrow
Lagenaria siceraria   Bottle gourd, calabash gourd
Luffa aegyptiaca   Smooth sponge gourd
Luffa acutangula   Ridged sponge gourd
Momordica charantia   Bitter melon
Sechium edule   Chayote

Euphorbiaceae – Spurge Family
Manihot esculenta   Cassava, yuca

Fabaceae (Leguminosae old name) – Pea or Bean Family
Cicer arietinum   Chickpea, garbanzo bean
Glycine max   Soybean
Lens culinaris   Lentil
Phaseolus coccineus   Scarlet runner bean
Phaseolus lunatus   Lima bean, sieva bean (butter bean)
Phaseolus vulgaris   Common bean (green, dry), snap bean, kidney bean
Pisum sativum   Garden pea, field pea, edible-pod pea
Psophocarpus tetragonolobus   Winged bean
Vicia faba   Fava bean (broad bean)
Vigna mungo   Urad, urd, black gram
Vigna radiata   Mung bean
Vigna unguiculata   Black-eyed pea, cowpea, asparagus bean, yard-long bean

Continued

Table 1.2. Continued.

Malvaceae – Mallow or Cotton Family
Abelmoschus esculentus   Okra, gumbo
Polygonaceae – Buckwheat Family
Rheum rhabarbarum   Rhubarb
Rumex acetosa   Sorrel
Rumex patientia   Dock, patience or monk's rhubarb
Solanaceae – Potato or Nightshade Family
Capsicum annuum   Pepper (bell, cayenne chili)
Capsicum frutescens   Tabasco pepper
Solanum lycopersicon   Tomato, cherry tomato
Solanum pimpinellifolium   Currant tomato
Physalis pruinosa   Strawberry ground cherry
Physalis philadelphica   Tomatillo
Physalis peruviana   Cape gooseberry
Solanum melongena   Eggplant
Solanum tuberosum   Irish potato
Tetragoniaceae – Carpetweed Family
Tetragonia tetragoniodes   New Zealand spinach
Apiaceae (formerly Umbelliferae) – Parsley Family
Anthriscus cerefolium   Chervil
Apium graveolens (Dulce group)   Celery
Apium graveolens (Rapaceum group)   Celeriac
Daucus carota   Carrot
Foeniculum vulgare   Fennel
Pastinaca sativa   Parsnip
Petroselinum crispum   Parsley, turnip-rooted parsley
Valerianaceae – Valerian Family
Valerianella locusta   Corn salad, mâche

The complete Latin binomial name includes a third element, the naming authority, which is not shown above. The name of the individual who first described the species is usually included after the Latin binomial. For example, the "L" that follows some vegetable species' names is an abbreviation for C. Linnaeus, who is considered to be the father of the Latin binomial classification system and was first to name a number of the vegetable species. For brevity, attached authorships for species mentioned in this publication are omitted.

this gene is expressed in broccoli or cabbage, the plants also produce a curd like cauliflower (Franco-Zorrilla *et al.*, 1999). Because of this and other genetic comparison studies, there has been a growing consensus among plant biologists that it is invalid to subdivide *Brassica* species into distinct botanical varieties. However, cauliflower is obviously different from cabbage and the other crops in *Brassica oleracea*. To make light of these horticultural differences, other references classify cauliflower as *Brassica oleracea* Group Botrytis rather than the botanical variety botrytis, which is more common in older literature (Griffiths, 1994). Therefore, the term Group (gp.) is used to show horticultural differences within a species that were previously classified into separate botanical varieties or subspecies.

Taking the cauliflower example a step further, 'Snowball' is a particular cultivar of cauliflower, Y is a particular strain of 'Snowball', and different production fields of 'Snowball' could be designated as individual lots. A Snowball type would refer to cultivars with the same basic characteristics as 'Snowball' (e.g. an early maturing cultivar that does not require vernalization to develop a curd) and would include all the various strains of 'Snowball'. For example, a cauliflower grower may ask a seed salesman if a cultivar is a "Snowball type" even though it may have a different cultivar name.

Classification of cantaloupe as an example

The following illustration shows the main divisions in plant classification and demonstrates how cantaloupe would be classified from general information to more specific:

- Vegetable Community – Plants
- Subcommunity – Spermatophyta (seed plants)
- Division – Angiospermae (angiosperms)
- Class – Dicotyledoneae (dicotyledons)
- Family – Cucurbitaceae
- Genus – *Cucumis*
- Species – *melo* L.
- Botanical Variety, Subspecies, or Group – Cantaloupensis (in this example, note this is not italicized and is capitalized.)
- Cultivated Variety (Cultivar) – 'Top Mark'
- Horticultural Strain – Seed company selection
- Stock or Lot Number – 1476

Generally speaking, genus and species names are italicized as a standard convention, especially when used together. Please note the difference between cultivar and variety. Notice in the example below that Top Mark is a cultivar and not a variety. Also, single quotes are used to signify a cultivar, e.g. 'Top Mark'.

Classification terminology

Family: an assemblage of genera that closely and uniformly resemble one another in general appearance and technical characters.

Genus: a more or less closely related and definable collection of plants that may include one or more species. The species in the genus are usually structurally or phylogenetically related, but do not routinely intercross.

Species: a group of similar organisms capable of interbreeding and are more or less distinctly different in morphological or other characteristics, usually reproductive parts, from other species in the genus.

Botanical Variety: a subdivision of a species consisting of a population with morphological characteristics distinct from other species' forms and is given a Latin name according to the rules of the International Code of Botanical Nomenclature. Variety was and continues to be used erroneously when the correct term is intended to be cultivar.

Group: a category of cultivated plants at the subspecies level that have the same Latin binomial, but have one or more characteristics sufficiently unique to merit a name that distinguishes them from another category. The term is used for horticultural convenience and has no botanical significance.

Thus, botanical variety, subspecies, and group are similar and therefore often used interchangeably.

Cultivar: contraction for "cultivated variety"; a plant that is clearly distinguished by identical physical characteristics and maintains these characteristics through proper propagation means.

Cultigen: a plant or group of plants known only in cultivation without a determined nativity, presumably having originated in its presently known form under domestication (Rubatzky and Yamaguchi, 1997). A plant whose origin or selection is due primarily to intentional human activity is called a cultigen. (A cultivated crop species that has evolved from wild populations due to selective pressures from traditional farmers is called a landrace.)

Clone: a population derived from a single individual and maintained by vegetative propagation. Individual members of a cloned population are genetically identical and can be maintained essentially uniformly with relatively little selection.

Landraces: plants that are ideally suited to a particular region or environment as the result of natural forces.

Line: a uniform sexually reproduced population, usually self-pollinated, that is seed propagated and maintained to the desired standard of uniformity by selection.

Strain: an improved selection of a cultivar that possesses similar characteristics, but differs in some minor attribute or quality.

Lot: a particular batch of seeds that were produced and processed together.

Type: a series of cultivars of a crop that have similar characteristics without specific reference to genetic or morphological characteristics.

References

Bailey, L.H. (1949) *Manual of Cultivated Plants*, revised edn. Macmillan, New York.

Balfour-Paul, H.G. (1996) Fertile crescent unity plans. In: Simon, R.S., Mattar, P. and Bullie, R.W. (eds) *Encyclopedia of the Modern Middle East*, Vol. 2. Macmillan, New York, pp. 654–656.

Bohs, L. (2005) Major clades in Solanum based in *ndh*F sequences. In: Keating, R.C., Hollowell, V.C. and Croat, T.B. (eds) *A Festschrift for William G. D'Arcy: The Legacy of a Taxonomist. Monographs in Systematic Botany from the Missouri Botanical Garden*, Vol. 104. Missouri Botanical Garden Press, St. Louis, Missouri, pp. 27–49.

Carlsson-Kanyama, A. (1997) Weighted average source points and distances for consumption origin-tools for environmental impact analysis. *Ecological Economics* 23, 15–23.

Cook, R. (2001) The Dynamic U.S. Fresh Produce Industry: An Industry in Transition. *Postharvest Technology of Horticultural Crops*, 3rd edn. University of California, Division of Agriculture and Natural Resources, Oakland, California.

ePIC (2011) Plant identification Database, KEW Royal Botanical Garden. Available at: http://epic.kew.org/searchepic/searchpage.do (accessed March 2008).

Erickson, D.L., Smith, B.D., Clarke, A.C., Sandweiss, D.H. and Tuross, N. (2005) An Asian origin for a 10,000-year-old domesticated plant in the Americas. *Proceedings of the National Academy of Sciences of the USA* 102(51), pp. 18315–18320.

FAO (2011) Food and Agriculture Organization of the United Nations. Available at: http://faostat.fao.org/site/339/default.aspx (accessed May 2011).

Franco-Zorrilla, J.M., Fernandez-Calvın, B., Madueño, F., Cruz-Alvarez, M., Salinas, J. and Martınez-Zapater, J.M. (1999) Identification of genes specifically expressed in cauliflower reproductive meristems. Molecular characterization of *BoREM1*. *Plant Molecular Biology* 39, 427–436.

Griffiths, M. (1994) *Index of Garden Plants*. Timber Press, Portland, Oregon.

GRIN (2010) Taxonomy for Plants, United States Department of Agriculture, Agricultural Research Service, Germplasm Resources Information Network. Available at: www.ars-grin.gov/cgi-bin/npgs/html/tax_search.pl (accessed June 2011).

Hillman, G., Hedges, R., Moore, A., Colledge, S. and Pettitt, P. (2001) New evidence of Lateglacial cereal cultivation at Abu Hureyra on the Euphrates. *Holocene* 11, 383–393.

Ladizinsky, G. (1998) *Plant Evolution under Domestication*. Kluwer Academic Publishers, the Netherlands.

Pirog, R. and Benjamin, A. (2003) Checking the food odometer: Comparing food miles for local versus conventional produce sales to Iowa institutions. *Report from the Leopold Center for Sustainable Agriculture*. Iowa State University, Ames, Iowa.

Poehlman, J.M. and Sleeper, D.A. (1995) *Breeding Field Crops*, 4th edn. ISU Press, Ames, Iowa.

Rubatzky, V.E. and Yamaguchi, M. (1997) *World Vegetables - Principles, Production, and Nutritive Values*, 2nd edn. AVI Publishing, Westport, Connecticut.

Schery, R.W. (1954) *Plants for Man* (Adapted from Vavilov). Prentice Hall, Englewood Cliffs, New Jersey.

Smartt, J. and Simmonds, N.W. (1995) *Evolution of Crop Plants*, 2nd edn. Longman Scientific & Technical, Harlow, UK.

USDA (2011) Plants Profile. Available at: http://plants.usda.gov (accessed 3 June 2011).

Welbaum, G.E. (1993) Water relations of seed development and germination in muskmelon (*Cucumis melo* L.). VIII. Development of osmotically distended seeds. *Journal of Experimental Botany* 44, 1245–1252.

Zohary, D. and Hopf, M. (1988) *Domestication of Plants in the Old World: The Origin and Spread of Cultivated Plants in West Asia, Europe, and the Nile Valley*, 3rd edn. Clarendon, Oxford, UK.

2 Tillage and Cropping Systems

Tillage

Soil preparation is an important aspect of vegetable preparation. There are different approaches to field preparation and the dealing with residue left behind by the previous crop. Plowing has been associated with crop production for much of recorded history. A plow is an agricultural implement with a sharp surface used for cutting and/or turning soil. Plows allow the soil to be broken so seeds can be planted. The plow may have first appeared around 1000 BC in the Near East and existed as early as 500 BC in China (Lal *et al.*, 2007). Moldboard plows were known in Britain after the late 6th century (Hill and Kucharski, 1990). The moldboard design consists of a curved plate with a sharp edge that turns over the soil so the top layers are buried and moist friable layers are brought the surface (Fig. 2.1).

Animals were initially used to pull these implements. Wooden plows remained the standard until Jethro Wood invented a cast-iron plow with interchangeable parts in the early 1800s. John Lane invented the steel plow shortly thereafter. In 1865, John Deere patented a steel plow in the USA similar in design to ones used today. Disc harrows were introduced in the 1860s to prepare soil with minimal residue or a plowed field for planting by breaking clods and incorporating residue into the top layer. Rotary tine tillers (RTT) that are powered by a tractor's power take-off became popular during the 1960s and 1970s because they pulverize soil close to the surface creating small soil particles well suited for the planting of small-seeded vegetables (Lal *et al.*, 2007).

Since its invention almost 170 years ago, the modern steel plow has been widely used for pre-plant soil preparation to establish vegetable crops. Plowing to prepare fields for vegetable production is popular because it produces a friable seedbed well suited for planting small-seeded vegetables that need good soil contact to germinate. Before planting, plowed soils are often further tilled with other implements like the disc harrow or RTT (Lal *et al.*, 2007). Tillage destroys existing weeds that may compete with emerging vegetable seedlings. The emergence of some vegetables is slowed by heavy residue on the soil surface and plowing buries the remains of the previous crop. Deep plowing also aids disease control by burying contaminated residue. Bare soil tends to warm faster in the early season compared to soil covered with residue, which leads to earlier maturity, an important goal for many growers in short-season areas (Table 2.1).

However, there are several negative factors associated with plowing. The bare soil produced is prone to erosion, particularly after fall plowing or on soils that are sloped. The heavy equipment used to plow and till compacts soil, uses energy and requires labor. Also, weed seeds are brought to the surface (Table 2.1).

Deep tillage with a moldboard plow and associated implements used to prepare a field before establishing the next crop is called conventional tillage. Conventional tillage is when <15% of previous crop residue remains on the soil surface following establishment. Conventional tillage can be thought of as "full-width" tillage because 100% of the topsoil is moved and mixed so that the majority of crop residue is incorporated into the soil.

An easy and effective way to obtain early-season weed control is by the stale bed technique (Riemens *et al.*, 2007). This technique requires that the seedbed be prepared several weeks before the intended planting date so weed seeds germinate before the crop is planted. To kill the first flush of weeds, herbicides or shallow cultivation are used prior to planting. If cultivation is used, care must be taken to avoid bringing new buried weed seeds to the surface (Riemens *et al.*, 2007). The stale bed technique is sometimes difficult to employ because a wet early season may prevent soil preparation in advance of planting.

Fig. 2.1. This "four bottom" steel moldboard plow is pulled by a tractor.

Table 2.1. The advantages and challenges of growing vegetable crops using conservation tillage techniques.

Conservation tillage limitations/challenges	Conservation tillage advantages
Lowers soil temperatures	Requires less machinery
Slows germination and emergence	Requires less labor
Slower early growth	Requires less fuel
May delay spring planting	Reduces soil erosion
May increase certain diseases	Reduces soil compaction
Heavier crop residue so planter operation is more difficult	Weed growth is delayed
Weed spectrum changes	More moisture is retained in soil
Increased insect pests	

In addition to the moldboard plow, less disruptive tillage implements have been developed. Graham-Hoeme Plow Company developed chisel plows in the 1930s. The chisel plow breaks the soil surface without inverting the top layer (Fig. 2.2; Lal *et al.*, 2007).

During the last 40 years, chisel and coulter-chisel plows have helped popularize conservation tillage practices. In the 1970s–1980s a wide range of conservation tillage systems were developed.

Conservation tillage is any method of soil cultivation that minimizes soil disturbance and leaves the previous year's crop residue in fields before and after planting the next crop (Derpsch, 1998). Crop residues reduce soil erosion, conserve moisture, inhibit weed growth and build soil organic matter. To be considered conservation tillage at least 30% of the soil surface must be covered with residue after planting the next crop. Some conservation tillage methods forgo traditional tillage entirely and leave 70% residue or more. Several types of conservation tillage are used as well as combinations of conservation and conventional tillage. A few of the more popular techniques include: no-till, strip-till, ridge-till, and mulch-till (Derpsch, 1998). Each method requires specialized or modified equipment and unique management practices.

No-till systems leave the soil undisturbed from the time the previous crop is harvested until the new crop is planted. Nutrients are precision placed in the soil rather than broadcast. Transplanting or seeding is accomplished in a narrow seedbed or slot created by coulters,

Fig. 2.2. This chisel plow mounts to the three-point hitch of a tractor and can be raised and lowered hydraulically.

disk openers, or in-row chisels (Derpsch, 1998). Weed control is accomplished through herbicides, by hand, or by mowing. Cultivation is only used for weed control in an emergency. No-till systems, when managed properly, frequently out-yield fields established with conventional tillage systems because of greater moisture retention especially in dry areas (Table 2.1).

Strip-till involves planting crops in a narrow tilled space with chisels or tillers cleared of residue with row sweepers while the rest of the field is left untilled. The tilled strip offers a more favorable soil environment for the rapid germination and emergence of small-seeded vegetables because a finer seedbed is prepared and mulch is removed above the seed to encourage rapid emergence.

Ridge-till involves planting row crops on permanent ridges about 10–15 cm (4–6 in) high. The previous crop's residue is cleared from ridge-tops into adjacent row middle furrows. Maintaining the ridges is essential and requires modified or specialized equipment such as cleaners, sweeps, disk openers, or coulters. Nutrients are precision-placed into the ridges rather than broadcast over the entire field. Weed control is accomplished by hand, herbicides, or cultivation.

Mulch-till disturbs the soil with chisels, field cultivators, disks, sweeps, or blades before planting. Mulch-till leaves at least one-third of the soil surface covered with crop residue (Derpsch, 1998). Weed control is accomplished by hand, herbicides, and/or cultivation.

Conservation tillage methods are advantageous because they reduce labor, fuel costs, field preparation time, erosion, and soil compaction (Table 2.1). Conservation tillage has become the predominant method for establishing agronomic crops in much of North America and is extensively used in other parts of the world as well. The transition to conservation tillage has been slower for vegetable crops because of the extensive use of plastic mulch, which requires a smooth seedbed free of residue. Successful stand establishment of direct-seeded vegetables also requires small soil aggregates to ensure good seed-to-soil contact. A well-tilled seedbed promotes uniformity and earliness, priorities for many vegetable growers (Table 2.1). However, conservation tillage is used for vegetable crops that are not grown on plastic mulch, where earliness is not a priority, are transplanted, and are produced on highly erodible soils.

Adjustments can be made to successfully deal with some of the challenges presented by conservation tillage while retaining the many advantages. For example, adjustments for planting in cooler soils that often result from conservation tillage include shallow planting (2.5 cm, 1 in or less), good seed–soil contact, slow planting (pull the planter at 8 km/h, 5 mph or less) and use of high-quality seed. Vegetable growers who use no-till often use pelleted seed to ensure good soil contact.

Techniques for decreasing root rot disease may include: planting when the field is not excessively wet, using ridge-till or raised beds to promote drainage, install tile drainage systems in fields, treating seeds with fungicide or biological controls that reduce disease, and using disease-resistant cultivars if possible.

In conservation tillage systems, growers must adapt to planting in heavier residues compared to conventional tillage. Techniques to mitigate problems include: spreading residue widely and evenly across a field, removing wheat straw or other cover-crop materials that are slow to decompose and may interfere with emergence, using row wipers or sweepers, using planters and transplanters especially designed for planting through residue, and installing harrows on planters.

Conservation tillage may change the type of weeds that growers must control. An effective weed control program for a conservation tillage system may include eliminating perennial weeds, killing weeds preplant, using post-emergence herbicides, and planting narrow rows.

Drainage and Erosion Control

Preventing wind and water erosion are challenges for vegetable farmers. Erosion is of special concern on sloping land where water can quickly displace topsoil, especially when conventional tillage is used. A good rule of thumb is to never leave soil uncovered. This means that residue or a cover crop should always be present to preserve soil resources when a crop is not growing. There are several other management strategies that reduce or prevent erosion.

Waterways are natural or constructed outlets for water, protected from erosion by grass or other perennial cover that holds soil (Fig. 2.3).

Once established, the waterway is not tilled. Waterways serve as safe outlets for runoff water from contour rows, terraces, and other diversions. Natural drainage ways make the best locations for waterways and often require minimal shaping to produce a good channel. Natural drainage ways eventually divert water into a stream or other tributary near a field. Waterways should be designed to be wide and flat so farm machinery can easily cross and yet provide capacity to carry storm runoff safely from the surrounding areas. Weed control in the waterways is generally by mowing. Waterways should be mowed before potential weed seeds are produced. Herbicides or hand removal can be practiced for troublesome perennial weeds.

Contour cropping is the practice of tilling and orienting crop rows along lines of consistent elevation on sloped land in order to conserve rainwater and to reduce soil losses from surface erosion. In simple terms, contour cropping orients rows around a hill rather than up and down it. The crop rows act as small reservoirs to prevent rapid runoff that causes erosion and catch and retain rainwater, improving infiltration and more uniform distribution of water. Contour cropping reduces erosion and is most effective on deep, permeable soils and on gentler slopes of about 2–6% that are less than 91 m (300 ft) long. The effectiveness of contouring

Fig. 2.3. The grass waterway in the center of a harvested corn field in Ohio directs runoff through a natural drainage way that is left in sod and natural weed free vegetation to prevent erosion.

is reduced greatly on steeper or longer slopes because of possible washover of rows by runoff water. In general, contouring can reduce erosion losses up to 50% compared with up-and-down-hill tillage on slopes of from 2–6%. On steeper slopes (18–24%) contour cropping without supplementary practices reduces erosion losses by only about 10%. Grass waterways are necessary to carry the runoff water safely from the contour rows.

Strip-cropping, the practice of alternating contour strips of sod and row crops, is even more effective than contouring. The sod strips help slow runoff and filter out eroded soil. Strip-cropping reduces soil losses to about half that of contouring alone or one-fourth that of up-and-down-hill tillage. Strip widths are governed by the percent slope and vary from up to 30.5 m (100 ft) on 2–6% slopes to 18.3 m (60 ft) or less on 18% slopes and above.

Terraces are channels or ridges built across slopes to catch runoff water and shorten the length of a slope (Fig. 2.4). They are generally more effective than either contouring or strip-cropping alone and are designed especially for steeper slopes. Most terraces are designed with gradual slopes to lead water off safely into grass waterways or other suitable drainage outlets. The number and spacing of terraces depend on the soil type, slope, and cropping practices, and should be designed by soil conservation specialists. Diversion terraces are especially designed to divert larger flows of water away from buildings, gullies, farm ponds, or fields below long slopes.

For production on sloped land, drainage occurs naturally and managing the runoff is of primary concern. However, flat land with heavy soils presents a different challenge for vegetable growers. On level fields that retain moisture and drain slowly, excessive rain or irrigation can cause soil saturation and flooding for extended periods. Most vegetables prefer well-drained soils because root zone anoxia stunts growth and increases root disease and epinasty. Too much water may prevent the use of farm machinery and operating heavy machinery in excessively wet conditions will damage soil structure and cause soil compaction. Wet soils warm slowly, often limiting early season growth. Subsurface field drainage tube systems are installed where drainage is poor to reduce soil moisture to optimum levels for crop growth.

To improve drainage, vegetables are often planted on raised beds shaped after plowing and/or tillage. Raised bed culture is also necessary when plastic mulch is used (Fig. 2.5).

Mulch laying after plowing to improve drainage is popular in regions where heavy rainfalls may saturate soils for part of the growing season (Fig. 2.6).

Fig. 2.4. Terraces enable crop production on steeply sloped land in Taiwan.

Fig. 2.5. Straw mulch and raised beds to improve drainage on level ground.

Fig. 2.6. Mulch laying after plowing to improve drainage.

Surface water can be drained via pumping, ditches, or waterways, but subsurface drainage is often the best option for removing excess water. The goal of a subsurface system is to drain gravitational water from the root zone (Nwa and Twocock, 1969). Beginning in the 1800s, round ceramic tile conduits were placed end-to-end and buried under fields to remove excess water. The tile lines were oriented to feed water into a waterway or stream. Ceramic tile conduits were heavy, expensive, labor intensive to install, and easily broken. Today lines of perforated plastic tubing are laid under fields instead of ceramic or concrete tiles. The introduction of plastic tile served to reduce the cost of installation because a continuous section of lightweight, flexible line can be mechanically laid in a trench and covered relatively quickly. The spacing of depth of a subsurface drainage system will vary based on soil type, crops grown, and precipitation, but a typical flexible plastic line spacing would be 12 m (40 ft), a typical depth would be 0.75 m (29.5 in), while a minimum cover depth would be 0.6 m (23.6 in) (Nwa and Twocock, 1969). Drainage lines that are too shallow can be broken or damaged by cultivation. Roughly 25% of the cropland in the USA and Canada has a subsurface drainage system installed (Wright and Sands, 2001).

Crop Residue Management

A cover crop is planted primarily to manage soil fertility, soil quality, water, weeds, pests, diseases, biodiversity, and wildlife in an agroecosystem (Lu *et al.*, 2000; Hartwig and Ammon, 2002). Typically, a cover or green-manure crop is grown for a specific period of time, and incorporated into the soil while green by plowing or disking. Incorporating cover crops into the soil increases the microbial activity that decomposes the plant matter, releasing nutrients such as nitrogen (N), potassium (K), phosphorus (P), calcium (Ca), magnesium (Mg), and sulfur (S) for the next crop. Soil microbial activity also leads to the formation of viscous materials that improve soil structure (Welbaum *et al.*, 2004). Soil structure refers to how individual soil granules clump or bind together to form aggregates. The size of the aggregates also determines the arrangement of soil pores between them. Soil with good structure is better aerated, absorbs water more quickly, and is easier to prepare for seeding than a soil with poor structure. The amount of organic matter in soil also increases with cover-crop decomposition, which improves soil health for subsequent crops.

Legume cover crops, such as beans, alfalfa, and clover, have root systems that form a symbiosis with *Rhizobium* spp. bacterium that fix atmospheric N, making them desirable green manure crops because they require little N fertilizer and have a lower carbon to nitrogen (C:N) ratio (Hartwig and Ammon, 2002). The C:N ratio in a green manure crop is a crucial factor to consider, since it will impact the rate of decomposition, the nutrient content of the soil and N availability for the cash crop. The C:N ratio will differ among cover-crop species and the age of the plant. In the C:N ratio, the value of N is always 1, whereas the value of carbon or carbohydrates is expressed in a value of about 10 up to 90. The C:N ratio should be less than 30:1 to prevent the bacteria from decomposing the green manure crop after it is incorporated into the soil, which would deplete existing soil N. If N in cover crop residue is limiting, decomposing microbes will use soil N instead, decreasing availability for crop plants. Legumes such as crimson clover and hairy vetch may have C:N ratios of 8:1 to 15:1, compared with ratios of 30:1 to 60:1 for wheat (*Triticum aestivum* L.) and rye residues (Ranells and Wagger, 1996). A high C:N ratio is beneficial if a cover crop is used to provide mulch for a vegetable crop planted into residue in a no or minimum tillage production system. For example, rye may be grown as a winter cover crop to stabilize soil. In spring, the rye is rolled or mowed to stop growth before it develops seed so tomato or other vegetables can be transplanted into the straw residue, which forms natural mulch (Fig. 2.7; Hartwig and Ammon, 2002).

In this situation, slow decomposition (high C:N ratio) of the straw cover crop is advantageous because it helps the straw act as mulch for a longer portion of the season before it breaks down. Legumes do not make good mulch crops because they decompose before the growing season for tomato or peppers are over.

Many crops can be used as cover crops (Hartwig and Ammon, 2002). Summer green manure crops that can be used with vegetables include: Egyptian clover, hairy vetch, oats, and sorghum/Sudan grass. Popular overwintering green manure crops include: hairy vetch, crimson clover, winter rye, and winter rape. The use of green manures should be a component of an effective crop management system that uses crop residues to increase soil fertility and prevent soil erosion during periods of high risk. In addition to legumes, certain cultivars of winter rape are used as a green manure crop to reduce nematode populations because of their high glycosinolate content in the residue (Potter *et al.*, 1998; Vargas-Ayala *et al.*, 2000). Glucosinolates and other natural compounds in cover crops may also break disease cycles

Fig. 2.7. A cover crop of rye straw was rolled before seed heads developed to provide mulch for a no-till transplanted broccoli crop.

and reduce populations of bacterial and fungal diseases (Everts, 2002).

Cover crops are sometimes used as "trap crops", to attract pests away from the cash crop to a habitat that is more attractive to the pest (Shelton and Badenes-Perez, 2006). Trap crop areas can be established inside a field or at other locations removed from a field. The trap crop is often grown during the same season as the cash crop. To control insects, the trap crop can be treated with a pesticide or by using sticky traps or suction (Kuepper and Thomas, 2002).

Cover crops may also be used to create species diversity in or around a vegetable field. Certain plants can attract natural predators of pests by providing elements of their habitat. This form of biological control is known as habitat augmentation, and can be achieved with the use of cover crops (Bugg and Waddington, 1994).

Cropping Systems

Much of the large-scale vegetable production in many parts of the world is an intensive monoculture where a single crop is grown in a field and other crops or vegetation are excluded during the production cycle. Intensive monoculture has evolved in countries were farmland is generally plentiful and there is a need for large concentrated harvests using minimal labor. Crop monocultures produce great yields by utilizing a plant's genetic potential to maximize growth in a uniform growing environment with less pressure from other species. The production of uniform cultivars, and particularly F-1 hybrids, bred for a specific environment, are grown at optimal spacing to use light, space, and nutrients to maximize yields. Standardized management practices for pest control, fertilizer inputs and harvesting allow growers to enjoy economies of scale, use less labor, and increase harvesting efficiency. Over the past 60 years monoculture practices including the use of synthesized fertilizers have greatly increased crop yields. Annual crop monocultures tend to rely on pesticide usage, large equipment to reduce labor, concentrated mineral fertilizers, and mechanical harvesting. Vegetable producers tend to be highly specialized because of unique equipment and expertise required to grow specific crops. Crop monocultures tend to favor large enterprises and in turn these enterprises favor and promote the system to ensure their future existence.

The ability of crop monocultures to produce large yields with reduced labor is widely accepted. However, crop monocultures can lead to loss of species diversity, the rapid spread of diseases, greater susceptibility to pathogenic attack, herbicide-resistant weeds, insects resistant to pesticides, greater corporate influence in agriculture, and increased energy usage for vegetable production (Pimentel *et al.*, 2005). There has also been a tendency for less locally grown fruits and vegetables because vegetable production in the developed world tends to concentrate in areas best suited to producing a crop and then shipping the produce to distant markets. While this practice optimizes productivity, it is very energy intensive and sometimes results in commodities that are no longer fresh when consumed. For example, to reduce postharvest losses, cantaloupes are often harvested prematurely and may still be immature when purchased.

In other regions of the world multiple cropping is widely used. A few examples of multiple cropping systems are listed below. Polyculture is the production of multiple crops in the same space at the same time. Polycultures strive to imitate the diversity of natural ecosystems, and avoid large stands of single crops or monoculture. Examples of polyculture include multi-cropping, intercropping, companion planting, beneficial weeds, and alley cropping.

Sequential cropping

Sequential cropping is growing two or more crops in sequence in 1 year in the same field. An example of this would be growing corn and planting cucumbers in the row middles. After the corn is harvested, the corn stalks serve as a support for the cucumber vines. Another example would be seeding lettuce in plastic mulch after a summer vegetable like tomato is harvested and the plants removed from the field. Options for sequential cropping are increased where the growing season is long and there is sufficient time to mature multiple crops.

Intercropping

Intercropping is another polyculture technique where two or more crops are grown simultaneously on the same land (Fig. 2.8; Zhang and Li, 2003). An example of this would be interplanting pumpkins among agronomic or grain corn, which was a common practice many years ago in the USA before mechanical corn harvesters were developed. The corn uses vertical space while the pumpkin vines cover the soil surface. Both crops mature in the fall, so hand-harvesting corn did not damage the pumpkin vines. In this historically important system, pumpkins were collected and sold or left in the field for livestock food after they were cut open.

Relay intercropping occurs when two or more crops are grown simultaneously during part of the growing season of each other. Usually, the second crop is seeded or transplanted after the first crop has reached the reproductive stage or the later part of the growth period but before the first crop is ready for harvest. An example of relay intercropping would be seeding corn next to a developing radish crop. In many countries, intercropping is used to maximize use of arable land (Horwith, 1985).

Farmscaping

Farmscaping is a whole-farm ecological approach to pest management, particularly for insects, designed to attract natural predators to fields where cash crops are grown (Fig. 2.9).

Fig. 2.8. Intercropping of squash and Chinese cabbage in Taiwan.

Fig. 2.9. Farmscaping of a fall broccoli crop surrounded by buckwheat (bottom) and beneficial plants (left) to promote insect diversity and natural control of predators to reduce insecticide use.

Farmscaping is a polyculture system that simultaneously uses specific beneficial plants to attract insects in and around production fields to retain natural predators that are lost in monoculture production. The co-cultivation of cash crops and beneficial plants promote insect diversity and thus controls insect pests with natural predators. If implemented correctly, farmscaping provides a sustainable strategy for insect pest management with little or no pesticide usage. Critics cite the acreage lost to beneficial noncash crops as a negative aspect of farmscaping.

Managed Weeds

Some studies have suggested that overseeding (also called underseeding in some literature) a low growing plant that acts as a managed weed or living mulch in the row middles after a vegetable crop has been established is an effective alternative to conventional cultivation and herbicide use (Hartwig and Ammon, 2002). Overseeding creates species diversity and may help attract beneficial insects. This technique is commonly employed in orchard management and is used by some vegetable growers to reduce erosion and allow the overseeded plant to inhibit the proliferation of more invasive weeds (Hartwig and Ammon, 2002). White clover is an example of an overseeded plant that is low growing, has a deep root system and is a legume that can fix atmospheric N. A downside to overseeding is that the mulch crop competes with vegetables for water and nutrients. In a dry climate where irrigation must be used, the disadvantages of overseeding may outweigh the advantages.

Ratoon Cropping

Ratoon cropping is a method of harvesting a crop while leaving the roots and crown intact to grow back for additional harvests. Ratoon cropping decreases the cost of preparing the field for a new planting. The yield of the ratoon crop tends to decrease following each cycle. The term ratooning is often used with crops harvested over multiple years but can be applied to annual crops where several cuts are made during a season. Globe artichokes, spinach, and lettuce are examples of crops that can be harvested by ratooning.

Crop Rotation

Crop rotation is a simple and effective management method that should always be used for vegetable crop production. Rotation is the practice of growing unrelated crops in succession in the same field to preserve the productive capacity of the soil. Crop rotation helps control disease buildup in the soil, reduce insect pests, and preserve soil nutrition. Most growers rotate multiple crops with each crop performing a separate function to improve soil. For example, a rotation may include a deep-rooted crop to improve soil structure and legume green-manure crop to improve soil fertility. Crops from related families should not be planted in the same field in subsequent years, like cabbage and broccoli for example, because they may harbor the same disease and insect pests.

References

Bugg, R.L. and Waddington, C. (1994) Using Cover Crops to Manage Arthropod Pests of Orchards, a Review. *Agriculture, Ecosystems & Environment* 50, 11–28.

Derpsch, R. (1998) Historical review of no-tillage cultivation of crops. In: *FAO International Workshop Conservation Tillage for Sustainable Agriculture*. Food and Agriculture Organization of the United Nations, Rome, pp. 205–218.

Everts, K.L. (2002) Reduced fungicide applications and host resistance for managing three diseases in pumpkin grown on a no-till cover crop. *Plant Disease* 86, 1134–1141.

Hartwig, N.L. and Ammon, H.U. (2002) Cover crops and living mulches. *Weed Science* 50, 688–699.

Hill, P. and Kucharski, K. (1990) Early medieval ploughing at Whithorn and the chronology of plough pebbles. *Transactions of the Dumfriesshire and Galloway Natural History and Antiquarian Society* 65, 73–83.

Horwith, B. (1985) A role for intercropping in modern agriculture. *BioScience* 35, 286–291.

Kuepper, G. and Thomas, R. (2002) 'Bug vacuums' for organic crop protection. *ATTRA Pest Management Technical Note*. ATTRA NCAT, Fayetteville, Arkansas.

Lal, R., Reicosky, D.C. and Hanson, J.D. (2007) Evolution of the plow over 10,000 years and the rationale for no-till farming. *Soil and Tillage Research* 93, 1–12.

Lu, Y.C., Watkins, K.B., Teasdale, J.R. and Abdul-Baki, A.A. (2000) Cover crops in sustainable food production. *Food Reviews International* 16, 121–157.

Nwa, E.U. and Twocock, J.G. (1969) Drainage design theory and practice. *Journal of Hydrology* 9, 259–276.

Pimentel, D., Hepperly, P., Hanson, J., Douds, D. and Seidel, R. (2005) Environmental, energetic, and economic comparisons of organic and conventional farming systems. *BioScience* 55, 573–582.

Potter, M.J., Davies, K. and Rathjen, A.J. (1998) Suppressive impact of glucosinolates in Brassica vegetative tissues on root lesion nematode *Pratylenchus neglectus*. *Journal of Chemical Ecology* 24, 67–80.

Ranells, N.N. and Wagger, M.G. (1996) Nitrogen release from grass and legume cover crop monocultures and bicultures. *Agronomy Journal* 88, 777–882.

Riemens, M.M., Van Der Weide, R.Y., Bleeker, P.O. and Lotz, L.A.P. (2007) Effect of stale seedbed preparations and subsequent weed control in lettuce (cv. Iceboll) on weed densities. *Weed Research* 47, 149–156.

Shelton, A.M. and Badenes-Perez, E. (2006) Concepts and applications of trap cropping in pest management. *Annual Review of Entomology* 51, 285–308.

Vargas-Ayala, R., Rodriguez-Kabana, R., Morgan-Jones, G., McInroy, J.A. and Kloepper, J.W. (2000) Shifts in soil microflora induced by velvetbean (*Mucuna deeringiana*) in cropping systems to control root-knot nematodes. *Biological Control* 17, 11–22.

Welbaum, G.E., Sturz, A.V., Dong, Z. and Nowak, J. (2004) Managing soil microorganisms to improve productivity of agro-ecosystems. *Critical Reviews in Plant Sciences* 23, 175–193.

Wright, J. and Sands, G. (2001) Planning an Agricultural Subsurface Drainage System. University of Minnesota Cooperative Extension Bulletin BU-07685. Available at: www.extension.umn.edu/distribution/cropsystems/dc7685.html (accessed 1 August 2013).

Zhang, F. and Li, L. (2003) Using competitive and facilitative interactions in intercropping systems enhances crop productivity and nutrient-use efficiency. In: *Structure and Functioning of Cluster Roots and Plant Responses to Phosphate Deficiency*. Springer, the Netherlands, pp. 305–331.

3 Vegetable Seeds and Crop Establishment

Introduction

Most vegetable crops are grown from seeds and not vegetatively propagated. A seed can be defined as "an immature plant in an arrested state" produced through sexual reproduction. If a plant produces seeds that germinate "true-to-type" and grow rapidly, it is cheaper, more efficient, and usually faster to propagate the crop by seed. True-to-type simply means that the plant that results from a seed has the same traits and appearance as the plant that produced the seed.

Vegetables that do not grow true-to-type from seed or that are difficult to propagate from seed such as potato, sweetpotato, or globe artichoke are vegetatively propagated. Vegetative propagation is a form of asexual reproduction of a plant where the stems, leaves, and roots, or other tissue not involved in reproduction are rooted. With vegetative propagation, the new plant is a clone that is genetically identical to the parent.

Seeds produced through tissue culture are sometimes called synthetic seeds. Synthetic seed can be defined as the artificial encapsulation of somatic embryos, shoot buds, aggregates of cells, or any tissues that have the ability to form a plant (Fujii et al., 1987). Synthetic seeds have been produced commercially but make up a small percentage of the commercial vegetable seeds sold in the world.

Seeds are one of the most important inputs in vegetable production. High quality seed is a prerequisite for a successful crop. Cutting production costs by using lower quality seeds is usually counterproductive. While modern seeds may seem expensive, they are a relatively small production cost and an ever smaller portion of the gross returns. For example, a seedsman who produces hybrid tomato seeds could expect to receive about US$12 for 10 g (0.35 oz) of seed, a commercial seed company would sell these seeds for about 3 cents each or 3,300 seeds for US$100. The seed retailer would sell the same 3,300 seeds for US$200, while the commercial tomato grower could produce US$60,000 of tomato fruits from these same 3,300 seeds. A grocery store would sell these same tomato fruits for US$110,000. So in this example, the tomato seeds yield a 300% return on investment. So sacrificing seed quality to save on production costs is not a wise business practice.

Seed quality is assessed through seed testing (Elias et al., 2011). Most countries have official testing procedures that establish standards for commercial seed sales. Seed laws exist in most countries to set standards for the labeling of seeds for sale, minimum germination percentages, inert matter (nonseed matter), and contamination from weed seeds. Viability tests determine whether or not seeds are able to germinate and results are reported as germination percentages. Most germination tests are standardized and conducted under laboratory conditions optimal for the germination of a species. Most international commercial companies sell vegetable seeds that germinate to percentages that exceed legal requirements and approach 100% for many species.

Since vegetable seeds are traded internationally, the International Seed Testing Association (ISTA) was established to deal with issues surrounding global seed trade. The primary purpose of this association is to develop, adopt, and publish standard procedures for sampling and testing seeds, and to apply these procedures to the evaluation of internationally traded seeds. ISTA provides internationally accepted standard rules for seed sampling and testing, accredits laboratories, promotes research, provides international seed analysis certificates and training, and disseminates knowledge on seed science and technology.

Because commercial seed lots germinate to very high percentages, seed vigor is of interest. Seed vigor is a measure of how well seeds germinate rather than if they are simply viable or not. Most commercial vegetable seed companies compete with one another on the basis of the vigor. Unlike viability testing, there are many different ways to assess the vigor of seeds (Elias *et al.*, 2011). Germinating seeds under adverse conditions is often used to assess vigor. Some of the more popular vigor tests measure how fast seeds germinate, how the seeds germinate at adverse temperatures or how seeds germinate after aging during short-term storage at elevated temperatures and moisture contents (Elias *et al.*, 2011). Because of the variable nature of seed vigor, most seed laws do not specify minimum vigor standards of seeds for commercial sale.

Seed Treatments

Commercial vegetable growers rarely use "raw" untreated seed for propagation. Many growers pay a premium for commercial seed treatments designed to insure the best possible stand establishment. A plant seed contains all the genetic information needed for the growth of the plant and reserve materials, such as proteins, fats, carbohydrates, and vitamins that are stored in the endosperm or cotyledons. These reserves sustain the young seedling through emergence until photosynthesis can make the new compounds required for growth and development. Insect pests and diseases are attracted to seeds, particularly aged or damaged seeds, which leak substances that are food for other organisms. Seeds produce natural defense compounds in response to predation similar to those produced by the mother plant (Welbaum, 2006). However, disease and insects can overwhelm vulnerable seeds, particularly under soil conditions unfavorable for germination, so treatments are often applied to protect seeds. Chemicals or beneficial microbes can be applied prior to planting to protect seeds and seedlings from decay or damping-off disease caused by soil fungi. Insecticidal compounds are used as preplant seed treatments to control biting and sucking pests on roots, stems, and leaves. Fungi or bacteria that are carried on or with the seed often cause diseases of vegetable crops. Disinfection of contaminated seed is important for control of seed-borne diseases.

Treating seeds before planting has been a practice since ancient times. The ancient Egyptians and Romans attempted to protect seeds against soil-borne pests by dipping them in onion sap prior to planting (FIS, 1999). Recent research suggests that this practice may be at least partially effective (Morsy *et al.*, 2009). Many diverse treatments have been developed to improve seed performance over the years. Some commonly applied vegetable seed treatments are described below.

Biological seed treatments with *Rhizobium* spp. improve nitrogen fixation of legumes after germination. Seeds may also be inoculated with other beneficial organisms like *Trichoderma* spp. fungi to naturally protect again seedling diseases or increase tolerance to environmental stress. Biological seed treatments are an emerging trend in the vegetables seed industry. *Pseudomonas chlororaphis*, *Bacillus subtilis* strains, *Bacillus lichenformis*, *Bacillus megaterium*, *Burkholderia cepacia* type Wisconsin, *Agrobacterium radiobacter* Strain 84, *Bacillus pumilus* GB34, and *Agrobacterium radiobacter* K1026 have all been used commercially as seed treatments (McSpadden Gardener and Fravel, 2002).

Chemicals such as fungicide are added to seeds for conventional production to control damping off, a very destructive seedling disease caused by soil-borne fungi. A light coating of fungicide, usually with brightly colored dye added to remind users of the treatment, is often applied to the seed surface to protect the plant through the seedling stage. Seed coating materials sometimes contain a small amount of fertilizer to simulate early seedling growth. This practice varies with species as some seeds, like legumes, are very sensitive to fertilizer placed on the seed.

Coating and pelleting are two widely used seed treatments, particularly with small-seeded species. Coated seeds have layers of clay or diatomaceous earth with a water-soluble binding agent added to the surface to increase seed size but not their overall shape.

Pelleted seeds have been coated until they are round. This makes singulation easier, particularly when belt or vacuum seeders with holes of a specific size are used (Fig. 3.1).

To ensure proper identification, coated seeds are often color coded, so cultivars or types are not inadvertently mixed during planting. Pelleting treatments split open upon hydration so that the coating material does not pose a barrier to radicle growth or limit seed oxygen availability.

Film coating is another innovation that adds polymers to the seed surface. Greater quantities

Fig. 3.1. (a) Precision vegetable vacuum seeder can plant to a final stand. (b) Vacuum simulation and metering system for singulating seeds. Pelleted seeds are often used with this type of planter for precision placement (photos courtesy of MaterMacc S.p.A.).

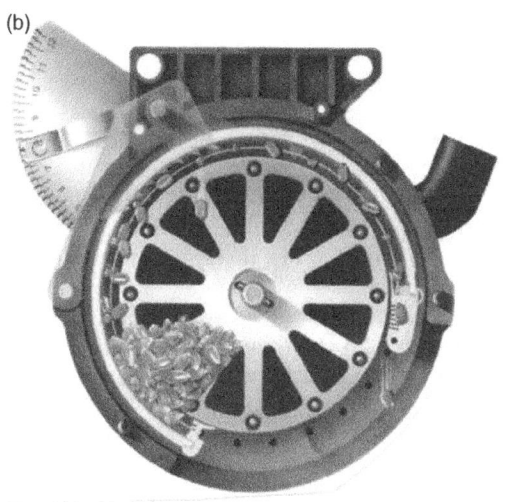

Fig. 3.1. Continued.

can be added to increase seed size. Many clay or diatomaceous earth coating materials produce dust when handled in bulk, posing a threat to worker safety. However, film coatings use the same water-soluble and dust-free polymers developed by the pharmaceutical and food industries to coat pills or candies.

Hot water soak treatments are used to treat brassicas and certain other species. If properly applied, the high temperature kills most seed-borne fungi and bacteria without affecting the seed. Prolonged exposure to hot water will damage seeds.

Pregerminated seeds have begun the germination process but have little or no root growth. Pregerminated seeds are all viable since they have begun germination. However, germinated seeds must be handled with extra care since they are in a transitional stage and more easily damaged than dry seeds.

Seeds and Stand Establishment

Primed seeds have been subjected to a controlled hydration process followed by redrying before planting. Priming is accomplished by imbibing seed in either osmotic solutions (osmoconditioning), moist solid carrier materials, such as moistened vermiculite (matriconditioning), or with water alone (drum priming) (Welbaum et al., 1998). These priming treatments permit early metabolic events associated with germination to proceed before radicle emergence commences (Welbaum et al., 1998). After a priming treatment has been applied, seeds are dried back to their original water content prior to planting.

Seed priming has become an important commercial seed treatment. Priming treatments are especially popular with vegetable seeds where rapid and uniform germination is required. Generally, priming reduces the time to germination and may improve germination under temperature or moisture stress. For example, priming increases the ability of lettuce seeds to germinate at high temperatures by removing thermodormancy and also advances seed maturity (Welbaum et al., 1998). Peppers, tomatoes, and lettuce respond well to priming, so the treatment is often applied to these seeds. Priming decreases the storage life of seeds, so treatments should be applied just prior to planting rather than long-term storage. However, priming does not restore viability to poor quality seed lots with a high percentage of dead or severely damaged seeds.

Storing Seed

Seeds are one of the true wonders of nature because of their desiccation tolerance, small size, and storability. Storage life is partially determined by genetic factors since some species are more long-lived than others (Copeland and McDonald, 2001). Even though seeds are desiccation tolerant, they do change in dry storage. Optimized seed storage is an important aspect of maintaining seed quality for long periods. Seeds are hygroscopic, which means they may gain or lose moisture from the air (Copeland and McDonald, 2001). The longevity of stored seeds is determined by the combination of seed moisture content and temperature. Harrington's Rule illustrates the relationship between seed moisture, temperature, and viability. This rule states that the potential storage life of seeds doubles for every 1% decrease in seed moisture content over the range from 5–14% or 5°C (41°F) decrease in storage temperature from 0–50°C (32–122°F) (Harrington, 1963). From this example you can see that seeds should be kept as cool and dry as possible. If seeds are stored at moisture contents greater than 18%, damage may occur due to heat buildup from high respiration. Between 10% and 18% moisture, fungi and mold may grow on seeds. Between 9% and 14% moisture content, insects may be active. For open storage, starchy seeds should be held at less than 12% moisture content, while oily seeds like watermelon should be stored at moisture contents less than 9%. Sealed storage requires moisture contents from 5–8%. In some seeds, storage at less than 4% moisture content may be damaging. Some legume seeds are more susceptible to mechanical damage when dried to less than 10% moisture content (Copeland and McDonald, 2001). As a general rule, the best moisture content for maintaining viability of most vegetable seeds is 5% and the best temperature below 5°C (40°F). Seed longevity may be increased for germplasm preservation using cryopreservation. Since sudden freezing and ice formation may damage seeds, moisture contents should be less than 14% before seeds are frozen (Copeland and McDonald, 2001).

Containers for storing seeds

Seed packaging is important for maintaining quality by protecting from moisture, predators, and mechanical damage. Good containers are moisture proof and may be resealed once opened to keep seed moisture low. Containers made of paper or cloth are less preferable because they do not provide a moisture barrier. Cloth and paper containers may be broken more easily, so they are ineffective against predation from rodents or other animals. Packaging made of self-sealing foil or plastic is preferred for handling small quantities of seeds because they keep moisture low after opening. Metal cans with resealable plastic lids are effective containers for larger quantities of vegetable seeds because they deter rodents and moisture is kept out after opening, protecting unused seeds. Quantities of large-seeded legumes are often shipped in large bags. Plastic bags prevent seeds from absorbing moisture until opened. If cloth bags are used, they should be stored in a cool dry place.

Seed Industry

Not long ago most countries with significant crop production industries had one or more of their own seed companies. Many larger countries had at least one vegetable seed company that would specialize in developing cultivars for the nation or specific production regions within nations.

The number of companies that actually produce their own seeds and develop new cultivars has been decreasing. Like many nonagricultural industries, the seed industry has seen much consolidation and increased specialization. The consolidation has occurred for several reasons such as economies of size, the need to compete globally, high labor costs in developed countries, the high cost of cultivar development, fewer vegetable farmers in some countries, and the expense of new technologies.

For vegetable growers, it is important to understand who is developing and producing seeds. The vegetable seed industry is global with increasing international trade (Fig. 3.2).

The vegetable seed industry is the primary developer and user of plant breeding technology. Vegetable seed companies are constantly developing high yielding cultivars to better withstand biotic and abiotic stresses with fewer inputs. Seeds serve as the delivery system not only for improved genetics, but also for new planting and production methods and crop protection strategies that improve the overall efficiency of agriculture (Halmer, 2004; Romeis *et al.*, 2008).

Large multinational companies tend to dominate vegetable seed sales for commercial production. Such companies specialize in seed production and cultivar development. Some companies do not actually produce seeds or develop new cultivars but specialize in retailing seeds produced by other companies or treating seeds. So there are many types of companies that comprise the international vegetable seed industry.

There are successful small vegetable-seed companies as well. Some small companies have been successful in satisfying the demand for organically produced seed. Certain companies have catered to regional markets to fill the demand for open-pollinated and heirloom cultivars that are popular in some areas.

Cultivar Selection

Cultivar selection is one of the most important duties of a vegetable grower. Selecting high quality, high yielding, and disease-resistant cultivars with consumer appeal that are adapted to a particular area is challenging. Some public or nonprofit institutions provide unbiased cultivar recommendations based on evaluation trials for specific regions that can help growers select the best cultivar.

The following section describes some of the diverse terminologies used to described vegetable cultivars.

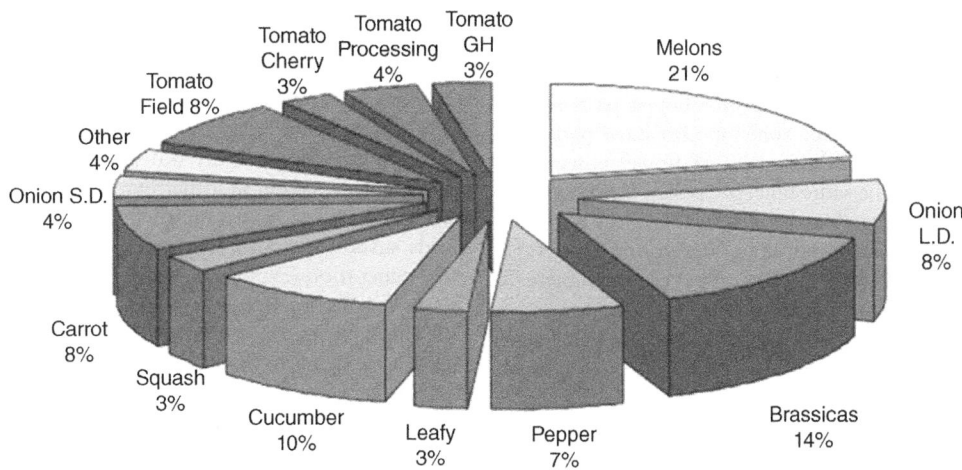

Fig. 3.2. Percentages of various vegetable seeds traded internationally (courtesy of Hazara Seed Company, Israel).

Heirloom vegetable

An heirloom cultivar is one that was commonly grown during earlier times in human history, but is not used for modern large-scale commercial production. Many heirloom vegetables have kept their traits through open-pollination and are not F-1 hybrids. The trend of growing heirloom plants in gardens has been growing in popularity in North America and Europe over the last decade. Some people grow heirlooms for historical interest, while others want to increase the available gene pool for a particular plant for future generations. Some grow heirloom cultivars due to an interest in traditional organic gardening. Some believe that the taste and quality is superior to modern commercial hybrid cultivars that were developed primarily to have good shipping qualities and high yield.

Open-pollinated

Many people now use this term rather loosely to simply refer to any cultivar that is a non-hybrid. A stricter definition is: "a heterogeneous cultivar of a cross-pollinated crop this is allowed to interpollinate freely during seed production". Open-pollinated cultivars will produce plants reasonably true-to-type but by nature there is more natural variation in an open-pollinated than a self-pollinated crop. Many heirloom cultivars of cross-pollinated vegetables, the cucurbits or brassicas for example, are open-pollinated.

Self- versus cross-pollinated crops

Many crops have perfect flowers that contain both male (anthers) and female (pistils) flower parts and therefore self-pollinate. Some species have perfect flowers but cross-pollinate because flower parts do not develop at the same time, the pollen is not compatible with stigmas on the same plant, or the pollen is sticky and most be moved by bees or other insects to adjacent flowers. Other crops cross-pollinate because they have separate male and female flowers or produce separate plants that are male or female.

Pure line

A self-pollinated crop that does not outcross is often referred to as a pure line cultivar because all the seeds were derived by self-pollination of a single flower, for example, lettuce or most beans. Pure line seeds are usually uniform and will grow true-to-type from seed.

Hybrids

The term hybrid has several meanings, each referring to the offspring of sexual reproduction (Rieger *et al.*, 1991). In terms of cultivar selection, F-1 hybrids are available for many vegetables. F-1 hybrids are made by sexually crossing two true-breeding homozygous plants to produce an F-1 hybrid generation (Fig. 3.3).

F-1 is short for Filial 1, which means "first offspring" (Rieger *et al.*, 1991). The F-1 hybrid produced is heterozygous, having two alleles, one contributed by each parent and typically one is dominant and the other recessive. The F-1 generation is also uniform and members of the population are very similar to one another. Increasingly the new cultivars developed by commercial seed companies are F-1 hybrids because growers prefer their superior vigor, uniformity and yield potential (Rieger *et al.*, 1991). F-1 hybrid vegetable cultivars express heterosis. Heterosis, or hybrid vigor, is the improved or increased function of any biological quality in hybrid offspring (Birchler *et al.*, 2003). There are two main theories as to why heterosis occurs. The dominance hypothesis attributes the superiority of hybrids to the suppression of undesirable (deleterious) genetic traits from one parent over the other (Birchler *et al.*, 2003). It attributes the poor performance of the inbred lines used to make F-1 hybrids to a loss of genetic diversity. The overdominance hypothesis says that some combinations of alleles (which can be obtained by crossing two inbred lines) are especially advantageous when paired together in one individual (Birchler *et al.*, 2003).

Seed companies also like F-1 hybrids because the seeds they produce are not suitable for producing another crop. In other words, commercial seeds grow true-to-type, but their offspring do not. Seeds saved from an F-1 hybrid plant will be the next generation (F-2) and usually produce variable plants that are inferior to their parents (Rieger *et al.*, 1991). Many of the F-1 hybrid seeds used today are more expensive than the open-pollinated seeds because they are more costly for seed companies to produce (Fig. 3.3). However, the productivity of an F-1 is often 30% or more greater from most vegetables so the return is worth the added investment.

Some crops are not amenable to mass production of hybrid seeds for commercial use. Lettuce, for

Fig. 3.3. Hand cross-pollination of two inbred cucumber lines to produce F-1 hybrid seed in Taiwan.

example, has perfect flowers and each floret produces a single seed, so emasculating and hand-pollinating each flower is not cost effective. Hybrid seeds are produced in some crops using "tricks" of nature to control the crosses that are made. For example, generic and cytoplasmic male sterility, genetic pollen incompatibility, and use of plant-growth regulating compounds to change floral sex expression are a few of the techniques used to control crosses for the production of F-1 hybrid seeds (Fig. 3.4; Rieger *et al.*, 1991).

In the case of lettuce, none of these techniques has been harnessed for mass production of hybrid seed, so all lettuce seeds are open-pollinated at this time.

Biotechnology

Biotechnology has become a cliché that is often used and means different things to different people. The term was first widely used in the 1980s to describe technologies that allowed foreign genes to be incorporated and expressed in organisms that were not closely related. Biotechnology in this sense was a significant development in plant science because, previous to its development, genes could only be incorporated through sexual crossing between relatively close relatives within the same genus. Some degree of the gene flow, often called horizontal gene transfer or lateral gene transfer, occurs naturally between plant species (Bock, 2010). Thus, biotechnology allows genes with beneficial traits to be systematically recruited from diverse plant species or even other organisms for expression in crop plants.

Many consumers falsely believe that most of the vegetables in the world today are transgenic. While it is technologically possible to transform vegetables with "foreign" genes from outside the genus or with synthetic genes using ballistics, *Agrobacterium*, or other biotechnologies (Khan, 2009), the reality is that relatively few of the commercial vegetable crops sold in the world are transgenic (Fig. 3.5).

Acceptance of genetically engineered crops varies with country. The discussion about the safety of genetically engineered foods is most intense in Japan and Europe, where public concern about transgenic food is higher than in other parts of the world such as the USA, Brazil, Australia, and China. In the USA, transgenic agronomic crops are commonly grown. However, despite government approval, transgenic vegetables are not widely grown because of a lack of consumer demand.

Fig. 3.4. A hybrid squash seed production field. Plants in the single male pollinator rows are larger. The ratio of female to male rows is six or eight to one. Plant sex expression is changed using plant hormones to ensure the female rows contain no male flowers that could self-pollinate and produce nonhybrid seed.

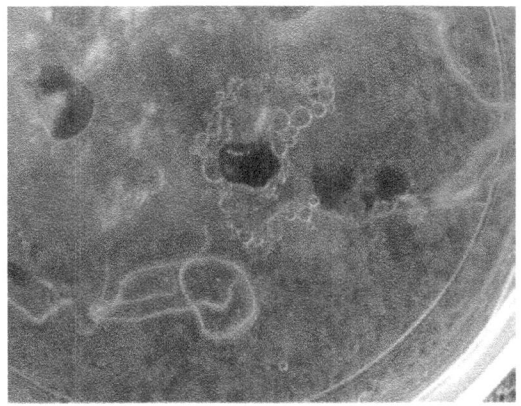

Fig. 3.5. Transgenic plantlet regeneration on selective tissue culture media. Genetic transformation of plants in the laboratory has become routine for many crop plants using *Agrobacterium* or ballistics, but government approval to grow transgenics for sale is a time-consuming and costly process in most countries. Very few transgenic vegetables are sold in the global marketplace.

What then is limiting the development and use of transgenic vegetables? Consumer acceptance is low in many areas because scientists have not convinced the public that transgenic crops are safe and do not pose an environmental risk. Also, some consumers believe that transgenic crops are "unnatural" while others do not like the large corporations that have developed transgenic crops. As consumers become more comfortable with transgenic crops they will likely appear in the marketplace. Also, it is expensive to develop, test, and gain government approval for the use of transgenic crops. Many companies are unwilling to incur developmental and approval costs until consumers are willing to purchase genetically engineered vegetables.

The concept of substantial equivalence was developed by the FAO and WHO to address the fears of consumers about the safety of transgenic crops. "Substantial equivalence embodies the concept that if a new food or food component is found to be substantially equivalent to an existing food or food component, it can be treated in the same manner with respect to safety (i.e. the food or food component can be concluded to be as safe as the conventional food or food component)" (UN-FAO, WHO, 1996). The statement of substantial equivalence basically acknowledges that it is unfeasible to test all the new crop cultivars released around the world every year for food safety. Only a few food crops on the market have been shown to cause adverse health effects and all of these were the result of conventional genetic modification and not from genetic engineering (National Research Council,

2004). Although there is scientific agreement that genetically engineered crops on the market are safe to eat (National Research Council, 2004), some scientists and advocacy groups call for additional and more rigorous testing before they feel marketing genetically engineered food is warranted.

Planting Vegetable Seeds

Many seeder types and models are available for direct-seeding vegetable crops. Smaller growers and gardeners typically use less expensive seeders, but larger growers need more sophisticated planters that precision-place seed. Precision placement eliminates the need for expensive thinning when seeds are drilled rather than precision planted.

Commercial seeders can be grouped into three categories: drills, plate planters, and precision planters. Drills can plant both small and large seeds, but cannot singulate seeds effectively because there is only a single orifice for seeds to pass through. These seeders can do a satisfactory job of planting when singulation is not required for crops such as radish, mustard, or turnip. Drills cannot deliver the low seeding rates required for some crops since there is little or no spacing adjustment in the row. Costly thinning would be required if close spacing is not desired.

Plate seeders offer more control over spacing because they pick up a single seed at a time in a cell on the plate and deposit it in the furrow. Most plate seeders work best when planting round seeds that are of medium to large size or pelleted. In-row seed spacing can be adjusted by changing spacing gears. Plate seeders work well for planting many types of raw or pelleted vegetable seeds. Horizontal, inclined and vertical plate designs are available. Selection of plate/cell size is critical to correct metering and singulation.

Precision seeders are the most accurate and do the best job of singulating and placing seeds accurately in a row. One design uses a revolving punched belt to accurately and uniformly space spherical or pelleted seeds. Belt seeders can also meter odd-shaped seeds effectively, but they will deliver multiple seed drops and/or skips more frequently if seeds do not accurately fit the belt holes. For maximum performance the belt hole size as well as the choke and base must be properly fitted to the seed or pellet size used.

Vacuum seeders are the other major type of precision seeders used for planting vegetable seeds. Vacuum seeders do a good job of singulating most vegetable seeds, especially oddly shaped seeds that cause problems for other seeder designs. Vacuum seeders have vertical plates with holes that are smaller than the seeds to be planted. The number of holes determines the seeding rate. The seeder has a blower that draws a vacuum on the side of the plate opposite the seed hopper. When the plate rotates through the seed hopper, the vacuum holds seeds against the holes and allows the rotating plate to pick up seeds (Fig. 3.1b). If the holes are the right size and the knock-off mechanism is adjusted correctly, one seed will be held on each hole. As the plate rotates the vacuum is broken, allowing the seeds to drop into the soil (Fig. 3.1). So-called air planters use the same basic principle except that the seed hopper is pressurized rather than pulling a vacuum on the other side of the plate.

Fluid drilling, sometimes referred to as fluid sowing or gel seeding, is a technique where germinated seeds are suspended in gel and transferred to a seedbed (Finch-Savage, 1981; Pill, 1991). This crop-establishment technique is used primarily in a field environment to overcome problems associated with conventional dry-seed sowing. For fluid drilling, seeds are germinated under ideal conditions, avoiding negative seedbed environmental effects such as insufficient moisture or high temperature, so seedling emergence is more rapid and synchronous. Faster seedling emergence should lessen the likelihood of soil crusting or pathogenic attack.

Fluid drilling is an integrated system that involves: (i) germination of seeds before sowing; (ii) separation of germinated and nongerminated seeds; (iii) storage of germinated seeds; (iv) preparation of the gel for suspending seeds; and (v) planting of germinated seeds suspended in gel using specialized equipment. Several types of gel materials are used: synthetic mineral clays, starch-polyacrylonitrile polymers, cellulose polymers, and copolymers of potassium acrylate and acrylamide, and natural gels of starch (Pill, 1991). A good gel should suspend seeds of various sizes for at least 24 h and yet be easily pumped through delivery tubing. Gels should be nonphytotoxic and mix easily with water of different pH and mineral content. Gel material should be relatively inexpensive, not dry to form a skin, and rapidly breakdown in the soil. Fungicides, herbicides, safening agents, beneficial organisms, and minerals are some of the common gel additives that have been reported to improve establishment. Various seed treatments are sometimes applied to increase the speed and uniformity of germination in preparation for fluid drilling (Pill, 1995).

Despite the benefits of fluid drilling, there are challenges as well. Specialized planting equipment is required, extra time is needed to prepare for planting, and seeds must be pregerminated. The use of primed seeds and plug transplants may deliver comparable performance to fluid drilling with less effort and expense (Pill, 1995). Once germinated, seeds have limited shelf life. If adverse weather prevents field establishment, germinated seeds must be stored with minimal radicle growth and loss of vigor. Germinated seeds have been successfully preserved in cold humid air, cold aerated water, aerated osmotic solutions, in plastic bags under partial vacuum, or nitrogen gas at 7°C (45°F), or in cold hydroxyethyl cellulose fluid drilling gel (Finch-Savage, 1981).

As the radicle of pregerminated seeds elongates, seeds progressively lose desiccation tolerance. Thus, the importance of adequate soil moisture at planting is essential for successful establishment. Seeds of low moisture content may be able to survive until soil conditions are favorable for germination and growth compared to pregerminated seeds whose desiccation tolerance has been lost and may perish if the soil is too dry. Although fluid drilling helps mitigate unfavorable soil conditions, planting pregerminated seeds into dry soils may result in poorer stands than if dry seeds were used.

Spacing seeds and transplants

It is increasingly difficult to provide stand recommendations for the ideal spacing of a particular vegetable crop because many variables are involved. The arrangement of plants in a field depends on how much space a crop needs to grow and develop properly, as well as the seeding, transplanting, and cultivation equipment available. Equidistant row spacing may give the highest yield for a particular crop but may not be suitable for accommodating the equipment needed to tend a crop or for promoting air circulation to minimize disease. Ideally, the arrangement of rows conforms not only to tractor wheel spacing, but also to equipment used to form beds, set transplants, control pests, and harvest the crop. Vegetables are increasingly grown on raised beds to ensure good drainage and to be compatible with certain irrigation systems. Spacing must be compatible with the raised bed format. Similarly, the use of plastic mulch may also impose spacing restrictions.

Plant genetics also play a major role in determining the appropriate spacing. For example, determinate, semi-determinate, and indeterminate tomato cultivars exist, so in-row and between-row spacing will vary depending on the cultivar selected as well as the training system used, i.e. ground culture, single stake, cage, string weave, etc. Tomato yields do not vary much on a field scale over a broad range of plant spacing because individual plants can compensate by producing fewer fruit at narrow spacing and more fruit per plant at wider spacing so the total yield remains the same. The same holds true for pickling cucumber production where the total yield may be similar for different plant populations, but early yield and fruits per plant vary.

Close spacing generally reduces commodity size. Exhibition pumpkins and squash need wide spacing to achieve their full genetic potential for fruit size. Pumpkins and squash for processing can be spaced closer to accommodate planting and harvest equipment since the largest possible fruit size is not the primary production goal. Some vegetables, such as asparagus, are more sensitive to spacing requirements and close spacing may reduce both yield and quality. For plants that produce a head like lettuce, close spacing may inhibit head formation and thus reduce marketable yields. Because so many factors impact plant spacing, no attempt is made to provide optimum spacing recommendations for individual vegetable crops. It is best to consider seed company recommendations based on cultivar characteristics and the production system used before determining the spacing requirements of a desired crop.

Transplanting vegetables

Transplanting, is another technique where seeds are sown outside a production field in a protective structure or special seedbed and the immature plants that develop are later replanted in a production field. This crop establishment technique is used to overcome problems associated with conventional field seeding, i.e. increased crop uniformity and turnover, increased use of marginal land, overcoming poor quality seed and seedling vigor, decreased pesticide use, decreased labor, and seed cost savings. For transplanting, seeds are germinated under ideal conditions, avoiding negative seedbed environmental effects such as insufficient moisture, high temperature, or predation, so seedling emergence is more rapid and synchronous. Use of transplants lessens the likelihood of soil crusting or pathogenic attack in the field and reduces the time the crop is in the field.

Vegetable transplants are often classified as being either bare root or having a ball of soil and roots attached. If soil remains around the roots, the chance of seeding damage and transplant shock is reduced. These transplants are increasingly grown in trays with individual cells so plants can be rapidly separated and planted. Transplant trays are also called plug trays and the transplants "plugs". Bare root transplants are historically grown in ground beds or cold frames and then pulled at the appropriate size for transplanting into a production field. In some areas, transplants are grown in warmer regions outdoors, shipped to northern areas, and field planted as early as possible to maximize the growing season. Bare root transplants can also be produced in cold frames or other protective structures in close proximity to a production field. Advantages of bare root transplants include lower production costs, increased root branching and, depending of the production system, hardier plants that are less succulent and more tolerant of environmental stress (Thompson and Kelly, 1957). Bare root plants are often hardier and more stress tolerant but are also subjected to greater stress due to root trauma during transplanting, which slows the resumption of growth in the field.

Many growers have made the transition to greenhouse-grown transplants using primarily plug trays (Fig. 3.6). With this system, each transplant grows in an individual cell so there is less competition among plants and greater uniformity.

Plug transplants often experience less shock during transplanting because there is less root disturbance and plants can often resume growth more rapidly in the field after transplanting. T-rail benches hold only the edges of trays in place so air circulates across the bottom of each tray. This system facilitates air pruning of roots so growth is confined to the cell, making it easier to pull plants for transplanting (Fig. 3.6). Water and nutrients are provided through overhead irrigation. Synthetic soils are frequently used to fill plug trays to improve drainage and root growth, and mitigate disease. Many synthetic soils used for transplants have small quantities of nutrients uniformly mixed in to ensure rapid seedling growth

Fig. 3.6. T-rail transplant production system in Israel. The spray boom is programmed to provide overhead irrigation by traveling above the beds. Edges of each tray rest on metal supports (i.e. T-rails) that allow air to circulate under benches to prevent root development below the tray.

after germination. Plug trays are sometimes sub-irrigated or floated on water or nutrient solutions in hydroponic production systems where root pruning is not required.

Any vegetable can be transplanted early in development if the root ball remains intact to avoid root damage and transplant shock caused by stresses in the new environment. As plants increase in size, transplanting is more difficult. Some plants are better adapted to transplanting than others (Table 3.1). Plants that form adventitious roots like tomato and cabbage can be bare-root transplanted, while other plants like cantaloupe and watermelon must be transplanted only as plugs or with the root ball intact. Other crops like sweet corn do not develop adventitious roots easily and transplanting is difficult even as plugs because of their sensitivity to transplant shock.

The reasons for transplanting vary with production area. The following crops are commonly transplanted to improve uniformity of harvest and reduce the time a crop occupies a field: the Brassicas (e.g. broccoli, cauliflower, cabbage) and lettuce, particularly heading types. Some vegetables are transplanted to accelerate maturity times so vegetables mature earlier in short-season areas: cantaloupe, mixed melons, pepper, tomato, and watermelon. Some plants are slow growing and compete poorly with weeds, so transplanting is more efficient and helps to ensure successful stands: celery, rhubarb, and globe artichoke.

Considerations for producing transplants

The cell size of transplant trays influences the field performance of plug transplants, especially earliness. When larger cells are used, the plant has more room to grow, so it is possible to produce an older, more mature transplant without it becoming spindly or root bound. In general, larger transplant trays result in earlier-maturing crops as long as they can be established without severe shock. Larger cells, however, take up more greenhouse space and are more expensive to grow. The optimum transplant age depends on the crop and tray cell size. In general, larger cells will enable production of larger, more mature transplants.

Tray effect on plant growth

Polystyrene transplant trays were once common but in many areas have been replaced with other materials. Polystyrene transplant trays are still used for float-bed systems (Fig. 3.7).

Polystyrene trays tend to be more expensive, stay cooler (thus slowing growth in some situations), less durable, promote algae growth, and harbor disease. Hard plastic trays are more popular because they last longer and are generally cheaper. Dark-colored trays absorb more heat and tend to produce faster growth than light-colored trays.

Deep-cell trays have larger cell volumes so more water and fertilizer are available to the plant. Deeper cells tend to promote faster growth and although they do not need watering as frequently as shallow cells, they will require more water to completely wet the media. With deep trays, it is important to water thoroughly and moisten all the media in each cell to promote deep root growth.

Growing media

Growing media, composed of various combinations of peat, vermiculite, and horticultural perlite, specifically formulated for germination is best suited for transplant production. Germination media tends to be well drained and does not provide

Table 3.1. A comparison of relative ease of vegetable transplanting.

	Transplant category	
Not well adapted to transplanting	Can be transplanted with care	Easily transplanted
Sweet corn, carrot, beans and peas, spinach, beet, turnip, radish	Watermelon, cantaloupe, mixed melons, lettuce, endive, celery	Potato, tomato, eggplant, pepper, cabbage, broccoli, cauliflower, onion, garlic, sweetpotato, asparagus
Transplanting is most successful when plants are small and soil is maintained around the root. Not adapted to bare root transplanting. Do not generate new roots easily. Damage to taproot may affect quality or marketable yield	Transplanting is most successful if plug transplants are used with the root ball intact. Severe stress should be avoided. Do not generate new roots prolifically	Plants generate new roots easily. Can be transplanted bare root. Tolerant of moderate stress

Fig. 3.7. Vegetable transplant production in a float bed production system.

a barrier to seedling emergence. Media containing coarser-textured (long-fibered) peat is free of disease, provides better drainage and aeration, and promotes transplant root development. Some soilless mixes contain small amounts of fertilizer often referred to as a "nutrient charge", which must be considered when designing a fertility program for transplants. A lower-nutrient charge in the media will allow more control over growth.

Seeding trays and seeding

Transplant trays are filled with premoistened growing media, before the media is compressed or "dibbled" to make a uniform surface for planting. The media should be compressed between 1–1.3 cm (0.4–0.5 in) deep for planting small-seeded vegetables and deeper for large-seeded vegetables like watermelon (Fig. 3.8).

One rule of thumb is that seeding depth should be at least three to five times the width of the seed. If seed is planted too shallow, certain crops, especially onions and brassicas, will tend to push up out of the cell without rooting properly.

The seeded trays should be uniformly covered with medium-grade vermiculite or the germination media used to fill trays. Seed of all vegetable species must be covered after seeding. Vermiculite works well as a tray covering because it can be evenly applied, is well aerated, is disease free, does not present a barrier to seedling emergence, does not support algae growth, and does not encourage roots to grow between cells.

After planting, trays should be watered to ensure the media is adequately hydrated and settled around the seed. Irrigation water should be heated during cold weather to raise the temperature of the soilless mix and start the germination process as soon as possible. If possible, water temperatures should be at least 21°C (70°F) when watering trays during germination and early seedling growth-stages. Good contact with soil is needed for rapid seed imbibition and emergence.

Temperatures should be maintained near the optimum for seed germination. Germination rooms or areas in a greenhouse should be set to optimize temperature and relative humidity, promote rapid germination in a confined area to minimize the cost of heating or cooling a large greenhouse to germination temperature. Heating pads or similar devices keep trays warm for rapid germination of warm-season vegetables. Air circulation is important to ensure uniform temperature and humidity throughout a germination room. Temperature should be regulated to minimize fluctuations that could delay germination and transplant development. Optimum temperature and time for germination varies among vegetable crops. Table 3.2 provides a range of germination and seedling growth temperatures for some commonly transplanted vegetables.

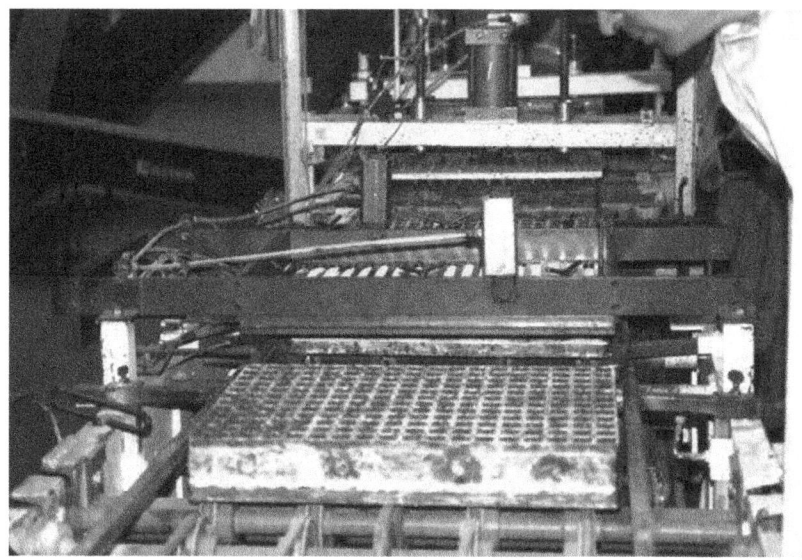

Fig. 3.8. Seed transplant trays for greenhouse production using automated equipment.

Table 3.2. Germination conditions for some of the major transplanted vegetables.

Crop	Optimum germination temperature °C (°F)	Approximate numbers of days to emergence[a]	Daytime growing temperature °C (°F)	Nighttime growing temperature °C (°F)
Brassicas	22–24 (72–75)	3–4	12–18 (54–64)	8–15 (46–59)
Cantaloupe	28–32 (82–90)	2–3	24–28 (75–82)	18–20 (65–68)
Celery	18–20 (64–68)	6–8	16–20 (61–68)	8–14 (46–57)
Cucumber	28–32 (82–90)	2–3	24–28 (75–82)	18–20 (64–68)
Lettuce	20–24 (68–75)	3–4	16–20 (61–68)	8–15 (46–59)
Pepper	26–30 (79–86)	4–6	20–24 (68–75)	18–20 (64–68)
Tomato	26–30 (79–86)	3–4	20–24 (68–75)	16–18 (61–64)
Onion	20–24 (68–75)	3–4	16–20 (61–68)	8–15 (46–59)
Watermelon	32–35 (90–95)	2–3	26–30 (79–86)	20–22 (68–72)

[a]Germination times vary widely among seed lots and with environmental conditions.

Warm-season vegetable crops (tomatoes, peppers, eggplant, and vine crops) are susceptible to chilling injury. Chilling occurs when transplants are exposed to temperatures above freezing but below 10°C (50°F) for an extended period. Chilling causes stunting of growth and may have a lasting effect on field establishment. For susceptible crops, it is important to maintain a minimum greenhouse temperature above 10°C (50°F).

Onion, cauliflower, and celery are biennial crops that are induced to flower if exposed to 13°C (55°F). Over-mature transplants that have advanced beyond the juvenile phase, which roughly corresponds to a pencil stem size, may be vernalized if exposed to 2–3 weeks of temperatures consistently below 13°C (55°F) during production, during transport, while waiting for transplanting, or in a field after transplanting. Vernalized plants flower prematurely when quite small, producing a low quality crop that yields poorly and is often unmarketable. Vernalization is generally not a problem because most high quality transplants are still juvenile at the time of transplanting and cannot be induced to flower.

Water quality

Good quality irrigation water is a key to successful transplant production. The pH of the water used for irrigating plug transplants should range from 5.5–6.5. At these values, micronutrients are more available. Water from ponds and wells may be alkaline (pH >7.0) and should be acidified to lower the pH.

Bicarbonates are a measure of water "hardness" and determine the amount of acid needed to lower pH. A water sample containing 100 ppm bicarbonates is considered "soft", while 400 ppm is considered to be very "hard". Although both samples may have the same pH, the 400 ppm sample would require more acid to "soften" the water. Many water sources have a bicarbonate level of 200–350 ppm and require treatment. The bicarbonate level of irrigation water is best in the 60–100 ppm range in order to avoid large changes in pH that occur when certain fertilizers such as ammonium are added (Bodnar and Garton, 1996).

Phosphoric and nitric acid are both used to acidify water. A volume of 7 l (1.9 gal) of phosphoric acid (85%) or nitric acid (67%) must be added per 100,000 l (26,417 gal) water to neutralize 60 ppm of bicarbonate (Bodnar and Garton, 1996). No more than 7 l (1.9 gal) of phosphoric acid should be used per 100,000 l (26,417 gal) of water to prevent excessive phosphorus fertilization. Nitric acid is the most common choice for neutralizing greater than 60 ppm bicarbonate. Each 7 l (1.9 gal) of nitric acid will also contribute 14 ppm of nitrate nitrogen to the water.

The EC (electrical conductivity) or total salts dissolved in water is another measurement of quality. Conductivities for untreated water should be less than 0.6 S/m (Siemens per meter) (0.6 mmho/cm) (milimhos per centimeter). Levels between 1.0 and 2.0 S/cm (1.0–2.0 mmho/cm) are good for transplant production (Bodnar and Garton, 1996). When fertilizer is dissolved in water the conductivity should be approximate 1.5 S/cm (1.5 mmho/cm). Lower conductivity readings may be an indication of nutrient levels that are too low to support plant growth, while higher conductivities may damage plants because of excessive amounts of salt.

It is important to know the nitrogen (N) content of irrigation water before extra N is added. Nitrate levels in untreated water are usually less than 20 ppm but may be as high as 50 ppm in some agricultural water sources. At levels >50 ppm, additional nitrogen should be used sparingly.

Watering transplants

The amount and frequency of watering will vary depending on the crop, cell type, growing media, greenhouse ventilation, and weather conditions. It is preferable to water plug transplants thoroughly in the morning, but not late in the day. If the plants remain wet overnight, disease problems may increase. If the plug is not watered thoroughly, root growth will be confined to the top of the plug. Trays should dry between watering to minimize disease, conserve water, and encourage hardy transplant development. Plants may wilt slightly but severe wilting should be avoided so that transplants are not damaged.

Fertility

A key to producing quality transplants is maintaining continuous vegetative growth from the greenhouse through field establishment. The fertility program used for vegetable transplants affects the quality of the finished transplant and its field establishment. A well-grown transplant will have adequate nutrient reserves to ensure rapid establishment under a wide variety of field conditions.

Vegetable transplants are usually fertilized through irrigation water. Fertilizers vary in their percent N, phosphate (P_2O_5), and potash (K_2O), and in micronutrient content (Table 3.3). Transplant fertilizers should have most of the N in nitrate form for rapid utilization and little in the form of ammonium or urea.

Phosphate is less water-soluble than other fertilizers and may precipitate out of solution. Many transplant fertilizers contain low- to medium-phosphate concentrations to keep phosphate soluble. Excessive phosphate (P_2O_5) may promote rapid seedling elongation under certain conditions. An alternative is to use a fertilizer with no phosphate (such as 14-0-14) for most feedings and apply a high-phosphate fertilizer periodically (once every four or five feedings) to promote growth. Withholding phosphate completely may produce transplants that have purple coloration along the stem and underside of leaves due to P deficiency. Fertilizer requirements vary depending on plant species, tray cell size (larger cells require less fertilizer), and the nutrient charge of the growing media (less fertilizer should be used if the media have a high nutrient charge).

Seeds and Stand Establishment

Table 3.3. Concentrations of N, P, K and electrical conductivity for 100 ppm solutions of some common water soluble fertilizers that are used for vegetable transplant fertility management (adapted from Bodnar and Garton, 1996).

Fertilizer analysis N-P-K	g/100 l water (oz/26 gal)	Parts per million			Conductivity S/cm (mmho/cm)[a]
		N	P	K	
20-20-20	50 (1.8)	100	43	83	0.40 (0.40)
20-10-20	50 (1.8)	100	21	83	0.60 (0.60)
20-8-20	50 (1.8)	100	17	83	0.75 (0.75)
15-5-15	67 (2.4)	100	14	83	0.70 (0.70)
14-0-14	71 (2.5)	100	0	83	0.85 (0.85)

[a]Electrical conductivity of a 100 ppm fertilizer solution made with distilled water in micromhos measured with a conductivity meter (Bodnar and Garton, 1996). The EC values vary depending on the background EC of the water source.

Fertility requirements of vegetable crops

Vegetable crops respond differently to fertilizer, so the feeding program must be tailored to a particular crop. For example, tomatoes are very responsive to fertilizer but excess fertilizer reduces transplant quality. If feeding at every watering, a common recommendation would be 50–100 ppm N, depending on the stage of plant development and environmental conditions. If feeding is weekly, concentrations of 250–350 ppm N would be more appropriate (Bodnar and Garton, 1996).

In contrast to tomato, Brassica crops require less fertilizer than other crops. One application per week of 100–150 ppm N should be sufficient under most conditions. Cucurbits have a relatively short growing cycle compared to other crops. Two to four applications of fertilizer at weekly intervals, at a 100–150 ppm N concentration, should be sufficient to produce good-quality cucurbit transplants (Bodnar and Garton, 1996).

With transplants, nutrient deficiency symptoms are more common than nutrient toxicities. Most growers are very cautious about over-fertilizing because of the cost of fertilizers and for fear of "burning" the plants. Many times, the application of fertilizer is delayed well beyond the time that transplants could benefit from fertilization. Plants with insufficient nitrogen have pale-green foliage. However, too much nitrogen will cause white stems, dark-green leaves, and tall, succulent, excessively vegetative plants.

Transplant height control

Height control is important because elongated transplants are more susceptible to damage during transplanting and stress after field planting. Height is important because some transplanter designs can only accommodate plants to a particular maximum size. Excessive stem elongation is caused by the following factors: too much heat, over fertilizing, overwatering, and low light conditions. High temperatures during the first 3–4 h after sunrise may cause rapid, succulent vegetative growth of vegetable transplants (Bodnar and Garton, 1996). Keeping the greenhouse cooler during the morning hours compared to the nighttime temperature can reduce poor quality growth.

There are a number of ways to control transplant height. Withholding water is commonly used to "harden" transplants to increase their dry matter and storage reserves, so they are more hardy and resistant to environmental stress during transplanting. Water can be withheld in the greenhouse or transplants can be moved outside for exposure to conditions similar to the field. Agitation from the wind promotes fiber development in the transplants. Transplant quality in a greenhouse can be improved using mechanical conditioning to substitute for air movement that stimulates fiber development (Latimer, 1998). Growth-retardant chemicals may be used for height control but plants may not resume rapid vegetative growth after transplanting, delaying crop development.

Cutting or clipping the tops with a mower or similar device can control the transplant height of some vegetables. However, cutting can cause unwanted branching that affects plant architecture. To discourage spread of disease, clipping transplants should be avoided. Transplants that were seeded at the right time and managed properly do not normally need to be clipped. Clipping is often done as a last resort to save transplants that have grown too tall to be planted by a carousel transplanter (Fig. 3.9).

Fig. 3.9. Carousel transplanter. Workers pull transplants from trays on racks and drop them into a revolving metal carousel that drops the transplants into a trench cut in the soil. Fixed shovels and press wheels under the worker's seat secure soil around the root ball.

Disease prevention

The primary means of controlling vegetable transplant diseases is by following good sanitation practices and by managing the greenhouse environment to suppress disease development (Jarvis, 1992). Controlling weeds inside and around greenhouses is important because they may also harbor disease organisms. If transplant trays are reused between crops, they should be washed to remove any soil or plug media that may adhere to the plastic and then disinfected to kill diseases and insects (Jarvis, 1992). A 1% solution of chlorine bleach is often used to disinfect trays although other commercial disinfectants are also widely used. Chlorine can be toxic to young seedlings, so bleach must be thoroughly rinsed off the trays before they are reused (Jarvis, 1992).

"Damping-off" control

"Damping off" is the term used for a number of different fungal diseases that can kill seeds or seedlings (Jarvis, 1992). Seeds are often sown in warm, wet conditions to speed germination and seedling growth for transplant production, but these conditions are also conducive to fungal attack that causes "damping off". Although "damping off" is often associated with *Pythium*, other diseases may also kill seedlings or transplants such as *Botrytis*, *Macrophomina phaseoli*, *Phytophthora*, *Rhizoctonia solani*, *Sclerotium rolfsii*, and *Thielaviopsis*. Different strategies can be used to control "damping off" including drier germination conditions, using sterilized/disease-free soil or potting mix, good air circulation, starting seedlings in sterilized soil, and use of seeds treated with a fungicide that prevents "damping off" such as metalaxyl, captan, or thiram (Jarvis, 1992).

Insects

Fungus gnats, shore flies, moth flies, and March flies live in damp, decaying vegetation, algae, and fungi that are often a problem in vegetable transplant greenhouses. Only fungus gnats commonly damage plants (Dreistadt, 2009). Larvae of these insects feed on roots, and since young seedlings have little root mass any feeding can stunt growth. In addition to larvae chewing on roots, both larvae and adults can spread plant pathogens such as *Pythium* and *Rhizoctonia* (Bodnar and Garton, 1996).

Avoiding overwatering, allowing surface soil to dry between watering and providing good drainage are important cultural practices for control. Keeping doors, vents, and windows closed or screened will prevent insects from flying into buildings. Using only pasteurized potting soil or treating potting soil with heat or steam kills flies as well as the algae and microorganisms they feed on (Dreistadt, 2009).

Natural predators, such as rove beetles (*Staphylinidae* family) and ground beetles (*Carabidae* family), help control fungus gnat larvae in greenhouses where broad-spectrum insecticides are not used. Commercially available *Steinernema* nematodes, *Hypoaspis* mites, or the biological insecticide *Bacillus thuringiensis* subsp. *israelensis* (Bti) can control fungus gnat larvae in container media (Wright and Chambers, 1994; Harris *et al.*, 1995).

Insect growth regulators (e.g. azadirachtin, kinoprene, diflubenzuron, cyromazine) applied to container media can be the most effective for controlling larvae (Dreistadt, 2009). Drenching media with an organophosphate (acephate, malathion) or carbamate (carbaryl) also kills larvae, but this can be hazardous and will kill many different organisms, including beneficial species (Dreistadt, 2009).

Algae

Algae can be a major problem in transplant production especially for transplants grown hydroponically or sub-irrigated when sunlight shines on the hydroponic solution. Shading hydroponic beds, using ebb-and-flood production systems and keeping nutrients out of the solution reduces algae growth.

In conventional transplant production with overhead irrigation, algae can become a problem on the surface of soilless mixes because it forms a crust, making it difficult to wet the medium. The use of medium-grade vermiculite spread over the surface of newly seeded trays will help prevent algal growth.

Algae tend to be a greater problem during cool, damp and overcast weather. The use of hydrated lime and certain registered materials like Agribrom (3-bromo-1-chloro-5,5-dimethylhydantoin) and copper hydroxide effectively control algae under and around container-grown plant benches, eliminating sites where fungus gnats and shore flies can live (Dreistadt, 2009).

Grafted transplants

Grafting has long been practiced by greenhouse growers and is increasingly used for field production by organic producers and as a methyl bromide alternative. Scions or tops of vegetable seeds are grafted on to a rootstock that differs genetically in order to: (i) provide soil-borne disease and nematode resistance; (ii) increase crop yields and quality; and (iii) improve the stress tolerance of plants (Kubota *et al.*, 2008).

The scion is the upper portion of the plant and is selected for its horticultural quality characteristics, while the rootstalk is selected for resistance to soil-borne disease, resistance to nematodes or stress tolerance. Rootstalks may be more stress tolerant because they possess an ability to mine the soil more efficiently for mineral nutrients or water. The tube grafting method (a small plastic tube or clip that holds the graft union together) is one of the most popular methods for creating vegetable transplants (Fig. 3.10). The grafting tube has a slit on one side so it falls off as the stem increases in diameter. Grafting is typically performed approximately 2–3 weeks after seeding.

Some rootstocks develop slower and may need to be sown earlier to achieve compatible stem sizes with the scion. Newly grafted seedlings are placed under high humidity (greater than 95%), ambient temperature (27–28°C/81–82°F) and low-light intensity for 4–7 days (Kubota *et al.*, 2008). After the graft union heals the transplants are acclimatized in a greenhouse prior to field planting.

Finishing and hardening transplants

Hardening transplants is important to minimize transplant shock, especially when planting under stressful conditions. Hardening acclimates plants by increasing dry matter accumulation and carbohydrate reserves. Vegetable transplants can be hardened using any of the following techniques either singly or in combination (Bodnar and Garton, 1996):

1. Reducing greenhouse temperatures through ventilation. However, the temperature should not be reduced below 10°C (50°F) on crops that are sensitive to chilling injury.
2. Expose transplants to air movement that agitates plants (Latimer, 1998).
3. Mild water stress, which causes the plants to wilt slightly but not excessively. Hardening transplants by withholding fertilizer can cause nutrient deficiencies that delay field establishment.
4. Holding plants outside for several days allows plants to become acclimated to the field conditions while they are still in the trays.
5. Treatments with ABA or other chemicals may harden transplants (Leskovar *et al.*, 2008).

Fig. 3.10. Grafted tomato transplants. The graft union is held together by a clip that is later removed before field planting.

References

Birchler, J.A., Auger, D.L. and Riddle, N.C. (2003) In search of the molecular basis of heterosis. *Plant Cell* 15, 2236–9.

Bock, R. (2010) The give-and-take of DNA: Horizontal gene transfer in plants. *Trends in Plant Science* 15, 11–22.

Bodnar, J. and Garton, R. (1996) *Growing Vegetable Transplants in Plug Trays*. Ontario Ministry of Food Agriculture and Rural Affairs, Harrow, Ontario.

Copeland, L.O. and McDonald, M.B. (2001) *Principles of Seed Science and Technology*, 4th edn. Kluwer Academic Publishers, Norwell, Massachusetts.

Dreistadt, S.H. (2009) Pest Notes: Fungus Gnats, Shore Flies, Moth Flies, and March Flies. *University of California Agriculture and Natural Resources Publication* 7448, 1–6.

Elias, S.G., Copeland, L.O., McDonald, M.B. and Baalbaki, R.Z. (2011) *Seed Testing Principles and Practices*. Michigan State University Press, East Lansing, Michigan.

Finch-Savage, W.E. (1981) Effects of cold-storage of germinated vegetable seeds prior to fluid drilling on emergence and yield of field crops. *Annals of Applied Biology* 97, 345–352.

FIS (1999) *Seed Treatment A Tool for Sustainable Agriculture*. International Seed Testing Federation, Nyon, Switzerland.

Fujii, J.A.A., Slade, D.T., Redenbaugh, K. and Walker, K.A. (1987) Artificial seeds for plant propagation. *Trends in Biotechnology* 5, 335–339.

Halmer, P. (2004) Methods to improve seed performance in the field. In: Bench-Arnold, R.L. and Sanchez, R.A. (eds) *Handbook of Seed Physiology. Applications to Agriculture*. Food Products Press, New York, pp. 125–166.

Harrington, J.F. (1963) Practical advice and instructions on seed storage. *Proceedings of the International Seed Testing Association* 28, 989–994.

Harris, M.A., Oetting, R.D. and Gardner, W.A. (1995) Use of entomopathogenic nematodes and a new monitoring technique for control of fungus gnats, *Bradysia coprophila* (Dipt.: Sciaridae), in floriculture. *Biological Control* 5, 412–418.

Jarvis, W.R. (1992) *Managing Diseases in Greenhouse Crops*. APS Press, St. Paul, Minnesota.

Khan, K.H. (2009) Gene Transfer Technologies in Plants: Roles in Improving Crops. *Recent Research in Science and Technology* 1, 116–123.

Kubota, C., McClure, M.A., Kokalis-Burelle, N., Bausher, M.G. and Rosskopf, E.N. (2008) Vegetable Grafting: History, Use and Current Technology Status in North America. *HortScience*, 43, 1664–1669.

Latimer, J.G. (1998) Mechanical Conditioning to Control Height. *HortTechnology* 8, 529–534.

Leskovar, D.I., Goreta, S., Jifon, J.L., Agehara, S., Shinohara, T. and Darrin Moore, D. (2008) ABA to enhance water stress tolerance of vegetable transplants. *Acta Horticulturae* 782, 253–264.

McSpadden Gardener, B.B. and Fravel, D. (2002) Biological control of plant pathogens: Research, commercialization, and application in the USA. Available at: www.plantmanagementnetwork.org/pub/php/review/biocontrol (accessed 28 September 2013).

Morsy, S.M., Drgham, E.A. and Mohamed, G.M. (2009) Effect of garlic and onion extracts on suppressing damping-off and powdery mildew diseases and growth characteristics of cucumber. *Egyptian Journal of Phytopathology* 37, 35–46.

National Research Council (2004) *Safety of Genetically Engineered Foods: Approaches to Assessing Unintended Health Effects.* National Academies Press, Washington, DC.

Pill, W.G. (1991) Advances in fluid drilling. *HortTechnology* 1, 59–65.

Pill, W.G. (1995) Low water potential and presowing germination treatments to improve seed quality: preplant germination and fluid drilling. In: Basra, A.S. (ed.) *Seed Quality: Basic Mechanisms and Agricultural Implications.* Food Products Press, New York.

Rieger, R., Michaelis, A. and Green, M.M. (1991) *Glossary of Genetics*, 5th edn. Springer-Verlag, Berlin.

Romeis, J., Shelton, A.M. and Kenney, G.G. (eds) (2008) *Integration of Insect-Resistant Genetically Modified Crops within IPM Programs.* Springer, New York.

Thompson, H.C. and Kelley, W.C. (1957) *Vegetable Crops*, 5th edn. McGraw-Hill, New York, 611 pp.

UN-FAO, WHO (1996) *Biotechnology and food safety: report of a joint FAO/WHO consultation.* Food and Agriculture Organization of the United Nations, Rome.

Welbaum, G.E. (2006) Natural defense mechanisms in seeds. In: Basra, A.S. (ed.) *Handbook of Seed Science and Technology.* Food Products Press, New York, pp. 451–473.

Welbaum, G.E., Shen, Z.-X., Oluoch, M.O. and Jett, L.W. (1998) The evolution and effects of priming vegetable seeds. *Seed Technology* 20, 209–235.

Wright, E.M. and Chambers, R.J. (1994) The biology of the predatory mite *Hypoaspis miles* (Acari: Laelapidae), a potential biological control agent of *Bradysia paupera* (Dipt.: Sciaridae). *Entomophaga* 39, 225–235.

4 Fertilization and Mineral Nutrition Requirements for Growing Vegetables

Introduction

Vegetables produce their own energy by photosynthesis using sunlight, carbon dioxide, and water, but require a fertile soil or growth media to supply mineral nutrients to live, grow, and reproduce successfully. Healthy, well-fed vegetable crops are better able to withstand diseases and insects and to compete with weeds. When the essential minerals required for plant growth and development are limited, vegetable quality suffers and yields decrease. Since vegetables are consumed directly, appearance is important and often times mineral deficiencies cause stunting, distortion, and color change that reduce quality and marketability. This is in contrast to most agronomic crops, which are harvested as grain and processed into foods. With agronomic crops mineral deficiencies affect yield but changes in appearance tend to be less critical.

Plant nutrients are classified into two categories: macronutrients and micronutrients. Macronutrients are those mineral elements that are needed in relatively large amounts and can be expressed as a percentage of the plant's dry weight. They include nitrogen (N), carbon (C), oxygen (O), hydrogen (H), potassium (K), sulfur (S), calcium (Ca), magnesium (Mg), and phosphorus (P) (Table 4.1). Micronutrients are those elements that plants need in small amounts, usually in the parts per million ranges, like iron (Fe), boron (B), manganese (Mn), nickel (Ni), zinc (Zn), copper (Cu), chlorine (Cl), and molybdenum (Mo) (Table 4.1).

Preferred Soils for Growing Vegetables

Except for certain aquatic vegetables like wasabi, watercress or Chinese water spinach, most vegetables prefer well-drained soils that are high in organic matter. As a general recommendation, well-drained loam is recommended for vegetable production. A loam soil is composed of sand, silt, and clay in relatively even concentrations (about 40-40-20% concentration, respectively; Kaufmann and Cutler, 2008). In actuality, many vegetables can be grown on a broad range of soil types, including sand, sandy loam, silty loam, clay loam, sandy clay loam, silty clay loam, and clay, as long as they are well drained and can be irrigated during dry periods.

Organic matter is an important soil constituent. Natural organic matter comes from decomposed plants and animals in the environment. For example, decomposing manure, animals, plants, and other living organisms and their remains collectively contribute to the pool of natural organic matter in soil. When organic matter decays to the point where it is no longer recognizable it is called soil organic matter (Kaufmann and Cutler, 2008). Therefore, soil organic matter comprises all of the organic matter in the soil exclusive of the material that has not yet decayed (Juma, 1999).

Natural organic matter enriches soils with minerals by binding to them for later release or by releasing them during decomposition. Once bound to organic matter, the minerals and metal ions are less likely to move through the soil during leaching rains or irrigation. Organic matter improves vegetable production, because it improves fertility by slowly releasing mineral nutrients to the crop, increases water retention and improves soil structure. In medium and light sandy soils, organic matter helps to hold moisture and nutrients.

When the organic matter has broken down into stable humic substances that resist further decomposition it is called humus. Humic substances are complex and heterogeneous mixtures of polydispersed materials formed by biochemical and chemical reactions during the decay and transformation of plant, animal, and microbial remains through a process called humification. Plant lignin

Table 4.1. General deficiency symptoms for the essential elements (except for C, H, O). Soil testing and plant analysis should be used to confirm infield diagnoses. Deficiency symptoms for a particular crop may vary dramatically depending on the crop investigated. For this reason, the deficiency symptoms listed here should only be used as a guide (McCauley et al., 2003).

Nutrient	Dry matter[a]	Deficiency symptom
Macronutrients		
Carbon (C)	C accounts for 40–60% of dry weight of plants depending on the species and tissues sampled.	C accounts for the majority of dry mass in plants. The source of C in plants is CO_2, which plants absorb from the atmosphere through their stomatal pores.
Oxygen (O)	C, O, H combine to account for 90% or more of plant dry weight.	O is used by respiring plants and given off as a byproduct of photosynthesis and therefore is not limiting.
Hydrogen (H)		H is derived from water and as an element does not limit plant growth.
Nitrogen (N)	2.0–6.0%[b]	Restricted growth of tops and roots; growth upright and spindly; leaves pale yellowish-green in early stages, more yellow and even orange or red in later stages; deficiency shows up first on lower leaves.
Phosphorus (P)	0.2–0.5%[b]	Restricted growth of tops and roots; growth is upright and spindly; leaves bluish-green in early stages with green color sometimes darker than plants supplied with adequate phosphorus; more purplish in later stages with occasional browning of leaf margins; defoliation is premature, starting at the older leaves.
Potassium (K)	1.2–5.0%[b]	Browning of leaf tips; marginal scorching of leaf edges; development of brown or light colored spots in some species, usually more numerous near the margins; deficiency shows up first on lower foliage.
Calcium (Ca)	0.5–2.0%[b]	Deficiency occurs mainly in younger leaves near the growing point; younger leaves distorted with tips hooked back and margins curled backward or forward; leaf margins may be irregular and display brown scorching or spotting.
Magnesium (Mg)	0.2–0.5%[b]	Interveinal chlorosis with chlorotic areas separated by green tissue in earlier stages giving a beaded streaking effect; deficiency occurs first on lower foliage.
Sulfur (S)	0.2–0.5%[b]	Younger foliage is pale yellowish green, similar to nitrogen deficiency; shoot growth somewhat restricted.
Micronutrients		
Zinc (Zn)	15–50 ppm[b]	Interveinal chlorosis followed by die-back of chlorotic areas.
Manganese (Mn)	20–200 ppm[b]	Light green to yellow leaves with distinctly green veins; in severe cases, brown spots appear on the leaves and the leaves are shed; usually begins with younger leaves.
Boron (B)	10–50 ppm[b]	Growing points severely affected; stems and leaves may show considerable distortion; upper leaves are often yellowish red and may be scorched or curled.
Copper (Cu)	2–20 ppm[b]	Younger leaves become pale green with some marginal chlorosis.
Iron (Fe)	25–150 ppm[b]	Interveinal chlorosis of younger leaves.
Molybdenum (Mo)	1–2 ppm[b]	Leaves become chlorotic, developing rolled or cupped margins; plants deficient in this element often become nitrogen deficient.
Nickel (Ni)	1–10 ppm[b]	Ni deficiency symptoms are not well established. Ni-deficient plants produce poor quality seeds that do not germinate well and have low vigor.
Chlorine (Cl)		Deficiency not observed under field conditions.

[a]Nondeficient range in normal plants
[b]Values for leaf tissue

and its transformation products, as well as polysaccharides, melanin, cutin, proteins, lipids, nucleic acids, fine char particles, etc., are parts of the humification process as well. One of the advantages of humus is that it holds water and nutrients that plants can use for growth. Another advantage of humus is that it improves soil structure (Juma, 1999). Soil structure describes the arrangement of the solid parts of the soil and of the pore spaces located between them (Marshall and Holmes, 1979). Healthy productive vegetable soils are often high in organic matter and humus. Managing vegetable soils so that they maintain or increase their humus and organic matter fraction is an important part of successful vegetable production.

There are techniques for increasing the amount of humus in production fields. Adding compost, plant or animal manures, or green manure to soil will increase the amount of humus in the soil. These three materials supply nematodes and bacteria with nutrients needed for them to thrive and convert organic matter into humus (Crow *et al.*, 2009).

Muck, peat, or organic soils are composed largely of plant material at various stages of decomposition (Fig. 4.1). These are soils that developed over many thousands of years as lake beds filled with moss and other organic debris, which gradually dried and decomposed to create a soil that has a very high percentage of organic matter and/or humus. The terms muck, peat, and organic are often used interchangeably to describe soils that are composed of at least 50% organic matter and humic substances (Thompson and Kelly, 1957). In some cases, muck and peat soils are differentiated by the structure of their organic fraction. If the structure of organic matter is evident they are called peat, but if the structure of the organic fraction is no longer evident then they are referred to as muck soils (Thompson and Kelly, 1957). Characteristics of muck and organic soils include the following. These soils have a brown or black color that grows darker with the stage of decomposition. They have high moisture-holding capacity and can absorb several times their dry weight of water. They are very high in N content: 2.5% or more of their dry weight is N. They contain low concentrations of some mineral nutrients such as potash, phosphate, and certain micronutrients. They oxidize over time at a rate of 2.5 cm (1 in) or more per year. Muck soils disappear over time and shallow deposits are eventually lost. Some muck production areas are noticeably lower than the surrounding mineral soils due to oxidation over time. Many salad and root vegetables, such as radish, potato, beets, carrot, celery, spinach, lettuce, etc., that do not go through a reproductive phase and have a high N requirement are productive on organic or muck soils.

Fig. 4.1. This is a muck soil production field near Celeryville, Ohio. Muck soils are flat black in color and composed of at least 50% organic matter. They are well suited for the production of many vegetables, such as celery, lettuce, spinach, onions, and radish. The windbreak on the left helps prevent erosion.

Soil Testing

One essential aspect of vegetable product is ensuring that the crop is adequately supplied with mineral nutrients throughout the growing season. A soil's fertility can be checked by chemical soil analysis to determine if any corrective action is needed. A soil test is the best tool for determining the lime and fertilizer needed for the best economic production of vegetables. Testing the soil removes guesswork and provides the information needed to select the right amount of fertilizer and lime to maximize productivity without wasting resources. Fertilizer is a substance that contains one or more recognized plant nutrients and promotes plant growth. Soil samples should be collected for testing prior to planting. Random soil samples should be taken at several locations throughout the field. Kits are available that enable growers to test their own soils. These are generally not as accurate as tests conducted by professional laboratories, however "home kits" give a general idea of the nutritional status.

A routine soil test normally provides a fertility evaluation and sufficient information to choose fertilizers to grow a crop. Fertilizer recommendations are based on potential crop yield. Since yields vary from soil to soil, information on your soils will enable the soil-testing lab to make a customized recommendation for your field. Soil information is available from government services in many countries. When soil classification maps are not available, an estimated yield can be provided as follows: average the three highest yields achieved over the last 5 crop years and the particular crop that was grown in the field (i.e. exclude the two lowest crop yields before calculating the average). Professional soil labs analyze soil samples, and computer recommendations on the fertility inputs needed to optimize production are returned to the grower in a matter of days for a reasonable cost. A standard soil test includes assessments of soil pH, P, K, Ca, Mg, Zn, Mn, Cu, Fe, B, and estimated cation exchange capacity plus a fertilizer and lime recommendation tailored to the crop to be grown and soil type. Note than an assessment of soil N is not provided because it is very difficult to estimate because of its dynamic nature and many different forms in the soil. If requested, two common additional tests include a soluble salts test to determine whether fertilizer salts are too high and an organic matter test that determines the percentage of organic material in soil.

In addition to the availability of mineral nutrients in the soil, other information is also provided by soil tests. Soil pH (or soil reaction) measures the "active" acidity in the soil's water (or hydrogen ion activity in the soil solution), which affects the availability of nutrients to plants. Maintaining proper soil pH is one of the most important vegetable production practices because it impacts both nutrient availability and nutrient-related toxicity.

How Acid Affects Soils

Vegetables grown in acidic soils exhibit a variety of symptoms including aluminum (Al), H, and/or Mn toxicity, as well as deficiencies of Ca and Mg (Nyle and Weil, 2008). Below pH 4, H ions damage root cell membranes. Aluminum is present in most soils, and toxicity is a widespread problem in acidic soils. Al^{3+} is most soluble at low pH and when dissolved in the soil solution is toxic, but above pH 5.2 little Al is soluble in soils, so plants are unaffected. Aluminum is not a nutrient but causes damage to plants by interfering with Ca uptake, binding with phosphate and interfering with production of adenosine triphosphate (ATP) and DNA, both of which contain phosphate.

In soils high in Mn, toxicity can become a problem at pH 5.6 and below. The solubility of Mn increases as pH drops, much like Al, so Mn toxicity symptoms are most common at pH 5.6. Classic symptoms of Mn toxicity are crinkling or cupping of leaves.

Nutrient Availability in Relation to Soil pH

Both trace and macronutrient availability is controlled by soil pH. In slightly to moderately alkaline (pH greater than 7.0) soils, Mo and macronutrient (except P) availability is increased, but P, Fe, Mn, Zn, Cu, and Co levels are greatly reduced and may limit plant growth (Havlin *et al.*, 1999). In acidic soils, micronutrient availability (except Mo and B) is increased. Nitrogen may be applied in fertilizer as ammonium (NH_4) or nitrate (NO_3) and dissolved N is most available in soil in a range of pH 6–8, but N availability is less sensitive to pH than other elements such as P. In order for P to be available, soil pH needs to be in the range 6.0–7.5, because at below pH 6.0, P starts forming insoluble compounds with Fe and Al (Havlin *et al.*, 1999). When pH is greater than 7.5, insoluble compounds are formed with Ca.

Liming Soils to Adjust pH

The majority of vegetable crops have a target pH range of 6.2–6.8. Lime applications, based on soil-test results, will adjust or maintain the proper pH. In agriculture, lime is usually defined as compounds containing Ca or Ca-Mg, capable of reducing harmful effects of an acidic soil by raising the soil pH (Mengel *et al.*, 2001). Lime benefits soils because it reduces concentrations of soluble Al and Mn that are harmful to plants and increases the availability of several essential plant nutrients.

Commercial soil labs provide fertilizer recommendations for specific crops based on soil-test results. However, there are different philosophies on how these recommendations should be made. It is not uncommon to receive vastly different fertilization recommendations from different labs for the same soil sample. For example, some labs assume that you need to build up the level of some nutrients like K to maintain a sufficient level in the soil, while other labs will only recommend the amount of fertilization needed to produce a crop for a single season.

Gaseous Plant Nutrients

Plants have been called the "lungs of the earth" because they produce the oxygen that humans and animals breathe. Animals, including humans, rely on plant-derived oxygen for life and exhale carbon dioxide (CO_2). Plants have a remarkable ability to take CO_2 from the atmosphere and convert it into dry matter (Fig. 4.2).

This process is called photosynthesis, which uses only light, water (H_2O), and CO_2 to make simple sugars (Taiz and Zeiger, 2010a). Water and CO_2 are nontraditional plant nutrients that are required for normal growth and development. They are nontraditional because, unlike most mineral nutrients that are derived from the soil, CO_2 is an atmospheric gas and H_2O is derived from soil primarily as a liquid (Fig. 4.2). During photosynthesis, H_2O is split into hydrogen and oxygen, which is released into the atmosphere as a gas, while hydrogen is chemically combined with CO_2 to make simple sugars, the basic building blocks of more complex compounds (Fig. 4.3; Taiz and Zeiger, 2010a). The energy to drive these chemical processes comes

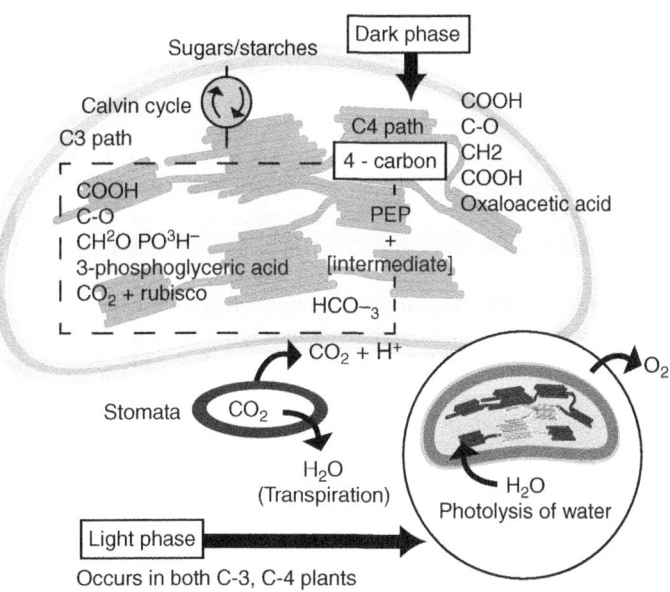

Fig. 4.2. Diagram of a chloroplast inside a green plant leaf. Atmospheric CO_2 enters through the stomatal pores on the surface of plant leaves. The CO_2 combines with H^+ produced when water is split by light to give off O_2 as part of the light phase of photosynthesis. The HCO molecules created are the basic building blocks of photosynthesis and are converted to more complex compounds.

Fertilization

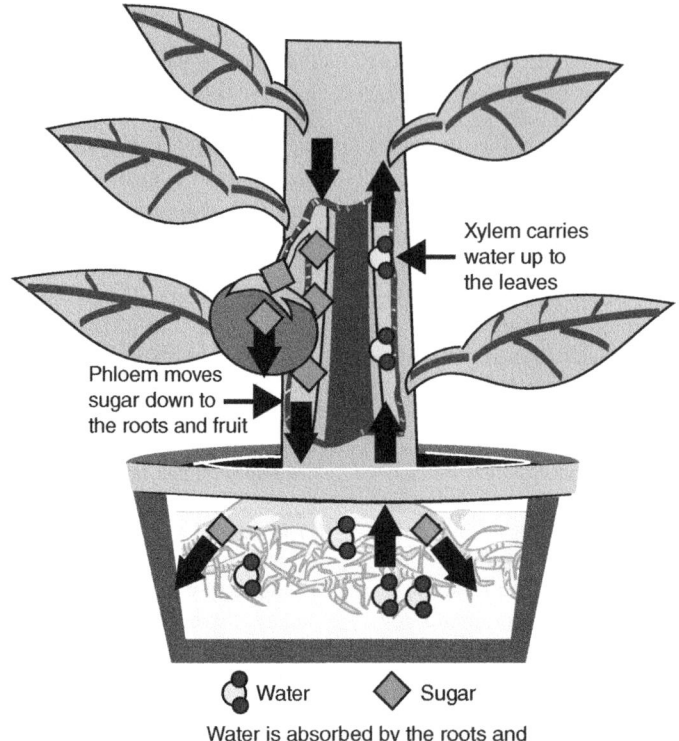

Fig. 4.3. Through the process of photosynthesis, the leaf chloroplasts manufacture sugars from CO_2 that are transported through phloem conduits to roots, growing tissues, and fruit. Water moves from the roots throughout the plant through other conduits called xylem. Plant macro- and micronutrients are dissolved in the water taken up through the roots.

from light (Fig. 4.3). Sugars are transported from the leaves to other organs throughout the plant through small veins called phloem.

The sugars are modified into tens of thousands of different compounds after translocation to destination tissues. These complex carbon compounds are used by plants for their own growth and development and, in the case of vegetables, form an important part of the human food chain.

The CO_2 global average concentration in Earth's atmosphere is roughly 0.039% by volume or 387 parts per million (ppm) by volume and rising. Plants acquire CO_2 from the air through tiny holes in the leaves called stomata (Fig. 4.2). The opening and closing of stomata is controlled by guard cells that modulate gas exchange between the atmosphere and open spaces inside the leaves. If water is limited, stomata close to conserve water, but as an unavoidable consequence photosynthesis is also reduced because insufficient CO_2 is acquired (Taiz and Zeiger, 2010a). In a field environment, there is sufficient CO_2 in the atmosphere so that its availability does not limit photosynthesis if the plant is well hydrated and stomata are fully open. However, in a closed greenhouse, CO_2 may be limited, so supplemental CO_2 is sometimes added as a gaseous nutrient to optimize photosynthesis.

About half of the sugars produced during photosynthesis are used to fuel a plant's growth. Scientists estimate that, globally, photosynthesis removes approximately 100 billion metric tonnes (110 billion short tons) of carbon from the atmosphere and sequesters it in plants each year. As ambient atmospheric CO_2 levels increase, a benefit could be increased photosynthesis and carbon sequestration by plants, if other factors like temperature and water availability remain the same. Higher atmospheric CO_2 could possibly result in increased yields of vegetable crops based on analyses of other crops (Long *et al.*, 2006; Ainsworth, 2008).

Rubisco, the protein C-3 plants use to "fix" CO_2, is the most abundant protein on earth (Fig. 4.2). The sugars produced by photosynthesis are translocated to other organs in the plant where they serve as food for respiration and plant growth and development. Most vegetables are C-3 plants. Corn and other grasses use a slightly different process for fixing carbon and are called C-4 plants (Fig. 4.2). The C-4 plants are generally more drought-tolerant because they use water more efficiently (Taiz and Zeiger, 2010b).

Respiration can be thought of as photosynthesis in reverse (Fig. 4.4). Respiration releases chemical energy stored in sugar (Fig. 4.4). Sugars are broken apart to harvest chemical energy that is used by plants for cell maintenance and growth. In this process, both CO_2 and H_2O are released as breakdown products and oxygen is consumed. Respiration always occurs both day and night (Fig. 4.4).

How do photosynthesis and respiration relate to vegetable production? The ability of vegetables to perform photosynthesis is a primary factor that determines both yield and quality. Carbon contributes from 40–60% and the elements C, H, and O combined account for over 90% of the dry weight of plants. Therefore, any factor that reduces photosynthesis also reduces dry weight accumulation, which is another way of saying yield. While some plants such as spinach tolerate shade, as a general rule vegetable crops are sun-loving plants that perform best in full sunlight because that is where photosynthesis is maximized. Some of the things that reduce photosynthesis include drought stress through stomal closure and smaller plants with less leaf area, plant-to-plant competition, nutrient deficiencies, reduced leaf area due to insect predation, plant disease, and reduced sunlight due to shading and cloudy weather.

Melons are a good example of how photosynthesis can affect vegetable quality. Sugar accumulation is the primary flavor component in melons. The soluble sugars that make fruits sweet come from photosynthesis, so any significant decrease will cause poor flavor. Drought stress (through stomatal closure), extended periods of cloudy weather, insect damage to leaves, leaf diseases, shading, and weed competition are some common factors that can reduce photosynthesis of a melon crop and consequently fruit quality. Since respiration consumes sugars, factors that increase respiration reduce sugars and decrease fruit quality (Fig. 4.4). Using cantaloupes as an example, high night temperatures increase respiration, causing loss of stored sugars and reductions in melon sugar content. Irrigated desert regions are often preferred for melon production because photosynthesis and sugar accumulation are maximized

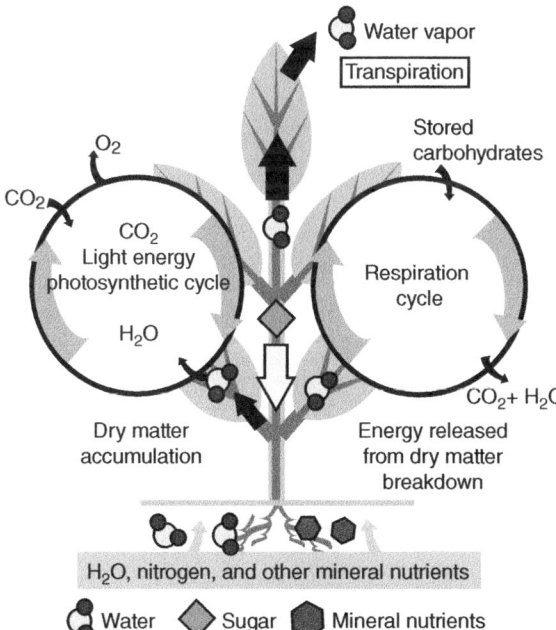

Fig. 4.4. Respiration breaks down stored molecules to provide energy. Respiration gives off CO_2 and water. So photosynthesis accumulates storage reserves including sugars while respiration consumes them. Humans and plants both respire. Humans must eat food while plants make their own through photosynthesis.

by sunny cloudless days and the cool night temperatures reduce respiration and favor sugar retention.

Plant Mineral Nutrients and Deficiencies

Mineral nutrients can be classified as mobile or immobile. Mobile nutrients move inside plants to where they are most needed. For example, a plant can reallocate mobile nutrients from its older leaves to younger ones. So when nutrients are mobile, nutrient deficiencies first appear on older leaves. However, not all nutrients are mobile. When immobile nutrients are lacking, the younger leaves show symptoms before older ones because the nutrients are fixed in the tissue and cannot move to where they are most needed.

Mineral nutrient mobility can be a valuable tool for diagnosing deficiency symptoms in vegetables (Figs 4.5 and 4.6).

However, visual symptoms sometimes vary between species and even cultivars within a species. Also, by the time visual symptoms appear a crop is often severely deficient, stressed, and damaged. Chemical analysis techniques discussed later in

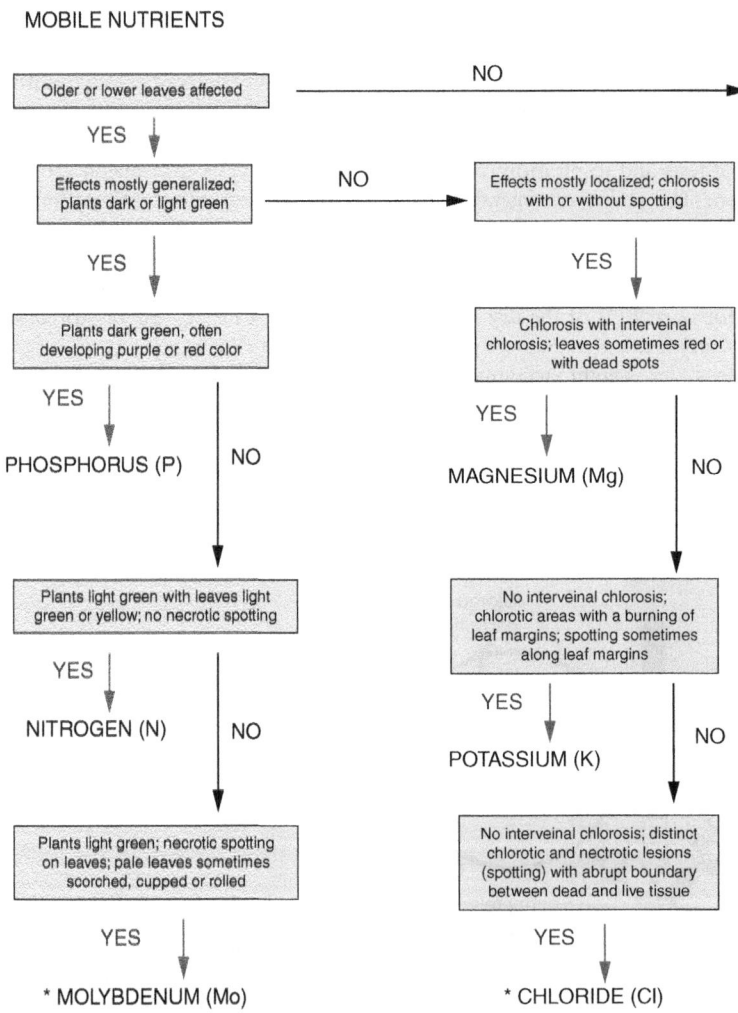

Fig. 4.5. Diagnosing deficiency symptoms of mobile plant nutrients (McCauley *et al.*, 2003 with permission).

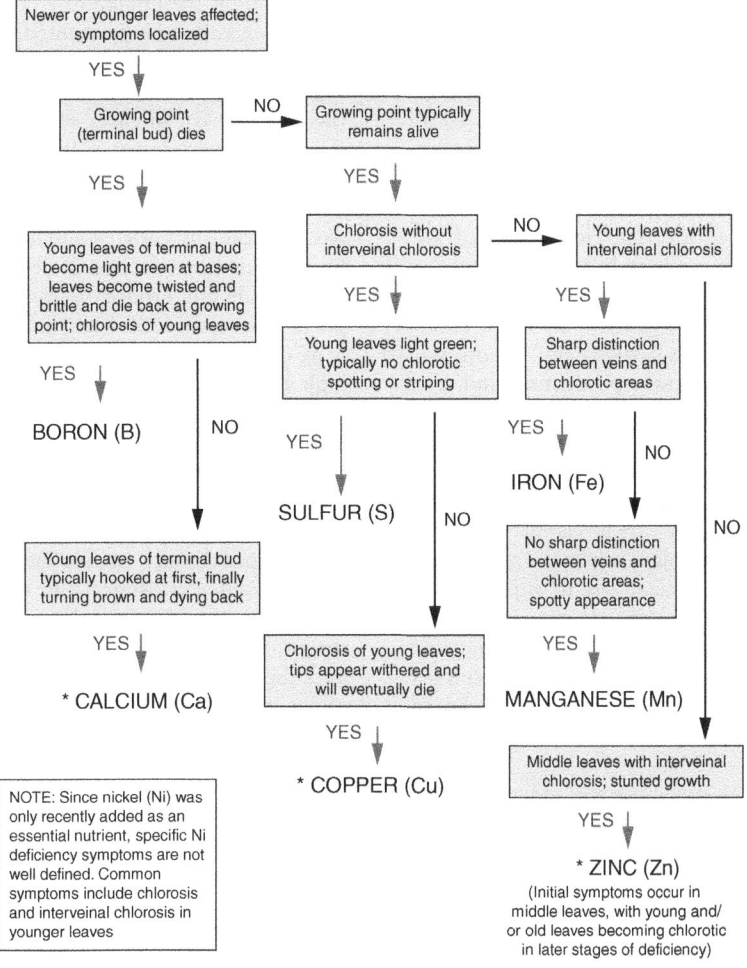

Fig. 4.6. Diagnosing deficiency symptoms of immobile plant nutrients (McCauley *et al.*, 2003 with permission).

this chapter are more accurate and can detect limiting nutrients before visual symptoms occur. Despite the limitations, visual identification is an important management skill that can help identify poorly nourished plants rapidly so corrective measures can be taken. Nitrogen, P, K, Mg, Mo, and Cl are considered mobile nutrients, while the other essential nutrients are immobile. While dividing nutrients into mobile and immobile categories is helpful for diagnosing symptoms, in actuality some minerals like Zn are semi-mobile and have intermediate properties between the two categories as will be discussed.

Mobile Nutrients

Nitrogen is a constituent of plant proteins, nucleic acids (DNA and RNA) and chlorophyll. Symptoms of N deficiency are general chlorosis of lower leaves (light green to yellow), stunted and slow growth, and necrosis of older leaves in severe cases. Nitrogen-deficient plants will mature early and crop quality and yield are reduced depending on the severity of the deficiency (Jones, 1998). Fields deficient in N can be either uniform or patchy in appearance, depending on conditions favoring the deficiency.

Nitrogen fertilization can also have a significant effect on plant growth and development. Excess N turns plants a deep green color and stimulates vegetative growth, often delaying maturity. Vegetables that consist primarily of vegetative organs, such as lettuce, cabbage, spinach, and celery, require relatively large amounts of N throughout the entire growing season. However, even with these vegetables excessive N fertilization may cause succulent growth and reduced storage life.

With vegetable crops that are harvested for reproductive tissues, such as broccoli, tomato, or melon, N also stimulates vegetative growth at the expense of reproductive growth. In these crops, excessive N fertilization early in the season may delay or even inhibit flower production and fruit set. Nitrogen status is often discussed in terms of the C-to-N ratio (C:N) of the plant tissue. The optimum nutrient status for some crops changes during the growing season. With tomato, for example, moderately low C:N levels early in development are best to stimulate vegetative growth, so the plant develops sufficient size to support fruit development during the reproductive cycle. Just prior to flowering, the C:N ratio should increase to promote flowering and fruit set. After fruit set or flower initiation, the C-to-N ratio should ideally fall to stimulate growth of the developing fruit.

High N can delay flowering and reduce fruit set in crops like tomato, pepper, cantaloupe, squash, and pumpkin. This results in delayed maturity and reduces early market yields, which are important for profits in short-season areas. Since N stimulates vegetative growth, excess N results in tall plants with weak stems and contributes to lodging (plants falling over). Excessive N causes new growth to be succulent with lower dry matter. Nitrogen toxicity is possible under dry conditions and may burn delicate tissues. Plants fertilized with ammonium (NH_4^+)-based fertilizers may exhibit ammonia toxicity, which is manifested in reduced growth, lesions on the stems and roots, and downward rolling of leaf margins (Mengel *et al.*, 2001).

Plants require P to produce ATP for energy storage, for sugar accumulation and as a component of nucleic acids (Jones, 1998). Phosphate deficiency symptoms are usually more noticeable in young plants, which have a greater relative demand for P than more mature ones (Grundon, 1987). Therefore, P deficiency is often seen on transplants in both the greenhouse and field because of insufficient fertilization and cool soils after transplanting, which inhibit the uptake of P. Phosphate-deficient plants generally appear stunted, with older leaves that turn purplish due to the accumulation of sugars, which favor anthocyanin synthesis (Bennett, 1993). In cases of severe deficiency leaf tips will brown and die. Plants suffering from P deficiency appear weak and maturity is delayed. Leaf expansion and leaf surface area may also be reduced, causing curled small leaves (Jones, 1998). From a field perspective, P deficiency is likely to occur on highly eroded or weathered soils, or in areas where subsoils high in calcium carbonate ($CaCO_3$) have been exposed. Crops grown in soils high in $CaCO_3$ are prone to P deficiency due to the precipitation of Ca-P insoluble minerals (Havlin *et al.*, 1999).

Plants need K to activate enzymes and coenzymes (specialized proteins serving as catalysts and co-factors). Potassium is a key nutrient for stress tolerance such as cold and hot temperatures and drought. Potassium is an important osmolite responsible for osmoregulation, maintaining cell turgor pressure, cell elongation during growth, and the opening and closing of stomata, which affect transpirational cooling and CO_2 accumulation for photosynthesis. Potassium deficiency does not immediately result in visible symptoms. Initially, there is only a reduction in growth rate, with chlorosis and necrosis occurring in later stages (Mengel *et al.*, 2001). Affected older leaves will show localized mottled or chlorotic areas with leaf burn at margins. Chlorotic symptoms typically begin on the leaf tip, and will advance along the leaf margins towards the base, usually leaving the midrib alive and green. As the deficiency progresses, the entire leaf will yellow. Small white or yellow necrotic spots may also develop, beginning along leaf margins.

Chlorine (Cl) is very abundant in soils, and reaches high concentrations in saline areas, but it can be deficient in highly leached inland areas. Until recently, little information was documented on Cl deficiencies. Chlorine deficiencies tend to be cultivar specific and can be easily misdiagnosed as physiological leaf spot (Engel *et al.*, 2001). Chloride is essential along with K for proper function of the stomata openings and plant water relations, splitting water into H and O during photosynthesis, cation balance, and transport of molecules within the plant.

The most common symptoms of Cl deficiency are chlorosis and marginal wilting of the young leaves (Berry, 2010). The chlorosis occurs on smooth flat depressions in the interveinal area of the leaf blade. In more advanced cases there often appears a characteristic bronzing on the upper side of the mature leaves and necrotic spotting along leaves with abrupt boundaries between dead and live tissue (Mengel et al., 2001).

Magnesium is the central molecule in chlorophyll and is an important cofactor for the production of energy storing compound ATP. Symptoms of Mg deficiency include interveinal chlorosis and yellow or reddish-purple leaf margins with green midribs. Leaves of Mg-deficient beets and potatoes become stiff and brittle while veins are often twisted. The Mg-deficient leaves show advanced interveinal chlorosis, with necrosis developing in the highly chlorotic tissue. The interveinal laminae tissue tends to expand proportionately more than the other leaf tissues, producing a raised puckered surface, with the top of the puckers progressively going from chlorotic to necrotic tissue (Berry, 2010). In the brassicas, tints of orange, yellow, and purple may also develop.

Molybdenum is needed to provide activity for specific enzymes and for N fixation in legumes. An early symptom for Mo deficiency is a general overall chlorosis, similar to N deficiency. Molybdenum is required for plants to utilize nitrates, so the initial symptoms of Mo deficiency are actually N deficiency. Molybdenum deficiency produces a classic symptom in cauliflower called "whiptail", which is often used to illustrate the deficiency. In the case of cauliflower, the lamina of new leaves fail to develop, resulting in a characteristic "whiptail" appearance (Berry, 2010). Other symptoms of Mo deficiency include mottled spots that develop into large interveinal chlorotic areas under severe deficiency and leaf scorching, cupping, or rolling. Molybdenum deficiency may develop when the pH is lower than 5.5. Raising the pH from 6.0 to 6.5 prevents Mo deficiencies in vegetable crops. At high concentrations, Mo has a very distinctive toxicity symptom of leaves turning brilliant orange.

Immobile Nutrients

Sulfur is an essential constituent of certain amino acids and proteins. Sulfur deficiency results in the inhibition of protein and chlorophyll synthesis, so deficient leaves are chlorotic. The veins and petioles often have a very distinct reddish color. The visual symptoms of S deficiency are very similar N and Mo deficiencies. In contrast to N or Mo deficiency, however, S deficiency symptoms initially occur in younger leaves, causing light green to yellow coloration. In later growth, the entire plant may be pale green. Additionally, plants deficient in S tend to be spindly and small and stems are often thin. With advanced S deficiency brown lesions and/or necrotic spots often develop along the petiole, and the leaves tend to become more erect and often twisted and brittle.

Boron is one of the most common deficient micronutrients in vegetable crops. Deficiencies of this element are most likely to occur in asparagus, most bulb crops, celery, beets, the brassicas and tomatoes where soils are low in B. The primary functions of B in plants are related to cell-wall formation, apical growth and reproductive tissues (Blevins and Lukaszewski, 1998). Plants suffering from B deficiency exhibit chlorotic young leaves and death of the terminal bud. In addition to chlorosis, leaves may develop dark brown, irregular lesions that will progress to leaf necrosis in severe cases. Poor cell-wall development causes leaves and stems of B-deficient plants to become brittle and distorted and leaf tips tend to thicken and curl (Blevins and Lukaszewski, 1998). Affected plants will grow slowly and appear stunted as a result of shortened internodes. Since B tends to accumulate in reproductive tissues, flower buds may fail to form or are misshapen, and pollination and seed viability is often poor in B-deficient plants (Blevins and Lukaszewski, 1998). Both the pith and the epidermis of stems may be affected, often resulting in hollow or roughened stems along with necrotic spots on the fruit. The leaf blades develop a pronounced crinkling and there is a darkening and crackling of the petioles often with exudation (Berry, 2010). A well-documented B classic deficiency in beets is crown rot, cavity spot, and heart rot (Mengel et al., 2001). Along with stunted growth, symptoms in beets include young leaves curling and turning brown or black in color. In later stages of the deficiency, the crown of the beet begins to rot and disease sets in, affecting the whole plant. The healthy part of the beet will be low in sugar. Boron at 200 ppm can cause marginal and tip necrosis in leaves as well as poor overall growth performance in sensitive crops (Blevins and Lukaszewski, 1998).

Plants deficient in Mn display interveinal chlorosis of young leaves since chloroplast development and thus chlorophyll accumulation is affected (Mengel et al., 2001). The early stages of Mn deficiency are somewhat similar to Fe deficiency because both start with a light chlorosis of the young leaves and netted veins of more mature leaves, especially when they are viewed through transmitted light. As the Mn stress increases, the leaves develop a gray metallic sheen with dark freckled and necrotic regions along the veins. However, unlike Fe, there is no sharp distinction between veins and interveinal areas, but rather a more diffuse chlorotic effect. A well-documented Mn deficiency symptom is marsh spot in peas. Manganese deficiency often occurs in plants growing on soils that have been over-limed.

Iron plays an important role in plant respiratory and photosynthetic reactions. Iron deficiency reduces chlorophyll production and is characterized by interveinal chlorosis with a sharp distinction between veins and chlorotic areas in young leaves since it is non-mobile (Berry, 2010). The symptoms evolve into an overall chlorosis, becoming totally bleached with extreme deficiencies. The bleached areas often develop necrotic spots. Until the leaves become nearly white, recovery is still possible with iron fertilization. Iron deficiency is strongly associated with calcareous soils and high pH, anaerobic conditions, and excess of heavy metals in soil (Berry, 2010).

Zinc is needed by plants for growth hormone production and is particularly important for internode elongation. Since Zn has limited mobility in the plant, symptoms are initially visible on the middle and younger leaves as interveinal chlorosis, especially midway between the margin and midrib, producing a striping effect or in some cases mottling. Chlorotic areas can be pale green, yellow, or even white. Severe Zn deficiencies will cause leaves to turn gray-white and abscise or die. Because Zn plays a prominent role in internode elongation, deficient plants are severely stunted. Flowering and seed set is also poor. Zinc deficiency generally does not affect fields uniformly and deficient areas commonly occur where topsoil has been removed or is very thin (Follett and Westfall, 2004).

Unlike many of the other immobile minerals, Ca is a macronutrient, like S, and may comprise as much as 2% of plant dry matter. Calcium is needed in larger quantities because it is a component of plant cell walls and regulates cell wall construction. Calcium deficiency is less common in calcareous soils, which are common around the world. Insufficient Ca can cause distortion of young leaves. Classic symptoms of Ca deficiency include blossom-end rot of tomato, pepper, and watermelon (necrotic tissues on the end of the fruit with a "burned" appearance), tip burn of lettuce, blackheart of celery, and death of the meristematic tissues in many plants (Ho et al., 1993). The occurrence of blossom-end rot tends to vary among cultivars and is made worse by drought stress, which can cause blossom-end rot to occur even when soil Ca is sufficient to support normal growth. All these symptoms show soft dead necrotic tissue in rapidly growing meristematic tissues, which is generally related to poor translocation of Ca to the tissue rather than low soil Ca (Berry, 2010). Very slow-growing plants with a deficient supply of Ca may re-translocate sufficient Ca from older leaves to maintain growth with only a marginal chlorosis of the leaves. This ultimately results in the margins of the leaves growing more slowly than the rest of the leaf, causing the leaf to cup downward. Plants under chronic Ca deficiency have a much greater tendency to wilt than non-stressed plants (Berry, 2010).

Copper is needed for chlorophyll synthesis, respiration and protein synthesis (Mengel et al., 2001). Plants with Cu deficiency have chlorotic younger leaves, stunted growth, delayed maturity and, in some cases, brown discoloration. Copper deficiency may be expressed as a light overall chlorosis, along with the permanent loss of turgor in young leaves. Recently matured leaves show netted green veining with areas bleaching to a whitish-gray. Some leaves develop sunken necrotic spots and have a tendency to bend downward (Berry, 2010).

Nickel is the metal component in urease, an enzyme that catalyzes the conversion of urea to ammonium (Havlin et al., 1999). Nickel is thought to be beneficial for N metabolism in legumes and other plants where ureides are important. Nickel deficiency symptoms include chlorosis and interveinal chlorosis in young leaves, which progresses to plant tissue necrosis. Other symptoms include poor seed germination and decreases in crop yield.

Fertilizer Formulations

Fertilizers come in various forms. Fertilizers are broadly divided into two categories. Organic fertilizers are composed of enriched organic matter, either plant or animal, in various stages of decomposition. The other category is inorganic fertilizers, which are composed of synthetic chemicals and/or minerals that are often mined from deposits.

The most typical form of inorganic fertilizer is granular fertilizer that often appears to be composed of small chips of stone or pellets. Some fertilizers come in a liquid form that can be fed through drip irrigation systems or sprayed in a foliar application. There are also slow-release fertilizers, which slowly release nutrients over time and reduce leaching and burning of plant tissues caused by excessive application. A wide range of organic fertilizers is available that are made from various natural materials.

The labeling of fertilizers varies among countries in terms of both analysis methodology and nutrient labeling. In many countries, the macronutrients are labeled as nitrogen, phosphorus, and potassium (N-P-K), always in that specific order of analysis, or N-P-K-S, adding sulfur or another nutrient after K. The three or four numbers on the fertilizer label represent an analysis of composition by weight. While the number for "N" represents the percentage weight of nitrogen, the number for "P" is actually the weight of an equivalent quantity of P_2O_5 and not elemental phosphorus. In order to calculate the weight of P in the formulation, the weight of P_2O_5 can be multiplied by 0.44 to compensate for the weight of the oxygen in the molecule. For example, a bag of 10-10-10 has 4.5 kg (10 lb) of N, 4.5 kg (10 lb) of P_2O_5, but only 2.0 kg (4.4 lb) of P. Similarly, the number for "K" is actually the weight of an equivalent quantity of K_2O, and not elemental potassium. In order to calculate the weight of K in the formulation, the weight of K_2O can be multiplied by 0.83 to account for the weight of the oxygen in the molecule. So, a bag of 10-10-10 has 4.5 kg (10 lb) of K_2O, but only 3.8 kg (8.3 lb) of K. A fertilizer that has an analysis containing all three of the nutrients N-P-K is often called a "complete analysis fertilizer".

Inorganic fertilizer (synthetic fertilizer)

Some fertilizers, such as urea, can be synthesized artificially. Inorganic fertilizer is often synthesized using the Haber-Bosch process, which produces ammonia as the end product (Smil, 2000). This ammonia is used as a feedstock for other N fertilizers, such as anhydrous ammonium nitrate and urea. Anhydrous ammonia is used as a liquid fertilizer because it provides an inexpensive source of N for some applications. Anhydrous ammonia is a very concentrated form of N (82%), so the amount applied is less than other types of N fertilizer. Unlike liquid fertilizers, anhydrous ammonia is unique and normally a gas at room temperature and must be stored under pressure to maintain its liquid state. Anhydrous ammonia has a high affinity for water and forms ammonium when mixed with water in the soil. Because of this, anhydrous ammonia is usually injected into the soil to improve the conversion to ammonium. However, if the soil is dry some of the ammonia will evaporate and be lost.

Some plant nutrients in fertilizers are mined from deposits containing high concentrations of minerals like phosphate or K. Ammonia can be combined with rock phosphate and K fertilizer in the "Odda process" to produce compound complete analysis fertilizer. Complete analysis inorganic fertilizer is often sold in a granular formulation in 23 kg (50 lb) bags or in bulk. The fertilizers should be stored in a dry place because this formulation is hygroscopic and may absorb moisture from the atmosphere and clump. The granules are a mixture of the actual fertilizer and an inert carrier and some types are supplemented with micronutrients such as boron that are often limited in some vegetable production regions. Granular fertilizers are a concentrated source of mineral nutrients and in some cases alter soil pH but do little else to improve long-term soil fertility.

The worldwide use of synthetic N fertilizers has increased steadily since the 1950s, rising almost 20-fold to a rate of 100 million metric tonnes (110 million short tons) of N per year in 2003 (Glass, 2003). The use of phosphate fertilizers has also increased from 9 million metric tonnes/year (2.2 million short tons/year) in 1960 to 40 million metric tonnes/year (44.1 million short tons/year) in 2000. Many inorganic fertilizers may not replace trace mineral elements in the soil, which become gradually depleted by crops over time.

Organic fertilizer

There is renewed interest in organic fertilizers for vegetable production. Because of reduced availability, low cost of other forms of fertilizer and handling issues, the use of animal manures and other solid wastes as fertilizers for vegetable crops declined for many years. Historically, animal manures were widely used to fertilize vegetable crops before inorganic fertilizers became popular. Many vegetables respond well to animal manures, because they supply nutrients to the soil and improve soil structure by adding organic matter.

Today there are many types of organic fertilizers available, including composted manure, worm castings, blood meal, seaweed, guano, composted natural wastes, or naturally occurring mineral deposits, to name just a few. The definition of organic varies somewhat from country to country so there is not universal agreement on what constitutes organic fertilizer. Organic fertilizers have been known to improve the biodiversity and long-term productivity of soil, and may prove a large depository for excess CO_2 (Pimentel et al., 2005).

Another source of organic fertilizer is cover crops planted as "green manures", which supply organic matter, and N if the crop is a legume. Green manure crops are planted in rotation with cash crops but are not allowed to reach harvest maturity. Instead, the green manure biomass is incorporated by plowing or left on the surface as mulch to decompose and recycle mineral nutrients back into the soil. Thus the decomposing plant tissue or "green manure" acts as a slow-release fertilizer, recycling mineral nutrients back into the soil, inhibiting erosion during nonproductive periods, and building soil organic matter reserves. Legumes, like hairy vetch and clover, and small grains, such as barley, rye or millet, are examples of overwintering green-manure crops. However, cover crops such as small grains that have a high C:N ratio must be supplemented with N to stimulate microbial decomposition of the plant matter that has been incorporated. If the green manure does not fulfill the nutritional requirements for the next crop in rotation, additional fertilizer inputs may be needed. It is important to prevent small-grain green-manure crops from becoming a weed problem by killing the plants before seed set.

Soil priming is an evolving trend, which uses special cover crops that promote beneficial soil bacteria. Cover crops can be selected to include "primer plants" that support establishment of soil microbial communities (Yunusa and Newton, 2003). Roots of plants used to "prime" soil penetrate deep into the subsoil and secrete sugars to create and sustain symbioses with microbes that benefit the growth and development of the vegetable crop planted in rotation (Wardle, 2002).

Fertilizer Application Methods

There are many factors to consider when determining the type of fertilizer to use, how much should be applied and the timing of application. Factors include organic or conventional production, the crop grown, the soil type, cultural practices used, such as plasticulture or conservation tillage, the irrigation system, potential environmental impact, the cost, and availability.

The higher cost of petrochemicals has dramatically increased inorganic fertilization costs for vegetable growers. In addition, environmental damage to drinking water and other natural resources caused by poor crop nutrient management has raised awareness about best nutrient management practices. Both of these factors have resulted in development of techniques to more efficiently apply nutrients to crop plants. This can be thought of as a type of precision agriculture where technology is used to minimize waste and create more precise management practices.

Today many vegetable farmers practice "precision" or "prescription" fertilization. The principle of prescription nutrient management is test–apply–test. Using this philosophy, no materials are applied unless soil or plant tissue analysis shows that the crop actually requires and would benefit from the input. Inputs are precisely calculated so only the amount of material is applied to correct the deficiency and reduce waste. This principle saves growers money by limiting inputs to only those that are needed to optimize production. It is also environmentally friendly because fertilizer inputs are restricted to those that are absolutely necessary, which reduces pollution.

On certain soil types and with some crops, a single application of fertilizer at planting is sufficient for an entire season. For example, radish production on organic soils may require only a single application of fertilizer at planting due to the fertile nature of the soil and the rapid maturation of the crop. However, for many vegetables, particularly those grown on sandy soils, split or multiple applications are usually necessary.

Common application methods include: broadcasting, row banding, top-dressing, side-dressing, liquid "starter" solutions, foliar application, and fertigation – the combined application of nutrients through irrigation water. Broadcasting is randomly scattering fertilizer on top of the soil, usually with mechanical spreaders. While this application method may be effective for grain crops that are planted at high density with narrow row spacing, it is inefficient for more widely spaced vegetable crops like melons or tomatoes.

Most solid fertilizers are not placed directly in the row with the seeds or transplants. With most seeds, especially beans, this practice is not recommended since the high salt concentration may inhibit seed germination or damage small seedlings. Row banding is a common technique used with granular fertilizer where a thin band of fertilizer is placed a few inches to the side and below the seed or transplant. Row banding confines the fertilizer to the root zone so it is not wasted on areas inaccessible to plant roots. Row banding works well for crops at wide-row spacing and when only a single application of fertilizer is needed.

Side-dressing applies nutrients to the soil after the crop is established. This technique places fertilizer either on top of or below the soil surface and to the side of the row. Side-dressing is beneficial for many vegetables, since it adds fertilizer when it is most needed by the crop. However, adverse weather conditions can sometimes interfere with timely fertilizer application and nutrients can be washed away from the soil surface. Top dressing is a similar to side dressing, but the fertilizer is applied to the top of the soil near the crop (Fig. 4.7). This technique is often used to supply nutrients to a perennial crop such as asparagus. It is also used on raised beds where furrow irrigation activates or dissolves the fertilizer band on the top of the row after a crop is transplanted.

Foliar feeding is application of liquid fertilizer formulations directly to the foliage rather than the root zone. Since the leaf cuticle of vegetables, particularly the brassicas, is impervious to water, minerals dissolved in aqueous solution on the foliage are predominately absorbed through the stomata. While foliar feeding has long been used to supply micronutrients, which are only needed in very small quantities, foliar feeding does not appear to be the most efficient way to apply N, P, K, and other macronutrients, which are required in much higher quantities.

Starter fertilizers are most beneficial when crops are planted into cold, wet soils in early spring or late fall, regardless of soil fertility. Starter fertilizer is important because it meets the demands of the seedling for readily available nutrients until the plant's root system develops. Using a starter is especially important in conservation tillage systems and transplanted vegetable crops. Crops planted in late spring or early fall generally do not require a starter fertilizer unless soil fertility is low. Starter fertilizers may be solid or liquid. A liquid starter fertilizer is often applied to the root zone of vegetable transplants. Solid starters are generally applied like row-banded fertilizer 5 cm (2 in) to the side of and below the seed. Liquid formulations supply fertilizer and water directly to the root zone where

Fig. 4.7. Dry fertilizer is placed on the soil surface at the edge of raised bed prior to planting. The fertilizer will be activated or dissolved by rain or irrigation water and be available to developing plants.

they can immediately be used by a fragile transplant to minimize shock.

Starter fertilizer composed of N and P provides the most favorable response. Unlike N, which is mobile and has a greater chance of root uptake, phosphorus compounds are easily bound and do not readily move within the soil. To be absorbed, the roots must essentially contact the phosphate in soil. Thus, pouring a liquid fertilizer high in P around the roots of transplants is beneficial. Phosphorus is important for promoting vigorous root growth that results in healthy, dark-green plants. Phosphorus deficiency causes stunted, purple-colored plants. Phosphorus has limited solubility in water so a formulation that will be dissolved in water should be selected for liquid starter solutions. Nitrogen in the starter may help plants overcome early season N deficiency due to the slow release of N from organic matter in cold soils. Also, it has been shown that some N in the ammonium form common in starter fertilizers will enhance P uptake from the starter and from the soil. Potassium is not as critical as N or P in a starter, but some response is likely when soil K levels are marginal, especially under cold, wet conditions. Therefore starter fertilizers containing N-P-K are commonly used for vegetable transplants where rapid growth and earlier maturity are particularly important. Starter fertilizers are routinely used for warm-season vegetable transplants such as watermelon, tomato, pepper, and eggplant grown in medium and short-season areas. Overall, the use of a starter fertilizer increases fertilizer efficiency and thus reduces fertilizer costs.

When fertilizer is applied in the irrigation water it is termed fertigation. This is most commonly done through a drip irrigation system but fertilizers can be added though other types of irrigation as well. It is important to use soluble forms of fertilizer for fertigation. Various types of fertilizer injectors may be used to add fertilizer to irrigation water. The most sophisticated types are computer-controlled, but less expensive types are also effective. Water-powered, non-electric chemical injectors precisely add liquid fertilizer to irrigation lines. Systems can be programmed to provide different nutrient levels based on the needs of the crop. Back-flow regulators are required by law in some areas to prevent the back-flow of fertilizer water from the drip-tubing into the water source.

Usually N and K are applied in this manner because they are water soluble. Nutrients may be applied entirely through drip irrigation or a split application may be employed. Some growers prefer to apply P at planting, since it is not very soluble in water and has low mobility in the soil. For split applications, most growers apply all P and 50% of the N and K to the soil at planting, although other combinations are sometimes used. Fertigation is one of the most efficient methods for feeding plants. Nutrients can be supplied directly to the root zone in precise quantities so there is very little waste. Nutrients are only provided as needed and can be altered throughout the season to meet the changing nutrient requirements of the crop.

For example, a crop of melons needs higher N early in the season to encourage vegetative growth and a good canopy to support fruit development. After the canopy is developed, N levels should decline to encourage flowering and fruit set. After fruit set, the demand for N again increases to support development. Later in the season as fruits ripen nutrient requirements subside again. Nutrient feeding through fertigation can be weekly or as needed to meet nutrient needs throughout the season.

Monitoring Plant Nutrient Status in the Field

As discussed earlier, soil testing is an important part of nutrient management but it tells only part of the story. Many times soil nutrient levels will be adequate but the plants may still be deficient in one or more nutrients. Cold soils, wet soils, root damage by disease or insects as well as other factors may prevent plants from taking up nutrients even though they are available in the soil. Therefore it is important to measure plant nutrient status throughout the season. Many of the same labs that test soil can also test foliar samples for mineral nutrients to determine whether or not a crop is adequately fed. However, commercial foliar analysis can be costly and it can often take a week or more to obtain test results. In many cases, deficiencies can occur before symptoms appear so it is wise to test routinely to monitor crop nutrient status.

Plant Tissue Analysis

Coupled with other data and observations, plant tissue analysis is an effective aid in evaluating the actual nutrient status of a crop and is an important part of prescription fertilization (test–apply–test). For example, soil tests might show that a field is properly fertilized while foliar analysis may show that a critical nutrient may be lacking in the tissue. This discrepancy suggests a root disorder that is preventing the nutrient from being taken up from the soil. Additional fertilization would be wasteful and would not solve the problem. So despite the additional cost, plant tissue analysis can determine the effectiveness of various fertilizer treatments and show whether the plant is properly fed. Petiole analysis was developed to take advantage of portable hand-held nutrient meters, which were developed for use by field practitioners (Fig. 4.8).

Samples taken several times during the season or over several growing seasons may be required to correctly evaluate the nutrient status of a particular soil–plant system. The desired concentration of an element should occur within the sufficiency range. The limits for this range vary depending on the crop, the tissue being analyzed and the stage of growth sampled. Commonly, fully expanded, recently matured leaf samples are used for analysis.

Portable hand-held nitrate and K meters are used to measure the nutrient content of leaf petiole sap that has been expressed using a garlic press or similar device. Although results may not be as accurate as laboratory results, they give growers a rapid real-time assessment of the mineral nutrient status of a crop.

Studies have shown that petiole nutrient levels for many vegetables are stable and petioles provide a reliable tissue to sample. For a real-time analysis of crop nutrient status, six to eight petioles are collected and the sap is expressed. A droplet is placed on the measuring junction of the meter. A value in parts per million is immediately registered and this value can be compared to published values to determine when crops are deficient for the nutrient measured. Corrections in nutrient management can be made immediately using fertigation or one of the other application techniques. The meters can be calibrated with standard solutions to ensure accuracy. This real-time assessment and feeding is a fast and accurate way of feeding crops and a good example of the test–apply–test prescription nutrient management technique currently used by progressive growers.

GPS and Crop Management

Another aspect of precision agriculture that can be applied to nutrient management is the use of Geographic Information System/Global Positioning System (GIS/GPS). The GPS is a space-based global navigation satellite system that provides location and time information in all weather, anywhere on or near the Earth, where there is an unobstructed line of sight to four or more GPS satellites. The GIS is designed to capture, store, manipulate, analyze, manage, and present all types of geographically referenced data. In the simplest terms, GIS is the merging of cartography, statistical analysis, and database technology. When applied to fertility management, data can be collected and precisely mapped to coordinates in a crop field. For example soil fertility data could be inputted for a field based on soil samples and/or aerial data to create a GIS database. Row-banding equipment with GPS units could vary fertilizer inputs to field requirements based on their precise location and the field fertility map developed. This technology can save money by placing fertilizer where it is needed the most.

Feeding the Microbes to Improve Soil Health and Crop Productivity

Most of this chapter has focused on the nutrient requirements of crops and how best to supply their needs. In addition to nutrients, a healthy soil is filled

Fig. 4.8. Portable hand-held meter provides real-time assessment of plant potassium. Droplets of petiole sap are expressed on to the sensor concealed on the left side of the device opposite the display. The meter is calibrated before use with solutions of known concentration and the read out is in parts per million.

with a diversity of microorganisms. While the tendency for agriculturists is to focus only on microbes that cause plant disease, there is an emerging realization that many of these soil organisms benefit crops in a variety of different ways and should be nurtured rather than destroyed. For example, microbes decompose crop residues and organic matter, slowly releasing mineral nutrients for crop use.

In many cases these beneficial soil organisms are sustained by sugars and amino acids that are released by decomposing plant material, such as green manure crops. The symbiosis that exists between plants and soil microbes is much more highly evolved and extensive than first realized. For decades, the use of amino acid supplements, sugars, humic acids, and various gaseous materials remained outside the mainstream of US agriculture because the scientific basis of such amendments was poorly understood. However, it is increasingly apparent that many of these materials and treatments directly benefit or stimulate microbial populations that in turn perform functions that benefit plants. For agricultural production on organic soils or mineral soils amended with animal manures, crop residues, and various organic waste materials, sufficient carbon sources and nutrients are available to adequately "feed" soil microbes so the soil remains "healthy" and alive with biological activity. However, in most modern crop production monocultures that rely on inorganic mineral fertilizers rather than organic matter, carbon sources for microbes may be limited. Populations of beneficial microbes may suffer when crop residues are removed from fields and soil organic matter is low (Welbaum *et al.*, 1994).

Tailoring inputs and cultural practices to promote beneficial soil microbes as well as the needs of the crop plants is an evolving area of crop nutrient management that could potentially increase agricultural productivity.

What is good for beneficial microbes is also good for the plant may not apply in every instance but it is generally true. It is increasingly apparent that practices that promote beneficial microbial activity, such as applying carbon amendments to the soils, also benefit crops indirectly through a broad range of mechanisms that include increased nutrient availability, improved soil characteristics, better protection against diseases, removal of waste materials from the plant, alterations in plant gene expression, and contributing compounds that stimulate growth.

References

Ainsworth, E.A. (2008) Rice production in a changing climate: a meta-analysis of responses to elevated carbon dioxide and elevated ozone concentration. *Global Change Biology* 14, 1642–1650.

Bennett, W.F. (ed.) (1993) *Nutrient Deficiencies and Toxicities in Crop Plants.* APS Press, St. Paul, Minnesota.

Berry, W. (2010) Symptoms of deficiency in essential minerals. In: Taiz, L. and Zeiger, E. (eds) *Plant Physiology*, 5th Edition. Online version available at: http://5e.plantphys.net/article.php?ch=5&id=289 (accessed 28 July 2011).

Blevins, D.G. and Lukaszewski, K.M. (1998) Functions of boron in plant nutrition. *Annual Review of Plant Physiology and Plant Molecular Biology* 49, 481–500.

Crow, S.E., Lajtha, K., Filley, T.R., Swanston, C.W., Bowden, R.D. and Caldwell, B.A. (2009) Sources of plant-derived carbon and stability of organic matter in soil: implications for global change. *Global Change Biology* 15, 2003–2019.

Engel, R., Bruebaker, L.J. and Ornberg, T.J. (2001) A chloride deficient leaf spot of WB881 Durum. *Soil Science Society of America Journal* 65, 1448–1454.

Follett, R.H. and Westfall, D.G. (2004) Zinc and Iron Deficiencies. Colorado State University Cooperative Extension Fact Sheet No. 0.545. Colorado State University, Ft. Collins, Colorado. Available at: www.ext.colostate.edu/pubs/crops/00545.html (accessed 28 July 2011).

Glass, A. (2003) Nitrogen use efficiency of crop plants: Physiological constraints upon nitrogen absorption. *Critical Reviews in Plant Sciences* 22, 453–470.

Grundon, N.J. (1987) *Hungry Crops: A Guide to Nutrient Deficiencies in Field Crops.* Queensland Government, Brisbane, Australia.

Havlin, J.L., Beaton, J.D., Tisdale, S.L. and Nelson, W.L. (1999) *Soil Fertility and Fertilizers*, 6th edn. Prentice-Hall, Inc., Upper Saddle River, New Jersey.

Ho, L.C., Belda, A.R., Brown, M., Andrews, J. and Adams, P. (1993) Uptake and transport of calcium and the possible causes of blossom-end rot in tomato. *Journal of Experimental Botany* 44, 509–518.

Jones, J.B. (1998) *Plant Nutrition Manual.* CRC Press, Boca Raton, Florida.

Juma, N.G. (1999) *Introduction to Soil Science and Soil Resources.* Vol. 1. *The Pedosphere and its Dynamics: A Systems Approach to Soil Science.* Salman Productions, Sherwood Park, Alberta, Canada, 335 pp.

Kaufmann, R.K. and Cutler, J.C. (2008) *Environmental Science.* McGraw-Hill, Debuke, Iowa.

Long, S.P., Ainsworth, E.A., Leakey, A.D., Nosberger, J. and Ort, D.R. (2006) Food for thought: lower-than-expected crop yield stimulation with rising CO_2 concentrations. *Science* 312, 1918–1921.

Marshall, T.J. and Holmes, J.W. (1979) *Soil Physics.* Cambridge University Press, Cambridge, UK.

McCauley, A., Jones, C. and Jacobsen, J. (2003) Plant Nutrient Functions and Deficiency and Toxicity Symptoms. Nutrient Management Module 9. Montana State University Cooperative Extension Service Publication No. 4449-9. MSU, Bozeman, Montana.

Mengel, K., Kosegarten, H., Kirkby, E.A. and Appel, T. (2001) *Principles of Plant Nutrition*, 5th edn. Kluwer Academic, Dordrecht, the Netherlands.

Nyle, B. and Weil, R. (2008) *The Nature and Properties of Soils*, 14th edn. Prentice Hall, Upper Saddle River, New Jersey.

Pimentel, D., Hepperly, P., Hanson, J., Douds, D. and Seidel, R. (2005) Environmental, energetic, and economic comparisons of organic and conventional farming systems. *BioScience* 55, 573–582.

Smil, V. (2000) *Enriching the Earth: Fritz Haber, Carl Bosch, and the Transformation of World Food Production.* MIT University Press, Boston, Massachusetts.

Taiz, L. and Zeiger, E. (2010a) Photosynthesis: The Light Reactions. In: *Plant Physiology*, 5th edn. Online version available at: http://5e.plantphys.net/article.php?ch=5&id=289 (accessed 29 July 2011).

Taiz, L. and Zeiger, E. (2010b) Photosynthesis: The Carbon Reactions. *Plant Physiology*, 5th edn. Online version available at: http://5e.plantphys.net/article.php?ch=5&id=289 (accessed 29 July 2011).

Thompson, H.C. and Kelley, W.C. (1957) *Vegetable Crops*, 5th edn. McGraw-Hill, New York, 611 pp.

Wardle, D.A. (2002) *Communities and Ecosystems; Linking the Aboveground and Belowground Components.* Princeton University Press, Princeton, New Jersey.

Welbaum, G.E., Sturz, A.V., Dong, Z. and Nowak, J. (2004) Managing soil microorganisms to improve productivity of agro-ecosystems. *Critical Reviews in Plant Sciences* 23, 175–193.

Yunusa, I.A.M. and Newton, P.J. (2003) Plants for amelioration of subsoil constraints and hydrological control: the primer-plant concept. *Plant Soil* 257, 261–281.

5 Irrigation of Vegetable Crops

Introduction

Irrigation may be defined as the science of applying water to the land or soil. Irrigation has many diverse uses, including the growing of agricultural crops, maintenance of landscapes, revegetation of disturbed soils in dry areas and during periods of inadequate rainfall. Additionally, irrigation also may provide frost protection for vegetable crops (Snyder and Melo-Abreu, 2005). In contrast, production that relies only on direct rainfall is referred to as rain-fed or dryland vegetable farming. Successful vegetable production in many regions is dependent upon farmers having sufficient water for irrigation. Water scarcity is a critical constraint to farming in many parts of the world.

Arid regions frequently suffer from physical water scarcity. Physical water scarcity is where there is insufficient water to meet all demands, including those needed for ecosystems to function effectively. Symptoms of physical water scarcity include environmental degradation and declining groundwater. Economic scarcity is caused by the inability to use existing water resources. Symptoms of economic water scarcity include a lack of infrastructure to deliver water to production farms where it is most needed for agricultural uses. Both physical and economic scarcity limit vegetable production in many parts of the world. Some 2.8 billion people currently live in water-scarce areas (FAO, 2007).

In the middle of the 20th century, the advent of diesel and electric motors allowed groundwater to be pumped out of major aquifers to irrigate crops. Unfortunately, the rate of depletion was often faster than aquifer recharge by rainwater and other natural sources. This can lead to permanent loss of aquifer capacity, decreased water quality, ground subsidence and other problems. Areas such as the North China Plain, the Punjab and the Great Plains of the USA are threatened because of aquifer depletion.

Globally, 2,788,000 km^2 (689 million acres) of agricultural land were equipped with irrigation infrastructure around the year 2000. About 68% of irrigated areas were located in Asia, 17% in America, 9% in Europe, 5% in Africa, and 1% in Oceania. The largest contiguous areas of high irrigation density are found: in north India and Pakistan along the rivers Ganges and Indus; in the Hai He, Huang He and Yangtze basins in China; along the Nile River in Egypt and Sudan; and in parts of California. Smaller irrigation areas are spread across almost all populated parts of the world (Siebert et al., 2006).

Irrigation is particularly essential for vegetable production. By definition, vegetable crops are intensively managed and irrigation management is a fundamental feature that distinguishes vegetables from lower-value dryland agronomic crops. Many vegetables have shallow root systems and are susceptible to drought stress. Even areas that receive regular rainfall experience some degree of drought stress nearly every growing season. In areas with seasonal rainfall, vegetable production must be irrigated to maximize profitability. Timely irrigation generally increases both yields and quality of most vegetable crops. Since most vegetables yield a high return per acre, an irrigation system can often pay for itself over the course of several seasons.

Why Water is So Important

Most plants are approximately 95% water, so maintaining full hydration is necessary to preserve appearance and prevent wilting. Full hydration enables developing plants to reach their full size (Fig. 5.1). This is especially critical for developing leaves, which when subjected to drought stress are stunted and smaller than plants that were well watered (Hsaio and Bradford, 1983). Large leaves create an extensive plant canopy, which maximizes the photosynthetic capacity of the plant. Since most dry matter comes from organic

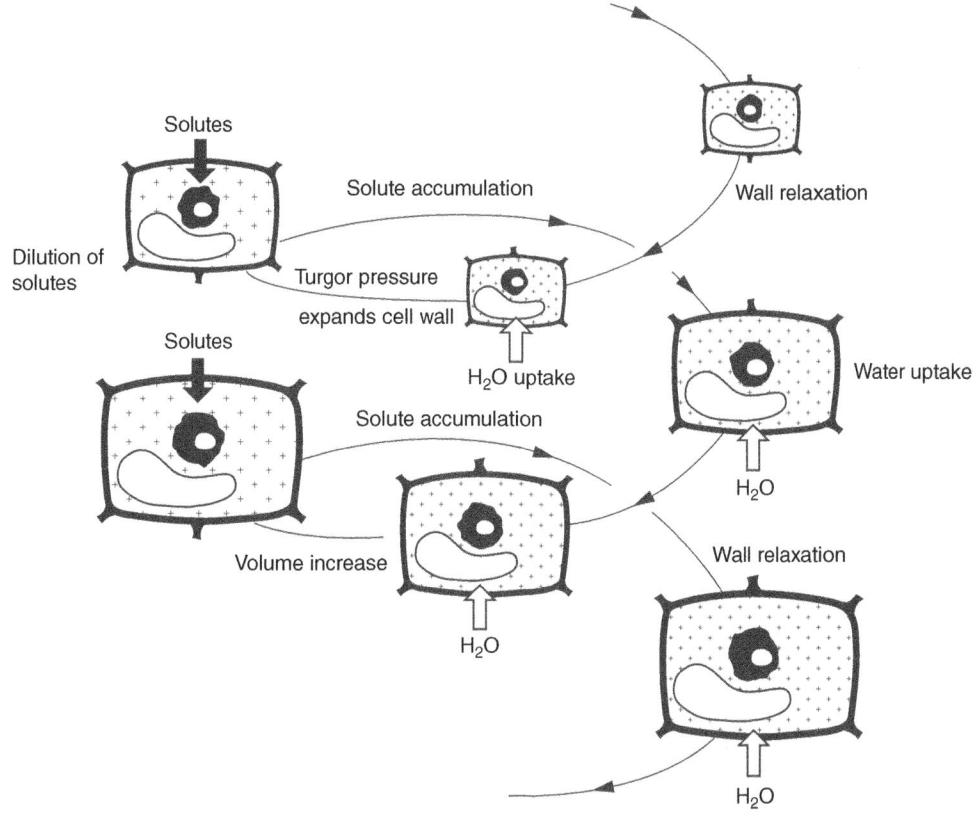

Fig. 5.1. Plant cells are initially small after cell division but rapidly increase in size by accumulating molecules that attract water called solutes, cell wall loosening, and water uptake. When water is limiting plant leaves and fruits are smaller because cell expansion is inhibited, so yields are reduced.

molecules created through photosynthesis, plants with greater photosynthetic capacity also have greater yield potential (discussed in Chapter 4).

Water availability impacts photosynthesis in other important ways. Even slight drought stress can close the stomatal pores on plant leaves. Closed stomata reduce access to the carbon dioxide in the atmosphere that plants use to make the compounds necessary for growth and development through photosynthesis (Hsaio and Bradford, 1983).

Drought stress during flowering may inhibit fruit set and yield and causes physiological disorders like blossom end rot. Mineral nutrients are not supplied fast enough from the roots to the leaves through transpirational water flow to support rapid cell division and enlargement resulting in blossom end rot. So keeping vegetables regularly watered is fundamental for optimizing yield and quality.

Determining When Plants Need to Be Irrigated

As a general rule, vegetables grown on a loam soil require approximately 2.5 cm (1 in) of water per week. If less rainfall is received, irrigation may be necessary to make up the difference. More water would be recommended for crops grown on well-drained sandy soils or in environments where evapotranspiration is high. Less water may be required for crops nearing maturity. For example, after crops such as tomatoes or melons have reached their maximum size and are nearing harvest, water requirements are less than early in development. Water applied late in development can cause fruit cracking, which may damage certain crops or reduce shelf life.

There are many techniques and devices for measuring soil moisture. If soils cannot be measured scientifically, some farmers determine when to irrigate by observing soil characteristics. A loam soil that is free of cracking may be adequately hydrated. Another primitive but useful method is called "hand feel" (Ross, 2004). To use this method, a handful of soil is obtained from just below the surface and rubbed between one's hands. If the soil is so dry that it cannot be rolled into strands or "worms", irrigation is needed. Another basic method for measuring moisture that does not require expensive instrumentation is the "gravimetric" method where soil-moisture content is determined by weighing the soil before and after oven drying.

There are more scientific methods for determining soil-moisture status. A tensiometer is a simple nonelectric device that operates on physical principles to determine the tension of water in soil (Fig. 5.2; Ross, 2004).

Tensiometers are sealed tubes filled with water that are inserted into the root zone. When the soil is dry, water flows through a porous cap at the end of the tensiometer into the soil, creating a tension inside the tube. This tension is related to the dryness of the soil and is recorded on a gauge built into the tensiometer that is often calibrated in centibars. Most crops should be watered when the tensiometer reads between 20 and 40 kPa (20 and 40 centibars), depending on the soil type (Ross, 2004). Tensiometers must be periodically filled with water to replenish losses into the soil and checked to ensure proper function.

Fig. 5.2. Tensiometers measure water tension in soil through a porous ceramic cone sealed to a plastic tube filled with water and connected to a meter. This tensiometer can automatically activate irrigation when a critical water tension is reached.

Some tensiometers can be hooked to the irrigation system so that water is applied automatically as soon as a threshold soil-water tension is reached. There are other types of "moisture meters" on the market that can be used to determine when irrigation is required. These devices vary widely in cost, accuracy, ease of use, and reliability.

A more advanced and sophisticated method for measuring soil-moisture status is Time-Domain Reflectometry (TDR). TDR is a relatively accurate and convenient method for measuring soil moisture at different depths through access tubes. Advantages of TDR include accuracy, an ability to make continuous measurements, freedom from calibration, and tolerance of high salt concentrations in soils. Disadvantages of TDR include complex electronics and the expense of some equipment.

Some of the other devices for measuring soil-water content include: Hydrotek's velocity differentiation domain (VDD), capacitance (or FDR), gypsum blocks/electrical resistance, granular matrix sensors, heat dissipation, soil conductivity sensors, psychrometry, and neutron probes (Ross, 2004).

Another approach is to monitor plant-water status rather than soil. This approach has the advantage of determining when plants are actually water stressed so water can be applied to restore full hydration. The following are techniques for field assessment of plant-water status. Measuring crop water stress in a field is challenging because of the size and number of plants involved. The Crop Water Stress Index (CWSI) was developed mainly from measuring the canopy temperature with an infrared thermometer. Infrared measurement of leaf temperature is based on the relationship between leaf temperature and transpiration, because as the rate of transpiration is reduced by drought stress, leaf temperature increases. CWSI is used mainly for determining when crops are water stressed and irrigation is needed.

Single representative plants can also be selected for direct measurements of water status in a field. There are many devices that can measure plant-water status, but the easiest, quickest and probably best adapted to field environments is the pressure chamber, which is also called a pressure bomb. Leaf water potential (LWP) can be measured by quickly placing detached leaves into a portable pressure chamber designed for fieldwork. When the leaf is cut off, the xylem sap is under tension and sucked back into the leaf. The xylem tension is roughly equal to LWP. To measure the tension, the detached leaf is sealed in a steel chamber with the cut end, usually the petiole in dicot leaves, protruding from the top and sealed by a grommet so that the chamber can be pressurized. Pressurized nitrogen gas or air is applied to the chamber and when the sap meniscus appears at the xylem surface, the pressure is recorded as the leaf water potential. Typical LWP values for transpiring leaves range from −0.3 MPa (−44 psi) (well hydrated) to −2.5 MPa (−360 psi) (severely drought stressed).

Determining How Much Water to Apply

Tensiometers and TDRs can assess soil dryness, and pressure bombs and other techniques can directly measure plant-water stress to determine when irrigation is required. However, there needs to be a way of determining how much water should be applied when irrigation is needed. An accurate way of determining the amount of water to apply to a field is the "pan" method of estimating evapotranspiration (Brouwer and Heibloem, 1986). This method is simple and does not require sophisticated equipment. An evaporation pan is necessary to hold water for determining the amount of evaporation at a given location. Evaporation pans are available in a variety of sizes and shapes, but are most commonly circular or square. The best-known methods are the "Class A" evaporation pan and the "Sunken Colorado Pan" (Brouwer and Heibloem, 1986). In Europe, India and South Africa, a Symon's Pan (or sometimes Symon's Tank) is also commonly used. Often the evaporation pans are automated with water-level sensors.

In the USA, the National Weather Service has standardized its measurements using the Class A evaporation pan, which is a cylinder 120.7 cm (47.5 in) in diameter with a depth of 25 cm (10 in). Other sizes of pans with straight sides can be used, but may create inaccuracies when making calculations based on Class A pans. The measurement period should begin with the pan filled to exactly 5 cm (2 in) from the top.

The Sunken Colorado pan is placed in the field so the top of the pan is near the soil surface. The rate of water loss from the pan is taken as an approximation of evapotranspiration. If 2.5 cm (1 in) of water is lost from the pan, the surrounding soil and crop has also lost a similar amount of water. When it rains, water is added to the pan, so additions as well as losses are recorded.

Pan evaporation is related to evapotranspiration by a crop coefficient that has been determined experimentally for many vegetables (Brouwer and Heibloem, 1986). Crop coefficients change with crop development, so the amount of water needed for irrigation changes during the season (Table 5.1).

For example, the crop coefficient for developing cantaloupes and other types of melons has been determined through research to be 0.75 (Table 5.1).

Table 5.1. Crop coefficients for some vegetables at various stages of development (Brouwer and Heibloem, 1986).

Crop	Planting to seedling establishment	Seedlings to flower formation or fully developed true leaves	Fruit development/first true leaf development to initial harvest	After first harvest
Green snap bean	0.35	0.70	1.10	0.90
Bean, dry	0.35	0.70	1.10	0.30
Cabbage/carrot	0.45	0.75	1.05	0.90
Cucumber/squash	0.45	0.70	0.90	0.75
Eggplant/tomato	0.45	0.75	1.15	0.80
Dried annual legumes	0.45	0.75	1.10	0.50
Lettuce/spinach	0.45	0.60	1.00	0.90
Sweet corn	0.40	0.80	1.15	1.00
Cantaloupe/ mixed melons	0.45	0.75	1.00	0.75
Onion, green	0.50	0.70	1.00	1.00
Onion, dry	0.50	0.75	1.05	0.85
Peanut	0.45	0.75	1.05	0.70
Pea, fresh	0.45	0.80	1.15	1.05
Pepper, fresh	0.35	0.70	1.05	0.90
Potato	0.45	0.75	1.15	0.85
Radish	0.45	0.60	0.90	0.90

The crop coefficient is multiplied by the amount of water lost from the pan to determine how much water should be applied back to the crop when irrigation is needed. So in this example, if 5 cm (2 in) of water were lost from the pan since the last irrigation, 3.8 cm (1.5 in) of water should be applied to the melon field when the tensiometer reading reaches 20 kPa (20 centibars). In summary, the tensiometer tells when it is time to irrigate, while pan evaporation and the crop coefficient enable a calculation of how much water should be applied.

Water Disposal and Recycling

Fifty years ago there were fewer than half the current number of people on the planet and a commonly held perception was that water was an infinite resource in many parts of the world. The world's population 50 years ago was not as wealthy as today and thus people ate more vegetables, grains, and less meat, so less water was used to produce food. Today, the competition for water resources is much more intense. Since agriculture is one of the largest users of water, there are many issues affecting its responsible use for vegetable production. Agriculture is in competition with industry and urbanization for water. Water demands will continue to grow with world population, so the competition is likely to worsen with the passage of time. To avoid a global water crisis, all parties must learn to share water resources. Farmers must use water more efficiently to meet the increasing demands for vegetables and other food, while industry and cities must also find ways to use water more efficiently.

Increasingly, agriculture water is being reclaimed or recycled for other uses, and recycled water is increasingly used for agricultural production as well. Reclaimed or recycled water is former wastewater or graywater treated to remove solids and certain impurities before it is reused for irrigation or to recharge groundwater aquifers. Recycled water improves sustainability and conserves water since it is reused rather than discharged into rivers and oceans.

There are many levels of treatment and many different methods, which leads to big differences in the quality of the reclaimed water. The best-quality reclaimed sewage water comes from adding a gravitational filtering step after the chemical and biological cleansing. Gravitational filtering allows water to seep from small ponds through the sand into an aquifer over a 400-day period before it is pumped out as purified water for reuse.

The cost of fresh potable water is considerably less than recycled water in parts of the world where water is plentiful. Recycled water is nonpotable and often sold cheaper to encourage agricultural uses. Recycled water sometimes contains higher levels of nutrients, such as nitrogen and phosphorus, which can benefit vegetable crops when used for irrigation.

Public health concerns persist over the use of reclaimed water for vegetable production. Treated wastewater effluent or reclaimed water may contain pathogens that could be transferred through water used to irrigate vegetables, depending on the degree of retreatment. Therefore, it is essential for human health that irrigation water for vegetable production be free of biological contamination, heavy metals, and chemicals. Water quality is especially critical when irrigation is applied directly to foliage that is later consumed, rather than to the soil through drip or subsurface irrigation systems.

Recycled water allows farmers to plan the next year's crop without worrying about limitations due to water shortages. Some countries effectively recycle a large portion of their water for agriculture use. In 2010, Israel treated 80% of its sewage (400 billion l/year (100 billion gal/year)); 100% of the sewage from the Tel Aviv metropolitan area was treated and reused as irrigation water for agriculture or public works. In Spain, approximately 12% of the nation's wastewater is recycled for agriculture and other uses.

Sources of Water

Sources of irrigation water can be subsurface groundwater or surface water. Groundwater may be extracted from springs or wells. Surface water is withdrawn from rivers, lakes or reservoirs or nonconventional sources like treated wastewater, desalinated water, or drainage water.

A source must be selected with sufficient volume to supply irrigation water during all conditions. For example, a stream or a well that provides an adequate supply of water during spring may lack adequate capacity later in the season or during drought. It is important to select a water source with sufficient capacity so that streams are not pumped dry or the water table is not lowered if a well is used (Ross, 2004). Irrigation practices that deplete water tables, alter the flow of streams and limit water availability to others may lead to legal action against the user.

Surface water contains more particulates (sand, algae, etc.), so proper filtration is a critical component, particularly for drip systems, to prevent clogging of the lines. It is wise to install valves to prevent the backflow of water from an irrigation system back into the water source (Hochmuth *et al.*, 2004; Ross, 2004). Before use, surface irrigation water should be tested for salt levels, mineral contamination, chemical pollutants, and biological contamination.

Irrigation pumps may be powered by electric motors or gasoline or diesel engines. In some less-developed areas, irrigation water is pumped or moved by human or animal power. The size of the pump and engine depends on the required flow rate. Sprinkler systems require greater flow rates than drip systems and therefore need larger pumps and motors. In areas where government-supplied irrigation water is delivered through underground conduits or surface canals, water administrators should be consulted for rules and procedures for access.

Types of Irrigation Systems

Irrigation systems distribute water from a source across a field. Many different types of irrigation systems exist. In general, the goal is to supply each plant with sufficient water to sustain growth and proper development without waste. Some systems can be described as "overhead" irrigation because water is applied through the air similar to rain. Overhead systems require relatively high water pressure, use large amounts of water and wet the foliage. Surface irrigation systems supply water directly to the soil without wetting the plants and require low-pressure pumps in many cases.

Overhead delivery systems

Sprinkler

There are many different designs and types of overhead irrigation for vegetable production. With sprinkler irrigation, water is piped to one or more locations within the field and distributed by overhead high-pressure sprinkler heads. Some sprinkler systems are mounted above the soil on permanent risers and are often referred to as "solid-set" irrigation. Other systems are portable and can be assembled and disassembled for transport. Portable sprinkler systems often consist of sections of aluminum pipe placed on the ground between rows with sprinkler heads elevated on risers above the canopy (Fig. 5.3).

Fig. 5.3. Portable sprinkler irrigation systems like this one near Watsonville, California, are often used for crop establishment.

Another sprinkler-system design, often referred to as a "lateral-move" system, uses a series of pipes elevated by fixed wheels about 1.5 m (5.0 ft) in diameter mounted along the length of the pipe. Each section can be coupled together to form a line of pipe across a field. Water is supplied at one end using a large hose. After sufficient water has been applied to a portion of the field, the hose is removed and the remaining assembly rotated, either by hand or with a motorized mechanism, so that the sprinklers move across the field. After repositioning, the hose is reconnected and the new spot in the field is irrigated. The process is repeated until the desired area is irrigated. Lateral-move systems are less expensive to install than the center-pivot system described below, but are much more labor intensive to operate. Most lateral-move systems utilize 10 or 13 cm (4 or 5 in) diameter aluminum pipes with water supplied using high-pressure pumps. The lateral-move system is most often used for small or oddly shaped fields, low growing crops or where labor is inexpensive.

Water "guns"

Water guns, also sometimes called water cannons, are higher-pressure sprinklers that rotate by a ball drive, gear drive, or impact mechanism (Fig. 5.4).

Guns rotate in a full or partial circle and operate at very high pressures of 275–900 kPa (40–130 lbf/in^2) and flows of 3–76 l/s (50–1,200 gal/min), usually with nozzle diameters ranging from 10–50 mm (0.5–1.9 in). The rotation of the gun is similar to that of a lawn sprinkler, only on a much larger scale. Fixed "guns" are mounted on a moveable platform connected to the water source by a hose and set to shoot a stream of water over a portion of a field. Water is usually pumped to the "gun" through a large pipe or plastic hose. A large water gun can be set to irrigate a large portion of a field before being pulled to a new location by truck or tractor.

Automated moving wheeled systems known as traveling guns, water guns, or water reel traveling irrigation are also available. These systems are capable of irrigating fields unattended (Fig. 5.5). Most traveling guns operate with polyethylene tubing. The tubing is wound on a drum, which pulls the gun on wheels across a field. The irrigation water or a small gas engine supplies power. The system shuts off when the sprinkler gun arrives back at the reel.

"Central-pivot" systems

Center-pivot is a form of sprinkler irrigation consisting of several segments of pipe (usually galvanized

Fig. 5.4. A water gun irrigates a full or partial circle in a field.

Fig. 5.5. A traveling gun in an onion field in Nova Scotia moves as an irrigation hose is wound on to a large reel.

steel or aluminum) joined together and supported by trusses, mounted on wheeled towers with sprinklers positioned along the length of pipe (Fig. 5.6).

Center-pivot systems sometimes have "drops" that lower the sprinkler heads to just above the crop canopy to reduce evaporative losses. "Drops" can also be connected to drag hoses to irrigate the ground between rows. Central-pivot systems are used throughout the world on many types of terrain. Crops are often planted in a circle to conform to the center pivot. The irrigated circle is large and can be 1 km (0.6 miles) or more in diameter. Water feeds the system at the pivot point in the center of the arc. Originally many center-pivot systems were water-powered, but more modern systems are driven by hydraulics or electric motors (Fig. 5.7).

A similar system to the central-pivot sprinkler irrigates a field in a rectangular pattern by moving back and forth along an open irrigation canal, which serves as the water source (Fig. 5.8).

Surface delivery systems

With open-surface irrigation systems, water moves across the land by gravity flow. Water is delivered to the crop through open ditches or furrows and enters the soil by infiltration. Flood irrigation covers cultivated land under a shallow layer of water. Historically, open-surface delivery has been the most common method of irrigating agricultural land.

More recently, irrigation water has been more precisely delivered through a series of plastic tubes laid on the soil surface or buried underground. This system efficiently delivers water under low pressure to the plant root zone, reducing the waste typically associated with other surface irrigation systems.

Furrow irrigation

In flat areas, irrigation water can be directed by gravity through fields using a series of earthen ditches and furrows. Crops are planted on raised beds of various widths and spacings. Water is released down the furrow between beds (Fig. 5.9). Small dikes, made of soil or plastic, control the water flow through the furrow. Over several hours, water infiltrates the soil across the bed. For furrow irrigation to work effectively, the ground must be perfectly flat and the soil a deep, porous loam. Furrow irrigation does not work well on sandy soils because water tends to move down the soil profile rather than across a bed. Furrow irrigation requires minimal equipment, does not require a

Fig. 5.6. A central-pivot irrigation system in a Texas spinach field. Sprinkler heads are mounted on extensions or drops that position them closer to the ground to reduce evaporative losses.

high-pressure source of water, and does not wet the foliage. However, furrow irrigation requires large quantities of water, leads to salt buildup on soil surfaces in arid regions, can be difficult to manage, and requires more labor than some other systems.

Seepage irrigation

Seepage irrigation was once widely used in parts of Florida in the USA where there is an impermeable hardpan beneath a permeable sandy soil. This system has been described as "farming in a soup bowl" because the water has restricted mobility vertically but not laterally. Water is conveyed down irrigation furrows between raised beds similar to furrow irrigation. Since water moves rapidly laterally, one irrigation furrow may water several beds. Furrows may be lined with plastic film to control lateral movement.

Controlling the water table is crucial with seepage irrigation. Since water movement into the subsoil is limited, it is easy to waterlog the soil if too much is applied. Water may also be pumped from the irrigation furrows to remove the excess. Seepage irrigation requires little equipment and does not wet the foliage. However, it does require large quantities of water and may be difficult to manage.

Seepage irrigation may also describe subirrigation in fields with naturally high water tables. Seepage in this case refers to the artificial raising of the water table to moisten the soil in the root zone. Often subsurface seepage irrigation is used on organic or muck soils that were created by the decomposition of peat-moss bogs in old lake beds. In muck soils, water can be pumped in or out to adjust the water table.

Spate water irrigation

A special form of irrigation using surface water is spate irrigation, also called floodwater harvesting. During a flood, water is diverted to normally dry rivers (wadis) or overflow diversion beds using a network of dams, gates, and channels spread over large areas. The moisture from floodwater is stored in the soil or held in reservoirs for later use as irrigation. Spate irrigation areas are often located in semi-arid or arid, mountainous regions.

While floodwater harvesting is an accepted irrigation method, rainwater harvesting is usually not considered irrigation unless it is transported and applied to a crop in a different location. Rainwater harvesting is the collection and consolidation of runoff water from roofs or unused land, which can be retained and reused for vegetable irrigation.

Manual irrigation

In areas with limited infrastructure and equipment, high labor inputs or animals must deliver irrigation water. Manual irrigation systems use a series of buckets and ropes to draw water from one source and deliver it to another. Cans are hand carried from a water source to irrigate vegetable fields around some large cities in parts of Africa.

Trickle or drip irrigation

Drip, also known as trickle irrigation, distributes water under low pressure through a network of

Fig. 5.7. A central-pivot irrigation system near Bakersfield, California. High-mounted sprinkler heads provide uniform irrigation. Electric motors mounted on the wheels move the system.

Fig. 5.8. This overhead irrigation system is similar to a central pivot but pulls water from an irrigation canal and moves across a field irrigating a rectangle rather than a circle.

tubes in a predetermined pattern and applies a small discharge at the base of each plant or the surrounding soil (Fig. 5.10; Ross, 2004).

In a drip system, water is slowly discharged from an emitter in the irrigation tube directly into the root zone. If managed properly, drip is a very water-efficient method of irrigation since water is delivered directly to the root zone, thus minimizing evaporation and runoff. In modern agriculture, drip irrigation is often combined with plastic mulch to further reduce evaporation and conserve moisture.

Fig. 5.9. Furrow irrigation of a tomato crop on raised beds near Davis, California.

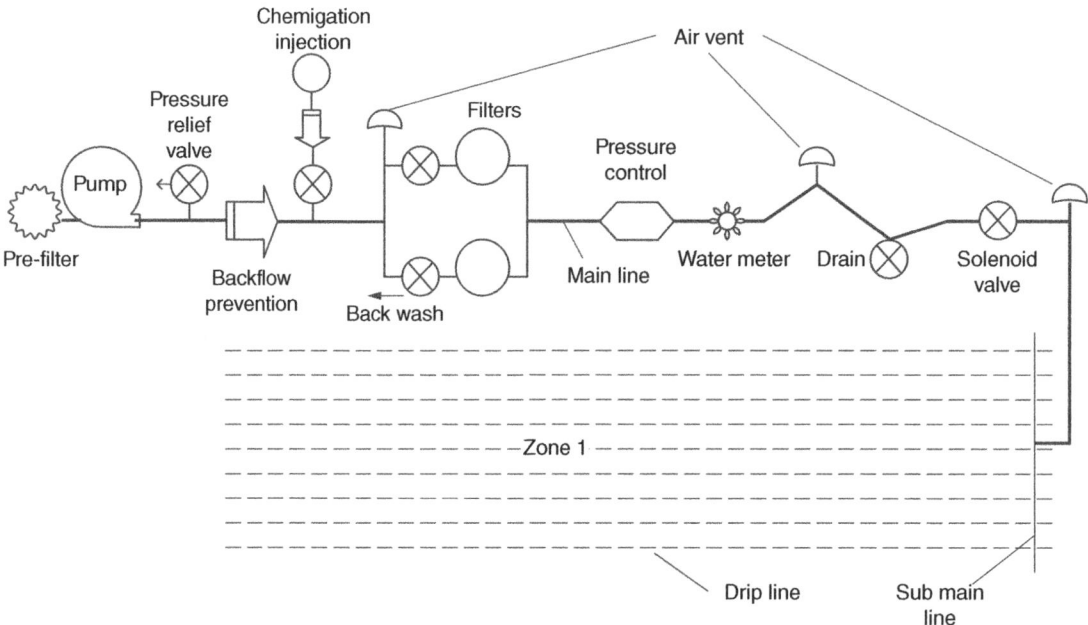

Fig. 5.10. Diagram of a drip irrigation system. (Source: *Production of Vegetables, Strawberries and Cut Flowers Using Plasticulture*, NRAES-133 (NRAES, 2004).)

There are several different drip-irrigation designs. One of the most common designs used for vegetable irrigation consists of a series of thin-walled plastic tubes that are laid on the soil surface in the row or sometimes beneath a layer of thin plastic mulch (Fig. 5.11).

These tubes are fragile, so care must be taken not to damage them during planting or cultivation. The tubes, or "tape", have premade small holes spaced uniformly along their length. The spacing of the holes can be ordered to match the in-row spacing

Fig. 5.11. Drip irrigation tubing placed on the soil surface between rows of a lettuce crop near Gilroy, California.

of the crop being grown. A larger diameter "lay-flat" tube that runs from the water source feeds the drip tape. The drip line screws into a small hole cut in the wall of the lay-flat feeder tube (Fig. 5.12).

Drip-irrigation management is very different from other types of irrigation. Since most drip systems only apply 12.7–25.4 cm/km^2 (0.05–0.10 in/h), it takes considerable time to supply water through the system. Drip irrigation is generally applied frequently to keep the soil in the root zone moist. If operated for too long or if the delivery rate is too high, water can be wasted through deep percolation as it moves below the root zone.

A pressure regulator maintains between 55 and 100 kPa (8 and 15 psi) in the system. These water pressures are significantly lower than for most other irrigation systems, with the exception of gravity-fed surface.Drip irrigation can deliver water uniformly throughout a field. Pressure-compensating emitters can help regulate pressure on steep slopes, so the field does not have to be level for drip irrigation. If trickle irrigation is used in conjunction with plastic mulch, applications can be limited to about three times a week, depending on the crop and soil type. Tape-style plastic tubing is generally used for just 1 year, because it breaks down in the sun, can be damaged during freezing weather and can clog over

Fig. 5.12. Drip irrigation components include: pressure regulator, filter, shut-off valve, and lay-flat water feeder line that attaches to drip tape that runs under the plastic mulch in each row.

time. Filters are required, particularly if surface water is used. To keep drip tubes operating at peak efficiency over longer periods, dilute bleach or other chemicals to control algae and bacterial growth are added to flush the drip to prevent emitters from clogging.

Subsurface drip irrigation (SDI) is similar to drip irrigation but the drip tubes or tape are buried in the root zone below the soil surface (Fig. 5.13).

Fig. 5.13. Subsurface drip irrigation near Watsonville, California. The irrigation lines are buried in the row middles below the soil surface. Special drip tape is used for subsurface use.

This method saves water in a number of ways. Surface-water evaporation is eliminated and the incidence of disease may be reduced. Water is applied directly to the crop's root zone and not to the soil surface where most weed seeds overwinter, so weed germination is reduced. Also, some crops may benefit from the additional heat provided by dry surface conditions. Since the water is applied below the soil surface, there is no crusting or ponding, and potential for surface runoff is eliminated during irrigation. However, a subsurface drip system may require higher initial investment and costs will vary due to water quality, filtration need, choice of material, soil characteristics, and the degree of automation desired.

Fertigation

Fertigation is the application of fertilizer through irrigation water. Fertigation is commonly used with drip irrigation. When done correctly, fertigation allows fertility inputs to change during the season to match the changing needs of a developing crop (Hochmuth *et al.*, 2004). Fertigation is often used in conjunction with petiole analysis, described in Chapter 4, to determine immediate nutriment needs. Based on petiole analysis results, fertigation can be tailored to meet the nutritional needs of a crop.

Nutrients may be applied entirely through drip irrigation during the production season or through a split application with some of the fertilizer applied to the soil before planting (Hochmuth *et al.*, 2004). For split applications, most growers apply all P and 50% of the N and K to the soil at planting, although other combinations are sometimes used. Some growers prefer to apply P at planting, since it is not particularly water soluble and has low mobility in the soil. It is important to use soluble forms of fertilizer for fertigation.

Various types of fertilizer injectors may be used to add fertilizer to irrigation water. The most sophisticated types are computer-controlled, but less expensive types are also effective. Back-flow regulators are required by law in some areas to prevent fertilizer water from draining out of the irrigation system and into the water source.

References

Brouwer, C. and Heibloem, M. (1986) Irrigation Water Management. FAO Training Manual No. 3. United Nations FAO, Rome. Available at: www.fao.org/docrep/S2022E/s2022e00.htm (accessed 9 September 2013).

FAO (2007) Coping with water scarcity - Challenge of the twenty-first century. Available at: www.plantstress.com/Articles/drought_i/drought_i.htm (accessed 11 September 2011).

Hochmuth, G.J., Paterson, J.W. and Garrison, S.A. (2004) Fertigation. In: Lamont, W. (ed.) *Using Plasticulture for Vegetables, Strawberries and Cut Flowers (NRAES-133)*. Natural Resource, Agriculture and Engineering Service (NRAES), Ithaca, New York, pp. 36–45.

Hsiao, T.C. and Bradford, K.J. (1983) Physiological Consequences of Cellular Water Deficits. In: Taylor, H.M., Jordan, W.R. and Sinclair, T.R. (eds) *Limitations to Efficient Water Use in Crop Production*. American Society of Agronomy, Madison, Wisconsin, pp. 227–265.

NRAES (Natural Resource, Agriculture and Engineering Service) (2004) *Production of Vegetables, Strawberries and Cut Flowers Using Plasticulture*. NRAES-133, Ithaca, New York. www.nraes.org

Ross, D.S. (2004) Drip Irrigation and Water Management. In: Lamont, W. (ed.) *Using Plasticulture for Vegetables, Strawberries and Cut Flowers (NRAES-133)*. Natural Resource, Agriculture and Engineering Service (NRAES), Ithaca, New York, pp. 15–35.

Siebert, S., Hoogeveen, J., Döll, P., Faurès, J.-M., Feick, S. and Frenken, K. (2006) The Digital Global Map of Irrigation Areas – Development and Validation of Map Version 4. Available at: www.tropentag.de/2006/abstracts/full/211.pdf (accessed 12 September 2011).

Snyder, R.L. and Melo-Abreu, J.P. (2005) *Frost Protection: Fundamentals, Practice, and Economics*, Vol. 1. Food and Agriculture Organization of the United Nations, Rome.

6 Mulches

Introduction and History

In this chapter, we will discuss the use of mulches in vegetable production. Irrigation was discussed in Chapter 5, although there is overlap with this chapter.

Plastic mulches were first used experimentally for vegetable production in the early 1950s (Lamont, 2004a). By the 1960s, plastic mulches were widely used because growers quickly recognized that they were affordable, easy to install, provided effective weed control, and increased early harvest conserved moisture. Early mulches were primarily black or clear. Black mulches were popular because they provided weed control and heated the soil to accelerate early season production of warm-season vegetables (see Fig. 6.2). Plastic mulch has been shown to benefit the production of many crops, but the cucurbits, pepper, tomato, and eggplant seem to show the greatest response. Approximately 6,500 km^2 (2,500 $miles^2$) of polyethylene mulch are used for crop production in the world today.

Plasticulture is defined as "a system for growing vegetable crops where significant benefit is obtained from using products derived from synthetic polymers" (Lamont, 1993). Typical plasticulture production consists of raised beds covered with plastic mulch, drip irrigation, delivery of chemicals through the drip irrigation (fertigation/chemigation), and pre-plant soil fumigation under the plastic mulch. Claimed benefits of the plasticulture system compared to conventional bare-soil production include: earlier production; higher yields per acre; cleaner, higher-quality product; more efficient use of water and fertilizer; reduced leaching of mineral nutrients; less soil erosion; fewer disease problems; fewer weed problems; better management of some insects; reduced soil compaction; less root pruning; and maximum efficiency through double- or triple-cropping (Lamont, 2004a).

Mulches from Organic Material

Other mulch materials were used before plastic mulches were created. Historically, mulches were composed of natural waste materials that could be easily placed in the field around crop plants and would naturally decompose over time. One popular mulch material was small-grain straw (Fig. 6.1).

Straw refers to the stalks of crops like wheat, barley, oats, or rye. After grain harvest, straw can be left in the field, plowed under, or burned in dry areas where there is insufficient moisture for timely decomposition. Straw can also be baled and used as animal bedding, landscaping, or agricultural mulch. Straw was applied by hand around established plants or sometimes uniformly spread or blown on to fields prior to transplanting. Straw was widely used as a vegetable mulch for many years because it was readily available, inexpensive, easy to spread, prevented compaction, decomposed naturally, helped keep crops clean, helped retain soil moisture, and provided some inhibition of weed growth. Problems included contamination by weed seed, cost, and inconsistent decomposition times. Straw mulch has a high carbon-to-nitrogen ratio, which means that it breaks down more slowly than leaf litter, yard wastes, or more succulent vegetative materials that have high nitrogen content. Microbes that decompose straw also use nitrogen and may compete with crops for plant-available nitrogen. Therefore, additional nitrogen fertilizer is sometimes warranted when heavy straw mulch is used.

Cover Crops and No-till

Over the past 30 years, straw has been largely replaced by plastic mulch for traditional large-scale commercial production operations in many regions of the world. Organic and sustainable vegetable growers have shown renewed interest in straw and other natural mulch materials (Lu *et al.*, 2000). One reason for the current interest in straw and renewable mulches is that straw mulch provides many of the same benefits as plastic but is compatible with organic production requirements, depending on how the straw was produced. Natural phytotoxic chemicals in wheat and rye straw, as well as wheat-straw mulches used in no-tillage (no-till) production systems, suppress certain broadleaf weeds (Shilling *et al.*, 1985). No-till production systems allow vegetables to be planted or transplanted directly into plant residues of the previous cover crop, which provide a mulch effect. In many cases, cover crops are grown in rotation with vegetables and serve as mulch crops when no-till planting is used. Planting directly into crop residue is more efficient because the need to bale straw in one location and spread it as mulch in another is eliminated. Annual grains are popular cover crops and can be killed by herbicide, chopping, or rolling to break stems before advanced grain development, to serve as mulch for the no-till planting of vegetables.

In addition to small grains, legume cover crops such as hairy vetch and crimson clover fix nitrogen (N) that is released upon decomposition and contributes to the N requirements of the vegetable crop planted in rotation. Cover crops suppress weeds, provide suitable habitat for beneficial predator insects and are non-hosts for nematodes and other pests in rotation. Fresh-market tomatoes transplanted into the residue of a hairy vetch cover crop used as mulch produced higher yields and more profits in comparison with the same crop grown using black polyethylene mulch or on bare soils without mulch (Lu *et al.*, 2000). A disadvantage of legume-based no-till mulches is that decomposition can be rapid and not timed to crop maturity. In other words, they may decompose prematurely before the cash crop is mature. Straw-based mulches, because of their higher carbon-to-nitrogen (C:N) ratio, tend to decompose slower, which can be a major advantage in long-season areas.

Fig. 6.1. Rice straw mulch of a Chinese cabbage crop in Taiwan.

Fig. 6.2. Comparison of early-season pumpkin growth with and without black plastic mulch. Each row was direct-seeded on the same date and drip irrigated. The plants on black plastic are larger because the plastic helps warm the soil and retain moisture.

Plastic Film Mulch

Many of the benefits of organic mulches also apply to plastic. However, plastic mulches have some additional advantages that make them very popular. These include consistent long-term weed control, an ability to fumigate soil and modification of soil temperature. Black mulch absorbs most UV, visible, and infrared wavelengths of solar radiation and heats the soil by conduction, so improving contact between the plastic mulch and the soil surface improves heating efficiency. Some of the absorbed energy heats the soil, but unfortunately much is re-radiated as thermal or long-wavelength infrared radiation and lost to the atmosphere. Soil temperatures under black plastic during the daytime are generally 2°C (36°F) higher at a 5 cm (2 in) depth and 1.3°C (34°F) higher at 10 cm (4 in) depth compared to bare soil. For warm-season vegetables like tomato, pepper, and melons, this increase in soil temperature translates into a marketable maturity that is 7–21 days earlier, depending on the crop and other environmental effects (Fig. 6.2).

Higher yields obtained from plasticulture production were attributed to increased water-use efficiency, better utilization of nutrients through fertigation, reduced fertilizer leaching, less soil erosion (wind and water), reduced soil compaction and root pruning, the ability to double-crop, cleaner/higher-quality produce, and a reduction in disease, insect pests, and weeds (Lamont, 1993).

Some of these advantages may require additional clarification. Plastic mulch helps retain fertilizer in the root zone by deflecting leaching rains off the bed and into the row middles. The weed control provided by plastic film eliminates cultivation with heavy equipment, reducing compaction and root damage that may occur when tilling close to the plant. The superior drainage on raised beds and more precise management of soil-moisture contents under plastic may reduce the incidence of certain soil-borne diseases. Without mulch, rain splashing soil on to the vegetables sometimes increases disease and necessitates washing before sale. Plastic mulch keeps crops cleaner and reduces fruit rot.

Components of the plasticulture system may be used individually or in combination. Plastic mulch can be installed by hand with or without using raised beds, but obtaining a tight installation is more difficult. If plastic is not spread tightly without wrinkles or creases over a smooth, solid surface that is highest in the center, then rainwater tends to collect on top of the plastic. The weight of the water can pull out the edges of the mulch, causing the mulch to dislodge and be blown away by the wind. Water that collects on plastic can become a breeding ground for insects and support weed growth, negating one of the prime reasons for using mulch.

Crops may be planted manually through plastic mulch by carefully cutting or punching clean holes in the plastic without damaging irrigation tubing. Holes cut in plastic should be kept as small as possible to prevent weed growth and loss of heat and water. Using plastic mulch without drip irrigation is not recommended. During the growing season, water will be depleted beneath the mulch by plant roots and transpiration. Rain or overhead irrigation cannot replenish water used by the crop, because the plastic prevents moisture from entering except through the small hole around the plant. If drip irrigation is not available, porous mulch should be used. Landscape fabric, which inhibits weed growth, and perforated plastic mulch both allow water to infiltrate.

Plastic mulch may be installed on raised beds with drip irrigation but without fumigation. Strategies like grafting allow vegetable production on soils with pest problems without fumigation. Some soils do not have pest problems that are severe enough to warrant fumigation. Crop rotation can be an effective soil-disease-management strategy if other profitable cash crops can be grown.

Site Preparation, Mulch Installation, and Planting

Proper site preparation is essential for the successful use of plastic mulch. The soil should be tested and amendments made before mulch is laid. In addition to testing the nutrient content of the soil, having soil tested for nematodes is also wise, particularly in known problem areas. A multipurpose fumigant is sometimes injected into the soil when the mulch is laid to control weeds, nematodes, soil-borne insects, and diseases. The field should be plowed early to ensure that organic matter and crop residues are properly incorporated into the soil. Some residues, such as wheat stubble, may puncture plastic mulch if exposed through the soil surface when the mulch is laid. Most problems encountered with plastic mulch are associated with poor soil and bed preparation. The soil must be friable, free of large stones, clods, sticks, and plant residue. The top of the bed should be firm and flat, so the plastic can fit tightly over the bed.

Most plastic mulch is installed on raised beds, which ensures good drainage. Raised beds are created using a bed shaper (Fig. 6.3a). A bed shaper is a three-sided metal-framed device that molds the soil when pulled behind a tractor to create a smooth, flat bed that is higher than the furrows on either side. Raised beds are usually 10–15 cm (4–6 in) high, 75 cm (30 in) wide and have a declining slope of 2–5 cm (0.8–2 in) from the center to the shoulder (Lamont, 2005). After the bed is created, a mulch layer pulled behind a tractor uniformly installs the plastic film (Fig. 6.3b). It is important to lay plastic immediately after bed shaping so that rain and wind do not damage the bed. Irrigation drip tubing is also placed under the plastic mulch as it is laid. The edges of the plastic that hang over the sides of the raised bed are mechanically buried under soil by a roller and coulters that securely cover the edges of the plastic film to hold it tightly across the top of the bed and keep it from becoming dislodged during storms (Fig. 6.3b). The film must be stretched tightly over the bed to ensure maximum heating if black or dark-colored plastic is used and to allow rainwater to drain into the furrows between rows. Tight installation is important for planting because the plastic must be punctured or burned to make a hole in order to place the transplant or seed in the soil beneath the mulch. Seeds or transplants should be set in the center of the hole so the developing plants do not emerge or grow beneath the plastic mulch. Plants should not be in direct contact with the plastic, because solar-heated plastic may burn or damage succulent transplants that have not been properly hardened.

Laying plastic mulch and drip tube and fumigating in the fall prior to spring planting is sometimes done in mild climates where winters are not severe. Fall installation of a plasticulture system offers several advantages, including avoidance of wet spring weather, more effective fumigation because of higher soil temperatures and higher nematode populations, and fall mulch installation enables earlier crop establishment in the spring.

Solid Fertilizer Application

The addition of solid fertilizer and mulch installation is often done in a single pass through a field. Solid fertilizer may also be applied as a separate process after the beds are formed but before plastic is laid (Fig. 6.4).

Fertilizer and lime should be applied according to soil test results. In some cases, 25–50% of the N and potassium (K) and all of the phosphorus (P) are applied at planting if fertigation is used. The remaining N and K can be applied through the drip tube using soluble fertilizers (e.g. calcium nitrate, sodium nitrate, or potassium nitrate). Some growers apply 100% of the fertilizer at planting if fertigation is not used. However, on some soils and with many crops, side-dressing late in the season is necessary. If fertigation is not used, approximately 75% of the fertilizer should be applied preplant and the rest side-dressed. This can be accomplished by applying granular fertilizer at the base of each plant.

Fumigation

Soil fumigants are sometimes injected under plastic mulch at installation to control insects, termites,

Fig. 6.3. Beds are shaped by molding soil (a) before the bed is covered with plastic film (b) (photos courtesy of Josh Freeman).

Fig. 6.4. The spreading of dry granular fertilizer on the tops of beds after shaping but before plastic film and drip tape are installed.

rodents, weeds, nematodes, and soil-borne diseases. Methyl bromide was used as a standard soil fumigant beginning in the early 1960s, but now has been removed from the market in much of the world because it is an ozone-depleting chemical and is toxic to humans. Farm workers must wear personal protective equipment when working with fumigants and other farm chemicals. Other chemicals can be used as fumigants; a partial list includes metham sodium, 1, 3-dichloropropene, chloropicrin, dimethyl disulfide, and methyl iodide (Gao *et al.*, 2011). Chloropicrin is commonly used in combination with other fumigants for increased potency and as a warning agent because, unlike other odorless gaseous fumigants, chloropicrin has a characteristic smell.

For effective fumigation, soil temperature at a depth of 15 cm (6 in) must be at least 10°C (50°F). Higher soil temperatures favor greater volatilization of fumigants and greater movement through soil spaces. Decomposed organic matter improves soil structure and generally helps fumigant dispersion in the soil. However, very high amounts of organic matter may absorb or tie up a fumigant, reducing its effectiveness. Undecomposed crop debris may hinder fumigant dispersion in the soil and may also harbor insects and pathogens that escape the fumigant. Organic matter is most beneficial when it is thoroughly decomposed. Fumigants move through air spaces in soil and dissolve in soil water. They must enter the soil solution to contact and kill pests. Moderate levels of soil moisture therefore aid in obtaining effective fumigation.

Gaseous fumigants are injected directly into the soil during plastic installation. Leakage of fumigant gas during application poses a threat to workers and the environment, and reduces the efficacy of the treatment. Some of the methyl bromide replacements are classified as emulsifiable concentrates (EC), which means they consist of an active ingredient and an emulsifying agent in an organic solvent. The solvent is usually not soluble in water, but when an EC product is mixed with water prior to application the resulting mix is a dispersion of fine, oily particles in water that can be delivered through drip irrigation lines. These fumigants, such as 1, 3-dichloropropene, can be applied through drip lines after the plastic is installed. For maximum effectiveness, two drip tapes are needed to achieve adequate coverage across a bed. With fumigant applications through the drip line, soil temperature and moisture do not dictate when plastic is laid and fumigation applied, losses of chemical into the atmosphere are reduced, farm-worker safety is increased, the need for worker training is reduced, the number of workers who must wear personal protective equipment is decreased and re-treating of existing rows is possible. Most fumigants require a waiting period of 7–60 days before crops can be safely planted in the treated bed, depending on the fumigant used and environmental conditions. After this period, the waiting fumigant is neutralized or leaks into the atmosphere.

Types and Colors of Plastic Mulch

Mulch technology has changed considerably over the last 50 years. Today there are many different textures and thicknesses available to growers, depending on the application. Mulches are available in almost every color of the rainbow (Fig. 6.5; Lamont, 2004b).

The following paragraphs summarize some of the many popular types of plastic mulches available for vegetable production. Black mulch remains one of the most popular types, and its characteristics were described above.

In the 1960s, clear plastic mulch was introduced. Clear plastic causes a greenhouse effect between the soil and plastic, increasing soil temperatures more than black mulch. Clear plastic mulch transmits 85–95% of sunlight, depending on the thickness and clarity of the film (Lamont, 2004b). Often the inner surface of clear plastic mulch is covered with condensed water droplets. This water does not absorb incoming shortwave radiation, but does collect long-wave infrared radiation from soil, so much of the heat lost to the atmosphere from a bare soil is retained by clear plastic mulch. Daytime soil temperatures under clear plastic mulch are generally 4.5–7.8°C (40–46°F) higher at a 5 cm (2 in) depth and 3–5°C (37–41°F) higher at a 10 cm (4 in) depth compared to bare soil, and higher than under black plastic. The down side of clear plastic mulch is the loss of weed control (Fig. 6.6). Under the right conditions, weed seeds can germinate and grow under the clear plastic, competing with crop plants and in some cases pushing the mulch out of the soil. Fumigants or herbicides are often used to provide weed control under clear plastic mulch.

Infrared transmitting mulch (IRT) provides the weed control of black mulch with the soil-warming ability of clear mulch. These mulches absorb

Fig. 6.5. This experimental plot compares effects of white, yellow and brown colored mulches on tomato development in Florida.

Fig. 6.6. Clear plastic warms the soil to increase early-season growth of warm-season vegetables under early-season cold conditions. Notice the weed growth under the plastic. If uncontrolled, weeds growing underneath may lift the plastic out of the soil. Selective-wavelength-transmission mulches have been developed to inhibit weed growth but warm soils.

photosynthetically active radiation, but transmit solar infrared wavelengths of radiation responsible for heating soil, providing a compromise between black and clear mulches. The IRT mulches are usually colored blue-green or brown (Fig. 6.5). Some weed seeds may germinate under IRT mulch, but the plants are weak and stunted and do not grow well.

In warmer climates, heating the soil is not required and plants benefit from cooling the soil instead. White mulches reduce soil temperature, which benefits certain vegetables grown in warm climates (Fig. 6.7).

Thin-white mulch can transmit enough light for weeds to grow beneath it. If thin-white film is used, chemical weed control with herbicides or fumigation may be required. Some mulches are coextruded white and black so that they are white on one side but black on the other. This design provides the cooling effect of white mulch with the weed control of black. The coextruded mulches are versatile because either side can be used facing up.

Highly reflective silver mulches have been developed primarily to repel aphids, which prevents or delays virus symptoms, particularly in squash crops (Lamont, 2004b). The soil temperatures under these mulches will be several degrees (3–4.5°C/37–40°F) cooler compared to black plastic mulch.

Red mulches have been tested on many different vegetables with varied results. Red mulch raises soil temperatures similar to black mulch, so crop maturity can be accelerated. The effect of some colored mulches is reportedly dependent on the wavelengths of light they reflect back on to the crop canopy. For example, red mulch increased the ratio of red:far red wavelengths (R:FR) in the reflected light. Red, blue, yellow, gray, and orange mulches reflect different radiation patterns back on to the lower side of the crop canopy, thereby affecting plant growth and development (Lamont, 2004b). Yet it is difficult to imagine how light reflected from mulches could have a consistent measureable effect on plant growth past the seedling stage, since the mulch directly below the crop is shaded and the reflected light has much lower intensity than full sunlight. Differences in reflected light of red mulch were minimal 40 cm (16 in) above the canopy on the shaded side of the row, which would seem to restrict effects of reflected light from color mulches to the early stage of plant development (Loy et al., 1998).

Colored mulches may affect insect populations on crops. Green peach aphid populations increased on squash and pepper plants grown on red mulch (Orzolek and Murphy, 1993). Yellow mulches attract cucumber beetles and potato beetles similar to the way insect traps work in a greenhouse (Fig. 6.5). Blue mulches reportedly attract thrips (Lamont, 2004b).

Fig. 6.7. White plastic mulch is used to cool this summer pepper crop in Florida.

In addition to the creation of new colors, plastic mulch designs have changed in other ways over the years as well. The thickness of plastic mulch is measured in mils where 1 mil is 0.025 mm (1/1,000 in). Today, plastic mulch is often made from either low- or high-density polyethylene (LDPE and HDPE), which ranges from 0.5–1.25 mil (12.7–31.75 µm) thick and 1.2–1.5 m (4.0–5.0 ft) wide (Lamont, 2005). Metalized mulches are usually LDPE films coated with a microscopically thin layer of aluminum. The amount of aluminum is so small that it is less than 1% of the weight of the roll. Metalized film has excellent adhesion qualities, strength characteristics and high reflectivity. Metallic film is available with metal over a white or black polyethylene backing.

Modern mulches have been downsized compared to the original mulches introduced in the 1960s. Mulches historically were several mils thick, but now some types of mulch are less than 1 mil in thickness with the same strength as thicker mulches thanks to the reformulation of mulch polymers. HDPE reduces mulch weight and cost, while increasing strength compared to LDPE.

Some mulches are embossed rather than smooth. Embossed mulches have a rough texture and small diamond pattern integrated into the film. Embossed mulches stretch, so when stepped on or driven over in the field, the plastic gives rather than pulling out of the soil.

Plastic Mulches for Fumigation

Traditional mulches leak gaseous fumigants into the atmosphere, which reduces the effectiveness of the treatment, wastes money, poses a hazard to farm workers, and may cause environmental damage. Mulches that have lower permeability have been developed to contain fumigant in the soil. The most advanced high-retention mulches are Virtually Impermeable Film (VIF) and Totally Impermeable Film (TIF) (Fig. 6.8). These can be used in conjunction with fumigation to prevent volatile losses of gaseous fumigants (Gao *et al.*, 2011). The properties, such as type of polymer, mulch thickness, and oxygen transmission levels, vary with each film because there are no clearly defined specifications for either type. They can be constructed with many of the same colors and characteristics of the conventional mulches described earlier in this chapter.

Virtually Impermeable Film is typically a multi-layer film composed a layer of ethylene vinyl alcohol (EVAL or EVOH) or nylon polyamide between two polyethylene layers, making it less permeable than

Fig. 6.8. VIF and TIF are multilayer films designed for use with fumigants.

LDPE and HDPE mulches (Gao *et al.*, 2011). For example, methyl bromide emissions in the field may be reduced from 60% to less than 5% when using a VIF compared to HDPE (Yates *et al.*, 2002). Laboratory tests have shown that some types of VIF are at least 75 times less permeable to methyl bromide and may be as much as 500–1,000 times less permeable compared to standard polyethylene films. Fumigant permeability is strongly affected by temperature. In general, permeability increases 1.5 to 2 times for every 10°C (50°F) increase in temperature.

Totally Impermeable Film utilizes a high barrier ethylene vinyl alcohol (EVAL) copolymer (Chow, 2009). Characteristics of TIF include good film-handling properties and extremely low fumigant-vapor permeability. TIF reduced drip-applied chloropicrin emissions by approximately 85% compared to HDPE film. Benefits of mulches with increased fumigant retention are a reduction in the amount of fumigant needed for effective pest control, lower emissions, and a reduced buffer-zone requirement in areas where fumigant use must be isolated.

Solarization

A much more sustainable alternative to chemical fumigation is solarization, which is the process of utilizing solar energy to warm the soil to a temperature hot enough to kill living organisms (Stapleton, 2000). Solarization is effective mainly in areas that have high temperatures and sunny weather, such as the Middle East, southern Europe, southern Asia, Australia, northern Africa, southern North America, Central America, and northern South America. To apply a solarization treatment, clear plastic mulch is laid over the soil and the edges buried (Fig. 6.7).

The covered soil is left in place for an extended period of several weeks as soil temperatures rise until they reach 50–60°C (120–140°F). Adding compost to the soil from certain types of crop residue, such as cabbage leaves, which contain isothiocyanates, may increase the effectiveness of solarization.

Solarization is less effective in northern latitudes. Other limitations include a drop in effective temperatures with increasing soil depth. Soil-treatment periods often coincide with the growing season and require time to be most effective.

Disposal and Recycling

Double- and even triple-cropping of vegetables can be achieved on the same plastic mulch by growing vegetables in succession in long-season areas. In northern areas, single-crop production is more the norm. A plasticulture system is generally renewed annually because drip lines fail to operate efficiently with time, plastic becomes damaged or loose, fumigation loses its effectiveness, or crop rotation needs to be practiced, to name a few reasons. Non-biodegradable plastic mulches must be removed from the field and disposed of properly. Mechanical aids that collect used plastic from the field help reduce labor. However, these devices are relatively slow and not cost-effective for small growers. In some areas plastic mulch is still removed by hand. Approximately 8 h labor is required to remove 4,000 m^2 (1 acre^2) of plastic mulch from a field (McCraw and Motes, 2007).

Disposal of plastic mulch is considered a serious environmental issue. Unfortunately, used plastic mulches are often disposed in landfills. This is not a sound environmental practice, since plastics degrade very slowly when buried. Plastic films may be burned as a concentrated form of fuel by power plants. However, the lack of a consistent supply of mulch throughout the year and a limited number of power plants capable of handling mulch for fuel limits this application.

The recycling of used/disposed plastic mulch into resins for reuse in the manufacturing industry is practiced in some areas (Hochmuth, 1998). Plastic mulches may be recycled, but generally this is not cost-effective. When recycling agricultural plastic, they must first be separated into different categories. Cleaning can often be a difficult task because the mulch may be covered with soil, manure, and plant residues that are difficult to remove. The costs of transporting, sorting, shredding, cleaning, and pelleting is often more than its market value, making plastic mulch less attractive than other recycling source materials such as plastic bottles.

Biodegradable Mulches

Researchers developing degradable plastic mulches to reduce labor and eliminate disposal problems. Films that are degraded by UV light have been available for many years. However, getting these products to decompose at crop maturity has been a problem that has not been resolved. Also, the buried edges of UV-degrading mulches do not break down because they are not exposed to sunlight and must be collected at the end of the season.

Scientists developed biodegradable mulches containing starch that are broken down slowly by soil microbes into carbon dioxide and water. The most widely available biodegradable mulch is made of a biopolymer from corn starch, plus proprietary biodegradable complexing agents derived from renewable, synthetic, or mixed sources. This mulch is approved for use on European and Canadian organic farms and performed comparably to black plastic in raising soil temperatures. Melon yields in comparative trials were also equivalent to those grown on black plastic mulch. Premature breakdown of plastic during the growing season occurred with clear and wavelength-selective products, which may necessitate supplemental weed control under certain circumstances (Waterer, 2010). Biodegradable mulch is thinner (0.5–0.8 mil (12.7–20.32 μm) thicknesses) than typical black polyethylene (1.25 mil (31.75 μm)), so extra care is required to prevent tears. Biodegradable mulches can range from two to three times the cost of standard black plastic, but end-of-season labor and disposal costs are avoided. Although biodegradable plastic mulches exist, non-biodegradable plastics are more widespread at the time of writing, making recycling and responsible disposal an important part of plasticulture usage.

References

Chow, E. (2009) An update on the development of TIF mulching films. *Proceedings of the 2009 Annual International Research Conference on Methyl Bromide Alternatives and Emissions Reductions.* Methyl Bromide Alternative Outreach, San Diego, California, pp. 50-1–50-3. Available at: www.mbao.org/2009/

Proceedings/050ChowEMBAO2009.pdf (accessed 28 September 2013).

Gao, S., Hanson, B.D., Wang, D., Browne, G.T., Qin, R., Ajwa, H. and Yates, S.R. (2011) Methods evaluated to minimize emissions from preplant soil fumigation. *California Agriculture* 65, 41–46.

Hochmuth, G. (1998) What to do with all that mulch? *American Vegetable Grower* 46(4), 45.

Lamont, W.J. (1993) Plastic Mulches for the Production of Vegetable Crops. *HortTechnology* 3, 35–39.

Lamont, W.J. (2004a) Plasticulture – An Overview. In: Lamont, W. (ed.) *Using Plasticulture for Vegetables, Strawberries and Cut Flowers (NRAES-133)*. Natural Resource, Agriculture and Engineering Service (NRAES), Ithaca, New York, pp. 1–8.

Lamont, W.J. (2004b) Plastic Mulches. In: Lamont, W. (ed.) *Using Plasticulture for Vegetables, Strawberries and Cut Flowers (NRAES-133)*. Natural Resource, Agriculture and Engineering Service (NRAES), Ithaca, New York, pp. 9–14.

Lamont, W.J. (2005) Plastics: Modifying the microclimate for the production of vegetable crops. *HortTechnology* 15, 477–481.

Loy, J.B., Wells, O.S., Karakoudas, S.M. and Milbert, K. (1998) Comparative effects of red and black polyethylene mulch on growth, assimilate partitioning and yield in trellised tomato. In: *Proceedings of the 27th National Agricultural Plastics Congress*, Tucson, Arizona, pp. 188–197.

Lu, Y.C., Watkins, K.B., Teasdale, J.R. and Abdul-Baki, A.A. (2000) Cover crops in sustainable food production. *Food Reviews International* 16, 121–157.

McCraw, D. and Motes, J.E. (2007) *Use of Plastic Mulch and Row Covers in Vegetable Production*. Oklahoma Cooperative Extension Service, Stillwater, Oklahoma.

Orzolek, M.D. and Murphy, J.H. (1993) The effect of colored polyethylene mulch on the yield of squash and pepper. In: *Proceedings of the 24th National Agricultural Plastics Congress*, Overland Park, Kansas, pp. 157–161.

Shilling, D.G., Liebl, R.A. and Worsham, A.D. (1985) Rye (*Secale cereale* L.) and wheat (*Triticum aestivum* L.) mulch: The suppression of certain broadleaved weeds and the isolation and identification of phytotoxins. In: Thompson, A.C. (ed.) *The Chemistry of Allelopathy*, Vol. 286. American Chemical Society, New York, pp. 243–271.

Stapleton, J.J. (2000) Soil solarization in various agricultural production systems. *Crop Protection* 19, 837–841.

Waterer, D. (2010) Evaluation of biodegradable mulches for production of warm-season vegetable crops. *Canadian Journal of Plant Science* 90, 737–743.

Yates, S.R., Gan, J., Papiernik, S.K., Dungan, R. and Wang, D. (2002) Reducing fumigant emissions after soil application. *Phytopathology* 92, 1344–1348.

7 Protected Culture

A Brief History of Protected Culture

A greenhouse is a building with clear or translucent walls and ceiling used for growing plants. Some of the very earliest greenhouses and conservatories were built in the 1500s, 1600s, and 1700s in Italy, France, Germany, England, Belgium, and the Netherlands, and were made from expensive blown glass flattened into small sheets. Royalty and the very wealthy commissioned construction of early greenhouses to house exotic plant materials from distant lands (Muijzenberg, 1980).

Although glass was discovered over 4,000 years ago, it was not until the 1800s that glass greenhouses were used for commercial enterprises. The creation of large metal-frame greenhouses for vegetable production coincides with the industrial production of flat glass in the 1800s. The first free-standing greenhouse was built in England in 1806 (Muijzenberg, 1980). The first commercial greenhouse in the USA was reported in 1820. Significant greenhouse production began around Boston, Massachusetts, with lettuce being produced in the 1860s, radishes and cucumbers in the 1870s, and tomatoes in 1883. By 1899, the greenhouse industry in the USA had rapidly grown to 930 ha (2,300 acres), although this figure includes both flowers and vegetable production. Similar greenhouse industry growth occurred in Europe. Between 1870 and the late 1890s, the greenhouse area in the Worthing district in the UK increased from 0.4–18 ha (1–44 acres) (Gras, 1940).

Since early glass was very expensive, vegetable growers built simple structures like glass-covered cold frames, hotbeds, and sash houses beginning in the 1800s and extending into the 1900s (Gras, 1940). A cold frame, also called sun box, is a transparent-roofed enclosure, built low to the ground and used to passively protect plants from cold weather. Essentially, a cold frame functions as a miniature greenhouse and a season-extension device. The height inside a cold frame is most commonly 0.6 m (2 ft) or less. Some low-growing vegetables like lettuce could be produced out of season to maturity in a cold frame. Larger vegetables like tomatoes and peppers could be started as seedlings in a cold frame, and as the plants developed the glass top or sash was removed after the frost-free date to allow the vegetables to grow to maturity.

The cold-frame roof is often sloped towards the winter sun to capture more light and improve water runoff, and is hinged for easy access. Sunlight heats cold frames to create a microclimate several degrees warmer than outside air and soil temperatures. The transparent top transmits sunlight and prevents heat escape via convection, particularly at night. Since the volume of a cold frame is small, direct sunlight can quickly cause overheating. The glass cover may be raised to ventilate the cold frame so that plants do not overheat. Cold frames were often used to produce transplants for field planting. For transplant production, seeds were sown in flats of soil and placed in cold frames where the modified environment favored germination and seedling development.

As greenhouses became more common in the 1900s, seeds were germinated in greenhouses and then moved to the cold frame to be hardened before transplanting. Cold frames are still used today primarily by home gardeners. Plastic sheets are often used as coverings instead of glass. Cold frames have largely been replaced by plastic row covers that are more commonly used by large-scale commercial growers and will be discussed later in this chapter.

Hotbeds, also called hotboxes, are similar to cold frames except that they have an internal heat source instead of relying on the sun for heat. Hotbeds were originally heated with steam pipes or fresh manure buried beneath the rooting zones of the plants. Microbial activity decomposed the manure, generating heat in the process. Recently, soil-heating cables are used to warm hotbeds. Depending on the weather conditions, heat source, and design of the hotbed, plants can be maintained even though outside temperatures fall below freezing for extended periods. Low-growing vegetables could be grown to maturity in hotbeds. Although hotbeds were once significant for out-of-season vegetable production, today they are mainly used by gardeners and have been largely replaced by greenhouses.

Sash houses were another adaptation of flat-glass technology. A sash house was a primitive greenhouse made of panels of class held together by wooden frames on a steep slope. Sash houses were typically taller than cold frames so that people could walk inside. Sash houses were often build against the side of another building or were partially buried beneath the ground. The ends were often made of solid material rather than glass panels. Most sash houses were unheated and designed primarily for starting young plants.

Cold frames and hotbeds have been replaced by row covers made of plastic film, which provide similar frost protection and season extension, but are more flexible, require less labor, and are cheaper to use. These row covers can be moved or installed in different locations each season. In Chapter 6 we discussed the use of plastic mulches to heat soil, retain moisture and control weeds. The coverings described in this section are different in that they often cover entire rows of plants, to generate a microclimate to enhance growth. Some covers also provide a barrier to insects. The plasticulture system discussed in Chapter 6 is compatible with row covers and the two systems are often used in combination during part of the growing season.

Modifying Field Environments for Vegetable Production

In areas with short growing seasons, air and soil temperatures increase gradually in spring, leading up to a frost-free date that starts the growing season. In the fall, the reverse occurs as the air and soil gradually cool before a killing frost occurs, terminating the season.

However, field environments can be modified to extend the growing season. For example, a simple technique used in desert areas where there is ample sunlight but cool temperatures early in the season, is to angle the tops of beds toward the sun to increase solar heating of the soil. This simple practice, without any type of row covering, can speed germination of direct-seeded crops and improve early-season growth of transplants.

Greenhouse structures can be built to allow vegetable production beyond the normal growing season in short-season areas. However, rather than covering an entire field with a permanent structure, it is often more practical, less expensive, and more flexible to cover individual rows to create a favorable microclimate around plants. Using row covers does not allow for year-round production in cold climates because the degree of environmental modification obtained is considerably less than in a greenhouse. However, in areas with slow seasonal temperature transitions, lightweight temporary covers can extend the season for weeks or even months. Season extension is important for vegetable growers in short-season areas because the highest market prices are paid for first-of-the-season, locally grown fresh vegetables. A premium is also paid for fresh vegetables late in the season after a killing frost has ended the season for other growers.

Temporary covers are primarily used to retain heat and protect against frost. In some cases, they reduce evapotranspiration and protect against insects. The negative aspects of these materials are that they reduce light intensity, may damage plants on sunny days due to overheating if not adequately ventilated, and they have to be disposed of responsibly at the end of their life expectancy. Also, most covers are lightweight and must be anchored to prevent wind damage. The following is a list of protective materials and structures, progressing from simple to complex, that are commonly used to protect vegetable crops against unfavorable conditions.

Hotcaps

Hotcaps are individual plant covers that protect against low-temperature injury. Some hotcaps are made of translucent water-resistant waxed paper pressed into a tepee shape. They are inexpensive and are generally used for only one season. A hotcap is placed over a transplant and the edges buried to hold the cover in place. The cap provides modest frost protection depending on environmental conditions (Welbaum, 1993). The top of the cap can be cut open to give the plant room to grow and provide ventilation. If not ventilated, hotcaps may overheat on warm sunny days. Unventilated hotcaps may create a high-humidity environment that increases disease and succulent growth due to reduced light intensities and excess water. New hotcaps made of plastic are reusable and may provide extra frost protection, particularly those that can be filled with water (Welbaum, 1993). Because of the labor involved in placement, adjustment and removal of individual hotcaps, plastic row covers have largely replaced hotcaps in many areas. Hotcaps are still used for small-scale production.

Floating row covers

Floating row covers are made of lightweight fabric of varying densities laid directly on top of the row. Various thicknesses and weights of row covers are available for different applications. Row covers are generally classified as woven or non-woven. The fabrics of non-woven row covers are made of spunbonded polyester or polypropylene fibers.

Light-transmitting, thin-mesh row covers are laid on top of the ground with the edges buried after planting. These materials may provide slight frost protection, allow water to infiltrate, reduce soil evaporation, and protect against insects that fly in from other areas to attack the crop. After seedling emergence, the covers are removed when the crop starts to push them out of the ground. If undamaged, thin row covers can be reused. In some cases, the cover is suspended above the crop using shallow stakes to provide room for crop growth so protection is provided over a longer period.

Heavier row covers are unrolled over a crop row early in the evening and rolled up the following morning to provide protection from isolated frost at night. These heavier floating row covers generally increase minimum night temperatures by 3–4°C (37–40°F) depending on the type of cover and environmental conditions. These thicker covering materials do not transmit as much light as thin fabric row covers and are removed from the field when the threat of frost has passed and stored for later use. Floating row covers can be placed in row middles on sunny days to improve ventilation when not needed.

Bed covers

Bed covers are similar to the plastic film mulches used for plasticulture production except than the bed is shaped to provide an open area to allow the crop room to grow beneath the plastic. During installation, a thin film of clear plastic is spread tightly across a V-shaped bed surface. The edges of the plastic are buried under soil to hold it in place. The bed is direct-seeded or transplanted near the center of the depression. The plastic covering traps heat, maintains moisture, and stimulates seed germination or transplant growth. As the seedlings emerge, the plastic is first slit to ventilate the row and later cut completely open to allow the crop to grow across the entire bed. Bed covers are most commonly used to produce warm-season vegetables in irrigated production areas that receive little rain, which could collect on the plastic and flood the depression in the uncovered bed. This bed cover is inexpensive and effective for warm-season vegetables, but requires deep soil.

Low tunnels

Low tunnels, also called low row covers, are constructed of sheets of clear plastic (Fig. 7.1). Tunnels can be used to protect warm-season crops from cold weather early in the season and extend the season past the killing frost. Plastic mulch can be used with row covers to modify soil temperatures, control weeds, and conserve moisture (Fig. 7.2).

Fig. 7.1. Low tunnels.

Fig. 7.2. Cucurbits grown under low tunnels and plastic mulch in Taiwan. The sides of the tunnels can be pulled to the top for ventilation.

The height of most tunnels varies from a few inches to a maximum of about 1.2 m (48 in). Because tunnels are relatively low, workers cannot get inside and can only access the crop by removing or adjusting the plastic cover. The taller covers are often held in place by a series of wire pieces bent into a semicircle. The ends of the wires are stuck securely in the soil at intervals to form a series of hoops over a row. Light-transmitting plastic film is stretched over the wire hoops and secured by wires spread over the top of the plastic. This design is sometimes called a hoop tunnel, although the supports are half hoops rather than complete circles. The edges are left unburied so they can be raised and lowered for ventilation on sunny days or lowered at night to provide maximum heat retention.

Other tunnel designs use two sheets of plastic on either side of the row (Fig. 7.3). The outer edge is buried or pulled close to the soil while the inner or upper edge of the plastic sheet is secured with a wire or string between hoops. When both sheets of plastic on either side of the row are fully extended, a tent is formed over the crop. The sides of the tunnel can be lowered or in some cases raised as well to provide ventilation.

High tunnels

High tunnels, sometimes called high row covers, are temporary or portable hoop-style houses covered with single or double-layered plastic film. Unlike low tunnels, high tunnels are tall enough to allow workers to stand inside and to accommodate small equipment (Fig. 7.4).

High tunnels are essentially temporary narrow portable plastic-covered greenhouses that cover sections of a field for a growing season. These tunnels are not designed to provide protection during the coldest months of the winter, and many contain no or minimal heating. The primary function of a high tunnel is to extend the season, allowing vegetable production earlier in the spring or later in the fall. The entire high tunnel can be disassembled at the end of the season and moved to a new location the following year. High tunnels are often used to grow warm-season crops, such as tomatoes and melons, for the early market. Because of their larger size and taller profile, high tunnels must be securely constructed to withstand high winds. The supports are often constructed of cross-linked plastic or aluminum pipe anchored in the soil (Fig. 7.5).

Fig. 7.3. Low tunnels in Japan composed of two sheets of plastic held in place by a series of semicircular hoops and tied together with string in the center protect low growing vegetable crops (foreground). The top and sides can be adjusted to increase or decrease ventilation. High tunnels, perpendicular to the low tunnels, are visible in the background.

Fig. 7.4. High row covers with black plastic mulch enable early season production of summer squash and red beets in the Annapolis Valley of Nova Scotia.

Fig. 7.5. A high tunnel in Japan with a cross-linked pipe frame protecting a Chinese cabbage crop.

This design gives the structure sufficient strength to withstand storms, but is simple enough to be disassembled and relocated. Horizontal straps laid across the outside secure the plastic film sheets to the tunnel frame. Polyethylene (PE) film and polyvinyl chloride (PVC) are two common types of plastic film used to cover tunnels. The vertical end panels of tunnels are removable to allow equipment to enter and to provide ventilation. The sides of many tunnel designs can also be raised for ventilation.

Permanent Greenhouses Used for Vegetable Production

Quonset-style greenhouses

One of the least-expensive permanent greenhouse designs is the Quonset. A Quonset greenhouse is shaped like a longitudinal half of a cylinder with a semicircular shape in cross section and no separate sidewalls and roof. The frame is often made of curved steel T-shaped supports or pipes that are cross-linked and mounted in the ground or held together by a rectangular metal frame. This simple design is less expensive to build than greenhouses with vertical sidewalls, but some of the space at the base of each sidewall is difficult to utilize. The Quonset greenhouse can be covered with a wide range of materials, including flexible plastic sheets or plastic film. In some areas, these structures are not taxed as permanent buildings when they are covered with plastic film. Air circulation is through a fan and vent system installed on the sides or ends of the house.

Vertical-wall permanent greenhouses

Tall greenhouses with metal frames, vertical sidewalls, and glass, polycarbonate, or acrylic glazing are the most expensive designs to build and maintain. Glasshouses are still widely used despite the high cost of construction (Fig. 7.6).

Glass greenhouses still have the advantage of higher light transmittance, which is important in northern latitudes where sunlight is a limiting factor during the winter. The size of glass panels used in greenhouse construction has increased dramatically over time.

Rigid polycarbonate plastic sheets are a popular greenhouse covering material with high light transmittance, durability, superior insulating properties, and reduced cost compared to glass. Double-walled plastic sheeting provides a dead air space for added insulation and is popular in cold-season areas to reduce heating costs (Fig. 7.7).

Greenhouses can also be covered with low-cost polyethylene (PE) film. Double-layer PE-coverings are more energy efficient than single-layer plastic or glass in some cases. Polyvinyl chloride (PVC)

Fig. 7.6. Greenhouses in the "Westland" region of the Netherlands.

Fig. 7.7. Double-walled polycarbonate covered greenhouse in Virginia, USA.

is another film used to cover greenhouses in some areas. Polyethylene film coverings are less expensive than glass, but must be replaced more frequently.

Originally, glass greenhouses were constructed of relatively small glass panes supported by wood and later steel or aluminum. The height of these early designs varied but was generally lower than the greenhouses built today. Some of the early designs were partially buried beneath the soil to reduce the profile of the house, used less glass, and utilized the temperature moderation provided by the ground. Such greenhouses also tended to be damp and poorly drained and were generally used to raise transplants rather than grow plants to maturity. Later high-roof greenhouses with straight sides like those used today became more prevalent. Modern glasshouses used for vegetable production have vertical sidewalls with both side and roof ventilation. Over time the volume of greenhouse space has increased as roofs have become taller to reduce the rapid temperature fluctuations and to allow access to larger equipment.

Greenhouse Cultural Systems

Originally, greenhouse crops were grown in the native soil inside the greenhouse. Greenhouses were not always built on the most productive soils, so animal manure or other organic matter was often added in an attempt to increase fertility. Compared to the field, a special challenge for greenhouse vegetable production was crop rotation to prevent the build-up of soil pathogens in a confined area. In a greenhouse, the same soil must be used to produce all crops and the number of profitable crops that can be grown in rotation was limited because low-value crops were not economically feasible.

When native soils were used, greenhouse soils were sterilized with steam generated by the heating system to remove pathogens that would accumulate over time. Steam sterilization enabled the same or related crops to be grown repeatedly without rotation. However, increased energy costs and the challenge of uniformly treating soil in large greenhouses made steam sterilization impractical. Chemical fumigants have been used in greenhouses, but using toxic chemical fumigants in a confined area is dangerous. Some fumigants pose environmental risks, so their use was not a good alternative to steam sterilization. Another approach to greenhouse soil-disease management was to graft the scions of vegetables on to disease-resistant rootstalks. Grafting created vegetables that could tolerate many diseases, so production in greenhouse soils was possible without sterilization. For example, susceptible cucumber cultivar scions can be grafted on to a *Lagenaria* species rootstalk that is highly resistant to many soil pathogens.

Another approach to greenhouse disease management is solarization. Solarization is a simple nonchemical technique to suppress greenhouse pests by exposing them to high temperatures generated by solar radiation. In areas with high sunlight, greenhouses can be sealed and the temperature increased to at least 45°C (110°F) during a fallow period (Scopaa *et al.*, 2008). Duration increases the efficacy. Solarization is effective against some pests more than others (Scopaa *et al.*, 2008). Solarization is most effective at lower latitudes that experience extended periods of intense sunlight with few clouds.

Today, much of the advanced greenhouse production utilizes hydroponic production systems to produce greenhouse vegetables. Hydroponics is a technology for growing plants in nutrient solutions with or without the use of an artificial media (sand, gravel, vermiculite, rockwool, perlite, lava rock, peat moss, coir, or sawdust) to provide mechanical support (Fig. 7.8; Jensen, 1997).

There are many types of controlled-environment/hydroponic systems that are used for vegetable production. Liquid hydroponic systems have no media to support root growth. Hydroponic systems are further categorized as open (the nutrient solution that is delivered to the plant roots is not reused) or closed (surplus solution is replenished and recycled) (Jensen, 1997). Hydroponic culture eliminates the need for soil sterilization, optimizes nutrient availability and can be adjusted to provide the proper amount of water.

Managing Nutrients in a Hydroponic System

Plants require 16 elements for growth and these nutrients can be supplied from air, water, and fertilizers as discussed in Chapter 4. The 16 elements are carbon (C), hydrogen (H), oxygen (O), phosphorus (P),

Fig. 7.8. Spinach plant growing in recirculating nutrient solution using porous lava rock as substrate.

potassium (K), nitrogen (N), sulfur (S), calcium (Ca), iron (Fe), magnesium (Mg), boron (B), manganese (Mn), copper (Cu), zinc (Zn), molybdenum (Mo), and chlorine (Cl). The amounts of these minerals that accumulate in plants vary with growing conditions and the type of vegetables grown (Table 7.1).

A hydroponic fertilizer program must ensure optimum concentrations of all nutrients are available through the crop production cycle, since inadequate or excessive amounts may reduce both crop yield and quality (Table 7.2). Excess fertilization wastes money, damages the crop, and may pollute the environment if fertilizer is released from the delivery system.

Carbon, H, and O are usually supplied in adequate amounts from air and water, but in cold-season areas where greenhouses are not ventilated when outside temperatures are low, fertilizing with gaseous carbon dioxide (CO_2) may be beneficial. Other nutrients are taken up though the roots and must be carefully managed. Nutrient-management programs should begin with an understanding of nutrient-solution concentrations in parts per million (ppm) for the various nutrients required by vegetables. By properly managing nutrient concentrations, growers can control the development and yield of the crop. There are different strategies for ensuring that hydroponic vegetables are adequately fertilized.

One approach is to vary the nutrient concentration of the hydroponic solution to match the demands of the developing plant. During the crop-production season, the demand for nutrients changes. Small amounts of nutrients are needed early, and then the demand increases with crop development, especially after fruit-set. A common problem occurs early in the season when too much N is added, causing excessively rapid succulent growth. This condition, which is also called stretching, results in plants with low dry matter, excessive internode length, and cracks and grooves in the stems of tomatoes and peppers. Stem cracks are entry points for decay-causing organisms such as soft rot. Succulent tomato plants often produce misshapen fruits with a high percentage of the physiological disorders blossom-end rot and cat-facing. To avoid poor-quality growth, N levels should be maintained from 60 to 70 ppm during the early stages of development (Hochmuth and Hochmuth, 2008). For example, N for a tomato crop grown in perlite can initially be 70 ppm from transplanting to first cluster formation. As the nutritional requirements increase, the N in solution increases sequentially from 80, 100, 120, and 150 ppm for plants with 2, 3, 4, and 5 fruit clusters, respectively. Nutrient-solution recommendations vary among vegetables. For example, cucumbers will need more N early in the season than tomatoes, so a hydroponic solution for cucumber seedlings from establishment to first flowering should be in the range 80–90 ppm.

An alternative method of hydroponic nutrient management is to monitor plant nutrition through foliar analysis and adjust the nutrient solution in response to plant needs. Plants would be grown in a hydroponic nutrient solution such as those listed in Table 7.2. Nutrients can be added to the hydroponic solution to satisfy the growth requirements of plants. Plants quickly remove their daily supply of some

Table 7.1. Approximate tissue composition for hydroponically grown lettuce, tomato, and cucumber.

Element	Butterhead lettuce[a]	Tomato[b]	Cucumber[b]
Nitrate N	1.3–1.7%	1.4–2.0%	1.0–2.0%
PO_4	0.7–0.9%	0.6–0.8%	0.8–1.0%
K	7–8%	5–8%	8–15%
Ca	1–3%	2–3%	1–3%
Mg	0.3–0.4%	0.4–1.0%	0.3–0.7%
Fe	200–300 ppm	40–100 ppm	90–120 ppm
Zn	25–50 ppm	15–25 ppm	40–50 ppm
Cu	5–10 ppm	4–6 ppm	5–10 ppm
Mn	30–55 ppm	25–50 ppm	50–150 ppm
Mo	N/A	1–3 ppm	1–3 ppm
B	15–30 ppm	20–60 ppm	40–60 ppm

[a]Siomos *et al.* (2001)
[b]Lorenz and Maynard (1988)

Table 7.2. Nutrient solutions for hydroponic vegetable production (modified from Lorenz and Maynard, 1988).

	Nutrient solution			
	Johnson's	Jensen's	Larsen's	Cooper's
Compound (g/100 l water)				
Potassium nitrate	25	20	18	61
Monopotassium phosphate	14	27	Not used	26
Phosphoric acid (75%)	Not used	Not used	40 ml	Not used
Potassium sulfate	Not used	Not used	34	Not used
Potassium magnesium sulfate	Not used	Not used	44	Not used
Magnesium sulfate	25	49	0.4	51
Calcium nitrate	46	50	95	100
Chelated iron (FeDTPA)	2	3	3	Not used
Chelated iron (FeEDTA)	Not used	Not used	Not used	8
Boric acid	0.1	0.3	0.6	0.2
Manganese sulfate	0.07	Not used	Not used	0.6
Manganese chloride	Not used	0.2	Not used	Not used
Zinc sulfate	0.01	0.04	0.1	0.04
Cupric chloride	Not used	0.01	Not used	Not used
Copper sulfate	0.003	Not used	0.1	0.04
Molybdic acid	0.001	0.005	0.01	Not used
Ammonium molybdate	Not used	Not used	Not used	0.04
Mineral (ppm)				
N	105	106	172	236
P	33	62	41	60
K	138	156	300	300
Ca	85	93	180	185
Mg	25	48	48	50
S	33	64	158	68
Fe	2.3	3.8	3	12
B	0.23	0.46	1	0.3
Mn	0.26	0.81	1.3	2
Zn	0.024	0.09	0.3	0.1
Cu	0.01	0.05	0.3	0.1
Mo	0.007	0.03	0.07	0.2

nutrients, while other nutrients tend to accumulate in solution. This means that the concentrations of N, P, and K can be at low levels in the solution (0.1 mM (millimolar) or a few ppm) because these nutrients are rapidly taken up and metabolized by the plant.

When water is removed from a hydroponic solution through transpiration, it must be replaced. For example it is often necessary to add about 0.5 mM P in the refill solution. If the refill solution is added daily, the dissolved P would rapidly be absorbed by the plant and the solution P concentration would again be close to zero. This does not indicate a deficiency; rather it indicates a healthy plant with rapid nutrient uptake. If the P level is continuously maintained at 0.5 mM in a recirculating solution, the P concentration in the plant can increase to over 1% of the dry mass, which is higher than the optimum in many plants (Table 7.1). Maintaining high concentrations of nutrients in the solution can be dangerous to the plants, resulting in excessive uptake that can lead to nutrient imbalances. This high P level can also induce Fe and Zn deficiency (Chaney and Coulombe, 1982).

Plant-tissue analysis can assess the content of essential minerals in hydroponically grown lettuce, tomato, and cucumber. The tissue measurements can be compared to published values to determine whether plant nutritional requirements are being met (Table 7.1). Once the plant reaches its optimum mineral content (Table 7.1), the concentrations of the minerals in the hydroponic solution can be adjusted by adding or subtracting components to maintain the desired plant-tissue percentages in Table 7.1. These recommendations are applicable to different production systems such as perlite, rockwool, and nutrient film.

Formulation methods

There are basically two methods to supply the fertilizer nutrients to the crop: (i) premixed products; or (ii) grower-formulated solutions (Hochmuth and Hochmuth, 2008). The two methods have different nutrient-use efficiencies. Fertilizer materials used for both methods are presented in Table 7.2 (Lorenz and Maynard, 1988). The recommendations are for creating solutions by mixing grams of minerals in 100 l (22 gal) water. When using proportioners installed in parallel on a one-water source line, the nutrients can be supplied through a more concentrated stock solution calculated to provide a final concentration after dilution similar to those given in Table 7.2. Final solution pH should be in the range 5.8–6.2.

Pre-mixed hydroponic solutions

There are several commercial premixed hydroponic fertilizer formulations available. Before selecting a premade fertilizer, the limitations and various formulations must be understood. Some of these preformulations contain Mg, while others do not. Those lacking Mg need to be supplemented with magnesium sulfate (Table 7.2).

Many formulations need supplemental Ca from calcium nitrate or calcium chloride and N. Supplemental formulations are listed in Table 7.2. The premixed materials often contain large amounts of K, making it difficult to achieve the desired K and Ca concentrations. Excessive K can interfere with Ca uptake by the root. Ca is essential for eliminating blossom-end rot in tomato and pepper and tip burn in lettuce (Hochmuth and Hochmuth, 2008). More Ca can be added from calcium chloride, but it is often better to have lower K concentrations. A related problem is that some of the premixed formulae have too much N to allow for providing adequate Ca by adding calcium nitrate. An option to increase the Ca in the solution is to supplement the calcium-nitrate stock with calcium chloride. The premixed products come fairly close to providing the desired concentrations of micronutrients, but may need to be adjusted for certain crops and production conditions.

Formulation recipe method

Individual ingredients for four different hydroponic solutions are provided in Table 7.2. Four formulae are presented to provide options for formulating nutrient solutions depending on grower preference (Lorenz and Maynard, 1988). Larsen's solution uses phosphoric acid to provide P and partially acidify the nutrient solution. If additional acidification is required, sulfuric acid can be used.

Greenhouse-grade potassium chloride can also be used as an alternative source of K. There is no problem using potassium chloride as a partial source because it provides K in the same form as potassium nitrate and potassium chloride, but is less expensive. The chloride ion is not toxic to plants in the amounts needed and some research shows it reduces soft rot (Hochmuth and Hochmuth, 2008).

Water–fertilization relationship

The nutrient solution formulations in Table 7.2 were designed for hydroponic production systems for vegetable culture. Solutions are often pulsed through the media or as a nutrient film at a rate of 10–20 cycles/day, depending on the weather and greenhouse environment, and the crop (Hochmuth and Hochmuth, 2008). It is impossible to have an exact formulation that will work for every production system under all environmental conditions. For example, when using a media of high-water-holding capacity, fewer irrigation cycles will be needed during the day. In this situation, the nutrient concentrations in the irrigation water might need to be greater so that adequate nutrition is supplied. A general recommendation is that nutrient concentrations need to be greater in production systems with media that require fewer irrigations, compared to systems requiring more frequent irrigations (Hochmuth and Hochmuth, 2008). The following are some examples of greenhouse media used for vegetable production.

Rockwool

Rockwool is a popular growing medium that was originally used as insulation and was later adapted for plant production in northern Europe. Rockwool is made from natural basalt rock and chalk that are mixed and melted at high temperatures and spun into fibers of different shapes and sizes. Rockwool has a high water-holding capacity and is initially disease free, providing an excellent support for root growth in hydroponic systems (Fig. 7.9).

Nutrient solution is added to blocks of rockwool using drip tubing or a trough that is carefully

Fig. 7.9. Rockwool media used for tomato growth in the Netherlands. (a) Rockwool promotes excellent root growth. (b) The plastic film cover excludes light and conserves moisture.

monitored to supply the right amount of water and nutrients (Tyson *et al.*, 2010). Recirculating nutrient solution through troughs may lead to the rapid spread of disease if one of the plants in the system becomes infected. Although this system is popular for growing vegetables, the disposal of the rockwool is a concern because the fiberglass decomposes very slowly and has a high salt content. Several recycling options have been developed (Bussell and McKennie, 2004).

Nutrient film

Nutrient film is a hydroponic system that is simple and can be used to grow a wide range of crops. In this system, lettuce or other plants are rooted in a trough or PVC pipe containing circulating nutrient solution (Tyson *et al.*, 2010). Troughs are covered to inhibit algae, which rapidly grows where light shines on the nutrient solution, and gently sloped so the nutrient solution will drain back to the reservoir (Fig. 7.10).

Cubes of rockwool or other synthetic media anchor roots and plants in place. There are many variations of this system. Various combinations of nutrients can be applied through the circulating solution or in growing media if plants are grown in containers that rest in troughs. Plants continuously fed nutrients and water may be overly succulent with limited shelf life, so to harden plants the recirculating solution may be pulsed to limit the availability of water. Nutrient concentrations must be continually monitored and adjusted to compensate for evapotranspiration and plant uptake of nutrients. Since a common nutrient solution bathes the roots of many plants, disease can spread rapidly through the circulating solution. Foliar diseases are inhibited because the leaves are not wetted in this system.

Ebb and Flow Systems

The ebb and flow (also called ebb and flood) system is used to produce bedding plants and container-grown vegetables. Plastic-lined beds are designed to hold water or nutrient solution. Containerized plants or flats of transplants are placed in beds. Polystyrene or other lightweight floating trays are often used in ebb and flow systems. A matrix of PVC pipe or other water-resistant material supports trays above the bottom of the bed during the ebb cycle so air circulates underneath to promote root pruning. During the hydration cycle, the bed is flooded with nutrient solution or water for a specific period of time, depending on crop needs. The bed is then drained during the ebb cycle by pumping the solution into an adjacent bed or a holding tank. This system controls the amount of hydration that each bed receives to prevent excessively rapid growth. Since beds are exposed to sunlight, algal growth can be a problem, especially if mineral nutrients are in solution. Nutrients can also be added to the containerized media and water pumped among beds. The ebb and flood system uses relatively large volumes of water, has significant evaporative losses, the nutrient concentration in the water is difficult to precisely control, and wastewater may have a high mineral content compared to other greenhouse-vegetable production systems. However, it is a relatively simple system that uses a minimum of hardware and equipment.

Aeroponics

Aeroponics is a minimalistic system that produces plants without media. Plant stalks are held in place by a series of braces. The roots are suspended in air and

Fig. 7.10. Butterhead lettuce growing in a recirculating nutrient film system of interconnected troughs.

bathed in a fine mist of nutrient solution. The roots are enclosed in the dark using sheets of dark plastic film to confine the nutrient mist and control the growth of algae. The nutrient fog condenses on the roots, producing a surface film. Plants absorb nutrients and water through the surface film. Plant-nutrient status must be continually monitored to ensure they are well fed. Aeroponic production is used in areas where water, soil, and space are limited, such as desert regions or densely populated urban areas where farmland is at a premium. Since there are no media in this system to act as a reservoir for water and nutrients, power outages or equipment failures for even a short duration can damage or kill the crop.

Bag Culture

A simple system called bag culture is also used to grow vegetables (Tyson *et al.*, 2010). Bags of commercially prepared soilless rooting media are placed in the greenhouse, and crops such as tomatoes, peppers, eggplants, or cucumbers are transplanted into a hole cut in the bag (Greer and Diver, 2000). Sometimes bags of perlite media are used. Perlite is an amorphous, naturally occurring volcanic material that has the unusual property of greatly expanding when heated at very high temperatures. Perlite drains well but has relatively high water content when hydrated. The plastic bag reduces evaporative losses and confines nutrients to the root zone. The plants are trained to grow vertically on strings suspended from overhead supports. Drip tubes provide water and/or nutrients to each bag. Although the media is not generally used long-term, a second unrelated crop may be grown. After use, the bagged media may be spread on outdoor production fields, sterilized, or composted for reuse. A negative aspect of this system is dealing with relatively large volumes of expended media.

Organic Production

Greenhouse vegetables can be produced organically (Greer and Diver, 2000). Natural composted media with no added chemicals are accepted by many organic certifying agencies. Composted animal manure with organic amendments, such as worm castings, enhances microbial diversity, increases fertility, and enhances drainage (Fig. 7.11).

Liquid organic fertilizer can be used in ebb and flood systems or to fertigate pot culture. Organic seeds of greenhouse cultivars are also available.

Fig. 7.11. Organic greenhouse production of basil near Harrisonburg, Virginia. Plants are grown in pots containing composted organic media and fertigated with liquid fertilizer through troughs beneath the plants.

Organically grown greenhouse crops often receive a premium market price.

Controlled-environment Production

Greenhouses are not the only permanent structures used for vegetable production. Controlled-environment agriculture (CEA) can be practiced in windowless buildings where all aspects of production are controlled, including light. Vegetables have been successfully grown using artificial light and hydroponic production in commercial buildings similar to warehouses in Canada and China. Some greenhouse crops such as butterhead lettuce grow well under artificial light at intensities lower than full sunlight. This so-called "warehouse" production of vegetables can take advantage of lower electricity rates at night. Traditional buildings can be cheaper to construct, easier to maintain, and cheaper to heat compared to traditional glass greenhouses. The economic success of these non-traditional production operations is dependent on a number of factors, including electrical costs and the market value of the crop, and may not be economically viable in all areas.

Changes to the Greenhouse Vegetable Industry

In the early 1970s, energy costs escalated rapidly, changing the economics of vegetable greenhouses in many short-season areas. Also during this time, transportation networks expanded, making the shipment of vegetables by truck or train possible to distant markets for a reasonable cost. Improved knowledge of postharvest handling techniques helped preserve the quality of vegetables shipped over long distances as well. As a consequence, greenhouse-grown vegetables were suddenly competing with field-grown vegetables produced at a fraction of the cost. For example, vegetables could be grown during the winter in south Florida, USA and shipped to commercial centers in the northeast USA and southern Canada in 2 or 3 days. Similarly, produce could be quickly shipped from production areas in the Mediterranean to northern Europe as well. Using controlled atmosphere storage, broccoli and cantaloupes grown in North America could also be shipped to Asia.

Urbanization also hurt the glasshouse vegetable industry in some areas. Originally, greenhouses were located at the edge of many large cities. Eventually, these areas were developed until the greenhouses were surrounded by urbanization. In some cases, greenhouses were taxed on their value like other permanent structures, which increased production costs. Some areas now offer tax relief for urban greenhouse producers. Also, urbanization brought increased pollution. Acid rain in particular caused "etching" of the glass, which reduced its light transmittance. Thus maintenance costs increased.

Some of the changes that occurred to the greenhouse vegetable industry in the 1970s, 1980s, and 1990s were in response to the challenges of higher energy costs and increased competition from vegetables produced outdoors in warm-season areas. These included shifting away from energy-intensive practices such as steam sterilization, growing fewer warm-season vegetables during the coldest seasons, relocating production greenhouses to more southern locations where production costs were less, and shifting from vegetable to flower production, which is considered to be more profitable in some areas.

By the 1980s and 1990s, trade agreements made exchange of vegetables more advantageous among countries. These events caused changes in the greenhouse vegetable industry in many parts of the world. In countries that lacked trading partners in long-season areas with winter vegetable production, greenhouse vegetable production continued to thrive and remains an important source of fresh vegetables when outdoor production is not possible. In some northern and eastern European countries, greenhouse vegetable production continues as an important industry due to changes in management practices, government subsidies, consumers' willingness to pay premium prices, and a lack of outdoor production alternatives. Greenhouse vegetables from the Netherlands, such as tomatoes and peppers, could be exported to markets in North America, where they compete well with locally grown vegetables (Cantliffe and Vansickle, 2003).

The Netherlands has long been the leader in greenhouse vegetable production, innovation, and research. Much of the Dutch greenhouse vegetable industry uses advanced technology and hydroponic culture to optimize production. Productivity in European greenhouses is nearly three-fold, and in some cases ten-fold greater than field production (Cantliffe and Vansickle, 2003). Product quality is generally much higher for greenhouses compared to field-produced vegetables (Cantliffe and Vansickle, 2003). In eastern Europe, countries such as Poland and Romania also have large acreages of greenhouse vegetable production.

Role of Greenhouses in Modern Vegetable Production

Modern hydroponic vegetable production in permanent greenhouses is both technology and capital intensive. These highly sophisticated production systems do more than provide a warm environment to prolong vegetable production in short-season areas. All aspects of production can be optimized, including air and root temperatures, light, water, and plant nutrition, thus mitigating adverse climatic effects often encountered during outdoor production (Fig. 7.12; Jensen, 1997).

Therefore, CEA/hydroponic systems can be used in a variety of environments that would normally be unfavorable for vegetable production, including regions where there is concern about the pollution of groundwater with nutrient wastes or soil sterilants (Jensen, 1997). For example, greenhouses are built in desert environments for CEA where soils lack fertility, water is scarce, and temperatures are often excessive. In areas of the Middle East where ambient temperatures are often too high for optimum vegetable production, temperatures can be reduced with cooling systems to enhance vegetable production. Also, humidity, water, and nutrients may be optimized in a hydroponic greenhouse where outdoor conditions are ill-suited for efficient production. Insects and other pests can be excluded from greenhouses as well. So greenhouses may be used for CEA in a broad range of environments and should not be thought of solely for production in short-season production areas.

The fact that greenhouse vegetables are grown in hydroponic systems rather than soil seems troubling and unnatural to some. The absence of native soils in modern greenhouses should not suggest that soil is bad or ill-suited for growing vegetables. Most vegetables are terrestrial plants and have evolved to grow in soil. Research on plant–microbial interactions is uncovering a complex mutualistic interdependence between soil microbes and plants that may be lost when production is shifted to synthetic media or solution culture (Welbaum *et al.*, 2004). Organic greenhouse production utilizes composted media with microbial diversity similar to agricultural soils.

However, greenhouse production provides a unique environment that is not well-suited for soil-based production systems. Hydroponic systems improve control of water and nutrients available to plants in an environment with limited root volume. Greenhouse production, particularly without

Fig. 7.12. Summer greenhouse tomato and pepper production using rockwool in Israel. This greenhouse is designed to exclude pests while maximizing light, water, and temperature for optimum vegetable growth.

Fig. 7.12. Continued.

adequate rotation, necessitates the use of hydroponics to reduce pathogens that tend to accumulate in soils when crops are grown repeatedly in a confined area.

References

Bussell, W.T. and McKennie, S. (2004) Rockwool in horticulture, and its importance and sustainable use in New Zealand. *New Zealand Journal of Crop and Horticultural Science* 32, 29–37.

Cantliffe, D.J. and Vansickle, J.J. (2003) *Competitiveness of the Spanish and Dutch Greenhouse Industries with the Florida Fresh Vegetable Industry, publication #HS918.* University of Florida IFAS Extension, Gainesville, Florida.

Chaney, R. and Coulombe, B. (1982) Effect of phosphate on regulation of Fe-stress in soybean and peanut. *Journal of Plant Nutrition* 5, 469–487.

Gras, N.S.B. (1940) *A History of Agriculture*, 2nd edn. F.S. Crofts, New York.

Greer, L. and Diver, S. (2000) *Organic Greenhouse Vegetable Production. Appropriate Technology Transfer for Rural Areas*. University of Arkansas, Fayetteville, Arkansas, pp. 1–19.

Hochmuth, R.C. and Hochmuth, G.J. (2008) Nutrient Solution Formulation for Hydroponic (Perlite, Rockwool, NFT) Tomatoes in Florida. Publication #HS796. Available at: http://edis.ifas.ufl.edu/cv216 (accessed 29 September 2013).

Jensen, M.H. (1997) Hydroponics. *Hortscience* 32, 1018–1021.

Lorenz, O.A. and Maynard D.N. (1988) *Knott's Handbook for Vegetable Growers*, 3rd edn. John Wiley and Sons, Hoboken, New Jersey, pp. 62–65.

Muijzenberg, E.W.B. (1980) *A History of Greenhouses*. Institute for Agricultural Engineering, Wageningen, the Netherlands.

Scopaa, A., Candido, V., Dumontet, S. and Miccolis, V. (2008) Greenhouse solarization: effects on soil microbiological parameters and agronomic aspects. *Scientia Horticulturae* 116, 98–103.

Siomos, A.S., Beis, G., Papanopoulou, P.P. and Barbayiannis, N. (2001) Quality and composition of lettuce (cv. 'Plenty') grown in soil and soilless culture. *Acta Hort* 548, 445–449.

Tyson, R., Hochmuth, R.C. and Cantliffe, D.J. (2010) *Hydroponic Vegetable Production in Florida. Publication #HS405*. University of Florida IFAS Extension, Gainesville, Florida, pp. 1–8.

Welbaum, G.E. (1993) Effects of three hotcap designs on temperature and tomato transplant development. *HortScience* 28, 878–881.

Welbaum, G.E., Sturz, A.V., Dong, Z. and Nowak, J. (2004) Fertilizing soil microorganisms to improve productivity of agroecosystems. *Critical Reviews in Plant Science* 23, 175–193.

8 Organic and Sustainable Vegetable Production

Introduction

Background of conventional and organic systems

Organic vegetable production is often considered as an alternative to what is variously called high input "conventional" farming, "modern" agriculture, or "traditional" farming. Actually, organic production pre-dates the advent of modern vegetable production. World War II caused many to realize that food was a strategic resource. Limited manpower and the need to maximize food production during the war lead to agricultural research and policies that accelerated the ascendancy of "modern" agrichemical systems of crop production that began in the early 1900s (Welbaum et al., 2004). The new technologies included synthetic concentrated fertilizers, mechanization and chemical weed control to increase production efficiencies. Another part of this system was the development of plant cultivars that were increasingly more dependent upon the support of agrichemistry in the subsequent post-war period (Welbaum et al., 2004).

This so-called conventional production system delivered impressive gains in crop productivity and efficiency over the past 80 years including the Green Revolution (Griffin, 1974). Consumers in most developed countries have come to rely on a diversity of high quality inexpensive vegetables because of conventional production systems. Food production worldwide has risen in the past 50 years and the World Bank estimates that between 70% and 90% of world food production over this period was the result of adaptation of conventional agricultural practices rather than greater acreage under cultivation. Modern agriculture has also dramatically reduced the numbers of people involved in production agriculture.

Concerns about conventional vegetable production

Despite the successes of conventional farming, some began to question the negative environmental impacts as far back as the 1960s. One of the initial reasons was degradation of soil quality that was not universally realized at first. The broad use of relatively inexpensive synthetic chemical fertilizers that were part of the "modern" system replaced the "natural" accumulation of soil nutrients from plant biomass and animal manures as well as the microflora that produced them. The concentrated industrially synthesized fertilizers increased yields, but at the same time the yield potential of agricultural soils declined as the biological processes that maintained their health and quality became over burdened (Greenland and Szabolcs, 1994; Pankhurst et al., 1997). Accordingly, an environmental price was paid for a food security system based upon technological and agrichemical innovation and the exploitation of non-renewable energy. In the 1980s and 1990s, the scientific literature contained reports of "soil-fatigue", "soil degradation", and "soil loss" (Welbaum et al., 2004). These reports along with a realization that environmental degradation was caused by modern chemically-based agriculture lead to an agricultural organic movement that began in the late 1960s and continues today (Kuepper and Gegner, 2004).

The negative consequences of conventional crop production have largely driven the debate about possible alternatives. In the minds of some, these negative attributes outweigh the positives. It is best to break these concerns into subcategories because they pertain to vastly different areas of the food chain continuum that extends from ecological concerns over production methods to the human health of consumers. The lengthy list of concerns is complex and involves a myriad of social, scientific, and environmental issues. While considering these individually, it should be remembered that: (i) interactions between farming systems and soil, water, biota, and atmosphere are complex and there is much to be learned about their dynamics and long term impacts; (ii) most environmental problems are complex and include economic, social, and political forces external to agriculture; (iii) some of these problems are global while others are experienced in specific locations; and (iv) many of these problems are being addressed through conventional, as well as alternative, agricultural channels (Stauber et al., 1995).

Ecological concerns

Vegetable production affects the environment in different ways. Negative consequences of conventional practices include ecological effects on soil quality, water quality, water availability, gases, crop pests, and air pollutants as some examples.

Conventional vegetable practices have been associated with a decline in soil productivity due to wind and water erosion of topsoil, increased soil compaction, loss of soil organic matter, decreased soil water-holding capacity, loss of biological activity, and salinization of soil and water on irrigated farms (Welbaum et al., 2004).

Conventional crop production practices may contribute to non-point source water pollutants that include sediments, salts, fertilizers, pesticides, and manures. Pesticides have been found in groundwater and surface water near agricultural land. Poor water quality linked to conventional production techniques impacts crop production, drinking water, recreational use, and aquaculture production. Scarcities in many locations can be traced to overuse and poor management of surface- and groundwater for irrigation in conventional systems (Stauber et al., 1995).

Other negative ecological effects associated with conventional production include: weeds, insects, mites, and fungal pathogens that are resistant to pesticides; detrimental effects on pollinator and beneficial insect species through pesticide use; loss of wetlands and wildlife habitat; and reduced genetic diversity due to reliance on relatively few F-1 hybrid vegetable cultivars.

Modern agriculture's link to global climate change is becoming clearer. Destruction of tropical forests and other deforestation for crop production leads to elevated carbon dioxide and other greenhouse gases in the atmosphere. Recent studies have found that soils may be important sources or sinks for greenhouse gases.

Economic and social concerns

Farming plays an important role in the development and identity of nations around the world. For example, the USA, Canada, and much of western Europe have evolved cultures with few farmers. Economic pressures have led to losses in the number of farms, including vegetable farms, in the USA and parts of Europe (Weiss, 1999). Less than 1.5% of Americans now produce food for the entire country and this statistic is similar for much of the developed world (Lobao and Meyer, 2001). More than 155,000 farms were lost between 1987 and 1997 in the USA (Lobao and Meyer, 2001). These losses contribute to the disintegration of rural communities and local marketing systems. What cultural values have changed and will continue to change with the decline of rural life and farmland ownership? Many Americans are totally unaware of the social aspects of small-town decline in rural America, which should also be considered a component of rural sustainable development.

Many agriculturalists feel that a growing anxiety between agrarian and urban citizens is because of a growing mistrust and in many cases misunderstanding about where the vegetables we eat come from and how they are grown. Is sustainable and equitable food production possible when most consumers have so little connection to the natural processes that produce their food? World population continues to grow and is predicted to stabilize at slightly less than 11 billion around 2200 (Guerrini, 2010). The rate of population increase is especially high in many developing countries. In these countries, the population factor, combined with rapid industrialization, poverty, political instability, and large food imports and debt burden, make long-term world food security a concern.

Market competition through wholesale supply chains for conventionally grown vegetable crops are limited. Farmers have little control over commodity prices, and they continue to receive a smaller portion of consumer dollars spent on vegetables. In the USA and many European countries, there is a history of government involvement in agricultural decisions; in many countries there is a widening disparity among farmer and farm worker incomes; and an increasing concentration of agribusiness companies/industries involved in production, processing, and distribution of vegetable crops (Stauber et al., 1995). Economically, it is very difficult for new conventional vegetable growers to enter the business today.

Impacts on food quality, supply, and human health

In many conventional production systems, vegetables are grown in areas with the best climate for production and then shipped to distant markets (Halweil, 2002). The best conditions usually mean a combination of some or all of the following: soil, climate, water, lower incidence of disease, accessibility, and sufficient labor. In the USA, the average vegetable purchased in a commercial chain store travels 2,400 km (1,500 miles) before reaching the consumer. This system often requires vegetables to be harvested at grossly immature stages of development to ensure they will withstand the rigors of shipping, distribution, and marketing with minimal losses. The result is vegetables are often not sold at peak consumer quality (Halweil, 2002). For example, tomatoes are often harvested at the mature green stage to handle the rigors of shipping and cantaloupes are harvested well before abscission to make sure they do not spoil during marketing. Consumers often associate the poor quality of store-bought vegetables with a failure by plant breeders to develop desirable cultivars. This situation has led some consumers to prefer heirloom cultivars instead of modern F-1 hybrids despite their lack of disease resistance and lower productivity. In actuality, their dissatisfaction is misplaced and often due to premature harvest coupled with long transportation times dictated by vegetable distribution systems. Although many modern cultivars have good flavor and quality if allowed to fully mature before harvest, some plant breeders have developed vegetables with traits such as greater fruit firmness and shelf life to meet the goals of industry rather than more flavorful and nutritious vegetables that consumers would prefer (Halweil, 2002).

Extensive use of pesticides

Human health problems can be directly linked to pesticide usage. Several high profile contamination events during the 1980s caused North American consumers to question chemical pest control for fruit and vegetable production. The use of unlabeled systemic insecticides on watermelons in California in the late 1980s caused severe illness for dozens of people who ate tainted fruit (Goldman et al., 1990). Reports that the plant growth regulator Alar may negatively affect human health in the 1980s heightened public concern about chemical residues on fruits and vegetables (Van Ravenswaay and Hoehn, 1991). The Bhopal India disaster, which killed thousands from exposure to methyl isocyanate gas that leaked from the Union Carbide pesticide plant in 1984, focused worldwide attention on the human health risks from exposure to agricultural chemicals (Dhara and Dhara, 2002). The presence of chemical residues on fruits and vegetables continues to be a major concern for consumers because of these and other incidents. One of the outcomes has been an increased consumer demand for pesticide-free vegetables.

Studies have tested the amount of pesticide residues on both fruits and vegetables to assess the risk. Many grocery store chains routinely check their conventionally grown produce for residues to make sure they are in tolerance with US law and safe for consumers. The FDA also spot-checks samples to make sure they are in compliance with federal laws. In one study, the FDA examined 18,113 non-organic food samples for pesticide residue as part of their routine pesticide monitoring program. Residues within legal tolerances were found on 31.2% of the non-organic samples while 2.5% contained residues that exceeded tolerance or occurred on crops for which they were not labeled (Beall et al., 1991).

The California Department of Food and Agriculture studied 9,403 samples of non-organically grown produce. In this study, roughly 88% contained no detectable residues, 21% contained pesticide residues within legal limits, and less than 1% contained illegal residues, most from chemicals that were not registered for use on the crop (Beall et al., 1991). The illegal residues may have been wrongfully used or inadvertently picked up from neighboring crops or soils. Not all the vegetables sold to consumers are tested for chemicals. The chance that illegal chemical residues on vegetables, applied either intentionally or unintentionally, is always possible (Beall et al., 1991). However, the long-term trend in vegetable production in the USA is lower use of agrichemicals because of increased organic production, fewer chemical registrations for vegetable use, and improved nonchemical alternatives for crop production (Chapter 9).

In many developed countries, vegetables are imported from countries where pesticide usage laws maybe less restrictive. Government inspectors test a relatively small percentage of imported vegetables before they enter a country. Food processing, transportation, and handling practices can decrease the amounts of residues. Some processes like concentration or dehydration can increase the amount of pesticides and pesticide breakdown products found in vegetable products.

There are several steps consumers can take to minimize their exposure to pesticide residues on vegetable crops. Washing vegetables with water (soap or detergent is not recommended) can effectively reduce the amount of residue present. One study showed that as much as 97% or as little as 9% of the residue may be removed by washing (Beall *et al.*, 1991).

In the USA and other developed countries, extensive testing is required to have a new pesticide registered for use in vegetable production. The cost of developing and registering a new pesticide has been estimated to exceed US$20 million in some cases. The development time until government approval for a new pesticide has been estimated to be 8–10 years. Much of the toxicology data needed for registering new agricultural chemicals is based on long-term animal studies, which are just one portion of the required tests, and take approximately 4–5 years to complete. As a result, many of the older chemicals are lost from the market rather than re-registered. Because of the time and cost, pesticide manufacturers have been slow to develop new chemical products, particularly for minor vegetable crops that are typically grown on small acreages. So farmers today have fewer synthetic chemical options for vegetable production compared to the past. The reduction in chemical pesticide options has caused many conventional growers to consider nonchemical pest control alternatives.

Contrasting organic and conventional systems

Conventional farming systems and practices also vary. However, they share many characteristics such as: rapid technological innovation; large capital investments in order to apply production and management technology; repeated use of intensive monoculture production systems; extensive use of uniform high-yielding hybrid crops; use of pesticides, fertilizers, and external energy inputs; high labor efficiency; and dependency on agribusiness partners (Stauber *et al.*, 1995).

The philosophy of some conventional vegetable producers may be summarized as follows: (i) nature is a competitor to be overcome; (ii) progress requires an unending evolution of larger farms using less labor; (iii) progress is measured primarily by increased productivity; (iv) efficiency and success are measured by looking at farming from a business stand point; and (v) science is an unbiased approach that can be used to produce social good (Stauber *et al.*, 1995).

Organic vegetable production is a natural approach that does not use synthetic pesticides, chemical fertilizers, or genetically engineered organisms. Organic vegetable producers minimize the use of external inputs in general and rely as much as possible on recycling (Kuepper and Gegner, 2004). Organic practices vary by country and group.

Although organic vegetable production is perceived by some as a primitive "back-to-nature" approach, in actuality it is a sophisticated scientifically based system with an interesting history. However, being a less mature sector within agriculture, organic production is less evolved with a considerable knowledge gap created by 170 years of intensive research devoted to conventional vegetable production (Kuepper and Gegner, 2004).

Throughout much of its early history, organic agriculture was treated with either hostility or apathy by the agricultural establishment consisting of universities, government agencies, and other proponents of conventional agriculture. Conventional agriculture was developed and promoted by these same establishments, so their hostile attitudes toward organic crop production were understandable. Defining and contrasting these production systems is filled with controversy and emotion. Some conventional growers view organic production as personal criticism or an attack on their methods. Most conventional farmers, particularly those who own land, take pride in their profession and want to preserve the environment and particularly soil to maintain their livelihood. So any suggestion that conventional production harms the environment and results in inferior vegetables sometimes causes emotional and defensive responses (Green, 1993).

However, attitudes toward organic and other alternatives to conventional production gradually improved with time. Many agricultural universities eventually developed organic research and educational programs. A number of factors are responsible for the change in attitudes toward organic agriculture, among them were consumer acceptance and the rapid growth of the organic industry. Also critical, especially from the perspective of the research community, were some high-quality scientific studies that added credibility to organic farming as a viable alternative.

For example, a series of studies funded by the US National Science Foundation and conducted by Washington University showed that some farms were able to maintain productivity without depending on the high-energy inputs of conventional farming (Kuepper and Gegner, 2004). These results focused the study on organic practices, including crop yields, attitudes, and the sustainability indices. The researchers concluded that organic farms could be competitive with conventional farms in the marketplace (Lockeretz et al., 1981). A national USDA study produced similar results, opening the way for a lengthy debate about organic production, which ultimately led to the National Organic Program (NOP) in 2002 in the USA (Kuepper and Gegner, 2004). The NOP facilitates the marketing of fresh and processed organic food sold in the USA to provide consumers with consistent, uniform standards. The NOP uses a federal accreditation program administered by the Agricultural Marketing Service through certifying agents rather than government officials (Kuepper and Gegner, 2004). Other countries and groups have similar organic programs. Third party organic certification is available in some parts of the world as well.

Organic Standards

One of the difficulties organic growers face is determining which products or materials are suitable for production. This is a very real problem that frustrates and complicates organic production. As a general rule, natural or nonsynthetic materials can be used for organic production but there are exceptions. In the USA, the NOP National List of Allowed and Prohibited Substances provides guidance on materials that can be used for certified organic crop production. Included are lists of synthetic materials that are allowed in organic crop production such as sulfur compounds, insecticidal soap, etc. Another list contains natural or nonsynthetic materials that are prohibited such as ashes derived from the burning of manure, nicotine sulfate, etc. (Kuepper and Gegner, 2004). Other countries and groups have similar listings of allowed and prohibited substances.

When considering commercial products to use for production, a grower may only use approved organic ingredients. If a full disclosure of ingredients is not found on a product label, details must be obtained from the distributor or manufacturer and retained on file. This level of detail even extends to inert product ingredients like fillers or carriers. Ultimately an agent from the certifying program must approve a list of all materials used for organic production.

The Organic Materials Review Institute (OMRI) plays an important role in determining which materials are suitable for certified organic production under the NOP in the USA. OMRI is a nonprofit, nongovernment organization that evaluates products for suitability in organic production and processing. OMRI is not a regulatory body. However, its decisions are highly respected and accepted by most certifiers of organic production. OMRI-listed products can be purchased and used with a high degree of confidence that they are suitable for organic production although there are many acceptable products in the marketplace that have not been evaluated by OMRI (Kuepper and Gegner, 2004).

Production Principles and Practices

Organic agriculture also shares management tools with conventional production such as crop rotation, raised beds, and maintaining soil pH by liming to name a few. However, many contemporary organic practices are used as well although they vary widely from farm to farm. There are several fundamental principles that characterize organic crop production. They include biodiversity, integration, sustainability, natural plant nutrition, natural pest management, and integrity (Kuepper and Gegner, 2004). Organic production includes a diversity of practices to maintain these principles including use of cover crops, green manures, animal manures, and crop rotations to fertilize the soil, maximize biological activity, and maintain long-term soil health. Another fundamental aspect is the use of biological controls, crop rotations, and other techniques to manage weeds, insects, and diseases. Organic principles include a reduction of external and off-farm inputs as well as the elimination of synthetic pesticides, fertilizers, and other unnatural materials. A focus is placed on renewable resources, soil, and water conservation, and management practices that restore, maintain, and enhance ecological balance (Kuepper and Gegner, 2004).

Individual practices can be thought of as tools in a toolbox that can be utilized differently to accomplish the same task in an organic system. There is no single cookbook recipe that works for all farmers. Examples of some of the many organic practices are to follow.

Biodiversity

Some insect pest outbreaks occur because of imbalances in the agroecosystem. In nature, massive pest outbreaks are relatively rare and short-lived, due to the presence of natural predators, parasites, and disease agents that quickly restore ecological balance. Diverse natural ecosystems are generally more stable than ones containing a limited number of species. The same principle is essentially true for agroecosystems as well. Organic farmers as well as some conventional growers rely heavily on biological pest control. Biological control relies heavily on an ecological balance with populations of beneficial insect predators and parasites, pest disease agents, insect-eating birds and bats, and other natural pest control strategies. Organic farmers sometimes release beneficial insects like ladybird beetles (ladybugs), lacewings, trichogramma wasps, tachinid flies, etc. to enhance biocontrol. Farmscaping is designing and maintaining both permanent and temporary habitats that support beneficial insects, spiders, and other species. The absence of pesticides also favors biocontrol by supporting species diversity. These biological control measures help keep pest levels in natural balance so that damage is below economic thresholds for crop losses. In many cases biological control can be so effective that no additional action is even needed.

Organic farms often grow a diverse mixture of vegetable crops and native plants to encourage proliferation of beneficial organisms that assist in pest management (Kuepper and Gegner, 2004). Good organic farmers mimic natural biodiversity through practices such as intercropping, companion planting, growing noncash crops to create beneficial habitats for insects, crop rotation, companion planting, and sequential cropping (Kuepper and Gegner, 2004). Interplanting two or more compatible crops in close proximity is one strategy for increasing biodiversity and more efficient use of resources (also see Chapter 2). When practiced on a large-scale, interplanting is called intercropping, while small-scale practice is often referred to as companion planting (Kuepper and Gegner, 2004). An example of companion planting would be the interplanting of sweet corn, indeterminate beans, and vining cucurbits. In this system, the beans provide some nitrogen for the corn, cornstalks support the beans, and the cucumber or squash vines suppress weeds and provide a barrier to predation by small animals. In such a system, there is greater balance so natural predators can prevent an explosion of a single pest that could cause crop damage.

The diversity of crops grown in rotation, the use of cover crops, good soil water management, and other practices contribute to diverse soil biology that resists pests. Diversity above ground also suggests biological diversity in the rhizosphere, providing greater nutrient cycling, disease suppression, soil structure, and nitrogen fixation (Kuepper and Gegner, 2004).

In contrast, in conventional farming systems like monocultures the broad diversity of plants and insects is lost so pest problems are continuous. Explosions of insect pests are common and typically worsen with time. Monoculture farmers become overly reliant on chemical methods to control insect pest imbalances.

Natural Plant Nutrition

Through the process of photosynthesis, plants use atmospheric carbon dioxide and water as elementary building blocks to produce more complex compounds like carbohydrates, proteins, and oils. Plants also need mineral nutrients to make complex molecules required for rapid healthy growth and development. Plants obtain minerals from the rhizosphere through their roots. The rhizosphere is a dynamic zone surrounding the roots where complex interactions occur among roots, soil microbes, and the soil itself.

The organic philosophy of crop nutrition begins with proper care and nourishment of the soil organisms (Kuepper and Gegner, 2004). These organisms are responsible for breaking down soil organic matter, releasing nutrients and forming complex mutualistic relationships that benefit both plants and microbes. Organic soil management strives to maintain a healthy rhizosphere by avoiding inputs such as unnatural chemicals and practices like excessive tillage that may harm soil organisms. Organic matter and natural rock minerals are added to benefit both microbes and provide an optimal environment for root development. From the organic perspective, the conventional approach has several flaws:

- Conventional vegetable production systems typically provide the minerals needed for plant growth directly in a soluble form with less emphasis on maintaining soil organic and microbial activity (Welbaum *et al.*, 2004).

- Applying large quantities of soluble fertilizer to a crop a limited number of times per season overwhelms the rhizosphere, causing imbalances that lead to water pollution, crop diseases, insect infestations, and rapid, poor-quality plant growth.
- Failure to support and care for soil biotic life ultimately leads to soil decline. As a result, soil organisms are unable to produce compounds that directly benefit vegetable crops, soil structure is poor and successful production is increasingly dependent on synthetic inputs (Welbaum et al., 2004).
- Conventional fertilization tends to concentrate on nitrogen, phosphate, and potassium, even though the need for at least 13 soil minerals is scientifically recognized. This skewed focus is also responsible for generating imbalances.
- Application of excessive soluble nutrients can stimulate problem weed species.
- Soluble nutrients, especially nitrate, are prone to leaching, which can cause a number of environmental and health problems.

It is the organic approach to soil building and plant fertilization that is the basis for a belief that organic vegetables have superior nutritional value, much more so than the absence of pesticide residues, which has drawn the spotlight ever since the 1960s (Kuepper and Gegner, 2004). However, most studies show no clear-cut nutritional advantage from eating organically grown vegetables (Bourn and Prescott, 2002).

Some farmers may incorrectly believe the nutrients released through organic soil management practices are in a chemically organic form. The natural digestion of soil releases minerals in a form similar to commercial fertilizers. However, plants can absorb some large organic molecules such as chelated minerals, certain pesticides, and hormones (Kuepper and Gegner, 2004).

Organic agriculture believes that pests like weeds, insects, or diseases, are indicators of how far a production system has strayed from its natural balance (Kuepper and Gegner, 2004). Using this philosophy, some weeds may predominate when soils are too acidic or basic, others may predominate in anaerobic soil conditions when structure is poor, and still others may thrive when excessive fertilizer or salts are present. Organic proponents also believe that insect pests are more successful in attacking poorly nourished weak plants and well-nourished healthy plants are more resistant (Kuepper and Gegner, 2004).

Planned Crop Rotation

Essentially a tool for annual cropping systems, crop rotation refers to the sequence of crops and cover crops grown on a specific field. Sequences confer particular benefits to long- and short-term soil fertility and to pest management. For example, the penetration of deep-rooted cover crops breaking apart large aggregates creates improved soil structure and "prime" soil for later cash crops while adding organic matter (Welbaum et al., 2004). Some benefits of crop rotation include:

- Legumes fix atmospheric nitrogen, which is used by crops that follow in rotation as legume residues decompose. Incorporating as much legume crop biomass in the soil as possible increases nitrogen recycling.
- Insect pest cycles are interrupted, such as the northern and western rootworm species that attack sweet corn for example.
- Diseases and pests may be suppressed, including nematodes.
- Weed control is often enhanced because perennial weeds are destroyed through cultivation of annual grains. Many annual weeds are outcompeted by mowing alfalfa or other dense forages grown in rotation.
- Livestock or composted manures are applied just in advance of crops like sweet corn that are heavy nitrogen feeders.

Soil Disease Management

Even with crop rotation, soil-borne diseases, nematodes, and insects may damage crops particularly when ecological balance is lost. Organic growers can use solarization for control of soil-borne diseases until balance is restored. Soil solarization is a hydrothermal process that occurs when moist soil is covered by plastic film and heated by sunlight during the warm months. Soil solarization at 37°C (99°F) for 2–4 weeks kills many annual weed seeds. Soil solarization is an environmentally friendly method for controlling pests such as soil-borne fungi, bacteria, nematodes, insects, mites, weed seed and seedlings (Stapleton, 2000). Solarization initially may reduce populations of beneficial microorganisms, but they quickly recolonize soil after treatment.

The addition of different types of organic matter to the soil before soil solarization increases efficacy. Conventional growers have increasingly used solarization as an alternative to gaseous fumigation to control soil-borne pathogens as well.

The use of cultivars resistant to soil-borne diseases and nematodes is another effective management tool for soil diseases. If disease-resistant cultivars are not available, grafting a scion of a susceptible cultivar on to resistant rootstock is a technique popular in parts of the Middle East and eastern Asia (King et al., 2008).

Often soil diseases are caused by excessive moisture and can be controlled by effective water management rather than chemical control measures. Calculating irrigation needs saves water and prevents waterlogged soils that increase root disease. Using tensiometers with pan evaporation is an accurate method for determining when irrigation is needed and how much to apply (Chapter 5). Raised beds improve drainage, particularly on heavy soils, prevent erosion on slopes and are a valuable management tool for both conventional and organic vegetable production (Parish, 2000).

Cover and Green Manure Crops

Cover crops are grown for soil and nutrient conservation. The combined benefits become economically feasible when the cover is grown during the offseason or interseeded with a primary cash crop (Kuepper and Gegner, 2004). A rule of thumb in sustainable crop management is that "soil should not be left bare for extended periods". Cover crops offer a variety of benefits including:

- improved soil quality;
- increased organic matter;
- reduced soil erosion;
- reduced soil compaction;
- increased nitrogen supply when legumes are grown and incorporated;
- improved water infiltration;
- decreased runoff;
- weed suppression;
- soil moisture conservation;
- reduced nitrate leaching; and
- increased yields of certain crops.

A crop grown for soil improvement is called a green manure (Cherr et al., 2006). Cover and green manure crops are related. In most cases, cover crops are used as green manures and incorporated prior to the planting of a cash crop. Legumes make good green manure crops because they fix atmospheric nitrogen (N), which is released for subsequent crops after the biomass is incorporated and thoroughly decomposed. However, if the green manure is also used for grazing or as hay some nutrients are removed as biomass and the fertilizer value is diminished (Cherr et al., 2006). For example, forage soybeans produce greater biomass and thus greater residual soil N than grain soybeans, because more N is removed from the field when the grain is harvested.

Green manure crops were under utilized in monoculture systems where cash crops and concentrated fertilizers were often emphasized. In recent years, there has been renewed interest in both conventional and organic agriculture in green manures as a technique for enhancing and preserving soil resources by maintaining year-round soil cover (Cherr et al., 2006).

Manure and Compost

Animal manures are traditional organic fertilizers. Before the advent of synthetically derived fertilizer in the last century, animal fertilizers predominated, so there is a long history of manure use for vegetable production. When livestock enterprises are integrated into a whole-farm operation, manure becomes an input in a closed system of nutrient recycling. However, with increased vegetable crop specialization, it is not always feasible to combine animal and vegetable production. When on-farm manure is not available it must be transported. Since the nutrient analysis of most manure is low relative to concentrated granular fertilizers used by conventional growers, transportation costs are higher by comparison. Large-scale intensive animal production enterprises generate considerable quantities of animal waste that are sometimes sold to vegetable producers. However, there are concerns about the quality of manures from these sources since they may be contaminated with heavy metals, antibiotics, pesticides, or hormones. Some organic growers object to partnering with industrial-scale livestock operations for fear that they would be promoting a less sustainable form of production that is at odds with their environmental and social values. Despite these concerns, the NOP does not differentiate between livestock manure sources in the USA. However, the NOP regulations require that livestock manure only contains compounds found on the National List of synthetic substances permitted for organic crop production.

Another issue of concern for using manure in organic vegetable farming involves food safety. Much of the human illness caused by eating uncooked produce can be traced to biological contamination (Chapter 9). At a time when concerns about microbial contamination are high, some have questioned the risks associated with manure use on food crops. One of the leading sources of biological contamination is bacteria transmitted through fecal matter to vegetables at any point in the supply chain. It has been argued that organic vegetables are more likely to become contaminated because of the greater use of animal manures in their production (Winter and Davis, 2006).

If proper safeguards like composting standards or application restrictions are not followed, manure can be a source of biological contamination. However, both GAPS and the NOP contain guidelines to ensure the safe handling and application of manures to reduce the risk of biological contamination on vegetables (Kuepper and Gegner, 2004). If these safety guidelines are followed, the risk of biological contamination on organic vegetables is minimal. For example, many organic regulations specify when raw manures may be applied relative to crop harvest. The NOP requires that raw animal manure be incorporated into the soil not less than 120 days prior to the harvest for vegetables that form in direct contact with soil and not less than 90 days prior to the harvest for vegetables that do not directly contact soil.

On-farm composting stabilizes the nutrients in manure, builds populations of beneficial organisms and has a greater beneficial effect on soils and crops. Composting also helps rid manure of biological contamination and improves the quality of the manure as fertilizer (Kuepper and Gegner, 2004). Minimum composting temperatures are often specified for organic programs to ensure that harmful bacteria are killed.

Additional products from composts, such as compost teas, have special applications in organic agriculture. Human manures, including composted sewage sludge commonly known as biosolids, are forbidden in many organic crop production programs around the world.

Field Sanitation

Sanitation is a simple but vitally important technique for pest control that requires discipline and diligence. Sanitation practices include the removal or deep plowing of crop residues that may harbor plant disease or insect pests, destroying weedy habitats that shelter pests, cleaning weed seeds from farm equipment between harvests, cleaning cultivation equipment between fields, and sterilizing pruning tools (Kuepper and Gegner, 2004).

Currently there is interest in flame or thermal weeding (Kuepper and Gegner, 2004). Torches mounted on a tractor's three-point hitch direct flames to the row middles. Tractor speed is adjusted so that weeds are seared while crop plants are unharmed. Searing uses less fuel and is sufficient to kill most weed seedlings. Liquid propane is the most common fuel, though alternatives such as alcohol and methane may be produced on-farm. The burning of crop residues to kill disease, of fence-rows to kill weeds and pest habitats, and of organic matter in dry climates has largely been abandoned in much of the world because of the air pollution created (Kuepper and Gegner, 2004). Also, deep plowing and the burning of crop residues can increase erosion and reduce biodiversity.

Tillage and Cultivation

Organic growers focus on maximizing the benefits of tillage instead of using herbicides (Kuepper and Gegner, 2004). Tillage and cultivation are important tools used for weed control, crop residue management, hardpan reduction, sanitation to destroy pests, and disease habitats to name a few. Plowing is intended to incorporate green manures, manures, and crop residues in the upper, biologically active zones of the soil, rather than burying them deeply.

Moldboard plowing is practiced only before planting, so soils are not left completely bare and vulnerable to erosion for extended periods. In larger-scale operations, tillage equipment includes rolling cultivators, rotovator tillers, finger-wheel cultivators, and torsion cultivators that enable tillage close to the row with minimal root damage while providing weed control. Small-scale farmers often rely on hand weeding, hand hoes, wheel hoes, stirrup hoes, and other relatively inexpensive equipment.

There are downsides to tillage, such as increased cost and damage if cultivation is too close to the plant. Determining the type, frequency and right amount of tillage while minimizing costs is sometimes challenging (Kuepper and Gegner, 2004). Tillage also aerates soil and speeds the decomposition and loss of organic matter. Excessive tillage may decrease earthworm activity. There is also the danger of compaction since every tractor pass compresses soil. Both organic and conventional farmers are concerned about minimizing production costs by reducing field operations. Reducing labor, fuel, and equipment costs has always been one of the major appeals of chemical weed control. But chemical weed control often has negative effects on the environment and this is one reason why organic growers do not use herbicides.

Conservation Tillage and Organic Farming

Conservation tillage is not considered a traditional organic practice. Organic agriculture is often characterized as being an intensive and extensive tillage production system (Kuepper and Gegner, 2004). This reputation has been advanced by the popularity of small-scale organic systems like the French Intensive and Biointensive Mini Farming models that encourage digging to create deep root zones (Jeavons, 2001). While appropriate to such intensive systems, this degree of cultivation is not characteristic of all organic agriculture (Kuepper and Gegner, 2004).

Some organic producers have adapted conservation tillage without using herbicides to kill the cover crop. Chisel plowing can incorporate light residues in the upper layers of the soil while leaving a significant cover to reduce erosion. Advances in seeding and transplanting equipment enable successful establishment of vegetable crops in light to moderate residue. Mechanically killing small-grain cover crops before they seed by rolling or flailing is an effective alternative to herbicides. The practice of ridge-tillage and strip tillage is widely used for small-seeded vegetable crops where the residue serves as mulch after planting (also see Chapter 2).

Conventional growers were quick to embrace conservation tillage because it saves fuel, and reduces labor and compaction. Conventional growers traditionally used herbicides to kill cover crops but are now transitioning to the alternative technologies (Abdul-Baki and Teasdale, 1997).

Mulching

Mulching is a practice often used by both organic and conventional growers. Historically, mulches for vegetable production consisted of straw, old hay, or other inexpensive biomass that was spread across a field to cover soil between plants. Mulches remain popular for vegetable production for a variety of reasons. They regulate soil moisture and temperature, suppress weeds, help keep vegetables clean, and increase soil organic matter (Chapter 6). Mulching has evolved so that cover crops are grown in place and killed by a variety of methods to generate biomass for no-till, strip-till, or ridge-till systems (Abdul-Baki and Teasdale, 1997).

During the past 40 years, plastic film has become the mulch of choice for conventional and some organic growers for many reasons, including longer life without breaking down, greater temperature modification, outstanding soil-moisture retention, reduced leaching of nutrients from the root zone following heavy rains, and weed control depending on the mulch used (Chapter 6). However, not all organic growers agree that plastic film is sustainable because most are petrochemically based. Despite these concerns, the NOP in the USA allows the use of plastic mulch as long as it is removed from the field at the end of the growing season. Some no-till organic farming systems use perennial deep-rooted living mulches that create minimal competition with shallow-rooted, annual, cash crops.

Supplemental Fertilization

Organic fertilizers are derived from animal sources or vegetable matter. These materials are sometimes characterized as either natural or processed. Examples of naturally occurring organic fertilizers are manure, plant tissue, worm castings, peat, seaweed, mined mineral deposits, and guano. Processed organic fertilizers include compost, blood meal, bone meal, humic acid, amino acids, and seaweed extracts. Other examples are natural enzyme-digested proteins as well as fish and feather meal (EPA, 2013).

Many organic farmers rely on animal and green manures as their primary fertilizers instead of concentrated industrially synthesized fertilizers. This is especially true for deep and rich soil at higher latitudes where organic matter decomposes slowly. Nutrient inputs for both organic and conventional growers are based on soil testing results so fertilizers are not wasted. To correct mineral deficiencies in organically managed soils, organic and some conventional growers often apply ground or powdered rock minerals.

A commonly used rock mineral is high-calcium aglime. Dolomitic limestone, natural sodium nitrate, various rock phosphates, gypsum, sulfate of potash-magnesia, and natural potassium sulfate are also common natural mineral fertilizers applied during vegetable production (Willer *et al.*, 2013; Fig. 8.1).

Fish emulsion and blood, feather, bone, alfalfa, and soybean meal are organic fertilizers that provide nitrogen. Some of these N fertilizers also supply organic matter, although that is not their primary function. When used in combination, organic mineral sources provide the essential plant macronutrients (N, P, K) and/or micronutrients (Ca, Mg, S). Natural minerals like glauconite (greensand), glacial gravel dust, lava sand, azomite, granite meal, and others are added to correct micronutrient deficiencies. A wide range of other products such as humates, humic acids, catalyst waters, bioactivators, and surfactants, have been approved for organic crop production although not everyone agrees on their effectiveness.

Biorational Pesticides

It is common belief that pesticides are not permitted for organic vegetable production. However, a rather wide range of natural or biorational pesticides is permitted by many organic standards. The frequency of pesticide use varies with both crop and location. The pesticides permitted in organic farming may be classified into several categories (Kuepper and Gegner, 2004).

Fig. 8.1. Sodium nitrate is a natural form of nitrogen that is mined from mineral deposits. This bag was approved for use in the NOP in the USA by OMRI.

Minerals

These include sulfur, copper, diatomaceous earth, and clay-based materials.

Botanicals

Botanicals include natural plant extracts such as rotenone, neem, and pyrethrum. Less common botanicals include quassia, equisetum, and ryania. Tobacco products like Black-Leaf 40 and strychnine are also botanicals but are prohibited in many organic programs due to their high toxicity. Strychnine is a naturally occurring alkaloid poison that was first identified by Carl Linnaeus in 1753 in plants from the genus *Strychnos*, family Loganiaceae, including trees and climbing shrubs (Buckingham, 2010). The case of strychnine serves as a reminder that some botanicals are highly toxic and that natural compounds are not necessarily safer than synthetic ones.

Soaps

A number of commercial soap-based products are effective as insecticides, herbicides, fungicides, and algicides (Weinzierl, 2000). Soaps in general are made from salts of fatty acids. Fatty acids are building blocks of natural fats and oils found in both animals and plants. Oleic acid, present in high quantities in olive oil and in lesser amounts in other vegetable oils, seems to have the greatest insecticidal activity. The potassium salt of oleic acid is the active ingredient in Safer's soaps, one of the more common currently available insecticidal soaps (Weinzierl, 2000). Soaps from synthetic fatty acids or detergents cannot be used for organic production (Kuepper and Gegner, 2004).

Even after many years of use, the mode of action of insecticidal soaps remains somewhat unresolved. Although the action of soaps involves physical disruption of the insect cuticle or outer body covering, additional toxic effects likely occur. Soaps may enter the insect's respiratory system and cause internal cell damage by breaking down membranes or disrupting metabolism. Soaps are also detrimental to the development of immature insects.

Insecticidal soaps are nonselective and may harm beneficial insects. For example, wet soap sprays may kill ladybird beetles (ladybugs) and green lacewing larvae if they are present on a plant when treated. There is a common misconception that any "soap" or "detergent" can be used as an insecticide or miticide. Only a few select soaps have insecticidal or miticidal properties. Many common household soaps and detergents have only limited activity on soft-bodied insect and mite pests when applied as a 1% or 2% aqueous solution. The reliability of cleaning products is less predictable than insecticidal soaps formulated as pesticides (Weinzierl, 2000).

Pheromones

A pheromone is a secreted or excreted chemical factor that triggers a social response in members of the same species. Pheromones are chemicals capable of acting outside the body of the secreting individual to impact the behavior of the receiving individual (Howard and Blomquist, 2005). Pheromones regulate many types of insect behavior and are used for trapping insects for monitoring populations or control. There are alarm pheromones, food trail pheromones, sex pheromones, and many others that affect behavior or physiology. Sex pheromones are produced by one sex (usually the female) to attract the other sex for mating. Mass insect attacks are coordinated by aggregation pheromones that attract others to a common location. Alarm pheromones are produced to help in colony defense. Ants produce trail pheromones to find food sources (Howard and Blomquist, 2005).

Insect pheromones, particularly the sex pheromones of moths, are among the most biologically active compounds known. A single molecule of pheromone can be detected by some species. Because of this sensitivity, sex pheromones are a common tool in integrated insect management programs. Both conventional and organic growers use pheromones to confuse and disrupt pests during their mating cycles or for trapping (Fig. 8.2).

Synthetic sex pheromones are available for insect pest detection and population monitoring. Pheromone traps allow the grower to make better application decisions based on threshold pest populations as part of an Integrated Pest Management (IPM) system (Howard and Blomquist, 2005).

Biologicals

One of the fastest-growing areas in natural pesticide development is biopesticides. Examples of this technology include the use of *Bacillus thuringiensis*, *Phlebiopsis gigantean*, and *Agrobacterium*

Fig. 8.2. Pheromone trap at the edge of a corn field.

radiobacter. Biopesticides are one aspect of biological control. Biocontrol is both a naturally occurring process and can be defined as the designed use of one organism to control another. Biocontrol can be achieved by three methods:

1. Inundative release (also termed "classical biocontrol") in which a natural enemy of a target pest, pathogen, or weed is introduced to a region from which it is absent, to give long-term pest control.
2. A biocontrol agent is applied as required (often repeatedly), in the same way that chemical pesticides are used.
3. Management and manipulation of the environment to favor the activities of naturally occurring control agents (Landis *et al.*, 2000). For example farmscaping, which provides host plants for natural predators in or near a production field.

One of the best-known biopesticides is *B. thuringiensis* (Bt), which is widely used to control lepidopterous pests. *Bacillus thuringiensis* is a gram-positive, soil-dwelling bacterium, commonly used as a biological pesticide by spraying on vegetable crops (Schnepf *et al.*, 1998). *Bacillus thuringiensis* also occurs naturally in the gut of caterpillars of various types of moths and butterflies, as well on leaf surfaces, aquatic environments, animal feces, and insect-rich environments such as grain storage facilities. During sporulation, many Bt strains produce crystal proteins (proteinaceous inclusions), called δ-endotoxins, which have insecticide action (Schnepf *et al.*, 1998). Not all crystal-producing Bt strains have insecticidal properties. Spores and the crystalline insecticidal proteins produced by *B. thuringiensis* have been used to control insect pests for years and are often applied as sprays under trade names such as DiPel and Thuricide. Because of their specificity and natural origin, these pesticides are regarded as environmentally friendly, with little or no effect on humans, wildlife, pollinators, and most other beneficial insects, and are widely used by organic farmers.

A few vegetable crops, such as sweet corn, have been genetically transformed to express the Bt δ-endotoxins genes that have insecticide action (Shelton *et al.*, 2002). The *Bt* gene was expressed in corn to control worms like the European corn borer that burrow into the ear through the husk and feed on developing kernels from the inside out. Control occurs when worms ingest the δ-endotoxin during feeding. Conventional spray treatments with either traditional insecticides or Bt sprays do not provide effective control because the husk protects the worms. Sweet corn expressing the *Bt* gene is one of a hand full of genetically engineered crops approved by the US government for commercial sale in the USA but very little transgenic sweet corn has ever been sold because of consumer concerns surrounding genetic engineering technology (Chapter 3). Genetically engineered crops are not allowed for organic production (Kuepper and Gegner, 2004).

Physical traps

Physical traps attract insects and capture them so they cannot escape. Sticky cards can be placed among or above plants to collect flying insects (Fig. 8.3). Sticky traps are usually yellow colored cards or strips to attract flying insects and are covered with strong waterproof adhesive. Insect populations can be identified and quantitated from tape or cards as part of IPM. The sticky traps are discarded when covered with insects.

Funnel traps, water pan traps, and cone traps may be used to catch certain species either to monitor

Fig. 8.3. The moving boom agitates the foliage, causing insects to fly into the yellow stick tape mounted above in this organic greenhouse in the eastern USA.

their populations or provide control. Blacklight traps may help determine when insects are flying as well as their relative abundance. This information allows pest managers to determine the timing of peak periods of activity and design pest management programs. Blacklight traps are useful for monitoring pests, but they are not designed to reduce populations. Blacklight traps are non-specific and require electricity, so pheromone traps are more easily used.

Another strategy is to repel flying insects away from plants by using silver or reflective material. Reflective mulch repels certain flying insects searching for plants, apparently because reflected ultraviolet light interferes with the insects' ability to navigate. Reflective mulch may be effective against winged aphids, leafhoppers, thrips, and whiteflies (Dreistadt *et al.*, 2007).

A trap crop attracts insect pests away from vegetable crops. This form of companion planting can provide pest control without the use of pesticides. Trap crops can be planted around the outside of a vegetable field to be protected or interspersed among rows. The trap crop with high insect population can be destroyed to disrupt reproduction (Landis *et al.*, 2000).

Hand removal of pests is one of the oldest and most basic of insect control methods. This technique is effective on small plots but is generally not cost effective for larger ones. Some organic growers have used tractor-mounted vacuum systems designed to suck and collect insects from the leaves of plants in vegetable fields. In some cases, individual plants in a vegetable field may become heavily infested with pests while the surrounding plants are not. In this situation the entire infested plant can be removed or "rogued" and destroyed as a control measure.

Buffers and barriers

Some use floating row covers as a physical barrier to exclude fly-in insects (Kuepper and Gegner, 2004; Fig. 8.4). Field buffers are strips of sod or other permanent vegetation that run through sections of fields or surround them. They may also consist of rows of trees or wood lots. There is considerable interest in both conventional and alternative agriculture in buffers, since they help reduce soil erosion and improve water quality. If managed properly by maintaining weed control and nurturing certain plants, some buffers also serve as beneficial insect habitats.

For organic production, field buffers may reduce genetic contamination from unwanted cross-pollination and chemical drift from adjacent land. Most agencies require a minimum 7.6 m (25 ft) buffer along "uncontrolled" borders where there is the potential for chemical use by a neighbor. Wider borders maybe required where hazards are great such as when an adjacent farm applies synthetic pesticides by aircraft (Kuepper and Gegner, 2004).

Fig. 8.4. Floating row covers provide a barrier to exclude fly-in insects.

Record Keeping

Because it is not feasible for representatives of an organic certifying agency to visit a farm regularly throughout a growing season, these organizations must rely on record keeping and production plans to monitor program compliance. Since vegetables are not routinely tested for residues by the certifying agency, most organic programs are based largely on record keeping and trust. A production plan consisting of detailed accounts of how and where a crop will be grown, specific products to be applied and storage facilities to be used is completed and reviewed before a crop is planted. To be organically certified a grower must provide the reviewing agent with extensive documentation to prove that the crop was organically grown, it was not inadvertently contaminated with chemicals, and it was not commingled with a similar conventional commodity produced on the same farm (Kuepper and Gegner, 2004).

Certification also comes at a financial cost. For middle size and larger growers, the cost of certification can be significant and is usually passed on to the consumer. Many programs have provisions for small growers to reduce or avoid the cost of certification.

High-input Organic Agriculture

Not all types of organic farming are the same. Organic farming was defined at the beginning of this chapter as a system that uses a minimum of off-farm inputs. This still describes the philosophy of many growers and the roots of the organic movement. However, in the USA the NOP has evolved so that some larger organic growers have a much greater reliance on off-farm inputs. These large growers who traditionally have been associated with conventional production, have shifted to the NOP to take advantage of the higher profit margins that sometimes exist for organic vegetables. Intensive organic vegetable production using plasticulture is an example of this phenomenon. In the organic plasticulture system, traditional rotations and soil-building practices are usually employed, followed by clean cultivation and the laying of plastic mulch and drip irrigation tape on shaped beds. During the season, significant quantities of soluble organic fertilizers, often fish-based, are fed to the crop through the drip system (i.e. organic fertigation). At the end of the season, all plastics are removed from the field, and it is returned to more standard organic management often including an off-season cover crop. Such systems are often exceptionally productive and economically attractive, when premium organic prices are received.

The labeling of such high-input systems as certified organic presents a paradox for many proponents of traditional organic agriculture. These technological advancements are unlikely to reflect the kind of farming most of the original practitioners and supporters think of as truly "organic". Of particular concern is the potential for soil erosion in an organic plasticulture system. Low-input organic farms produce less soil erosion, while fields grown using plasticulture are 15 times more erodible (Kuepper and Gegner, 2004). Traditional organic farms leach minimal amounts of N into runoff or groundwater, while losses from fields fertilized with high levels of soluble organic fertilizers would be predicted to be much greater. Also, of concern is the fossil fuel energy involved in plastic manufacture, transportation, and application that may or may not be compensated by reductions in the tractor fuel needed for greater cultivation. A further consideration when analyzing the efficiency of plastic mulch is the issue of disposal following removal. There traditionally are limited options in many areas for recycling plastic mulches at the end of each season. Newly developed biodegradable plastic films may reduce some of the environmental impact of the plasticulture system for vegetable production (Chapter 6).

Many organic farms perform well on many of the measurable indicators associated with sustainability, such as energy consumption and environmental protection when compared to conventional farms (Dalgaard *et al.*, 2001). Finally, the lowered capital investment required to produce a crop by traditional organic methods makes this form of farming more accessible to resource-poor farmers and entails less risk in years of crop failure or low prices. These factors are less certain in a high-input system.

Summary: Is Organic Sustainable?

The organic movement started as a small-grower initiative to produce vegetables with fewer off-farm inputs and minimal environmental impacts. It has evolved into a mainstream agriculture system that is now government controlled in many countries, embraced by large food-producing corporations because of its profit potential, and the subject of mainstream research. The growth of organic vegetable sales has been driven by strong consumer demand for healthy nutritious vegetables, developed through traditional plant breeding rather than bioengineering, and produced by environmentally responsible farmers rather than large faceless corporations whose primary concern is profit. Related to the organic movement is a consumer desire to purchase locally grown produce that is fresher and healthier than produce grown at distant locations, often harvested in an immature state, and transported great distances with little regard for quality or energy consumption. Studies have shown that consumers purchase organic produce for a variety of reasons. One is a consumer perception that organic produce is pesticide free and therefore safer, more nutritious, and produced in an environmentally sustainable manner (Raviv, 2010). Some consumers buy organic for the presumed environmental benefits of organic production including erosion and runoff control and environmentally safe pest management (Raviv, 2010). Other consumers buy organic because they are concerned about the effects of genetic modification on the food they eat. Some consumers have expressed a desire in helping small local producers with their organic purchases as well.

While the perceptions of organic farming have changed over the years, the growth of the industry and the introduction of certification standards have led to a new clearer definition. This definition describes organics as a viable production system, based on sound scientific practices, that does not include synthetic chemicals or transgenic crops and that is recognized outside agricultural circles.

As the organic industry continues to grow and evolve, it faces many challenges, including the consequences of its own success. Economic opportunities invite new players into the market who may have little interest in sustainability or the core values of the original organic movement. The term organic is no longer synonymous with small family farms. In fact, the availability of organic products will likely continue to grow driven by the involvement of large corporations given current organic standards (Fig. 8.5).

It will be of interest to see how the organic movement evolves in the future for several reasons. At least in the USA, attitudes appear to be changing somewhat as consumers gain a greater understanding of the NOP and who is growing the certified organic produce they buy. These changing attitudes may be the first signs that in the USA the NOP is no longer addressing all consumer needs. This shift in attitude is likely due to multiple reasons. Consumers are

Fig. 8.5. A large greenhouse operation producing pots of organic basil hydroponically for retail sale through supermarkets.

becoming aware that organic produce is not necessarily more flavorful, safe or nutritious than non-organic produce (Bourn and Prescott, 2002).

There is a growing awareness that biological contamination is one of the most critical issues in vegetable safety and both conventional and organically crops are susceptible (Chapter 9). Good Agricultural Practices (GAPs) certification, rather than organic, is the best way to ensure vegetables are free of biological contamination. Some argue that the present organic system does not adequately address the problems of true economic sustainability of smaller family farms or farm workers, particularly in foreign countries that export vegetables to developed countries. Many consumers now understand that the NOP is a government-run program that adds to the cost of their food. They understand that large corporations are increasingly involved in certified organic vegetable production rather than small local growers. Consumers are purchasing more locally grown vegetables from farmers who do not participate in the NOP but grow crops in an environmentally responsible fashion with minimal synthetic chemical inputs. Several successful large grocery chains are emphasizing sustainably grown crops in their advertisements rather than offering organic produce exclusively. Still other consumers are learning that some types of genetic modification, like traditional plant breeding, are safe, natural, and have been used for generations to feed the world. The Green Revolution, F-1 hybrid corn, bush tomatoes for mechanical harvesting, and bush soybeans are all examples of how plant breeding genetically improved crop productivity without compromising quality. While many remain skeptical of the more radical genetic engineering technologies, consumers are increasingly aware that some genetic transformation technologies are natural and can safely be applied to human foods. For example, the debate about golden rice has educated some consumers that genetic engineering can have a positive impact to help feed malnourished, at risk populations.

What is becoming clear is that conventional and organic vegetable production are no longer diametrically opposed as they once were (Fig. 8.6). Both conventional and organic producers alike will increasingly adapt effective and economically viable organic practices now that many of the ideological barriers of the past have been removed and the vegetable production systems continue to evolve (Fig. 8.6).

In present-day agriculture, both conventional and organic growers are concerned with environmental sustainability, largely through a raised awareness brought about by the organic movement and environmental research. In fact, some technologies that were mainly developed for organic growers are now widely used by conventional growers as well. Floating row covers were widely used by organic growers but now are used by some conventional growers. The many beneficial effects of organic soil management will continue to be integrated into conventional production. The environmental advances from the organic movement such as a judicious use of minimal amount of fertilizers and non-chemical pest control technologies will likely be retained. As agricultural chemical prices, chemical availability, and fuel prices increase, conventional growers are more likely to embrace nonchemical alternatives. In addition to continuing environmental concerns, both producers and consumers will demand socially responsible practices to ensure the health and prosperity of the agricultural sector.

The flow of innovation moves in both directions. No-till cultivation was initially used by conventional growers but with adaptations is now used by many organic growers as well. Conventional research has contributed advanced IPM strategies, highly selective biopesticides, and genetically improved crops, that have improved production with minimal environmental impact.

Whether certified organic farming will survive its own success and continue as a socially and environmentally responsible alternative or merely become a parallel production system based on minimal compliance to standards, remains to be seen (Raviv, 2010). It is likely that a hybrid between organic and conventional production systems will emerge in the coming years that combines the best features of each system (Fig. 8.6). Undoubtedly the advances and alternatives developed through the organic movement will be retained in an evolved hybrid vegetable production system. Organic certification may not be needed if consumers have confidence and knowledge of the sustainable techniques used to produce their food. GAPs will be implemented into all vegetable production in the future, as growers, handlers, and consumers are educated about the issues surrounding biological contamination. Profitability is a common bond that unites both types of growers and is needed to maintain a stable and viable agricultural sector.

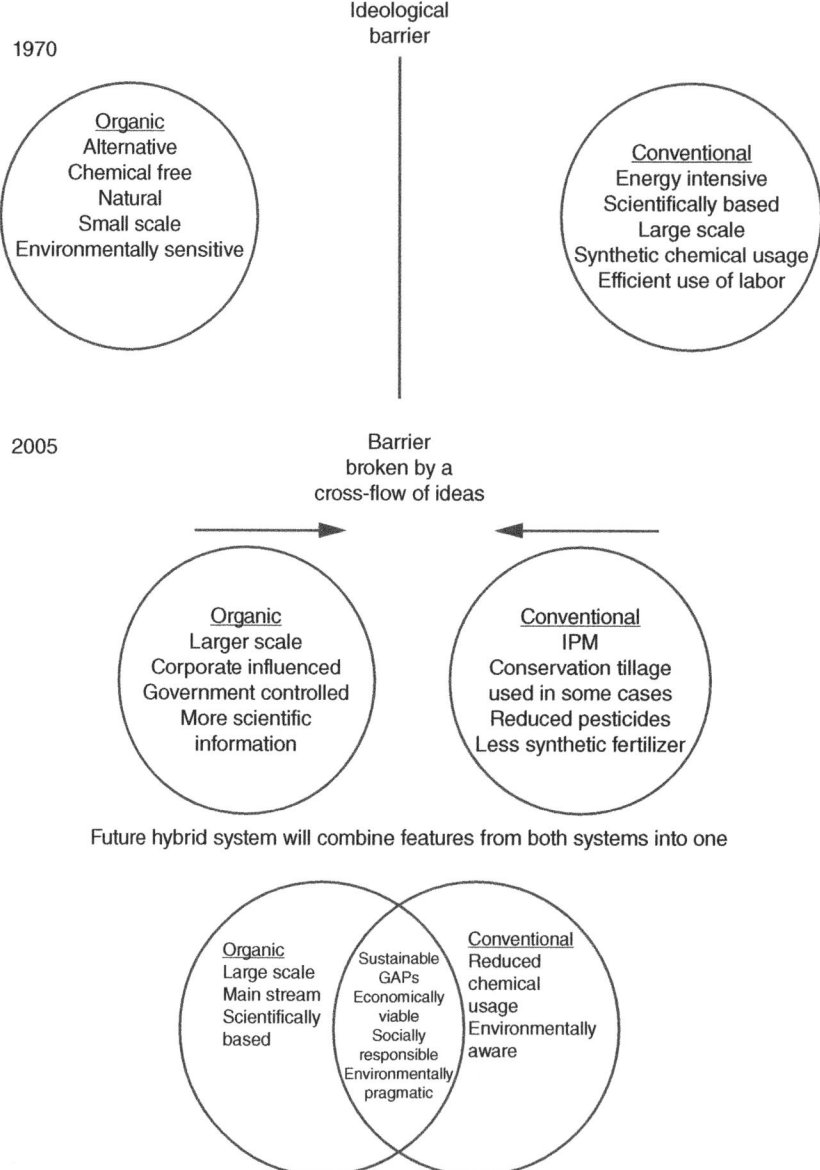

Fig. 8.6. Over the past the past 40+ years there has been convergence of organic and conventional vegetable production. While the two systems still differ in several important regards, it is likely that future production will evolve into a new type of production that is called neither conventional nor organic.

The organic production system may evolve into something even more broadly sustainable than the current programs through a greater awareness of the importance of small farmers and the need for locally grown food.

As the percentage of people directly involved in production agriculture in the developed world continues to decline, it is natural for consumers and producers to become disconnected from one another. One of the lessons from the organic movement is that consumers do care about the quality of their food, who grows it, and how production impacts their environment and health. Agriculturalists have not always done a good job explaining the logic and need behind technological advances in agriculture and how they may affect consumers. In the years ahead, agriculturalists must do a better job of informing the public about how their food is grown.

The production chapters of this book provide examples of both organic and conventional practices. Unfortunately, space does not allow a complete discussion of the practices used by both systems, particularly for the topics of pest control and fertility management. For these topics, many of the major crop pests are listed and the reader should consult references for more detailed information about various control measures available. For fertility management, the amount of the minerals removed by the crop is given in many cases and the reader can decide how these inputs can be best supplied from the many references cited.

References

Abdul-Baki, A.A. and Teasdale, J.R. (1997) Sustainable Production of Fresh-Market Tomatoes and Other Summer Vegetables With Organic Mulches. *Farmers' Bulletin* No. 2279 (Rev.). US Department of Agriculture–ARS, Beltsville, Maryland.

Beall, G., Bruhn, C., Craigmill, A. and Winter, C. (1991) Pesticides and your food: How safe is "safe"? *California Agriculture* 45, 4–11.

Bourn, D. and Prescott, J. (2002) A comparison of the nutritional value, sensory qualities, and food safety of organically and conventionally produced foods. *Critical Reviews in Food Science and Nutrition* 42, 1–34.

Buckingham, J. (2010) *Bitter Nemesis: The Intimate History of Strychnine*. CRC Press, Boca Raton, Florida.

Cherr, C.M., Scholberg, J.M.S. and McSorley, R. (2006) Green manure approaches to crop production. *Agronomy Journal* 98, 302–319.

Dalgaard, T., Halberg, N. and Porter, J.R. (2001) A model for fossil energy use in Danish agriculture used to compare organic and conventional farming. *Agriculture, Ecosystems & Environment* 87, 51–65.

Dhara, V.R. and Dhara, R. (2002) The Union Carbide disaster in Bhopal: a review of health effects. *Archives of Environmental Health: An International Journal* 57, 391–404.

Dreistadt, S.H., Phillips, P.A. and O'Donnell, C.A. (2007) UC Davis Pest Notes: Thrips. UC ANR Publication 7429. Available at: www.ipm.ucdavis.edu/PMG/PESTNOTES/pn7429.html (accessed 18 October 2013).

EPA (2013) United States Environmental Protection Agency: Organic Farming. Available at: www.epa.gov/oecaagct/torg.html (accessed 16 October 2013).

Goldman, L.R., Beller, M., Oregon, H. and Jackson, R.J. (1990) Aldicarb food poisonings in California, 1985–1988: toxicity estimates for humans. *Archives of Environmental Health: An International Journal* 45, 141–147.

Green, J. (1993) Sustainable agriculture: why green ideas raise a red flag. *Grower: Vegetable and Small Fruit Newsletter* 93, 7.

Greenland, D.J. and Szabolcs, I. (eds) (1994) *Proceedings of the Symposium Soil Resilience and Sustainable Land Use and the Second Workshop on the Ecological Foundations of Sustainable Agriculture (WEFSA II)*. CAB International, Wallingford, UK.

Griffin, K. (1974) *The Political Economy Of Agrarian Change, An Essay On The Green Revolution*. CAB International, Wallingford, UK.

Guerrini, L. (2010) The Ramsey model with AK technology and a bounded population growth rate. *Journal of Macroeconomics* 32, 1178–1183.

Halweil, B. (2002) Home grown: The case for local food in a global market. *Worldwatch Institute* 163, 1–83.

Howard, R.W. and Blomquist, G.J. (2005) Ecological, behavioral, and biochemical aspects of insect hydrocarbons. *Annual Review of Entomology* 50, 371–393.

Jeavons, J.C. (2001) Biointensive Sustainable Mini-Farming: I. The Challenge. *Journal of Sustainable Agriculture* 19, 49–63.

King, S.R., Davis, A.R., Liu, W. and Levi, A. (2008) Grafting for disease resistance. *HortScience* 43, 1673–1676.

Kuepper, G. and Gegner, L. (2004) Organic Crop Production Overview. Fundamentals of Sustainable Agriculture. Available at: http://attra.ncat.org/attra-pub/organiccrop.html (accessed 10 October 2013).

Landis, D.A., Wratten, S.D. and Gurr, G.M. (2000) Habitat management to conserve natural enemies of arthropod pests in agriculture. *Annual Review of Entomology* 45, 175–201.

Lobao, L. and Meyer, K. (2001) The great agricultural transition: crisis, change, and social consequences of twentieth century USA farming. *Annual Review of Sociology* 27, 103–124.

Lockeretz, W., Shearer, G. and Kohl, D. (1981) Organic farming in the Corn Belt. *Science* 211, 540–547.

Pankhurst, C.E., Doube, B.M. and Gupta, V.V.S.R. (1997) *Biological Indicators of Soil Health*. CAB International, Wallingford, UK.

Parish, R.L. (2000) Stand of cabbage and broccoli in single-and double-drill plantings on beds subject to erosion. *Journal of Vegetable Crop Production* 6, 87–96.

Raviv, M. (2010) Is Organic Horticulture Sustainable? *Chronica Horticulturae* 50, 7–14.

Schnepf, E., Crickmore, N., Van Rie, J., Lereclus, D., Baum, J., Feitelson, J., Zeigler, D.R. and Dean, D.H.

(1998) *Bacillus thuringiensis* and its pesticidal crystal proteins. *Microbiology and Molecular Biology Reviews* 62, 775–806.

Shelton, A.M., Zhao, J.Z. and Roush, R.T. (2002) Economic, ecological, food safety, and social consequences of the deployment of Bt transgenic plants. *Annual Review of Entomology* 47, 845–881.

Stapleton, J.J. (2000) Soil solarization in various agricultural production systems. *Crop Protection* 19, 837–841.

Stauber, K.N., Hassebrook, C., Bird, E.A.R., Bultena, G.L., Hoiberg, E.O., MacCormack, H. and Menanteau-Horta, D. (1995) The promise of sustainable agriculture. In: Bird, E.A.R., Bultena, G.L. and Gardner, J.C. (eds) *Planting the Future: Developing an Agriculture that Sustains Land and Community*. Iowa State University Press, Ames, Iowa, pp. 3–15.

Van Ravenswaay, E.O. and Hoehn, J.P. (1991) The impact of health risk information on food demand: a case study of Alar and apples. In: Caswell, J.A. (ed.) *Economics of Food Safety*. Springer, the Netherlands, pp. 155–174.

Weinzierl, R.A. (2000) Botanical insecticides, soaps, and oils. In: Rechcigl, J.E. and Rechcigl, N.A. (eds) *Biological and Biotechnological Control of Insect Pests*. CRC Press, Boca Raton, Florida, pp.101–118.

Weiss, C.R. (1999) Farm growth and survival: econometric evidence for individual farms in Upper Austria. *American Journal of Agricultural Economics* 81, 103–116.

Welbaum, G.E., Sturz, A.V., Dong, Z. and Nowak, J. (2004) Managing soil microorganisms to improve productivity of agro-ecosystems. *Critical Reviews in Plant Sciences* 23, 175–193.

Willer, H., Lernoud, J. and Kilcher, L. (eds) (2013) *The World of Organic Agriculture, Statistics and Emerging Trends*. FiBL-IFOAM Report. Bonn, Germany, 340 pp.

Winter, C.K. and Davis, S.F. (2006) Organic foods. *Journal of Food Science* 71, R117–R124.

9 Vegetable Safety

Pesticide Residues

Monoculture production systems were established over 100 years ago to allow greater mechanization to help farmers grow larger acreages with less labor. Monoculture systems are more prone to pest outbreaks because the ecological balance is disturbed (Altieri, 1999). To counter outbreaks of harmful insects, plant diseases, and weeds that occur in monoculture production systems, pesticides were developed to provide control by spraying with minimal labor. Pesticide usage increased steadily from the 1940s through the 1970s as new chemicals were developed for pest control for a wide variety of reasons (MacIntyre, 1987). However, by the 1970s and 1980s, there was increasing public concern about pesticide residues on vegetables and their negative effects on human health. A series of high profile cases that resulted in serious illness and death were traced to pesticide residues on vegetables and focused public attention on this issue. For example, poisoning caused by improper use of the chemical aldicarb (Temik) 2-methyl-2-(methylthio) propionaldehyde-O-(methylcarbomoyl)oxime caused illness after eating contaminated watermelons and cucumbers (Goes et al., 1980; Green and Wehr, 1987). The largest pesticide-related foodborne outbreak in the USA occurred in 1985 when 1,373 persons reported becoming ill after eating watermelons grown in soil treated with aldicarb; 78% of these persons had probable or possible pesticide-related illnesses (Green and Wehr, 1987). The pesticide manufacturing disaster in Bhopal, India also helped focus attention on agricultural pesticides and their usage in the 1980s (Broughton, 2005). These and other incidents of pesticide poisoning fueled public interest in organic and naturally grown produce without synthetic chemical inputs (Heckman, 2006).

After decades of increasing usage, there are signs that the adoption of sustainable production practices is reducing pesticide usage. Pesticide use in California, the leading vegetable crop production state in the USA, dropped for a fourth consecutive year in 2009 (Brooks, 2010). In developed countries, the better educational programs, greater use of biological control measures, reduced use of soil fumigants, adoption of integrated pest management strategies, loss of pesticides approved for use on vegetables, increased pesticide costs, and more effective pesticides are combining to decrease pesticide usage for crop production and increase awareness of the dangers from improper pesticide use (Brooks, 2010). While pesticide awareness is increasing in developing countries, cases of misuse remain and concerns about overuse for crop production persist (Gunnell et al., 2007).

Biological Contamination of Vegetables

Increased availability combined with a realization of the health benefits from eating a diet rich in fresh fruits and vegetables has increased demand for fresh and minimally processed vegetables in many countries. Consumer demand has increased vegetable quantity and resulted in a variety of convenience-packaged products such as bagged ready-to-eat salad mixes and cut and peeled vegetables that are now available to consumers (USDA, 2002). However, eating fresh vegetables is not without risk.

Beginning in the 1990s, headlines of illness caused by biological contamination of vegetables informed the public of a new threat to vegetable safety (Rangel et al., 2005). Biological contamination can be defined as the presence of living organisms such as viruses, bacteria (Fig. 9.1), fungi, or agents derived from these organisms, and mammal or bird antigens that negatively affect human health (Sivapalasingam et al., 2004).

Outbreaks of biological contamination are often associated with *Cyclospora*, Hepatitis A, *Salmonella enterica*, *Escherichia coli* O157:H7, and *Listeria monocytogenes*, particularly on salad greens such as spinach and lettuce (Fig. 9.1; Table 9.1; Sivapalasingam et al., 2004).

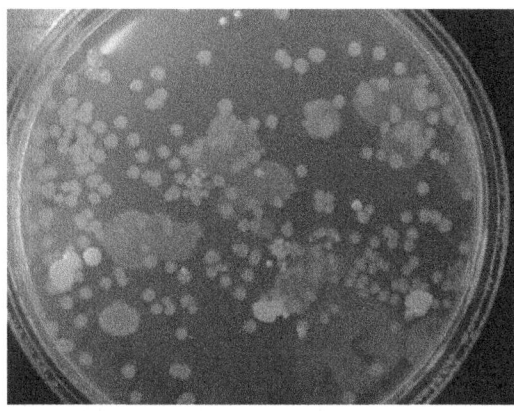

Fig. 9.1. Bacterial colonies cultured from lettuce leaves. Most bacteria found on vegetables do not cause sickness.

Biological contamination of vegetables is a significant health risk for consumers (Altekruse et al., 1997). Producers, and others who work in the vegetable supply chain, must diligently work to prevent infection by biological contaminants. *Escherichia coli* O157:H7 and *S. enterica* are enteric (intestinal) pathogens that infect a broad range of hosts, including humans (Fig. 9.2). Due to its severity and the apparent low infective dose, less than a few hundred cells, *E. coli* O157:H7 is considered one of the most serious foodborne pathogens (Strachan et al., 2005).

It is likely that biological contamination of fresh vegetables has always existed, but was only viewed as "food poisoning" when modern sophisticated detection techniques allowed for specific identification and classification. This point is brought out by the fact that produce-related illnesses accounted for only 0.7% of all reported foodborne illnesses in the 1970s compared to 6% in the 1990s (Sivapalasingam et al., 2004). During this 20-year period, 54% of the produce-related illnesses were associated with known pathogens. Of the pathogens, 60% were bacteria, of which 48% were *Salmonella* spp., while *Cyclospora* and *E. coli* O157:H7 were newly recognized as causes of foodborne illness at the time. Since 1982, *E. coli* O157:H7 has been identified in many countries as a cause of hemorrhagic colitis and other serious conditions that result in sickness or even death (Wells et al., 1983). Between 1991 and 2004, a US study showed that outbreaks were most commonly associated with lettuce (34%), salad mixes (11%), coleslaw (11%), melons (8%), and sprouts (2%), while fruit crops accounted for 44% of reported contamination events (Sivapalasingam et al., 2004).

Cases of biological contamination of vegetables have continued to be numerous and widespread during the decade from 2000 to 2010 (Rangel et al., 2005). For example, a serious *E. coli* O157:H7 outbreak in spinach in the USA caused 171 illnesses and three deaths in 2006 (CDC, 2006a; Table 9.1). The spinach contamination event was related to a smaller outbreak in lettuce shortly thereafter (CDC, 2006b; Todd, 2007).

An understanding of how biological contaminants survive in nature and become established on vegetables is important to anyone involved in vegetable production. Native benign bacteria are common on vegetables and have adapted to the conditions specific to the microenvironment on vegetables leaves (Lindow and Brandl, 2003). This environment provides low nutrient and water availability, high exposure to UV radiation, and rapid changes in environmental conditions. The interactions between the plant and these native microorganisms, as well as microbe–microbe interactions, are pivotal for establishment of pathogenic bacteria on vegetables (Lindow and Leveau, 2002; Lindow and Brandl, 2003). Enteric human pathogens also interact with other phyllosphere bacteria and this interaction may be important in allowing pathogens to establish themselves on vegetable leaves and spread disease to humans (Brandl, 2006).

Once a production field is contaminated by *E. coli* O157:H7 or other human pathogens, attachment is the first step in colonizing a vegetable surface. *Escherichia coli* O157:H7 may use the same attachment mechanism to attach to vegetables that it uses to attach to animal and human epithelial cells (Brandl, 2006).

Vegetables become biologically contaminated with *E. coli* O157:H7 when they come in contact with contaminated manure, tainted irrigation waters, contaminated sewage sludge, or farm workers who practice poor sanitation (Fig. 9.2; Franz and van Bruggen, 2008). *Escherichia coli* and other pathogenic organisms may find niches in the phyllosphere where they can proliferate, complicating their removal from fresh produce (Aruscavage et al., 2006).

Many microbes cannot survive boiling water, so thorough cooking effectively inactivates most biological contaminants. Fresh consumption of

Table 9.1. Produce-associated outbreaks that occurred in the USA from 2000 to 2007[a], 2008[b], and 2009[c].

Microorganism involved	State(s)	Month	Number of cases	Vehicle
		Year 2000		
Salmonella enterica Thompson	Multistate	Nov	43	Tomato
		Year 2001		
S. enterica Newport	California	May	8	Cilantro
Escherichia coli O157:H7	Texas	Nov	20	Lettuce
		Year 2002		
E. coli O157:H7	California	July	5	Alfalfa sprouts
S. enterica Newport	Colorado	July	13	Cilantro
S. enterica Newport	Multistate	July	510	Tomato
E. coli O157:H7	Illinois	Nov	13	Lettuce
		Year 2003		
S. enterica Saintpaul	Multistate	Feb	16	Alfalfa sprouts
E. coli O157:H7	California	Sep	51	Lettuce
E. coli O157:H7	California	Oct	16	Spinach
S. enterica Chester	Multistate	Nov	26	Alfalfa sprouts
S. enterica Enteriditis	California	Nov	14	Lettuce
S. enterica Saintpaul	Multistate	Nov	33	Iceberg lettuce/tomato
		Year 2004		
E. coli O157:H7	Georgia	Apr	2	Alfalfa sprouts
S. enterica Bovismorbificans	Multistate	Apr	35	Alfalfa sprouts
S. enterica Braenderup	Multistate	June	137	Roma tomato
S. enterica Newport	Multistate	July	97	Iceberg lettuce
S. enterica Javiana	Florida	Sep	24	Green salad
E. coli O157:H7	New Jersey	Nov	6	Lettuce
		Year 2005		
E. coli O157:H7	Multistate	Sep	12	Pre-packaged lettuce
E. coli O157:H7	Multistate	Oct	34	Pre-packaged lettuce
		Year 2006		
Shigella sonnei	Oregon	Jan	35	Lettuce-based salad
S. enterica Berta	Multistate	Jan	16	Tomato
Salmonella typhimurium	Maryland	June	18	Lettuce/tomato
S. enterica Newport	Multistate	June	115	Tomato
E. coli O121	Utah	July	3	Lettuce-based salad
E. coli O157:H7	Multistate	Aug	205	Spinach
E. coli O157:H7	New Mexico	Aug	5	Spinach
E. coli O157:H7	Multistate	Nov	77	Lettuce
E. coli O157:H7	Multistate	Nov	65	Lettuce
E. coli O157:H7	New York	Nov	20	Lettuce
E. coli O157:H7	Oregon	Nov	3	Vegetable-based salad
		Year 2007		
E. coli O157:H7	Florida	Jan	2	Caesar salad
S. typhimurium	Maine	Feb	76	Lettuce/spinach
E. coli O157:H7	Alabama	June	26	Lettuce-based salad
S. enterica Newport	DC	June	46	Tomato
S. enterica Newport	New York	July	10	Tomato
S. sonnei	California	July	72	Lettuce-based salad
S. typhimurium	Minnesota	Oct	23	Tomato
		Year 2008		
S. enterica Saintpaul	Multistate	Nov	1,442	Jalapeño peppers[b]
		Year 2009		
S. enterica Saintpaul	Multistate	May	228	Alfalfa sprouts[c]

[a]Data collected from CDC: Bacterial foodborne and diarrheal disease. National case surveillance annual reports, years 2000–2007
[b]CDC, 2008
[c]www.cdc.gov

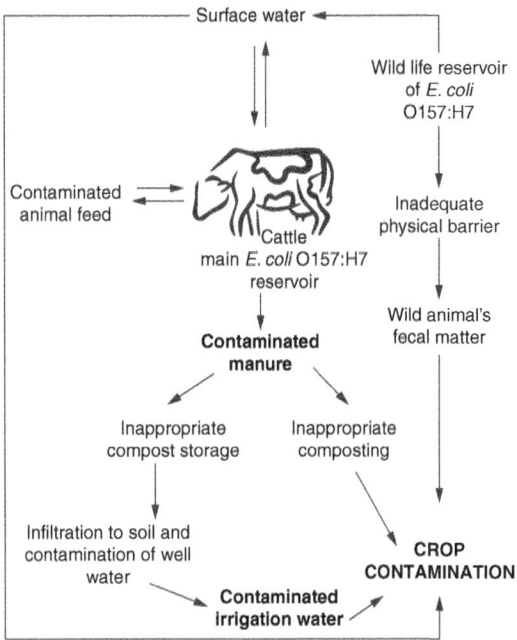

Fig. 9.2. Possible routes taken by *E. coli* O157:H7 to contaminate vegetables (Lopez-Velasco, 2010).

vegetables poses the greatest threat to human health from biological contamination. Once contaminated, fresh vegetables and minimally processed ready-to-eat vegetables become likely vehicles for foodborne disease. The food industry has developed its own strategies to reduce the risk of contamination in minimally processed vegetables, including the use of gamma irradiation, chlorination, and modified atmosphere packaging, all in combination with storage at low temperatures, usually 4°C (39°F) (Luning *et al.*, 2006).

Decontamination techniques for minimally processed bagged vegetables, such as chlorine disinfection or modified atmosphere packaging, do not always adequately control microbial populations (Lee and Baek, 2008). Refrigeration, shrink-wrapping, or bagging may multiply human pathogens under certain conditions. In addition, bacteria may colonize around cut vegetable surfaces or damaged plant tissues using leaked nutrients to feed their growth (Aruscavage *et al.*, 2006). Furthermore, *E. coli* O157:H7 is able to live inside lettuce plant tissues, not only on leaf surfaces, as was shown after its recovery from surface-sterilized leaves (Franz *et al.*, 2007). *Escherichia coli* O157:H7 inoculated on to lettuce persisted for up to 20 days (Solomon *et al.*, 2003). Other studies showed that when human pathogens were introduced on to spinach seeds they concentrated on root junctions shortly after germination, even though *E. coli* O157:H7 was not necessarily detected in mature leaves (Jablasone *et al.*, 2005; Johannessen *et al.*, 2005). Leafy vegetables, such as spinach or lettuce, are capable of supporting biological contaminants because they have a pH greater than 4.6 and high water content, so contamination avoidance, improved sanitation, and preservation treatments are the primary strategies for avoiding human disease transmission (Foley *et al.*, 2002).

Sources of Biological Contamination of Vegetables

Livestock (in particular cattle), are considered to be natural reservoirs for bacteria that cause human disease, such as *E. coli*. Infected cattle can spread *E. coli* O157:H7 to humans through their feces (Franz and van Bruggen, 2008). Although cattle are believed to be the primary reservoir of *E. coli* O157:H7, other species can also asymptomatically carry and shed this bacterium in fecal material, potentially contaminating irrigation water, soil, and crops. Other sources of *E. coli* O157:H7 contamination may include animal excreta (birds, insects, rodents, and reptiles) or workers (Pirovani *et al.*, 2000). *Escherichia coli* O157:H7 has been isolated from deer, dogs, ducks, kangaroos, and wild pigs (Ahmed *et al.*, 2007). Feces from feral swine may have caused contamination of spinach by *E. coli* O157:H7. Feral swine, which contained the same strain of *E. coli* as cattle from a nearby dairy, likely contaminated spinach fields while they passed through (Jay *et al.*, 2007). *Escherichia coli* O157:H7 has persisted in non-aerated sheep manure for over 1 year and dairy cattle manure for 49 days at 22°C (72°F) (Kudva *et al.*, 1998). Insects like the lesser mealworm (*Alphitobius diaperinus*), fruit fly (*Drosophila melanogaster*), and housefly (*Musca domestica*) can transmit *E. coli* O157:H7 (Dingman, 2000).

Animal manure is sometimes used as an organic fertilizer in vegetable crop production (Franz and van Bruggen, 2008). Vegetable contamination by human pathogens may occur following application of poorly composted or raw manure (Bihn and Gravani, 2006). Biological contamination may occur in either conventional or organic vegetable production systems. Since animal feces may serve as a reservoir for *E. coli* O157:H7, proper handling of manure for vegetable production is critical for preventing the transmission of bacterial contamination through fresh vegetables (Franz and van Bruggen, 2008). Proper composting reduces or eliminates bacterial pathogens, parasites, fly larvae, and weed seeds while converting manure and other materials into safe fertilizer, soil amendments, or potting media (Kashmanian and Rynk, 1996). Various organic production guidelines used internationally call for composted manure to reach a minimum critical temperature to kill harmful organisms and also provide restrictions on the application of uncomposted manure to vegetable crops (Luning et al., 2006). However, if these procedures are not carefully followed, the risk of contamination is great (Luning et al., 2006). During composting, temperatures increase from 25°C–58°C (77–136°F) due to bacteria and yeast metabolism, followed by gradual cooling to ambient temperatures (Ishii et al., 2000). During the heating phase of composting, human pathogens are inactivated, but if the composting process fails to reach the 58°C (136°F) threshold temperature for inactivation or if the duration of composting is insufficient, the compost may harbor biological contaminants (Ishii et al., 2000).

Contaminated livestock and poultry manure may degrade water quality, particularly if manure is over applied and enters irrigation water supplies through runoff or leaching. Large-scale commercial livestock and poultry operations in areas where fresh vegetables are grown have increased risk of microbial contamination due to improperly handled animal wastes polluting irrigation water (Ribaudo et al., 2003; Ibekwe et al., 2004). Irrigation water may be contaminated by the leaching of pathogens through soil into ground water (Gagliardi and Karns, 2000).

Irrigation method affects the risk of vegetable contamination. Drip or trickle irrigation systems have lower risk of crop contamination because water is applied directly to the soil and not vegetable tissues (Bastos and Mara, 1995). Sprinkler irrigation wets the entire vegetable surface, increasing the risk of crop contamination (Keraita et al., 2007). *Escherichia coli* O157:H7 applied with sprinkler irrigation for 14 days increased *E. coli* detection on leaves 30 days later, demonstrating the potential for vegetable contamination when overhead irrigation is used (Solomon et al., 2003).

Soil is directly contaminated by application of animal waste or indirectly through rainwater or irrigation runoff, which then percolates through soil layers. *Escherichia coli* O157:H7 has been isolated from beneath the soil surface following application of contaminated manure (Gagliardi and Karns, 2000). Furthermore, *E. coli* and other coliforms exhibited a high growth rate in soil (Gagliardi and Karns, 2000). Soil type, tillage practice (tilled or no-till), and rainfall amounts determined the degree of leaching, while different tillage practices failed to prevent vertical transport of bacteria in soil (Gagliardi and Karns, 2000). Ammonia and nitrate concentrations in soil were positively correlated with the leaching of *E. coli* O157:H7, which persisted in soils for more than 5 months after application of contaminated compost or irrigation water (Islam et al., 2004). When lettuce plants were grown on contaminated soil, *E. coli* O157:H7 were present 36 days after planting and a four-fold increase was observed between days 7 and 36 (Ibekwe et al., 2004).

Soil microcosms inoculated with *E. coli* O157:H7 showed long-term persistence in the phyllosphere, rhizosphere, and non-rhizosphere soils. In the case of lettuce leaves, the bacterium persisted for over 45 days on leaf surfaces, in non-rhizosphere soil, and in the rhizosphere (Ibekwe et al., 2004). The persistence of this bacterium for long periods in agricultural fields suggests that the contamination of vegetables is possible from soil under certain circumstances.

Good Agricultural Practices/Good Handling Practices

The terms Good Agricultural Practices (GAP) and Good Handling Practices (GHP) can refer to any collection of specific methods, which when applied to agriculture, produce the objectives of the proponents of those practices. There are numerous competing definitions of what production and handling methods constitute GAPs and GHPs, so whether a practice can be considered "good" will depend on the standards being applied (Gravani, 2009). There are many GAPs that apply to vegetable as well as agronomic crop production practices. The concept of GAPs has been changed in recent years by some groups because of a rapidly changing agriculture, globalization of world trade, food crisis (e.g. mad cow disease), nitrate pollution of water, appearance of pesticide resistance, soil erosion, etc.

There are many different GAP programs around the world with different goals and objectives. One of the primary objectives of most GAP programs is to help preserve water and soil resources from the risk of biological contamination. Third-party certifications such as GlobalGAP, EurepGAP, and ISOGAP are designed to improve food security, particularly of internationally traded commodities. Some governments have developed their own national GAP standards. Anyone involved in vegetable production or handling should be familiar with the applicable national and local GAP/GHPs.

"Good Agricultural Practices" as defined by the Food and Agriculture Organization (FAO) of the UN pertain to broad-based principles applying to on-farm production and post-production processes, resulting in safe and healthy food and non-food agricultural products, while taking into account economic, social, and environmental sustainability. These broad-based principles apply to many different aspects of agriculture including soil, water, and animal production, vegetable crop production and human health and welfare (FAO, 2005).

FAO GAPs may be applied to a wide range of diverse farming systems, including both organic and traditional, and can be applied to both large and small farms. The four overarching principles of the FAO GAP program are as follows:

1. Economically and efficiently produce sufficient (food security), safe (food safety) and nutritious food (food quality).
2. Sustain and enhance natural resources as a component of food production.
3. Maintain viable farming enterprises and contribute to sustainable livelihoods of people involved in agricultural production.
4. Meet cultural and social demands of society.

As an example, FAO GAPs that apply to soil include reducing erosion by wind and water through hedging and ditching, application of fertilizers at appropriate moments and in adequate doses (i.e. when the plant needs the fertilizer), avoiding runoff (see nitrogen balance method), maintaining or restoring soil organic content by manure application, use of grazing, crop rotation, reducing soil compaction issues (by avoiding using heavy mechanical devices), maintaining soil structure by limiting heavy tillage practices, and *in situ* green manure applications by growing legumes such as cowpea, clover, sunhemp, etc. (FAO, 2007b).

Examples of GAPs that apply to agricultural water management would include scheduling irrigation based on actual plant needs and soil water status to reduce runoff water, reducing soil salinization by limiting water input to crop needs, and recycling water when possible, avoiding crops with high water requirements in low availability regions, avoiding drainage and fertilizer runoff, maintaining permanent soil covering, especially during winter to avoid nitrogen runoff, managing water tables by limiting excess water output, restoring or maintaining wetlands or marshlands, providing water points for livestock away from irrigation water sources, preventing erosion and harvest runoff water by digging catch pits, and strip-cropping across slopes (FAO, 2007a).

The United States Department of Agriculture (USDA) has developed an audit/certification program to verify that farms use GAPs and/or GHPs that do not harm the environment and control the spread of biological contaminants in our food supply, including vegetable crops (Fig. 9.3; Bihn and Gravani, 2006).

Unlike the FAO guidelines, the USDA guidelines focus on food safety, and do not address topics such as animal welfare, social sustainability, biodiversity, or the use of antibiotics and hormones. The USDA GAP/GHP guidelines and principles grew out of the 1998 Food and Drug Administration publication entitled, "Guide to Minimize Microbial Food Safety Hazards for Fresh Fruits and Vegetables." The program was developed after state governments petitioned the USDA to create a GAP and GHP audit program. The need for an audit program arose from wholesale buyers asking farmers to demonstrate their adherence to GAP and GHP recommendations. "GAPs metrics" allow those who monitor GAPs to determine their effectiveness, by using threshold values for indicator organisms in water, for example, along with other measurables to gauge the effectiveness of the practices implemented.

Fig. 9.3. An example of Good Handling Practices is wearing protective clothing to prevent the spread of harmful organisms from humans on to fresh vegetables.

References

Ahmed, W., Tucker, J., Bettelheim, K.A., Neller, R. and Katouli, M. (2007) Detection of virulence genes in *Escherichia coli* of an existing metabolic fingerprint database to predict the sources of pathogenic *E. coli* in surface waters. *Water Research* 41, 3785–3791.

Altekruse, S.F., Cohen, M.L. and Swerdlow D.L. (1997) Emerging foodborne diseases. *Emerging Infectious Diseases* 3, 285–293.

Altieri, M.A. (1999) The ecological role of biodiversity in agroecosystems. *Agriculture, Ecosystems and Environment* 74, 19–31.

Aruscavage, D., Lee, K., Miller, S. and LeJeune, J.T. (2006) Interactions affecting the proliferation and control of human pathogens on edible plants. *Journal of Food Science* 71, 11.

Bastos, R.K.X. and Mara, D.D. (1995) The bacterial quality of salad crops drip and furrow irrigated with waste stabilization pond effluent: an evaluation of the WHO guidelines. *Water Science and Technology* 31(12), 425–430.

Bihn, E.A. and Gravani, R.B. (2006) *Role of Good Agricultural Practices in Fruit and Vegetable Safety.* ASM Press, Washington, DC.

Brandl, M. (2006) Fitness of human enteric pathogens on plants and implications for food safety. *Annual Review of Phytopathology* 44, 367–392.

Brooks, L. (2010) DPR Reports Pesticide Use Declined Again in 2009. Available at: www.cdpr.ca.gov/docs/pressrls/archive/2010/101229.htm (accessed 28 September 2013).

Broughton, E. (2005) The Bhopal disaster and its aftermath: a review. *Environmental Health: A Global Access Science Source* 2005, 4–6.

CDC (2006a) Ongoing multistate outbreak of *Escherichia coli* serotype O157:H7 infections associated with consumption of fresh spinach. Available at: www.cdc.gov (accessed 16 May 2011) Published online 10 May 2005, doi: 10.1186/1476-069x-4-6.

CDC (2006b) Multistate outbreak of *E. coli* O157 infections. Available at: www.cdc.gov/ecoli/2006/december/121006.htm (accessed 16 May 2011).

CDC (2008) Outbreak of *Salmonella* Serotype Saintpaul Infections Associated with Multiple Raw Produce Items – United States, 2008. Available at: www.cdc.gov/mmwr/preview/mmwrhtml/mm5734a1.htm (accessed 28 September 2013).

Dingman, D.W. (2000) Growth of *Escherichia coli* O157:H7 in bruised apple (*Malus domestica*) tissue as influenced by cultivar, date of harvest, and source. *Applied and Environmental Microbiology* 66, 1077–1083.

FAO (2005) Sustainable Agriculture and Rural Development (SARD) and Good Agricultural Practices (GAP). *Nineteenth Session.* Committee on Agriculture, Rome, 31 pp.

FAO (2007a) Good Agricultural Practices; water. Committee on Agriculture (COAG). Available at: http://www.fao.org/prods/GAP/home/principles_en.htm (accessed 28 September 2013).

FAO (2007b) Good Agricultural Practices, Soil. Committee on Agriculture (COAG). http://www.fao.org/prods/GAP/home/principles_en.htm (accessed 28 September 2013).

Foley, D.M., Dufour, A., Rodriguez, L., Caporaso, F. and Prakash, A. (2002) Reduction of *Escherichia coli* O157:H7 in shredded iceberg lettuce by chlorination and gamma irradiation. *Radiation Physics and Chemistry* 63, 391–396.

Franz, E. and van Bruggen, A.H. (2008) Ecology of *E. coli* O157:H7 and *Salmonella enterica* in the primary vegetable production chain. *Critical Reviews in Microbiology* 34, 18.

Franz, E., Visser, A.A., Van Diepeningen, A.D., Klerks, M.M., Termorshuizen, A.J. and van Bruggen, A.H. (2007) Quantification of contamination of lettuce by GFP-expressing *Escherichia coli* O157:H7 and *Salmonella enterica* serovar Typhimurium. *Food Microbiology* 24(1), 106–112.

Gagliardi, J.V. and Karns, J.S. (2000) Leaching of *Escherichia coli* O157:H7 in diverse soils under various agricultural management practices. *Applied Environmental Microbiology* 66, 7.

Goes, E.A., Gibbons, S.E., Aaronson, M., Ford, S.A. and Wheeler, H.W. (1980) Suspected foodborne carbamate pesticide intoxications associated with ingestion of hydroponic cucumbers. *American Journal of Epidemiology* 111, 254–260.

Gravani, R.B. (2009) *The Role of Good Agricultural Practices in Produce Safety*. Wiley-Blackwell, Oxford, UK.

Green, M.A. and Wehr, H.M. (1987) An outbreak of watermelon-borne pesticide toxicity. *American Journal of Public Health* 77, 1431–1434.

Gunnell, D., Phillips, M.R. and Konradsen, F. (2007) The global distribution of fatal pesticide self-poisoning: systematic review. *BMC Public Health* 7, 357.

Heckman, J. (2006) A history of organic farming: Transitions from Sir Albert Howard's War in the Soil to USDA National Organic Program. *Renewable Agriculture and Food Systems* 21, 143–150.

Ibekwe, A.M., Watt, P.M., Shouse, P.J. and Grieve, C.M. (2004) Fate of *Escherichia coli* O157:H7 in irrigation water on soils and plants as validated by culture method and real-time PCR. *Canadian Journal of Microbiology* 50, 8.

Ishii, K., Fukui, M. and Takii, S. (2000) Microbial succession during a composting process as evaluated by denaturing gradient gel electrophoresis analysis. *Journal of Applied Microbiology* 89, 768–777.

Islam, M., Doyle, M.P., Phatak, S.C., Millner, P. and Jiang, X. (2004) Persistence of enterohemorrhagic *Escherichia coli* O157:H7 in soil and on leaf lettuce and parsley grown in fields treated with contaminated manure composts or irrigation water. *Journal of Food Protection* 67, 1365–1370.

Jablasone, J., Warriner, K. and Griffiths, M. (2005) Interactions of *Escherichia coli* O157:H7, *Salmonella typhimurium* and *Listeria monocytogenes* plants cultivated in a gnotobiotic system. *International Journal of Food Microbiology* 99, 7–18.

Jay, M.T., Cooley, M.D., Carychao, D., Wiscomb, G.W., Sweitzer, R.A., Crawford-Miksza, L., Farrar, J.A., Lau, D.K., O'Connell, J., Millington, A., Asmundson, R.V., Atwill, E.R. and Mandrell, R.E. (2007) *Escherichia coli* O157:H7 in feral swine near spinach fields and cattle, central California coast. *Emerging Infectious Diseases Journal* 13, 1908–1911.

Johannessen, G.S., Bengtsson, G.B., Heier, B.T., Bredholt, S., Wasteson, Y. and Rorvik, L.M. (2005) Potential uptake of *Escherichia coli* O157:H7 from organic manure into crisphead lettuce. *Applied and Environmental Microbiology* 71, 2221–2225.

Kashmanian, R.M. and Rynk, R.F. (1996) Agricultural composting in the United States: trends and driving forces. *Journal of Soil and Water Conservation* 51, 194–294.

Keraita, B., Konradsen, F., Drechsel, P. and Abaidoo, R. (2007) Effect of low-cost irrigation methods on microbial contamination of lettuce irrigated with untreated wastewater. *Tropical Medicine and International Health* 12, 8.

Kudva, I.T., Blanch, K. and Hovde, C.J. (1998) Analysis of *Escherichia coli* O157:H7 survival in ovine or bovine manure and manure slurry. *Applied and Environmental Microbiology* 64, 9.

Lee, S.Y. and Baek, S.Y. (2008) Effect of chemical sanitizer combined with modified atmosphere packaging on inhibiting *Escherichia coli* O157:H7 in commercial spinach. *Food Microbiology* 25, 582–587.

Lindow, S.E. and Brandl, M.T. (2003) Microbiology of the phyllosphere. *Applied Environmental Microbiology* 69, 9.

Lindow, S.E. and Leveau, J.H. (2002) Phyllosphere Microbiology. *Current Opinion in Biotechnology* 13, 6.

Lopez-Velasco, G. (2010) Molecular characterization of spinach (*Spinacia oleracea*) microbial community structure and its interaction with microbial community structure and its interaction with *Escherichia coli* O157:H7 in modified atmosphere conditions. Ph.D. dissertation, Virginia Tech, Blacksburg, Virginia.

Luning, P.A., Devlieghere, F. and Verhé, R. (eds) (2006) *Safety in the Agri-Food Chain*. Wageningen Academic Publishers, the Netherlands.

MacIntyre, A.A. (1987) Why pesticides received extensive use in America: A political economy of agricultural pest management of 1970. *Journal of Natural Resources* 27, 53.

Pirovani, M.E., Di Pentima, J.H. and Tessi, M.A. (2000) Survival of Salmonella hadar after washing disinfection of minimally processed spinach. *Letters in Applied Microbiology* 31, 6.

Rangel, J.M., Sparling, P.H., Crowe, C., Griffin, P.M. and Swerdlow, D.L. (2005) Epidemiology of *Escherichia coli* O157:H7 outbreaks, United States, 1982–2002. *Emerging Infectious Diseases* 11, 7.

Ribaudo, M., Kaplan, J., Christensen, L., Gollehon, N., Johansson, N., Breneman, V., Aillery, M., Agapoff, J. and Peters, M. (2003) Manure management for water quality. Costs to animal feeding operations of applying manure nutrients to land. *USDA-ERS Agricultural Economic Report (824)*. Agricultural Research Service, US Department of Agriculture.

Sivapalasingam, S., Friedman, C.R., Cohen, L. and Tauxe, R.V. (2004) Fresh produce: a growing cause of outbreaks of foodborne illness in the United States, 1973 through 1997. *Journal of Food Protection* 67, 9.

Solomon, E.B., Pang, H.J. and Matthews, K.R. (2003) Persistence of *Escherichia coli* O157:H7 on lettuce plants following spray irrigation with contaminated water. *Journal of Food Protection* 66, 2198–2202.

Strachan N.J., Kasuga, F., Rotariu, O. and Ogden, I.D. (2005) Dose response modeling of *Escherichia coli* O157 incorporating data from foodborne and environmental outbreaks. *International Journal of Food Microbiology* 103, 13.

Todd, B. (2007) Outbreak: *E. coli* O157:H7. *The American Journal of Nursing* 107, 4.

USDA (2002) Profiling food consumption in America. *The USDA Fact Book (2001–2002)*. United States Department of Agriculture.

Wells, J.G., Wachsmuth, I.K., Riley, L.W., Remis, R.S., Sokolow, R. and Morris, G.K. (1983) Laboratory investigation of hemorrhagic colitis outbreaks associated with a rare *Escherichia coli* serotype. *Journal of Clinical Microbiology* 18, 9.

10 Family Cucurbitaceae

Origin and History

The cucurbits are largely tropical in origin with different genera originating in Africa, tropical America, and Southeast Asia. Commercial cucurbits are primarily herbaceous annuals that produce distinctive tendril-bearing vines and are commonly grown in temperate regions with long growing seasons. Some are adapted to humid conditions while others are found in arid regions. Most are frost-intolerant although some species are more tolerant of low temperature than others.

Taxonomy

The Cucurbitaceae family is well defined but taxonomically isolated from other plant families. The family Cucurbitaceae consists of about 120 genera and more than 800 species. Two subfamilies, Zanonioideae and Cucurbitoideae, are well characterized: the former by small, striate pollen grains and the latter by styles united into a single column. The food plants all fall within the subfamily Cucurbitoideae and belong to two tribes: the Cucurbiteae and Sicyoideae (Maynard and Maynard, 2000). The genera *Cucurbita*, *Cyclanthera*, and *Sechium* are of New World origin. All other genera originated from the African or Asian tropics. In this chapter, the production of cucumber (*Cucumis sativus*), netted and non-netted melons (*Cucumis melo*), watermelon (*Citrullus lanatus*), and squash and pumpkin (*Cucurbita* spp.) will be discussed. Whitaker and Davis (1962), Robinson and Decker-Walters (1997) and Rubatzky and Yamaguchi (1997) provide information about other crops in the family Cucurbitaceae.

CUCUMBER

Origin and History

The center of origin of cucumber is believed to be India where it has been grown for thousands of years (Zeven and Zhukovsky, 1975). The ancient Egyptians also cultivated cucumbers. *Cucumis sativus* var. *hardwickii*, a wild taxon native to India, has been proposed as the wild progenitor of the domesticated forms of *C. sativus*. Cucumbers spread to China and Greece from India about 2,000 years ago (Whitaker and Davis, 1962; Robinson and Decker-Walters, 1997). The Sikkim cucumber has been grown in the Himalayas as food for centuries.

The cucumber also spread to Italy, and was a significant crop during the Roman Empire. In classical Rome, Pliny reported greenhouse production of cucumbers by the 1st century, and the Emperor Tiberius was said to have eaten them throughout the year (Sauer, 1993). Cucumbers were probably spread to the rest of Europe by the Romans. The earliest records of cucumber cultivation appear in France by the 9th century, Great Britain by the 14th century, and the Caribbean at the end of the 15th century. Colonists introduced the cucumber to North America by the mid-16th century (Hedrick, 1919). Interaction between Europeans and Native Americans spread cucumbers throughout North America. Less than a century later, European explorers observed a wide range of Native American peoples cultivating cucumbers along the east coast of North America from Montreal to Florida. Slaves also introduced cucumber directly to the Americas from Africa in the early 1600s. By the 17th century, Native Americans living in the Great Plains region were also cultivating cucumbers (Wolf, 1982).

Botany

Cucumber (*C. sativus*) has a chromosome number of n = 7, West Indian Gherkin (*C. anguira*), n = 12, and Armenian cucumber (*C. melo*), n = 12. Commercial cucumber cultivars are warm-season, frost-sensitive annuals that are cross-pollinated by bees.

Plants commonly have a trailing or climbing growth habit, although some bush cultivars with shortened internodes also exist. Root systems are extensive but shallow. Stems are square with stiff bristle hairs, unbranched tendrils and generally range in length from 0.4–3 m (15.8–118.1 in). Tendrils at each node help anchor plants and allow climbing on supports. Cucumber petioles are 3–15 cm (1.2–5.9 in) long. Rough leaf blades have a triangular ovate shape from 5–25 cm (2.0–9.8 in) wide with three- to five-angled regions or shallow-lobed sinuses and a pointed apex (Rubatzky and Yamaguchi, 1997).

Cucumber plants exhibit monoecious (separate male and female flowers on a plant), andromonoecious (separate perfect and male flowers on the same plant), or gynoecious (all female) sex expression. Sex expression is genetically controlled but modified by the environment and chemicals. Application of auxin or ethephon favors both earlier and a higher percentage of female flowers, while gibberellin delays female flower formation and percentages. Auxin and ethephon favor female flower development while gibberellin stimulates the formation of male flowers. In monoecious cultivars, staminate flowers appear first and are several times more abundant than pistillate flowers. Flowers occur at the nodes, staminate in clusters or singly with only one flower in a cluster opening at a time. Pistillate flowers are borne singly on the main stem and lateral branches in monoecious types and singly or in clusters on the main stem and lateral branches on gynoecious types. The large inferior ovary of pistillate flowers resembles a miniature fruit. Both staminate and pistillate flowers are 1–3 cm (0.4–1.2 in) in diameter with a yellow, showy five-lobed corolla. Flowers are open for a single day and if not adequately pollinated rapidly abort.

Flowering is influenced by photoperiod with regard to the number and sex of the flowers formed. During short days, there is a tendency for earlier and more frequent pistillate flowering. Low temperatures may cause a similar response. Conversely, high temperature and long days promote male flower formation (Rubatzky and Yamaguchi, 1997). High nitrogen promotes vegetative growth and inhibits flower formation. Environmental stress (water, nutrient, etc.) promotes female flower formation.

Fruits can be spherical, blocky, oblong, or elongated in shape and variable in size. Fruit surfaces vary in the number and size of spiny warts, which are usually more apparent on young fruit. Cucumber fruit are consumed immature when their flavor is mild and seeds are small and underdeveloped. Cucumbers are used for different purposes depending on their characteristics (Fig. 10.1).

Cultivars are often divided into the general categories of "fresh market" or "processing" to describe their intended uses. Cultivars grown for fresh market usually have a fruit length-to-width ratio of at least 6:1, dark-green color when immature, thick skin, slightly tapered stem and blossom ends. The seed cavity of fresh market types is usually larger than processing cultivars.

Cultivars developed for processing often have smaller vines, fruit with a length-to-width ratio of about 3:1, bicolored pale or darker green immature color, thin skins, blocky shape, and more prominent warts. The warts bear white or black loosely attached spines (spicules) in their centers. Fruit with white spines turn from green to white or yellow at full physiological maturity when seeds have fully developed. The color of black-spined fruit changes from green to orange at maturity. Black-spined fruit have thinner skin and lighter fruit color compared to white-spined fruit and for many years were preferred for processing because they have a more attractive appearance after pickling. However, black-spined cucumbers tend to prematurely develop their mature orange color at the stem and blossoms ends under environmental stress or when over mature, so more white-spined cultivars have been developed for processing in recent years.

West Indian gherkin

Small-fruited cultivars of pickling cucumbers are sometimes sold as gherkins. However, the West Indian or true gherkin (*Cucumis anguria*) belongs to a different species and has a different appearance and characteristics than *C. sativus*. West Indian gherkin is an annual, monoecious climbing vine with flowers, leaves, tendrils, and fruit smaller than cucumber. Fruits, which are spiny, yellow, oval, and about 5 cm (2 in) long, are eaten fresh, cooked, or pickled. The plant may self-seed, escape from cultivation, and become an aggressive weed (Whitaker and Davis, 1962).

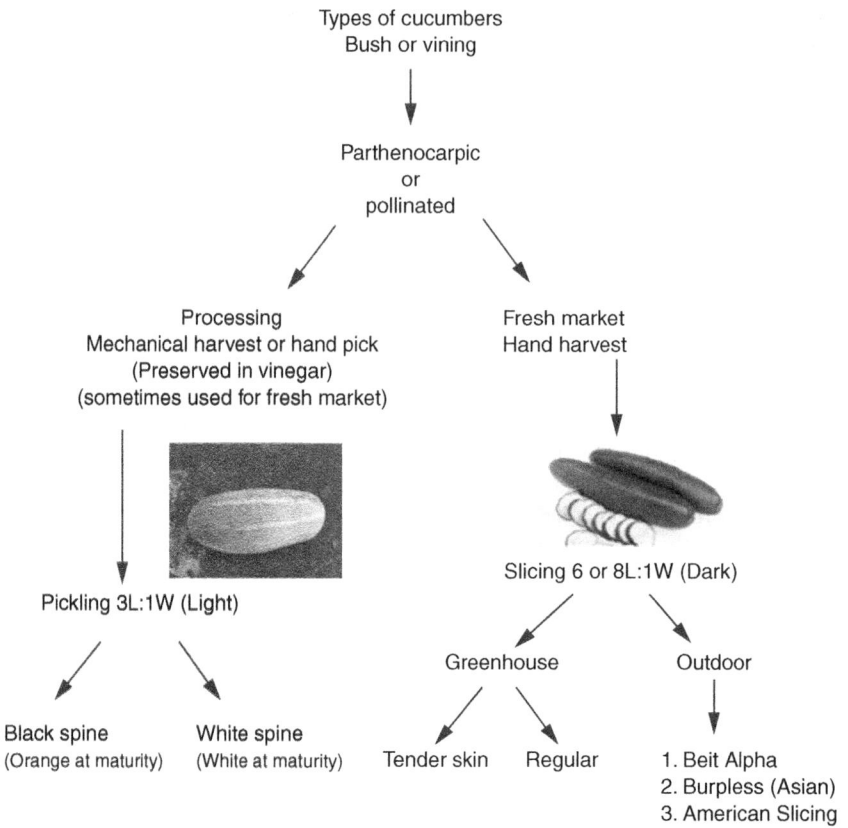

Fig. 10.1. Cucumber cultivars come in many different shapes and sizes that were developed for specific uses and markets. This figure summarizes some of the major commercial types. L:W, length to width ratio.

Types and Cultivars

There are many diverse cultivars of cucumber grown throughout the world. Cultivars differ in their degree of bitterness caused by cucurbitacins that accumulate in tissues of plants in the family Cucurbitaceae. The function of cucurbitacins is believed to be defense against herbivores. Cucurbitacins are chemically classified as steroids but often occur as glycosides (Chen et al., 2005). Non-bitterness in fruit is inherited as a simple recessive trait. Since plant breeders have been selecting against bitter flavor for many years, fruits of most modern cultivars are free of bitterness. Some older heirloom cultivars may develop bitter taste, particularly when grown under environmental stress.

Pickling

Immature pickling cucumber cultivars are processed with vinegar, spices, and herbs (such as dill seed) to make pickles and other products (Fig. 10.2).

Pickling cultivars have special characteristics that make them suitable for processing in vinegar (Motes, 1975). Resistance to carpel separation and bloating allow pickling cucumbers to retain their shape and integrity during the brining process. Fruit are harvested for processing while immature and not more than 12.5 cm (5 in) long. Cucumbers for processing may be open-pollinated or F-1 hybrids and are either hand or mechanically harvested, maturing 55–60 days from seeding. To ensure a concentrated set for mechanical harvesting, gynoecious cultivars are often preferred. The first flowers developing on a gynoecious cultivar are female, so every flower has the potential to produce fruit resulting in a concentrated early set and uniform crop maturation. However, to provide a source of pollen, a monoecious cultivar must be added so

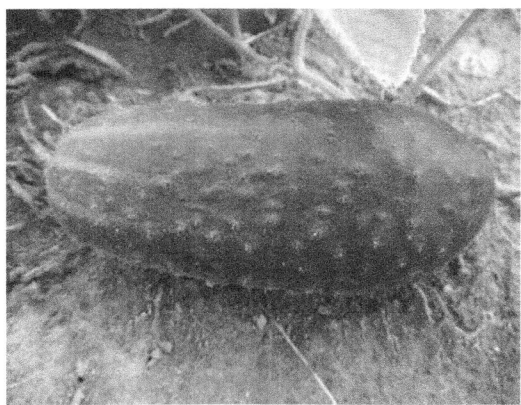

Fig. 10.2. Pickling-type cucumber fruit at harvest maturity.

that 12–18% of the plants are pollinators. The male flowers of the monoecious cultivar are sufficient to pollinate gynoecious plants (Rubatzky and Yamaguchi, 1997).

Total yields of gynoecious and monoecious cultivars are comparable because not all of the female flowers on the gynoecious plants develop into fruit. The gynoecious plant cannot produce enough photosynthates, so some female flowers abort. Applying gibberellic acid to small plants induces some male flower formation on gynoecious cultivars for seed production. The gynoecious character may not always be stable, so some plants may be monoecious in a gynoecious cultivar. Crowding may increase the percentage of monoecious plants in a gynoecious field depending on the stability of the cultivar. Pickling cultivars are sometimes marketed fresh for use in salads or direct consumption without pickling.

American slicing

Fresh market cultivars popular for slicing as a side dish or in salads are sometimes referred to as American types. Cultivars are usually monoecious F-1 hybrids that require bee pollination and mature in about 80–85 days from seeding. These cultivars have comparatively solid dark-green skin and a length-to-width ratio of 6:1 or greater, 8–10 cm (3–4 in) in diameter and 20–25 cm (8–10 in) long (Fig. 10.3).

The fruits are harvested immature well before physiological maturity, which occurs at approximately 120 days when the fruit are white or creamy yellow in color with fully mature seeds.

Plant breeding has improved the quality of the American slicing cucumber. Modern cultivars are uniformly dark green over the entire fruit without white striped ends. The fruit shape is nearly cylindrical with blunt ends and fruits appear circular in cross section. The pericarp (fruit) walls are thicker and the seed cavity greatly reduced. Modern cultivars tend to have fewer warts and spines compared to older cultivars (Rubatzky and Yamaguchi, 1997). Many modern cultivars are resistant to virus diseases and mildew.

Parthenocarpic types

Parthenocarpic slicing cultivars are also called seedless, English, or greenhouse cucumbers. Seedless or parthenocarpic cucumbers have traditionally been grown in greenhouses to prevent bees from introducing foreign pollen, which would cause seed development (Fig. 10.4).

Many cultivars are gynoecious, producing long, straight, smooth fruit at each axil, with thin-skins and medium to dark-green color. A slightly restricted "neck" at the stem end of the fruit serves to readily identify this unique type. Parthenocarpic cucumber fruit have a length-to-width ratio of about 6:1. The skin is so thin that greenhouse cucumber cultivars are often film wrapped to protect against physical damage and prevent shriveling (Rubatzky and Yamaguchi, 1997).

More recently, parthenocarpic cultivars have been developed for field production of both processing and fresh market cucumbers. Parthenocarpic pickling cucumber cultivars are not dependent on bee pollination and produce uniform fruit for mechanical harvest when conditions are unfavorable for pollination. For the processing of larger whole fruit, parthenocarpic cultivars are less desirable because they lack seed development, which contributes to texture and flavor components associated with quality.

'Beit Alpha' cucumbers were developed in Israel but are gaining popularity in other parts of the world as a fresh-market type. 'Beit Alpha', also known as 'Beta Alpha', are a type of parthenocarpic, all female, multi-fruited, dark-green hybrid that do not require pollination and offer high yield potential. The seedless fruit is 14–19 cm (5.5–7.5 in) long, with a thin smooth skin with few warts that does not need to be peeled before eating or film wrapping to prevent postharvest dehydration. Fruits can be harvested when 3–4 cm (1.2–1.6 in)

Fig. 10.3. Slicing cucumbers after harvest waiting to be washed and graded.

Fig. 10.4. Greenhouse-grown parthenocarpic cucumbers from the Netherlands.

in diameter or for the specialty markets with the flowers still attached.

Oriental types

Oriental cultivars are usually long, contain seeds, have thin dark-green skins, and considerable warts and spines. They are popular in Asian markets and are grown to a limited extent in other parts of the world. Oriental cultivars are often divided into two groups: the day-neutral North China group and the short-day South China group that is grown primarily for winter production. The South China group produces primarily pistillate flowers under short-day conditions and many staminate flowers under long-day conditions.

The Sikkim cucumber, also called the Concombre apple, is popular in India and is unique because the fruit has reddish brown skins. The Sikkim cucumber is a fat, large fruit reaching 38 cm (15 in) long and 15 cm (6 in) wide. The ripe fruit is eaten cooked, fermented, or raw and has a mild flavor (Tamang *et al.*, 1988).

Armenian

The Armenian cucumber, also called the serpent or snake cucumber, is a type of melon (*C. melo*) that is eaten fresh like a slicing cucumber when immature. The annual vine is creeping, with slender pubescent stems and rounded leaves with five lobes. The sex expression is monoecious with small and pale yellow flowers with five petals similar to other melons. The Armenian cucumber is long and slender with pale to dark green skin that is wrinkled longitudinally without spines. The slender fruit have their best flavor when 25–35 cm (10–15 in) in length but if allowed to grow will reach a length of about 0.9 m (36 in) and up to 15 cm (6 in) in diameter. Fruit are rarely straight and often bent and twisted. The fruit color changes to yellow when ripe with a mature odor similar to cantaloupe (Rubatzky and Yamaguchi, 1997).

Fig. 10.5. 'Lemon' cucumber fruit.

Novelty types

Novelty cultivars have a unique appearance that appeals to some consumers but generally have limited marketability. 'Lemon' or 'Apple' is a novelty or heirloom open-pollinated cultivar that has been passed along for many generations without significant genetic improvement. The fruits are round, have black-spines, and pale yellow-green color at the immature eating stage before turning bright orange at physiological maturity when the seeds are fully developed (Rubatzky and Yamaguchi, 1997; Fig. 10.5).

'Lemon' is andromonoecious (separate male and perfect flowers on the same plant) like cantaloupe. A hundred years ago, some botanists incorrectly identified this cultivar as a cantaloupe. The seed area is large and the pericarp wall is very thin compared to modern cucumber cultivars.

'White Wonder' is a white novelty pickling cucumber cultivar that lacks chlorophyll in the fruit epidermis during immature and mature stages of development. The color is the primary characteristic that makes 'White Wonder' unique. The shape, texture, and flavor are similar to other pickling cultivars.

Economic Importance and Production Statistics

Total world production of cucumbers and gherkins in 2011 was estimated to be 1,958,000 ha (4,836,260 acres). The average world yield was 30,927 kg/ha (27,613 lb/acre) in 2011. Compared with other vegetables, cucumber occupies fourth place in importance in the world, following tomato, cole crops, and onion. In 2011 total production of cucumbers and gherkins, according to FAO, was 37.6 million metric tonnes (41.3 million short tons); Africa producing 847,737 metric tonnes (934,470 short tons), North America 1.2 million metric tonnes (1.3 million short tons), South America 78,570 metric tonnes (86,609 short tons), Asia 30.8 million metric tonnes (34.0 million short tons) and Europe 3.9 million metric tonnes (4.3 million short tons). China alone produced 23.6 million metric tonnes (26.0 million short tons). For the most current information on cucumber production, please consult the latest FAO Crop Production Statistics.

Nutritional Values

Cucumbers are mainly water and do not contribute a great deal to the human diet. Cucumbers are a good source of certain minerals like potassium but are a poor source of protein, carbohydrates, fat, fiber, or most vitamins (Table 10.1).

Production and Culture

Temperature requirements and crop management

Cucumbers are a warm-season, frost-sensitive crop requiring warm soil temperatures for germination and

Table 10.1. Nutritional composition of cucumber (USDA, 2011).

Nutrient	Amount/100 g (3.5 oz) edible portion
Water (%)	96
P (mg)	17
Energy (kcal)	13
Fe (mg)	0.3
Protein (g)	0.5
Na (mg)	2
Fat (g)	0.1
K (mg)	149
Carbohydrate (g)	2.9
Ascorbic acid (mg)	4.7
Fiber (g)	0.6
Vitamin A (mg)	45
Ca (mg)	14

reliable field emergence. The minimum, optimum, and maximum soil temperatures for germination are 15.5°C (60°F), 35°C (95°F), and 40.5°C (105°F), respectively. The optimum range of soil temperatures for stand establishment is 15.5–35°C (60–95°F) (Masabni et al., 2011).

Little vine growth occurs when temperatures fall below 15.5°C (60°F). The growth rate increases steadily as temperatures rise above 21.1°C (70°F) (Curwen et al., 1975). Crops planted in early spring when soils are cool may take 8–9 weeks or more to reach harvest-stage, while later plantings may require only 6 weeks. Under optimal conditions, pickling cultivars mature approximately 55–60 days after seeding depending on the cultivar and the fruit size required. Slicing cultivars require an average of 80 days to harvest from seeding depending on the cultivar and environmental conditions.

Cucumbers should not be planted in a rotation after related cucurbit species like melons, squash and pumpkins because of the build-up of various insects and diseases, which live on crop residues and are common among these related species. Cucumbers are very sensitive to some herbicides, so carryover from previous crops must be considered when selecting fields for conventional production.

Soil requirements

Cucumbers can be grown profitably on most fertile, well-drained soils. Loam-textured soils with organic matter are well suited for cucumber production. Light, sandy soils are acceptable provided they are adequately fertilized and irrigated. A pH of 6.0–7.0 is suggested for optimum cucumber growth. Acid soils with pH < 5.5 should be avoided (Lorenz and Maynard, 1988).

Fertilizer and nutrition

Cucumbers respond favorably to fertilizer, so proper fertility management is a key to maximizing productivity. Fertilizer requirements vary, depending on the soil type, the environment, and cultivar. There is no single recommendation that will fit all production situations. Fertilization is most efficient and effective if based on preplant soil test recommendations and foliar analyses made throughout the season. Most soil testing labs will rate essential minerals as being high, medium or low and recommend how much fertilizer should be applied to successfully grow the crop. It is important to determine how many nutrients a cucumber crop is likely to remove during a season. The nutrients removed from a field by a typical cucumber crop would be approximately 196 kg/ha (175 lb/acre) of nitrogen (N), 28 kg/ha (25 lb/acre) of phosphorus (P) and 196 kg/ha (175 lb/acre) of potassium (K) (Masabni et al., 2011). Soil test recommendation for N and K often are less than this amount to account for release of nutrients for organic matter and inputs from the nitrogen cycle. Recommendations for P additions often exceed the amount removed by the crop because some P fertilizer will be fixed or immobilized in soil. An adequate supply of micronutrients should also be provided, especially for hydroponic production.

Nitrogen is a very important nutrient for cucumber production. Early in the season, adequate nitrogen promotes vegetative growth and canopy formation that is needed to maximize photosynthesis to support later fruit development. Lower available nitrogen a little later in the season promotes flower formation and fruit set, while excessive nitrogen fertilization may delay flowering and reduce fruit set.

Petiole or plant tissue analysis can be used to assess crop nutrient status during the growing season to know if adjustments in the fertility program are required. Later in the season, side dressing or fertigation with nitrogen, and possibly other nutrients, may correct deficiencies and maximum yields on sandy soils, especially when crops are handpicked over an extended period. Additional nitrogen applications are often needed following heavy, leaching rains on light, sandy soils. Side dressing is usually not necessary for crops grown for once-over mechanical harvesting because of the relatively short crop maturation time.

Seedbed preparation

Field production of cucumbers may be with or without mulch. Plasticulture is the most popular mulch in areas with short growing seasons to enhance early production of slicing cucumbers for fresh market. Plasticulture is rarely used for the production of earlier maturing pickling cucumbers because the added cost is not warranted and the mulch interferes with mechanical harvesting.

If cucumbers are not grown on plastic, production on raised beds is recommended to improve drainage and to reduce belly rot of the fruit. If fertigation is not used, mineral nutrients should be incorporated into the bed prior to planting. The soil should be tilled so that it is friable and free of clods to provide good seed to soil contact for direct-seeded crops.

Field establishment

There are approximately 1,100 cucumber seeds/oz (30–35/g) (Lorenz and Maynard, 1988). Precision planting of cucumber seeds to the right spacing and depth will encourage uniform emergence of seedlings. A uniform stand will lead to concentrated fruit set for efficient hand or mechanical harvesting. Precision seed spacing produces the desired stand of plants, saves seed, and eliminates the need for thinning after establishment. This is particularly important for once-over destructive machine harvests of pickling cucumbers. Cucumbers are normally planted at a depth of 1.3–5 cm (0.5–2 in), depending on soil type, time of season, and soil moisture availability. Many growers use precision planters and coated seed helps deliver fungicides, biologicals, or small amounts of fertilizer.

In short season areas, cucumbers particularly for fresh market are transplanted. Transplanting may reduce the time to first harvest by 10 days to 2 weeks depending on conditions. Cucumbers are sensitive to transplant shock and are best transplanted from plug transplants rather than with bare roots. To reduce transplant shock, plants should be hardened before transplanting and irrigated more frequently until properly established.

Spacing

Spacing is dependent on cultivar, environments, and cultural practices. Generally recommendations for fresh-market cucumbers are in-row spacing of 23–31 cm (9–12 in) with row spacing of 0.9–1.8 m (36–72 in). This results in populations of 18,000–29,600 plants/ha (7,300–21,780 plants/acre) (Fig. 10.6).

Pickling cucumbers usually have smaller vines than the slicer type and are grown at high populations. Depending on environmental conditions, maximum yields for hand-harvest may be obtained with in-row spacing of 15–31 cm (6–12 in) with 0.9–1.8 m (36–72 in) between rows. This spacing results in populations of 18,000–71,700 plants/ha

Fig. 10.6. A field of slicing cucumbers grown without plastic mulch, destined for fresh-market sale.

(7,300–29,040 plants/acre). Populations of 50 plants/m² gave higher yields compared to lower plant populations of pickling cucumbers (O'Sullivan, 1980). High yields of destructively harvested pickling cucumbers are obtained from plant populations of 123,500–222,300 plants/ha (50,000–90,000 plants/acre) (Curwen *et al.*, 1975). These populations are achieved with in-row spacing of 10–15 cm (4–6 in) with 31–71 cm (12–28 in) between rows.

Irrigation

Cucumber crops require large amounts of water. In many locations where the average rainfall during the growing season is appreciably less than 38 cm (15 in) or erratic during the growing season, irrigation is necessary to maximize both fruit quality and yields. Fruits contain about 95% water. The rule of thumb for at least 2.5 cm (1 in) of water per week applies to cucumber production. Even this amount may be insufficient to sustain growth during prolonged drought and/or high temperatures. In areas where environmental stress is common, 5 cm/week (2 in/week) may be required. Wilting decreases yields and may reduce fruit quality depending on the severity. The most critical need for water is during fruiting when drought may cause "nubbing" (small fruit that are shriveled at one end). In crops grown for once-over mechanical harvest, the last irrigation should be timed to allow the soil to dry sufficiently to allow equipment into the field.

Where possible, irrigation should be applied early in the day to permit the soil and leaf surfaces to dry before nightfall. Prolonged periods of wet soil and leaves promote infection by mildew, *Alternaria*, angular leaf spot, and other various fruit-rotting and foliage diseases. Overhead irrigation will prohibit bee activity and should not be used in the early morning hours when bees are most active.

Cucumbers can be successfully irrigated by furrow, trickle, and overhead sprinkler methods. Furrow irrigation is limited to areas where the land is level. Surface irrigation is preferable to overhead, which tends to spread various diseases to the foliage and fruit. Trickle or drip irrigation is commonly used with plasticulture but can be used on bare soil as well.

Pollination

Bees are extremely important for field production of non-parthenocarpic cultivars. In most areas, native bee populations are insufficient to pollinate a commercial crop. Motes (1975) recommends at least one colony per 50,000 plants. Most cucumber flowers are open for only 1 day and a successful pollination event requires multiple bee visits. Cool wet weather significantly reduces bee activity and pollination resulting in low yields and/or a high percentage of misshapen fruit. Pesticides that harm bees should not be used and when necessary spray applications should be scheduled for late afternoon or early evening.

Greenhouse production

Cucumbers are important greenhouse crops in North America, western Europe, and Japan. Cucumbers are grown during the winter in unheated greenhouses in temperate regions. Parthenocarpic cultivars are usually grown in greenhouses to avoid the use of bees since standard field cultivars must be cross-pollinated. Most parthenocarpic greenhouse cucumbers are gynoecious F-1 hybrids. The cost of parthenocarpic greenhouse cucumber seed is significantly higher than seed of standard field cultivars (Hochmuth, 2013).

Plants are transplanted into a wide range of synthetic rooting media including rockwool, gravel, sand, or bagged media to avoid the buildup of disease in native greenhouse soils. Greenhouse soils may be sterilized with steam or chemical fumigants. Cucumbers may also be grafted on to special disease-resistant rootstocks such as *Lagenaria siceraria* (Molina) Standl. and a *Cucurbita moschata* (Duchesne ex. Pow) × *C. maxima* (Duchense ex. Lam.) hybrid for greenhouse production in native soils if soil fumigation/sterilization is not used (Hochmuth, 2013).

Temperature control is important with an optimum recommended temperature of 25–28°C (77–82°F) and night temperatures of 17–18°C (63–64°F). Lower temperatures tend to favor vegetative growth while higher temperatures promote flowering. High light intensities are necessary to obtain maximum yields. Greenhouse CO_2 levels should be maintained between 400 and 1500 ppm to maximize photosynthesis (Hochmuth, 2013).

Harvesting and marketing

Fields of slicing cucumbers are repeatedly hand-harvested during the season. Fruit are harvested by carefully cutting or snapping fruit from the vine by hand to avoid damaging the plant. Cucumbers develop rapidly and must be harvested every other day or daily under favorable growing conditions.

Delayed harvest results in excessively large fruit and inhibits the production of new fruit. Fruit are carefully handled and graded after harvest.

Cucumbers for fresh use are often individually film wrapped or grouped in a plastic tray and wrapped to preserve freshness and prevent shriveling. Nontoxic water-soluble wax is sometimes added to the wash water to provide a thin coating on the fruit surface to reduce desiccation and give fruit an attractive glossy appearance. The parthenocarpic greenhouse slicing cucumbers are usually wrapped in plastic film because of their very tender delicate skin that is easily damaged.

Cucumbers are generally stored at 13–15°C (55–59°F) and 90–95% relative humidity (RH) and have a postharvest shelf life of 1–2 weeks. Like all cucurbits, storage at temperatures less than 13°C (55°F) causes chilling injury. Fruits injured by chilling will deteriorate more rapidly when returned to room temperature and have off-flavor. Short-term chilling before serving in salads or for fresh use improves crispness. In contrast, freezing injury will be initiated at −0.5°C (31°F). Symptoms of freezing injury include a water-soaked pulp becoming brown and gelatinous in appearance over time (Suslow and Cantwell, 2013).

Cucumbers are highly sensitive to exogenous ethylene. Accelerated yellowing and decay will result from low levels (1–5 ppm) of ethylene during distribution and short-term storage. Cucumber should not be mixed with commodities such as melons and tomatoes (Suslow and Cantwell, 2013).

Controlled or modified atmosphere storage and/or shipping offer moderate to little benefit for maintaining cucumber quality. Low O_2 levels (3–5%) delay yellowing and the onset of decay for a few days. Cucumber tolerates elevated CO_2 up to 10% in controlled atmosphere storage but storage life is not extended beyond the benefit of reduced levels of O_2 (Suslow and Cantwell, 2013).

Pickling cucumbers may also be repeatedly hand-harvested but the trend has been toward greater use of once-over destructive mechanical harvesting, particularly for processing, to cut labor costs. A mechanical harvester cuts the vines at the soil surface and conveys the plants up a conveyer to a set of rollers that pinch the fruit from the vine as the plants are fed between the rollers. The vines and immature fruit are returned to the field and fruit of the desired size are conveyed to an adjacent vehicle or wagon and taken to the processing facility for washing, grading, and processing.

Diseases

There are many diseases that affect cucumbers. Cucumber mosaic virus, watermelon mosaic 1 potyvirus, and zucchini yellow mosaic potyvirus are some of the most important viral diseases (Zitter *et al.*, 1996). The use of disease-free seed and sprayings to control aphids may help prevent the spread of virus diseases. Some cultivars are resistant to viruses.

Some of the most important fungal diseases are downy mildew (*Peronospora cubensis*), powdery mildew (*Erysiphe cichoracearum*), and damping-off (*Pythium*, *Rhizoctonia*) (Fig. 10.7).

Treating seed with thiram, fungicide applications to soil at planting, and spraying emergent seedlings with metalaxyl controls damping-off.

Mildew is common on cucumber leaves, particularly late in the season and in areas with high humidity. Cultural control measures include: removing plant debris at the end of the season to reduce overwintering of the fungus, reducing plant populations to improve air circulation, and avoiding excess nitrogen fertilization. Several fungicides, such as trifloxystrobin, azoxystrobin, and chlorothalonil + myclobutanil control the mildew diseases. Some cultivars are resistant to mildew. Biorational compounds can reduce the incidence of powdery mildew on cucurbits. Biorational materials include natural and mineral oils, peroxigens, cow's milk, silicon, and salts of monovalent cations such as sodium, potassium and ammonium (Bélanger and Labbe, 2002).

Stripped and spotted cucumber beetles transmit bacterial (*Erwinia tracheiphila*) wilt among plants during feeding. Cucumber beetle feeding does little direct damage to plants but the bacteria introduced during feeding proliferate in the xylem reducing water flow between the roots to leaves. Bacterial wilt symptoms

Fig. 10.7. Cucumber leaf with powdery mildew infection.

are similar to drought stress and in severe cases will kill plants. Some cucumber cultivars are more resistant to bacterial wilt than others. Other cucumber pathogens include: angular leaf spot (*Pseudomonas syringae* pv. *lachrymans*), gummy stem blight (*Didymella bryoniae*), common anthracnose (*Colletotrichum* spp.), fruit wet-rot (*Choanephora cucurbitarum*), Fusarium wilt, Cladosporium scab, and bacterial soft rot (*Erwinia* spp.) (Zitter *et al.*, 1996). Cucumber growers should consult local recommendations for tested and approved control measures as well as disease-resistant cultivars.

Insect Pests

Some of the most troublesome insects of cucumber include cucumber beetles (*Diabrotica* and *Acalymma* spp., vector of bacterial wilt), *Epilachna* spp. beetles, greasy worm (*Agrotis ipsilon*), the melon fruit fly (*Dacus* spp.), and aphids. These pests can be controlled with insecticides. Control with natural enemies is highly developed in greenhouses but less so for field cultivation. Please consult local recommendations for effective control measures.

NETTED AND MIXED MELONS

Origin and History

Netted melons, also called muskmelons or cantaloupes, appear indigenous to Africa (Zeven and Zhukovsky, 1975). The oldest record appears to come from Egypt where muskmelon cultivation occurred as early as 2400 BC. Netted melons were known to Greeks about 300 BC. Truly wild forms are found only in eastern tropical sub-Saharan Africa. The related wild forms reported in India are likely undomesticated escapes derived from local cultivars. Once domesticated, melon diversity expanded into numerous cultivars, especially in India, which is considered a secondary center of origin (Zeven and Zhukovsky, 1975). Cultivars were rapidly dispersed throughout Europe and fairly early into the Americas. Columbus reportedly carried seed to the New World in 1494 and the Spaniards introduced melons into the Americas in the late 1500s. North American natives were growing muskmelons during the 1660s (Rubatzky and Yamaguchi, 1997).

The term muskmelon is derived from musk, which is a Persian word for a kind of perfume. The name cantaloupe is believed to have arisen from the city of Cantaluppi in Italy or from the estate and castle of Cantalupo also in Italy (Rubatzky and Yamaguchi, 1997).

Modern-day cultivars share the main characteristics of earlier cultivars but are greatly improved for uniformity, size, shape, flesh thickness, sugar content, and quality. 'Netted Gem' introduced by W. Atlee Burpee Seed Company from France in about 1880 made an important contribution to cultivar development in North America. Shipping and storage characteristics were of no importance until after 1900. Today melons are most commonly eaten fresh, as a salad or as a dessert (Rubatzky and Yamaguchi, 1997).

Botany and Life Cycle

Melon is a general term for fruit produced by various members of the family Cucurbitaceae. In some areas the term is applied to both netted and non-netted members of *C. melo* but not watermelon. In other regions of the world, the terms cantaloupe and/or muskmelon are used to describe the netted melon while other types of *C. melo* are called "melon", "mixed melon" or sometimes "winter melon". Cucumber and melon are both members of the genus *Cucumis* but different species, so they do not cross-pollinate. Since cucumber and melons are related, some of the information on cucumber also applies to melon.

Naudin classified different forms of *C. melo* into separate botanical varieties in 1859. Although Naudin's classification was not recognized taxonomically, it was useful for many years to delineate obvious horticultural differences among fruit types within *C. melo*. Many of the botanical variety designations are now classified into eight horticultural groups under *C. melo* (Table 10.2; Rubatzky and Yamaguchi, 1997).

Table 10.2. Groups of *Cucumis melo* (Rubatzky and Yamaguchi, 1997).

Group name	Common name/characteristics
Cantaloupensis (formerly Reticulatus)	Netted melons (muskmelon and cantaloupe) as well as lightly or non-netted cultivars. Perishable
Inodorus	Winter melons or mixed melons. Some may be stored for several weeks
Flexuous	Armenian cucumber, snake or serpent melons
Comonon	Oriental pickling melon
Chito	Mango melon or vine peach melon
Dudaim	Queen Anne's pocket melon
Momordica	Snap melon
Agrestis	Wild types

Fruit size and characteristics vary widely among groups and cultivars (Table 10.2). Fruit shape varies from spherical to oblong. Many fruits are covered with corky reticulate netting while others are smooth. The vein tracts or sutures are longitudinal indentations or stripes on the fruit surface associated with vascular bundles, which may not be visible in fruits that are heavily netted. The edible melon flesh is actually the fruit pericarp, which varies in thickness, color, and texture. Pericarp colors may be white, green, pink, or orange. Aroma varies from strong to odorless and depends on the quantity and quality of volatile compounds produced. Melon fruits produce from 300 to 500 seeds each arising from separate fertilization events. The seeds are smooth, range from 5–15 mm (0.2–0.6 in) in length and have yellow or beige coloration. On average there are approximately 30 seeds/g.

Types and Cultivars

This section focuses on two of the most widespread and economically important Groups Cantaloupensis and Indorous. Cultivars in the Inodorous and Cantaloupensis Groups can be intercrossed. The Group Inodorus contains the so called "winter melons" such as 'Casaba', 'Canary', 'Honeydew', 'Crenshaw', 'Canary', and 'Santa Claus' that do not abscise from the vine at maturity, lack heavy netting, are generally less aromatic, and have longer storage life than the Cantaloupensis group. One exception to this description is the 'Crenshaw' melon, which has a short shelf-life and can be very odoriferous. Winter melons generally require a growing season that is several weeks longer than most members of the Cantaloupensis Group. Best fruit quality is obtained at high temperatures from 30–35°C (86–95°F) in semi-arid environments with low humidity.

'Honeydew' fruit have pale green rinds that turn creamy white when mature (Fig. 10.8). The firm, thick, juicy flesh is pale green to white at maturity. Fruit do not abscise when ripe so judging maturity can be more difficult than for netted melons that slip. 'Honeydew' fruit have moderate storage life of up to several weeks under the right conditions. Honeydew can accumulate high sugar contents and when fully ripe may have a soluble solids content in excess of 16%, among the highest of any melon. 'Honeydew' was selected from the French cultivar 'White Antibes' in the state of Colorado in the USA (Rubatzky and Yamaguchi, 1997).

'Crenshaw' originated by chance in a muskmelon field (Rubatzky and Yamaguchi, 1997). The 'Crenshaw' fruit is smooth and oblong with some ribbing at the tapered stem end (Fig. 10.9). 'Crenshaw' fruit exhibit intermediate stem abscission, are very susceptible to sunburn, and have poor shipping and storage qualities. The immature fruit rind is dark green.

Fig. 10.8. 'Honeydew' fruit in a California production field.

Family Cucurbitaceae

Rind color change is not uniform and ripe fruit often have sections of both green and pale yellow. The ripe fruit have moderately firm, salmon-colored flesh with a mild perfume-like aroma. The flesh is sweet with 12–14% soluble solids (Rubatzky and Yamaguchi, 1997).

'Casaba' fruit have a greenish yellow, thick, relatively hard rind with shallow ribs and no netting. Fruits are round but tapered at the stem end (Fig. 10.10). The white flesh is moderately firm with an aroma reminiscent of cucumber. The fruit do not abscise at maturity and have a long storage life of up to a couple of months at cool temperatures. The flesh has a soluble solids content of 6–8%.

'Juan Canary' fruit are oblong, smooth, and intense yellow. The flesh has little aroma and is whitish green. The flesh has a soluble solids content of 6–8%. The fruit do not abscise at maturity,

are well adapted to shipping and can be stored for several weeks.

'Santa Claus' is an oblong slightly warted melon with a hard rind that is green and yellow, vaguely striped, and has little aroma. The flesh is firm and whitish green with a soluble solids content of 6–8%. The fruit are suitable for shipping and under cool temperatures can be stored for approximately 3 months if the rind is not damaged.

The netted fruit of group Cantaloupensis (formerly Reticulatus) are called muskmelon and/or cantaloupe in many markets. A unique feature of some muskmelon/cantaloupe cultivars is that they detach from the vine at maturity in a series of stages known as "slip". The slip stages are useful for characterizing maturity. Fruits progress from complete stem attachment to one quarter, one half, three quarters, and finally full slip during ripening, becoming increasingly aromatic and yellowish orange in color (Fig. 10.11; Rubatzky and Yamaguchi, 1997).

The name cantaloupe is often associated with fruit produced in North and Central America in arid regions with low humidity and high temperatures. Cantaloupe fruit characteristics are desirable for long-distance shipping and include: round and small shape, weight of approximately 1.5–2 kg (3.3–4.5 lb), uniform heavy netting, indistinct ribbing, firm thick flesh, and small, relatively dry seed cavity (Fig. 10.12).

In contrast, muskmelon cultivars grown for local markets in the eastern USA and Canada are adapted for production under humid conditions, have spherical or elongated fruit shape, prominent vein tracts and ribbing, larger size (2.3–3.6 kg; 5–8 lb), strong flesh, strong aroma, large moist seed cavities, and relatively soft flesh. Most muskmelons have relative poor shipping and storage characteristics and are better adapted to local or regional markets (Fig. 10.13).

Many modern muskmelon cultivars developed for the eastern USA and Canada have incorporated the beneficial fruit characteristics of western cantaloupes to improve shipping and storage life but with larger fruit size and tolerance of high humidity. This trend has blurred the distinction between muskmelon and cantaloupe cultivars. Today, older large-fruited aromatic muskmelon cultivars are popular in some markets as heirlooms. Melon retailers are making a conscious effort to eliminate the word "muskmelon" from marketing because of the negative connotations associated with the term "musk" and the strong aromas the name implies.

Fig. 10.9. 'Crenshaw' fruit in a California field. The yellow fruit on the left is ripe and the green fruit is immature. The rind is very sensitive to sunburn and fruits not protected by the canopy are sometimes protected with bags.

Fig. 10.10. Mature bright yellow 'Casaba' fruit in a California production field.

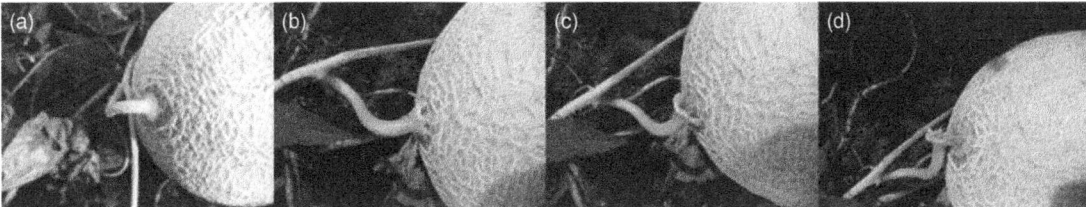

Fig. 10.11. Cantaloupe fruit naturally separate from the vine in a series of stages described by the percentage of stem attachment. Before the fruit begins to ripen, the stem is securely attached to the fruit (a). As ripening commences, the stem begins to separate so that when a quarter of the stem is free of the fruit it is often called quarter slip stage (b), half separation is half slip (c), and total separation is full slip (d).

Fig. 10.12. A western shipping-type melon commonly grown in California, Arizona, Texas, Mexico, or Central America for distant markets.

Increasingly the word cantaloupe is used exclusively for marketing netted melons in the USA.

Not all netted melons abscise at maturity. Some Japanese cultivars have netting and vein tracts like western cantaloupe cultivars but ripen without abscission like members of the Inodorus Group. Fruit of these cultivars are often sold with stems attached to the fruit in a T configuration (Fig. 10.14).

Persian melon cultivars produce large-round netted fruit that resemble western shipping-type cantaloupes. Persian fruit do not detach at maturity and are usually classified in the Inodorus Group despite their shallow netting, mild aroma, small moist seed cavity, and firm salmon-colored flesh. Persian melons can be crossed with other Cantaloupensis types to create melons with good shipping characteristics and longer shelf life.

Cultivars of other Cantaloupensis melons have either sparse netting or lack netting entirely but have vein tracts and thin rinds. Examples include 'Charentais', 'Ananas', 'Ha-ogen', and 'Valencia'.

Plant characteristics

Commercial cultivars of *C. melo* types are warm-season, frost-sensitive annuals that are cross-pollinated by bees. Sex expression is andromonoecious or sometimes monoecious. Commercial cultivars are day-neutral. Plant habit is vining with varied internode length. Bush cultivars have been developed for some types. Many cultivars have tendrils at each node and flowers have five yellow petals that are open for a single day, like cucumber. However, leaves differ from cucumber in being circular, oval, or kidney shaped with five to seven rounded lobes. Many commercial cultivars are F-1 hybrids. The main advantages of F-1 hybrids are uniformity and hybrid vigor. Most commercial cultivars are andromonoecious so F-1 hybrid seed production requires emasculation of the hermaphrodite flowers and cross-pollination by hand. Manual labor increases the cost of hybrid seed by several times compared to open-pollinated. It is estimated that at least 80% of the netted melons grown in the USA are F-1 hybrids. A lower percentage of winter melon cultivars are F-1 hybrids.

Production and Culture

According to FAO, in 2011 the total annual world production of melons was about 25.5 million metric tonnes (28.1 million short tons) from 1,008,700 ha (2,492,599 acres). Major melon-producing countries were China (12.2 million metric tonnes (13.5 million short tons)), followed by Turkey (1.7 million metric tonnes (1.9 million short tons)), Iran (1.2 million metric tonnes (1.3 million short tons)), the USA (1.1 million metric tonnes (1.2 million short tons)), and Spain (1.0 million metric tonnes (1.1 million short tons)). In the USA, California was the leading producer of all melons, accounting for 33% of total acreage in 2005, followed by Texas, Georgia, and Arizona. By acreage and weight, California leads the USA in cantaloupe and honeydew production.

Fig. 10.13. A classic muskmelon cultivar traditionally grown for local markets in the eastern USA and now as heirlooms.

Fig. 10.14. Japanese melons grown in tunnels early in the season and then uncovered later so the structure supports vine growth. These fruit resemble cantaloupe but do not abscise at maturity and have characteristics similar to the Winter Melon Group (Table 10.2).

In the past, melons were considered "seasonal delights" because of their limited availability throughout the year. Production for export has developed in Mediterranean countries, Australia, Costa Rica, Guatemala, Taiwan, and Japan, using F-1 hybrid cultivars with improved shelf life. Today, not only is the USA a net importer of melons (imports minus exports) but it is the largest importer of cantaloupes and mixed melons in the world. Many of the US imports come from Costa Rica and Guatemala. France and the UK are also large import markets for melons. In tropical Asia, melons are more of a luxury crop for urban markets, grown in the drier lowlands and highlands.

Site selection

Melons grow well in a wide range of fertile and well-drained soils. Maximum yields are achieved on medium-textured soils with high water-holding capacity and good internal drainage. Peat and heavy clay soils should be avoided because their poor aeration and restricted drainage inhibit root growth and fruit quality. Melons are sensitive to saline and acid soils but will grow well on slightly acidic (pH 6.8) to moderately alkaline (pH 8.0) conditions.

A crop rotation cycle of several years without growing other members of the family Cucurbitaceae (cucumber, pumpkin, watermelon, etc.) is recommended if pathogen populations are high and fumigation is not used. When possible, grasses, corn, or sorghum are good rotation crops. Carryover from herbicides, such as atrazine (2-chloro-4-(ethylamino)-6-(isopropylamino)-s-triazine), may inhibit melon growth.

Seedbed preparation

In many regions, melons are generally grown on raised beds 15–20 cm (6–8 in) high. Final spacing of the beds in furrow-irrigated culture is generally 203 cm (80 in) center-to-center. Beds can be created and shaped to final spacing before planting or to one-half final spacing 102 cm (40 in). In the latter procedure, every other bed is seeded at planting. When plants begin to vine, unplanted beds are broken open and the field is reworked to position the plant row in the middle of the final 203 cm (80 in) bed.

Plastic mulch

In short-season areas, plastic mulches heat cold soil to stimulate early season growth. In other areas, mulch is popular for fumigation, moisture conservation, reducing fruit rot, preventing the leaching of nutrients following heavy rains, and providing weed control. Plastic film is installed over a 203 cm (80 in) bed and secured at the edges with soil. Planting is by direct seeding or transplanting through holes punched or burned in the mulch. Water is applied under the plastic by drip irrigation or fertigation.

Trenches

For early-season plantings in some areas, seeds are planted in the bottom or on the north slope of a plastic-covered trench dug approximately 46 cm (18 in) wide and 15–20 cm (6–8 in) deep in the center of a standard east-west, 203 cm (80 in) raised bed. After planting, the trench is covered with clear plastic and secured at the edges with soil. The plastic cover is cut open or removed entirely when the melon plants begin to vine and temperature control is no longer required. The plastic cover protects from fly-in insects while intact.

Low row covers

Plastic tunnels are suspended over melon beds by tightly stretching clear plastic film over wire hoops for temperature modification in short-season areas. Row covers can be used longer in the season than with trenches but are more expensive and may require ventilation on sunny days. Beds under clear plastic tunnels may be covered with plastic mulch to provide weed control and to conserve water.

Fertilizer and nutrition

Melons develop extensive root systems that efficiently explore the soil for water and available nutrients. For this reason, fertilizer requirements are moderate compared with many other vegetable crops. An effective nutrient management program provides sufficient quantities of N, P, K, S, Ca, Mg, B, Cu, Fe, Mn, Mo, and Zn so that plants develop and maintain a large canopy of healthy leaves throughout the season. Fertilizer should be applied in accordance with both soil and foliar test results when possible. Nitrogen is the most commonly required fertilizer, although P is sometimes needed to promote good seedling vigor, maximum productivity, and high fruit quality, especially in alkaline soils. Mineral soils may contain adequate K but losses through leaching necessitate annual additions, especially in areas with light soils. The approximate absorption of N-P-K by a melon crop is 174 kg/ha (155 lb/acre), 28 kg/ha (25 lb/acre), and 174 kg/ha (155 lb/acre), respectively. Soil test recommendation for N and K often are less than these amounts to account for nutrient release for organic matter and inputs from the nitrogen cycle. Recommendations for P additions often exceed the amount removed a melon crop because some P fertilizer will be fixed or immobilized by certain soil types (Lorenz and Maynard, 1988).

For conventional production, granular phosphorus fertilizer may be applied preplant, as 10-34-0 or 0-52-0 for example, in twin bands 15 cm (6 in) deep and 10–15 cm (4–6 in) to either side of the planted row or in drip irrigation lines as part of a fertigation program. Phosphorus can also be banded and lightly incorporated before beds are formed, but alkaline mineral soils chemically tie up phosphorus, making it unavailable to the plants. Thus, banding the fertilizer near the seed is preferred over broadcast application (Lorenz and Maynard, 1988).

Nitrogen is often applied in two side-dressings or through fertigation, the first at the two- to four-leaf stage and the second at vine proliferation. Melon leaf petiole tissue analysis provides an effective diagnosis of crop nutrient status and allows continuous monitoring of crop fertility throughout the season so inputs can be applied only if required to maximize fruit yields and quality. Over-fertilization, especially with N, favors vegetative growth over reproductive growth and may inhibit fruit set. The N level, especially ammoniacal N, must be relatively low at flowering to encourage fruit set and development over the production of new leaves and vines. Moderate N tissue levels favor sugar accumulation in the fruit rather than additional vegetative development.

Field establishment

The minimum, optimum, and maximum soil temperatures for seed germination are 16°, 35°, and 38°C (60°, 95°, and 100°F), respectively. Germination is slow and erratic when soil temperature is below 20°C (68°F). Fruit maturing when daily mean air temperatures are below 21°C (70°F) have poorer quality. Melons grow best in hot weather but very high temperatures (43–46°C/110–115°F) may cause temporary vine wilting and softer ripe fruit with reduced shelf life (Lorenz and Maynard, 1988).

Melons are often planted in single rows spaced 2 m (80 in) apart (Fig. 10.15). Seeds are planted from 1.25 cm (0.5 in) to 2.5 cm (1.0 in) deep. In-row spacings of 5–25 cm (3–10 in) are common. After plants reach the two- to four-leaf stage, they are sometimes thinned to an in-row spacing of approximately 20 cm (8 in) for a final plant stand of 9,800 seeds/ha (3,920 seeds/acre). In some areas on small plots, melons are planted in hills 1.2–1.8 m (48–72 in) apart with three or four seeds per hill (Lorenz and Maynard, 1988).

Vacuum planters optimize use of expensive hybrid seed by precision placement. Under ideal conditions, seeds are planted to a final stand at 15–20 cm (6–8 in) spacing and not thinned (Table 10.3).

Narrow spacing reduces fruit size while wider spacing increases it. Hybrids usually are less affected by close spacing than open-pollinated cultivars. Melon seeds are often film coated. Film coating offers several advantages, such as less abrasion and clogging of planter parts, easier cleanup and maintenance of equipment, addition of fungicide or small quantities of fertilizer to seeds, and checking of seed placement in soil.

Fig. 10.15. Cantaloupe field on raised beds in the Central Valley of California during harvest season.

Table 10.3. Number of seeds needed to plant 0.4 ha (1 acre) when rows are spaced 2 m (80 in) apart (Lorenz and Maynard, 1988).

In-row spacing cm (in)	Seeds/ha (seeds/acre)
5 (3)	26,150 (10,460)
10 (4)	19,600 (7,840)
15 (6)	13,050 (5,220)
20 (8)	9,800 (3,920)
25 (10)	7,850 (3,140)

Transplants are used primarily in areas with short growing seasons to reduce development time. Melons do not easily produce adventitious roots and are considered difficult to transplant, especially with bare roots. Melons are most easily transplanted from plug trays with soil around the roots. Transplants are grown in protective structures until they have two to four true leaves and conditions are sufficiently warm for rapid field growth. Prior to field planting, transplants should be hardened to increase dry matter and stress tolerance by withholding water, mechanical conditioning, or exposure to suboptimal temperatures. Plants should be kept well watered to minimize stress after field planting. A liquid fertilizer or "starter" solution is often applied to the root zone at transplanting to encourage rapid growth.

Irrigation

Muskmelon have extensive, moderately deep, root systems that efficiently remove water from soil, so irrigations may be less frequent than with some other vegetables. However, the general recommendation for at least 2.5 cm (1 in) of water per week still applies particularly during pollination and rapid fruit development. Post-plant irrigations ensure seed germination, emergence, and stand establishment. However, when possible, irrigation may be withheld or carefully managed until the pre-vining stage to avoid cooling the soil, to discourage damping-off diseases and to encourage deep rooting. The final irrigation is typically 7–10 days prior to harvest, but may vary depending on environmental factors such as soil type, air temperature, and humidity.

Drip irrigation, either with or without plastic mulch, is common in the eastern and increasingly the western USA as well as other areas where irrigation is required. Drip conserves water and provides uniform application to the root zone. Melon fields are often established with sprinkler irrigation but this may increase foliar and fruit disease in mature plants.

Furrow irrigation is a relatively inefficient method that is still used in some flat desert areas. When using furrow irrigation, care must be taken to ensure that water reaches the seed or root zone, and that after vines start to run, the top of the bed does not become wet and encourage fruit rot. Poor drainage, flooding, or standing irrigation water may damage melon crops. Common problems associated with excessive water are weak root systems that may lead to vine collapse as the crop reaches maturity, poor fruit netting, and low fruit sugar content.

Flowering and pollination

Melon sex expression is most commonly andromonoecious because separate male and perfect flowers occur on the same plant. Both male and perfect flowers are open for only 1 day. Male flowers rapidly senesce and dehisce after closing, but fruiting flowers often remain attached for several days. If pollination is successful, the ovary rapidly enlarges and "set" fruit, while the remaining fruiting flowers senesce and abscise. Most vines only support a few developing fruit at once and most later fruiting flowers will not set.

Melon pollen is sticky and must be transferred from stamens to the pistil by insects. The most effective pollinators are bees. Several hundred pollen grains must be uniformly transferred to each lobe of the stigma of each fruiting flower. Poor or inadequate pollination of each lobe will result in misshapen and/or small fruit. Adequate pollination requires 10–15 bee visits during the single day a flower is open. The grower and beekeeper must time colony placement with the start of flowering. If colonies are placed prematurely before flowering, bees will migrate to another source of pollen and will not work the intended field. Conversely, if bees are placed in the field too late, the first flowers will not be pollinated. One or two strong colonies per acre are required to ensure sufficient bee visits to adequately pollinate a melon field and maximize yield and quality.

Bee colonies are often placed around the periphery of the field, but placement within a field may increase the frequency of flower visits with the same number of colonies. Communication between the grower and beekeeper is essential to minimize bee kill if harmful chemicals are used and colony disruption by cultural practices.

Environmental and disease factors may significantly influence flowering, pollination, and fruit set.

Plants under stress will have fewer flowers and will not set as many fruit as healthy plants. Rain, fog, strong winds, and extreme temperatures will reduce bee activity and consequently yields.

Fruit quality

Soluble solids (sugars) accumulation depends upon the plant's ability to produce sufficient quantities of sugars by photosynthesis to meet metabolic needs plus excess for fruit storage. For high fruit sugar content, it is important to have a large leaf canopy prior to fruit set to maximize photosynthesis and support fruit growth. Any factor that limits photosynthesis will also decrease sugar accumulation in fruit. Factors that limit fruit sugar content include: reduced leaf area (leaf size and number, disease, insects, mechanical damage); reduced photosynthesis (insufficient leaf area due to viral infection, foliage disease, insect damage, mechanical damage, chemical damage or air pollutants such as nitrogen oxides or ozone, cloudy weather, dust, shading by other plants, opaque sprays); water stress (dry soil, restricted root growth); and competing needs for sugar compounds within the plant (vegetative growth, repair of damaged tissue, combating disease). The highest melon sugar content is produced during warm sunny days and cool clear nights when plants are not diseased or stressed. Sugar content of melons are easily measured by expressing a few drop of fruit juice on to a refractometer. By some standards, 9% soluble solids, reported as °Brix, is the minimum industry standard allowed for shipment to preserve quality and reputation. Soluble solids of 9% are needed to achieve a grade of USA No.1, and USA Fancy Grade requires 11% along with other criteria. A good tasting netted melon contains about 14% soluble solids (Suslow et al., 2013).

Besides sugar accumulation, other characteristics associated with cantaloupe fruit quality are netting, flesh color, flesh thickness, flesh texture, aroma, and cavity size. During netted fruit enlargement, certain cells in the fruit epidermis divide to form a layer of corky tissue under the fruit surface. This corky tissue eventually protrudes through fissures on the rind surface that forms the fruit net. Favorable temperatures, plant nutrition, and moisture as well as freedom from disease, insects, or other stresses help produce a rugged attractive net that also protects the fruit from abrasion during shipment. Examples of stress conditions that reduce net formation are: insufficient or excess nutrients or moisture; poor root development; reduced photosynthetic capacity; and unfavorable growing conditions such as extreme temperatures, high humidity, excess moisture or competition from weeds. When selecting a fruit, consumers should avoid "slick" cantaloupes with poor netting. These are considered off-types because they have an abnormal appearance. The best quality cantaloupe fruit are true-to-type with heavy netting, uniform shape, and good size.

Fruit color is more stable under environmental changes than soluble solids, especially when stress occurs close to harvest. However, color intensity is decreased by plant stress during the earlier stages of fruit development. Disease, poor nutrition, water-logged root systems, extensive insect or mechanical damage, and strong weed competition (shading) reduce color. Interestingly, mild water stress, as might be present under semi-desert or desert conditions, seems to enhance fruit flesh color intensity (Rubatzky and Yamaguchi, 1997).

Flesh thickness is largely genetically controlled and is one of the most stable fruit quality characteristics. Fruit cavity size is a factor in shipping and holding ability of fruit. Fruit with a small tight cavity ships better. Rough handling of fruit, especially throwing, causes loose seed cavities, sometimes caused "shakers" because of the sound the dislodged seeds make when the fruit is rapidly moved back and forth. Large cavities may result in softer fruit that does not ship or hold as well as fruit with a small solid seed cavity.

Harvesting and marketing

The time to harvest netted melons is easily determined because the vine and the fruit naturally separate or "slip" as ripening progresses (Fig. 10.11). For local markets, netted melons are harvested at one quarter or half "slip". In other words, the fruit is harvested when the stem is only one quarter or half attached to the fruit. This varies somewhat because certain cultivars are riper at a particular slip stage than others (Suslow et al., 2013).

For distant markets, melons are sometimes harvested preslip, which is before full maturity. Consumers should avoid buying cantaloupe fruit with the stems still attached because they usually have lower sugar content and crunchy texture due to premature harvest. A "wet nose" (moist stem end of the fruit) is a good indication that the fruit was recently harvested.

Optimum harvest is more difficult to determine for melons that do not "slip" from the vine at maturity.

For example, honeydew fruit are often harvested as they reach mature size. The plant hormone ethylene is sometimes applied to ripen honeydew after harvest. Ethylene is a natural gaseous plant hormone that triggers ripening in many fruit including both netted and winter melons. Rates of ethylene production vary from 40 to 80 µl/kg/h (0.004–0.007 tsp/lb/h) at 20°C (68°F) for intact fruit and 7–10 µl/kg/h (0.006–0.009 tsp/lb/h) at 5°C (41°F) for fresh cut fruit (Suslow et al., 2013).

Fruits that are signaled to ripen by ethylene are called climacteric. The climacteric response includes increased fruit respiration, fruit softening, a change from immature to mature fruit color, and increased aroma. In the USA, cantaloupes with altered ethylene response have been genetically engineered to ripen more slowly to prolong shelf life and decrease storage losses. These cultivars were tested but never brought to market.

Controlled atmosphere storage or shipping offer only moderate benefits for cantaloupes under most conditions. With extended transit times of 14–21 days, cantaloupes are reported to benefit from delayed ripening, reduced respiration and associated sugar loss, and inhibition of surface molds and decay. Modified atmospheres of 3% O_2 and 10% CO_2 at 3°C (37.4°F) are effective. Elevated CO_2 at 10–20% is tolerated but will cause effervescence in the fruit flesh. This carbonated flavor is lost on transfer to air. Low O_2 (<1%) or high CO_2 (>20%) will cause impaired ripening, off-flavors and odors and other condition defects (Suslow et al., 2013).

Postharvest cooling

Postharvest cooling is very important to maintain fruit quality by slowing the rate of fruit respiration and ripening after harvest. In some areas, melons are harvested at night or in the early morning when fruit temperatures are naturally low to reduce field heat. After harvest, melons are forced-air cooled to 10–13°C (50–55°F) but no lower since they are chilling sensitive.

Because melons, and particularly netted melons, can harbor harmful bacteria in the corky tissues of the rind, melon fruit should be washed thoroughly before cutting. This is especially important for crops produced using animal manures, which can be a source of biological contamination along with poor worker sanitation. Fruit should be stored for less than 3 days after cutting to reduce the risk from *Salmonella* or other bacterial pathogens. A chlorinated postharvest dip treatment can provide some control of fruit rot diseases as well as human pathogens such as *Listeria*, *E. coli*, and *Salmonella*.

Diseases

There are a number of important diseases that affect melons. Growing resistant cultivars effectively prevents Fusarium wilt (*Fusarium oxysporum* f.sp. *melonis*) (races 0, 1, 2 and 1-2). Powdery mildew (*Sphaerotheca fuliginea* and *Erysiphe cichoracearum*) is controlled by fungicide applications, but modern F-1 hybrids are tolerant of most races. Downy mildew (*Peronospora cubensis*) is important in hot and humid climates and is controlled by fungicides. Gummy stem blight (*Didymella bryoniae*) is also a disease in humid and hot conditions. Seed treatment, crop rotation, and fungicides control anthracnose (*Glomerella cingulata*). Damping-off (*Pythium* sp. and *Rhizoctonia* sp.) is prevented by treating seed with fungicides, such as thiram. Bacterial wilt (*Erwinia tracheiphila*) causes symptoms similar to drought stress because bacteria clog the xylem, reducing the flow of water through the plant. Bacterial wilt is controlled by removing affected plants and by controlling the striped and spotted cucumber beetles, which vector the bacteria (Fig. 10.16).

Other diseases include: angular leaf spot (*Pseudomonas syringae* pv. *lachrymans*), Cercospora leaf spot (*Cercospora citrullina*), Alternaria leaf spot (*Alternaria cucumerina*), and scab (*Cladosporium cucumerinum*) fruit and foliage disease (Zitter et al., 1996).

Cucumber mosaic virus cucumovirus, watermelon mosaic 2 potyvirus, and zucchini yellow mosaic potyvirus all affect melon and are transmitted by aphids, in particular *Aphis gossypii*. Sources of resistance to these three viruses and also

Fig. 10.16. Bacterial wilt of cantaloupe.

to the vector *A. gossypii* are available. Other virus diseases in melon are papaya ringspot potyvirus (aphid transmitted), melon necrotic spot carmovirus (transmitted by the soil fungus *Olpidium* sp.), and beet curly top hybrigeminivirus (BCTV) (transmitted by leafhoppers) (Zitter *et al.*, 1996).

Insect Pests

Pests in melon are thrips (*Thrips palmi* and *Frankliniella* spp.), spider mite (*Tetranychus urticae*), melon aphids (*Aphis gossypii*), melon fruit fly (*Bactrocera cucurbitae*), cucumber beetles (*Diabrotica* spp.), leaf folder (*Diaphania indica*), and the leaf feeder (*Aulacophora indica*). Root knot nematodes (*Meloidogyne* spp.) can be a serious problem when melons are grown without proper crop rotation. Control by wide-spectrum soil fumigants can be effective, but these are expensive and hazardous to the environment. Other pests include squash bugs (*Anasa tristis*), leaf miners (*Liriomyza sativae*), melon worm (*Diaphania hyalinata*), and whiteflies (*Trialeurodes vaporariorum*). Please consult local recommendations for effective pest control measures in your area (Cornell, 2004).

Nutritional Values

Netted melons are a good source of carbohydrate in the form of sugar. They are also a reasonably good source of potassium and vitamin A (Table 10.4; USDA, 2011).

WATERMELON

Origin and History

The origin of watermelon is unclear but it was cultivated at least as early as 2000 BC based on evidence from the Nile Valley (Zohary *et al.*, 2012). Watermelon seeds were found in the tomb of Pharaoh Tutankhamen, as well as in other sites of the 12th Egyptian Dynasty, even though no literature or hieroglyphics depict or describe the eating of watermelons. One theory is that watermelon was derived from a perennial relative, *C. colocynthis*, which is endemic to Africa and was found in early archaeological sites before watermelon (Zohary *et al.*, 2012). However, some believe that watermelon was domesticated in Africa from putative wild forms of *C. lanatus*. The Kalahari Desert in south central Africa is a region where wild forms are still found, some of which do not have bitter fruit. The related species, *C. lanatus* var.

Table 10.4. Nutritional composition of netted melon pericarp tissue (USDA, 2011).

Nutritional value	Amount/100 g (3.5 oz) edible portion
Energy	141 kJ (34 kcal)
Carbohydrates	8.16 g
– Sugars	7.86 g
– Dietary fiber	0.9 g
Fat	0.19 g
Protein	1.84 g
Water	90.15 g
Vitamin A equivalent	169 µg (21%)
– β-carotene	2020 µg (19%)
Thiamine (vitamin B1)	0.041 mg (4%)
Riboflavin (vitamin B2)	0.019 mg (2%)
Niacin (vitamin B3)	0.734 mg (5%)
Pantothenic acid (vitamin B5)	0.105 mg (2%)
Vitamin B6	0.072 mg (6%)
Folate (vitamin B9)	21 µg (5%)
Vitamin B12	0.00 µg (0%)
Vitamin C	36.7 mg (44%)
Vitamin E	0.05 mg (0%)
Vitamin K	2.5 µg (2%)
Ca	9 mg (1%)
Fe	0.21 mg (2%)
Mg	12 mg (3%)
P	15 mg (2%)
Zn	0.18 mg (2%)
K	267 mg (6%)

Percentages are relative to USDA daily recommendations for adults.

citroides, known as citron, tsamma melon, or preserving melon, is apparently native to the Kalahari Desert region as well. David Livingstone found extensive areas covered with wild watermelon vines in central Africa during his explorations in the mid-1800s. Watermelons were also cultivated in China as early as the end of the 9th century AD (Zohary *et al.*, 2012).

Watermelon was largely unknown in Mediterranean countries until introduction by the Moors in the 13th century. The word "watermelon" first appeared in English dictionaries in the early 1600s (Mariani, 1994). Watermelons were apparently introduced to North America in 1500s, because Native Americans were found cultivating them by French explorers in the Mississippi valley.

Botany and Life Cycle

Watermelon is in the family Cucurbitaceae (Gourd Family) and the genus and species are *Citullus lanatus* (Thunb.) Matsum & Nakai, having formerly been classified as *C. vulgaris* or *C. citrullus* L. Watermelons are annual with monoecious sex expression although some andromonoecious forms exist. The long-trailing stems or vines may exceed 5–6 m (16–20 ft) in length and short internode "bush" cultivars require less space. Stems are thin, angular, grooved, and hairy, with branched tendrils that anchor the vines. Root systems are extensive, deeper than some other cucurbits, although still relatively shallow compared to deeply rooted crops such as tomato. The majority of the roots are within 60 cm (24 in) of the surface (Rubatzky and Yamaguchi, 1997).

Leaves are large, 5–20 cm (2–8 in) long, and deeply and prominently lobed. Solitary, yellow flowers range in size from 2 to 5 cm (0.8–2 in) in diameter and are open for only 1 day (Rubatzky and Yamaguchi, 1997).

Relatively large and flat, smooth watermelon seed may be colored white, tan, green, red, or black and 10–15 seeds weigh about 1 g (0.04 oz) (Lorenz and Maynard, 1988). Large-fruited cultivars may contain 500 seeds. The large cotyledons make up much of the seed volume. Watermelon seedlings like other cucurbits are epigeal with the cotyledons becoming photosynthetic organs after germination. Cotyledons are oblong in shape with an inconspicuous epicotyl developing between them.

Types and Cultivars

Watermelon fruit vary in size, shape, flesh color, and surface color among cultivars (Fig. 10.17).

Sizes range from 1–3 kg (2.2–5.5 lb) for small-fruited cultivars, sometimes referred to as "icebox" or "midget", to more than 24 kg (55 lb) for large-fruited types. Fruits of some cultivars, like 'Tom Watson', may weigh as much as 60 kg (132 lb). Most icebox cultivars are similar in size to cantaloupe and most have a thin rind that is not well suited for long-distance shipping. Birds can also peck seeds through the thin rind of some small-fruited cultivars. Icebox types are popular with home gardeners and in short-season areas because of their quick maturity. Many commercial markets prefer fruit in the 15–25 lb (7–11 kg) range, that are easier to handle and refrigerate. Fruit shapes range from round, oblong, or elongated with blocky or relatively pointed ends. The rind of mature fruit is usually smooth and can vary in thickness from less than 1 cm (0.4 in) to 4 cm (1.6 in). Exterior rind colors range from blackish green to yellow with solid, striped, or mottled coloring. Wax accumulation on the outer rind surface increases with maturity. The seed are scattered and imbedded in the edible placental tissue and there is no central cavity.

Fig. 10.17. Diverse sizes, shapes, and colors of watermelon fruit at the Taiwan Watermelon Festival.

Cultivars differ with regard to flesh texture, color, and sugar content. The placental tissue of modern cultivars is sweet and watery. The flesh texture of some fruit is fairly stringy or fibrous, particularly when over ripe. Flesh colors range from red, orange, pink, yellow, to white (Rubatzky and Yamaguchi, 1997). Red flesh color is due to the pigment lycopene while yellow color mostly comes from β-carotene and xanthophylls (Maness *et al.*, 2003). The flavor and sugar content of yellow-fleshed and red-fleshed melons are usually the same. Occasionally, flesh bitterness occurs in some heirloom fruit due to the presence of cucurbitacins, but seldom in modern cultivars. Most modern cultivars have black seeds because some consumers think a fruit with white seeds is immature. Black seeds make an attractive contrast with red or yellow flesh.

Usage

Watermelons are grown primarily for the sweet juicy fruit pericarp tissue that is chilled and eaten as slices or chunks, added to fruit salads or juiced. The rind may be preserved in vinegar as a sweet or brined pickled product. Some watermelon cultivars and *Citrullus colocynthis*, a related species, are grown exclusively for their abundant and large seeds that are roasted or boiled with flavorings such as licorice. In both Asia and the Middle East, roasted or boiled watermelon seeds are a popular snack food (Fig. 10.18).

Some watermelons are used for livestock feed and extracted juice is fermented into an alcoholic beverage. Watermelon fruit are an important source of water in some desert areas, during drought, or where drinking water is contaminated.

Citron is also called preserving melon and is a distinct inedible flesh type of watermelon that resembles small edible cultivars. 'Green Citron' *Citrullus lanatus* var. citroides has a hard tough rind that can be used for pickling or animal feed. Citron melon flesh is white, light green, or slightly pink but generally inedible due to its hard flesh and bitterness. The seed color is greenish tan. Citron melons grow wild in some sections of North America and can be weeds in commercial watermelon fields. Citron watermelons should not be confused with the candied diced citron that is an ingredient in fruitcake and made from the peel of *Citrus paradisi*, a tropical tree fruit also called citron.

Fig. 10.18. This small-fruited watermelon in Israel is grown for its edible seeds.

Production and Culture

Watermelon is a warm-season, frost-sensitive crop that requires a relatively long growing season from 75 to 120 days depending on the cultivar and environment. The optimum day and night temperatures are 32°C and 20°C (90°F and 68°F), respectively. Crop rotation helps control soil-borne diseases like *Fusarium* or nematodes. Plastic mulch, row covers, or low tunnels may increase soil and air temperatures in short-season areas. Some cultivars, like 'Klondike' and 'Peacock', are tolerant of low humidity, are better suited for production in desert areas, and have moderate drought tolerance. F-1 hybrid cultivars are more uniform, more productive, and popular despite their extra cost, although open-pollinated cultivars are still widely grown commercially in many areas.

Site selection and field preparation

Soil compaction restricts root growth, so friable, deep and well-drained sandy loam or loam soils are preferred for watermelon production. Watermelons are often grown on raised beds to improve drainage and support plasticulture production. Beds are formed so they are 15–20 cm (6–8 in) above the bottom of the adjacent furrow. Final spacing of the beds is generally 203 cm (80 in) center-to-center although other bed sizes may also be used.

Watermelons can be grown without irrigation or mulch but this is not recommended for commercial production because most areas experience water deficits sometime during the growing season that lower yields and reduce fruit quality. Watermelon is often produced on plastic mulch for weed control, to warm soil in short-season areas, to preserve soil moisture, and to reduce the leaching of fertilizer that occurs following heavy rainfall. Plasticulture can significantly increase watermelon yields. Establishment is by direct seeding or plug transplanting, especially for seedless production, through holes punched or burned in the plastic mulch. Drip irrigation or fertigation is used with plastic mulch to get sufficient water to the seed and root zone. Watermelon production with conservation tillage is technically feasible but not widely practiced in North America.

Fertilizer and nutrition

Watermelons tend to develop extensive root systems in the upper profile of the soil allowing efficient extraction of nutrients. For this reason, fertilizer requirements are moderate compared with many other vegetables. A fertility program should be developed with input from preplant soil tests and foliar analysis during the season. Nitrogen is the most commonly required fertilizer, although P is sometimes needed to promote seedling growth and maximum productivity and fruit quality. Potassium increases rind thickness and cracking resistance and fertilization is often needed on lighter soils. A watermelon crop will remove approximately 196, 28, and 196 kg/ha (175, 25, and 175 lb/acre) of N, P, and K, respectively, from the soil. General fertilizer recommendations for watermelon production range from 67 to 168, 45 to 168, and 112 to 224 kg/ha (60–150, 40–150 and 100–200 lb/acre) for N, P, and K, respectively (Lorenz and Maynard, 1988). The actual fertilization requirement depends on soil test results and plant utilization during the season. The difference between the amount applied and the actual crop need is due to leaching losses for N and K, soil binding which renders a portion of the P unavailable, and N gained from organic matter decomposition, the nitrogen cycle, and microbial fixation. The ideal soil pH is between 6.0 and 6.5 although a range from 5 to 7 is acceptable.

Fertilizer, as liquid or granular, is commonly applied as twin bands 15 cm (6 in) deep and 10–15 cm (4–6 in) to either side of the seed lines before planting, at planting, or in drip irrigation lines as part of a fertility program. Banding of fertilizer near the seed is preferred over broadcasting since the rows are far apart making broadcast applications inefficient. Nitrogen is commonly applied as a side dressing or as fertigation, particularly when vines start to spread. Petiole is the most effective means of diagnosing the nutrient status of the crop during the season. Sufficient values for N in the petioles sampled from the sixth leaf from the growing tip during early fruit development range from 5,000 to 7,500 ppm, P values range from 1,500 to 2,500 ppm, and K values range from 3 to 5%.

The goal for optimized production is to grow plants with a large canopy, while maintaining healthy leaves as long as possible. This strategy maximizes the photosynthetic capacity of the plant, which maximizes yield potential and fruit sugar content. Canopy development and health is aided by maintaining sufficient levels of N, P, K, S, Ca, Mg, B, Cu, Fe, Mn, Mo, and Zn throughout the growing season. Watermelons should not be over-fertilized especially with N. Excessive N increases production costs, may damage the environment by leaching into the water table, and may inhibit fruit set and delay harvest. The N level must be low enough at flowering that the plant will form fewer new leaves once fruit development begins. This allows more sugars to accumulate in the fruit, rather than being used for new vegetative growth.

Field establishment

Seeded watermelon cultivars are often direct-seeded in warm-season areas. Seedless watermelons are transplanted from plugs because they germinate poorly in the field under suboptimal conditions and seeds are very expensive. Bare-rooted watermelons do not survive well in the field because they are sensitive to shock, so most are established with plug transplants. However, both seeded and seedless cultivars are often transplanted in short-season areas. Watermelon transplant scions may be grafted on to squash or gourd rootstocks to increase resistance to soil-borne diseases and nematodes, improve drought resistance, or enhance plant growth (Fig. 10.19).

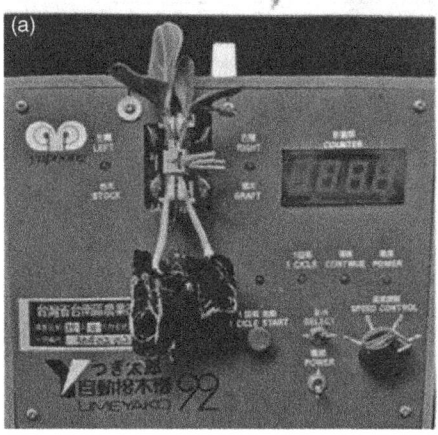 

Fig. 10.19. Simple machines can join rootstocks and scions to make a successful graft union and reduce labor (a). A watermelon seedling grafted just above the cotyledons by removing the apical meristem of the rootstock and replacing it with a scion of a production cultivar (b).

Grafting has long been used as a disease control strategy for greenhouse production of cucurbits and in the Middle East and parts of Asia where acreages are relatively small. With greater emphasis on sustainable production using fewer synthetic chemicals and decreased availability of soil fumigants, grafting of cucurbits is rapidly gaining popularity in other countries as well. The use of machines to aid the grafting process reduces labor (Fig. 10.19).

Watermelon seeds are planted 2–4 cm (0.8–1.6 in) deep. Most cultivars have a germination optimum between 25–35°C (77°–95°F), which will result in emergence in less than 1 week, but between 12°–20°C (54–68°F) germination is slow and erratic and may require 2 weeks or more depending on seed vigor. Seed fungicide treatments are recommended to control damping-off disease, especially when planting into cool wet soils (Lorenz and Maynard, 1988).

Watermelon vines spread rapidly once established so wide spacing is typical for watermelon production (Fig. 10.20). Except for brush-type cultivars, spacing ranges from 2.5–5.0 cm (1–2 in) in-row and from 2–3 m (6.5–10 ft) between rows. Seeds can also be planted in hills rather than rows. Hills of three or four seeds can be planted at equidistant spacing 2–3 m (6.5–10 ft) apart. Cultivars with long vines require wide row spacing and are susceptible to wind damage. Short internode or bush-type cultivars can be spaced closer at half the distance of vining cultivars. Fields may be seeded at higher density to ensure a good stand before thinning to a final spacing at the two- to four-leaf stage. Closer spacing generally produces smaller fruit and wider spacing larger fruit. Plant populations can range from 3,200–8,000/ha (1,296–3,239/acre) (Lorenz and Maynard, 1988). Vines are sometimes trained toward row centers to facilitate cultivation and harvesting.

Flowering and pollination

Watermelon is "day neutral" and flowers when plants are sufficiently large. Most commercial watermelon cultivars have monoecious sex expression producing separate male and female flowers on a plant. The most effective pollinators are bees. Hundreds of pollen grains must be equally deposited on all lobes of the stigma of each female flower to ensure set and full fruit development. Environmental and disease factors can significantly influence flowering, pollination, and fruit set. Plants under stress will have fewer flowers and will not set as many fruit as healthy plants. Rain, strong winds, and high or low temperature extremes also will reduce bee activity, fruit set, and yields. To transfer sufficient pollen, about ten bee visits are required on the day the flower is open. If pollination is inadequate or if fruit load is excessive, flowers will abort. Poorly pollinated fruit may sometimes set but are often misshapen.

If pollination is successful, the ovary of the female flowers will enlarge rapidly. Early developing fruit have an inhibitory influence on the development of late-formed fruit. Most vines of large-fruited cultivars can only support two or three fruit at a time and female flowers forming

Fig. 10.20. Watermelon production field with full vine cover and developing fruit.

after fruit set will abort. Fruit thinning is sometimes practiced to improve the size and sugar accumulation of remaining fruit. Small-fruited cultivars may support more fruit per vine.

The bee population needed to maintain that frequency of flower visitations is normally four to five strong colonies per hectare (1–2/acre). Placement of bee colonies within a field rather than along the periphery may double the flower visitation frequency with the same number of colonies. Good cooperation between the grower and beekeeper is essential for making sure the bees are present in time for flowering and to minimize bee kill from insecticide applications if used.

Seedless watermelon

Kihara (1951) first developed seedless watermelons in Japan during the early 1950s. Tetraploid (4n) and diploid (2n) plants are crossed to produce triploid (3n) seed (Kihara, 1951). Diploid watermelons are treated with colchicine, a natural product from autumn crocus that doubles the chromosome number, to create tetraploid plants (Deppe, 1993).

When performing the 4n × 2n cross, pistillate flowers of the 4n plant are pollinated with 2n pollen. Fertilization produces fruit with 3n seeds that are planted for seedless crop production. Fruits of watermelon are seedless because they are triploid and normally sterile so fertilization does not occur. On the rare occasions that fertilization does occur, the embryo usually aborts. Small, aborted seeds may be found in the placental tissue of seedless fruit, but they are soft with no embryo, and can be eaten like the seeds in an immature cucumber.

Pollination is still required with a diploid cultivar to stimulate ovary growth for parthenocarpic fruit development. About 12–18% of the field must be interplanted with a diploid cultivar that serves as a pollen source for the seedless triploid cultivar. Sometimes a standard diploid cultivar is planted to pollinate two to four rows of triploid plants. When this approach is used, a pollinator cultivar should be of a contrasting color or shape to distinguish the seedless fruit from the selfed diploid that are sold as seeded watermelon. Recently, diploid pollinator lines that only produce male flowers have been developed for seedless watermelon production (Freeman *et al.*, 2007).

Since the male pollinator only produces staminate flowers, fewer plants are required for adequate pollination. These pollinator lines make seedless production more efficient because a portion of the field does not need to be devoted to a lower-value seeded fruit pollinator cultivar.

The chromosome imbalance caused by triploidy does not significantly affect plant or fruit development. Early seedless cultivars had a higher percentage of fruit defects such as triangular shape, large blossom scars, hollow heart, light flesh color, off flavors, and delayed maturity. However, modern seedless cultivars have higher quality and greater uniformity with few if any defects.

Because triploid seeds germinate poorly, particularly at low temperatures, the seeds are more expensive and seedlings are not vigorous, seedless fields are often established by plug transplants to minimize shock. Seedless transplants are often germinated at a high temperature (32°C; 90°F) and the temperature is then lowered to 22–23°C (72–73°F) for seedling growth. The coats of triploid seed tend to adhere to the seed coats of emerging cotyledons, which can distort and cause poorly developed seedlings. Orienting seed with the radicle end up when planted may help seedlings lose their coats during emergence.

Irrigation

Despite the fact that watermelon has an extensive root system, a consistent supply of water particularly during fruit development maximizes both yield and fruit quality. Preplant or post-plant irrigations will ensure rapid seed germination, emergence, and stand establishment. Irrigation is often applied sparingly until the pre-vining stage to avoid cooling the soil, inhibit damping-off diseases, and encourage deep-root formation. Watermelon crops require from 400–700 mm (16–28 in) of rainfall or irrigation for maximum productivity. The general recommendation for at least 2.5 cm (1 in) water/week either from rain, irrigation, or both applies and more water is required on sandy soils or when mulch is not used.

Drip irrigation, either with or without the use of plastic mulch, conserves water and provides uniform application directly to the root zone at a slow rate. Sprinkler irrigation is often used for stand establishment, but this is not the best system once the vine canopy has developed because it may increase vine and fruit disease. Level production fields in the southwestern USA, western USA, and Mexico are furrow irrigated but this traditional inexpensive system uses water inefficiently and may lead to salt accumulation in soil.

Harvesting and marketing

Cultivars differ greatly in their times to fruit maturity. Some small-fruited cultivars mature 75 days after planting while others may require 140 days or longer depending on the environment. Ideally, fruit are harvested when the sugar concentration is highest. Sugar is the primary determinate of quality along with flesh texture. There are no other well-defined flavor components other than sugar. Soluble solids (sugar) readings should be taken from the center because sugars are not always evenly distributed within the fruit. Higher sugar accumulation often occurs on the side of the fruit facing the sun away from the ground. Soluble solids of 12% to more than 13% can be achieved by some cultivars. Fruits with less than 7% soluble solids generally do not have good flavor. Fruit maturing when daily mean air temperatures are below 21°C (70°F) has poor quality. Sugar content does not increase after harvest. Watermelon do not ripen once harvested so it is important to allow fruits to fully mature on the vine to achieve their maximum sugar content.

Watermelons do not abscise from the vine when ripe like cantaloupe. In commercial fields, harvesting crews determine maturity by examining the color of the fruit's ground spot. The ground spot is white where the fruit rests on the soil and the rind does not develop chlorophyll. As the melon ripens, the ground spot changes from white to pale yellow. When the fruit is ready for harvest the ground spot will be pale yellow, and will turn dark yellow when overly ripe (Fig. 10.21).

The ground spot color of ripe fruit varies somewhat with cultivar. Also during ripening the color of the rind will fade from glossy to dull green and the tendril dies at the node where the fruit is attached to the vine. These are the most reliable indications of maturity. Tapping or thumping to hear a dull sound is subjective and not a reliable method for determining fruit maturity. A dull sound may be the result of "hollow heart", a common disorder that occurs when an open space forms in the fruit tissue. Soluble solids measurements or tasting, although destructive, are the best ways to evaluate fruit maturity and quality. When sampled fruit are sufficiently ripe, it is assumed that fruit of similar size and age in the same field are of equivalent maturity and ready for harvest.

Fruit tissue becomes "grainy" around the seed when overripe. Delayed harvest or prolonged storage causes the flesh texture to become mealy and stringy as cells collapse when over mature.

Postharvest handling

Fruit should be cleanly cut at the stem end rather than pulled from the vine. Fruit rinds may give the appearance of strength but actually are susceptible to cracking from compression or impact shock and should not be stacked on the stem or blossom ends. Watermelons may be shipped in bulk or in cardboard or plastic pallets. Small-fruited cultivars can be packed in cardboard pallets to prevent fruit damage from compression.

Fig. 10.21. Observing the change of ground spot color from white to yellow is a reliable way of determining ripeness.

Fruit are most turgid and susceptible to cracking during early morning hours. Flesh firmness and rind toughness are important cultivar characteristics for shipping without damage. To maintain harvest quality, fruit should be quickly cooled and stored at 13–16°C (55–61°F). Watermelons are not suited to long-term storage, but can be held at 80% RH for 2–3 weeks with little loss of quality if optimum temperatures are maintained. Extended storage at 10°C (50°F) or less results in quality loss due to chilling injury. The waxy rind limits desiccation. Watermelons are not climacteric, and immature fruit cannot be ripened off the vine with ethylene treatment. Watermelons should, ideally, not be stored or transported together with climacteric fruit that give off ethylene at harvest such as apple, cantaloupe, or tomato because ethylene will cause pitting of the skin, flesh breakdown, and black rot, shortening the storage life.

Sale of precut watermelons has increased dramatically in some markets. Precut-watermelons offer greater convenience and an appropriate quantity for single people and couples. Precut watermelons are sold ready-to-eat in resealable containers with the rind removed. Precutting reduces preparation time and waste that the consumer must deal with. Precut products also allow the consumer to evaluate the flesh color and texture prior to purchase.

In Japan, watermelons are sometimes grown in square molds (Fig. 10.22a). Molded watermelons command a premium price and are often given as presents for special occasions. Watermelons are also used for carving to make table decorations. Very complex designs and figures can be sculpted using the different colored layers of a watermelon fruit (Fig. 10.22b).

Fig. 10.22. Watermelons can be molded into different shapes by placing rigid containers around the developing fruit (a). Watermelon fruit may become elaborate art objects when carved (b).

Diseases

A number of different diseases affect watermelon, many of which are most prevalent under humid conditions. Damping off is favored by cold wet conditions and is caused by *Phythium* spp., *Rhizoctonia* spp. or *Fusarium* spp. Damping off can infect seedlings in the field as well as greenhouse transplants. Fungicide seed treatments and limiting moisture can effectively control damping off. Phytophthora root and fruit rot (*Phytophthora capsici*) can infect maturing fruit (Noh *et al.*, 2007). Gummy stem blight (*Didymella bryoniae*) is a major fungal disease that can cause a number of disorders in watermelon fields or greenhouse transplants including damping off, crown rot, leaf spot, stem canker, and fruit rot (Gusmini *et al.*, 2005). Gummy stem blight causes round or irregular, brown lesions on the leaves and may also attack the stem causing elongated water-soaked areas that become gray. Gum may ooze from the stem cracks and plant dieback is another identifying feature.

Anthracnose (*Glomerella cingulata* var. *orbiculare*) attacks watermelon foliage and fruit. In severe cases, leaves die back leaving only bare stems. Anthracnose is most common under warm rainy conditions.

Downy mildew (*Pseudoperonospora cubensis*) occurs on watermelons worldwide and attacks the foliage giving rise to yellow spots, which coalesce into brown areas causing leaf curl and resulting in a scorched appearance of the crop. Powdery mildew (*Sphaerotheca fuliginea* and *Erysiphe cichoracearum*) is a widespread watermelon foliage disease (Tomason and Gibson, 2006). Fusarium wilt (*Fusarium oxysporum*) causes infected vines to wilt. In severe cases, the entire root system may become brown and a soft rot may develop near the crown.

Watermelon is also susceptible to virus diseases that are often transmitted by aphids. Watermelon mosaic virus causes leaf mottling and stunted plant growth. Other virus diseases that affect watermelon include: cucumber green mottle mosaic, cucumber mosaic, cucumber vein yellowing, cucurbit aphid-borne yellows, cucurbit chlorotic yellows, squash leaf curl, squash mosaic, tomato spotted wilt, watermelon chlorotic stunt, watermelon silver mottle, zucchini yellow mosaic, and cucurbit yellow stunting disorder.

Fruit blotch is a particularly devastating disease throughout the world caused by the bacterium *Acidovorax avenae* subsp. *citrulli*. There are several symptoms, including leaf lesions and fruit spots that may cover the entire fruit surface in severe cases. Fruit blotch bacteria may eventually cause the fruit to rot and can be transferred by infected seed, transplants or weeds (Lessl *et al.*, 2007). Another bacterium that may affect watermelon is *Xanthomonas cucurbitae*, which causes bacterial pumpkin spot. Other significant watermelon diseases include black root rot (*Chalara elegans*), fruit rot (*Choanephora cucurbitarum*), Alternaria leaf spot (*Alternaria alternata*), and cucumber blight (*Alternaria cucumerina*) (Zitter *et al.*, 1996).

Insect Pests

Several species of cucumber beetle attack watermelon, including *Diabrotica undecimpunctata*. Both the beetles and larvae damage watermelon; the beetles feed on the stems and foliage and larvae feed on roots. Cucumber beetles are vectors of bacterial wilt but watermelon is not as susceptible to the disease as *C. melo* or *C. sativum*. In India, the red pumpkin beetle (*Aulacophora foveicollis*) is a significant pest, which feeds on the immature leaves and flowers. *Epilachna* beetles (*Epilachna* spp.) are common pests of watermelon in Africa. The larvae and adults feed on the leaves and chew holes in the stems and fruit (Cornell, 2004).

The melon aphid, *A. gossypii*, and the green peach aphid, *Myzus persicae*, may infest watermelon crops. Heavy aphid populations seriously weaken the plant and cause yellowing and wilting. Aphids also vector several viruses, which reduce plant growth and fruit quality. Several species of thrips may also infect watermelon crops (Cornell, 2004).

Melon fruit fly (*Bactrocera cucurbitae*) is a pest of watermelon in some countries. Adult flies lay eggs, which hatch into larvae within 1 week and feed on the fruit, causing damage and also allowing entry of fruit-rot pathogens. Watermelon crops may be covered with netting to protect against fruit flies. Mites (*Tetranychus urticae* and related species) are pests of watermelon, particularly when conditions are warm and dry. Mites cause distortion of new growth and chlorotic spotting as well as fine webbing on the foliage. Armyworms, sometimes referred to as melonworm or rindworms, feed on the rind of developing fruit. There are several types of armyworms that feed on watermelon fruit including southern (*Spodoptera eridania*), beet (*Spodoptera exigua*), and fall (*Spodoptera frugiperda*) (Cornell, 2004).

Seed corn maggots (*Delia platura*, formerly *Hylemya platura*) are a pest favored by early planting dates, heavy cover crops, and cool-wet weather. Seed corn maggots attack a wide range of horticultural crops including beans, peas, cucumber, melon, onion, corn, pepper, potato, and watermelon. Although these maggots feed primarily on decaying organic matter, they will feed on seeds and seedlings of watermelon and other crops. Similarly cutworms (*Agrotis segetum* and *A. ipsilon*) reduce watermelon stands by feeding on young seedlings (Cornell, 2004).

Economic Importance and Production Statistics

According to the United Nations Food and Agricultural Organization (FAO), watermelon is the most widely grown cucurbit with an estimated 3,413,750 ha (8,435,560 acres) worldwide in 2011 and production totaling 98,047,947 metric tonnes (108,079,361 short tons) of fruit (FAO, 2011). An estimated 1,200 watermelon cultivars are grown in at least 96 countries. Asia leads the world in watermelon production with two-thirds of the volume, followed well behind by European (13%) and African (6%) production. China, with 23% of the world's watermelon production, provides most of the Asian volume. In 2009, China produced 65,002,319 metric tonnes (71,652,792 short tons) of watermelon on 1,776,579 ha (4,390,022 acres). Other major producing countries were Turkey (3,810,210 metric tonnes (4,200,038 short tons)), Iran (3,074,580 metric tonnes (3,389,144 short tons)), the USA (1,819,890 metric tonnes (2,006,085 short tons)), and Egypt (1,500,000 metric tonnes (1,653,467 short tons)).

Watermelon consumption per person in the USA for 2011 was 6.8 kg (15 lb)/year. In the USA, per capita watermelon consumption has steadily declined over the past 50 years likely due to competition from a wide range of convenience-packaged beverages and frozen snacks. The per capita consumption of watermelon remains highest in countries of the Middle East. In Egypt for example, consumption per person is nearly 45 kg (100 lb)/year. Of the watermelons used in the USA, 22% were imported primarily from countries in Central America. Texas produces the largest acreage of watermelons in the USA followed by Florida, while California, Georgia, and Arizona are other major producing states.

Nutritional Values

As the name suggests, watermelon consists mostly of water (Table 10.5). Watermelon fruit also contain soluble sugar and significant concentrations of some minerals. Lycopene, a pigment response for red flesh color, is also an antioxidant.

Table 10.5. Nutrient composition of watermelon fruit tissue (USDA, 2011).

Nutrient	Amount/100g (3.5 oz) edible portion
Water (g)	92.6
Protein (g)	0.5
Fat (g)	0.2
Carbohydrate (g)	6.4
Fiber (g)	0.3
Ca (mg)	0.7
P (mg)	10
Fe (mg)	0.5
Na (mg)	1.0
K (mg)	100
Ascorbic acid (mg)	7.0
Vitamin A (IU)	590
Thiamine (mg)	0.03
Riboflavin (mg)	0.03
Niacin (mg)	0.2

PUMPKINS AND SQUASH
Origin and History

The genus *Cucurbita* is believed to be native to tropical America. Archaeological evidence suggests squash may have been first cultivated in Mesoamerica some 8,000 to 10,000 years ago (Roush, 1997; Smith, 1997). *Cucurbita pepo* likely originated in what is today the southwest USA and Mexico. *Cucurbita argyrosperma* may have originated in Central America and southern Mexico, while *C. moschata* may have developed further south in Central America and northern South America. Similarly, *C. maxima* apparently originated in central and southern regions of South America (Zeven and Zhukovsky, 1975).

Species of *Cucurbita* are among the most ancient crops cultivated in the Americas. The four species *C. pepo*, *C. argyrosperma*, *C. moschata*, and *C. maxima* only exist in cultivation. The close relationship with humans along with evidence of these species in the ruins of ancient American civilizations, indicate their importance in the development of Native American agriculture. Squash was one of the "Three Sisters", maize (corn), beans, and squash, planted by Native Americans.

Botany and Life Cycle

Cucurbita spp. are generally warm-season, frost-sensitive annuals, although *Cucurbita foetidissima* HBK is a perennial. *Cucurbita foetidissima* includes buffalo gourd, which has a variety of common names including calabazilla, chilicote, coyote gourd, fetid gourd, Missouri gourd, stinking gourd, wild gourd, and wild pumpkin. The buffalo gourd is a xerophytic tuberous plant found in the southwestern USA and northwestern Mexico. The fruit is eaten cooked like a squash when very young but becomes too bitter for vegetable use at maturity. The buffalo gourd grows fast, requires little water, and produces a large massive underground tuber that has medicinal properties and has been proposed as a feedstock for biofuel production (Curtin *et al.*, 1997). The seeds can be used as food and are high in both lipid and protein (Berry *et al.*, 1976).

Cucurbita ficifolia, known as Bouche, malabar, or figleaf gourd, is a climbing vine that reaches from 5–15 m (16–49 ft) in length. The figleaf gourd is a frost-sensitive annual in temperate climates and a perennial in tropical zones (Andrés, 1990). In nature, the figleaf gourd grows in moist regions at altitudes from 1,000–3,000 m (3,281–9,843 ft), but it can be cultivated in other climates because of its hardy root system and disease resistance. The figleaf gourd is used as a disease-resistant rootstalk for grafting to cucurbits that are susceptible to root disease. The flowers and tender shoots are used in Mexico as vegetables. The most nutritional part of *C. ficifolia* is its fat- and protein-rich seeds (Andrés, 1990).

Other species of *Cucurbita* have been identified growing wild. These are not commercially important on an international scale, although they may be grown and consumed locally and may cross with other species. Three of the most significant species include *C. lundelliana* Bailey, which is native to Central America, *C. andreana* Millan, which is cross-fertile with *C. maxima* and is believed to be its wild progenitor from South America, and *C. texana* Gr. Millan, which is cross-fertile with *C. pepo* and native to southern USA and northern Mexico (Robinson and Decker-Walters, 1997).

The primary species of economic importance include: *C. pepo* L., which includes many diverse types including pumpkin, winter squash, summer squash, gourds; *C. moschata* Duch., which includes pumpkin and winter squash; *C. maxima* Duch., which includes pumpkin and winter squash; and *C. argyrosperma* Pang., which includes pumpkin, winter squash, and gourds (Robinson and Decker-Walters, 1997).

Summer squash are defined as fruits harvested in an immature stage before the rind becomes hard (Fig. 10.23).

Winter squash are physiologically mature fruit with a hard rind that cannot be penetrated with a fingernail, and contains viable seeds at harvest. They can be stored for several months at room temperature if properly cured and are free of mechanical damage and disease. Pumpkin is another term for winter squash in much of the world. However, in North America pumpkins are essentially a winter squash with a bright orange rind and stringy flesh grown as a decoration for the fall festival Halloween or made into pies. For Halloween celebrations, orange or white mature pumpkin fruits of predominately *C. pepo* and *C. maxima* are decorated or carved (Fig. 10.24).

Differences among the various species of *Cucurbita* are subtle based on seed anatomy and leaf and stem characteristics (Table 10.6). Although leaf shape and surface marking can vary within a species, a combination of stem, androecium, peduncle, flesh texture, and seed features is used to differentiate the species. Seed colors can be white, tan,

Fig. 10.23. 'Zucchini', an example of a bush type summer squash, grown on plastic mulch and drip irrigation in Nova Scotia. Fruits are classified as summer squash because they are harvested when immature a few days after anthesis.

Fig. 10.24. In parts of North America, pumpkin refers to orange cultivars of winter squash that are decorated or carved like this jack-o'-lantern to celebrate the Halloween holiday on 31 October. A jack-o'-lantern is a pumpkin whose insides have been removed and replaced with a light or candle.

brown, or black depending on the species. Examples of characteristics that differentiate the major cultivated species of *Curcurbita* are listed in Table 10.6.

The cutivated *Cucurbita* described in Table 10.6 are monoecious and most have long trailing vines and a prostrate growth habit unless supported. Certain summer squash cultivars of *C. pepo* have short internodes and a bush growth habit (Fig. 10.23). Taproots are moderate to deep with extensive shallow horizontal development. Flowers are bright yellow, borne singly in leaf axils, and seldom open for more than 1 day. Most *Cucurbita* species are day neutral, although in a few flowering is affected by day length.

Uses

Cucurbita fruit may be very large, particularly fruit of *C. maxima* and *C. argyrosperma*. Large-fruited cultivars of *C. maxima* bred for exhibition have weighed greater than 800 kg (1,700 lb) (Fig. 10.25).

Cucurbita pepo is probably the most versatile and widely used species. Cultivars of both summer and winter squash exist in *C. pepo*. Some bush cultivars of squash are grown for their immature fruits (summer squash, courgettes, vegetable marrow) or at anthesis (baby squash) that are steamed, boiled, baked, or fried. The time-to-harvest for this stage of development depends on the environment and cultivar but generally ranges from 35 to 50 days.

Table 10.6. Characteristics that differentiate species of *Curcurbita* (Rubatzky and Yamaguchi, 1997).

Cucurbita species	Leaves	Fruit stems (peduncle)	Seeds
C. pepo	Prickly, deep sinuses between lobes	Not noticeably flaring or enlarged at attachment to fruit	Tan colored Seed scar horizontal or rounded
C. moschata	Not prickly, sinuses indistinct or absent, lobes pointed; with rare exceptions, leaves soft hairy, with white spots at the intersections of veins	Distinctly five-sided, regularly grooved, hard Flaring at attachment to fruit Roughly cylindrical, not definitely, irregular grooves, not flaring or noticeably enlarged at attachment to fruit; hard	Color, grayish white to tan; margin thickened deeper in color and different texture from body of seed; seed scar slanting, rounded, or horizontal
C. maxima	Lobes rounded; rough hairy, kidney shaped, white spots never present	Cylindrical, soft and spongy, yielding readily to thumbnail	Margin, when present, identical in color and texture with body of seed; white or brown to bronze, seed scar slanting

Family Cucurbitaceae

Fig. 10.25. Prize-winning *C. maxima* fruit in a large pumpkin competition in Bradford, Ohio. The fruit in the background weighed 413 kg (910 lb).

Cucurbita pepo includes many cultivars of pumpkin and gourd that are widely used as ornamentals. Gourds have distinctive shapes and color, but are not edible because of their very hard, thin rinds. Ornamental gourds of *C. pepo* with considerable variation in color and shape are also used as decorations to celebrate Halloween and Thanksgiving holidays in North America. The flesh of physiologically mature pumpkin and squash fruits may be boiled, steamed, or baked for consumption (winter squash). The flesh may also be creamed, mixed with spices and eaten as a pudding or as pie filling. Pumpkin pies are traditionally made for fall and winter holidays in North America. While pumpkin pies are often made from small and medium-sized fruit of *C. pepo* or *C. maxima* some prefer using butternut squash (*C. moschata*) because they have less fiber, small seed cavity, mild flavor, and intense orange color (Fig. 10.26).

Some *C. pepo* cultivars such as 'Lady Godiva', 'Streaker', 'Triple Treat', 'Eat All', 'Sweetnut' and 'Hull-less' produce "naked seed". The naked-seeded pumpkin was derived from natural mutants whose seed coats include all tissue layers, but secondary wall thickening is reduced in the outer tissues (epidermis, hypodermis, and sclerenchyma). As mature hull-less seed dry, the outer tissues collapse, producing a thin seed coat that can be eaten without decoating

Fig. 10.26. Butternut is a popular and distinctive winter squash type that has excellent qualities for making pies and baking.

(Stuart and Loy, 1983). Because the seed coat is formed from maternal tissue, cross-pollination of naked-seeded cultivars does not affect coat development. Although the feature improves seed edibility, the characteristic makes seed more susceptible to mechanical damage and decay after planting.

'Spaghetti squash' also called "vegetable spaghetti" is another unique edible cultivar of *C. pepo*. After cooking, the edible pericarp tissue can be divided into loose strands resembling spaghetti but with crisp texture different from pasta.

Economic Importance and Production Statistics

World production of pumpkin, squashes, and gourds in 2009 was estimated at 22.1 million metric tonnes (24.4 million short tons) produced on 1.6 million ha (3,953,686 acres) (FAO, 2011). The greatest production was in Asia (14.4 million metric tonnes (15.9 million short tons)), with China being the leading producer (6.5 million metric tonnes (7.2 million short tons)). Total production in Europe was about 2.8 million metric tonnes (3.1 million short tons), with the Russian Federation, Ukraine, Italy, and Spain as the top four producers (FAO, 2011). Africa produced 1.9 million metric tonnes (2.1 million short tons), South America 0.7 million metric tonnes (0.8 million short tons), and North and Central America 1.4 million metric tonnes (1.5 million short tons), respectively. In North America, the USA is a significant producer with 0.7 million metric tonnes (0.8 million short tons). Winter squash and pumpkin are grown primarily in the northern USA. Illinois, New Jersey, California, Indiana, New York, Ohio, Michigan, and Pennsylvania are leading producers. Illinois has significant processing industry for producing pumpkin pie filling and other products totaling about 3,600 ha (8,000 acres) (Fig. 10.27).

Summer squash are grown throughout the USA with Georgia, Florida, and California being the leading producers.

Nutritional Values

Except for vitamin C, winter squash are generally more nutritious than summer squash (Table 10.7). Winter squash are a typically a good source of carbohydrates and vitamin A.

Production and Culture

Growth and development

Winter and summer squash and pumpkins are warm-season crops that are sensitive to cool temperatures and frost intolerant. Most cultivated *Cucurbita* are well adapted for growth at temperatures from 18–30°C (64–86°F) and are damaged by chilling at temperatures below 13°C (55°F). Although most are day neutral, *Cucurbita* spp. generally do not grow well in the wet tropics, although

Fig. 10.27. A pumpkin field windrowed for mechanical collection and processing into pumpkin pie filling.

certain forms of *C. moschata* are adapted to tropical conditions. Summer squash production is more widely dispersed than winter squash (Rubatzky and Yamaguchi, 1997).

Pumpkins and squash can be grown on a wide range of moderately fertile and well-drained soils. Peat and heavy clay soils are not recommended. Clay soils generally have poor aeration and restricted drainage, which inhibit root growth and increase fruit rot. Maximum yields are achieved on medium-textured soils with high water-holding capacity.

A crop rotation cycle of several years between planting members from the family Cucurbitaceae is recommended if pathogen populations are high and soils are not fumigated between crops. When possible, grasses, corn, or sorghum are good rotation crops. However, care must be taken that there are no carryover herbicide residues, which might inhibit growth. Pumpkins and squash are sensitive to acid conditions and salinity. A pH range from 6.5 to 7.5 is ideal for good growth and yields (Rubatzky and Yamaguchi, 1997).

Seed are planted about 2.5 cm (1 in) deep in heavy soils and about 5 cm (2 in) deep in sandy soils. Soil temperatures should be above the minimum of 15°C (59°F) for seed germination and at 30–35°C (86–95°F) emergence can occur within 1 week for high vigor seeds. In some tropical areas, plants occasionally are propagated from cuttings (Rubatzky and Yamaguchi, 1997). Pumpkins and squash are sometimes transplanted from plug trays with a root ball intact in short-season areas.

Plastic mulch is sometimes used for pumpkins and winter squash production primarily for weed control, to keep fruit clean and for moisture conservation since earliness is not a big issue except for summer squash production. Clear or infrared transmitting IRT is used in northern areas to increase soil temperatures especially for early season planting of summer squash (Fig. 10.28). Planting through mulch is by direct seeding or occasionally transplanting through holes punched or burned through the plastic. Drip irrigation or fertigation is used with plasticulture.

No-till pumpkin production works well and is rapidly gaining popularity. No-till pumpkins are direct seeded or transplanted when soil is sufficiently warm in the early summer directly into killed cover-crop residue, such as small grain-straw mulch, which serves as a weed barrier, conserves moisture, and keeps fruit clean.

Spacing depends on whether cultivars are bush or vining. Wide plant spacing accommodates spreading vine growth. Spacing within rows varies greatly depending on plant growth habit and the desired fruit size, number, and yield. Thus, spacing from 50–150 cm (20–59 in) in rows and from 2–3 m (6.6–9.8 ft) between rows is common. Wide spacing enables intercropping cultivation, which is used

Table 10.7. Nutritional composition of summer and winter squash (amount/100 g (3.5 oz) edible portion; USDA, 2011).

Nutrient	Summer squash	Winter squash
Water (%)	94	89
Energy (kcal)	20	37
Protein (g)	1.2	1.5
Fat (g)	0.2	0.2
Carbohydrate (g)	4.4	8.8
Fiber (g)	0.6	1.4
Ca (mg)	20	31
P (mg)	35	32
Fe (mg)	0.5	0.6
Na (mg)	2	4
K (mg)	195	350
Ascorbic acid (mg)	14.8	12.3
Vitamin A (mg)	196	4,060

Fig. 10.28. A maturing pumpkin production field using black plastic mulch and drip irrigation.

with cucurbits in some areas. Pumpkins and squash can also be planted in "hills" of three to five seeds each spaced 2–3 m (6.6–9.8 ft) apart. Bush cultivars are spaced closer with populations as much as two to three times greater than for vining cultivars.

The large leaf area of *Cucurbita* results in high evapotranspiration. However, many cultivars are drought tolerant due to their moderately deep root system and extensive horizontal root proliferation near the surface. Nevertheless, because of the high moisture requirements of these crops, soils with high water-holding capacity supplied with at least 2.5 cm (1 in) of moisture per week during active periods of growth is needed. Vining *Cucurbita* crops require from 500–900 mm (20–35 in) of water to produce a high-yielding crop. Summer squash with less extensive root systems are more easily stressed during periods of drought.

Fertilizer and nutrition

Since pumpkins and squash efficiently explore the soil for water and available nutrients, fertilizer requirements are moderate compared to many other vegetable crops. A high yielding crop of winter squash will remove roughly 168N-28P-168K kg/ha (150N-25P-150K lb/acre) from the soil depending on the cultivar, soil and environment (Lorenz and Maynard, 1988). A summer squash crop may remove slightly less nutrients. Fertilizer should be applied in accordance with soil tests before planting and foliar test during the season. Plant tissue analysis of petioles is the most effective means of diagnosing the nutrient status of the crop in the field.

Nitrogen is the most commonly required fertilizer, although P is sometimes needed to promote early-season growth particularly in cool soils, maximize production and achieve high fruit quality, especially in alkaline soils. Nitrogen is applied as fertigation in response to tissue analysis or as two side-dressings, the first at the two- to four-leaf stage and the second at vine spread. Care must be taken to not over-fertilize with N, particularly early in the season prior to flowering and fruit set. Excessive N favors vegetative growth over reproductive growth, which can inhibit fruit set. The N level must be low enough by the time of flowering so that the plant will form fewer new leaves after fruit set and growth begins. This allows more sugars to go to the fruit, rather than excessive vegetative growth.

Some mineral soils contain adequate K, but deficiencies are common on light or infertile soils. Potassium should be applied in conjunction with soil and foliar test results and can be made by row banding or through fertigation during the season.

Phosphorus fertilizer, as liquid or granular, is commonly injected in twin bands 10 cm (6 in) deep and 10–15 cm (4–6 in) to the side of a row before planting or as liquid by fertigation. Banding of fertilizer near the seed is more efficient than broadcasting since the rows are very far apart with vining cultivars.

Flowering and pollination

Squash and pumpkin have monoecious sex expression with separate male and female flowers on the same plant. Both kinds of flowers are open a single day. Male flowers rapidly senesce and abscise the day after closing. If female flowers are successfully pollinated, the ovaries will rapidly develop into fruit, but if not they will slowly wither and senesce after a few days. Most vines of large-fruited cultivars can only support two or three developing fruit at a time and later flowers will fail to set. Small-fruited cultivars can support many developing fruit on a single vine.

The most effective pollinators are bees. Several hundred pollen grains must be deposited on the pistil of each female flower to produce a fully developed fruit of marketable size and uniform shape. Full pollination requires 10–15 bee visits during the 1 day the flower is open. The bee population needed to maintain that frequency of flower visitations is normally four to five strong colonies per hectare (one to two per acre). Plants under stress will have fewer flowers and will not set as many fruit as healthy plants. Reduced photosynthetic capacity, rain, strong winds, and high or low temperature extremes also will reduce bee activity and consequently reduce yields.

Growing giant pumpkins

In some parts of the world, the size of winter squash and pumpkins is important especially when growing fruit for exhibition (Fig. 10.25). Growing giant pumpkins and squash also illustrates some of the important principles of fruit development. The first step is to select a cultivar that has the genetic potential to grow large fruit such as *C. maxima* cv. Atlantic Giant. Cultural practices help fruit reach their maximum genetic potential. Plants must be grown at extra wide spacing to reduce competition. After the first fruit sets, all other developing fruit and flowers should be removed to eliminate competing "sinks" that draw photosynthetic resources from the single selected exhibition fruit. Minimizing disease and insect attacks on the plant is important because these stresses reduce photosynthetic capacity. Plants should be watered regularly to optimize expansive growth and eliminate water stress that would reduce photosynthesis by causing stomatal closure. Plants should be fertilized weekly with a complete analysis fertilizer after the fruit has set to ensure that essential mineral nutrients are not limiting. Some pumpkin growers inject carbohydrates into the stem with the hope that they will be absorbed, translocated into the fruit, and metabolized to supplement natural dry matter accumulation through photosynthesis. Rotating the developing fruit periodically will help prevent the development of a flat side but this is for cosmetic purposes and will not affect final size or weight.

Harvesting and marketing

Summer squash, such as 'Yellow Straight Neck', 'Zucchini', and 'Patty Pan' are harvested as immature fruit often as soon as 40–50 days after planting. Summer squash should be harvested when fruits are immature and glossy in appearance before significant seed development begins. Preferred size varies among markets. Some summer squash are harvested very early in their development, often just a few days after anthesis with the flower corolla still attached. Other markets prefer larger fruit (Fig. 10.29).

It is important to harvest summer squash regularly to maximize plant productivity because larger developing fruit tend to suppress the development of new pistillate flowers. In some markets, clusters of open male flowers are harvested and sold as a delicacy (Fig. 10.30). The flowers are prepared by cooking in a number of different ways, usually with ingredients like eggs, flour, or meat.

Fig. 10.29. A succession of developing fruit with senescing flowers on a bush zucchini summer squash plant. Flowers are only open for a single day and then rapidly senesce.

Fig. 10.30. Squash flowers for sale at a farmers' market in Hania, Crete, Greece.

The development of most pumpkins and winter squash cultivars falls in the range of 80–150 days although the exact time is very cultivar and environmentally dependent. Pumpkins and winter squashes should fully mature before harvesting. Essentially, all *Cucurbita* fruit are hand harvested, and except for summer squash, rind hardness is a usual indication of maturity sometimes accompanied by vine senescence. Fruits for storage should be harvested after the skin is sufficiently hard that a fingernail or similarly sharp object cannot penetrate it. When ready for harvest, fruit are carefully cut off the vine with a sharp knife or clippers to minimize peduncle injury, a possible site for disease entry. Pumpkins for ornamental display are usually cut well above the fruit attachment point so the large stem can act as a fruit handle for carrying. Crop yields for winter squash range from 20–30 metric tonnes/ha (9–14 short tons/acre) (Lorenz and Maynard, 1988).

Harvesting of bush cultivars of summer squashes is complicated by short internodes that cause fruits to be closely spaced, which interferes with removal. Fruits are harvested by cutting with clippers, a sharp knife or by twisting the soft pedicle (fruit stem), by hand. Having soft skins, the fruit are easily scratched by the stiff foliar trichomes on the leaf petioles and are very susceptible to physical damage and rapid moisture loss. Summer squash are sometimes sold in plastic-wrapped trays to protect fruit. Care must be taken to protect the harvester's arms and hands from abrasion by foliar trichomes as well. Crop yields for summer squash range from 7–15 metric tonnes/ha (3–7 short tons/acre). Summer squash has a marketable shelf life of only about 7–10 days at 13°C (55°F), is chilling sensitive and should not be stored at lower temperatures (Cantwell and Suslow, 2013).

Rough handling damages pumpkins and winter squash despite their relatively hard rind. Fruit should not be exposed to bright sunlight or hard freezes. After harvest, winter squash are often cured at temperatures between 27–30°C (81–86°F) at 80% RH for about 10 days to heal wounds that occurred during harvest. Curing helps increase storage life by healing (suberize) cuts and bruises thus inhibiting entry of disease pathogens that cause fruit rot. Undamaged or cured, disease-free winter squash and pumpkins can be stored for several months at 13–15°C (55–59°F) and 55–60% RH depending on the genotype. Squash are chilling sensitive and should not be stored below 13°C (55°F), which can cause off flavors and rapid deterioration, particularly when fruit are returned to room temperature (Cantwell and Suslow, 2013).

Diseases

Anthracnose (*Colletotrichum orbiculare*) is a destructive disease that causes defoliation and lesions on the fruits. Angular leaf spot (*Pseudomonas syringae* pv. *lachrymans*) is caused by a bacterial pathogen. The bacterium can attack leaves, stems, and fruit. Leaf symptoms begin as small, water-soaked lesions that expand to fill the area between veins, giving an angular appearance (Cornell, 2004). Powdery mildew (*Erysiphe cichoracearum*), downy mildew (*Peronospora cubensis*), scab (*Cladosporium cucumerinum*), and leaf-spot (*Alternaria cucumerina*) primarily affect leaves and stems (Fig. 10.31).

The same fungus, *Didymella bryonia*, that causes gummy stem blight in other cucurbits causes black rot (Cornell, 2004). Black rot is the fruit-infecting phase of the disease, and is most common on butternut squash and pumpkins, while gummy stem blight refers to the foliar and stem-infecting phase of the disease. Choanephora wet rot (*Choanephora cucurbitarum*) causes a soft rot of squash fruit. Serious virus diseases include: cucumber mosaic cucumovirus (CMV), watermelon mosaic 2 potyvirus (WMV-2), watermelon mosaic 1 potyvirus, zucchini yellow mosaic potyvirus (ZYMV), and squash leaf curl bigeminivirus (SLCV). Damping off is favored by cold wet conditions and is caused by *Phythium* spp., *Rhizoctonia* spp. or *Fusarium* spp., and can infect seedlings and transplants in the field or greenhouse if disease-free media is not used. Phytophthora blight, caused by the fungal-like organism *Phytophthora capsici*, causes a sudden wilt of infected plants and/or white yeast-like growth on affected fruit. Fusarium wilt and crown rot are diseases caused by several different members of the genus *Fusarium*, which has many subspecies that are host-specific (Cornell, 2004). *Fusarium* species can be seed-borne, but also persist in the soil as spores for many years with no host. The spread of this pathogen often occurs through movement of infested soil and/or plant debris. Bacterial wilt is not as serious a problem with pumpkin and squash as with cucumber and muskmelon because the vascular elements are larger and less susceptible to clogging. Still, this disease, which is spread by cucumber beetles, has been reported in *Cucurbita* and appears cultivar specific.

Insect Pests

Squash vine borer is a major pest. This insect bores small holes that are visible on the outside of the stem to access the interior where if feeds and lives.

Fig. 10.31. Powdery mildew on squash leaves.

The debris surrounding each hole looks as though the stem had been drilled with a small-diameter bit. Once inside the stem the borers are sheltered and difficult to control by conventional means.

Cucumber beetles feed on flowers and young seedlings causing mainly superficial damage. The leaf-feeding *Epilachna* beetles are a serious problem for *Cucurbita* growers. The adult squash bug (*Anasa tristis*) is dark gray and about 16 mm (0.625 in) long and sucks sap from the leaves and stems. In severe cases, the leaf first wilts, turns black, and dies. Squash bugs can also feed directly and damage the fruit. Squash bugs live through the winter in protected areas both under debris in the fields and in buildings and lay eggs on the underside of leaves in the spring and summer. Aphids, primarily *A. gossypii*, do not cause serious injury to cucurbits. Aphid feeding may distort leaves. However, some species of aphids transmit virus disease. Resistant cultivars provide the most reliable control of virus diseases. Other insects that may affect *Cucurbita* spp. to varying degrees include cutworms (*Agrotis segetum* and *A. ipsilon*), leafminers (*Liriomyza sativae*), and rindworms (*Spodoptera* spp.).

References

Andrés, T.C. (1990) Biosystematics, theories on the origin and breeding potential of *Cucurbita ficifolia*. In: Bates, D.M., Robinson, R.W. and Jeffrey, C. (eds) *Biology and Utilization of the Cucurbitaceae*. Cornell University Press, Ithaca, New York, pp. 102–199.

Bélanger, R. and Labbe, C. (2002) Control of powdery mildew without chemicals: prophylactic and biological alternatives for horticultural crops. In: Belanger, R.R., Bushnell, W.R., Dik, A.J. and Carver, T.L.W. (eds) *The Powdery Mildews. A Comprehensive Treatise*. The American Phytopathological Society Press, St. Paul, Minnesota, pp. 256–267.

Berry, J., Weber, C., Dreher, M. and Bemis, W.E. (1976) Chemical composition of Buffalo Gourd, a potential food source. *Journal of Food Science* 41, 465–466.

Cantwell, M. and Suslow, T.V. (2013) Pumpkin and Winter Squash: Recommendations for Maintaining Postharvest Quality. Available at: http://postharvest.ucdavis.edu/pfvegetable/PumpkinWinterSquash (accessed 18 November 2013).

Chen, J.C., Chiu, M.C., Nie, R.L., Cordell, G.A. and Qiu, S.X. (2005) Cucurbitacins and cucurbitane glycosides: structures and biological activities. *Natural Product Reports* 22, 386–399.

Cornell (2004) Cornell Pest Management Guidelines for Vegetables 2004. Available at: www.nysaes.cornell.edu/recommends (accessed 31 December 2013).

Curtin, L.S., Moore, M., Kamp, M. and Austin, M. (1997) *Healing Herbs of the Upper Rio Grande: Traditional Medicine of the Southwest*, Revised Edition. Western Edge Press, Santa Fe, New Mexico.

Curwen, D., Powell, R.D. and Schulte, E.E. (1975) *Vine crops*. Cooperative Extension Programs, University of Wisconsin–Extension, Madison, Wisconsin.

Deppe, C. (1993) *Breed Your own Vegetable Varieties*. Little, Brown & Company Publishing, Boston, Massachusetts.

FAO (2011) FAOSTAT Production Crops. Available at: http://faostat.fao.org/site/567/default.aspx#ancor (accessed 12 June 2012).

Freeman, J.H., Miller, G.A., Olson, S.M. and Stall, W.M. (2007) Diploid watermelon pollenizer cultivars exhibit varying degrees of performance with respect to triploid watermelon yield. *HortTechnology* 17, 518–522.

Gusmini, G., Song, R. and Wehner, T.C. (2005) New sources of resistance to gummy stem blight in watermelon. *Crop Science* 45, 582–588.

Hedrick, U.P. (1919) Sturtevant's notes on cultivated plants. *New York Department of Agriculture Annual Report* 27, 1–686.

Hochmuth, R.C. (2013) Greenhouse Cucumber Production - Florida Greenhouse Vegetable Production Handbook, Vol. 3. Publication #HS790. Available at: http://edis.ifas.ufl.edu/cv268 (accessed 18 November 2013).

Kihara, H. (1951) Triploid watermelons. *Proceedings American Society of Horticultural Science* 58, 217–230.

Lessl, J.T., Fessehaie, A. and Walcott, R.R. (2007) Colonization of female watermelon blossoms by *Acidovorax avenae* spp. *citrulli* and the relationship between blossom inoculum dosage and seed infestation. *Journal of Phytopathology* 155, 114–121.

Lorenz, O.A. and Maynard, D.N. (1988) *Knott's Handbook for Vegetable Growers*, 3rd edn. Wiley-Interscience, New York.

Maness, N., Mcglynn, W., Scott, D. and Perkins-Veazie, P. (2003) Alternative uses of watermelons: Progress towards on-farm lycopene production. *Proceedings of Horticultural Industry Show* 2003, 77–80.

Mariani, J.F. (1994) *The Dictionary of American Food and Drink*. Hearst Books, Charlotte, North Carolina.

Masabni, J., Dainello, F. and Cotner, S. (2011) Texas Vegetable Growers' Handbook. Available at: http://aggie-horticulture.tamu.edu/publications/veghandbook/index.html (accessed 27 September 2011).

Maynard, D. and Maynard, D.N. (2000) Cucumbers, Melons, and Watermelons. In: Kniple, K.F. and Orneles, K.C. (eds) *The Cambridge World History of Food*. University Press, Cambridge, UK, pp. 298–313.

Motes, J.E. (1975) Pickling cucumber production-harvesting. Bulletin E837. Available at: http://archive.lib.msu.edu/DMC/Ag.%20Ext.%202007-Chelsie/PDF/e837.pdf (accessed 17 November 2013).

Noh, J., Kim, W., Lee, K., So, S., Ko, B. and Kim, D. (2007) Effect of furrow mulching with PE black film and dripping of phosphorous acid on control of Phytophthora root and fruit rot (*Phytophthora capsici*) occurred in field-grown watermelon. *Korean Journal of Horticultural Science and Technology* 25, 24–28.

O'Sullivan, J.N. (1980) Irrigation, spacing and nitrogen effects on yield and quality of pickling cucumbers grown for mechanical harvesting. *Canadian Journal of Plant Science* 60, 923–928.

Robinson, R.W. and Decker-Walters, D.S. (1997) *Cucurbits*. CAB International, Wallingford, UK.

Roush, W. (1997) Squash Seeds Yield New View of Early American Farming. *Science* 276, 894–895.

Rubatzky, V.R. and Yamaguchi, M. (1997) *World vegetables: Principles, production, and nutritive value*, 2nd edn. Chapman and Hall, New York.

Sauer, J.D. (1993) *Historical Geography of Crop Plants: A select roster*. CRC Press, Boca Raton, Florida.

Smith, B.D. (1997) The initial domestication of *Cucurbita pepo* in the Americas 10,000 years ago. *Science* 276(5314), 932–934.

Stuart, S.G. and Loy, J.B. (1983) Comparison of testa development in normal and hull-less seeded strains of *Cucurbita pepo* L. *Botanical Gazette* 144, 491–500.

Suslow, T.V. and Cantwell, M. (2013) Cucumber: Recommendations for Maintaining Postharvest Quality. Available at: http://postharvest.ucdavis.edu/pfvegetable/Cucumber (accessed 18 November 2013).

Suslow, T.V., Cantwell, M. and Mitchell, J. (2013) Cantaloupe: Recommendations for Maintaining Postharvest Quality. Available at: http://postharvest.ucdavis.edu/pfvegetable/Cantaloupe (accessed 18 November 2013).

Tamang, J.P., Sarkar, P.K. and Hesseltine, C.W. (1988) Traditional fermented foods and beverages of Darjeeling and Sikkim – a review. *Journal of the Science of Food and Agriculture* 44, 375–385.

Tomason, Y. and Gibson, P.T. (2006) Fungal characteristics and varietal reactions of powdery mildew species on cucurbits in the steppes of Ukraine. *Agronomy Research* 4, 549–562.

USDA (2011) National Nutrient Database for Standard Reference 2011. Available at: www.nal.usda.gov/fnic/foodcomp/search (accessed 7 October 2011).

Whitaker, T.W. and Davis, G.N. (1962) *Cucurbits. Botany, cultivation, and utilization*. Interscience Publishers, New York.

Wolf, E.R. (1982) *Europe and the People without History*. University of California Press, Berkeley, California.

Zeven, A.C. and Zhukovsky, P.M. (1975) *Dictionary of Cultivated Plants and their Centers of Diversity*, 2nd edn. Centre for Agricultural Publishing and Documentation, Wageningen, the Netherlands.

Zitter, T.A., Hopkins, D.L. and Thomas, C.E. (1996) *Compendium of Cucurbit Diseases*. APS Press, St. Paul, Minnesota.

Zohary, D., Hopf, M. and Weiss, E. (2012) *Domestication of Plants in the Old World*, 4th edn. Oxford University Press, Oxford, UK.

11 Family Solanaceae

POTATO

Origin and History

The potato is an ancient crop. Potatoes were used as food at least 8,000 years ago according to carbon dating of starch grains found in archaeological excavations in the Andean regions of Peru and Bolivia (Brown, 1993). The potato was unknown to the outside world until the Spanish explorer and conqueror Gonzalo Jiminez de Quesada (1499–1579) and his men took it to Spain. The Spanish thought the potato was a kind of truffle and called them "tartuffo". However, potatoes soon became a standard supply item on the Spanish ships because sailors who ate them did not suffer from scurvy (Brown, 1993).

Both wild and cultivated potato plants survive well in soil because of their high moisture content and starch and other nutrient reserves, which enable repeated regeneration of shoots. Unharvested tubers remain dormant in the soil but sprout under favorable conditions, enabling continued survival without replanting. The Inca's ability to preserve harvested potato tubers as *chuño*, a product made by the mashing and naturally drying tubers during repeated freezing and thawing cycles at high-elevations, increased their versatility as a food crop.

Following its introduction into Spain in about 1570, the potato rapidly spread to other parts of the world such as India in about 1610, China in 1700, and Japan in about 1766 (Brown, 1993). Scottish-Irish immigrants introduced the potato into North America in the early 1700s. Before wide adoption and acceptance of the potato in Europe, there was considerable skepticism concerning its suitability for human consumption. When the potato was first introduced into Europe, it was thought to be poisonous because the leaves resemble nightshade (*Solanum* sp.) (Hornfeldt and Collins, 1990). Because potato tubers developed underground, many considered them unfit for human consumption or suitable only as animal feed. Productivity was low, which also slowed acceptance, because Andean introductions (*Solanum tuberosum* subsp. *andigena*) from equatorial South America were not adapted and performed poorly in Europe. The more productive Chilean species (*S. tuberosum* subsp. *tuberosum*) were not introduced to Europe until the 19th century (Brown, 1993).

French military chemist and botanist Antoine-Augustin Parmentier (1737–1813), saw the potential of the potato as a new food crop because during his time many peasants in Europe subsisted on gruel and famine was common. Parmentier won a contest to find a new food with his study entitled "Chemical Examination of the Potato" (Brown, 1993). In 1785, Parmentier convinced King Louis XVI of France to encourage cultivation of potato as a food crop by allowing the planting of 45 ha (100 acres) of potatoes outside Paris, France under heavy guard. This succeeded in making the commoners believe that something valuable was being protected. One night Parmentier removed the guards so the local farmers would, as he had hoped, take potato tubers for planting on their own farms. Its value as a human food soon was recognized, along with the potential to produce more calories at a lower cost than grain crops. This incident helped establish potato as a subsistence staple to meet the growing food needs of the European population. The Irish in particular enthusiastically adopted potato as their primary food crop (Woodham-Smith, 1991).

The cool moist climate and rich soil in Ireland was well suited for potato. The Irish became overly dependent on the potato because it was so successful. However, when the late blight (*Phytophthora infestans*) disease arrived, the consequences were disastrous, killing millions and causing massive emigration during the Irish famine of 1845–1846 (Woodham-Smith, 1991). Limited crop rotation, the high percentage of land devoted to potato production and the limited genetic diversity of the potatoes grown all contributed to the disease and subsequent crop failure that caused the famine. The disease caused tubers to rot in storage and spread to subsequent crops because the potato is vegetatively propagated. Late blight remains a major disease problem in many parts of the world, especially where humidity is high.

The association with Irish history is likely responsible for the name "Irish potato", which is retained even though the crop is native to South America and is now grown throughout the world. White potato and tuber potato are other common names. Although some cultivars are white fleshed and have light skins (periderm), the name "white potato" does not adequately describe the many internal and external variations in color that exist among cultivars. Despite the inaccuracy, the terms "white" and "Irish" potato still persist in some areas.

Botany and Life Cycle

Potato (*S. tuberosum* L.) is a member of the family Solanaceae. Potatoes are dicotyledonous, short-lived perennials that are typically cultivated as annuals. Above-ground stems are erect and initially smooth and then branch with continued growth. Plant growth habit ranges from compact to sprawling depending on cultivar, stage of development, and environmental conditions. Compound pinnate leaves with leaflets vary in size, shape, and texture. Potato plants develop enlarged tubers on the end of underground rhizomes (stems) called stolons. The growth of stolons and tuber development is favored by decreasing daylength (short days) and cool temperatures in most cultivars (Miller and McGoldrick, 1941; Gregory, 1965).

Carbohydrates from photosynthesis in the leaves are translocated as sucrose, which accumulates largely as starch in tubers that develop at the end of the stolons. Tuber development continues throughout the season due to cell division and enlargement. Time to crop maturity depends on cultivar and is strongly dependent on the interaction of temperature and daylength (Miller and McGoldrick, 1941). The combined effects of both high temperature and long days will completely inhibit tuber formation in most cultivars (Gregory, 1965). Tuber development is optimized with a phasic pattern of high temperature during early growth and low temperature during later growth (Cao and Tibbitts, 1994).

Under suitable production conditions a potato crop can mature in the range of 90 to 120 days.

Potato is a cool-season crop but only tolerant of light frosts. Plants can be damaged by prolonged temperatures below −2.5°C (27.7°F). Plant acclimation history along with genetic make-up determines the degree of frost tolerance that plants exhibit (Chen and Li, 1980). The optimum temperatures for potato growth and development are generally listed as 15.5–18.3°C (60–65°F).

Individual flowers are perfect with a five-lobed fused corolla ranging in color from white to pink to bluish purple and clustered in a primary inflorescence (Fig. 11.1).

Flowers lack nectaries and are sometimes sterile. In some cultivars, most flowers abort and few fruits develop (Burton, 1966). Most flowers produce dry dusty pollen that is released through vibration from tube-like poricidal anthers as insects collect and distribute pollen (Batra, 1993; Harder and Barclay, 1994). Fertilized flowers produce small spherical berries that range in color from green to yellow to purple. The fruit contain glycoalkaloids and should not be eaten (Burton, 1966). Few to many small, flat-oval or kidney-shaped, yellow or yellowish-brown seeds are embedded in the gelatinous fruit pulp. Most potatoes do not grow true-to-type from seed and must be asexually propagated from tubers that develop finely branched, shallow, fibrous, spreading, adventitious roots (Stevenson, 1951; Burton, 1966). Some true-to-type diploid cultivars have been developed for seed propagation. Plants developing from true seed form a taproot with many laterals.

The potato tuber is a shortened, thickened, fleshy stem with leaves reduced to scales or scars subtending axillary buds known as "eyes" (Harris, 1978). The eyes are positioned in leaf axil scars and remain dormant during tuber enlargement. Each eye is a multiple bud cluster, and each bud may produce a stem. Bud numbers differ among cultivars. Tubers buds exhibit polarity, so buds near the stolon attachment or apical end exert dominance over distal buds at the opposite end (Harris, 1978). Such apical dominance can be overcome by cutting tubers into pieces. Seed pieces, containing at least one eye, provide more uniform field emergence. Low temperatures and aging of tubers also reduce apical dominance.

The primary tuber tissues include the periderm (skin), cortex, vascular cylinder of phloem and xylem, and pith tissues (Burton, 1966). The cambium

Fig. 11.1. A blooming potato field in western Ohio.

produces little secondary tissue. The tuber periderm may be smooth or rough because of netting or russeting with colors ranging from brown to light tan, red, or dark purple (Clark and Lombard, 1951). Flesh color usually is either light yellow or white; however, some cultivars have either deep yellow, orange red, or purple colored flesh. Tuber shapes of commercial cultivars vary from elongated, blocky, round to flattened (Clark and Lombard, 1951).

Types and Cultivars

Potato breeding is challenging because the plants are tetraploid, containing four sets of chromosomes instead of the more normal two. This genetic complexity prevents them from breeding true-to-type from seed. Since cultivar development is problematic, potato cultivars tend to predominate for much longer than other vegetables. For example, 'White Rose', introduced in 1881, is a widely adapted, heat-tolerant cultivar that is still grown in parts of the western USA. Four cultivars, 'Russet Burbank', 'Kennebec', 'Katadin', and 'Sebago', account for much of the US acreage. 'Russet Burbank', also called 'Idaho Baker', arose from a natural mutation discovered by Luther Burbank in the early 1900s and remains a leading cultivar in the USA. 'Russet Burbank' and 'Atlantic', a cultivar widely grown in the eastern USA and Canada, are successful because they have desirable characteristics for baking and processing such as high starch, low sugar, high specific gravity, and small tuber-cell size (Table 11.1). The sugar in tubers often caramelizes when deep-frying potato chips or French fries causing an undesirable, dark-colored product. The "russet" characteristic refers to the rough, mottled, corky periderm that is attractive and resists injury from abrasion during harvesting and transport (Clark and Lombard, 1951).

Not all cultivars are suitable for baking or processing. Some cultivars such as 'Red LaSoda' are best pan fried, boiled, or used in salads (Table 11.1). Cultivars suited for boiling have lower starch content and less dry matter (Table 11.1; Clark and Lombard, 1951). Tubers suited for boiling often have large rather than the small cells common in processing cultivars. Small cells slough when boiled, causing an unsuitable texture for some foods (Table 11.1). Specific gravity and dry matter values vary among locations and growing conditions.

Potatoes have been genetically engineered to express different transgenes. Potatoes expressing the Cry (crystal proteinaceous inclusions) toxin (δ-endotoxins) gene, from the bacterium *Bacillus thuringiensis* (or Bt), a gram-positive, soil-dwelling bacterium, have been developed and grown in some areas. The Bt bacterium occurs naturally in the gut of caterpillars of various types of moths and

Table 11.1. Potato cultivar characteristics and usage (modified from Curwen et al., 1982).

Representative cultivar	Specific gravity	Total dry matter (%)	Texture after cooking	Best use	Relative tuber cell size
White Rose	>1.06%	>16.2	Very soggy	Pan frying, salads	Large
Red LaSoda	1.06–1.07	16.2–18.1	Soggy	Pan frying, salads, boiling	Large
Goldrush	1.07–1.08	18.2–20.2	Waxy	Boiling, mashing, fair for chipping	Medium
Atlantic	1.08–1.09	20.3–22.3	Mealy, dry	Baking, processing	Small
Russet Burbank	<1.09	<22.3	Very mealy or dry	Baking, processing	Small

butterflies, as well as on the dark surface of plants (Roh et al., 2007). In the USA, for example, Bt potatoes were approved by the government and grown commercially in the late 1990s. However, consumer concerns over transgenic crops have resulted in little production of transgenic potatoes in recent years in the USA and western European countries. However, this is not the case in all countries. Development of transgenic potatoes is seen by some as a way to benefit resource-poor farmers in developing countries (Collins et al., 2000).

Production and Culture

Site selection and field preparation

Well-aerated and drained, deep, medium fine (silty clay) to medium coarse (sandy) textured friable, slightly acid (pH 5.5–6.5) soils are preferred for potato production (Davis, 1949). Soil texture and the degree of compaction strongly influence tuber shape, yield, and quality. A soil pH less than 5.5 helps to naturally control the potato disease scab (*Streptomyces scabies*) without affecting tuber quality or yield (Harris, 1978). Rotating potato with unrelated crops is important for disease and pest management.

Fertilization and nutrition

To maximize yields, potato fields must be supplied with sufficient mineral nutrients. Preplant soil testing helps assess soil fertility prior to establishment. Potato is a heavy feeder and a crop removes approximately 235 kg/ha (209 lb/acre) nitrogen (N), 34 kg/ha (30.3 lb/acre) phosphorus (P) and 308 kg/ha (275 lb/acre) potassium (K) (210 N, 30 P, 275 K lb/acre) from the soil. Of this total, approximately 168 kg/ha (150 lb/acre) N, 21 kg/ha (18.75 lb/acre) P, and 224 kg/ha (200 lb/acre) K (50 N, 19 P, and 200 K lb/acre) are sequestered in tubers alone (Masabni et al., 2011). When used, solid fertilizers should be banded near seed pieces to maximize uptake by shallow root systems. All fertilizer may be applied at planting, but the best nutrient management program matches inputs to the requirements of critical growth stages. Nutrient availability for crop establishment is important but the highest demand occurs during tuber enlargement. Excessive fertilizer applications, and in particular nitrogen, may delay tuber formation and maturity, and reduce the accumulation of tuber solids (Harris, 1978).

As the crop develops, tissue analysis can identify critical deficiencies. Sufficient potato leaf petiole nutrient levels for good growth vary with the stage of plant development. Local recommendations should be consulted if petiole analysis is used, but typical levels for dry petiole tissue at the early stages of growth should be approximately 12,000 ppm N, 2,000 ppm P and 11% K. During tuber development petiole levels decline to 5,000 ppm N, 1,000 ppm P, and 6% K as nutrients are remobilized to the below-ground organs (Rubatzky and Yamaguchi, 1997).

Vegetative propagation

For commercial potato production, "seed" refers to tubers used as propagules rather than true seed produced in fruits. The term "seed potato(es)" describes tubers used for crop establishment (Burton, 1966). Although vegetative propagation produces true-to-type clonal progeny, disadvantages include tuber bulk, potential for decay, and possible disease transmission. For successful vegetative propagation, an adequate supply of disease-free and pest-free seed potatoes is required (Davis, 1949).

Tissue culture techniques such as heat therapy and/or meristem micropropagation can produce plants that are initially disease-free. Plants regenerated from tissue culture are then used for seed potato production. Governments or other organizational programs certify seed potatoes as disease-free in many of the major potato-producing countries (Rubatzky and Yamaguchi, 1997).

Production and storage of seed potatoes

Seed potato production is commonly performed in cool, dry regions with less disease occurrence and little or no virus-vectoring insect activity. Under cool growing conditions, some foliar virus disease symptoms are more visible, allowing for inspection and removal of infected plants. In addition to field inspections, laboratory tests can identify virus diseases in tuber samples.

Some diseases, such as bacterial wilt and early blight, occur at higher temperatures. High temperatures during crop development may advance the physiological maturity of seed potatoes affecting subsequent plant growth and yield. For example, certain cultivars grown from seed potatoes produced under cool (13–14°C, 55–57°F) conditions may produce higher yields than those grown from tubers maturing under warm (26°C, 79°F) temperatures (Rubatzky and Yamaguchi, 1997).

The time of harvest, storage duration, and temperature influence the "physiological maturity" of seed potatoes. Late planting and/or early harvested potatoes have less time to develop and are physiologically immature, while tuber age is advanced by delayed harvest. Physiologically immature seed potatoes tend to produce plants having fewer stems with fewer but larger tubers. Physiologically mature seed potatoes tend to produce plants having many stems and a higher percentage of small tubers (Rubatzky and Yamaguchi, 1997). Accelerated aging can advance maturity when seed potatoes are subjected to stress during growth and/or physical injury during postharvest handling and storage.

Freshly harvested potatoes are usually physiologically immature and will not sprout until their rest requirement is satisfied during storage. The duration of the rest period is related to the physiological maturity of the tuber (Rubatzky and Yamaguchi, 1997). Sprout emergence and subsequent growth are influenced by storage period and temperature and there is significant variation among cultivars. Therefore, seed potatoes tubers must be stored prior to planting. A storage period of 6–8 weeks will usually satisfy the rest requirement for most cultivars, although some may require a longer period (Harris, 1978).

Seed potato tubers are stored at low temperatures, 3–4°C (37–39°F), and high, 90%, RH with proper ventilation to maximize storage life and minimize premature sprouting. Temperatures lower than 2°C (36°F) can cause injury and reduce sprouting. Tubers stored at 12–22°C (54–72°F) have shorter rest periods but exhibit stronger apical dominance compared to those stored at lower temperatures.

Exposing seed potatoes to light during low-temperature storage results in plants that produce many small tubers, whereas dark storage at high temperatures results in plants that produce fewer but larger tubers. In warm climates, the use of natural diffused light is a low-cost alternative to temperature-controlled storage because the diffused light has a similar effect as low temperature in inhibiting sprout growth and reducing apical dominance. In some European countries, this practice is called "chitting" or green sprouting (Rubatzky and Yamaguchi, 1997).

True potato seed

Potato is a polyploid and highly heterozygous with multiple copies of each chromosome, causing true potato seed (TPS) to produce both plants and tubers that are highly variable and not true-to-type. In other words, the progeny are unlike the parent plant, so TPS is used primarily for potato breeding.

Research, primarily at the International Potato Center in Peru, has led to the development of diploid cultivars that can be grown from TPS. Cultivars grown from TPS yield less and often lack uniformity of tuber size, shape, color, and quality, but are of interest in areas lacking facilities to store seed potatoes. Therefore, TPS offers great promise for some developing countries (Almekinders et al., 2009).

TPS usage offers freedom from propagation transmission of many viruses and circumvents the disadvantages associated with the handling, storage, and transport of bulky seed tubers since 100 g (3.5 oz) of TPS is the propagation equivalent of 2 or 3 metric tonnes (2.2 or 3.3 short tons) of seed potatoes (Almekinders et al., 2009). While the use of true seed for propagation has increased in some developing countries, vegetative propagation, using whole or tuber portions, remains the principal propagation method in most developed countries.

Field establishment

Prior to planting, seed tubers should be removed from low-temperature storage and placed at 10–13°C (50–55°F) for several days if their rest period has been satisfied (Burton, 1966). Acclimation at these higher temperatures improves sprout emergence after planting.

Small tubers can be used whole if they are disease free and produce more stems compared to an equivalent weight of cut seed-potato pieces. Whole seed tubers are easier to handle and reduce the spread of diseases associated with cutting (Davis, 1949).

Larger tubers are cut into uniform pieces containing at least one eye but preferably two or three (Fig. 11.2). Cut seed-tuber pieces should be blocky and weigh between 40 and 60 g (1.4 and 2.1 oz). Seed-potato size influences plant stem size and vigor because they contain more nutrient reserves for the developing plant. Large seed pieces produce large stems that grow faster and produce more leaf area and high yield, although with pieces larger than 60 g (2.1 oz) the advantage diminishes (Harris, 1978). Roots develop from the base of the stems and not seed potato tissues (Fig. 11.2).

Seed potato cutting is by hand or with automated equipment using disease-free tubers and sanitary practices. After cutting, seed pieces are usually placed at a warm temperature (15–21°C, 59–70°F) and high humidity for several days to heal cut surfaces by suberization, which is rapid at warm temperatures (Burton, 1966). Fungicides can also be applied to protect seed pieces from decay.

Planting methods range from hand to highly automated equipment. Minimizing damage to seed pieces during planting is important to reduce decay (Davis, 1949). Planting presprouted whole or cut seed can hasten field emergence and is practiced where growing seasons are short. To encourage sprouting, tubers and/or seed pieces are spread out in light at warm temperatures, 20–25°C (68–77°F) (Burton, 1966). After exposure to light, sprouts tend to be short, thick, and green, and do not interfere with planting unless they are elongated.

The quantity of seed tubers planted depends in part on production objectives, cultivar characteristics,

Fig. 11.2. A large cut seed piece, seen toward the bottom, gave rise to this young potato plant. A small developing tuber is just above and to the left of the seed-potato piece.

seed availability, and cost. Size of tubers, whether whole or cut, determines plant spacing and the amount required (Burton, 1966). Planting from 2,240–3,360 kg/ha (2,000–3,000 lb/acre) of seed pieces is typical for many areas. Close spacing tends to yield smaller tubers, whereas wide spacing results in fewer large ones. Cultivars with a heavy tuber set should be given wider spacing.

Field placement of seed potato pieces depends on several factors, including their maturity, storage conditions, cultivar, and length of the growing season (Burton, 1966). Spacing is increased in anticipation of greater tuber yields when physiologically mature (old) seed potatoes are planted. Less mature seed potatoes are spaced more closely to adjust for lower tuber production. When properly managed, equivalent yields can be obtained using seed potatoes of different physiological ages. The in-row spacing of seed pieces typically ranges from 15–38 cm (6–15 in) with between-row spacing ranging from 76–102 cm (30–40 in). Planting depth varies with soil type, temperature, and cultivar and generally ranges between 5–15 cm (2–6 in). Adequately covering seed potatoes with soil avoids desiccation, greening, sunburn, and other injuries to developing tubers caused by air and sunlight (Davis, 1949).

Seed potatoes may be planted as soon as soil temperatures reach at least 15°C (59°F), because lower temperatures delay emergence. Shoot emergence at 12°C (54°F) may require 30–35 days, while emergence may take less than 10 days at temperatures between 22°C and 30°C (72°F and 86°F) (Rubatzky and Yamaguchi, 1997).

Irrigation

Approximately 90% of potato roots occur within 50 cm (20 in) of the surface, which increases susceptibility to water stress. Soils should be regularly monitored with tensiometers or other devices to determine when irrigation is required. Water should be supplied continuously throughout the season, although highest moisture demand occurs during tuber initiation and enlargement. Water stress during enlargement may cause irregularly shaped, knobby tubers. Crop water requirements range from 250 to >500 mm (10–20 in) depending on soil type and environmental conditions. Soil moisture should not fall below 60% field capacity. Between 500 and 1,500 l of water are required to produce 1 kg of potato tubers (63–180 gal/lb) depending on climate and cultivar (Gleick, 2000). Excessive surface water and high humidity increases foliar diseases, while excessive soil moisture enlarges tuber lenticels.

Production and culture

For potato production, soil must be finely tilled and free of clods, stones, and other impediments. Fields are generally plowed and disked or rotovated to create friable soil for planting. Winter cover crops are often planted in preparation for potato production. However, no-till production of potato is less widely practiced than with some other crops.

Although potatoes can be planted on flat ground, forming hills over each row provides room for developing tubers to grow without being pushed out of the ground causing greening due to exposure to sunlight (Davis, 1949; Fig. 11.3). Hilling is also important for good drainage. Potato hills can be formed by mounding soil over seed pieces at planting or within 4 weeks of planting after initial emergence.

Harvesting and marketing

Unlike many other vegetables, potatoes are not extremely perishable so harvest scheduling is flexible. Potato tubers "hold" in the field for relatively long periods unless temperatures are extremely cold or soils waterlogged. Tubers are sometimes harvested before they are fully mature and sold as "new" (freshly harvested and often immature) potatoes to take advantage of favorable market prices. "New" potatoes are intentionally sold before the periderm is completely formed and usually have a lighter skin color. Many consumers prefer these slightly immature potatoes because they are fresh, easy to peel, and have slightly higher sugar content suitable for boiling or pan-frying.

The vines should mature and die before harvest so the periderm "skins" of the tubers are well suberized to minimize skinning and bruising during harvest. However, in shorter-season areas, potatoes are often harvested near the end of the growing season while the plants are still green. In this situation, artificial vine killing is necessary to improve tuber storage quality and for ease of harvest (Burton, 1966). Mechanical beaters or chemical desiccant sprays may be applied to hasten vine senescence. Most cultivars require 10–14 days after vine killing to ensure

Fig. 11.3. This hill of potatoes was opened approximately 2 months after planting to reveal the developing tubers inside. These "new" potatoes do not have a fully mature periderm at this stage of development.

that the tuber periderm is sufficiently suberized for safe harvest (Burton, 1966).

For small-scale production, potatoes may also be dug by hand or mechanically plowed for manual or mechanical collection. For large-scale commercial production, self-propelled or pull-type potato harvesters dig and separate the tubers from the soil and vines during one pass through a field, depositing the tubers in a wagon or truck driven alongside (Fig. 11.4).

Harvesting or other handling operations may cause external injury or internal bruising. Handling should be performed with great care to minimize damage to the highly sensitive, thin-skinned, turgid tubers especially when temperatures are near freezing. Crushing and bruising are common defects that lead to rapid water loss, shriveling, and decay in storage.

Postharvest handling

Potatoes are sorted and graded in a packinghouse. Truckloads of newly harvested potatoes are gently dumped on a conveyor belt for transport into the house. Impact areas are sometimes lined with soft material to reduce damage. It is important that the tubers be kept dry after harvest so water baths or flumes are not used in potato packinghouses. Tubers are sometimes cleaned with a brief spray wash but must be rapidly dried to reduce disease.

Wounds in the periderm should be healed by curing for about 1–2 weeks at 10–20°C (50–68°F) and 95–99% RH (Harris, 1978). During curing, the periderm hardens, wounds heal, adhering soil dries, and disease symptoms appear, which facilitates the removal of the infected tubers. To prevent tuber damage, grading and long-term storage should not occur until curing is completed.

Once in the packinghouse, the potatoes are sorted on the basis of size, graded, and bagged (Fig. 11.5). During the sorting process, damaged potatoes or inert matter are removed to prevent cross-contamination.

Depending on the situation, the potatoes can be marketed, processed, or stored. There are problems associated with storing potatoes such as sprouting, increasing sugar content, and weight loss due to evaporation and respiration. Tubers can be stored for several months at 1–2°C (34–36°F) and 95% RH. Storage at these low temperatures causes starch to be converted to sugar (Burton, 1966). Tubers stored at low temperature for long periods are unsuitable for processing into potato chips or French fries, because their sugar content is too high and will burn when deep-fried. Storage at 10–13°C (50–55°F) and 95% RH prevents starch from converting to sugar. Tubers stored at low temperature can be reconditioned at 18–20°C (64–68°F) and 85–90% RH for several days to convert the sugar back to starch prior to processing. For retail use,

Fig. 11.4. Mechanically harvesting potatoes in eastern Colorado.

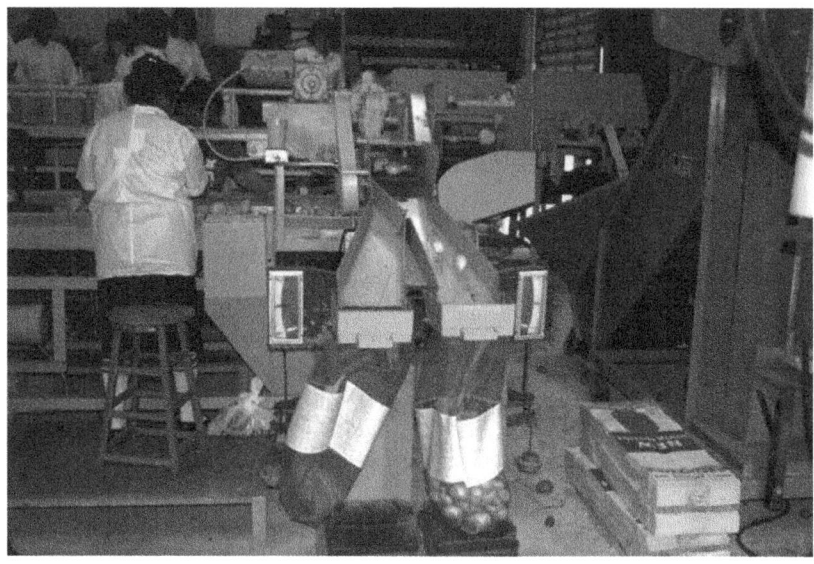

Fig. 11.5. Grading newly harvested winter potatoes near Homestead, Florida.

most tubers are stored at 1–2°C (34–36°F) and high humidity until marketed. Several days in a grocery store at room temperature converts most of the sugar back to starch so consumers rarely experience tubers with high sugar content. Potato tubers are usually marketed in bags or displayed in bulk.

Potato tubers are not very sensitive to external ethylene. Low levels of external ethylene elevate respiration, especially in immature potatoes, and result in weight loss and mild shriveling. After aging for 2–3 months at temperatures above 5°C (41°F) and in the absence of sprouting inhibitor application, low levels of ethylene may retard sprouting. High concentrations of external ethylene may induce sprouting (Suslow and Voss, 2013).

Controlled or modified atmospheres offer little benefit to potato. Periderm development and wound healing is delayed at atmospheres below 5% O_2. Injury from low O_2 (<1.5%) or elevated CO_2 (>10%) will induce off-odors, off-flavors, internal discoloration, and increased decay (Ma *et al.*, 2010; Suslow and Voss, 2013).

Greening can occur before or after harvest and is caused by the exposure of tubers to light, which causes the production of both chlorophyll and the alkaloid solanine (Burton, 1966). Solanine can also be produced in response to mechanical injury. Heat does not destroy solanine although some may be leached during boiling. Most of the solanine in tubers is removed during peeling. High concentrations can make humans sick. The acceptable concentration is less than 20 mg/100 g (0.003 oz/lb). Above this concentration, the flavor is noticeably bitter. The best advice is to not eat green potatoes.

Nutritional Values

Nutritional composition

Potatoes are 78% water and higher in dry matter than many other vegetables (Table 11.2). Tubers contain high amounts of starch and are a good energy source. The protein is of good quality but low in the essential amino acid methionine. Much of the vitamin C is in the outer layers near the periderm, encouraging food preparations without peeling (Table 11.2).

Economic Importance and Production Statistics

Potatoes are the world's fourth largest food crop, after rice, wheat, and corn. Potato production in developing countries has increased considerably since the early 1960s. More than one-third of all potatoes are grown in developing countries. Total world harvested area for potato was 18,651,838 ha (46,089,695 acres) with total production of

Table 11.2. Nutritional value of potato tubers (USDA Nutrient Database, 2011).

Potato, uncooked, with skin
Nutritional value per 100 g (3.5 oz)[a]

Energy	321 kJ (77 kcal)
Carbohydrates	17.5 g
– Starch	15.4 g
– Dietary fiber	2.2 g
Fat	0.1 g
Protein	2 g
Water	75 g
Thiamine (vitamin B1)	0.08 mg (7%)
Riboflavin (vitamin B2)	0.03 mg (3%)
Niacin (vitamin B3)	1.1 mg (7%)
Pantothenic acid (vitamin B5)	0.3 mg (6%)
Vitamin B6	0.3 mg (23%)
Folate (vitamin B9)	16 µg (4%)
Vitamin C	19.7 mg (24%)
Vitamin E	0.01 mg (0%)
Vitamin K	1.9 µg (2%)
Ca	12 mg (1%)
Fe	0.8 mg (6%)
Mg	23 mg (6%)
Mn	0.15 mg (7%)
P	57 mg (8%)
K	421 mg (9%)
Na	6 mg (0%)
Zn	0.3 mg (3%)

[a]Percentages are approximated using USDA daily recommendations for adults.

329,581,307 metric tonnes (363,302,202 short tons). The leading countries include: China 5,083,034 ha (12,560,451 acres) producing 73,281,890 metric tonnes (80,779,456 short tons), the Russian Federation 2,182,400 ha (5,392,828 acres) producing 31,134,000 metric tonnes (34,319,360 short tons), India 1,828,000 ha (4,517,086 acres) producing 34,391,000 metric tonnes (37,909,588 short tons), and Ukraine 1,411,800 ha (3,488,634 acres) producing 19,666,100 metric tonnes (21,678,164 short tons). In comparison, the USA production consisted of 422,901 ha (1,045,011 acres) producing 19,569,100 metric tonnes (21,571,240 short tons) and cumulatively European countries had 6,275,139 ha (15,506,206 acres) producing 123,755,681 metric tonnes (136,417,287 short tons) (FAOSTAT, 2011).

The area planted and potato production declined in Europe in the 1970s, 1980s, and 1990s (Scott et al., 2000). The per capita consumption of fresh potatoes and in the use of potatoes for animal feed also declined. Production also doubled in the Netherlands between 1963 and 1993, but has since been declining similar to other European countries. Potato output in the USA and Australia increased dramatically from 1963 to 2003 (FAOSTAT, 2011). However, per capita potato consumption in the USA has fallen by some 18% over the past 10 years while total production has declined by 12%, mimicking the European trend (USDA, 2009).

However, in the rest of the world, potato production has increased, especially in developing countries (Scott et al., 2000). Potato typically plays the role of a complementary vegetable or seasonal staple in developing countries rather than a principal source of carbohydrates. As incomes improve, consumers in developing countries often prefer to eat more potatoes as a way of bringing greater diversity to their cereal-based diets. Globally, potato production in developing countries rose steadily from 28 million metric tonnes (31 million short tons) in 1963 to 149 million metric tonnes (164 million short tons) in 2005, which more than offset the drop in production in industrialized countries (FAOSTAT, 2011). Annual per capita potato consumption in developing countries rose dramatically during the past 30+ years but remains lower in Asia (16 kg, 35 lb), Africa (7 kg, 15 lb) and Latin America (19 kg, 42 lb) than in Europe (86 kg, 190 lb) (FAOSTAT, 2011). Worldwide, potatoes are used fresh for direct human consumption (48%), processing (11% of which 2% was for starch production), vegetative propagation (13%), stock feed (20%), and the remaining 8% waste (FAOSTAT, 2011).

The consumption of processed potato products has increased dramatically in the USA and other developed countries in recent years. In the USA, approximately 88% of the US crop was used for human consumption, of which 68% was consumed as processed products. The per capita consumption total of 53 kg (113 lb)/person/year breaks down as follows: 32% fresh, 45% frozen, 10% dehydrated, 8% processed into chips, >1% canned, and ~3% processed into starch and flour etc. (USDA, 2009). A similar trend toward processing has occurred in western and eastern Europe, Russia, Argentina, Colombia, China, and Egypt (Scott et al., 2000). Worldwide alcohol production from potatoes for biofuels and human beverages is statistically negligible but important in some locations.

Diseases

Diseases of potatoes are numerous and widespread, affecting both yield and quality. Potato diseases may be bacterial, fungal, or viral (Harris, 1978). Some major bacterial diseases include bacterial wilt (*Ralstonia solanacearum*), bacterial soft rot (*Erwinia carotivora*), and common scab (*S. scabies*).

Fungal diseases are promoted by high humidity and moist conditions. A partial list includes late blight (*Ph. infestans*), early blight or target spot (*Alternaria solani*), black scurf (*Rhizoctonia solani*), and pink rot (*Phytophthora erythroseptica*).

Important viral diseases include potato leafroll luteovirus (PLRV), the mosaic viruses, notably potato X potexvirus (PVX), and potato Y potyvirus (PVY).

Insect Pests

Insect pests of potato are particularly destructive in many regions. Some insects feed on potato foliage while others feed on tubers. Some of the common foliage feeders include: Colorado potato beetle (*Leptinotarsa decemlineata*), variegated cutworm (*Peridroma saucia*), western yellow striped armyworm (*Spodoptera praefica*), redbacked cutworm (*Euxoa ochrogaster*), alfalfa looper (*Autographa californica*), grasshopper (*Melanoplus* spp.), blister beetle (*Epicauta* spp.), tuber flea beetle (*Epitrix tuberis*), and spotted garden slug (*Agriolimax reticulatum*) (Berry *et al.*, 2000).

Insects that feed on tubers include: wireworm (*Limonius* spp.), tuber flea beetle larve (*Epitrix tuberis*), white grubs (*Polyphylla* spp.), leatherjacket (cranefly) larvae (*Tipula dorsimacula*), larvae of the western spotted cucumber beetle (*Diabrotica undecimpunctata*), variegated cutworm (*Peridroma saucia*), redbacked cutworm larvae (*Euxoa ochrogaster*), symphylan (*Scutigerella immaculata*), and slugs (Berry *et al.*, 2000).

Sucking insects that attack potato include green peach aphid (*Myzus persicae*), potato aphid (*Macrosiphum euphorbiae*), twospotted spider mite (*Tetranychus urticae*), potato leafhopper (*Empoasca filament*), lygus bugs (*Lygus* spp.), thrips (*Frankiniella occidentalis*), and whitefly adults (Homoptera: Aleyrodidae) (Berry *et al.*, 2000). The population of aphids should be strictly controlled, particularly in seed-potato production fields, because they vector viruses. Consult local recommendations for biological control measures and treatments available to control pests in your area.

Weed Management Strategies

Weed control is an important aspect of potato production to reduce competition for nutrients, water, and sunlight. When properly grown, the potato plants normally cover the ground, smothering weed competition. However, early-season control may be required to stop broadleaved weeds or annual grasses from competing for nutrients and moisture during tuber enlargement. Weeds may also harbor insect pests and may interfere with harvesting equipment (Holm *et al.*, 1991). Many of the major weeds in potato fields are annuals although some perennials may be problematic as well. Nutsedge (*Cyperus* spp.) is a troublesome perennial weed whose rhizomes can penetrate potato tubers and lower both yield and quality (Boldt, 1976). There are many weeds that can populate potato fields, but according to Holm *et al.* (1991), the most important ones worldwide are *Chenopodium album*, *Portulaca oleracea* (South America and Asia), *Galinsoga parviflora*, *Stellaria media* (Europe, Chile and New Zealand), and *Echinochloa crus-galli* (Bulgaria, Poland, and the USA).

Physiological Disorders

Physiological disorders are nonpathogenic tuber defects that reduce quality and the percentage of marketable tubers if the condition is severe. The most common physiological disorders are hollow heart and internal spotting. Hollow heart is somewhat cultivar dependent and is associated with rapid growth and uneven water supply.

Blackheart is rare in early-crop potatoes due to typical marketing practices and more common in late season production or storage. In conditions of restricted airflow, low oxygen, and high respiration, tubers held above 15–20°C (59–68°F) develop brown discoloration, which eventually becomes deep black in the center of the tuber (Suslow and Voss, 2013).

Black spot and brown spot are caused by the same factors and are responsible for significant postharvest losses, particularly in response to over-fertilization with N, low soil K availability, irregular irrigation, and other pre-harvest practices. Non-pigmented compounds form in the vascular bundle tissue just under the skin during storage. Following severe bruising or cutting, the affected tuber tissue turns reddish, then blue, and finally black in 24–72 h. Severity increases with time. Cultivars differ significantly in their susceptibility and symptom expression (Suslow and Voss, 2013).

Chilling injury occurs at temperatures near 0°C (32°F) for a few weeks and may result in a mahogany discoloration of internal tissue in some cultivars. Much longer periods of low temperature storage are generally required for chilling injury to occur (Suslow and Voss, 2013).

Internal brown spot symptoms are internal dry, corky reddish-brown or black spots or sectors. Uneven water management and/or widely fluctuating temperatures induce this calcium (Ca) uptake deficiency, usually early in tuber development (Suslow and Voss, 2013).

Freezing injury occurs at temperatures less then −0.8°C (30.5°F). Symptoms include a water-soaked appearance, glassiness, and tissue breakdown upon thawing. Mild freezing may also result in chilling injury (Suslow and Voss, 2013). Common external tuber defects include greening, growth cracking, and tuber deformation (Harris, 1978). Cracking and irregular-shaped tubers are associated with uneven moisture.

PEPPERS

Origin and History

Peppers, *Capsicum* spp., are endemic to tropical and subtropical America and have been grown as a food for centuries. Evidence of early cultivation was found in Peruvian burial sites, and seed remnants dated older than 5000 BC in caves near Tehuacan, Mexico. The major species were likely domesticated in different areas: *C. annuum*, central to southern Mexico; *C. frutescens*, Central America, probably Guatemala; *C. chinense*, Ecuador and southern Columbia; *C. baccatum*, eastern Bolivia; and *C. pubescens*, mountainous Bolivia and southern Peru (Pickersgill, 1997).

Columbus was likely the first to take specimens to Europe where they were eventually adopted as food crops and quickly moved into usage through much of the world. The non-pungent types became more popular in Europe. Spanish and Portuguese traders were largely responsible for initiating the worldwide dispersal of peppers (Bosland, 1996).

Peppers have long been used in folk medicine, most notably in Africa and by indigenous peoples of Latin America, to treat a wide variety of conditions such as arthritis pain, herpes zoster-related pain, diabetic neuropathy, mastectomy pain, dyspepsia, flatulence, headaches, dropsy, colic, toothache, and cholera as examples. Pepper extracts are also used as counter-irritants for rheumatism and as a constituent of some throat gargles and lozenges (Bosland, 1996). Peppers are used medicinally by many folk cultures. Chili peppers have antimicrobial properties and this is a possible reason why the consumption of pungent peppers has traditionally been greater in tropical regions to help fight disease (Cichewicz and Thorpe, 1996).

Peppers have become a major vegetable crop in many countries. The term chili pepper (also spelled chile or chilli pepper) refers to the fruit of plants from the genus *Capsicum*. In some cultures, chili is used alone without the word pepper. Also the name chili implies elongated fruits that are often but not always pungent rather than blocky bell types. The most notable feature of peppers is flavor and whether sweet and mild or strongly pungent. *Capsium* sp. peppers should not be confused with black pepper (*Piper nigrum*), a flowering tropical vine native to Asia in the family Piperaceae, cultivated for its fruit, which are dried, ground, and widely used as a spice and seasoning (McGee, 2004).

Botany and Life Cycle

Peppers belong to the family Solanaceae and like tomato and eggplant are warm-season, frost-sensitive, tropical perennials grown commercially as annuals particularly in short-season areas. Peppers are herbaceous but eventually become woody and shrublike with advanced maturity. Generally, plants are erect and highly branched becoming 0.5–1.5 m (19.7–59.1 in) tall at maturity. Root systems tend to be well developed and anchored by a taproot. The relatively simple, smooth leaves are broadly lanceolate to ovate with few trichomes (Rubatzky and Yamaguchi, 1997).

Among the different species, corolla colors vary from white to greenish white, and lavender to purple. Anther colors are blue, purple, or yellow. The bell-shaped calyx usually enlarges along with the fruit and covers a part or much of the fruit. All domesticated cultivars are predominately self-pollinated, although outcrossing may occur (Rubatzky and Yamaguchi, 1997).

Botanically, the fruit is an indehiscent, either pendent or erect, many seeded berry. Fruit are frequently borne singularly at each node for *C. annuum* cultivars, and with multiple fruit (typically two or three) per node for some other species. Fruit of domesticated species are variable in pungency and do not abscise. During development, the pericarp grows faster than placental issues, resulting in a fruit cavity. Carpel walls are fused with the placenta at the base of the fruit and may or may not continue that attachment to the tip. With maturity, the texture of the outside pericarp wall becomes smooth and glossy, while the inside has a blistered rough texture composed of large cells (Rubatzky and Yamaguchi, 1997).

Capsicum annuum is the most widely cultivated and economically important species and includes both sweet- and pungent-fruited cultivars of various shapes and sizes. Domesticated types are sometimes classified as *C. annuum* var. *annuum* and wild members as *C. annuum* var. *auiculare*. *Capsicum frutescens* is a semi-domesticated species found in the lowlands of tropical America. Southeast Asia is recognized as a secondary area of diversity. *Capsicum frutescens* are perennials with blue anthers, greenish-white corollas and commonly have two or more fruit developed per node. Wide variations in flavor exist among the many accessions. *Capsicum chinense* cultivars produce some of the most pungent fruit. Except for a ring-like constriction at the base of the calyx, *C. chinense* resembles *C. frutescens* and *C. annuum*. Fruit are smooth and shapes are as diverse as those of *C. annuum*, but often with puckered fruit walls. *Capsicum chinense* has a unique citrus-like aroma. Both *C. baccatum* and *C. pubescens* are grown primarily in South America and will not be discussed further (Pickersgill, 1997).

Types and Cultivars

Pepper fruit colors are highly variable: green, yellow, or even purple when young and later turning to red, orange, yellow, or a mixture of these colors with advancing age. Green color is due to chlorophyll, red and purple due to carotenoids, and purple is due to the pigment anthocyanin. As with color, fruit shapes vary greatly and may be blocky, conical, round, pencil shaped, or combinations. Fruit may also have thick or thin walls and range from 1 cm (0.4 in) to more than 30 cm (11.8 in) in length and from 1 cm (0.4 in) to about 15 cm (5.9 in) in diameter.

The systematic classification of the many types of pepper is challenging because of the wide range of fruit and plant types that fit multiple groupings. Fruit shape is frequently used as a horticultural classification method and peppers are often divided into pungent and nonpungent types. However, fruit shapes are often falsely associated with pungency level. Although this association is commonly used to differentiate sweet and pungent pepper types, it does not apply in all cases since pungency is not genetically linked to fruit size, shape, maturity, or other horticultural traits.

Nonpungent

Historically, large, blocky, four-lobed "**bell peppers**" are the most important type grown in the USA and Europe accounting for the largest percentage of commercial fresh market volume (Fig. 11.6).

Bell peppers are used fresh in salads or stuffed with meat. They can also be stir-fried and added to various dishes for color, texture, and flavor. The highest quality bells, like cv. 'California Wonder', have a thick fruit wall, four locules (fruit chambers) and both ends of the fruit are generally flattened. 'California Wonder' reaches green maturity in about 95–100 days and red physiological maturity, when the seeds are fully developed, after approximately 125 days. Similar cultivars are often referred to as 'California Wonder' types. Cultivars that mature in 65–75 days are available but have poorer quality with three locules rather than four, thinner fruit walls, less blocky shape, and smaller size. Immature fruit colors range from green to orange to purple. As fruits mature, they obtain their red or yellow mature color.

Other nonpungent types/products include **paprika**, which is a spice made by grinding dried, usually red, pepper fruits from a wide range of cultivars. **Pimento** (or pimiento) is processed from cultivars that have conical or tomato-shaped fruit with thick walls. The fruit are sliced, canned, and used as a condiment to add color to dishes or for stuffing olives.

Fig. 11.6. High-quality, thick-wall, four-lobe green bell pepper at the time of harvest.

Pungent

Pungent cultivars tend to be smaller and elongated but this is a general rule that does not always apply. There is no genetic relationship between fruit shape and pungency. Although pungency may increase with pepper maturity, the increase is gradual with development and does not dramatically increase with color change or physiological maturity. In other words, these peppers have a degree of pungency throughout development and do not transition from sweet to pungent at a particular stage. A few of the many types of pungent peppers are described below.

'**Tabasco**' is probably the best-known cultivar of *C. frutescens*, and is widely grown in warm-temperate as well as tropical regions. Tabasco fruit are small and elongated with an irregular shape, 5 cm (2 in) long, and are moderately pungent (Table 11.3). Fruit are pickled or processed into a spicy red seasoning sauce.

'**Cayenne**' (*C. annuum*) is also known as the Guinea spice and is named for the city of Cayenne in French Guiana. It is also called the cow-horn pepper because of its elongated curved shape. Cayenne is a red, hot chili pepper that is often dried and made into powder for flavoring and for medicinal purposes (Table 11.3), i.e. cv. Long Red Cayenne.

'**Banana**', with a shape and immature color similar to a banana, is an elongated type of "wax" pepper that turns red at maturity. "Wax" pepper is a general term for any cultivar that produces yellow fruit. Wax peppers may be either pungent or sweet. For some cultivars, yellow is the immature color but for other cultivars it may be the mature color. Cultivars of banana peppers can be either mild or slightly pungent (0–500 Scoville units; Table 11.3). They are often processed by pickling. i.e. cvs Hungarian and Anaheim (yellow cultivars).

'**Jalapeño**' (*C. annuum*) are bullet shaped, 5–9 cm (2–3.5 in) long, very dark green when immature, red at maturity, and somewhat pungent (Table 11.3). The name is derived from Xalapa, Veracruz, Mexico where it is traditionally grown. Jalapeños have a wide range of uses including fresh, grilling, baking, stuffing, and processing into green sauce. Mild cultivars, such as 'TAM Mild Jalapeño II', have been developed for those who enjoy the distinctive flavor without pungency.

Table 11.3. Scoville ratings of capsaicins, pepper spray, and various types of peppers (Tainter and Grenis, 2001; Lopez, 2007; Roberts, 2008; The Scoville Scale, 2012).

	Scoville rating
Compound name	
Pure capsaicin	16,000,000
Nordihydrocapsaicin	9,100,000
Homodihydrocapsaicin	8,600,000
Product	
Police grade pepper spray	5,300,000
Common pepper spray	2,000,000
Pepper type or cultivar	
Red Savina habanero	350,000–580,000
Habanero	100,000–350,000
Scotch bonnet	100,000–325,000
Birds eye	100,000–225,000
Jamaican hot	100,000–200,000
Carolina cayenne	100,000–125,000
Bahamian	95,000–110,000
Tabiche	85,000–115,000
Thai chili	50,000–100,000
Tepin (chiltepin)	50,000–100,000
Piquin	40,000–58,000
Cayenne	30,000–50,000
Tabasco	30,000–50,000
de Arbol	15,000–30,000
Manzano	12,000–30,000
Serrano	5,000–23,000
Hot wax	5,000–10,000
Chipotle	5,000–10,000
Jalapeño	2,500–8,000
Ancho	1,000–2,000
Coronado	700–1,000
Anaheim	500–2,500
Pepperoncini	100–500
Pimento	100–500
Cherry	0–3,500
Banana pepper (Hungarian wax)	0–500
Sweet bell pepper, Cubanelle, Aji dulce	0

'**Cherry**' peppers (*C. annuum*), named for their resemblance to the tree fruit, are round green when immature and turn intense red at maturity. Cherry peppers are small and round and may be pungent or sweet depending on the cultivar (Table 11.3). Cherry peppers are consumed fresh or pickled in vinegar.

'**Scotch Bonnet**' (*C. chinense*) is a pungent cultivar popular in the Caribbean islands, Guyana, the Maldives, and western Africa (Table 11.3). The name is derived for its resemblance to a Tam

O'Shanter hat (Andrews, 1998). 'Scotch bonnet' fruit change from green to colors ranging from pumpkin orange to scarlet red as they advance from immature to mature. 'Scotch bonnet' is one ingredient that gives jerk (pork/chicken) and other Caribbean dishes their distinctive flavor.

'Habanero', habañero in Spanish, describes a *C. chinense* pepper with similar appearance, pungency, and flavor to 'Scotch bonnet' (Table 11.3). Although the names "habanero" and "Scotch bonnet" are sometimes used interchangeably, they are distinct cultivars. 'Habanero' fruit are slightly elongated and pointed compared to the compressed appearance of 'Scotch bonnet'. A typical mature 'habanero' fruit is 2–6 cm (0.8–2.4 in) long. There are subtle differences in flavor as well. Immature habaneros are green and transition to orange-red with maturity.

'Tepin', also called 'chiltepin', 'chiltepe', and 'chile tepin', is the botanical variety *glabriusculum* of *C. annuum* that is native to southern North America and northern South America (Singh, 2006). The name tepin is derived from a Nahuatl word meaning "flea". The plant is a shrub that attains a height of approximately 1 m (59 in), but sometimes reaches 3 m (118 in) (Richardson, 1995). The tiny tepin fruits are extremely pungent, red to orange-red, round or slightly ellipsoidal, and about 0.8 cm (0.31 in) in diameter (Table 11.3). When tepin fruit are dried for preserving, they become round even if they were slightly ellipsoidal when fresh.

Thai chili, also called bird's eye chili, is a very pungent pepper used in traditional dishes in Malaysian, Indonesian, Filipino, and other Southeast Asian cuisines (Table 11.3). The name bird's eye chili is generic and describes several small-fruited types including the "tepin" that are spread by birds. They are perennial producing small, tapered fruits, 2–3 cm (0.8–1.2 in) in length, similar in appearance to tabasco, with two to three fruits per node. Red is the most common mature fruit color. Taxonomically, bird's eye chili types were traditionally classified as *C. frutescens* L. but they are increasingly listed as *C. chinense* (DeWitt and Bosland, 1996).

Peppers are versatile and can be used in many food preparations. Both sweet and pungent types are often pickled in vinegar. Fresh peppers add color, texture, nutrients, and flavor to salads. Much like bell peppers, other chili peppers can be eaten raw in salads or cooked (boiled, baked, fried, stuffed) as main dishes or side dishes usually using less-pungent types. Ornamental cultivars have green or variegated foliage and many small, multicolored fruit that develop above the canopy exposing them to view. The fruit are edible allowing them to be used in container gardens in urban areas as both ornamentals and vegetables/spices. Ornamental fruit may be pungent or nonpungent (Bosland, 1996).

There is considerable variation in plant characteristics among cultivars. Like eggplant and tomato, pepper plants can be classified as determinate, semi-determinate, or indeterminate depending on how many nodes are produced before vertical growth is terminated with a flower cluster. Determinate plants are small and bushy, while indeterminate plants are tall with continuous vegetative growth. Semi-determinate plants are intermediate between the two types. Other important characteristics that distinguish cultivars include leaf cover, which is an important characteristic to protect fruits from sunburn, flowers per node, single or multiple, which helps determine fruit size, and position of fruit bearing, e.g. some cultivars have fruit that develop in the center of the plant, complicating harvest.

Genetic resistance has been incorporated into various cultivars of *C. annuum* by traditional plant breeding for the following diseases: bacterial leaf spot, cucumber mosaic virus, *Phytophthora*, pea enation mosaic virus, pepper mottle virus, potato virus Y, tobacco etch, tobacco mosaic virus, tabamovirus O, and tomato spotted wilt. Many of the commercial cultivars of *C. annuum* are F-1 hybrids. Most growers feel that the extra cost of F-1 hybrid seed pays for itself in greater productivity and fruit uniformity. Transgenic cultivars have been created experimentally but are not widely grown commercially at this time.

Source of pungency

Capsaicin (pronounced cap-say-i-sin), an odorless, colorless, and flavorless chemical, is responsible for the pungency in peppers. Capsaicin stimulates chemoreceptor nerve endings in the skin and especially mucous membranes. Human taste buds can detect as little capsaicin as 10 ppm. A single drop diluted in 100,000 drops of water will produce a persistent burning of the tongue. Biting into a pepper stimulates nerve receptors in the mouth, sending a pain signal to the brain, inducing sweating, salivation, and gastric flow in an attempt to rid the body of the irritation. Capsaicin is a stable alkaloid ($C_{18}H_{27}NO_3$) with a chemical structure similar to peperin ($C_{17}H_{19}NO_3$) that gives the unrelated black pepper seasoning its bite (Bosland, 1996). Capsaicin is a vanillyl amide of isodecylanic acid chemically related to vanilla and very acrid (Guzman et al., 2011).

Capsaicin is concentrated in glands located in the pepper's placenta, the inner white tissue that supports the seeds. Cutting or chewing disrupts these glands, spreading capsaicin throughout the fruit tissue giving the impression that the whole pepper is pungent. However, if the fruit wall is isolated without disrupting the placenta where the capsaicin glands are located, the tissue will not taste spicy. The key is to break the fruit in half by pulling or snapping so the capsaicin glands are not ruptured.

There is a scientific way to assess the pungency of peppers based on Scoville units (Peter, 2001). The scale is named after its creator Wilber Scoville, an American pharmacist. Scoville heat units (SHU) indicate the amount of capsaicin present in peppers and is therefore a quantitation of the spicy heat or pungency (Table 11.3). The greater number of SHU the more pungent a pepper will taste.

Pungency is usually determined by high performance liquid chromatography (HPLC), which identifies and measures the concentration of heat-producing chemicals like capsaicin and its relatives. A measurement of one part capsaicin per million corresponds to about 15 Scoville units. However, this technique under estimates pungency compared to the actual Scoville method and results may vary up to 50% among laboratories.

Purified capsaicin has its own commercial uses. Capsaicin induces a warm feeling when applied in concentrated form to the skin and is used medicinally in sore-muscle remedies. It is also used to give the "bite" to some commercial ginger ale brands and is the powerful irritant in some "pepper" sprays.

Production and Culture

Field establishment

Peppers are established by either direct seeding or transplanting. Plants can also be propagated vegetatively by rooting cuttings, but this practice is rarely used for large-scale commercial production.

Direct seeding was the standard method of establishment in long-season areas before modern transplant technologies and F-1 hybrids seeds were developed. Direct seeding is more feasible when less expensive open-pollinated cultivars are grown. Today, direct seeding is still used for large-scale production of mechanically harvested peppers for processing (Bevacqua and VanLeeuwen, 2003). Direct seeding makes for higher density plantings that are preferred for pimiento pepper production in Spain (Gil Ortega et al., 2004).

Seed of *C. annuum* cultivars are flat, smooth, typically pale yellow, ovoid, and 3–5 mm (0.08–0.12 in) long. One gram contains about 150–160 seeds (approximately 4,245 seeds/oz). Newly harvested seeds may require after-ripening to achieve full viability and vigor depending on the cultivar and production practices. Even high vigor seeds tend to germinate slower than tomato by comparison. High quality pepper seeds germinate in 6–10 days at 15.6–29°C (60–85°F) but very slowly at 15°C (59°F). Seed enhancement technologies such as seed coating (pelleting and filmcoating), controlled hydration (priming) or gel coatings containing growth-promoting substances such as gibberellic acid can improve the speed and synchrony of pepper seedling emergence (Halmer, 2008). Seed priming in particular can effectively reduce the mean time to germination, particularly at low temperatures, and is a commonly used commercial seed treatment (Bradford et al., 1990; Khan et al., 1992).

Plant density varies among cultivars but frequently range from 25,000 to 30,000 plants/ha (10,416–12,500 plants/acre). Plant spacing ranges from about 40 to 50 cm (16–20 in) within rows to approximately 75 cm (30 in) between rows and depends on the cultivar grown. Closer spacing tends to reduce fruit size, although high densities have the advantage of providing shade that may reduce sunburn.

Transplanting

Transplanting is the standard method of establishing pepper plants in short-season areas and when expensive F-1 hybrid seeds are grown. Pepper transplants are considered "moderate" in their ability to survive transplanting because root and shoot growth are slower compared to crops that are considered easy to transplant like tomato (*Solanum lycopersicon*) or cabbage (*Brassica oleracea* L. Capitata group) (Loomis, 1925). Containerized transplants grown in shallow plastic trays with individual cells for each plant are preferred for the establishment of peppers (Cantliffe, 2009). Also known as plugs, containerized transplants minimize root disturbance, shock, and seedling diseases compared to field-grown, bare-root transplants (Styer and Koranski, 1997). Transplants enable a late-maturing crop such as habanero peppers to be grown in short-season areas because field time is reduced. Since pepper seeds are slow to germinate and seedlings grow slowly early in the season, transplanting can improve weed management without additional thinning costs needed after direct seeding. Furthermore, substantial water is required to establish a direct-seeded pepper crop compared to transplanting, which is an important consideration when growing irrigated peppers.

Grafting

Grafting is a common practice in Asian countries (China, Korea, Japan, Taiwan) and the Mediterranean area (Spain, Italy, Israel, Tunisia, and Turkey). The protected cultivation of pepper relies heavily on grafted plants in many parts of the world (Lee and Oda, 2003). However, the demand for grafted transplants is anticipated to increase worldwide due to loss of methyl bromide fumigant and the adoption of more sustainable production practices.

The primary reason for grafting pepper scions on special rootstocks is to increase plant vigor, uniformity, and disease tolerance. Several commercial pepper rootstocks show resistance to major soil-borne pathogens such as *Phytopthora capsici*, *Verticillium dahliae*, *Fusarium oxysporum*, and *Meloidogyne* spp. (Oka *et al.*, 2004; Santos and Goto, 2004; Saccardo *et al.*, 2006). In addition, grafted plants may have higher tolerance to abiotic stresses such as salinity, root hypoxia, and extreme temperatures.

Cleft, approach, micro- and tube grafting are all techniques that can reliably join pepper, eggplant, and tomato scions with compatible rootstocks. Unlike some other species where dissimilar rootstalks give the best performance, pepper scions are most compatible with rootstocks from the same genus. In Taiwan, several lines from *C. baccatum*, *C. frutescens*, and *C. chacoense* were identified as suitable rootstocks for sweet pepper production during hot-wet (summer) and hot-dry (fall) seasons (Palada and Wu, 2008). In an Italian study, two bell pepper hybrid cultivars (scions) were cleft-grafted on to five commercial rootstocks and on some combinations growth promotion, improvements in fruit quality, and a 25% increase in yield were reported compared to ungrafted plants (Colla *et al.*, 2008).

Site selection and field preparation

Peppers can be grown on a wide range of soil types from sandy to fine textured clays with a pH range from 6.5 to 7.0. Peppers are very sensitive of waterlogged conditions so poorly drained soils of any kind must be avoided. Waterlogged plants tend to defoliate and are more susceptible to root diseases.

The soil can be tilled in a number of ways for pepper production. Rotovators can create a smooth bed if there is little residue. In some cases, fields are plowed, particularly if heavy residue must be incorporated, tilled, and the beds reshaped. Most commercial pepper production in the USA occurs on 15–20 cm (6–8 in) raised beds to promote drainage. Beds can be various sizes but are commonly 122 cm (48 in) wide and 183 cm (72 in) center-to-center. No-till pepper production is possible as well, particularly for later plantings (Fig. 11.7). Soil temperatures are cooler in no-till production systems, which can be problematic for early plantings in short-season areas.

Fig. 11.7. A blooming bell pepper plant grown no-till in southwest Virginia. A small-grain cover crop was killed before planting to provide mulch for moisture conservation and to control weeds.

Growth and development

Peppers do not grow well under cold conditions when night temperatures fall below 10°C (50°F). Low temperatures tend to limit flavor and color development, and plants and fruit are susceptible to chilling injury. Since early season growth is slow in cold weather, planting should be delayed until soil and air temperatures warm in spring. Mean day temperatures of 20–25°C (68–77°F) are ideal.

Peppers are more tolerant of high temperatures than tomatoes. Pungent types tend to be more high-temperature tolerant and set fruit better at high night temperatures compared to bell peppers. Flowers are not properly fertilized at temperatures below 15°C (59°F) or above 32°C (90°F) because of poor pollen production. Peppers pollinate better at higher temperatures than tomato. Cool temperatures can sometimes result in parthenocarpic fruit development. Pollination and fertilization are optimum between 20–25°C (68–77°F). In general, small-fruited pepper cultivars are more tolerant of both high and low temperature extremes. In outdoor plantings, wind movement and insect activity is sufficient for self-pollination to occur. Plants are not photoperiod sensitive. Flowering usually begins between 1 and 2 months after planting with fruit achieving desired or full size about 1 month after anthesis (Rubatzky and Yamaguchi, 1997).

Pepper plants have extensive root systems that are moderately deep-rooted with taproots reaching a depth of more than 1 m (39.4 in). Although peppers are generally drought resistant, intermittent periods of moisture, and/or nutritional stress can dramatically reduce plant growth, fruit size, and yield. Flowers and young fruit may abscise if water stress occurs during flowering.

The water requirements for pepper production range from 400–1,000 mm (16–39 in). The broad range reflects differences in cultivar and environmental conditions. The general irrigation recommendation of a minimum of 2.5 cm (1 in)/week applies to pepper. On sandy soils, more water may be required. Drip irrigation provides uniform and efficient water application directly to the root zone. Furrow irrigation may be an option in areas where water is plentiful and the infrastructure supports it.

Fertilizer and nutrition

Peppers do not remove as many nutrients from the soil as tomatoes. When both the plants and fruit are considered, a pepper crop will remove approximately 134 kg N, 13.4 kg P, 134 kg K/ha (120 lb of N, 12 lb of P, and 120 lb of K/acre) (Maynard and Hochmuth, 1996). Fertilization should be performed in response to both soil and foliar testing to ensure that only the nutrients needed by the plant are applied. Peppers respond to N fertilization, which is often applied prior to transplanting by row banding. Depending on testing results, a second application of N may be applied again at first bloom. However, excessive N wastes costly fertilizer, may damage the environment and favors vegetative growth while inhibiting reproductive growth thus delaying maturity and decreasing marketable yields.

Fertigation is common with plasticulture pepper production. Phosphorus fertilizer may be banded at transplanting because of limited solubility and the rest of the nutrients fertigated during the season in response to tissue or petiole sap mineral analysis. A liquid starter solution high in P is often applied to the root zone at transplanting to stimulate early growth. A standard starter solution recipe may consist of 1.5 kg/200 l (3 lb/50 gal water) of 8N-24P-8K or 10N-52P-17K with 118 ml (0.03 oz) applied to each transplant (Maynard and Hochmuth, 1996).

Culture

Peppers should not be grown in the same field in consecutive years to prevent a buildup of disease. Rotating peppers with other crops outside the family Solanaceae is necessary, especially when pathogen populations are very high.

Peppers are often grown on plastic mulch to conserve moisture, modify soil temperatures, and in some cases control weeds. Peppers are grown under white or reflective mulch when soil temperatures are excessive (Fig. 11.8).

In short-season areas, clear or black plastic can heat the soil and accelerate development of early season plantings (Waterer, 2000). In the western USA, production is often on unmulched raised beds with furrow or drip irrigation.

Peppers are commonly planted in double rows on top of 122 cm (48 in) wide beds although many other widths are possible. The in-row spacing varies among cultivars but generally ranges from 46–71 cm (18–28 in). For outdoor production, determinate or semideterminate cultivars are preferred. Bell pepper plants can be very top heavy when loaded with fruit. Lodging can occur particularly when

Fig. 11.8. Double row pepper production in Florida on raised beds with white mulch to reduce soil temperature.

soils are saturated. The string-weave system, common for tomato culture, may be employed to keep pepper plants with heavy fruit loads upright in long-season areas where mature plants set fruit near the top of the plant. Pepper plants grown in short-season areas are usually not staked.

Greenhouse production

Peppers production in glasshouses and other protective structures is significant primarily in areas were crops cannot be matured outdoors. Indeterminate cultivars developed for greenhouse production are carefully trained vertically on strings and pruned to efficiently use greenhouse space (Fig. 11.9).

Usual plant spacing is approximately 50 × 90 cm (20 × 30 in) and depends on the cultivar grown. Well-grown and managed crops on average can produce yields of approximately 15 kg/m^2 (3 lb/ft^2).

The Dutch greenhouse system uses fertigated rockwool media similar to the system used for greenhouse tomato production. Night temperatures should be maintained <15°C (59°F) for best fruit development. Bees or mechanical agitation are used to improve greenhouse pollination. A bumblebee hive containing approximately 60 bees per 1,440 m^2 (16,000 ft^2) provides sufficient greenhouse pollination. Manual pollination by workers is generally less efficient and considerably more expensive.

Harvesting and marketing

Fresh market

The harvesting of peppers is rather flexible and based on color, fruit size, and/or wall thickness. Generally, the surface of older fruit is firmer, shinier, and waxier than younger fruit. More-mature fruit tend to resist shrinkage after harvest more than less-mature fruit whose cuticle is not fully developed. Green fruit are harvested after reaching their mature size but before physiological maturity and color change occur. Fruit are physiologically mature when the seed they contain are fully developed and they obtain their mature color, which is often red or yellow. Some growers use accumulated heat units for scheduling harvests.

Bell and other peppers for fresh market are hand harvested. There is no abscission zone, so the pedicle of fruits must be carefully cut or expertly snapped by hand without damaging the brittle branches and surrounding leaves. Destructive mechanical harvesting of fresh market peppers can reduce labor costs but is not widely used because fruit damage is increased and yields reduced. Mechanical harvesting is more common for processing crops because fruit damage is less critical if no storage is needed before processing (Fig. 11.10).

Postharvest handling

Fruits are susceptible to mechanical injury and may be easily damaged during shipment. After harvest, fruit for fresh market should be rapidly cooled to about 10°C (50°F) to remove field heat. Peppers are washed in ambient or warm chlorinated (300 ppm) water to reduce postharvest diseases before storage or shipment. Bell peppers in particular are chilling sensitive, which causes off-flavors, reduced

Fig. 11.9. Summer greenhouse production of yellow bell peppers in Israel. In this case, the greenhouse is used to produce peppers because outdoor temperatures are too low. The greenhouse provides an optimized, protected environment with efficient water use, improved disease-free rooting medium, temperature control, humidity control, and freedom from fly-in insect pests.

Fig. 11.10. Mechanically harvesting green peppers for processing.

storage life, pitting, decay, discoloration of the seed cavity, and excessive softening (Fig. 11.11). As a general rule, bell peppers should not be stored below 10–13°C (50–55°F) for extended periods, although brief exposure to 7–8°C (45–46°F) may not damage fruit (Cantwell, 2013b). Chili peppers are not as chilling sensitive as bell peppers. Chili peppers stored above 7.5°C (45°F) suffer more water loss, shrivel, color change, and decay. Storage at 7.5°C (45°F) is considered the best for maximum shelf-life of 3–5 weeks. Chilis can be stored at 5°C (41°F) for at least 2 weeks without visible signs of injury. Storage at 5°C (41°F) reduces water loss and shrivel, but after 2–3 weeks, chilling injury is mostly detected as discoloration of the seeds. Ripe or colored chilis are less chilling sensitive than mature-green chilis (Cantwell, 2013a).

Unlike tomatoes, peppers are not climacteric. Green bell and many chili peppers do not ripen after harvest when treated with the gaseous plant hormone ethylene. Bell pepper ethylene production rates are very low in the range of 0.1–0.2 μl/kg/h at 10–20°C (50–68°F) (Cantwell, 2013b).

Some chilis, such as habaneros, show increased ethylene production during ripening and may produce over 1 μl/kg/h at 20–25°C (68–77°F). Responses to added ethylene depend on the particular type of chili. Chili 'Poblanos' for example may respond to ethylene treatment, while 'Jalapeño' peppers do not.

As with bell peppers, holding partially colored chili peppers at warmer temperatures of 20–25°C (68–77°F) with high humidity (>95%) is effective to complete color development. Adding ethylene may further enhance ripening but the response is cultivar dependent (Cantwell, 2013a).

Red and yellow physiologically mature bell peppers are more expensive to produce than green because fruits must remain on the plant longer to develop their mature color. This reduces total yields because mature peppers inhibit the set and growth of new fruit. Fruit are also susceptible to disease and insect attack during their extra time on the plant. Mature red fruit also have ten times more provitamin A content, softer texture, and higher soluble sugar content, which also adds to their value (Rubatzky and Yamaguchi, 1997).

The red color develops because of increased carotenoid synthesis and a breakdown of chlorophyll. As green bell peppers begin to turn to red, their color first changes to brown during the transition because both the red carotenoids and green chlorophyll are in the fruit wall simultaneously. This is often called the "chocolate" stage. Once harvested, chocolate peppers will remain brown because they do not easily ripen after harvest. In some markets, chocolate peppers have less value because the color is unappealing to consumers.

Fig. 11.11. Grading and packing red bell peppers outdoors in Florida.

High relative humidity is necessary to limit desiccation. Water-soluble wax is sometimes used to coat peppers to decrease shriveling and increase storage life. However, waxing may increase the possible incidence of bacterial soft rot. Bell peppers are also wrapped with a thin layer of plastic film individually or grouped in a plastic tray to preserve freshness and reduce mechanical damage from handling. Waxed or film-wrapped peppers can be stored from 10–14 days at 13°C (55°F). Shelf life varies among cultivars. Deterioration is often due to moisture loss and some fruit types are more prone to desiccation than others.

Peppers generally do not respond well to controlled atmosphere storage. Low O_2 atmospheres (2–5% O_2) alone have little effect on quality and high CO_2 atmospheres (>5%) can cause pitting, discoloration, and softening, especially at temperatures >10°C (50°F). Atmospheres of 3% O_2 + 5% CO_2 are more beneficial for red than green peppers stored at 5–10°C (41–50°F) (Cantwell, 2013b).

Processing

Peppers may be processed by canning or drying. Total solids usually increase with fruit age as well as carotene and ascorbic acid content. Processed peppers, such as pimientos and paprikas and many chilis, are usually harvested at physiological maturity when fruit are fully red or yellow. Jalapeño and other types are processed in the immature green stage. Fruit for canning are detached by carefully breaking or by cutting through the pedicel (stem) to minimize damage. It is important that fruit have an intact pedicel to reduce desiccation and pathogenic attack prior to canning.

Peppers for processing are destructively harvested either by hand or machine. For small scale production of chili powder, paprika, and other dried pepper products, entire plants with fully colored fruit are cut and field dried when environmental conditions are favorable. For field dehydration, peppers remain attached to the plant until desiccated. Field drying reduces the time and energy required for oven drying if needed, but may result in a product that is less clean. The cut plants are often laid on plastic sheets for drying. Ethephon, an ethylene-releasing compound, is occasionally applied to accelerate fruit coloration in the production of some types of dry chili and paprikas. The dried peppers are separated from the plants by hand so that crushing or grinding equipment can process the dried peppers into various products (Walker, 2013).

For large-scale production, mechanical harvest of the red chili and paprika crops has been widely adopted in recent years, with more than 80% of the commercial crop in the USA harvested by machine (Walker, 2013).

There are several different types of mechanical harvesters currently used for red chili and paprika. The three most common picking heads are the finger-type, the belt-type, and the double helix. The finger-type head is designed with a series of counter-rotating bars with fingers. The fingers comb the plants to strip off the pods on to conveyor belts. The mechanism is aggressive and may harvest unwanted plant debris. A belt-type harvester has two sets of counter-rotating vertical belts imbedded with fingers that comb both sides of a plant. The most widely used picking head is the double-helix design (Marshall, 1997). The helices may be vertical or oriented at an angle. The helices rotate in opposite directions to each other, snapping pods off of the plants and flipping them on to conveyor belts on either side.

Larger, commercial operations employ either tunnel or belt dryers rather than use sun drying. Both systems are more reliable and sanitary than sun drying (Wall, 1994). These methods also use more energy, especially early in the harvest season when succulent pods are harvested and must be dried. Another drawback to tunnel and belt dryers is their limited capacity (Walker, 2013).

Nutritional Values

Red bell peppers are one of the very best sources of vitamin A. Vitamin A content increases dramatically as the fruit ripen and turn red (Table 11.4). Vitamin C and sugar content increase with color change as well. Cultivars that turn yellow when mature do not contain as much vitamin A as red peppers.

Economic Importance and Production Statistics

World production of dried peppers exceeds fresh by 2 million ha (5 million acres) to 1.8 million ha (4.5 million acres) (FAOSTAT, 2011). Total world production of fresh peppers including chilis was 29,601,175 metric tonnes (32,629,710 short tons)

Table 11.4. Nutritional comparison of red and green bell peppers (USDA Nutrient Database, 2011).

Constituent	Green bell	Red bell
	Amount/100g (3.5 oz) edible portion	
Water (%)	93.0	91.0
Protein (g)	0.9	0.8
Fat (g)	0.3	0.6
Carbohydrate (g)	4.4	5.3
– Sugars	2.4	3.5
Thiamine (mg)	0.06	0.11
Riboflavin (mg)	0.02	0.08
Niacin (mg)	0.4	0.7
Vitamin C (mg)	160	220
Vitamin A (IU)	530	5,700
Vitamin E (mg)	0.4	1.6
Vitamin K (µg)	7.4	4.9
Dietary fiber (g)	1.7	2.1
Ca (mg)	10	7
Fe (mg)	0.34	0.43
Mg (mg)	10	12
Mn (mg)	0.12	0.11
P (mg)	20	26
K (mg)	175	211
Na (mg)	3	4
Zn (mg)	0.13	0.25

in 2011, while total dried production was 3,457,533 metric tonnes (3,811,278 short tons) in 2011. Leading producers of fresh peppers include: China 705,000 ha (1,742,093 acres), Indonesia 239,770 ha (592,484 acres), Mexico 144,391 ha (356,798 acres), Turkey 93,826 ha (231,849 acres), Ethiopia 89,205 ha (220,430 acres), Nigeria 57,382 ha (141,794 acres), South Korea 47,388 ha (117,098 acres), Egypt 39,666 ha (98,017 acres), the USA 30,110 ha (74,403 acres), and North Korea 22,500 ha (55,598 acres). The leading producers of fresh peppers in western Europe are Spain 17,193 ha (42,485 acres) and Italy 10,327 ha (25,519 acres).

Leading producers of dried peppers are: India 869,467 ha (2,148,500 acres), Ethiopia 330,000 ha (815,448 acres), Myanmar 131,783 ha (325,643 acres), Bangladesh 104,967 ha (259,379 acres), Pakistan 68,370 ha (168,945 acres), Thailand 64,341 ha (158,990 acres), Vietnam 63,538 ha (157,005 acres), Romania 54,403 ha (134,433 acres), China 42,773 ha (105,694 acres), Nigeria 35,000 ha (86,487 acres), Mexico 31,471 ha (77,767 acres), and Egypt 17,327 ha (42,816 acres).

As a general rule, nonpungent types of pepper are most popular in northern Europe and northern North America, while the more pungent chili peppers are widely consumed in the tropics and subtropics (Cichewicz and Thorpe, 1996). In the USA, per capita pepper consumption has steadily increased over the past 80 years. Americans consumed 0.7, 1.1, and 1.6 kg (1.5, 2.4, 3.6 lb) per person in 1930, 1960, and 1980, respectively, and the average increased again from 6.9 kg (15.3 lb) per person in 2005 to 7.5 kg (16.6 lb) in 2010. This steady climb can be attributed to Hispanic influences on US cuisine, greater availability of all types, and perceived health benefits from consuming peppers. Analysis of the types consumed shows that bell peppers grew from 4.2–4.8 kg (9.2–9.8 lb), while chili pepper consumption grew from 2.8–3.0 kg (6.1–6.8 lb) per person per year (Burden, 2012). In 2011, California led the USA in bell pepper production and was followed by Florida. California was also the leader in chili pepper production with New Mexico second (Burden, 2012).

Diseases and Physiological Disorders

There are a number of serious disease problems that affect pepper (Pernezny and Momol, 2006). Please consult local recommendations for control measures. Anthracnose (*Colletotrichum acutatum*, *C. gloeosporiodes*, *Colletotrichum* spp.) spores may infect fruit any time during development, but symptoms are usually expressed on mature fruit as small, water-soaked lesions that quickly develop into larger sunken areas. A dark growth of the fungus may be visible in these lesions, with tan to pink concentric circles of spores evident in some cases. Occasionally, leaf spots and stem dieback may occur as well. Anthracnose can be controlled by using pathogen-free seed and avoiding overhead irrigation whenever possible (Pernezny and Momol, 2006).

Bacterial spot (*Xanthomonas euvesicatoria*) begins as small, water-soaked spots on leaves that may reach 0.6 cm (0.25 in) in diameter, turn dark brown and appear greasy. Scabby lesions may also appear on the fruit. During periods of heavy rainfall or humidity, spots on leaves may coalesce causing "blight" symptoms and abscission. Bacterial spot can be seed-borne and spread rapidly among transplants.

Bacterial soft rot (*E. carotovora* pv. *carotovora*) occurs primarily after harvest and during shipment. Infected fruits become soft and turn to mush. The rot often begins in the stem and advances quickly into the fruit during transit and storage. Field symptoms are quite obvious as fruit soften and sag from the pedicel like a balloon filled with water. Softened areas usually turn gray. The invasion by numerous soft rot organisms may cause a characteristically foul odor. Insect vectors spread the bacteria. To control soft rot, avoid bruising and wounding, harvesting wet fruit, and exposing fruit to the sun, which causes sunscald and secondary infection (Pernezny and Momol, 2006).

Frogeye spot (*Cercospora capsici*) causes circular leaf lesions (approximately 0.6 cm (0.25 in) in diameter) with light tan to white centers and narrow dark borders. Heavy infection may cause leaf abscission and reduced yield.

Gray leaf spot (*Stemphylium solani*) infection causes circular spots on leaves. These spots are at first brown but later turn light tan to white with sunken centers and reddish-brown margins. Spots can also appear on stems, petioles, and fruit pedicels, but not on fruit or flower petals.

Phytophthora blight (*Ph. capsici*) can affect all parts of the plant, causing seedling death as well as root rot, stem canker, leaf blight, and fruit rot in older plants. Stem infection at the soil line is common, causing sudden wilting and death. When mature plants become infected at the nodes, whole branches die. Individually infected leaves exhibit small circular to irregular leaf spots that appear scalded, dry, and bleached to a light tan color with a papery consistency. Establishing pepper fields in well drained fields while practicing crop rotation will help prevent the disease (Pernezny and Momol, 2006).

Sclerotinia stem rot (*Sclerotinia sclerotiorum*) is most common under cool, damp conditions. The causal fungus infects the stem at the soil line, individual petioles of leaves and occasionally fruit close to the soil surface. Stem infections frequently girdle the stem causing wilting and death. When weather is moist, the white mycelium will often grow up the stem surface several inches (approximately 17.5 cm (6.9 in)) above ground. Petiole or bud infections proceed rapidly downward in the plant. Entire branches may be girdled in this manner. Fruit infected directly from the soil surface or downward through the pedicel rot quickly into a watery mass. The fungus survives as sclerotia formed in stems and lesions associated with diseased fruit. Rotations involving other susceptible crops such as cabbage, celery, lettuce, potatoes, or tomatoes should be avoided. Deep-plowing fields with a history of this disease to bury fallen sclerotia benefits control. Flooding fields in the off-season with water for 6 weeks may also kill sclerotia.

Southern blight (*Sclerotium rolfsii*) occurs in warm humid weather causing root and stem infection, wilting, and ultimately death. A white collar of coarse white fungal mycelium develops on the stem at the soil line. Tan sclerotia are the size of mustard seed and provide an over-seasoning mechanism in soil. Wind, water, or equipment may spread the fungus in a field. Rotating with a grass crop and deep plowing to bury the sclerotia may provide control.

Wet rot (*Choanephora cucurbitarum*) produces a blossom blight, fruit rot, and occasionally leaf blight. Fungal hyphae capped with black heads (sporangia) grow on petals as they senesce, causing blossom drop. Young fruit may become infected, soften, and abort shortly after pollination. Maintaining good air circulation and fungicides may provide control.

Many virus diseases affect pepper. Viruses (e.g. cucumber mosaic, pepper mottle, potato Y, tobacco etch, tobacco mosaic, tomato spotted wilt) are difficult to distinguish as single or multiple viruses may infect a field. Most of these viruses induce degrees of mosaic, mottle, vein banding, and plant stunting. Malformation, leaf cupping, and fruit distortion may also occur. Accurate diagnosis is dependent on laboratory testing (Pernezny and Momol, 2006).

Tobacco mosaic virus (TMV) is commonly mechanically transmitted during transplant production, harvesting, and packing. Pepper mottle, potato Y, and tobacco etch are primarily transmitted by aphid feeding. Tomato spotted wilt virus (TSWV) is transmitted by thrips. These viruses are known to survive in numerous weed hosts such as ground cherries (*Physalis* spp.), nightshades (*Solanum* spp.), common groundsel (*Senecio* sp.), wild tobacco (*Nicotiana* sp.), toadflax (*Linaria* sp.), sicklepod (*Cassia* sp.), and jimson (*Datura* sp.).

Cultivars resistant to TMV should be grown when possible. Workers, especially those who use tobacco products, should wash hands with strong soap and water or 70% alcohol before handling plants to prevent the spread of TMV. To reduce insect transmission of viruses (e.g. tobacco etch virus, potato Y virus, and cucumber mosaic virus) wild host plants around fields should be eradicated. Old infected crops should be destroyed well before planting subsequent crops. A 15 m (50 ft) barrier strip of a non-susceptible crop (corn, wheat, etc.) can trap flying insects before they infect a field.

Physiological disorders include blossom-end rot, which appears as a dry black lesion at the blossom end of fruit. Blossom-end rot is caused by a localized Ca deficiency in rapidly growing fruit tissue induced by drought stress. Some cultivars are more susceptible to this disorder than others. Pepper speck is a disorder of unknown cause that appears as spot-like lesions that penetrate the fruit wall. Some cultivars are more susceptible than others (Cantwell, 2013b).

Direct sunlight can kill or severely damage exposed fruit tissue, resulting in sunscald. Scalded areas of the fruit have a bleached appearance and are more susceptible to bacterial soft rot and secondary infection by other pathogens.

Insect Pests

A number of arthropod pests can seriously damage peppers. Bell peppers for fresh market must be essentially blemish free for successful marketing. Insect/mite pests reduce yield and quality by directly feeding on the plant and/or its fruit and by vectoring destructive agents such fungi and viruses, especially tobacco etch and TSWV.

Caterpillar-like pests including fall armyworm (*Spodoptera frugiperda*), beet armyworm (*S. exigua*), southern armyworm (*S. eridania*), yellowstriped armyworm (*S. ornithogalli*), cutworms (various species), and loopers (various species) damage foliage and fruiting structures. Young larvae feed under leaf surfaces of leaflets leaving the upper epidermis intact creating a "windowpane" effect. Older larvae, which may grow to 7.5 cm (3 in) in length for some species, consume foliage and eat large holes on the fruit surface, ruining its appearance and creating avenues for secondary infection. The beet armyworm is a major pest of bell pepper in Florida. Cutworm larva damage plants at night when they climb the plants and feed on the foliage, or they cut seedlings off at the soil surface (Olson et al., 2005).

If possible, these pests should be managed in the first several larval instars with repeated applications of Bt every 3 or 4 days during peak hatching periods. Older larvae may be controlled by hand removal on small plots, approved insecticides or biological control measures.

Pepper weevils (*Anthonomus eugenii*) are small shiny, brown- or gray-colored snout beetles, about 3.5 mm (0.13 in) long. Adults feed on leaf and flower buds. Females bore a small hole in developing fruit or flower buds. The hole is plugged with fecal matter (frass) after an egg is deposited. A tiny, legless grub hatches from the egg and eats its way toward the core of the fruit where it feeds on seeds and pulp. Both adult and larval feeding causes bud/fruit drop. Punctures in the fruit, caused by worms, frequently enable fungal colonization and destruction of the tissue. Pepper weevil populations are monitored visually and with sticky yellow card traps. The action threshold for this pest is one adult per 400 terminal buds or 1% of buds infested (Capinera, 2005).

Broad mite (*Polyphagotarsonemus latus*) adults are tiny white and usually found on the underside of young, developing leaves. Broad mite feeding distorts leaves, causing them to become thickened and narrow. Flower abortion and dark, smooth russeting of fruit occur when feeding is heavy. Broad mite is mostly an early- or late-season pest and may be effectively managed with sulfur or insecticidal soap.

Adult aphids (superfamily Aphidoidea), also known as plant lice, are delicate small sap-sucking insects and may be among the most destructive pests on pepper plants. Winged and nonwinged forms are all female and give birth to nymphs. Nymphs are smaller but otherwise adult-like in appearance; they quickly mature in only 7–10 days. Heavy aphid infestations may result in plant debilitation, sooty mold growth on honeydew, and leaf distortion. Aphids also spread plant viruses such as tobacco etch virus. Acquisition and transmission of these viruses is rapid but the virus does not persist in the aphid for more than several minutes. Most transmission results from winged aphids probing several different plants in succession rather than by colonized feeding. Aphids maybe managed with insecticidal soaps or neonicotinoids, which work like the natural insecticide nicotine.

Both nymph and adult melon thrips (*Thrips palmi*) abrade pepper buds, flowers and/or leaf tissues, and then suck the exuding sap. When present in high numbers, melon thrips produce silvering, yellowing and bronzing of affected tissues. Leaves may crinkle and die, growing tips may become stunted, discolored, and deformed, and fruits may abort or develop scar tissue. The overall effect is a loss of plant vigor and a reduction in marketable yield. Adults are quite mobile and can move into new plantings quickly from old fields. In addition pepper, eggplant, cucumber, potato, beans, and watermelon are also susceptible to melon thrip damage. New plants should be isolated from old infested fields. Some conventional insecticides seem to stimulate melon thrip populations (Funderburk et al., 2004).

Western flower thrips (*Frankliniella* sp.) insert an egg in flower parts and very small fruit causing flower abortion, poor fruit set, and impaired fruit development. The larvae then feed on the flowers and fruit after hatching. This pest is also a vector of TSWV (Funderburk *et al.*, 2004).

Thrips (order Thysanoptera) are early season pests and populations decline greatly in the summer and fall as natural enemies provide control. Pepper flowers can tolerate up to 15 thrips per flower for short periods with little damage. A biological control program integrating pirate bug (family Anthocoridae) predation with reflective mulch can control this pest and slow the spread of tomato spotted wilt disease (Funderburk *et al.*, 2004).

Adult leafminers are small flies (*Symphyta*), approximately 2.4 mm (0.1 in) long, with black heads, yellow between the eyes, and black thorax. Females have a tube ovipositor at the end of the abdomen used to puncture the upper leaf surface for egg laying. The white, oval egg is inserted in the leaf tissue, but many punctures (called stipples) are used by the adult for feeding and do not contain eggs. The larva, a yellow maggot with black, sicklemouth hooks, feeds between the upper and lower leaf surface for approximately seven days, leaving a serpentine mine containing a black string of frass. The mature larva exits from the mine and falls to the ground (or plastic mulch) where it pupates. The adult emerges in 7–14 days. Serpentine mines in leaves reduce photosynthetic area and may provide entry points for foliar pathogens. Heavily damaged leaves become necrotic, predisposing fruit to sunscald.

EGGPLANT

Origin and History

In China, India, Japan, and many Mediterranean countries, eggplants are a popular and widely grown vegetable. The name "brinjal" is of Indian or Arabic derivation, while the English name "eggplant" is because plants produce white fruit resembling a chicken egg (Rubatzky and Yamaguchi, 1997). Aubergine is another common name for eggplant used in some English-speaking countries.

Eggplant is believed to be native to the Indian subcontinent where large-fruited cultivars were likely domesticated and many wild forms exist. From India, domesticated eggplant spread eastward into China, which became a secondary center of origin. By the 5th century BC Arabic traders were responsible for subsequent movement of eggplant to Africa and Spain (Rubatzky and Yamaguchi, 1997). Eggplant cultivation in the Mediterranean region and the Americas is comparatively recent.

Botany and Life Cycle

There are more than 1,000 species of eggplant in the genus *Solanum*, but only a few are grown commercially as vegetables. The fruit of *Solanum melongena* cultivars vary from egg-shaped to long and slender and are grown commercially in many regions of the world. *Solanum macrocarpon* is known as Gboma eggplant in West Africa. The Chinese scarlet eggplant, *S. integrifolium*, is grown for food and as an ornamental. Wild species of eggplant generally have bitter fruit and sharp spines that cover much of the plant, including the calyx. The bitterness is due to glycoalkaloids. However, most domesticated cultivars are not bitter and have few or no spines. Fruit tissues contain high levels of phenolics and turn brown when cut or damaged due to rapid oxidation by polyphenol oxidase.

Eggplants are short-lived perennials in the tropics but are cultivated as annuals in temperate zones. Some cultivars are determinate (Fig. 11.12) and grow as a bush to a height of 0.5 m (19.7 in) while indeterminate cultivars may reach a height of 2.5 m (98.4 in). Root systems have strong taproots and are moderately deep and spreading. Stems are erect and branching, becoming woody with maturity. Leaves are generally large, alternate, and simple and

Fig. 11.12. A determinate eggplant cultivar that produces long, green fruit.

may have thorns on the stems and leaves depending on the cultivar. Leaf blades are ovate to ovate-oblong with wavy lobed margins. The leaf base is usually rounded.

Perfect flowers, solitary or multiple in a determinate simple inflorescence, usually form opposite or nearly opposite leaves rather than in leaf axils. Flowers are 2–3 cm (0.8–1.2 in) in diameter with purplish pubescent corollas that are predominantly self-pollinated although a small percentage of outcrossing may occur. Flowers generally remain open for 2 or 3 days and are most receptive during the morning. The fruit is a large, solid berry.

Seed are small and light brown and imbedded in the placental tissue. There are about 225 eggplant seeds/g (6,367/oz). Fruit shapes are diverse and may be round, pear shaped, oblong, or elongated, with lengths ranging from 4–5 cm (1.6–2.0 in) to >30 cm (11.8 in). The calyx is deeply lobed and may contain thorns. Skin surfaces are smooth and usually shiny. Fruit colors, whether solid or streaked, are varied and may be white, yellow, green, brown, and red, purple, black, or mixtures of these colors.

F-1 hybrids are increasingly grown because of their improved productivity and uniformity compared to open-pollinated cultivars. Parthenocarpic cultivars are available. Transgenic cultivars have been developed through genetic engineering to express the *Bacillus thuringiensis* gene that confers resistance to certain types of insects. However, these cultivars are not of commercial importance at this time.

Economic Importance and Production Statistics

Total world production of eggplant is 1,817,798 ha (4,489,961 acres) with a total yield of 46,825,331 metric tonnes (51,616,092 short tons) for an average yield of 25.8 metric tonnes/ha (10.4 tons/acre) (FAOSTAT, 2011). Production is concentrated in a relatively few countries primarily in Asia and the Middle East. China (787,000 ha/1,943,890 acres) and India (680,000 ha/1,679,600 acres) dominate world production. Other important producing countries include Indonesia (52,233 ha/129,016 acres), Egypt (45,020 ha/111,199 acres), Iran (38,785 ha/95,799 acres), Turkey (25,355 ha/62,627 acres), the Philippines (21,377 ha/52,801 acres), and Iraq (19,917 ha/49,195 acres).

Romania (10,020 ha/24,749 acres) and Italy (9,423 ha/23,275 acres) are the leading European producers. Eggplant is a minor crop in the USA with only 1,922 ha (4,747 acres) produced. Florida has year-round production except in August and September and produces more than 30% of the US crop. New Jersey is the second largest US producer with harvests from July–October, while California produces 19% from April to December. Mexico (1,091 ha/2,695 acres) exports eggplant primarily to the USA and Canada from January to March.

Nutritional Values

Eggplant is a relatively good source of vitamin C, vitamin K, thiamine, niacin, vitamin B6, dietary fiber, folate, pantothenic acid, magnesium, P, K, and manganese (Table 11.5).

Production and Culture

Eggplants are well adapted to tropical conditions and mid-temperate regions that provide a long growing season with continuous warm temperatures. Eggplants benefit more from warm temperatures and

Table 11.5. Nutritional value of raw uncooked eggplant (USDA Nutritional Database, 2011).

Nutrient	Unit	Value per 100 g (3.5 oz)
Water	g	92.3
Energy	kcal	25
Protein	g	0.98
Total lipid (fat)	g	0.18
Carbohydrate	g	5.88
Fiber, total dietary	g	3.0
Sugars, total	g	3.53
Ca	mg	9
Fe	mg	0.23
Mg	mg	14
P	mg	24
K	mg	229
Na	mg	2
Zn	mg	0.16
Vitamin C (ascorbic acid total)	mg	2.2
Thiamine	mg	0.039
Riboflavin	mg	0.037
Niacin	mg	0.649
Vitamin B-6	mg	0.084
Folate, DPE	µg	22
Vitamin B-12	µg	0.00
Vitamin A	IU	23
Vitamin E (α-tocopherol)	mg	0.30

are more sensitive to low temperatures than either tomatoes or peppers. In temperate regions with cold early-season soils, eggplants are produced on raised beds in a plasticulture system with either clear or black mulch to warm the soil (Fig. 11.13). Indeterminate cultivars are also used for greenhouse production. Favorable daytime temperatures are between 22°C (72°F) and 30°C (86°F) and optimal when coupled with warm nights preferably between 18°C (64°F) and 24°C (75°F). At temperatures less than 17°C (63°F) or greater than 35°C (95°F), growth is negligible and pollen dysfunction increases. Cultivars producing elongated fruit tend to be more resistant to high temperatures than small egg- or oval-shaped ones.

Flowering is considered day neutral. Early cultivars may begin flowering when plants have as few as six leaves, while others may not flower until 14 or 15 leaves have developed.

Roots are moderately deep and extensive. Most soils are acceptable unless they impede root development or are poorly drained. Eggplants have better drought tolerance than tomato or pepper, but for maximum yields, an adequate supply of moisture is required. Typical eggplant crop moisture requirements are approximately 900–1,000 mm (35–39 in). A soil pH between 5.5 and 7.5 is preferred. Eggplants have a fairly high nutrient demand and supplemental fertilization is commonly provided.

Fig. 11.13. A determinate eggplant cultivar with purple stems, calyx, and elongated fruit grown on black plastic mulch with drip irrigation to warm the soil and control weeds. Such cultivars are referred to as Asian types in some Western markets.

Propagation and establishment

Propagation is either by direct seeding or transplants. Optimum temperatures for seed germination are 24–32°C (75–90°F). Below 15°C (59°F) and above 35°C (95°F), germination is poor. Bare root or plug transplants at the two or three true-leaf stage are suitable. For production in disease-infested soils, eggplants can be grafted to resistant rootstocks of *Solanum torvum* or *S. integrifolium*.

Cultivars and cultural practices determine field spacing. Common spacing for most non-supported cultivars varies from 20 to 30 cm (8–12 in) within rows and about 90 cm (35 in) between rows. Determinate cultivars with large bushy growth are spaced further apart than dwarf types, and trellised plantings are also given more space. Trellis plantings are also spaced 20–30 cm (8–12 in) within rows but 100–120 cm (39–47 in) between rows.

Harvesting and marketing

Generally, 3–4 months of growth after seed germination are required to mature fruit. The period from anthesis to harvest maturity varies among cultivars and temperatures and may be as little as 10 days or as long as 40 days. Under favorable conditions, flowering and fruit production is continuous. For best edible quality, fruit are consumed when immature, glossy, and before the seeds have developed. Cultivars with elongated fruit may be harvested after achieving about one-half their fully mature size. Market or edible maturity of some eggplants may be determined by gently pressing a thumb against the side of the fruit. If the indentation retracts, the fruit is immature. If the indentation does not retract, the fruit is likely physiologically mature and beyond best eating quality. A loss of skin glossiness is another indication of advanced maturity. The flesh of overly mature fruit becomes dry, bitter, and pithy, and the seeds hard. Delayed or failure to harvest mature fruit reduces flowering and subsequent fruit production.

Eggplants are repeatedly hand harvested and should be cut free rather than torn from the plant to prevent damage, since there is no abscission zone on the pedicle. Growth is rapid under optimum conditions, so frequent harvests are necessary to prevent excessive fruit size. The foliage is mildly abrasive and some cultivars have thorns, so the harvester's hands, legs and arms should be protected. Fruit are very susceptible to postharvest physical injury. Cotton gloves are often worn during harvest. Bruising and compression injury are very common when attention to careful harvest and handling practices is not followed. Eggplant cannot withstand stacking in bulk containers.

Rapid cooling soon after harvest is essential for preserving quality and shelf life. Forced-air cooling is effective to preserve quality following washing or hydrocooling to remove field heat. Storing eggplant in plastic bags or polymeric film overwraps may reduce chilling injury and water loss. The storage life of eggplant is generally less than 14 days because visual and sensory qualities deteriorate rapidly. The optimum storage conditions are 10–12°C (50–54°F) and 95% RH. Storage at low humidity will result in visible signs of desiccation such as a reduction of surface sheen, skin wrinkling, spongy flesh, and browning of the calyx. Short-term storage or transit temperatures below the optimum range reduce weight loss, but will result in chilling injury after several days (Cantwell and Suslow, 2013). Fruit are chilling sensitive and injured by prolonged storage below 10°C (50°F). At 5°C (41°F) chilling injury will occur more quickly in 6–8 days (Cantwell and Suslow, 2013). Symptoms include surface pitting, surface bronzing, and seed and tissue browning. Chilling injury is cumulative and may be initiated in the field prior to harvest. Freezing injury may occur at −0.8°C (30.6°F), depending on the soluble solids content. Symptoms of freezing injury include water-soaked pulp that becomes brown and desiccated over time (Cantwell and Suslow, 2013).

Eggplant fruit have a moderate to high sensitivity to exogenous ethylene. Calyx abscission and increased deterioration, particularly browning, may be a problem if eggplants are exposed to >1 ppm ethylene during distribution and short-term storage.

Controlled or modified atmosphere storage or shipping does not benefit eggplant quality. Low O_2 levels (3–5%) delay deterioration and the onset of decay by a few days. Eggplant tolerates up to 10% CO_2, but storage life is not extended beyond the benefit of reduced levels of O_2.

Diseases

Circular brownish spots on the fruit and leaves are characteristic of leaf spot and fruit rots caused by *Phomopsis vexans*. Fruit rot may occur during postharvest transport even though no symptoms were apparent at harvest. Early blight (*A. solani*) causes dieback known as collar rot in seedlings, but the plant can be infected at all growth stages including fruiting. The pathogen produces distinctive patterned leaf spots, stem lesions, and fruit rot. *Alternaria* is favored by temperatures between 16–32°C (60–90°F). Stressed plants are more susceptible than healthy plants. Anthracnose fruit rot from *Colletotrichum melongenae* causes sunken spots and lesions on the fruit surface. Although temperatures between 13–35°C (55–95°F) support growth, the optimum is 27°C (80°F) and humidity at 93% or higher. Wilt caused by *Verticillium albo-atrum* attacks the vascular system resulting in stunted growth, yellow discoloration, and eventually defoliation of lower foliage and plant death. This fungus is favored by temperatures between 13–30°C (55–86°F). Tobacco ring spot virus (TRSV) is characterized by yellowing foliage and plant death if the infection is severe. Crop rotation can lessen the effects of this disease. The dagger nematode (*Xiphinema* spp.) is a known vector of TRSV. The most serious postharvest diseases that affect fruit include: *Alternaria* spp. (black mold rot), *Botrytis* spp. (gray mold rot), *Rhizopus* spp. (hairy rot), and *Phomopsis* rots (Aguiar *et al.*, 2013).

Insect Pests

Many insect pests attack eggplant. Some of the most destructive ones include: spider mites (*Tetranychus* spp.), green peach aphids (*M. persicae*), lygus (*Lygus* spp.), flea beetles (*Chrysomelidae*), and wireworms (*Elateridae*). Spider mites are especially harmful at warmer temperatures. Flea beetles are usually only a problem early in the season when plants are young. Lygus bugs will feed on flowers causing them to drop during the blooming period. Root knot nematodes (*Meloidogyne* spp.) cause wilting and leaf yellowing (Aguiar *et al.*, 2013).

TOMATO

With the exception or the potato, tomato is the most widely grown crop in the family Solanaceae. The popularity of tomato is likely due to its acid sweet taste and unique flavors. Although indigenous to western South America, its dietary and economic importance in the region of origin has lagged in comparison to other parts of the world where is has been enthusiastically adapted after introduction.

Origin and History

As judged by the distribution of wild species, the progenitor of the cultivated tomato is considered to have originated in the narrow, dry, tropical, coastal areas of Ecuador and Peru and portions of northern Chile. The wild cherry tomato, *Solanum lycopersicum* var. *cerasiforme*, is considered the most likely immediate ancestor of the cultivated tomato. This wild form spread from Ecuador and Peru throughout tropical America. Native people around Vera Cruz and Puebla Mexico were likely responsible for domestication (Peralta and Spooner, 2007).

The tomato was unknown to Europeans until samples were collected by early explorers in Mexico and returned to Europe. The initial introduction of tomato to Europe appears to have been from Mexico rather than the Andean regions. It is suggested that the name "tomato" comes from the Nahuatl language of Mexico (Peralta and Spooner, 2007).

As early as 1554, tomato was grown as a food crop in Italy were it was called "porni di oro" or golden apple, suggesting those first introduced were yellow fruited. The tomato was grown in France, the American colonies, and other European countries as an ornamental curiosity called "love apple", but was not widely accepted as a food crop. Tomato was considered poisonous in English-speaking countries possibly because it is in the same family as deadly nightshade. Tomato leaves and immature fruit contain the toxic alkaloid tomatine. Following the realization after 1800 that it was not poisonous, tomato quickly became a food crop and cultivation quickly spread throughout the world, especially during the 20th century (Peralta and Spooner, 2007).

Botany and Life Cycle

Solanum lycopersicum cultivars are short-lived perennials in the tropics, but in temperate regions are grown as annuals. Among the cultivars, growth habit ranges from highly indeterminate to strongly determinate. Plants grow from 0.5–2.0 m (19.7–78.7 in) tall with solid, thick stems. Some dwarf cultivars, grown as novelties, reach less than 30 cm (11.8 in) tall. Growth habit may vary from erect, to semi-prostrate, and to vining. Taproots usually are strong and deep, occasionally reaching depths of 1.83 m (72 in). Small glandular hairs that appear on stems, leaves, and peduncles produce a noticeable odor. Leaves are compound pinnate, coarsely toothed, and curled or flat.

Flowers are borne opposite and between leaves. Although some cultivars have 30 or more flowers per cluster, usually 4–12 develop on a broad, flat raceme. Flowers are perfect, about 2 cm (0.8 in) in diameter, and yellow anthers are fused to form a pendent-shaped tube when viewed longitudinally, with a yellow, star-shaped corolla. Self-pollination is most common. Flowers do not produce nectar, although cross-pollination, usually by bees, occurs with varying frequencies.

Pedicels, or fruit stems, typically have an abscission zone a few centimeters (less than an inch) above the fruit, depending on the cultivar. Many recent cultivar introductions have been selected to have a "jointless" characteristic where the abscission layer never develops. Therefore, fruit must separate without the pedicel or stem attached. When the pedicel portion remains attached to the fruit, puncture of other fruit can occur during bulk handling, leading to undesirable postharvest losses (Rubatzky and Yamaguchi, 1997).

The tomato fruit is a fleshy berry with a slightly hairy surface when very young but smooth at maturity. Fruit of most cultivars are globose but other shapes such as elongated, plum, and pear-like exist as well. Noticeable lobes are present with some cultivars, an indication that the fruit has multiple ovaries. Ripe fruit colors range from red, pink, tangerine, orange, yellow, purple, to colorless. Most fruit are solid colors but striped cultivars also exist. Red color is due to the linear carotene pigment lycopene. Other carotenoid pigments cause orange color, and only trace amounts of the yellow pigment xanthophyll (oxygenated carotenoid) is present in ripe fruit. Intermediate colors are due to differing ratios of these pigments, plus anthocyanin (purple) in rare cases, in combination with skin color. Red tomatoes have a yellow skin and red flesh (pericarp), while pink cultivars also have red flesh, but because of a recessive gene, their skin is colorless. Yellow flesh, the result of another recessive gene, when overlaid by yellow skin produces bright yellow fruit. If yellow flesh is combined with colorless skin, the fruit is pale yellow (Rubatzky and Yamaguchi, 1997).

At fruit maturity, the locules are normally filled with gelatinous material that surrounds the seeds. Fruit usually contain many seeds, which are flat and a light cream to brown color. Seed are typically

2–3 mm (0.08–0.1 in) long and there are approximately 300–350 seeds/g (8,500–10,000/oz).

Types and Cultivars

Tomato genetics are well understood due in part to pioneering work of the late Dr. Charlie Rick at the University of California, Davis, who spent 60 years collecting and characterizing tomato germplasm from South and Central America. More recently, the tomato genome has been sequenced and extensively studied using modern scientific techniques. Beneficial genetic traits that have been transferred to modern fresh market and processing cultivars through classical plant breeding include: uniform ripening, improved pollination and fruit set, crack resistance, improved stress tolerance, high pigmentation, greater disease resistance, uniform ripening, bush character, and improved fruit solids.

In 1960, both processing and fresh market tomatoes were produced from the same or similar large-fruited indeterminate cultivars that were hand-harvested. However, over the past 50 years processing and fresh market tomatoes have evolved into two dramatically different crops. Starting in the mid-1960s, cultivars were developed through plant breeding to accommodate once-over destructive mechanical processing in anticipation of shortages in farm labor (Schmitz and Seckler, 1970). For machine harvest adaptation, tomato plants were specifically bred by Jack Hanna and others for a determinate growth habit with small vines that produce a concentrated fruit set of many small, thick-skinned, uniformly ripened fruit. Fruit for processing into paste and sauces are firm and thick walled with few and small locules. Small fruit with these characteristics are less subject to impact and compression damage. Fruit intended for processing into juice tend to have larger locules. In addition, red ripe fruit remain attached to the vine until harvested. Such vine storage is critical for high yields, and although fruit do not abscise naturally, they are easily detached by harvest equipment. Processing cultivars also have small pedicle attachment and the jointless character, so the fruit breaks clean with no stem attachment (Grandillo *et al.*, 1999).

The transition to mechanical harvesting increased average yields from about 16 metric tonnes/ha (7 short tons/acre) for hand-picked, large-fruited cultivars in the 1960s to 110 metric tonnes/ha (50 short tons/acre) or more for the destructive mechanical harvesting system of small-fruited cultivars used today.

Fresh-market tomatoes are large-fruited, predominately hand-harvested, and grown on semi-determinate or indeterminate plants often trained to a support system. In some areas, unsupported determinate cultivars that produce large fruit are grown for fresh-market use. Mechanical harvesting of mature green medium- to large-fruited fresh-market tomatoes is used in some markets although hand-harvesting continues to predominate in most markets.

Consumers in developed countries of North America and western Europe sometimes complain about the poor quality of tomatoes purchased in supermarkets (Bland, 2005). This dissatisfaction is often blamed on plant breeders who are perceived as having sacrificed eating quality for better shipping and storage characteristics. What is lost in this conversation is the fact that tomatoes are often produced in distant locations where they grow best and shipped great distances to market. On average, the typical vegetable in a US supermarket travels 2,400 km (1,500 miles) before it reaches the shelf. Since tomatoes are soft at maturity and cannot stand the rigors of handling, they are frequently shipped at the firm mature-green stage. Ripening occurs during shipment or upon arrival, sometimes after treatment with the plant hormone ethylene, which accelerates color change and fruit softening. However, this practice has a negative effect on eating quality. Mature green fruit do not have the flavor and aroma of ones that develop to a red color on the vine (Rubatzky and Yamaguchi, 1997). Best eating quality is associated with the red ripe stage when both high acid and sugar contents are achieved, the two characteristics most associated with tomato-like flavor (Kader *et al.*, 1977). So blame for poor quality of some supermarket tomatoes has much to do with the shipping and distribution system rather than poor genetics.

Dissatisfaction with tomato quality has driven some commercial growers and gardeners to turn back to heirloom cultivars that comparatively speaking have received less genetic improvement. However, heirlooms as a general rule, depending on the cultivar selected, have little disease resistance, yield a low percentage of marketable fruit, have lower solids, less crack resistance, less intense fruit color, and may pollinate poorly. In a blind comparison of vine-ripened fruit of the heirloom cultivar 'Brandywine' and the F-1 hybrid 'Mountain Spring', developed for commercial production in the region, 42 college students preferred the taste and appearance of 'Mountain Spring' 2 to 1 over 'Brandywine' (unpublished results). So local production of red-ripe fruit from modern adapted cultivars may be a solution to poor fresh-market tomato quality in some areas.

In an attempt to increase the firmness of vine-ripe fruit, the 'Flavr Savr' genetically engineered tomato was the first transgenic to be granted a license by the government for commercial production and sale in the USA in 1994. 'Flavr Savr' contained a second copy of the tomato polygalacturonase gene inserted backwards (antisense direction) into the genome. The polygalacturonase enzyme produced by this gene degrades pectin, a component of the tomato cell wall, which causes fruit softening. When the antisense gene is expressed, it interferes with the production of the polygalacturonase enzyme, delaying the ripening process. The 'Flavr Savr' failed to achieve commercial success for various reasons and was withdrawn from the market in 1997. Similar technology, but using a truncated version of the polygalacturonase gene, has also been used to slow fruit softening.

DNA Plant Technology (DNAP), Agritope, and Monsanto each developed slow-ripening tomatoes by preventing the production of ethylene, the gaseous hormone that naturally triggers fruit ripening. All three tomatoes inhibited ethylene production by reducing the amount of ethylene precursor. DNAP's tomato, called 'Endless Summer', inserted a truncated version of the ACC synthase gene into the tomato, which interfered with the endogenous ACC synthase. 'Endless Summer' was briefly tested in the marketplace, but patent arguments and lack of consumer interest caused it to be withdrawn. Tomato plants are relatively easy to genetically engineer, but as long as consumers have little interest in buying transgenic crops genetic improvement is likely to be limited to traditional plant-breeding approaches at least in North America and Europe. However, the technology exists to genetically engineer tomatoes for greater environmental stress tolerance, pest resistance, improved nutrition, improved taste, and to deliver vaccines (Ruf et al., 2001; Goyal et al., 2007).

Naturally occurring genetic mutations that slow fruit ripening have been characterized and used by plant breeding programs to extend storage life (Matas et al., 2009). The ripening inhibitor mutant gene *rin* limits ethylene production, and the nonripening *nor* gene prevents ethylene development. Another mutant gene, *ale*, prolongs the storage of harvested fruit, and has less adverse effects on color development. These mutants have been bred into cultivars such as the Extended Shelf Life (ESL) group that ripen slowly for long-distance shipping of fresh market tomatoes (Plunkett, 1996; Matas et al., 2009). The ESL cultivars are another approach to developing tomatoes that can withstand the rigors of shipping at more advanced stage of ripeness. Since the *rin*, *nor*, and *ale* mutants occurred naturally and were not engineered by scientists, consumers are more accepting of their use.

The use of hybrid (F-1) cultivars has increased over the past 30 years. Crossing two inbred parental lines by hand produces hybrid seeds. This makes hybrid tomato seeds more expensive to produce compared to open-pollinated. There is debate over whether this extra cost is worth the benefits, especially for processing-tomato production. Tomatoes are naturally self-pollinated and do not exhibit the same degree of heterosis as crops that cross naturally. Hybrid cultivars are more widely used for fresh-market production and increasingly for processing as well, because of their extra earliness, improved uniformity, vigor, and increased yield compared to open-pollinated cultivars.

Fresh-market tomato cultivars are often divided into large fruit and small fruit categories. There has been renewed interest in small-fruited cultivars in some markets because they are less prone to damage during harvesting and transport and can be shipped at a more advanced stage of maturity. Small fruit are sometimes marketed as healthy snacks and come in a variety of shapes and colors.

Economic Importance and Production Statistics

In 2011, tomato was ranked at the eighth most valuable agricultural commodity in the world with a value of over US$58 billion dollars with production of 159 million metric tonnes (175 million short tons) produced on 4,751,530 ha (11,741,285 acres). Leading tomato-producing countries in 2011 were: China (981,000 ha/2,424,104 acres), India (865,000 ha/2,137,462 acres), the USA (146,510 ha/362,056 acres), Turkey (328,000 ha/810,506 acres), Nigeria (264,430 ha/653,421 acres), Egypt (212,446 ha/524,966 acres), Iran (183,931 ha/454,303 acres), Russia (117,000 ha/289,113 acres), and Italy (103,858 ha/256,639 acres) (FAOSTAT, 2011). China (48 million metric tonnes/53 million short tons), India (17 million metric tonnes/19 million short tons), and the USA (13 million metric tonnes/14 million short tons) are the leading producing countries (FAOSTAT, 2011). Much of the world trade in tomatoes comes in processed products grown in the Mediterranean region, the USA, and South and Central America. Fresh market tomatoes are freely traded among North American and European Union countries.

In the USA, the fresh-market tomato industry is distinctly separate from the processing industry. Fresh-market tomatoes are responsible for a larger share of US total crop value than processed tomatoes because they command comparatively higher prices. In the USA, the fresh tomato market consists of a relatively small number of firms competing in the market, which enables individual firms to affect prices. There are less than an estimated 1,000 farms that produce for the wholesale fresh tomato market and fewer than 50 shipping companies controlled the movement of fresh tomatoes into wholesale, retail, and food service sectors (Thompson and Wilson, 1998). In addition, much of the industry is vertically integrated, with companies owning the entire line of grower, packer, and shipping firms.

According to the US Economic Research Service, the USA produced 39,845 ha (98,400 acres) of fresh market tomatoes in 2012 with a farm value of US$0.9 billion, the highest of any fresh-market vegetable. Of this total, California harvested 12,758 ha (31,500 acres), compared to 12,150 ha (30,000 acres) for Florida. California had average yields of 37.2 metric tonnes/ha (16.6 short tons/acre) and accounted for 41% of total US fresh market tomato production (Thornsbury, 2013).

Canada and Mexico are the largest export destinations for US fresh tomatoes. In value, over 66% of US fresh-tomato exports were shipped to Canada in 2012 and exports to Canada and Mexico accounted for 98% of the export total. US imports of fresh tomatoes reached over 1.4 million metric tonnes (1.5 million short tons) in 2012. Mexico and Canada, in that order, were the main suppliers (Thornsbury, 2013).

US per capita consumption of fresh tomatoes has been increasing over the past 20 years. Average annual per capita consumption has increased from 5.6 kg (12.3 lb) in 1981 to 9.3 kg (20.6 lb) in 2012 (Thornsbury, 2013). Part of the increase in consumption has been due to the increasing popularity of fresh-market tomato use in salads and sandwiches, year-round availability, and a growing awareness of the health benefits from eating fresh vegetables.

Nutritional Values

Tomato is also a good source of vitamin E (α-tocopherol), thiamine, niacin, vitamin B6, folate, Mg, P, K, dietary fiber, vitamin A, vitamin C, vitamin K, and Mn (Table 11.6).

Table 11.6. Nutritional value of red ripe uncooked tomatoes (USDA Nutrient Database, 2011).

Constituent	Amount per 100 g (3.5 oz) sample
Energy	74 kJ (18 kcal)
Carbohydrates	3.9 g
– Sugars	2.6 g
– Dietary fiber	1.2 g
Fat	0.2 g
Protein	0.9 g
Water	94.5 g
Vitamin A equiv.	42 µg (5%)
– β-carotene	449 µg (4%)
– Lutein and zeaxanthin	123 µg
Thiamine (vitamin B1)	0.037 mg (3%)
Niacin (vitamin B3)	0.594 mg (4%)
Vitamin B6	0.08 mg (6%)
Vitamin C	14 mg (17%)
Vitamin E	0.54 mg (4%)
Vitamin K	7.9 µg (8%)
Mg	11 mg (3%)
Mn	0.114 mg (5%)
P	24 mg (3%)
K	237 mg (5%)
Lycopene	2,573 µg

Percentages are approximated using USDA daily recommendations for adults.

Production and Culture

Tomato cultivation is adaptable to many environments. Production ranges from high-elevation regions near the equator to temperate regions at high latitudes. Tomatoes do not grow well in the humid tropics because of high disease incidence, poor pollination, and regions where low temperatures or short growing seasons limit growth.

Site selection and field preparation

Tomatoes are grown successfully on a wide range of soil types, from sandy to fine-textured clays, as well as soils with high organic matter content. A soil pH range from 5.5 to 7 is usually satisfactory. Plants grow best with a uniform supply of moisture and good drainage. Tomatoes are intolerant of waterlogging, especially during emergence and near fruit maturation. Excessive moisture promotes damping-off and root rot diseases. Raised-bed culture is recommended on moist heavy soils to improve drainage. Resistant cultivars should be grown on soils with a history of Fusarium or Verticillium wilt diseases if other fields are not available. Crop rotation is strongly recommended to help minimize disease (Le Strange *et al.*, 2000).

Root systems

Tomato plants generally have extensive root systems, but most develop within the upper 60 cm (23.6 in) of soil. Taproots may extend to a depth of 1.8 m (6 ft) when not restricted by hard pans or high water tables. The deep-rooted system provides the plant some drought tolerance. However, when rainfall is insufficient, irrigation is needed to maximize yields (Rubatzky and Yamaguchi, 1997).

Irrigation

Both subsurface and surface trickle or drip irrigation are effective and have been adopted in many areas. The initial cost of installation is high but once installed, drip systems often require less labor, utilize less water, reduce the occurrence of soil diseases, such as Phytophthora root rot, and also restrict between-row weed growth. In addition to water conservation, other advantages include the ability to use drip systems to fertigate. Other irrigation methods sometimes used include furrow, subbing (raising subsurface water level) and overhead sprinkler. Sprinkler irrigation uses water less efficiently and wets the plants, which may increase foliar disease.

Water usage commonly is about 25–30 mm (1–1.2 in) weekly, and on a hot dry day, evapotranspiration may exceed 10 mm (0.4 in). Mid-season crop coefficients that relate crop-water needs to evapotranspiration generally range from 1.05 to 1.25 for processing tomato (Hanson and May, 2006). Although irrigation frequency and amounts vary, processing tomato crops grown in California are usually supplied with 600–900 mm (24–35 in) of water. In order to increase the fruit soluble solids content, irrigation is sometimes curtailed during the latter stage of fruit development for processing crops.

Temperature

Tomatoes can be grown in most open-field locations where there is a minimum of 3–4 months of warm, frost-free weather, with an average temperature above 16°C (61°F). Vegetative and reproductive growth at lower temperatures is slow, and an extended period of exposure to 12°C (54°F) or less can result in chilling injury. Chilling injury increases membrane leakage and other metabolic abnormalities in affected fruit (Zhao *et al.*, 2009). Although frost sensitive, tomatoes are hardier than peppers or eggplants. Day temperatures of 25–30°C (68–86°F) with night temperatures between 16–20°C (61–68°F) are optimal for growth and flowering. A wide diurnal difference between day and night temperatures tends to improve flowering, growth, and fruit quality. Fruit set is best between 18–24°C (64–75°F) and is poor below 15°C (59°F) or above 30°C (86°F). Night temperatures are more critical than day temperatures for good fruit set (Rubatzky and Yamaguchi, 1997).

Fertilization and nutrition

Nitrogen is very important for vegetative growth. Fertilization with a mixture of NO_3^- and NH_4^+ with a higher ratio of NO_3^- compared to NH_4^+ generally gives best results. For high yields, sufficient plant and canopy size must be maximized prior to flowering to support fruit development. However, excessive vegetative growth, stimulated by excess nitrogen (N) fertilization, can reduce early and subsequent fruit set (Rubatzky and Yamaguchi, 1997). Some N is often applied preplant or at transplanting based on soil testing. Later additions are made based on petiole or foliar analysis, so tissue values are initially high to stimulate vegetative growth, lower at flowering, and increase during fruit enlargement after fruit set. Fertigation is the most efficient application method for nitrogen but side dressing can also be used (Le Strange *et al.*, 2000).

Adequate P is also important for early plant development and flowering. Attaining high fruit-soluble solids relies upon adequate K. Calcium is important for cell wall development and helps avoid blossom-end rot disorder. Usually a starter fertilizer is applied prior to or at planting to stimulate early seedling growth. Both P and K are commonly applied preplant (Le Strange *et al.*, 2000).

Propagation

Tomato establishment is by direct-field seeding or transplants. It is also possible to propagate indeterminate cultivars from stem cuttings although this is not a common practice. The minimum soil temperature for seed germination is 10°C (50°F) while the maximum is about 40°C (104°F). Between 25–30°C (77–86°C), seedling emergence occurs within 6–9 days if moderate to high vigor seeds are planted. Seed priming treatments can improve the speed of emergence. Seed are sometimes coated or pelleted to deliver chemicals or biologicals, increase uniformity, improve seed handling, and enable precision placing of seeds to a final stand at planting. Seed is commonly treated with fungicides or biologicals to reduce damping-off diseases.

Processing tomatoes are often direct seeded at relatively close spacing to ensure a complete stand and sometimes are planted in "clumps", which involves sowing several seed at each location. This procedure improves emergence in soils prone to crusting. Plants are not thinned from individual clumps. Crowded plants produce less fruit per plant, but the additional plants produce more total fruit, so total yields are higher or comparable to other planting practices. Another method sometimes used for processing tomatoes is called plug planting. Dry or imbibed seeds are mixed into medium containing vermiculite, peat moss, or other material, such as small amounts of fertilizer, biologicals, or chemicals. A portion of this mixture containing three to seven seeds is precision placed into the seedbed. The mulch-like medium is helpful to prevent crusting and supporting early seedling growth. In situations with high quality seed and favorable seedbed conditions, processing tomatoes may be precision seeded to a final stand without thinning.

Where earliness is a priority and environmental conditions restrict direct seeding, fresh market and tomatoes for processing are grown from transplants (Fig. 11.14). Many fresh market and processing cultivars are now F-1 hybrids and the increased commercial use of expensive hybrid seed has resulted in a return to transplant practices to minimize seed costs. Advances in transplant production technology have reduced the cost. Reduced field production times, earliness, weed control, and a general reduction of inputs are some of the advantages of transplanting.

For large-scale commercial production, plug-tray plants grown in disease-free soilless media, have replaced bare-root transplants grown in fields or outdoor beds. Plug-tray production is often in greenhouses or protected structures with uniform environment. Plug plants have a small root ball that reduces transplant shock. Transplants are established in the field with machines, some of which are fully mechanized. To advance early

Fig. 11.14. Early season transplanting of fresh-market tomatoes into black-plastic mulch on Virginia's Easternshore.

production, row covers or plastic tunnels are commonly employed.

Grafted tomato plants are used in much of the world. Grafting procedures were described previously in the pepper section of this chapter and most apply to tomato as well. The demand for grafted tomato transplants is anticipated to rise in North America due to the decreased use of chemical fumigants. The primary reason for using grafted tomato plants is improved disease tolerance. Several commercial rootstocks are resistant to major soil-borne pathogens. In addition, grafted plants may have greater vigor, uniformity, and tolerance of abiotic stresses such as salinity, root hypoxia, and extreme temperatures (Oka *et al.*, 2004).

Spacing

Plant spacing is determined by cultivar growth habit, end use, and harvest method. Determinate plants, like those grown for processing, are spaced closer than indeterminate cultivars. Mechanically harvested processing crops are planted on wide beds ranging from 150–180 cm (59–70 in), with in-row spacing from 30–60 cm (12–24 in) to provide populations of 10,000–20,000 plants/ha (4,049–8,097/acre). High populations are used for processing because total yield rather than fruit size is the objective (Rubatzky and Yamaguchi, 1997).

Plant density of field-grown fresh-market tomatoes range from 8,000 to 14,000 plants/ha (3,239–5,668/acre) for ground culture and about 6,000–8,000/ha (2,429–3,239/acre) if supported. Typical plant spacings range from 60–75 cm (24–30 in) within rows and from 120–150 cm (47–59 in) between rows (Le Strange et al., 2000). Tying plants to individual stakes or using the string-weave system are popular techniques for supporting indeterminate or semi-determinate cultivars (Fig. 11.15). Stakes are spaced between every plant or every other plant. The string-weave is relatively easy to install by running a string the length of the row and looping it around each wooden stake and tomato stalk when the plants are about 60 cm (24 in) tall. A total of three strings are typically placed at increasing heights through the row during the season.

This system is sufficient to keep fruits off the soil, make harvest easier, improve airflow, and increase spray coverage. Supported tomato production significantly increases labor costs but results in higher yields, less disease, and higher quality fruits. Determinate fresh-market cultivars are sometimes grown on plastic mulch without support in a system called "ground culture", but the yield potential is generally less than for indeterminate plants.

Flowering and fruit set

Fertilization of flowers is generally favored by day temperatures between 21–30°C (70–86°F) and night temperatures between 15–21°C (59–70°F). For many cultivars, day temperatures above 32°C (90°F) reduce fruit set and it is negligible at 40°C (104°F). High temperatures interfere with viable pollen production and its dispersion and negatively affect ovule viability (Ho and Hewitt, 1986). Hot, drying winds can have similar effects. Cool temperatures and high humidity and/or low light intensity limit pollen shedding. At or below 15°C (54°F) pollen formation and function is greatly inhibited. Cold temperatures can also reduce ovule viability.

Fig. 11.15. Placing tomato stakes in a string-weave plasticulture field for summer fresh-market production in Virginia. Air hammers set wooden stakes between transplants. Later, string will be run the length of the row and looped around plants and stakes to keep plants upright. White plastic mulch is used to cool soils.

Stigmas are receptive to pollination for 4–7 days. Style elongation occurs within the anther cone and usually coincides with pollen release from dehiscing anthers. Depending on temperature, fertilization occurs within 48 h after pollination. The most temperature-sensitive periods for fruit set are about 5–10 days before anthesis and 2–3 days following pollination. High light intensity tends to accelerate flowering in many cultivars, whereas low light intensity limits vegetative growth and may also delay flowering.

The appropriate level of vegetative growth should be achieved before flowering begins so the plant can support maximum fruit development with minimal drop. Once reproductive growth begins, the fruit becomes the major photosynthetic sink with proportionally less directed to vegetative growth. Underdeveloped plants typically yield poorly. Low light intensity along with high night temperatures encourages excessive vegetative growth, which competes with the fruit for assimilates. Also, low light intensities and night temperatures less than 10°C (50°F) or greater than 27°C (81°F) can cause green-fruit drop (Rubatzky and Yamaguchi, 1997). Some cultivars, such as 'Solar Set', are better adapted to setting fruit during adverse temperatures.

During cool weather, hormones such as indoleacetic acid (IAA), para-chlorophenoxyacetic acid (4-CPA), or naphthalene acetic acid (NAA) may increase fruit set if sprayed on flowers. However, the resulting fruit, may be fully or partly parthenocarpic, often appear puffy because the carpels are poorly filled. To help fruit develop normally, gibberellins are applied in combination with fruit-setting hormones.

Fruit ripening

Tomatoes, like cantaloupes, are climacteric fruit. Ripening is naturally triggered by the gaseous hormone ethylene (Brady, 1987). Depending on the cultivar, most tomato fruit mature 32–60 days after anthesis. Temperature greatly influences the rate of fruit ripening. The degree-day calculation, described elsewhere, is sometimes used to schedule planting and harvests of processing tomatoes. Optimum temperatures for fruit maturation and color development are between 20–24°C (68–74°F). At favorable temperatures, the red pigment lycopene develops during ripening even in the dark. However, light accelerates color development and intensity.

Lycopene synthesis is inhibited to a greater extent than other carotenoids above 32°C (90°F) or below 10°C (50°F). At these extremes, mature fruit tend to develop a yellowish to an orange-red color instead of deep red. At temperatures greater than 40°C (104°F), fruit tend to remain green because chlorophyll is not broken down. Below 10°C (50°F) ripening essentially stops and chilling injury may occur (Brady, 1987).

For most cultivars, ripening is accompanied with increases in sugars and organic acids in addition to color change and fruit softening as part of the climacteric response. The accumulation of sugars and aromatic compounds in the presence of acids gives the fruit its characteristic flavor and aroma (Kader et al., 1977).

Harvesting and marketing

The time from field planting until first fruit harvest is dependent on cultivar and growing conditions and usually ranges from as few as 55 to as many as 125 days. Currently in the USA, essentially all processing tomatoes are machine harvested, as are some green fruit for fresh market (Fig. 11.16).

The transition to mechanical harvesting has occurred in other major tomato-producing countries as well. Electronic color-sorting equipment is incorporated into modern harvesters to remove culls, unripe fruit, and inert material. Machine-harvested fruit are typically handled in bulk and processed usually within 24 h so that physical injury, unless severe, has little effect on processed products.

Fresh market fruit are harvested when physiologically mature and range from mature green to the full red-colored stage. At many fresh-market tomato-packaging facilities, fruit are washed in chlorinated water to remove dirt and to limit postharvest diseases. When wash water is cold, postharvest decay often increases because of contaminated water entering the fruit though the stem scar. However, when the wash water is equal to or warmer than the fruit temperature, contaminated water is not drawn into the fruit. Rapid cooling soon after harvest is essential for optimal postharvest keeping quality. The precooling endpoint is typically 12.5°C (55°F). Forced-air refrigeration is the most effective but room cooling may be more common (Le Strange et al., 2000).

Fresh-market fruit are graded for uniform size and quality before packaging. Standard tomato

Fig. 11.16. Mechanically harvesting small-fruited tomatoes for processing.

quality is primarily based on uniform shape and freedom from growth or handling defects. In some facilities, size grading is accomplished by machine and color sorting is performed electronically. Occasionally, fruit are waxed or wrapped in plastic film to reduce moisture loss. US grades are No. 1, Combination, No. 2, and No. 3, while color designations are green, breakers, turning, pink, light red, and red. Distinction among grades is based predominantly on external appearance, bruising, and firmness. High quality tomatoes should have a shape that is true-to-type. Colors range from orange-red to deep red and should be uniform and intense over the entire fruit with no green shoulders or other ripening disorders. The appearance of a quality tomato should be smooth, with small blossom- and stem-end scars. In addition, fruit should be free from growth cracks, catfacing, zippering, sunscald, insect injury, and mechanical injuries. Top-graded fruit should be firm and not easily deformed from being overripe (Le Strange et al., 2000).

Although of different firmness, both mature green and fully colored fruit are physiologically mature because seed are capable of germinating. Full color and fruit softening are usually highly correlated, and the terms "red ripe" and "full color development" are often used interchangeably. Determination of color stage when fruit are harvested depends on how they will be handled and used. Fruit to be transported to distant markets or not intended for immediate use are harvested at the mature green or breaker stage, and further color development occurs during holding or in transit. Mature green or breaker stage fruit are firm and better able to withstand postharvest handling than fully colored soft fruit. Breaker or pink stage fruit are harvested when extended postharvest life is less essential (Le Strange et al., 2000). ESL cultivars are better suited for long distance shipment after the onset of color change because they soften more slowly.

Postharvest handling and storage

Postharvest fruit deterioration because of excessive softening is a major reason for marketing losses. Rough handling, poorly designed containers, and exposure to hot and dry conditions also contribute to fruit losses. Tomatoes are chilling sensitive at temperatures below 10°C (50°F) if held for longer than 2 weeks or at 5°C (41°F) for longer than 6–8 days. Consequences of chilling injury are failure to ripen and develop full color and flavor, irregular (blotchy) color development, premature softening, surface

Family Solanaceae

pitting, browning of seeds, and increased decay (especially black mold caused by *Alternaria* spp.) (Snowdon, 2010). Chilling injury is cumulative and may be initiated in the field prior to harvest. The optimum relative humidity for storage is 90–95% to maximize postharvest quality and prevent water loss (desiccation). Extended periods of higher humidity or condensation may encourage the growth of stem-scar and surface molds.

Mature-green tomatoes can be stored up to 14 days prior to ripening at 12.5–15°C (55–60°F) without significant loss of sensory quality and color development. Decay of mature green fruit is likely to increase if they are stored for longer than 2 weeks. After reaching the firm-ripe stage, shelf life is from 8 to 10 days at 7–10°C (44–50°F). Lower temperatures are sometimes used for short-term storage or transit but will cause chilling injury after several days. Standard ripening temperatures are 18–21°C (65–70°F) at 90–95% RH, while slow ripening is used during transit and occurs from 14–16°C (57–61°F) (Rubatzky and Yamaguchi, 1997).

Color development and fruit softening occur in response to ethylene (Brady, 1987). Ethylene application can accelerate ripening when supplied at or slightly beyond the mature green stage. Fruit are treated with 100–150 ppm ethylene in a sealed ripening room. The best ripening response to ethylene occurs at 20–21°C (68–70°F) and 85–90% RH for 12–24 h.

Controlled atmosphere storage or shipping offers some benefits. Mature green fruit have been stored for 6 weeks at 13°C (55°F) in an atmosphere of 3% O_2 and 97% N_2 and had no noticeable flavor or other quality impairment when ripe. Low O_2 levels of 3–5% delayed ripening and the development of surface and stem-scar molds without severely impacting sensory quality. Storage times of up to 7 weeks have been reported for tomatoes in 4% O_2, 2% CO_2, and 5% CO. More typically, 3% O_2 and 0–3% CO_2 can maintain acceptable quality for up to 6 weeks prior to ripening. Elevated CO_2 above 3–5% will cause injury in most cultivars. Low O_2 (1%) will cause off-flavors, objectionable odors, internal browning, and other defects.

Glasshouse production

Protected structures such as greenhouses or tunnels allow fresh market tomato production in a modified environment (Fig. 11.17). Tomato production

Fig. 11.17. Summer greenhouse tomato production in Israel. The greenhouse optimizes the growing environment for the crop. Individual plants are trained on a support system suspended from the top of the house.

in glasshouses or other protective structures is important in many regions where climate limits field production. In areas with poor soils and inadequate rainfall, greenhouses create an optimized environment even during the growing season. Tomato production in glasshouses often results in higher yield and better quality than field production (Cook and Calvin, 2005).

Greenhouse tomato production is an important industry in some European countries, China, and Japan (Cook and Calvin, 2005). Some of the leading producers of greenhouse tomatoes in Europe and the Americas include Spain 12,146 ha (30,000 acres), the Netherlands 4,615 ha (11,400 acres), England/Wales 1,215 ha (3,000 acres), Canada 344 ha (850 acres), Mexico 304 ha (750 acres), and the USA 263 ha (650 acres) (Peet and Welles, 2005). It has been estimated that 90% of Canadian fresh tomato production is in greenhouses. Mexican greenhouse tomato production continues to increase, accounting for an 8% share of total Mexican production (Cook and Calvin, 2005).

Protected production requires intensive cultural practices unique from those used in the field. Plants are grown in a broad range of growing media ranging from soil, peat, synthetic soil, composted matter, straw bales, gravel, sand, or rockwool slabs. The media can be fertigated with drip tubes. It is important that adequate aeration of the root zone is always provided.

In the nutrient film technique (NFT), tomato roots are bathed in shallow recirculating troughs of nutrient solution, while the tops are trained to a support. The flow need not be continuous, but is frequent enough to provide nutrients and water. Fertilizers and/or nutrient solutions are carefully monitored for mineral content, pH, and possible pathogen contamination. Certified organic greenhouse tomato production is possible using approve media and inputs.

Tomatoes can be grown aeroponically with roots suspended in air misted with nutrient solution (Peterson and Krueger, 1988). The nutrient solution condenses on the shielded roots providing water and nutrients to the plant. This system is efficient where space and water are limited. Mechanical failures can quickly damage plants because there is no soil to buffer rapid change in moisture or nutrients.

Special indeterminate cultivars capable of growth, fruit-set, and development at extreme temperatures and light conditions have been developed by plant breeders especially for greenhouse environments. In the North American market, greenhouse tomato clusters are often marketed as tomatoes-on-the-vine, which gives them a fresh, natural appearance and special market identity.

Greenhouse establishment is almost exclusively with transplants. Growth from seed requires too much valuable glasshouse space and time. Another consideration is the high cost of F-1 hybrid seed for greenhouse production. The optimum plant population is 3–4 plants/m^2. During early development, greenhouse day temperatures are maintained between 15–21°C (59–70°F) with nights between 14–17°C (57–63°F). During flowering and fruit development, day temperatures can be elevated from 18–30°C (64–86°F), whereas night temperatures are usually maintained at 14–17°C (57–63°F). Night temperatures should never fall below 13°C (55°F) to avoid chilling injury and soil temperatures should be maintained above 14–15°C (57–59°F) (Rubatzky and Yamaguchi, 1997).

Supplemental light is provided when tomatoes are grown in greenhouses when intensity is low and day lengths are short. Heating and lighting energy are major production costs. Various practices have been developed to save energy such as double-wall polymer glazing, thermal blankets to reduce radiation losses at night, LED lights, direct-air heating, and reflective mulches.

When air temperatures are high, ventilation is critical to lower the temperature and, if necessary, the relative humidity. Low light intensity frequently limits production, so light transmission into structures should not be restricted. Low CO_2 concentration can reduce photosynthesis. If temperature and/or light is not limiting, CO_2 atmospheric enrichment will increase photosynthesis. With enrichment, CO_2 concentrations are raised two to three times greater than the normal 300 ppm ambient level.

Because of a lack of wind and insect activity, greenhouse environments are not conducive to self-pollination. Plants or flower clusters must be shaken to disburse pollen. The shaking is accomplished using hand-held electric pollinator wands, vibrating support systems, leaf blowers, or bumblebees (Greenleaf and Kremen, 2006). Midday is the best time for pollination. Fruit-set hormones are normally not used in greenhouses because fruit quality is often compromised.

Vine training and pruning are important management practices to enhance the ratio of foliage to fruit production and also to improve light penetration, aeration, disease management, and ease of harvesting. Pruning also allows for some regulation of fruit size and flowering. The typical glasshouse tomato production period is from 3 to 5 months or more. Therefore, vine training is an ongoing process. Plants are often supported by wires and strings suspended from the top of the structure to maximize plant use of available light and space.

During the tomato production cycle, scouting for disease and other pests is important in order to control an outbreak that could spread rapidly through a confined greenhouse. Rather than native soil, permanent greenhouses use disease-free media or hydroponics for tomato production.

Insects can be as troublesome as diseases and may require rapid application of control measures. The restricted environment of the glasshouse presents opportunities for biological control such as the use of parasitic and predator insects. Insecticidal soaps, traps, IPM programs, and good sanitation practices can control glasshouse pests such as whitefly, thrips, mites, and aphids.

Physiological Disorders

The incidence of nonpathogenic fruit disorders is largely determined by cultivar characteristics and usually induced by environmental cues such as moisture stress, nutrient stress, and temperature extremes. Disorders include blossom-end rot, which is the formation of a necrotic black sunken area on the fruit. Blossom-end rot is a water stress-induced localized Ca deficiency that is cultivar dependent (Snowdon, 2010).

Blotchy ripening is the failure of areas of the fruit wall to color uniformly. The tissue tends to remain firm and may be white, pink, or yellow and occasionally turns brown. The disorder seems to be associated with mineral stress and cool conditions. Internal browning is caused by discoloration of vascular tissues in the fruit wall. Gray wall is observed as areas of discoloration within the fruit wall. One form of irregular fruit coloring is attributed to a toxin produced by whitefly feeding (Snowdon, 2010).

Freezing injury is initiated at −1°C (30°F), depending on the soluble solids content. Symptoms of freezing injury include a water-soaked appearance, excessive softening, and desiccation of the locular gel.

Additional disorders are zipper, which occurs when floral parts adhere to the surface of the fruit as it expands, fruit puffiness caused by inadequate pollination resulting in partially filled locules, and catface, also caused by poor pollination, and often the result of low or high temperatures. Symptoms of catfacing include non-uniform locule formation, an asymmetric extension of the locules and a large irregular corklike blossom scar that resembles a cat's face to some. Some cultivars like 'Beefsteak' and 'Ponderosa' have a natural tendency to produce a high percentage of catfaced fruit because their stigmas naturally extend beyond the flower pollen tube, reducing pollination efficiency. Low light intensity may also contribute to catfacing in some cultivars. Radial cracking is a genetically controlled trait that has been eliminated from most modern cultivars but is still a problem in certain heirlooms. Concentric fruit cracking is caused by water collecting around the fruit stem. Genetic resistance to cracking disorders has been bred into many modern cultivars. Green shoulders is a condition where the top of the fruit fails to turn red and remains green and hard even in ripe fruits. The uniform ripening gene prevents green shoulders and has been bred into many modern cultivars. Sunscald occurs when fruits are exposed to direct sunlight, which kills fruit tissues due to heat buildup.

Weed Management Strategies

Weed competition is most damaging early in the growing season. Popular weed-control strategies include plastic mulch, organic mulch, solarization, mechanical and hand cultivation, fumigation, and selective herbicides (Stall and Gilreath, 2002). Two parasitic plants dodder (*Cuscuta* spp.) and broomrape (*Orobanche ramosa*) are major pests of tomato in some areas of the world, which are difficult to control. Weed management is easier with transplanted crops, because clean cultivation can be practiced before transplanting. Crop rotation can effectively reduce difficult weed problems by altering the environmental conditions that favor a particular species or by allowing alternative control methods. Alfalfa (*Medicago sativa*) is a good rotation crop with tomato because its frequent cutting cycle reduces some weeds. Corn (*Zea mays*) is also effective in rotation with tomato because some corn herbicides control *Solanum nigrum* and field bindweed (*Convolvulus arvensis*) and corn is not a host for dodder (*Cuscuta* spp.) (Stall and Gilreath, 2002).

Soil solarization controls many soil-borne diseases, nematodes, and weed pests in some areas. Preventing weeds from seeding in or around production fields reduces populations in subsequent crops. A stale bed prepared 10 days to 2 weeks before seeding or transplanting often allows the first flush of weeds to be killed with shallow cultivation or nonselective herbicides. When tomato seedlings are about 10 cm (4 in) tall, shallow cultivation can create a dry layer of soil (dry mulch) along the seed line to prevent weed seeds from germinating and to smother small seedlings (Stall and Gilreath, 2002).

Diseases

Many diseases attack tomato. Bacterial wilt (*R. solanacearum*) is a serious disease in hot and humid areas. Long-term crop rotation, wilt-resistant cultivars, and sanitation practices are recommended for control. There are no effective chemical options. Bacterial spot (*X. vesicatoria*, *X. campestris*) are serious seed-transmitted diseases occurring primarily during cold, rainy weather. These diseases damage blossoms, foliage, and stems, but may be difficult to detect on mature fruit. Control measures include copper fungicides, resistant cultivars, disease-free seed, and healthy transplants. Cultivar resistance is not available for some strains of these pathogens. Other bacterial diseases of tomato include canker (*Clavibacter michiganensis* subsp. *michiganensis*), a seed-borne disease that spreads rapidly within greenhouses. Growing disease-free transplants is the most important control measure. Other diseases include bacterial speck (*Pseudomonas syringae* pv. *tomato*), bacterial soft rot (*E. carotovora* subsp. *carotovora*), bacterial wilt (*R. solanacearum*), pith necrosis (*Pseudomonas corrugata*), and syringae leaf spot (*Ps. syringae* pv. *syringae*) (American Phytopathological Society, 1991).

Some of the most important fungal diseases of tomatoes are early blight (*A. solani*), black leaf mold (*Pseudocercospora fuligena*), leaf mold (*Mycovellosiella fulva*), powdery mildew (*Leveillula taurica*), southern blight (*Corticium rolfsii*), Botrytis blight (*Botrytis cinerea*), and target spot (*Corynespora casiicola*). Fusarium wilt (*F. oxysporum*) and Verticillium wilt (*Verticillium dahlia*) are soil-borne diseases that attack vascular tissue. They can be controlled by crop rotation and fumigation, and disease-resistant cultivars are available for some races. Phytophthora root rot (*Phytophthora parasitica* and *Ph. capsici*) may infect plants throughout the season.

Preventing soil saturation for extended periods is the most useful control. Corky root (*Pyrenochaeta lycopersici*) is problematic in some areas, and a typical control strategy is to avoid early planting in cold soils. Late blight (*Ph. infestans*) is a concern particularly in high humidity and occurs during rainy periods and in fall. Fruit-protectant chemicals, applied by ground application equipment for optimum coverage, may provide control. On fall fields, fungicides are applied to minimize damage from black mold (*Alternaria alternata*) (American Phytopathological Society, 1991).

There are many virus diseases that infect tomato. Virus damage can cause negligible to substantial yield loss. Depending on the virus, transmission is through direct contact or through insect vectors such as aphids, whiteflies, and thrips. Some of the most common viruses are curly top, cucumber mosaic, tobacco mosaic, tobacco etch, and potato Y. Tomato spotted wilt and tobacco streak are common in some areas. Other viruses affecting tomato may include whitefly-transmitted gemini viruses, tobacco leaf curl virus, and tobacco ringspot virus. Growing virus-resistant cultivars, if available, is the best control method. Early control of insect vectors and general field sanitation is beneficial (American Phytopathological Society, 1991).

Diseases are an important source of postharvest loss depending on season, region, and handling practices. Decay or surface lesions result from the fungal pathogens black mold rot (*A. alternata*), gray mold rot (*B. cinerea*), sour rot (*Geotrichum candidum*), and Rhizopus rot (*Rhizopus stolonifer*). Bacterial soft rot, caused by *Erwinia* spp., can be a serious problem particularly if harvest and packinghouse sanitation is poor. Hot air or hot water immersion at 55°C (131°F) for 0.5–1.0 min often prevents surface mold. Controlled atmospheres can delay fungal growth on the stem-end and fruit surface. Greenhouse fruit marketed on-the-vine as cluster tomatoes are susceptible to Botrytis gray mold, especially if film-wrapped in a tray. General postharvest control measures include proper sanitation (leaving diseased fruits in the field or removing before shipment or storage), chlorinated wash water, UV light treatment, fungicide treatments, and controlling the temperature of the wash water (Snowdon, 2010).

Insect and Nematode Pests

Insect pests of tomato are numerous and vary with location (Kennedy, 2003). The following is a partial listing of some of the major pests of tomato. The tomato fruitworm (*Helicoverpa armigera* and *Helicoverpa zea*) and various armyworms (*Spodoptera* spp.) are destructive insect pests, causing significant losses due to fruit boring. Naturally occurring beneficial insects *Trichogramma* spp. egg parasites, the larval parasite *Hyposoter exiguae*, and predators such as bigeyed bug and minute pirate bug are important for the biological control of tomato fruitworm. Sprays of Bt and the Entrust formulation of spinosad are effective on many worms that feed on tomato and are generally acceptable for organic certification. Pyrethroids and/or insect growth regulators may provide effective control for armyworms and stink bugs (*Euschistus conspersus* and *Nezara viridula*). However, pyrethroids are a broader-spectrum chemistry and may inadvertently kill beneficial insects. IPM monitoring programs have been developed in many production regions for determining treatment thresholds for initiating fruitworm and armyworm control programs. Tomatoes should not be planted near other alternative insect hosts, such as corn and cotton that serve as a refuge.

Cotton aphid (*Aphis gossypii*) is a major pest during the dry season and vectors cucumber mosaic virus. The silverleaf whitefly and sweetpotato whitefly (*Bemisia argentifolii* and *Bemisia tabaci*) are both serious pests themselves and also vector tomato yellow leaf curl virus. Thrips are also a problem pest in glasshouses, especially *Frankliniella occidentalis*, which vectors tomato spotted wilt virus. Other important tomato fruit pests include tarnished plant bug (*Lygus hesperus*), potato tuber moth (*Phthorimaea operculella*), and tomato pinworm (*Keiferia lycopersicella*). Some important foliage pests are beet leafhopper (*Neoaliturus tenellus*), green peach aphid (*M. persicae*), tobacco hornworm (*Manduca sexta*), tomato hornworm (*Manduca quinquemaculata*), leafminers (*Liriomyza sativae* and other species), looper caterpillars (*Autographa californica* and *Trichoplusia ni*), potato aphids (*Macrosiphum euphorbiae*), tomato russet mite (*Aculops lycopersici*), flea beetles (*Epitrix hirtipennis* and others), and wireworms (*Limonius* spp.) (Kennedy, 2003).

The primary insect pests of tomato seedlings are flea beetles (*Epitrix* spp.), darkling ground beetles (*Blapstinus* spp.), and cutworms (*Peridroma* and *Agrotis* spp.), but they do not typically cause major problems during transplant production or after field establishment. Sometimes the garden centipede (*Scutigerella immaculata*) can damage young transplants. Root knot nematodes (*Meloidogyne incognita* and other species) will cause gall formation. Crop rotation, resistant cultivars, or fumigation will control root knot nematodes.

References

Aguiar, J., Molinar, R. and Valencia, J. (2013) Eggplant Production in California. Publication #7235. Available at: http://anrcatalog.ucdavis.edu/pdf/7235.pdf (accessed 14 November 2013).

Almekinders, C.J.M., Chujoy, E. and Thiele, G. (2009) The use of true potato seed as pro-poor technology: the efforts of an international agricultural research institute to innovating potato production. *Potato Research* 52, 275–293.

American Phytopathological Society (1991) *Compendium of Tomato Diseases*. APS Press, St. Paul, Minnesota.

Andrews, J. (1998) *The Pepper Lady's Pocket Pepper Primer*. University of Texas Press, Austin, Texas.

Batra, S.W. (1993) Male-fertile potato flowers are selectively buzz-pollinated only by *Bombus terricola* Kirby in Upstate New York. *Journal of the Kansas Entomological Society* 66, 252–254.

Berry, R.E., Reed, G.L. and Coop, L.B. (2000) Identification & Management of Major Pest & Beneficial Insects in Potato. Publication No. IPPC E.04-00-1. Available at: http://ippc2.orst.edu/potato (accessed 29 December 2011).

Bevacqua, R.F. and VanLeeuwen, D.M. (2003) Planting date effects on stand establishment and yield of chile pepper. *HortScience* 38, 357–360.

Bland, S.E. (2005) Consumer acceptability of heirloom tomatoes. MS thesis. The University of Georgia, Athens, Georgia.

Boldt, P.F. (1976) Factors influencing the selectivity of U-compounds on yellow nutsedge. MS thesis. Cornell University, Ithaca, New York.

Bosland, P.W. (1996) Capsicums: Innovative uses of an ancient crop. In: Janick, J. (ed.) *Progress in New Crops*. ASHS Press, Arlington, Virginia, pp. 479–487.

Bradford, K.J., Steiner, J.J. and Trawatha, S.E. (1990) Seed priming influence on germination and emergence of pepper seed lots. *Crop Science* 30, 718–721.

Brady, C.J. (1987) Fruit ripening. *Annual Review of Plant Physiology* 38, 155–178.

Brown, C.R. (1993) Proceedings of the symposium past, present and future uses of potatoes origin and history of the potato. *American Journal of Potato Research* 70, 363–373.

Burden, D. (2012) Bell and chili peppers profile. Available at: www.agmrc.org/commodities_products/vegetables/bell_and_chili_peppers_profile.cfm (accessed 15 December 2013).

Burton W.G. (1966) *The Potato*, 2nd edn. Wageningen Veenman, Rotterdam, the Netherlands.

Cantliffe, D.J. (2009) Plug transplant technology. *Horticultural Reviews* 35, 397–436.

Cantwell, M. (2013a) Chile Pepper: Recommendations for Maintaining Postharvest Quality. Available at: http://postharvest.ucdavis.edu/pfvegetable/chilepeppers (accessed 18 November 2013).

Cantwell, M. (2013b) Bell Pepper: Recommendations for Maintaining Postharvest Quality. Available at: http://postharvest.ucdavis.edu/pfvegetable/bellpepper (accessed 18 November 2013).

Cantwell, M. and Suslow, T.V. (2013) Eggplant, Recommendations for Maintaining Postharvest Quality. Available at: http://postharvest.ucdavis.edu/pfvegetable/eggplant (accessed 18 November 2013).

Cao, W. and Tibbitts, T.W. (1994) Phasic temperature change patterns affect growth and tuberization in potatoes. *Journal of the American Society for Horticultural Science* 119, 775–778.

Capinera, J.L. (2005) Pepper Weevil, *Anthonomus eugenii* Cano (Insecta: Coleoptera: Curculionidae). *Entomology and Nematology Department Document ENY-278*, Florida Cooperative Extension Service, Institute of Food and Agricultural Sciences, University of Florida, Gainesville, Florida.

Chen, H.H. and Li, P.H. (1980) Biochemical changes in tuber-bearing *Solanum* species in relation to frost hardiness during cold acclimation. *Plant Physiology* 66, 414–421.

Cichewicz, R.H. and Thorpe, P.A. (1996) The antimicrobial properties of chile peppers (*Capsicum* species) and their uses in Mayan medicine. *Journal of Ethnopharmacology* 52, 61–70.

Clark, C.F. and Lombard, P.M. (1951) Descriptions of and key to American potato varieties. *United States Department of Agriculture Circulars* 741, 50.

Colla, G., Rouphael, Y., Cardarelli, M., Temperini, O., Rea, E., Salerno, A. and Pierandrei, F. (2008) Influence of grafting on yield and fruit quality of pepper (*Capsicum annuum* L.) grown under greenhouse conditions. *Acta Horticulturae* 782, 359–363.

Collins, W., Witcombe, J., Lenne, J. and Eden-Green, S. (2000) Workshop on Transgenic potatoes for the benefit of resource-poor farmers in developing countries. In: Lizárraga, C. and Hollister, A. (eds) *Proceedings of the International Workshop on Transgenic Potatoes for the Benefit of Resource-poor Farmers in Developing Countries*. International Potato Center (CIP), Lima, Peru, pp. 5–8.

Cook, R.L. and Calvin, L. (2005) *Greenhouse tomatoes change the dynamics of the North American fresh tomato industry* (No. 3). United States Department of Agriculture, Economic Research Service, Washington, DC.

Curwen, D., Kelling, K.A., Schoenemann, J.A., Stevenson, W.R. and Wyman, J.A. (1982) *Commercial Potato Production and Storage*. Publication A2257. University of Wisconsin Cooperative Extension, Madison, Wisconsin.

Davis, G.N. (1949) *Growing Potatoes in California*. College of Agriculture, University of California, California.

DeWitt, D. and Bosland, P.W. (1996) *Peppers of the World. An identification guide*. Ten Speed Press, Crown Publishing, New York, 219 pp.

FAOSTAT (2011) Online Database of Crop Production Statistics 2011. Available at: http://aostat.fao.org/site/567/DesktopDefault.aspx?PageID=567 (accessed 27 December 2011).

Funderburk, J., Olson, S., Stavisky, J. and Avila, Y. (2004) Managing Thrips and Tomato Spotted Wilt in Pepper. Available at: http://edis.ifas.ufl.edu/in401 (accessed 31 December 2013).

Gil Ortega, R., Gutierrez, M. and Cavero, J. (2004) Plant density influences marketable yield of directly seeded 'Piquillo' pimiento pepper. *HortScience* 39, 1584–1587.

Gleick, P.H. (2000) Water for Food: How Much Will Be Needed? In: Gleick, P.H. (ed.) *The World's Water*. Island Press, Washington, DC, pp. 63–91.

Goyal, R., Ramachandran, R., Goyal, P. and Sharma, V. (2007) Edible vaccines: Current status and future. *Indian Journal of Medical Microbiology* 25, 93–102.

Grandillo, S., Zamir, D. and Tanksley, S.D. (1999) Genetic improvement of processing tomatoes: A 20 years perspective. *Euphytica* 110, 85–97.

Greenleaf, S.S. and Kremen, C. (2006) Wild bee species increase tomato production and respond differently to surrounding land use in Northern California. *Biological Conservation* 133, 81–87.

Gregory, L.E. (1965) Physiology of tuberization in plants. *Encylopedia Plant Physiology* 15, 1328–1354.

Guzman, I., Bosland, P.W. and O'Connell, M.A. (2011) Heat, Colour, and Flavour Compounds in Capsicum Fruit. In: Gang, D.R. (ed.) *Recent Advances in Phytochemistry 41: The Biological Activity of Phytochemicals*. Springer, New York, pp. 117–118.

Halmer, P. (2008) Seed technology and seed enhancement. *Acta Horticulturae* 771, 17–26.

Hanson, B.R. and May, D.M. (2006) Crop coefficients for drip-irrigated processing tomato. *Agricultural Water Management* 81, 381–399.

Harder, L.D. and Barclay, R.M.R. (1994) The functional significance of poricidal anthers and buzz pollination: Controlled pollen removal from dodecatheon. *Functional Ecology* 8, 509–517.

Harris P.M. (ed.) (1978) *The Potato Crop. The Scientific Basis for Improvement*. Chapman & Hall, London.

Ho, L.C. and Hewitt, J.D. (1986) Fruit development. In: Atherton, J.G. (ed.) *The Tomato Crop*. Chapman Hall, London, pp. 201–239.

Holm, L.G., Pancho, J.V., Herberger, J.P. and Plucknett, D.L. (1991) *A Geographic Atlas of World Weeds*. Krieger Publishing Company, Malabar, Florida.

Hornfeldt, C.S. and Collins, J.E. (1990) Toxicity of nightshade berries (*Solanum dulcamara*) in mice. *Journal of Toxicology - Clinical Toxicology* 28, 185–192.

Kader, A.A., Stevens, M.A., Albright-Holton, M., Morris, L.L. and Algazi, M. (1977) Effect of fruit ripeness when picked on flavor and composition in fresh market tomatoes. *Journal of the American Society for Horticultural Science* 102, 724–731.

Kennedy, G.G. (2003) Tomato, pests, parasitoids, and predators: tritrophic interactions involving the genus *Lycopersicon*. *Annual Review of Entomology* 48, 51–72.

Khan, A.A., Maguire, J.D., Abawi, G.S. and Ilyas, S. (1992) Matriconditioning of vegetable seeds to improve stand establishment in early field plantings. *Journal of the American Society for Horticultural Science* 117, 41–47.

Lee, J. and Oda, M. (2003) Grafting of herbaceous vegetable and ornamental crops. *Horticultural Reviews* 28, 61–124.

Le Strange, M., Schrader, W. and Hartz, T. (2000) Fresh-Market Tomato Production in California. Available at: http://anrcatalog.ucdavis.edu/pdf/8017.pdf (accessed 31 December 2013).

Loomis, W.E. (1925) Studies in the transplanting of vegetable plants. *Cornell Agricultural Experiment Station Memoirs* 87, 1–63.

Lopez, S.L. (2007) NMSU is home to the world's hottest chile pepper. Available at: http://web.archive.org/web/20070219124128/http://www.nmsu.edu/~ucomm/Releases/2007/february/hottest_chile.htm (accessed 21 February 2007).

Ma, Y., Hong, G., Wang, Q. and Cantwell, M. (2010) Reassessment of treatments to retard browning of fresh-cut Russet potato with emphasis on controlled atmospheres and low concentrations of bisulfite. *International Journal of Food Science & Technology* 45, 1486–1494.

Marshall, D.E. (1997) Designing a pepper for mechanical harvesting. *Capsicum and Eggplant Newsletter* 16, 15–27.

Masabni, J., Anciso, J., Lillard, P. and Dainello, F.J. (2011) *Texas Commercial Vegetable Production Guide*. Publication No. B-6159. Texas A&M Cooperative Extension Service, State College, Texas.

Matas, A.J., Gapper, N.E., Chung, M.Y., Giovannoni, J.J. and Rose, J.K. (2009) Biology and genetic engineering of fruit maturation for enhanced quality and shelf-life. *Current Opinion in Biotechnology* 20, 197–203.

Maynard, D.M. and Hochmuth, G. (1996) *Knott's Handbook for Vegetable Growers*, 4th edn. Wiley-Interscience, New York.

McGee, H. (2004) *On Food and Cooking*, revised edn. Scribner Publishing, New York.

Miller, J.C. and McGoldrick, F. (1941) Effect of day length upon vegetative growth, maturity and tuber characteristics of the Irish potato. *American Potato Journal* 18, 261–265.

Oka, Y., Offenbach, R. and Pivonia, S. (2004) Pepper rootstock graft compatibility and response to *Meloidogyne javanica* and *M. incognita*. *Journal of Nematology* 36, 137–141.

Olson, S.M., Simmone, E.H., Maynard, D.N., Hochmuth, G.J., Varina, C.S., Stall, W.M., Pernezny, K.L., Webb, S.E., Taylor, T.G. and Smith, S.A. (2005) *Pepper Production in Florida*. Horticultural Department Document HS-732, Florida Cooperative Extension Service, Institute of Food and Agricultural Sciences, University of Florida, Gainesville, Florida.

Palada, M.C. and Wu, D.L. (2008) Evaluation of chili rootstocks for grafted sweet pepper production during the hot-wet and hot-dry seasons in Taiwan. *Acta Horticulturae* 767, 151–157.

Peet, M.M. and Welles, G. (2005) Greenhouse tomato production. In: Heuvelink, E. (ed.) *Tomatoes*. CAB International, Wallingford, UK, pp. 257–304.

Peralta, I.E. and Spooner, D.M. (2007) History, origin and early cultivation of tomato (Solanaceae). *Genetic Improvement of Solanaceous Crops* 2, 1–27.

Pernezny, K. and Momol, T. (2006) Florida Plant Disease Management Guide: Pepper. Available at: http://edis.ifas.ufl.edu/pg052 (accessed 5 January 2012).

Peter, K.V. (ed.) (2001) *Handbook of Herbs and Spices*, Vol. 1. CRC Press, Boca Raton, Florida.

Peterson, L.A. and Krueger, A.R. (1988) An intermittent aeroponics system. *Crop Science* 28, 712–713.

Pickersgill, B. (1997) Genetic resources and breeding of Capsicum spp. *Euphytica* 96, 129–133.

Plunkett, D.J. (1996) Mexican Tomatoes – Fruit of New Technology. *Vegetables and Specialties S&O/VGS* 268, 26–30.

Richardson, A. (1995) *Plants of the Rio Grande Delta*. University of Texas Press, Austin, Texas.

Roberts, S. (2008) Scoville Scale. Available at: www.scottrobertsweb.com/scoville-scale (accessed 31 December 2013).

Roh, J.Y., Choi, J.Y., Li, M.S., Jin, B.R. and Je, Y.H. (2007) *Bacillus thuringiensis* as a specific, safe, and effective tool for insect pest control. *Journal of Microbiology and Biotechnology* 17, 547–59.

Rubatzky, V.E. and Yamaguchi, M. (1997) *World Vegetables, Principles, Production, and Nutritive Values*, 2nd edn. Chapman and Hall, New York.

Ruf, S., Hermann, M., Berger, I.J., Carrer, H. and Bock, R. (2001) Stable genetic transformation of tomato plastids and expression of a foreign protein in fruit. *Nature Biotechnology* 19, 870–875.

Saccardo, F., Colla, G., Crino, P., Paratore, A., Cassaniti, C. and Temperini, O. (2006) Genetic and physiological aspects of grafting in vegetable crop production. *Italus Hortus* 13, 71–84.

Santos, H.S. and Goto, R. (2004) Sweet pepper grafting to control phytophthora blight under protected cultivation. *Horticultura Brasilera* 22, 45–49.

Schmitz, A. and Seckler, D. (1970) Mechanized agriculture and social welfare: The case of the tomato harvester.

American Journal of Agricultural Economics 52, 569–577.

Scott, G.J., Rosegrant, M.W. and Ringler, C. (2000) *Roots and Tubers for the 21st Century: Trends, Projections, and Policy Options*. Food, agriculture, and the environment discussion paper. Vol. 31. Intl Food Policy Res Inst, 64 pp.

Singh, R.J. (2006) *Genetic Resources, Chromosome Engineering, and Crop Improvement: Vegetable Crops*. CRC Press, Boca Raton, Florida.

Snowdon, A.L. (2010) *Post-harvest Diseases and Disorders of Fruits and Vegetables: Vegetables*, Vol. 2. Manson Publishing, London.

Stall, W.M. and Gilreath, J.P. (2002) Weed control in tomato. In: *Weed Management in Florida Fruits and Vegetables*. IFAS Extension Publication #HS200, University of Florida Cooperative Extension Service, Gainesville, Florida, pp. 5–58.

Stevenson, F.J. (1951) The potato – its origin, cytogenetic relationships, production, uses and food value. *Economic Botany* 5, 153–171.

Styer, R.C. and Koranski, D.S. (1997) *Plug and Transplant Production: A Grower's Guide*. Ball Publishing, Batavia, Illinois.

Suslow, T.V. and Voss, R. (2013) Potato, Early Crop: Recommendations for Maintaining Postharvest Quality. Available at: http://postharvest.ucdavis.edu/pfvegetable/PotatoesEarly (accessed 18 November, 2013).

Tainter, D.R. and Grenis, A.T. (2001) *Spices and Seasonings*. Wiley-IEEE, New York.

The Scoville Scale (2012) Available at: www.happystove.com/recipe/32/The+Scoville+Scale (accessed 1 January 2012).

Thompson, G.D. and Wilson P.N. (1998) The organizational structure of the North American fresh tomato market: Implications for seasonal trade disputes. *Agribusiness* 13, 533–547.

Thornsbury, S. (2013) North American Fresh-Tomato Market. Available at: www.ers.usda.gov/topics/in-the-news/north-american-fresh-tomato-market.aspx#backgroundstatistics (accessed 16 December 2013).

USDA (2009) Economic Research Service. 2009. US Potato Utilization. Available at: www.ers.usda.gov/data/foodconsumption/Spreadsheets/potatoes.xls (accessed 29 December 2011).

USDA Nutrient Database (2011) The USDA Nutritional Database for Standard Reference. Available at: http://ndb.nal.usda.gov (accessed 11 July 2011).

Walker, S.J. (2013) Red chile and paprika production in New Mexico. Publication Guide H-257. Available at: http://aces.nmsu.edu/pubs/_h/h-257/welcome.html (accessed 15 December 2013).

Wall, M.M. (1994) *Postharvest Handling of Dehydrated Red Chiles*. Guide H-236. New Mexico State University Cooperative Extension Service, Las Cruces, New Mexico.

Waterer, D.R. (2000) Effect of soil mulches and herbicides on production economics of warm season vegetable crops in a cool climate. *HortTechnology* 10, 154–158.

Woodham-Smith, C. (1991) *The Great Hunger: Ireland 1845–1849*. Penguin Publishing, London.

Zhao, D.Y., Shen, L., Fan, B., Liu, K.L., Yu, M.M., Zheng, Y., Ding, Y. and Sheng, J.P. (2009) Physiological and genetic properties of tomato fruits from 2 cultivars differing in chilling tolerance at cold storage. *Journal of Food Science* 74, 348–352.

12 Family Asteraceae

Origin and History

Asteraceae is a very large and widespread family with more than 23,000 species, spread across 1,620 genera (Jeffrey, 2007). Family members are annual or perennial herbs, many are weeds or wild flowers, and a few are woody but are not usually classified as trees. Asteraceae contains many familiar ornamental plants including aster, marigold, calendula, daisy, chrysanthemum, dahlia, and zinnia and medicinal plants including grindelia, echinacea, yarrow, and many others (Duke, 2013).

The Latin name "Asteraceae" is derived from the Greek word for "star". Compositae is an older family name that still appears in the literature and is derived from the word composite, which refers to the characteristic inflorescence found in only a few angiosperm families.

A characteristic of many Asteraceae species is milk-like latex contained in its tissues. Latex from dandelion roots can be used as a source of rubber. During World War II, some European nations grew dandelions for rubber production when tropical sources were unavailable. Today, several species of dandelion, most particularly Russian dandelion, are again being investigated for commercial rubber production (van Beilen and Poirier, 2007). Lactucarium, the dried latex produced from *Lactuca virosa*, has medicinal use (Duke, 2013). Dried latex has some narcotic properties and has been used as a sedative. Dandelion is a common weed in North America that can be invasive. Interestingly, European settlers originally brought dandelion to North America as a green vegetable before it escaped. Young dandelion plants are still used as a vegetable by some cultures today. The vegetable crops we will consider in some detail in this chapter include lettuce, chicory, and endive.

LETTUCE, ENDIVE, AND CHICORY
Origin and History

Lettuce (*Lactuca sativa* L.) is a major world salad crop. It is a cool-season vegetable adapted to temperate regions. Cultivated lettuce has an uncertain origin since it no longer exists in the wild. Modern lettuce is possibly derived from its wild relative *L. serriola*. The *L. sativa-serriola* complex is large, polymorphic and capable of cross-breeding, so *L. sativa* may have been derived directly from *L. serriola* through selection (Jeffrey, 2007).

The Egyptians are credited with domesticating lettuce and using not only its leaves but also seeds to produce oil. Lettuce was apparently cultivated as far back as 4500 BC judging from paintings of putative romaine lettuce leaves identified in Egyptian tombs. These paintings suggest that lettuce was a widely known and appreciated crop. At the same time, the Egyptians also cultivated endive and chicory, which are also native to the Mediterranean region. Lettuce spread to the Greeks and Romans and then throughout the Mediterranean basin. By AD 50 multiple types were described in the region (Zohary *et al.*, 2012). Lettuce was mentioned in medieval writings as a medicinal herb. The 16th–18th centuries saw the development of many types in Europe, and by the mid-18th century cultivars were described that can still be found today. Europe and North America originally dominated the market for lettuce, but by the late 1900s the consumption of lettuce had spread throughout the world (Zohary *et al.*, 2012).

Botany and Life Cycle
Lettuce

Lettuce (*L. sativa* L.) is an erect smooth herbaceous annual plant grown for its crisp edible leaves that form continuously during the vegetative stage of development. Leaves are arranged spirally in a dense rosette. Plants have considerable diversity of leaf shape, texture, and margin among the many different forms. For example, glabrous leaves can be smooth, savoy, or crumpled. Leaf margins may be lobed, smooth, or finely divided, and leaf colors can vary from light to dark green with some cultivars having red or purple coloration (Fig. 12.1).

Interior leaves of leafy cultivars tend to be lighter in color, whereas those of heading types are white. With the exception of stem lettuce, the cylindrical stem is short and compressed. Upon bolting, the stem elongates, becoming erect, tall and branched (Rubatzky and Yamaguchi, 1997).

The inflorescence or flower head of Asteraceae is a structure called a capitulum (McKenzie *et al.*, 2005). What may appear as a single flower is actually many individual florets that share a common receptacle to form a composite inflorescence. The capitulum is a contracted raceme composed of numerous individual sessile flowers, called florets. The inflorescence is a panicle consisting of 10–25 florets, each opening in the morning. The florets are self-pollinating, although insect pollination occasionally occurs. Flowering may continue for 1–2 months after induction. Asteraceae have achene-like fruit, called a cypsela, with fused carpels and one locule, that contains a single small, dry, ribbed seed topped with a pappus. The pappus is a feathery mast-like appendage that acts as a parachute to disperse seeds through the wind. The pappus is a modified calyx that is milled off during processing to produce seed for commercial planting. The achenes are oblong, broadest towards the apex, and range in color between white and shades of yellow, brown, gray, and black (Rubatzky and Yamaguchi, 1997).

Most newly harvested seed exhibit a short postharvest dormancy that disappears during dry storage or after leaching. A few older heirloom cultivars exhibit photodormancy and require light to germinate. Many modern cultivars have varying levels of thermodormancy, which prevents germination above 24°C (75°F). On average, 1,000 lettuce seeds weigh 1 g (28,571/oz). In deep soils, a taproot rapidly develops after germination along with an extensive lateral root system. Although the taproot can grow up to 1 m (3.3 ft) deep in some situations, lateral roots near the surface absorb the most moisture and nutrients (Rubatzky and Yamaguchi, 1997).

Fig. 12.1. Red and green leaf lettuce production near Salinas, California.

Endive

Endive (*Cichorium endiva* L.) is a cool-season, hardy, annual or biennial plant initially producing a dense rosette of leaves on a short compressed stem. The foliage resembles loose-leaf lettuces, but tends to be prostrate. Favorable growing temperatures are between 18–20°C (64–68°F) with harvest maturity occurring about 60–80 days after seeding under optimal conditions. Endives are more heat tolerant than lettuce. High temperatures and long days cause bolting. Vernalization is required for biennial types (Rubatzky and Yamaguchi, 1997).

Chicory

Chicories (*Cichorium intybus* L.) are deep-rooted frost-tolerant perennials grown as annuals or biennials. Upon bolting, the stem elongates and branches. The inflorescence bears many capitula, usually with pale blue or white flowers. Flowers are self-pollinating and each floret produces an individual seed. About 800 seeds weigh 1 g (22,800/oz). Chicories are cool-season plants that grow best at temperatures between 15–18°C (59°–64°F) with harvest maturity occurring about 70–100 days after planting. Chicory is grown for its foliage, which is used in salads or as a cooked vegetable. The roots of some cultivars are also used. Foliage, both green and red, is strongly flavored and exhibits different levels of bitterness. Old and dark-colored leaves have greater bitterness compared to immature or pale foliage (Rubatzky and Yamaguchi, 1997).

Types and Cultivars

Lettuce

There are four widely recognized morphological forms of lettuce: crisphead, butterhead, romaine or cos, and loose leaf (Table 12.1). Stem, oil seed, and Latin lettuce are additional forms (Rubatzky and Yamaguchi, 1997). The Batavian lettuce is another crisphead type that pre-dates iceberg with a smaller and less firm head (Miles, 2013).

Crisphead forms are often referred to as "iceberg" or simply head lettuce (Table 12.1). Following early rosette development, additional leaf growth begins overlapping, eventually entrapping newly formed young leaves inside a dense nearly spherical head that weighs 0.7–1.0 kg (1.5–2.2 lb) (Fig. 12.2).

Iceberg may become very firm and heads may burst if harvest is delayed. Outer leaves usually are dark green, while inner leaves are progressively lighter in color. The leaves are somewhat brittle but with a crisp texture and high water content. Crisphead lettuce does not grow well under warm conditions but prefers cool conditions and full sunlight. One of the outstanding attributes of crisphead lettuce is its long-distance shipping and storage ability. It withstands handling well and has

Table 12.1. Major forms of lettuce (Doležalová *et al.*, 2004).

Botanical variety or Group name	Common names	Description
L. sativum bot. var. *asparagina*	Stem or asparagus lettuce, celtuce	Grown for its erect thickened edible stem 30–40 cm (12–16 in) long
L. sativum bot. var. *capitata*	Crisphead, iceberg	Very dense rosette with cabbage-like structure and crisp texture
L. sativum bot. var. *capitata*	Batavian, summer crisp, French crisp	Smaller, less firm more open head than crisphead
L. sativum bot. var. *capitata*	Butterhead, Boston, bibb	Smaller, flattened, less compact head, leaves are broad and tender, with a soft oily texture
L. sativum bot. var. *crispa*	Loose-leaf, bunching	Loose rosette of cut, fringed or crisped leaves varying in color
L. sativum bot. var. *longifolia*	Cos or romaine	Upright head and narrow or columnar, leaves ovate to oblong, obtuse with broad midrib
L. sativum	Latin lettuce	A rosette of loose, elongated and soft cos-like leaves that may form a partly closed head
L. serriola	Oil seed	Leaves are not eaten, seeds contain 35% oil historically used for cooking

high density, which allows it to be shipped efficiently. Crisphead is used as the main ingredient for tossed salads and on sandwiches. Increasingly it is chopped in bags for food service companies, restaurants and consumers as prewashed ready-to-eat items or premixed with other vegetables as salad (Turini et al., 2011).

Butterhead is another heading type with smaller tender leaves (Table 12.1). Leaves overlap to form a smooth soft head with an almost oily texture and delicate flavor. Butterhead is more fragile than iceberg and is not as well suited for mass handling or long-distance shipping. Plants are easily bruised, susceptible to leaf breakage and wilting, so extra care must be exercised to preserve quality. Batavia-type cultivars have characteristics intermediate between crisphead and butterhead types (Miles, 2013).

Cos or romaine heads have elongated, coarse, crisp-textured leaves with prominent broad midveins (Table 12.1). The long relatively narrow leaves tend to grow upright and may loosely overlap each other but do not form solid heads (Fig. 12.3). Postharvest handing is similar to crisphead types with a relatively long shelf life. Romaine is more tolerant of warm weather and many consider the flavor superior to crisphead.

Loose-leaf or bunching cultivars are easy to grow and a common spring crop for home gardeners (Table 12.1). Loose-leaf cultivars exhibit considerable variation in leaf size, margin shape, color, and texture. Each plant develops leaves in a tight rosette cluster (Fig. 12.1). Many have crispy tender foliage with smooth leaves. The texture varies from crisp to oily depending on the cultivar. Edible leaves mature in as little as 50 days from seeding and 30 days for small leaf or baby cultivars. Postharvest handling is more critical with

Fig. 12.2. Double row of crisphead lettuce a few weeks prior to harvest.

Fig. 12.3. Romaine lettuce field near Gilroy, California. The field is drip irrigated with four rows per raised bed.

leaf lettuces because leaves break and wilt easily even with refrigeration and good handling. Under cool short-day conditions, successive leaf cuttings can be made until bolting occurs in summer (Miles, 2013).

Stem lettuce is an important crop in Egypt, Japan, China, and other countries in Southeast Asia but it is not widely grown in other areas (Table 12.1). Plants are nonheading and are produced mainly for their erect, thickened, edible stem. Stem lettuce is peeled and the soft green core is eaten raw or cooked. Except for the youngest foliage, leaves are not palatable because of their high latex content and bitterness (Rubatzky and Yamaguchi, 1997).

Since lettuce is predominately a self-pollinated crop with tiny flowers that produce only one seed (achene), no F-1 hybrid cultivars have been developed. Therefore, lettuce cultivars are non-hybrid, self-pollinated purelines. Seed can be saved from season to season with minimal variability. Genetically engineered transgenic lettuce has not had an economic impact.

Chicory

Distinctive forms of chicory include leafy salad, heading, forcing heads, and root types. The green leafy vegetable that resembles dandelion is called foliage chicory and is a popular vegetable in parts of Europe (Fig. 12.4).

Seeds are generally direct-seeded 2.5 cm (1 in) apart in rows 46 cm (18 in) wide. For microgreen production seeds can be sown at closer spacing.

Foliage colors vary from light green to dark purple. In general, smooth leaf forms are usually cooked, while crispy leaf types most often are eaten fresh in salads. Leafy forms of chicory tend to have a bitter flavor that is not liked by all. Foliage chicory cultivars mature in about 50–90 days from seeding and 30–40 days if harvested in the baby stage as microgreens. Successive leaf harvests can be conducted or the whole plant can be destructively harvested. Foliage types are less sensitive to high temperatures and can be grown at 24°C (75°F) with cool night temperatures, although high temperatures and drought stress increases bitterness and fiber. The postharvest life of both chicory and endive is about 15 days if stored at 1°C (34°F) and 95% RH (Rubatzky and Yamaguchi, 1997).

Chicory cultivars that produce small red or green-colored heads are called radicchio (Fig. 12.5). The head resembles a small head of cabbage but with prominent white leaf veins and weigh from 0.25 to 1 kg (0.5–2.2 lb). Cultivars are divided into round 'Chiaggia' and upright 'Trevisio' types that mature from 55 to 70 days after transplanting. The most popular cultivars in the USA have red with white veins. Plants can be direct-seeded or transplanted with a final spacing of 15 cm (6 in) apart in rows 46–61 cm (18–24 in) wide. Radicchio adds color, texture, and bitter flavor to salads and vegetable dishes. Radicchio is susceptible to leaf-tip burn and head rot when grown at high temperatures. Optimum daytime growth occurs at <21°C (70°F) and night temperatures <15°C (60°F) (Rubatzky and Yamaguchi, 1997).

Chicory forced to produce narrow leafy heads called chicons are known as witloof, witloof chicory,

Fig. 12.4. A large foliage chicory plant of the cultivar 'Catalogna'.

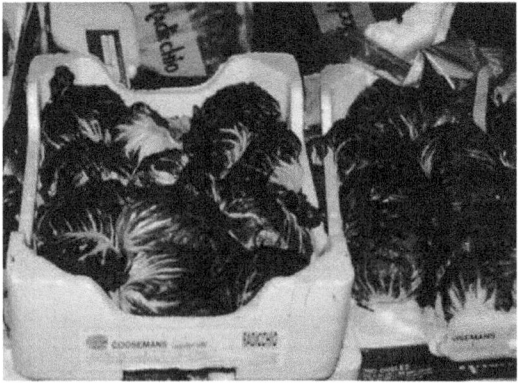

Fig. 12.5. Heads of red radicchio at a vegetable market.

French endive, or Belgium endive. Witloof types are grown vegetatively during the first year and roots are harvested and stored for chicon production at a later time (Fig. 12.6).

Cultivation and crop care of witloof chicory is similar to parsnips. For the production of forcing roots, the crop is direct-seeded in late spring and allowed to grow through the summer, usually outdoors. First-year growth results in production of a dense rosette of leaves. The tops are cut off in early fall. About 5 cm (2 in) of the top foliage is retained above the root crown so the growing point for forcing is not damaged during handling. The roots are stored for a 20–30 day induction period below 1.7°C (35°F) to assure uniform bud growth during forcing (Rubatzky and Yamaguchi, 1997). Individual induced roots are "forced" to produce the small upright chicons indoors in hydroponic culture or in a densely planted trench covered with soil at 10°C (50°F) in the dark. A soil cover is sometimes applied to exclude light and maintain compactness during chicon development. Other plastic or cloth covering materials may be used to exclude light and keep the chicons free of dirt. Chicons mature in approximately 110–120 days from root planting indoors (Rubatzky and Yamaguchi, 1997). At harvest, the white to pale yellow chicons are snapped from the roots and excess tissue is trimmed away. Chicons are expensive because they have a long production cycle and are labor intensive. Chicons are used as an additive to salads or cooked with meat or other vegetables to add a slightly bitter flavor and unique texture (Rubatzky and Yamaguchi, 1997).

Some chicories are grown for the high carbohydrate content of their fleshy enlarged root. Roots are processed after drying for use as a coffee adulterant or substitute. 'Magdeburg' and 'Brunswick' are two important cultivars of root chicory. Root chicory foliage is considered too coarse for salads. Production practices for root chicories are similar to those used for sugarbeets.

Endive

Endives are hardy annual or biennial plants, which produce a dense rosette of leaves on a short compressed stem. The foliage resembles that of some loose-leaf lettuces, but it is not a type of lettuce, as commonly believed in some circles. It is a distinct species of vegetable from *Cichorium*, the same genus as chicory. Therefore, much of the information described above for chicory also applies to endive. Two main types of endive are grown. The escarole type produces broad, coarse, crumpled leaves, whereas the endive type produces narrow, deeply cut, curled leaves (Fig. 12.7).

Upon bolting, the stem elongates and branches. The inflorescence produces many pale blue or sometimes white florets. Flowers are self-pollinating and each floret produces an individual seed similar to lettuce. There are about 800 seeds/g (22,850/oz) (Rubatzky and Yamaguchi, 1997).

Optimum growing temperatures are between 15–18°C (59–64°F). Harvest maturity occurs about 70–100 days after seeding depending on the cultivar and environmental conditions. The crop is grown much like lettuce and rapid continuous growth is

Fig. 12.6. Witloof chicory chicons packed for sale at a Dutch vegetable auction.

Fig. 12.7. An endive plant cultivar 'Salad King'.

encouraged to assure tender crisp leaves. Many people dislike the slight bitterness of endive, while others enjoy it (Rubatzky and Yamaguchi, 1997). High temperatures and drought stress tend to increase bitterness and fiber. High temperatures and long days are conducive to bolting for annual cultivars and high light accelerates bolting. Vernalization is required for biennial cultivars. The plants are sometimes blanched, by tying outer leaves together to exclude light or by covering the plants with straw or other material to reduce bitterness, increase tenderness, and improve appearance. Most cultural, spacing, harvest, and handling practices are similar to those for lettuce. Endive is used as a garnish and as a salad ingredient that adds unique flavor and texture when added to and/or substituted for lettuce. The broad-leaved escarole cultivars are occasionally used as potherbs in stews or soups. Endive is relatively easy to grow and more popular in Europe than the USA, where production is quite limited (Rubatzky and Yamaguchi, 1997).

Economic Importance and Production Statistics

Statistically China dominates world lettuce production. China is the leading country in both lettuce production and consumption. In 2011, China produced over 13.4 million metric tonnes (14.8 million short tons) of lettuce and chicory on 570,000 ha (1,425,000 acres) (FAOSTAT, 2011). The USA is the second largest lettuce-producing country with over 3.9 million metric tonnes (4.3 million short tons) and India ranks third with 1.1 million metric tonnes (1.2 million short tons). Other leading producers in descending order include: Spain, the largest EU producer with over 864,000 metric tonnes (952,397 short tons), Japan, Turkey, and Mexico (FAOSTAT, 2011).

In terms of production value, lettuce is the leading vegetable crop in the USA. Nearly all (98%) of the lettuce consumed in the USA is produced domestically. Total acreage of USA lettuce planted in 2010 was 121,500 ha (267,300 acres) (ERS, 2011). More than 90% of US lettuce production occurs in California and Arizona. The majority of production from April to October occurs in the Salinas Valley of California, while production from November to March occurs in Yuma, Arizona, and California's Imperial Valley (Boriss et al., 2012). The main lettuce types grown in the USA include iceberg, butterhead, romaine, and leaf. Iceberg lettuce consumption in the USA has decreased steadily from 12.6 kg (27.8 lb) per capita in 1990 to 7.3 kg (16 lb) in 2011 (ERS, 2011). The growing popularity of lightly processed, ready-to-eat, packaged salad greens, introduced in the late 1980s, has contributed to the decline in iceberg popularity combined with dramatic growth in romaine and leaf lettuce and spinach consumption. Estimates suggest that about one-third of all iceberg lettuce in the USA is lightly processed into prewashed, ready-to-eat, bagged salads (Boriss et al., 2012).

Spain is the leading exporter of lettuce in the world, primarily to other EU countries, followed by the USA (FAOSTAT, 2011). The USA exported 296,893 metric tonnes (327,268 short tons) of lettuce in 2010 valued at US$439.3 million (Boriss et al., 2012). The largest share of US lettuce exports go to Canada, Taiwan, and Mexico. Canada receives approximately 86% of the total US lettuce exports. In 2010, the USA imported 133,191 metric tonnes (146,818 short tons) of mainly romaine and leaf lettuce valued at US$134,688 million primarily from Canada and Mexico (Boriss et al., 2012).

Nutritional Values

The leafy tissues of lettuce, chicory, and endive are high in water and low in dry matter. Most of these crops are eaten for their texture and flavor and not because they are outstanding nutritionally since they contain little protein, fat, or carbohydrate. However, especially romaine contains significant quantities of vitamins A and C and minerals, although the amounts vary widely among crops (Table 12.2). Crisphead lettuce is very popular largely because of its unique texture, long-distance shipping ability, and long shelf life.

Production and Culture – Lettuce

Lettuce is a cool-season, frost-tolerant crop. Some seedlings can survive as low as 4.0°C (25°F). Minimum survival temperatures of immature plants vary widely with genotype and degree of acclimation. However, as lettuce approaches harvest maturity it is more easily damaged by frost. For example, light frost may cause surface blistering of the epidermis on the outermost leaves, particularly on crisphead types (Rubatzky and Yamaguchi, 1997).

Growth is extremely slow at temperatures below 7.2°C (45°F). Optimum daily temperatures for growth are 18.3–21°C (65–70°F) with night temperatures in the range of 7.2–10.0°C (45–50°F). Crop maturation is strongly temperature dependent and harvests

Table 12.2. Nutrient composition of fresh witloof chicory, endive, and butterhead, crisphead, loose leaf and romaine lettuce (amount/100 g (3.5 oz) edible portion) (USDA ARS, 2013).

	Chicory, witloof	Endive	Lettuce, butterhead	Lettuce, crisphead	Lettuce, loose leaf	Lettuce, romaine
Water (%)	95	94	96	96	94	95
Energy (kcal)	15	17	13	13	18	16
Protein (g)	1.0	1.3	1.3	1.0	1.3	1.6
Fat (g)	0.1	0.2	0.2	0.2	0.3	0.2
Carbohydrate (g)	3.2	3.4	2.3	2.1	3.5	2.4
Fiber (g)	N/A	0.9	N/A	0.5	0.7	0.7
Ca (mg)	N/A	52	N/A	19	68	39
P (mg)	21	28	25	20	25	45
Fe (mg)	0.5	0.8	0.3	0.5	1.4	1.1
Na (mg)	7	22	5	9	9	8
K (mg)	128	314	257	158	264	290
Vitamin A (IU)	0	2,050	970	330	1,900	2,600
Thiamine (mg)	0.07	0.08	0.06	0.05	0.05	0.10
Riboflavin (mg)	0.14	0.08	0.06	0.03	0.08	0.10
Niacin (mg)	0.50	0.40	0.30	0.19	0.40	0.50
Ascorbic acid (mg)	10.0	6.5	8.0	3.9	18.0	24.0
Vitamin B6 (mg)	0.05	0.02	0.06	0.04	0.06	0.06

can occur in as few as 60 days from transplanting with warmer weather, whereas winter-grown crisphead cultivars may require more than 120 days. Nonheading cultivars mature more rapidly than crisphead types. Temperatures >30°C (86°F) usually stunt growth, and result in bitterness and poor quality loose-head formation. High temperatures cause tip burn in some cultivars, a necrosis of the leaf margins, which leads to head rot through secondary infections. Romaine and leaf lettuce are slightly more tolerant of high temperatures than iceberg and butterhead types (Turini et al., 2011).

Generally, high light intensity and long days increase growth rate, hastens leaf development, produces broader leaves and promotes rapid head formation. However, long days induce lettuce to bolt and high temperatures accelerate the process. Some cultivars flower prematurely when planted in late spring or early summer because of the combination of high temperatures and long days. After bolting is initiated, flower stalks rapidly grow to a height of 0.6–1.2 m (2–4 ft) with little vegetative growth after flower-stalk development is initiated. Because the crop is ruined by flower-stalk formation, most modern lettuce cultivars have been bred to be extremely resistant to bolting. Some cultivars are so resistant to bolting that seed production is difficult (Rubatzky and Yamaguchi, 1997).

Site selection and field preparation

Lettuce can be grown on a wide range of well-drained soils such as organic, sandy, or silt loams with good moisture-holding characteristics. Lettuce is sensitive to soil compaction, crusting, and acidity, so the pH should be >5.5 and a range from 6 to 8 is preferred. Seedlings are sensitive to salinity, but tolerance increases with age. Electrical conductivity of soil extracts >2.5 dS/m may reduce seed germination and/or growth. Irrigation water with salinity levels <1 dS/m are considered best suited for lettuce production (Turini et al., 2011). Crusted soils interfere with seedling emergence and may reduce plant populations and stand uniformity. Crust formation can be minimized by irrigations that do not exceed infiltration rates. Shallow cultivations with a rotary hoe or other implements break crusts. Mulches also help in preventing crusting. Uniform, well-formed smooth seedbeds, either flat or raised, are important for uniform seedling establishment and growth. Lettuce should be rotated with unrelated crops to prevent the buildup of soil-borne diseases and pests (Turini et al., 2011).

Propagation and stand establishment

Stand establishment is a major production challenge for lettuce growers. Lettuce is established by either transplanting or direct seeding. Where the soil environment is uniform lettuce production is direct-seeded using precision seeders. Coated or pelleted seeds help ensure precision placement, good soil contact, and singulation, because lettuce seeds are small, thin, and elongated. For direct-seeded crops, the soil is finely tilled using rotovators or disks to be free of clods, stones, and other impediments to rapid germination and uniform emergence. Weed control is important in the seedling stage because lettuce does not compete well with weeds. In many cases, the crop can be precision planted to a final stand with high quality seed when soil conditions are favorable. However, in cases where soil conditions are less favorable, fields are densely seeded and thinned to the desired plant population.

To further complicate establishment, lettuce seeds may exhibit different types of dormancy. The 'Grand Rapids' cultivar has a well-characterized, phytochrome-mediated photodormancy and requires light to germinate (Borthwick et al., 1954). However, this dormancy has been bred out of most modern cultivars but may affect certain heirloom cultivars. Some lettuce exhibits primary dormancy at harvest and should be stored until the seeds have after-ripened and are fully germinable and vigorous before planting.

In addition, exposure of hydrated seeds to temperatures above 24°C (75°F) for as little as 24 h may induce thermodormancy that blocks germination. This is a particular problem in desert production areas where lettuce is direct seeded into warm soils in late summer for fall and winter crops. Generally the higher the temperature the greater number of seeds in the population are affected and the deeper the dormancy. Recently progress has been made in understanding the genetic basis of thermodormancy in lettuce seed. Seed priming, controlled hydration followed by redrying, can increase the ceiling temperatures for lettuce germination (Schwember and Bradford, 2010).

In many other locations with variable soil conditions or short seasons lettuce is grown from transplants. Transplants are also used when market prices are favorable to speed maturity and reduce the time a crop remains in the field. Lettuce seedlings are sensitive to stress and not suited to bare-root transplanting. Lettuce transplants are grown in plug-transplant trays started in greenhouses or other protective enclosures and transplanted into production fields after several weeks' growth. Carousel transplanters are often used to set lettuce plugs in the field. Workers hand-pull individual plants from the trays and place them into a revolving carousel that drops each plant into a trench before press wheels secure soil around the root ball or plug as the planter slowly moves through the field. Mechanized transplanters that remove plants from trays and place them in soil mechanically are used to establish large acreages of lettuce in California and Arizona (Fig. 12.8).

Fig. 12.8. Highly automated self-propelled mechanical transplanter planting lettuce in the Salinas Valley of California.

No-till production of lettuce is usually by transplanting and not yet widely practiced by large growers in the USA.

Lettuce is often grown on raised beds because drainage and aeration is improved, reducing disease. A common configuration is raised beds that are centered 102 cm (40 in) apart. The top of the bed is typically 61 cm (24 in) wide and the rows are spaced at the edges of each bed. Depending on the type of lettuce and cultivar grown, double rows are spaced 30–51 cm (12–20 in) apart. Leaf lettuces are spaced closer together than heading types. In-row spacing varies widely among cultivars and ranges from about 5 cm (2 in) for leaf types to 30–46 cm (12–18 in) for crisphead cultivars (Turini et al., 2011).

Fertilizer and nutrition

Lettuce is a heavy feeder with a shallow fibrous root system, so precision timing and placement of applications avoids waste. Areas under continuous intensive crop production may have elevated levels of nitrate nitrogen (N) and/or phosphorus (P), so new applications should only be applied after soil testing.

Phosphorus fertilization should be based on the soil test amounts of bicarbonate-extractable P. Levels above 60 ppm are adequate for lettuce growth. If test results are below 60 ppm, preplant applications of 45–90 kg/ha (40–80 lb/acre) of P_2O_5 or at-planting applications of 22 kg/ha (20 lb/acre) of P_2O_5 are recommended. Zinc (Zn) fertilization is recommended if extractable soil level is less than 1.5 ppm. Zinc is sometimes needed in areas where phosphate levels are high, because phosphate inhibits Zn uptake (Turini et al., 2011).

A potassium (K) fertilization program should also be based on the ammonium acetate-exchangeable test results. Soils containing >150 ppm K are sufficient for growing lettuce. Potassium fertilizer presents less of an environmental risk than N or P so growers may routinely apply K in fields with adequate exchangeable soil K. A lettuce crop will remove approximately 134 kg/ha (120 lb/acre), so fertilizing to replace more K than what the crop removes is unnecessary and wasteful (Turini et al., 2011).

Nitrogen applications in advance of crop planting risk losses due to nitrate-N leaching. Small amounts, such as 22 kg/ha (20 lb/acre) N, are often applied just prior to or during planting. At thinning, if required, 56–90 kg/ha (50–80 lb/acre) N is side-dressed into the bed (Turini et al., 2011). Additional side-dressings are common several weeks apart and may be based on foliar analysis or by pre-side-dress soil nitrate testing. If foliar analysis is used to determine fertilization applications, 6,000, 3,000, and 30,000 ppm N, P, and K, respectively, are considered sufficient on a dry weight basis (Rubatzky and Yamaguchi, 1997). Soil nitrate levels greater than 20 ppm in the top 12 in (30 cm) of soil are adequate for lettuce growth. From 11–17 kg/ha (10–15 lb/acre) of N is often applied 7–10 days before harvest to ensure adequate final growth and good head color. In drip-irrigated fields, N can be applied through the drip system as needed. Typically, drip systems deliver both N fertilizer and water more efficiently, so fertilizer additions are 20% to 30% lower than for conventionally irrigated fields. A seasonal total of roughly 169 kg/ha (150 lb N/acre) is typical for early season crops, while 224–280 kg/ha (200–250 lb/acre) is applied during cold weather (Turini et al., 2011).

Lettuce is sensitive to high levels of ammonium that sometimes accumulate in cold soils early in the season or in heavy soils when the transformation from ammonium to nitrate is slow. Ammonium injury causes root tips to turn brown and cavities to form inside older roots. Due to concerns about biological contamination, uncomposted manures should not be used to fertilizer lettuce crops. Thoroughly composted manures and organic waste are sometimes applied to improve soil structure. Quantities of compost applied are typically in the range of 9 metric tonnes/ha (4 short tons/acre) (Turini et al., 2011).

Weed management strategies

Lettuce seedlings compete poorly with weeds so effective management is essential for successful production. Several effective herbicides have been developed for lettuce weed control. Pre-emergence herbicides are typically applied in a band 12.7–15.2 cm (5–6 in) wide over rows after planting. The stale-bed technique, which is irrigation to encourage weed emergence followed by shallow cultivation, is an effective nonchemical, preplant control strategy. Burn-down herbicides or propane flaming may provide effective weed control. Additional strategies include mechanical cultivation, mowing, hand weeding, and eliminating weeds surrounding production fields. It is important to prevent weed seeds from blowing into production fields (Turini et al., 2011).

Irrigation

Since lettuce has a shallow root system, water must be consistently supplied throughout the season. Irrigation is always needed to supply water in arid and semi-arid areas. In some other locations, rainfall may be inadequate to produce lettuce. However, most temperate areas experience drought stress at some point during the growing season so supplemental irrigation is needed to optimize production. Irrigation methods include flooding, subbing, furrow, sprinkle, and drip. Combinations of various systems may be used as well. For both direct-seeded and transplanted crops, sprinkler irrigation is often applied every 2–3 days until seedlings emergence or transplants are established, which usually takes 6–10 days. After establishment, irrigation is less frequent until 2–3 weeks after seeding. Sprinkler irrigation is often applied before thinning, if required, and when fertilizer is side-dressed. Depending on terrain and soil type, fields may be sprinkled to maturity using manually moved, linear move or buried pipe systems, then a switch is made to furrow irrigation or drip irrigation for the remainder of the season. The use of surface drip irrigation, installed after cultivation and side-dressing, has markedly increased because water use efficiency and productivity are improved. Drip systems also have appeal because low levels of nutrients can be delivered on demand throughout the season, minimizing pollution and groundwater contamination. Typically one drip line is installed between two rows on 1 m (40 in) beds or three drip lines are installed among five or six plant rows on 2 m (80 in) beds. In some cases, drip system water lines are collected before harvest and reused for subsequent crops. Buried 2.5 cm or 7.6 cm (2 or 3 in) drip tape is increasingly used before planting to supply water for an entire season, including germination or transplant establishment (Turini et al., 2011). Acid is injected periodically to clear drip emitters of bicarbonate and iron precipitates.

Whichever irrigation system is used, the goal is to provide a uniform moisture supply to optimize growth. Prolonged soil saturation promotes bottom rot disease. As a crop matures, excess water and fertilizer causes heads to split, formation of excessively large heads, low head density, and reduced storage life. About 40 cm (16 in) of water, well distributed during growth, is usually adequate for growing lettuce. However, the amount varies depending on the type of lettuce, season, climate, and soil type. To give a general idea of the quantities of water required based on California production recommendations, sprinkler irrigation requires 750–1,000 m^3/ha (1.5–2.0 acre-ft/acre), 1,000–1,250 m^3/ha (2–2.5 acre-ft/acre) for furrow irrigation and 500–750 m^3/ha (1–1.5 acre-ft/acre) for drip irrigation (Turini et al., 2011). Water requirements are highest the last 30 days' prior to harvest.

Determining when and how much to irrigate is an important consideration for lettuce to preserve water and ensure efficient utilization of fertilizer. A combination of soil moisture monitoring and weather-based scheduling can effectively predict the amount of water needed and the timing of applications. Soil tensions for sustained growth should be maintained at <25–30 centibars (25–30 kPa). The amount of water needed to irrigate a lettuce crop can be estimated using reference evapotranspiration data adjusted with a crop coefficient, which is related to the percentage of ground covered by canopy. For lettuce, at a maximum canopy cover of 85%, the crop coefficient is nearly 1.0. So if evapotranspiration data show that 5 cm (2 in) of water has been lost since the previous irrigation, applying a crop coefficient of 0.95 would mean that 4.8 cm (1.9 in) of water should be added back to the field to replace losses when the soil tension reaches 25 kPa (25 centibars). The crop coefficient multiplied by the evaporative losses gives the amount of water that should be added back to the crop though irrigation when the soil tension threshold is reached. Because evaporation represents a majority of water loss during the early stages of growth, a smaller crop coefficient of between 0.3 and 0.7 should be used until the canopy has covered 30% of the field (Turini et al., 2011). In some areas, government agencies provide daily estimates of evapotranspiration to calculate the amount of water needed for irrigation. If this information is not available, evapotranspiration can be estimated using the pan evaporation technique described previously. Crop coefficients are determined from research and are available in irrigation references for many areas.

Harvesting and marketing

Although the technology is available to mechanically harvest lettuce, most head lettuce is still harvested by hand. When possible, lettuce is field packed. Field packing of lettuce is more efficient because no packinghouse is need, the crop is subjected to less handling, and crop residues are left in the field. Most crisphead, butterhead, and romaine lettuce grow in California and Arizona are field packed. Lettuce grown where the weather is frequently inclement is still field cut, but graded and packed at a facility outside the field. Unless field temperatures are extremely cold, harvested lettuce is cooled to remove field heat before shipment to market.

For field packing, crews of 20–30 workers typically divide into groups of three with two cutters and a packer (Fig. 12.9). Mobile field harvesting and packing aids move very slowly along with the harvesting crew through the field to provide a work platform with conveyer belts to collect packed cartons and distribute empty ones to workers. Only the solid heads are hand cut and trimmed to a head with a few wrapper leaves. Each cut and trimmed head, with outer leaves removed, is placed on a board by the cutter so the packer can wrap and seal each head in plastic film or bag before placing it in a carton (Turini et al., 2011). Lettuce can be packed in cartons with or without plastic wrapping to preserve moisture and reduce mechanical damage.

The marketing trend in lettuce is toward more convenience-packaged, chopped-lettuce products rather than the sale of whole heads. Lettuces harvested for processing are placed into large bulk containers for transport to processing facilities where lettuce is lightly processed by washing, cutting, and centrifuging to remove excess water so the lettuce can be bagged for marketing as a value-added, ready-to-eat, convenience product. After chopping, the lettuce is treated with citric acid or other antioxidant to reduce discoloration of the cut surfaces, extend the shelf life, and provide a fresh appearance. Chopped lettuce, packaged alone or mixed with other salad ingredients, such as cabbage, carrots, etc., has been very successful and resulted in substantial market expansion. Lettuce produced for value-added processing has fewer size restrictions or head-shape requirements. Thus, it is possible to conduct non-selective destructive harvests using mechanized harvesting equipment for lightly processed lettuce crops, although this has yet to become a standard practice. Because of the many types of lettuces grown and harvested at various maturity states, it is difficult to summarize yields. Yields range from 40 metric tonnes/ha (18 short tons/acre) for baby lettuce products to more than 75 metric tonnes/ha (34 short tons/acre).

Field-packed lettuce should be rapidly transported to cooling facilities to remove field heat,

Fig. 12.9. Crisphead lettuce being cut, wrapped, and boxed in one operation near Salinas, California. Uniformity is important because it allows all heads to be harvested in a single pass through the field.

Family Asteraceae

reduce respiration, and maximize storage life. Vacuum cooling is an excellent method for quickly lowering commodity temperature, with in-field packing possibly eliminating the need for packing sheds and cooling with slurry ice (Fig. 12.10).

Vacuum cooling removes field heat in about 15 min in a sealed container at reduced atmospheric pressure. Where vacuum-cooling is not available, forced-air cooling is used to remove field heat. Hydrocooling can be used for nonheading lettuces but is not suitable for head lettuce because excess water is retained. Reliance on good postharvest handling and temperature management throughout the marketing chain is critical for extending shelf life. After cooling, lettuce must be stored at 1.1°C (34°F) and 98% relative humidity to maximize shelf life. Crisphead lettuce can last for 2–3 weeks while other types have a shorter shelf life. Storage at 3.3°C (38°F) reduces shelf-life to only 1–2 weeks (Turini *et al.*, 2011). Lettuce is sensitive to ethylene gas, the natural gaseous plant hormone responsible for the ripening of climacteric fruit. Russet spotting is a disorder that occurs when lettuce is stored in the presence of ethylene generated by ripening fruit or engine exhaust (Lipton *et al.*, 1972).

In California, some growers began successfully growing and marketing organic, ready-to-eat, salad mixes composed of specialty lettuces and other salad greens in the 1980s and early 1990s. Some of these innovative operations were successful in meeting the growing demand for high-quality, organic, bagged salad mixes, but many have now been purchased by larger traditional national vegetable-producing corporations who wished to take advantage of the rapidly growing organic market in North America (Fig. 12. 11).

Lettuce production in protected culture

Lettuce is one of the most commonly grown greenhouse vegetables. A significant amount of lettuce production occurs in glasshouses and simple protective structures such as plastic tunnels. Protected structures are especially useful during periods when low temperatures limit outdoor growth. Butterhead lettuce is commonly grown in protected structures, especially permanent greenhouses. Off-season lettuce production in a greenhouse is possible with managed temperature, light environment, and hydroponics. However, the relatively high cost of production associated with the structure and heating costs may make competing on price with lettuces grown outdoors and shipped from distant markets difficult. In harsh environments, like desert areas of the Middle East with unfavorable soils or excessive temperatures, greenhouses can modify the environment to make lettuce production possible at any time of the year (Shah, 2010).

Fig. 12.10. Vacuum cooling chamber near Salinas, California.

Fig. 12.11. Organic endive (right) and butterhead lettuce (left) production near Watsonville, California for ready-to-eat bagged salad-mix sale. Plots are relatively weed free because of high plant populations and hand weeding.

Temperatures of 13°C (55°F) at night and 16°C (61°F) during the day promote early seedling growth. When temperatures exceed 21°C (70°F), structures usually are ventilated. During the rosette stage, a 10°C (50°F) night and 13°C (55°F) day regime is recommended with ventilation when temperatures exceed 18°C (64°F). To minimize heating costs, a lower temperature regime of 7°C (45°F) night and 13°C (55°F) day with ventilation at 16°C (61°F) or above is often practiced. In addition to directly increasing greenhouse air temperatures, heating the soil is sometimes practiced. For hydroponic production, heating the circulating nutrient solution improves growth (Rubatzky and Yamaguchi, 1997).

Light is a commonly limiting factor in most off-season protected culture. Unfortunately, light supplementation is often more expensive than supplying heat. In eastern Canada and certain other locations, growers can take advantage of lower electrical rates at off-peak hours to cut costs of supplemental lighting. When light intensity is sufficient, additional heating and CO_2 enrichment can further increase plant growth. Some lettuce cultivars have been bred specifically to perform in low light and low-temperature conditions (Yamaguchi and Rubatzky, 1997).

Crops grown in protected structures are often established using transplants to help speed maturity in valuable greenhouse space. Transplants are produced more efficiently in specialized rooms that accommodate high populations and provide optimum light, temperature, and moisture for seedling growth. Transplanting occurs 3–5 weeks after seeding. Plant spacing is as close as possible to maximize available greenhouse space.

Pest management is a major concern for any protected cultivation. When greenhouse crops are grown in native soils diseases must be managed through crop rotation, fumigation, solarization, or steam sterilization. Hydroponic systems are preferred for greenhouse lettuce production to avoid costly soil fumigation/sterilization needed to control disease. A hydroponic system also provides an alternative to unproductive native greenhouse soils, better utilization of valuable vertical space, and rapid lettuce growth without water stress. A recirculating nutrient film system (NFS) with shallow gutters or troughs is popular for greenhouse production of butterhead lettuce. In the NFS, individual plants are rooted in a porous well-drained substrate like sponge or rockwool that holds the plant in place in a trough. Other hydroponic systems using gravel, ebb and flood, straw bales, or other alternative media can also be used to grow lettuce. Lettuce is grown outdoors in native soils with temporary covers such as tunnels that are mobile, so crop rotations can be practiced. The integration of good sanitation, biological and/or chemical pesticide control is critical for effective pest management in a greenhouse. Identifying diseased plants and removing them from the greenhouse is especially important for hydroponic production because disease can spread rapidly through the nutrient solution to adjacent plants, particularly in the NFS. Cultural practices that limit plant access to water, reduce relative humidity, and provide good ventilation help to reduce greenhouse pest problems.

Diseases – Physiological Disorders

Lettuce is affected by several physiological disorders that can greatly diminish quality. Cultivars vary widely in their susceptibility to the following disorders. Tipburn is associated with rapid growth and restricted transpiration during abrupt temperature increase (CABI, 2013). These conditions reduce calcium transport to rapidly growing new tissues resulting in necrosis of leaf margins on young leaves. Secondary bacterial infection of tip-burned plants often leads to head rot.

Russet spotting is a result of ethylene exposure that occurs more frequently in older plants and is worse during transit, high storage temperatures, and low O_2. Russet spots may occur anywhere in a head, except on the heart leaves. The spots are small, tan, russet-brown, or olive. They are mostly on the midrib, but may develop on other parts of a leaf as well. On the midrib, the spots are pit-like, while on the blade, they are shallow, rounded, and diffuse (Lipton *et al.*, 1972).

The cause of internal rib necrosis, also called rib blight, is unclear but appears to be associated with specific cultivars. Internal rib necrosis appears as a diffuse, dark, gray-green, or occasionally coal-black discoloration of the lower midrib. Symptoms normally are most distinct on the outer head leaves and some of the smaller inner leaves but occasionally appear on the wrapper leaves (Lipton *et al.*, 1972).

Brown stain disorder is a form of CO_2 injury that results in distinct lesions. Typical lesions average about 2.5 cm (1 in) wide and have distinct margins that often are darker than the slightly sunken centers. The dark margins produce a halo effect. Lesions commonly develop on several head leaves just under the cap leaves, but also may develop on leaves deeper in the head. The heart and wrapper leaves are not affected by brown stain (Lipton *et al.*, 1972).

Pink rib occurs most commonly in over-mature lettuce, but may also affect less-mature heads. Pink rib is characterized by a diffuse, pink discoloration near the bases of midribs of the outer head leaves. The discoloration usually is most intense on the inner (adaxial) surface, but it is often visible on the outer (abaxial) surface. In heads with severe symptoms, all but the youngest head leaves may be pink, and the discoloration may reach into the large veins (Lipton *et al.*, 1972). Causes of pink rib are not clear; however, unfavorably high temperatures in transit or storage accelerate its development and O_2 availability also seems to be involved (Martínez and Artes, 1999).

Glassiness disorder occurs when transpiration is restricted during high humidity and soils are waterlogged. Glassiness creates patches with water-soaked areas that are almost translucent on lettuce leaves. Glassiness often has no lasting effect and once good ventilation and airflow are introduced in a greenhouse or tunnel, the symptoms disappear. If glassiness persists for long periods, leaf cells may die.

Pathogens

Lettuce infectious yellows, lettuce necrotic yellows virus, beet western yellows, turnip mosaic, and lettuce mosaic are some of the most important virus diseases of lettuce (CABI, 2013). The spread of lettuce mosaic virus (LMV), as well as other viral diseases, may be stopped by targeting the aphid vectors, the immediate removal of diseased plants from the field or greenhouse, and the use of healthy virus-free seed. Production and certification of seed that is free of LMV greatly limits potential disease and crop losses. Methods for the detection of LMV-infected seed from sampled seed lots include: (i) leaf symptom expression within the population of 30,000 seedlings grown from a sample of the seed lot; (ii) the reaction of LMV-sensitive plants to extracts of ground-up seed obtained from a 30,000-seed sample of the seed lot; and (iii) a less intensive method tests the seed sample for a positive antibody response in an enzyme-linked immunosorbent assay (ELISA).

Big vein (Mirafiori lettuce virus) disease causes leaf veins to become deformed, enlarged, and clear. The enlarged veins cause the rest of the leaf to be ruffled and deformed and head lettuce cultivars may fail to form a head. However, infected plants showing less severe symptoms can still be harvested and marketed (UC IPM, 2013).

The virus that causes big vein is soil-borne and is introduced into lettuce plants by the soil fungus

Olpidium brassicae that attaches itself to lettuce roots. Even though the virus is soil-borne and always present in infested fields, the severity of the disease varies greatly from season to season and is most prevalent during cool weather (UC IPM, 2013). In fields having chronic and severe big-vein histories, planting susceptible cultivars during spring should be avoided. Crop rotations are not effective in controlling big vein. Some big-vein-resistant cultivars are now available.

Bottom rot (*Rhizoctonia solani*) commonly occurs under wet field conditions. The symptoms are a slimy rotting of the underside of the plant progressing upward and into the head. Growing lettuce on well-drained soils and/or using raised beds, crop rotation, good sanitation, and fungicidal sprays can control bottom rot (UC IPM, 2013).

Lettuce drop (*Sclerotinia sclerotiorum*) is a soil-borne fungal disease that affects crops from the rosette stage until harvest. The disease causes a wet rot of the entire plant, beginning at the stem base. Lettuce drop is worse during wet and cool conditions. The best control of lettuce drop is good sanitation, crop rotation, fungicide applications, and good drainage (UC IPM, 2013).

Downy mildew (*Bremia lactucae*) is one of the most serious diseases of lettuce in temperate areas. Downy mildew causes light green to yellow angular spots on the upper surfaces of leaves. White fluffy growth of the pathogen develops on the lower sides of these spots. With time these lesions turn brown and dry. Older leaves are attacked first. Severely infected leaves may die. If downy mildew infects the cotyledons of young seedlings, the plants can die. Greenhouse-grown lettuce transplants can also be infected. Downy mildew is controlled by using cultivars with resistance to the relevant race of the fungus or by spraying with fungicides (UC IPM, 2013).

Bacterial leaf spot (*Xanthomonas campestris* pv. *vitians*), varnish spot (*Pseudomonas cichorii*), and anthracnose (*Microdochium panattonianum*) are foliar diseases that can affect developing lettuce. Bacterial leaf spot is favored by cool wet conditions and can be controlled by fungicides, although other control options are limited. The varnish spot bacterium is found in reservoir water and can be spread by sprinkler irrigation, so not using contaminated water for irrigation can control this disease. Anthracnose only occurs in fields where the resting fungal structure resides in wet soil. Planting lettuce in fields with a disease history should be avoided; fungicide applications can control this disease. Corky root is caused by the soil-borne bacterium *Rhizomonas suberifaciens*. Crop rotations, resistant cultivars, and lower soil N levels provide control (UC IPM, 2013).

Verticillium wilt (*Verticillium dahliae*) is a problem in California and affects all types of lettuce. Severe infestations can destroy an entire crop. The fungus produces microsclerotia that are long-term resting structures that often lie dormant in soil for 10 years or more. Equipment used in infected fields should be cleaned after use to prevent transfer to clean fields. Fumigation, rotations with unrelated crops, and disease-resistant cultivars are control options (UC IPM, 2013).

Avoiding movement of infected soil to clean fields is the best control for Fusarium wilt of lettuce. The disease affects lettuce more at higher soil temperatures. Therefore, planting lettuce in contaminated fields should be avoided when soil temperatures are high.

Damping-off (*Pythium* sp.), gray mold (*Botrytis* sp.), and leaf-spot (*Cercospora* sp.) are also reported on lettuce. Aster yellows is a phytoplasma (mycoplasma) transmitted by the six-spotted leafhopper (*Macrosteles fascifrons*) (CABI, 2013).

Insect Pests

One of the most serious pests is the lettuce aphid (*Nasonovia ribisnigri*), which infests the inner leaves. The foxglove aphid (*Aulocorthum solani*) also infects inner leaves of lettuce. Since the lettuce and foxglove aphids are protected within the head as it develops, timely detection and treatment are essential for control. The lettuce and foxglove aphid may not be adequately controlled with neonicotinoid insecticide treatments via soil injection or foliar sprays. The green peach aphid (*Myzus persicae*) and the potato aphid (*Macrosiphum euporbiae*) tend to build up on outer leaves, making insecticide treatment more effective. Natural enemies such as parasitic wasps help suppress aphids that colonize the outer leaves of lettuce. Fungal pathogens of whiteflies are common during cool wet spring weather. Hoverfly larvae (Syrphidae, *Baccha* spp., *Criorhina* spp.) and other aphid predators such as ladybugs, also called ladybird beetles, or lady beetles (Coccinellidae) can suppress aphids on outer leaves. *Nasonovia*-resistant romaine lettuce has been developed but it is not resistant to other aphid species (Turini *et al.*, 2011).

Leafminers (*Liriomyza* spp.) form tunnels inside leaves while feeding on mesophyll tissue between the upper and lower surfaces. Leafminer female flies cause stippling damage by feeding on fluids obtained by puncturing leaf surfaces with their ovipositor. Parasitic wasps, particularly from the genus *Diglyphus*, can help suppress leafminer populations. Insecticides can manage larvae but are largely ineffective on the mobile, insecticide-resistant adults (Turini *et al.*, 2011).

Beet armyworms (*Spodoptera exigua*), cabbage loopers (*Trichoplusia ni*), and other adult caterpillars may sometimes damage lettuce. The beet armyworm and cabbage looper larvae are susceptible to natural enemies including diseases, predators, and parasitoids. Caterpillars can be treated with selective insecticides. *Bacillus thuringiensis* (or Bt) is a gram-positive, soil-dwelling bacterium, commonly used as a biological pesticide. The Cry toxin produced by Bt may be extracted and used as a pesticide (UC IPM, 2013).

Several species of whiteflies may infest lettuce in warmer climates, including the silverleaf whitefly (*Bemisia argentifolii*), greenhouse whitefly (*Trialeurodes vaporariorum*), and banded whitefly (*Trialeurodes abutilonia*) to name a few. Whitefly adults vary in size but most are tiny, approximately 1.5 mm (0.06 in) long, with white wings and variation in body color (UC IPM, 2013).

Whiteflies are found mostly on the undersides of leaves and fly when plants are disturbed. This propensity to fly when perturbed can be used as a whitefly control. Mechanical agitators periodically pass through a greenhouse and disturb lettuce leaves, catching whiteflies in sticky traps mounted behind the mobile device. The tiny, oval eggs hatch into a mobile first larval stage that has legs and antennae. The last nymphal stage, often called the pupa or the red-eye nymph, is the stage that is easiest to identify. Silverleaf whitefly pupae are oval, whitish, and soft. The edge of the pupa tapers down to the leaf surface and has few to no long waxy filaments around the edge. In contrast, greenhouse whitefly pupae have many long waxy filaments around the edge. Whiteflies feed on lettuce tissue, producing sticky honeydew on the leaves. A black, sooty mold often grows on the excreted honeydew. Silverleaf whitefly feeding can cause a stunting and yellowing of head lettuce (UC IPM, 2013).

Several wasps, including species in the *Encarsia* and *Eretmocerus* genera, parasitize whiteflies. Bigeyed bugs, lacewing larvae, and lady beetles also prey upon whitefly nymphs. Silverleaf whitefly is an introduced pest that has escaped its natural enemies in much of the western USA. Indigenous native parasites and predators do attack it, but do not provide adequate control. The lady beetle (*Delphastus pusillus*) may assist in biological control. Insecticidal sprays may control whiteflies as well (UC IPM, 2013).

The adult western flower thrip (*Frankliniella occidentalis*) is a minute, slender-bodied insect possessing two pairs of long, narrow wings, fringed with long hairs at the margins. The bodies of adult thrips can be yellow, orange, brown, or black. The larvae are white, yellow, or orange. In spring, populations build up on weeds and other vegetation and move into lettuce fields when the weeds decline. On lettuce plants, adults reproduce and rapidly colonize into large populations (UC IPM, 2013).

Western flower thrip feeds on lettuce and vector plant viruses. Thrips damage lettuce by puncturing leaves and sucking sap. Punctured leaves develop a silvery appearance that eventually turns to brown scarring and can be confused with windburn or blown sand damage. The presence of small, black fecal specks in the damaged area confirms thrip damage (UC IPM, 2013).

The western flower thrip is the most important vector of tomato spotted wilt virus and the only known thrip species to vector impatiens necrotic spot virus. Western flower thrip feeds on virus-infected weeds, ornamentals, or other vegetation surrounding the field before they migrate to the lettuce field and introduce virus disease. Only the larval stage can acquire these tospoviruses, but they remain infected throughout their life.

Managing vegetation in and around lettuce fields, biological control and cultural practices are important for minimizing damage from western flower thrips. When thrips are present on the lettuce crop, insecticides are often used for control. Natural enemies that feed on thrips include predaceous mites, minute pirate bugs, and lacewings. These beneficials are very susceptible to insecticide sprays, however, and may not be present in fields when insecticides are used (Turini *et al.*, 2011).

Plant parasitic nematodes are microscopic roundworms that feed on lettuce roots, and include root knot nematode (*Meloidogyne incognita*, *M. javanica*, *M. arenaria*, and *M. hapla*), needle nematode (*Longidorus africanus*), stunt nematode (*Merlinius* sp.), and spiral nematode (*Rotylenchus* sp.).

Nematodes live in soil and plant tissues and several species may occur in a field. The host range varies according to the nematode species. Some nematodes are able to infest a wide variety of crops and others are limited to a narrow crop range. Symptoms of nematode infestation also vary according to the nematode species and crop type, and are often non-specific such as yellowing or stunting. Root knot nematode species, however, cause typical galling on roots of infested plants. The geographical distribution of the different species is highly dependent on temperature, soil type, and cropping history. Plants infested as seedlings may be stunted, with patches of affected plants becoming evident by midseason. Fumigation, sanitation, cultural methods, and crop rotation are possible nematode controls (UC IPM, 2013).

GLOBE ARTICHOKE AND CARDOON

Origin and History

The globe artichoke is native to the Mediterranean region of southern Europe and northwestern Africa. The crop has ancient origins. Southern Italy is a possible site of domestication (Pignone and Sonnante, 2004). The Greek philosopher Theophrastus, 371–266 BC, referenced the cultivation of globe artichoke in Sicily (De Candolle, 1959). At the beginning of the Christian Era, the Roman scholars Pliny and Galen wrote about therapeutic properties of globe artichoke. In the Middle Ages, 5th to 15th century, the globe artichoke was associated with the Arabs, who spread production from Sicily throughout the Mediterranean region (Pignone and Sonnante, 2004). Since the 15th century, globe artichoke has been grown throughout southern Europe.

Europeans introduced globe artichoke to the USA during colonial times (Ryder *et al.*, 1983). The American statesman, agriculturalist, and president Thomas Jefferson raised globe artichokes on his farm in central Virginia as early as 1767 (Welbaum, 1994).

Cardoon (*Cynara cardunculus* L.) is a close relative of globe artichoke and has been cultivated for thousands of years in the central and western Mediterranean region. The thistle-like cardoon is in the same genus as the globe artichoke, and is a minor vegetable crop in much of the world except Italy where it is a significant vegetable (Rubatzky and Yamaguchi, 1997).

The Jerusalem artichoke (*Helianthus tuberosus*) is also in the family Asteraceae and is sometimes confused with the globe artichoke. Jerusalem artichoke, also called sunchoke, is a relative of sunflower (*Helianthus annuus*) and native to North America. The Native Americans were cultivating Jerusalem artichoke when the Europeans arrived. Today Jerusalem artichoke is grown on a small scale throughout the world for its small, irregularly shaped tuber, which stores the carbohydrate inulin, a macromolecule composed of fructose, which can be safely eaten by diabetics (Rubatzky and Yamaguchi, 1997).

Botany

The globe artichoke (*Cynara scolymus* L.) is a member of the family Asteraceae and is also known as green artichoke or French artichoke. Globe artichoke is a cool-season thistle-like herbaceous perennial dicot grown for its immature flower buds. The "globe" refers to the immature flower head called a capitulum, which includes the fleshy bases of the outer bracts, the inner bracts, the receptacle and portions of the floral stem (Rubatzky and Yamaguchi, 1997). The edible portions of the bud are the tender bases of the bracts and the fleshy receptacle or heart (De Vos, 1992). At full maturity, an artichoke plant may reach 1.2–1.5 m (4–5 ft) tall and 1.5–1.8 m (5–6 ft) across. At maturity, buds fully open to expose brilliant blue thread-like styles that give the flowers an ornamental appearance and are attractive to bees. At this stage of development the bracts and other flower parts are fibrous and inedible but the flowers are sometimes used fresh or dried as ornamentals (Fig. 12.12).

Fig. 12.12. This unharvested globe artichoke plant produced beautiful blue flowers that are inedible but have ornamental value.

Globe artichoke is a cool-season crop that grows best in 24°C (75°F) days and 13°C (55°F) nights. Artichoke is day-neutral but vernalization plays a role in flower induction. The percentage of annual plants producing buds the first season after transplants were established in May was greater when a cold treatment was applied (Rangarajan *et al.*, 2000). 'Imperial Star', developed for annual culture, produced a higher percentage of buds the first year compared to 'Green Globe' following planting in late spring without vernalization (Welbaum, 1994). The vegetative parts of the plant can tolerate much higher temperatures but bud development is excessively fast and of poorer quality if temperatures exceed 29°C (85°F). Cool days >13°C (55°F) and nights >7°C (45°F) induce buds and extend the production period. Less of the bud tissue can be eaten because of increased fiber development as the flower tissues develop inside. The highest bud quality is achieved when the crop is grown under cool conditions that result in slow growth and bud development.

Because of winter-kill, it is difficult to overwinter globe artichokes in northern areas with harsh climates. Reliable commercial culture is limited to USDA hardiness zone 7 or above. Acclimated plants can survive low temperatures of −6.7°C (20°F) with little damage, but flower buds can be damaged at −1.1°C (30°F). Prolonged exposure to temperatures below −9.4°C (15°F) will kill leaves and eventually the crown. Light freezes 0–−2.2°C (~32–28°F), depending on the duration, damages bud bracts causing blistering of the outer bud tissue and a whitish appearance but will not affect eating quality. Prolonged exposure to temperatures below −2.2°C (28°F) will damage the bud so it is not edible (Smith *et al.*, 2008).

Cardoon (*C. cardunculus* L.), like globe artichoke, is a member of the family Asteraceae. Cardoon produces flowers prolifically but the flowers are smaller than globe artichoke and are sometimes used as ornamentals. Although these less fleshy flower buds can be eaten, the immature cardoon leaves and leaf bases are most commonly consumed after boiling, braising, or sautéing and have an artichoke-like flavor (Fig. 12.13; Rubatzky and Yamaguchi, 1997).

Types and Cultivars

Artichoke cultivars can be grouped as annual or perennial. Annual cultivars were developed by plant breeding to be propagated from seed for a single year of production. Cultivars that are grown from seed include 'Imperial Star', 'Desert Globe',

Fig. 12.13. The fleshly leaf bases of these cardoon plants can be eaten after cooking.

'Talpiot', and 'Emerald' (Schrader and Mayberry, 1992; Smith *et al.*, 2008). Annual cultivars have an advantage because they can be grown in short-season areas where plants cannot be overwintered and they can be grown in rotation with other annual crops. Some annual cultivars, such as 'Imperial Star', produce more marketable buds in the first year compared with perennial cultivars when both are grown from seed (Welbaum, 1994).

While perennial cultivars may also produce seed, they generally speaking do not grow true-to-type and are propagated vegetatively from crown division or tissue culture. The most important perennial cultivars grown in the USA are 'Green Globe' and 'Green Globe Improved', which account for about 50% of California production (Smith *et al.*, 2008).

Globe artichoke cultivars vary in bud color. In some cultivars the capitulum is entirely green while others have red or purple coloration at the tip of the bract. 'Magnifico' has totally red bracts. Although 'Imperial Star' and 'Talpiot' were developed to grow true-to-type from seed, there is plant-to-plant variability associated with these cultivars. However, the variability is considerably less than when 'Green Globe' is propagated from seed (Schrader and Mayberry, 1992).

In southern Italy, globe artichoke is grown for autumn, winter, and early spring harvest using the cultivars 'Violetto di Sicilia', 'Spinoso Sardo', and 'Brindisino', while in central Italy globe artichokes are raised for early spring harvest using 'Campagnano', 'Castellamare', 'C3', 'Terom', and 'Violetto di Toscana' (CABI, 2011).

Globe artichoke cultivars are defined by many important characteristics, which affect their quality and marketability. Some important characteristics that differentiate cultivars include: plant height and width, disease resistance, insect resistance, bud size, bud shape, rate of bud development, number of buds per plant, number of bracts, size of bracts, rate of fiber development, degree of thorniness, and flavor.

Cardoon has very pubescent foliage, more so than globe artichoke. The degree of pubescence varies among cultivars. 'Long Spanish' and 'Ivy White' are two common cardoon cultivars. 'Ivy White' is very pubescent and has a white rather than green appearance.

Economic Importance and Production Statistics

In 2010, just less than 125,000 ha (307,542 acres) of globe artichokes were grown in the world, producing 1,440,903 metric tonnes (1,588,324 short tons). Table 12.3 shows major artichoke-producing countries around the world. Overall Europe is the most important producer with Italy as the leading country. Important production occurs outside of Europe as well with Egypt, Peru, and Argentina all ranking in the top five. Production in China has increased dramatically in recent years.

The major Italian production areas are located in Sicily and the central regions of Latium, Sardinia, Campania, and Tuscany, where globe artichokes are generally planted from July to September and grown year-round as perennials. In Italy and Spain, approximately 80–75% of harvested globe artichokes are sold for fresh market and the remaining 30–25% are processed as canned hearts and crowns or as frozen artichokes (CABI, 2011).

In the USA, California accounts for 99% of commercial production. About half of the California acreage is annual culture and half perennial. California exports fresh artichokes to Canada, Mexico, Japan, and Europe. The vast majority of California artichokes are sold fresh (Smith *et al.*, 2008).

Nutritional Values

Globe artichoke is considered a luxury vegetable by many because it takes a lot of space to grow, it is not easy to grow, there is relatively little to eat from each capitulum, it does not provide many nutrients to the diet, and it is relatively expensive (Table 12.4). Many people like globe artichokes simply because of their unique flavor and because they are fun to eat.

Production and Culture

Soil requirements

The pH optimum for both globe artichoke and cardoon is 6.0–6.8. Globe artichokes develop deep

Table 12.3. Leading globe artichoke-producing countries by continent/region (FAOSTAT, 2010).

Country	Production (t)	World rank
Europe	711,432	
Italy	480,112	1
Spain	166,700	3
France	42,153	8
Greece	20,400	13
South America	247,303	
Peru	127,503	4
Argentina	84,000	5
Chile	35,000	11
Middle East	N/A	
Turkey	29,070	12
Iran	15,800	15
Syria	6,100	16
Northern Africa	319,194	
Egypt	215,534	2
Algeria	39,200	10
Morocco	45,460	7
Tunisia	19,000	14
Other		
USA	40,820	9
China	59,900	6

Table 12.4. Globe artichoke heart nutritional composition (Rubatzky and Yamaguchi, 1997).

Nutrient	Amount per 100 g (3.5 oz) edible portion
Water (%)	84
Energy (kcal)	51
Protein (g)	2.7
Fat (g)	0.2
Carbohydrate (g)	11.9
Fiber (g)	1.1
Ca (mg)	48
P (mg)	77
Fe (mg)	1.6
Na (mg)	80
K (mg)	339
Vitamin A (IU)	185
Thiamine (mg)	0.08
Riboflavin (mg)	0.06
Niacin (mg)	0.76
Ascorbic acid (mg)	10.8
Vitamin B6	0.11

root systems that may grow from 0.9–1.2 m (3–4 ft) deep. Artichokes can be grown on a wide range of soils but perform best on deep fertile well-drained loams. Raised beds are sometimes used, especially on heavier soils, to improve drainage during rainy periods. Well-drained soils with poor moisture retention properties, such as sandy soils, should be avoided. Globe artichokes are moderately salt tolerant. They can be grown in up to 6 dS/m (ECe in mmho/cm) at 25°C (77°F) with no reduction in yield. Yield losses of roughly 11% have been calculated for every 6 dS/m increase in soil salinity above this threshold (Smith et al., 2008).

Perennial planting

Perennial artichokes are propagated from crown divisions. Rooted sections of crowns or stumps selected from commercial fields are planted by hand in trenches 10–15 cm (4–6 in) deep with 1.0–1.1 m (3.3–3.5 ft) in-row spacing and 2.7–4 m (9–10 ft) between rows. Plants are often configured in a grid pattern to make weeding and other management practices easier. The fields are generally reestablished every 5–10 years, because with crown expansion, competition among plants increases, decreasing bud production. The cropping cycle for perennial artichokes begins when plants are cut back to the ground. For fall, winter, and spring harvests, the plants are cut back from mid-April to mid-June. For summer harvest plants are cut back in late August or September. The plants are cut at ground level to stimulate new shoot development. During the harvest season, old bearing stalks are often removed after harvest to encourage new shoot development. This process called stumping (similar to ratooning) consists of harvesters chopping out the stalk by hand just below the ground using an axe or stalk knife. Stalks are removed at 3- to 4-week intervals throughout the year depending on the growth of new bud-bearing stalks. Stumping is believed to increase yield and extend the productivity of the field.

Annual planting

Artichoke seeds germinate slowly, compete poorly with weeds, and are difficult to establish by direct seeding; they are usually started in plug trays 4–6 weeks prior to field establishment. Artichoke seedlings produce a large central taproot and plants do not perform well in small-celled, shallow, transplant trays. Artichoke plants are not as easy to establish as many other vegetable crops. Extra care should be taken to ensure survival in the field such as avoiding transplanting on hot windy days and irrigating transplants regularly until establishment.

Artichokes produce a long taproot, so the soil should be deeply tilled in advance of planting. Transplants are generally set in a single row per 1.8–2.0 m (72–80 in) wide bed with in-row spacing of 76 cm (30 in). Annual plantings can mature at different times to take advantage of marketing opportunities. In short-season areas, annual crops are planted after the time for hard freezes has passed. Annual artichokes may be direct seeded in areas with long growing seasons but the majority of production is transplanted to reduce the time the crop is in the field, weed competition, and diseases.

Globe artichokes will grow under a wide range of temperatures, but for best bud quality, cool sunny conditions result in slow even growth. Optimum production temperatures are 13–24°C (55–75°F). Frost can damage buds but the above-ground tissues can survive temperatures down to about −5.5°C (22°F).

Perennial artichokes are often treated with gibberellic acid (GA3 or GA4+7) to increase earliness and uniformity of bud development. Gibberellic acid treatments are often sprayed on a field 6 weeks before the predicted first harvest. A standard application is 10 g (0.04 oz) GA added to 380 l (100 gal) water and applied per 0.4 ha (1 acre) (Smith et al., 2008).

Production recommendations for cardoon are similar to globe artichoke. Cardoon is more commonly planted from seed than artichoke. Leaves are frequently tied 0.3 m (1 ft) above the ground and soil piled around the base of the plant to blanch the leaves and stems to reduce bitterness.

Fertilizer and nutrition

Perennial globe artichoke crops require moderate amounts of N. Many growers need to apply from 112–224 kg/ha (100–200 lb/acre) N based on soil test results. For annual artichoke production, single applications of N are not recommended since nitrate may leach beyond the root zone, particularly on sandy soils that receive heavy rainfall. Small applications of N 22–34 kg/ha (20–30 lb/acre) applied preplant provide transplants with sufficient N for the first month's growth. Crop N requirements increase with maturity, so at the beginning of the season, 6 kg/week/ha N (5 lb/week/acre) may be sufficient. With further plant growth and the onset of bud formation, the amount of N needed to sustain growth doubles to 12 kg/week/ha (10 lb/week/acre). A cumulative seasonal total application of 144–180 kg/ha N (120–150 lb/acre) is often sufficient to mature a high-yielding crop depending on soil characteristics. If annual artichokes follow other crops in rotation, such as lettuce or broccoli, a benefit may be realized from residual nutrients in the soil. Nitrate levels of 20 ppm or greater can sustain early growth of artichoke transplants and save on fertilizer costs. The soil should always be analyzed prior to any nutrient applications to assess residual levels and prevent unnecessary applications that could harm the environment and waste money.

Soils containing 60 ppm P or more are adequate. For soils with less P or plantings during the winter when soils are cold and uptake reduced, applications of 45–90 kg/ha (40–80 lb/acre) of P_2O_5 are recommended based on test results. The requirement for K can also be determined through soil testing. Soils with <150 ppm K are sufficient for growing globe artichoke.

Irrigation

Water stress will slow growth and result in fibrous buds of poor quality that will begin to open prematurely at a small size before fully enlarging. Moisture stress likely contributes to black tip physiological bud disorder, which causes bracts to turn dark brown and unmarketable.

Irrigation regulates globe artichoke production cycles and relates directly to timing of crop harvest, so soil moisture management is particularly critical. Uniform soil moisture is essential during all phases of globe artichoke production. Irrigation begins about 1 month after plants are cut back at the beginning of a new production cycle for perennial production in areas with seasonal rainfall. Subsurface and overhead sprinkler irrigation are two of the most common types of irrigation used for globe artichoke production. Annual plantings are often established with overhead sprinkler irrigation to ensure the root zone is uniformly moist. Subsurface irrigation is used for annual production after the crop is well established. Subsurface drip irrigation lines are buried 30–35 cm (12–14 in) below the surface on one or sometimes both sides of the plant row, providing efficient distribution of water directly to the root zone. On uneven ground, pressure-compensated drip lines help distribute water uniformly. Durable drip tape that can be reused at the end of the annual artichoke production season is popular in some areas.

With sprinkler irrigation, the crop is often watered at 2- or 3-week intervals during the summer months depending on the soil type. Approximately 506–706 m^3/ha (2–3 acre-inches) of water are applied at each irrigation by sprinklers. In comparison, drip irrigation is scheduled more frequently at 1 week intervals depending on the soil type and weather conditions to maintain a higher soil moisture content and reduce evaporative losses common to sprinkler systems. Drip irrigation may reduce water usage by 25% on heavy clay soils, increase yields on sandy soils, and has largely replaced furrow irrigation because of its greater efficiency. Approximately 2,539–6,093 m^3/ha (10–24 acre-inches) of water are needed per year to mature a perennial artichoke crop. Annual crops often need from 5,078–6,093 m^3/ha (20 to 24 acre-inches) per season due to their denser leaf canopy, higher transpiration and shorter season (Smith *et al.*, 2008).

Since moisture is a key element in globe artichoke management and production often occurs in areas where water is limited, the combination of soil moisture monitoring and weather-based irrigation scheduling often determines when water should be applied and how much is needed. Water use by artichoke plants is greatest during summer months when leaf canopy cover is greatest. Perennial globe artichoke crops have an extensive deep root system capable of supporting growth and development early in the season with little rain or irrigation. However, a consistent water supply is critical during bud formation and development to maximize quality and yield. Water use by globe artichoke crops can be approximated using pan evaporation or estimated from reference evapotranspiration data adjusted by a crop coefficient that is closely related to the percentage of ground covered by the leaf canopy. At full canopy cover, the crop coefficient is nearly 1.0 for annual artichoke crops (Smith *et al.*, 2008).

Harvesting

Perennial globe artichokes are harvested year-round, but the highest volume in the northern hemisphere usually occurs between March and May (Fig. 12.14).

Seeded artichokes may be harvested any time of the year by adjusting planting times. Depending on weather conditions, globe artichokes are generally

Fig. 12.14. Artichoke field near Castroville, California in late July. One plant in the foreground has bolted and is producing buds. The peak harvest period for this perennial field will be in October and November.

hand-harvested twice each week, more frequently during hot weather and less frequently during cold. During cold winter conditions, perennial fields may go 2 weeks between harvests. Perennial artichokes are commonly harvested 30 or more times during a season. Annual artichokes have a shorter, more concentrated production period. Globe artichoke should be harvested when the buds have reached their maximum size, are still tight, before the bracts open, and the internal flower tissue grows above the receptacle (Fig. 12.12; Rubatzky and Yamaguchi, 1997).

Artichoke production is labor intensive because the crop is hand-harvested. Harvesters walk between rows repeatedly over several weeks inspecting candidate buds and harvesting those deemed ready based on their size and compactness. The plants are very similar to thistles and some cultivars have thorns, which complicates harvest. The terminal or primary bud is harvested first and is usually the largest. The secondary and tertiary buds are harvested later as they reach their maximum size. Lower order buds may be too small to harvest. Globe artichokes are harvested by cutting the stem 7.6 cm (3 in) below the base of the bud. For some markets, the stem is cut much lower to include the first leaf below the bud. Generally a healthy perennial plant should produce about 5–8 marketable buds per season.

The bracts of some seeded artichoke cultivars have the tendency not to spread open with increasing maturity as readily as perennial cultivars such as 'Green Globe', so over-mature buds may have a compact appearance if harvest is delayed. Over-mature buds have well-developed flower tissues, are fibrous and bitter, and have less fleshy edible tissue (Fig. 12.12). Sampling a few buds to determine the interior characteristics can help determine the optimum harvest time based on exterior appearance. Buds are placed in cloth sacks that are held open by metal frames that move through the field to support the harvest crew. Yields for annual artichokes of 19,552 kg/ha (17,600 lb/acre) are considered good, although this number can vary greatly based on both environmental and market factors. Yields of 14,652 kg/ha (13,200 lb/acre) are considered good for perennial artichoke fields (Smith *et al.*, 2008).

Postharvest handling

Both annual and perennial globe artichokes are visually inspected in the field after harvest for defects and damage from insects or disease. Marketable buds are graded by size and quality and field packed in cartons or bags. Unmarketable buds are not packed and are discarded in the field. In some areas, artichokes may be harvested in bulk and transported to a packing facility outside the field where they are graded, boxed, and cooled (Fig. 12.15).

However, field harvesting and packing is becoming increasingly popular when feasible to cut costs and reduce handling. Cartons of field-packaged buds are transported to a cooling facility where field heat is removed and they are held until transport to market. Bud size classifications represent the number of buds that can be packed in a standard carton, so for example 18 buds (18 buds per carton or 18s) are larger than 11.3 cm (4.5 in) in diameter; 24s are 10–11.3 cm (4.0–4.5 in); 36s are 8.8–10 cm (3.5–4.0 in); 48s are 7.5–8.8 cm (3.0–3.5 in); and 60s are 6.9–7.5 cm (2.75–3.0 in). Random small buds between 2.5–6.9 cm (1–2.75 in) are jumble packed with an average of 100–175 buds per carton. The fresh market in the USA and Canada prefer 24s and 36s although some retailers and markets in other parts of the world prefer 36s and 48s because globe artichokes are priced by the bud rather than by the pound. There is a perception among some consumers that smaller buds are less developed and therefore have less fiber and flower development although this is often untrue (Smith *et al.*, 2008).

Field-packed globe artichokes are often cooled by forced air. Globe artichokes should be held at or near 1°C (33°F) at 90–95% relative humidity during storage and shipping to prevent desiccation. Bud quality can be maintained up to 2 weeks if stored at 1°C (33°F) at 90–95% relative humidity. For long-distance transport, artichokes are often shipped in refrigerated trucks (Smith *et al.*, 2008).

Diseases

Black tip is a physiological disorder of the bracts caused by calcium deficiency that is worse in some cultivars than others. Black tip seems to be a greater problem for production at higher temperatures and on sandy soils. Powdery mildew (*Leveillula taurica*) and Ramularia leaf spot (*Ramularia cynarae*) infest bracts and foliage, resulting in leaf drop and damaged buds. Verticillium wilt (*V. dahliae*) causes chlorosis, stunting, and wilting. Affected plants produce small buds and plants collapse and die in severe cases. All cultivars seem to be susceptible and annual artichokes can be rotated with broccoli or cauliflower to manage the disease (Smith *et al.*, 2008). Botrytis (*Botrytis cinerea*) causes a gray or brown fungal growth that develops on damaged

Fig. 12.15. Globe artichokes at this packing house are washed, graded by size, packed, and cooled prior to shipment to market.

tissues. This disease is worse under rainy weather and moderate temperatures.

Curly dwarf is a virus that stunts and eventually kills plants. Symptoms include leaf curl, dwarfing, and reduced bud production. The only known control measures are to use noninfected planting material and remove and destroy infected plants immediately. Milkthistle (*Silybum marianum*) is an alternate host for the virus.

Other virus diseases include: artichoke latent, artichoke mottled crinkle, artichoke yellow ringspot, autographa gamma (silver-Y moth), broad bean wilt (lamium mild mosaic), tobacco streak (stunt of asparagus), tomato black ring (ring spot of beet), tomato infectious chlorosis, and tomato spotted wilt (CABI, 2011).

Insect Pests

The globe artichoke plume moth (*Platyptilia carduidactyla*) lays eggs on the underside of leaves or the stem below the bud. The larvae tunnel into the stem, foliage, and buds damaging the receptacle and bracts while distorting and stunting bud development. Losses of 20–50% of the crop are not uncommon. The plume moth is not a problem in desert production areas. Control depends on strict sanitation practices including the removal of infested buds found during harvest and the immediate incorporation of plant debris back into the soil after a planting is cut back to the soil line. Effective management combines sanitation with Integrated Pest Management (IPM) strategies such as the use of insect growth regulators, pheromone mating disruption, biological control agents, and mass trapping with the reduced usage of conventional pesticides. Plume moth was one of the first pests where effective IPM strategies were developed for large-scale commercial production beginning in the 1970s. Before effective IPM control measures were implemented, massive amounts of insecticide were sprayed, which led to insect resistance to the pesticides and public health concerns in urban areas surrounding artichoke fields.

Aphids, including the bean aphid (*Aphis fabae*), green peach aphid (*M. persicae*), and artichoke aphid (*Capitophorus elaeagni*), can all attack artichokes at different times of the year. In addition to affecting growth, the artichoke aphid may cause sooty mold on the buds that reduces yield and marketability. Cribate weevil (*Oyiorhynchus cribricollis*) larvae feed on roots while adults feed on foliage and buds. Caterpillars, including the salt marsh caterpillar (*Estigmene acrea*) and cutworms (*Peridroma saucia*) and others, feed on both artichoke foliage and buds. Caterpillars are a particular problem for transplanted annual production, since they may kill the growing point of developing seedlings. The proba bug (*Proba californica*) has a life cycle and feeding habits similar to the lygus (*Lygus hesperus*) bug. Proba nymphs and adults feed mainly on young leaves and inject a toxin into the plant that causes stunting. The developing buds are deformed by the phytotoxin. The two-spotted spider mite (*Tetranychus urticae*) causes serious infestations that may reduce plant vigor and yield. Larvae of the chrysanthemum leaf miner (*Phytomyza syngenesiae*) may seriously damage foliage.

Other pests may include *Agrotis ipsilon* (black cutworm), *Cossus cossus* (carpenter moth), *Cynthia cardui* (painted lady butterfly), *F. occidentalis* (western flower thrip), *Peridroma saucia* (pearly underwing moth), *Pratylenchus penetrans* (nematode, northern root lesion), *Rhizobium radiobacter* (crown gall), *Rhizobium rhizogenes* (gall), *Spodoptera littoralis* (cotton leafworm), *Spodoptera litura* (taro caterpillar), and *Trichodorus* (stubby root nematodes) (CABI, 2011).

References

Boriss, H., Brunke, H., Geisler, M. and Jore, L. (2012) Lettuce profile. Available at: www.agmrc.org/commodities__products/vegetables/lettuce-profile (accessed 13 June 2013).

Borthwick, H.A., Hendricks, S.B., Toole, E.H. and Toole, V.K. (1954) Action of light on lettuce-seed germination. *Botanical Gazette*, 205–225.

CABI (2011) Crop protection compendium. Available at: www.cabi.org/cpc/?compid=1&dsid=17585&loadmodule=datasheet&page=868&site=161 (accessed 12 July 2012).

CABI (2013) Datasheets: *Lactuca sativa* (lettuce) production and trade. Available at: www.cabi.org/cpc (accessed 28 May 2013).

De Candolle, A. (1959) *Origin of Cultivated Plants*. Hafner Publishing Co., New York.

De Vos, N.E. (1992) Artichoke production in California. *HortTechnology* 2, 438–444.

Doležalová, I., Lebeda, A., Tiefenbachová, I. and Křístková, E. (2004) Taxonomic reconsideration of some *Lactuca* spp. germplasm maintained in world genebank collections. *Acta Horticulturae* 634, 193–201.

Duke, J. (2013) Dr. Duke's phytochemical and ethnobotanical databases. Available at: www.ars-grin.gov/duke (accessed 19 June 2013).

ERS (2011) Vegetables and melons yearbook. Available at: http://usda.mannlib.cornell.edu/MannUsda/viewDocumentInfo.do?documentID=1212 (accessed 11 June 2013).

FAOSTAT (2010) Food and Agriculture Organization of the United Nations. Available at: http://faostat.fao.org/site/567/DesktopDefault.aspx?PageID=567 (accessed 16 July 2012).

FAOSTAT (2011) Lettuce and Chicory production statistics. Available at: http://faostat.fao.org/site/567/Desktop_Default.aspx?PageID=567 (accessed 3 June 2013).

Jeffrey, C. (2007) Compositae: Introduction with key to tribes. In: Kadereit, J.W. and Jeffrey, C. (eds) *Families and Genera of Vascular Plants,* Vol. VIII: *Flowering Plants, Eudicots, Asterales.* Springer-Verlag, Berlin, pp. 61–87.

Lipton, W.J., Stewart, J.K. and Whitaker, T.W. (1972) *An illustrated guide to the identification of some market disorders of head lettuce* (No. 950). Agricultural Research Service, US Department of Agriculture. Washington, DC, pp. 1–7.

Martínez, J.A. and Artes, F. (1999) Effect of packaging treatments and vacuum-cooling on quality of winter harvested iceberg lettuce. *Food Research International* 32, 621–627.

McKenzie, R.J., Samuel, J., Muller, E.M., Skinner, A.K.W. and Barker, N.P. (2005) Morphology of cypselae in subtribe arctotidinae (compositae–arctotideae) and its taxonomic implications. *Annals of the Missouri Botanical Garden* 92, 569–594.

Miles, C. (2013) Winter Lettuce. Available at: http://agsyst.wsu.edu/WinterLettuce.html (accessed 28 June 2013).

Pignone, D. and Sonnante, G. (2004) Wild artichokes of south Italy: did the story begin here? *Genetic Resources and Crop Evolution* 51, 577–580.

Rangarajan, A., Ingall, B.A. and Zeppelin, V.C. (2000) Vernalization strategies to enhance production of annual globe artichoke. *HortTechnology* 10, 585–588.

Rubatzky, V.E. and Yamaguchi, M. (1997) *World Vegetables: Principles, Production and Nutritive Values,* 2nd edn. Chapman & Hall, New York.

Ryder, E.J., De Vos, N.E. and Bari, M.A. (1983) The globe artichoke (*Cynara scolymus* L.) production in California. *HortScience* 18, 646–653.

Schrader, W.L. and Mayberry, K.S. (1992) 'Imperial Star' artichoke. *HortScience* 27, 646–653.

Schwember, A.R. and Bradford, K.J. (2010) A genetic locus and gene expression patterns associated with the priming effect on lettuce seed germination at elevated temperatures. *Plant Molecular Biology* 73, 105–118.

Shah, M. (2010) Gulf Cooperation Council food security: Balancing the equation. Nature Middle East. Available at: www.nature.com/nmiddleeast/2010/100425/full/nmiddleeast.2010.141.html (accessed 8 July 2013).

Smith, R., Baameur, A., Bari, M., Cahn, M., Giraud, D., Natwick, E. and Takele, E. (2008) *Artichoke production in California.* University of California, Division of Agriculture and Natural Resources, Publication 7221, Oakland, California, pp. 1–6.

Turini, T., Cahn, M., Cantwell, M., Jackson, L., Koike, S., Natwick, E., Smith, R., Subbarao, K. and Takele, E. (2011) Iceberg Lettuce Production in California. Publication 7215. Available at: http://anrcatalog.ucdavis.edu (accessed 29 June 2013).

UC IPM (2013) How to manage pests: Lettuce. Available at: www.ipm.ucdavis.edu/PMG/selectnewpest.lettuce.html (accessed 28 June 2013).

USDA ARS (2013) National nutrient database for standard reference, release 25. Available at: http://ndb.nal.usda.gov/ndb/search/list (accessed 1 July 2013).

van Beilen, J.B. and Poirier, Y. (2007) Guayule and Russian dandelion as alternative sources of natural rubber. *Critical Reviews in Biotechnology* 27, 217–231.

Welbaum, G.E. (1994) Annual culture of globe artichoke from seed in Virginia. *HortTechnology* 4, 147–150.

Zohary, D., Hopf, M. and Weiss, E. (2012) *Domestication of Plants in the Old World: The Origin and Spread of Domesticated Plants in Southwest Asia, Europe, and the Mediterranean Basin.* Oxford University Press, Oxford, UK.

13 Family Poaceae

SWEET CORN, POPCORN, AND ORNAMENTAL CORN

Origin and History

Sweet corn, also known as green maize or sweet maize in many parts of the world, is a crop of New World origin. Scientists believe that sweet corn was domesticated in southern Mexico very long ago (Ranere, 2009). The progenitor of modern corn was a wild, annual grass, perhaps with a terminal flowering structure with male flowers above and female flowers below. Another theory suggests that the original plant had a terminal male spikelet with several small female spikelets at the nodes immediately below the male flower cluster (Goodman, 1988). Pollen samples collected near Mexico City were estimated to be 60–70,000 years old, illustrating how old corn is (Beadle, 1981; Sears, 1982).

Deliberate cultivation of corn began approximately 7,000 years ago. Teosinte (*Zea mays* spp. *mexicana*) may be similar to the wild plant from which corn was developed (Matsuoka *et al.*, 2002). From the original 2.5 cm (1 in) wild pod, human selection has created a pod many times larger than the wild form (Galinat, 1992). Modern cultivars of corn bear little resemblance to their wild progenitors (Goodman, 1988). Corn spread throughout the Americas as a basic food crop hundreds of years before the arrival of Europeans (Ranere, 2009). Both sweet corn, as well as popcorn, sources in North America originated with native Americans. Sweet corn has become a major commercial vegetable crop in North America during the past 200 years. The popularity of sweet corn in other parts of the world is even more recent. Sweet corn is not widely consumed in some cultures because of its close relationship with agronomic corn.

Botany and Life Cycle

Sweet corn (*Z. mays* var. *saccharata*) is a warm-season, frost sensitive, annual monocot from the family Poaceae (formerly Gramineae). The sex expression is monoecious and pollination is by wind. The pollen moves in the wind and by gravity from the tassel to the female flower located lower on the stalk (Fig. 13.1).

Sweet corn kernels (a fruit called a caryopsis) are eaten when immature while either on the cob or detached. In contrast, field or agronomic corn is harvested when the plants are physiologically mature as dry grain. The plants of most sweet corn cultivars are smaller than field corn. Also, the tassels of some sweet corn cultivars are distinct with larger size and lighter color than field corn (Fonseca *et al.*, 2003).

After germination the primary seed root anchors the seedling. Seminal (seed) roots originate from the scutellar node located within the seed embryo and are composed of the radicle and lateral seminal roots (Gardner *et al.*, 1985). The lateral seminal roots emerge later from behind the coleoptile. The seminal root system helps sustain seedling development by virtue of water uptake from the soil, but a young corn seedling depends primarily on the energy reserves of the kernel's starchy endosperm for nourishment until the nodal root system develops later. Once a seedling has emerged, the rate of new growth of the seminal root systems slows as the nodal root system begins to develop from above the mesocotyl (Gardner *et al.*, 1985).

The nodal root system is composed of whorls of secondary adventitious roots that develop at the basal nodes of the stem and grow laterally. These lateral roots branch profusely, spread widely, and penetrate deeply into the soil, though depth of penetration depends on soil characteristics. These relatively shallow branched secondary roots absorb nutrients from the soil for the plant. Besides nodal roots, "prop or brace" roots help to anchor the plants in the soil and provide some nutrient absorption. Brace roots are produced on the first two or three nodes of the stalk above the soil line (Gardner *et al.*, 1985).

Fig. 13.1. Corn is monoecious with separate male (tassel, a) and female (ear, b) flowers at the top and middle of the plant, respectively. Pollen from the tassel must land on a silk, germinate, and grow the length of the silk to pollinate the ovule that will become the edible kernel. This sequence must occur repeatedly for normal ear development.

The rigid stem (axis or culm) ranges in height from 1.5–2.5 m (5.0–8.2 ft) depending on cultivar and is enveloped by alternating leaf sheathes originating from prominent nodes (Fig. 13.2). The leaf sheaths form at the node and closely enclose a

Fig. 13.2. The corn stem is sturdy and erect with the tassel at the terminus, the ear in the middle. The long alternating opposed leaves with parallel venation occur along the length of the stem.

length of the main stem, often covering the node above. At the ligule, each leaf sheath angles away from the stem as a long broad curved leaf with prominent midrib. The large leaf blades are alternate and grasslike in appearance. The long leaves are fairly uniform in width and have many parallel veins running along their length (Rubatzky and Yamaguchi, 1997).

Branching or suckering may occur at the base of the plant in some genotypes (Fig. 13.3). Suckers are secondary shoots or tillers that develop in the axils of lower leaves near the soil surface. Ears formed on the secondary shoots are late to develop and much less productive than the main stalk. There is significant genetic variation with regard to branching. Many modern cultivars are predominantly single stalked without suckers.

Fig. 13.3. Suckers or tillers are smaller plants that develop at the base of the corn plant. Wide spacing supports sucker development. Modern cultivars lack suckers.

Plant and flower characteristics

Male flowers

The male inflorescence is a terminal panicle (called the tassel) borne on the top of the plant, consisting of the spikelike central axis with lateral branching (Fig. 13.1a). The central axis usually has four or more rows of paired spikelets, while the lateral branches are in paired rows. Pairs of spikelets consist of a sessile and pedicelled flower. Tassel florets contain stamens and a rudimentary pistil that degenerates early although under some circumstances the pistil may develop. As male flowers mature, anthesis starts near the center of tassel spike and then proceeds simultaneously in both directions. Anthers first exert from the upper spikelet, followed rapidly by anthers from the sessile spikelet, resulting in an extended period of pollen release. The pollen exits through a pore at the tip of each anther. Sweet corn hybrids are estimated to produce from 8,200,000 to 14,960,000 pollen grains per tassel (Fonseca *et al.*, 2003). The amount of pollen shed is affected by temperature, wind, and cultivar and usually occurs over 3–10 days. Pollen shed begins before the stigmas (silks) of the female flowers emerge from the husks, thus forcing cross-pollination (Gardner *et al.*, 1985).

Female flowers

Female flowers are in terminal inflorescences enclosed within husks (Fig. 13.1b). The ear of the corn plant is the female flower. The silks are the styles of the female flower and are attached to each ovule. The flower is a spike-like carpellate inflorescence with the fruit (caryopsis) paired in vertical rows along the cob. The pollen grain must germinate and grow the length of the style to pollinate each ovary on the inflorescence. The cob and developing fruit are protected by a husk (Fig. 13.1b). If husk development is incomplete and the tip of the cob exposed, kernel development near the tip of the ear may be impaired, insect damage may occur, and the ear is more susceptible to diseases (Gardner *et al.*, 1985).

If properly pollinated, the ovules develop into fruit that is harvested in an immature stage of development as sweet corn for human consumption. The outer layer of the fruit is the pericarp and is fused to the testa beneath. The embryo occupies the lower one-quarter corner of the fruit. The majority of the kernel (roughly 75% of the volume) is endosperm surrounded by the aleurone layer. In flour, dent, and flint corn the endosperm is predominately starch and oil while in sweet corn the endosperm contains sugar and starch, the proportions of each depending on the genotype of the cultivar as well as the stage of development.

Each of the different types of sweet corn genotypes disrupt starch synthesis, resulting in the accumulation of sugar in the endosperm that is not rapidly converted into starch. This is what makes sweet corn sweet instead of starchy. Sugar molecules are shorter than starch molecules, which are actually long, complex chains of sugar molecules. These shorter sugar molecules pack more tightly when corn is dried and this is what gives sugary seed its characteristic wrinkled, glassy appearance (Rubatzky and Yamaguchi, 1997).

As the young axis of the cob develops, flower primordia develop on its periphery. Each of these primordia forms two equal lobes by division that lie side-by-side, resulting in the paired appearance of ovules in cross section. Each lobe develops into a two-flowered spikelet. The flowers are arranged one above the other, and in most corn cultivars only the upper of these flowers develops into a mature kernel (Rubatzky and Yamaguchi, 1997).

An exception to his ordered flower arrangement is the cultivar 'Country Gentlemen' (also called 'Shoe peg'). 'Country Gentleman' is an heirloom cultivar that produces kernels randomly spread across the ear rather than in paired rows (Fig. 13.4).

With 'Country Gentlemen' both flowers produce kernels. The kernels are crowded during development, which prevents the formation of paired straight rows of kernels (Rubatzky and Yamaguchi, 1997).

Fig. 13.4. 'Country Gentleman' is a novelty white heirloom sweet corn cultivar, also called 'Shoe Peg', that does not form kernels in straight rows, making it more difficult to eat on the cob.

Pollination

Corn is wind pollinated and must be planted in block configuration to ensure effective transfer of pollen from the male to female flowers.

Different types of corn should be separated by at least 210 m (700 ft), and preferably further, to ensure that unwanted cross-pollination and xenia effects are avoided. If adequate space is not available, different types of corn can be separated in time by altering planting dates to avoid cross-pollination (Gardner et al., 1985).

Sweet corn genetics

Corn was one of the first crops sold commercially as F-1 hybrids dating back to the 1930s. Corn is easily hybridized because its monoecious sex expression can easily be manipulated by detasseling to control pollination and produce the inbred lines needed for F-1 hybrid production. Today, almost all commercial sweet corn and field corn cultivars are F-1 hybrids. However, there still are a few novelty and specialty non hybrid, open pollinated cultivars available (Darrah et al., 2003).

Sweet corn genetics is complex with a number of different alleles affecting sugar content. Sweet corn was not created through genetic engineering but by naturally genetic mutations that have occurred and been selected and maintained as new cultivars. Sweet corn genes can be divided into three groups. The oldest type is due to the recessive su gene that causes sugar rather than starch to accumulate in the endosperm of each kernel (Tracy et al., 2006). This is "standard" sweet corn (Table 13.1). After the harvest of standard sweet corn, sugar is very rapidly converted to starch, so the ears must be rapidly cooled to slow the conversion in order to

Table 13.1. Effects of pollination on sugar content of su, se, and sh_2 sweet corn cultivars.

Genotype	Standard sugar (su)	Sugar enhanced (se)	Super sweet (sh_2)
su	OK	OK	Starchy
se	OK	OK	Starchy
sh_2	Starchy	Starchy	OK

OK, indicates that cross-fertilization will either have no effect on flavor or the kernels may taste like standard sweet corn. Starchy means that isolation is required or the endosperm will be like field corn.

Family Poaceae

preserve the sugar content. The *su* gene probably arose as a naturally occurring mutation several different times throughout history and was maintained by indigenous peoples.

In addition to the *su* gene there are two additional major sweetness genes of sweet corn: the sugary enhancer gene (*se*) results in a class of sweet corn that has a higher sugar content than the *su* cultivars (Table 13.1). In these cultivars, the conversion of sugar to starch at harvest occurs more slowly. Sugary enhancer cultivars also lose moisture at a slower rate than normal sweet corn, which helps extend their harvest period. 'Illini Extra Sweet' was the first cultivar released in the 1970s (Tracy, 2000).

Supersweets are a class of sweet corn cultivars that carry the shrunken-2 gene (sh_2) (Table 13.1). This gene restricts the conversion of sugar to starch, causing the supersweets to have higher sugar contents than other sweet corn types (Darrah *et al.*, 2003). Studies show that supersweet cultivars can maintain their sweetness on the stalk for up to 7 days and off the stalk for up to 30 days in cold storage (Brecht, 2004). Supersweet cultivars have a tender pericarp in addition to a higher sugar content compared to *su* and *se* types. The supersweet characteristic results in extremely shriveled kernels because of the reduced size of the endosperm that allows seed corn to be easily identified (Darrah *et al.*, 2003). Some supersweet cultivars display poor seed germination and low seedling vigor, making this type of corn more difficult to establish, particularly in cold soils.

Xenia refers to the immediate and recognizable effect of a pollen parent gene on the phenotype of the developing kernel (Rubatzky and Yamaguchi, 1997). Corn is one of the few vegetable crops where the identity of the pollen source determines the appearance and characteristics of the fruit, which in the case of sweet corn is the developing kernel. The two most common xenia effects are kernel color and starch content. Understanding xenia is important for maintaining quality because unintended cross-pollination from another cultivar with different characteristics is visually apparent on the developing ear almost immediately after pollination. Controlling pollination of sweet corn is especially important when sweet corn is grown next to field or agronomic corn because many of the field-corn pollen traits are dominant and expressed over the desirable traits of sweet corn, including starch content. The starchy gene in field corn is dominant to *su* and *se*, so if a sweet corn plant is pollinated by field corn the developing endosperm will be starchy rather than sweet (Fig. 13.5; Darrah *et al.*, 2003). For example, if white sweet corn is unintentionally cross-pollinated with high-starch yellow dent corn, the kernels on a sweet corn ear pollinated by pollen from a field corn plant will be yellow, dented, and recognizable during development because of xenia (Fig. 13.5; Darrah *et al.*, 2003).

It is also important to isolate supersweet (sh_2) corn from *su* and *se* types of sweet corn. If isolation is not maintained and cross-pollination occurs, then both the supersweet and the *su* and *se* types of sweet corn may produce starchy kernels. Both the *su* and *se* types can cross-pollinate and maintain their normal sugar contents (Darrah *et al.*, 2003).

In determining sweet corn color, the gene for yellow is dominant while white is recessive. When pollen from a plant with yellow pericarp fertilizes a plant with white pericarp, the kernels have a yellow pericarp because of the Mendelian dominance of yellow over white. When yellow corn is pollinated by white corn, the kernels are yellow but lighter in color and often capped with white. There is no genetic linkage between kernel color and sweetness because each trait is independent. An isolation distance of at least 213 m (700 ft) should be maintained between sweet and other types of corn, including field corn (Darrah *et al.*, 2003).

Fig. 13.5. Sweet corn with some lighter colored full kernels pollinated by field or dent corn. Sweet corn must be isolated to prevent unwanted cross-pollination. The starchy genetic trait carried by field-corn pollen has an immediate effect on the appearance and development of sweet corn. The immature kernels that result from outcrossing are edible but not sweet.

Types and Cultivars

There are many types of corn produced around the world. It is important to understand these different types because most can be used for human food albeit in very different forms (Table 13.2). Corn is often divided into two major categories: corn that is harvested when mature as a dried grain and corn that is harvested immature as a vegetable. Both are used for human food. Vegetable corn includes sweet corn, baby corn, and roasting ears (Rubatzky and Yamaguchi, 1997). Grain corn includes popcorn, ornamental corn, dent corn, and flint corn (Hallauer, 2001). In addition to the types of sweet corn described above in the section entitled "Sweet Corn Genetics", several other commercially important types of corn are described below (Table 13.2).

Dent corn (field or agronomic corn)

Dent corn is a one of the more common types of field or agronomic corn produced in the world (Fig. 13.6). For example, over 99% of the corn production in the USA is dent corn and less than 1% is sweet corn.

In some areas waxy, flint, or flour corn may also be grown on a commercial scale as a grain. Dent corn, and other types of agronomic corn, is mechanically harvested in the fall as a mature dry grain and is distinguishable from other types of corn by the distinctive dent on the top of each mature kernel. Dent corn is very versatile and has many different uses, which is one of the reasons it is so popular. A kernel of dent corn is roughly 61% starch, 19.2% feedstock (gluten and hull), 3.8% oil, and 16% water (Barker and Beuerlein, 2005). Dent corn is used for animal feed, starch conversion to ethanol for vehicle fuel, vegetable oil, high fructose corn syrup, and human food. Human foods derived from field corn include breakfast cereals, corn starch, corn meal, hominy, and grits.

One of the major industrial uses of agronomic corn is for production of high-fructose corn syrup (HFCS). HFCS is not made from sweet corn, which is a common misconception. HFCS is also called

Table 13.2. Common types of corn and their uses.

Common name	Scientific name	When harvested	Uses
Sweet corn	*Zea mays* var. *saccharata*	Eaten when immature as a vegetable	Boiled or sometimes roasted
Popcorn	*Z. mays* var. *everta*	A type of flint corn harvested as a grain when mature	Used for popping when heated to high temperatures and as an ornamental
Ornamental corn "Indian corn"	*Z. mays* – several cultivars harvested mature as a whole ear for ornamental usage	As mature ears	Decoration, often displayed as whole ears with colorful kernels exposed
Dent corn	*Z. mays* var. *indentata*	The most common type of corn, also called field, agronomic, or grain corn, widely traded internationally	Cooking oil, flour, ethanol, animal feed, starch
Flint corn	*Z. mays* var. *indurata*	The kernel is hard but smooth at maturity. It has little soft starch. Grown commercially in India and some other countries	Decoration, many Indian corn cultivars are flint types, and cooking e.g. polenta and hominy
Flour corn	*Z. mays* var. *amalacea*	Resembles flint corn but kernels contain soft starch and have a little or no dent. Flour corn is one of the oldest types	Processed into flour when mature
Waxy corn	*Z. mays* var. *ceratina*	Mature kernels have a waxy appearance and contain gummy starch high in amylopectin	Grown commercially as a source of starch
Baby corn	*Z. mays*	The unfertilized young ear 45–50 days after sowing obtained from sweet corn or other corn harvested within 2 days of tasselling	Used as a vegetable in salads or cooking. It can be preserved in 1.5% saline solution, canned or frozen
Pod corn	*Z. mays* var. *tunicata*	Progenitor of modern corn with each kernel in an individual husk	Not grown commercially at present

Family Poaceae

Fig. 13.6. Dent (also called field or agronomic corn) is the most common type grown in the world. The plants are mechanically harvested as a grain when dry and fully mature. Field corn can be eaten as a vegetable at the immature milk stage of development when the pericarp is still tender.

glucose-fructose syrup in the UK, glucose/fructose in Canada, and high-fructose maize syrup in other countries, and comprises any of a group of corn syrups that has undergone enzymatic processing to convert some of its glucose into fructose to produce a desired sweetness (Marshall and Kooi, 1957). In the USA, consumer foods and products typically use high-fructose corn syrup as a sweetener. It has become very common in processed foods and beverages in the USA, including breads, cereals, breakfast bars, lunch meats, yogurts, soups, and condiments. HFCS became popular in the late 1970s. According to the USDA, HFCS consists of 24% water and the rest sugars. The most widely used varieties of high-fructose corn syrup are: HFCS 55, which is commonly used to sweeten soft drinks, approximately 55% fructose and 42% glucose; and HFCS 42, which is used in cereals and baked goods, approximately 42% fructose and 53% glucose. HFCS 90, approximately 90% fructose and 10% glucose, is used in small quantities for specialty applications, but primarily is used to blend with HFCS 42 to make HFCS 55 (USDA, 2012).

In the USA and some other countries, HFCS is among the sweeteners that have primarily replaced table sugar in many foods because of government production quotas of domestic sugar, subsidies that favor domestic corn production, and an import tariff on foreign sugar, which combine to raise the price of sugar, making HFCS cheaper for many applications (USDA, 2012).

Popcorn

Popcorn is a special kind of flint corn, also known by some as "Indian corn" and sometimes "calico corn". Flint corn is named because of its recognizable hard outer pericarp and a resemblance to the mineral flint, a type of quartz (Hallauer, 2001).

Popcorn is harvested as a mature dried grain. The kernel is smaller than sweet or field corn and distinctly pointed. The endosperm of popcorn is hard and the pericarp is very thick, sealing moisture inside. The kernel moisture content must be less than 13.5% for popping to occur. During heating, the moisture trapped in the kernel becomes superheated pressurized steam. Under these conditions, the starch inside the kernel gelatinizes, softens, and becomes pliable. The pressure continues to increase until about 930 kPa (135 psi) when the breaking point of the pericarp is reached at a temperature of roughly 180°C (356°F) (Lusas and Rooney, 2001). The hull ruptures rapidly, causing a sudden drop in pressure inside the kernel and a corresponding rapid expansion of the steam, which expands the starch and proteins of the endosperm into airy foam. As the foam rapidly cools, the starch and protein set into the familiar crispy puff. Most popcorn is white or yellow although other colors do exist. Ornamental types have different colored pericarps ranging from red to blue. Hulled and hulless types are available.

Ornamental corn

Ornamental corn, also called in Indian or decorative corn, may take many forms (Fig. 13.7). Some types accumulate starch and have a vividly colored pericarp ranging in color from white to black. An ear may have kernels with the same color or each may be colored differently. Whole ears are often displayed as decorations after drying with the husks tied back to display the brightly colored kernels. Ornamental corn may have tan, white, or red husks when dried (Fig. 13.7). Ornamental popcorn is also available, including the cultivars 'Cutie Pops', 'Feather Mixed', and 'Strawberry'. Ornamental corn and sweet corn will cross-pollinate, so they must be isolated from each other (Hallauer, 2001).

Baby corn

Baby corn, also sometimes called candle corn, is harvested when ears are very small and immature. It typically is eaten whole-cob before ovules are pollinated. Baby corn is eaten both raw and cooked and is common in Asian cuisine and as a relish (Chutkaew and Paroda, 1994).

Baby corn can be produced either as a primary crop or as a secondary crop in a planting of sweet corn or field corn. When produced as a primary crop, a cultivar is chosen that is well suited for baby corn production (Miles and Zenz, 2000). While many cultivars may be suitable, those developed specifically for baby corn production tend to produce more ears per plant (Bar-Zur and Saadi, 1990). In the second production method, the cultivar is selected to produce sweet or field corn. The second ear from the top of the plant is harvested for baby corn, while the top ear is allowed to mature.

Baby corn ears are hand harvested as soon as the corn silks emerge or a few days after. Harvest requires about the same amount of time and labor as hand-harvesting sweet corn. Corn generally develops very quickly, so the harvest of baby corn must be timed carefully to avoid overly mature ears. Baby corn ears are typically 4.5–10 cm (1.8–4 in) in length and 7–17 mm (0.3–0.6 in) in diameter. After harvest, the baby corn ears should be immediately refrigerated with the husks intact to conserve ear moisture and preserve quality (Miles and Zenz, 2000). Baby corn is canned or frozen often in countries were labor is plentiful.

Roasting ears

Roasting ears describe cultivars of corn with slow pericarp development and strong corn flavor

Fig. 13.7. Plants in an ornamental corn cultivar evaluation trial. There are many types of ornamental corn. Ornamental corn shows variation in kernel characteristics and colors. The size and color of the plants and husks vary greatly as well.

harvested for human consumption when immature. Typically, roasting ears are harvested in the early dough stage and roasted or baked. Since roasting-ear cultivars are starchy dent corn, the dried kernels can be ground for flour or used as livestock feed just like other types of field or dent corn when mature. Examples of roasting ear cultivars are 'Truckers Favorite' and 'Asgrow Favorite'. Today, sweet corn is more commonly used for roasting or baking since true starchy roasting ears are difficult to obtain in many markets. Because of its higher sugar content compared to roasting ears, sweet corn has a greater tendency to burn when roasted or baked.

Sweet corn

There are many important characteristics that differentiate cultivars. Corn plants should have sufficient stalk size and durability to resist lodging during wind storms or under heavy fruit loads. Suckers are small weak plants that grow from the base of the main plant and that seldom produce marketable ears. Most modern cultivars do not produce suckers but some heirloom and Indian corn cultivars do (Rubatzky and Yamaguchi, 1997).

The following are examples of the important traits that differentiate cultivars and affect quality and sweet corn production. For both mechanical harvest and pick-your-own operations, uniform maturity is important for destructive mechanical harvesting to reduce the number of unmarketable ears and to improve harvest efficiency. The position of the ear on the plant and number of ears per plant affects maturity and harvest efficiency as well. The ears of some cultivars are difficult to pull by hand because they are so well attached to the plant and this is undesirable for pick-your-own operations.

The general appearance of sweet corn husks varies widely among cultivars ranging from light green, dark green, green and purple to mostly purple. Husk cover of ears improves resistance to bird predation, worm damage, and resistance to certain diseases like corn smut. Adequate ear-tip cover by the husks helps prevent ovules at the open end of the ear from aborting or from being damaged during development. The tip of each ear should be well covered by the husk to prevent against insect damage. Flag leaves are small leaves that protrude perpendicular to the ear and should be dark green and free of tears or blemishes. Increasingly, corn is sold without the husk so the presence and appearance of flag leaves is only important where corn is sold in the husk. Some markets prefer that sweet corn be sold with the shank (stock below the ear) attached to the ear. However, selling husked ears is decreasing in popularity in some markets because consumers prefer ears of fresh and frozen ear corn that are convenience packaged after being dehusked, desilked, and prewashed so the corn is ready to eat after cooking.

The ear characteristics are the most important for determining sweet corn quality. Ear size is desirable and tends to increase with maturity class such that earlier maturing cultivars have very small ears while later maturing cultivars have large ones. Small cobs are preferred to large and ear shape should be long rather than short and stubby. Kernel size and number creates a visual texture. Early-maturing cultivars with fewer rows (12–14 rows per ear) and large kernels appear coarse textured compared to the more refined look of a mid-season or main-season 18-row corn with smaller kernels. Rows are paired and occur in even numbers. Ears with a higher number of rows are preferred and deemed to be of higher quality. High quality ears contain straight rows of kernels, which make eating corn on the cob easier.

Kernel characteristics are an important determinant of eating quality. Deeper kernels are preferred to shallow kernels to reduce biting into the cob. Kernels that are excessively wide must be punctured while biting, so medium sized kernels are preferred to ones that are excessively wide or narrow. Light or clear silks detract less from the ear appearance and are preferred when eating corn on the cob. Ear tip fill is a sign of complete pollination and an important quality attribute of whole ear corn. Kernels should extend to the tip; poor pollination and drought stress cause poor tip fill. Husk cover of the entire ear will help protect the terminal kernels from desiccation and insect predation (Rubatzky and Yamaguchi, 1997).

Sugar content is one of the most important quality characteristics. Brix readings range from 10% to 15% for su corn, 13–28% for se corn, and 25–35% for sh_2 corn. Sugars are converted to starch as the corn matures and after harvest. In sh_2 corn, this conversion is much reduced, leading to greater retention of sugar content after harvest (Rubatzky and Yamaguchi, 1997).

The texture of the sweet corn also depends on the type of corn, the cultivar, and the harvest maturity. Sweet corn kernels contain phytoglycogen, which provides the characteristic creamy texture. The covering or pericarp of the kernel contributes to the texture. The pericarp of *su* and *se* corn is similar, while the pericarp of sh_2 corn often is described as having a crunchy texture or firmness, depending on the cultivar. Within each sweet corn type there is a range of firmness associated with the pericarp. As a general rule, *se* types tend to have a more tender pericarp than *su* corn (Rubatzky and Yamaguchi, 1997).

Sweet corn cultivars are often divided into three categories based on kernel color: yellow, white, and bicolor. Bicolor simply means that the kernel colors are a mixture of white and yellow. Yellow corn is characterized as having a stronger flavor, while the flavor in white corn is sometimes described as mild.

Maturity classes

Sweet corn is often divided into different maturity classes based on time to harvest from direct seeding. For example, first-early cultivars mature from 65 to 74 days after planting and usually produce small plants, small ears, and have poorer quality compared to later maturing cultivars. These cultivars have only 12–14 rows per ear and shallow kernels. When growing first-early cultivars, ear quality is sacrificed for earliness. Early cultivars mature in a range from 75 to 80 days. These cultivars have more rows and deeper kernels compared to first-early cultivars. Quality is satisfactory but not outstanding. Main-season cultivars are often seen in the marketplace during midsummer and are used extensively for processing. They mature in 81–90 days and produce large ears with 18 rows of deep kernels with excellent eating quality. Late cultivars take more than 90 days to mature and have very high quality and produce large ears. Later cultivars are often used to separate cultivars in "time" to prevent unwanted crossing. Planting a range of maturity classes early in the season helps to ensure a continuous supply of fresh sweet corn throughout the summer (Tracy, 2000).

Economic Importance and Production Statistics

Sweet corn production

The USA is the leading producer of sweet corn with 236,860 ha (585,294 acres) with production totaling 3,788,030 metric tonnes (4,175,588 short tons). Other important producers of sweet corn include: Nigeria 705,700 metric tonnes (777,901 short tons), Mexico 627,092 metric tonnes (691,250 short tons), Indonesia 458,200 metric tonnes (505,079 short tons), Peru 408,181 metric tonnes (449,943 short tons), South Africa 402,100 metric tonnes (443,239 short tons), and France 351,184 metric tonnes (387,114 short tons). In the USA in 2009, 255,438 ha (613,050 acres) of sweet corn were harvested including 94,514 ha (379,500 acres) destined for fresh market and 153,578 ha (379,500 acres) for processing (FAO, 2011). Florida and California are the largest producers of fresh-market sweet corn, while Minnesota, Wisconsin, and Washington are the largest producers of sweet corn for processing. The value of the US sweet corn crop was US$1,171,396,000 for both processing and fresh market combined. Interestingly sweet corn only accounts for less than 1% of the total corn (maize) production in the USA. The USA is the leading agronomic corn-producing country in the world with around 40 million ha (100 million, acres) and a value of roughly US$15 billion. Most of the agronomic or grain corn grown in the USA is used primarily as animal feed or processed into sweetener and ethanol. The average American adult consumes about 68 kg (150 lb) of caloric sweeteners per year and over half of this is derived from HFCS made from field corn by enzymatically converting starch to HFCS (USDA, 2012).

World sweet corn production increased 14% on about 6% additional land area from 1995 to 2005 and continues to grow. The popularity of sweet corn increases with disposable income. Large increases in sweet corn production and consumption are occurring in China, India and other countries, coinciding with their rapid economic growth (FAO, 2011).

Nutritional Values

Sweet corn is a good source of the minerals phosphorus, magnesium, iron, zinc, vitamins, particularly yellow corn, and antioxidants (Table 13.3). White corn is typically poorer nutritionally particularly in vitamin and antioxidant content. One ear of sweet corn supplies 88 calories. Most of these calories come from the 19.1 g (0.67 oz) of carbohydrates in an ear. The sweet flavor of corn comes from its 6.4 g (0.23 oz) of sugar. The majority of this sugar is in the form of glucose although fructose and sucrose are also present.

Production and Culture

Corn is a warm-season crop, with cultivars adapted to areas from about 50° latitude to the equator. Corn can survive brief exposures to adverse temperatures as low as 0°C (32°F) and as high as 44°C (112°F), but growth decreases once temperatures drop to 5°C (41°F) or exceed 35°C (95°F) (Barker and Beuerlein, 2005). Planting before the average soil temperature reaches 13°C (55°F) at seed depth may lead to poor

Table 13.3. Nutritional value of yellow sweet corn (uncooked kernels)[a] (USDA Nutrient Database, 2011).

Nutritional value	100 g (3.5 oz)
Energy	360 kJ (86 kcal)
Carbohydrates	19.02 g
– Sugars	3.22 g
– Dietary fiber	2.7 g
Fat	1.18 g
Protein	3.2 g
– Tryptophan	0.023 g
– Threonine	0.129 g
– Isoleucine	0.129 g
– Leucine	0.348 g
– Lysine	0.137 g
– Methionine	0.067 g
– Cysteine	0.026 g
– Phenylalanine	0.150 g
– Tyrosine	0.123 g
– Valine	0.185 g
– Arginine	0.131 g
– Histidine	0.089 g
– Alanine	0.295 g
– Aspartic acid	0.244 g
– Glutamic acid	0.636 g
– Glycine	0.127 g
– Proline	0.292 g
– Serine	0.153 g
Water	75.96 g
Vitamin A equivalent	9 µg (1%)
Thiamine (vitamin B1)	0.200 mg (17%)
Niacin (vitamin B3)	1.700 mg (11%)
Folate (vitamin B9)	46 µg (12%)
Vitamin C	6.8 mg (8%)
Fe	0.52 mg (4%)
Mg	37 mg (10%)
K	270 mg (6%)

[a]One ear of medium size (6¾–7½ in (17.1–19 cm) long) corn has 90 g (3.2 oz) of kernels.
Percentages are relative to USDA daily recommendations for adults.

higher market prices. However, early plants can be damaged if excessively low temperatures or prolonged periods below freezing occur. Seedling vigor and cold temperature are very important characteristics for early-season plantings and vary among genotypes (Welbaum et al., 2001). Many of the early sh_2 cultivars suffered from low seedling vigor because of their reduced storage reserves in the endosperm. Shrunken-2 seed corn leaked more sugars compared to standard sweet corn, which attracted soil-borne diseases, further reducing stands in some cases. Difficulties establishing early cultivars of sh_2 delayed production for processing and under cool-season conditions for fresh market. Increasing seed vigor as a breeding priority and preplant seed treatments have improved establishment (Westgate and Hazzard, 2005).

Optimal temperatures for growth vary between day and night and during the growing season. For example, optimal daytime temperatures range between 25–33°C (77–91°F), while optimal nighttime temperatures range between 17–23°C (62–74°F). The optimal average temperatures for an entire crop growing season, however, ranges between 20–23°C (68–73°F) (Barker and Beuerlein, 2005).

High temperature stress during ear formation, reproduction, and grain-fill can reduce yield, but temperatures less than 38°C (100°F) usually do not cause much injury if soil moisture is adequate. For nonirrigated corn, stress usually begins when air temperatures exceed 32°C (90°F) during the tasseling-silking (pollination). Corn yield may be significantly reduced each day the temperature exceeds 35°C (95°F) or higher during pollination because hot, dry winds may cause tassel blasting and loss of pollen. Pollen shed usually occurs during cooler morning hours when conditions are less severe (Barker and Beuerlein, 2005).

Corn is apparently a short-day plant, because maturity for many cultivars is accelerated under 12–14 h of light and delayed under longer photoperiods. Cultivars developed for different latitudes have differing photoperiod requirements.

Site selection and field preparation

Corn can grow on a wide range of soil types ranging from sandy loams to clay loams. Lighter soils are often selected for earlier plantings because they warm up faster in the spring, while plantings for late market are often placed on heavier soils. Some sweet corn production occurs on muck soils

germination and reduced stands. The growing point of germinating seedlings remains near or below the soil surface and is usually not damaged by frost until plants reach the five- to six-leaf node (collar) stage. By this time, corn is approximately 25 cm (10 in) tall, and the probability of freezing temperatures greatly decreases (Barker and Beuerlein, 2005). The loss of leaves from frost generally does not seriously injure small plants, but will delay plant development. The fact that newly germinated seeds and small seedlings can tolerate frost has led to the increased early planting of corn before the frost-free date in northern areas in order to sell product as early as possible to obtain because of the favorable response to organic matter and nitrogen (N).

The best yields take place at a pH of 6.0–6.8 although the crop grows fairly well over a wider pH range of 5.5–7.5. Acid soils with pH below 5.5 should be avoided. Corn should be rotated with other unrelated crops to help control diseases and insects (Barker and Beuerlein, 2005).

Fertilizer and nutrition

Soil test levels determine the fertilizer requirements for the crop. Sweet corn has a relatively high nutrient requirement. Sufficient nutrients should be available early in the season to ensure rapid seedling growth and good stand establishment. The crop responds well to N, and N fertilization varies widely depending on soil type. Generally more N is used on light sandy soils than heavier silt or clay loam soils. Side-dressing with N is common with corn. Many large growers row band N at planting to support early seedling growth and side-dress with anhydrous ammonia when the corn is knee high to simulate reproductive growth. Anhydrous ammonia is a pressurized gas that contains 82% N. It is injected into the soil where it reacts with water to form the positively charged ammonium ion, which is held by negatively charged particles in the soil. It is very important to use anhydrous ammonia in moist soils to reduce losses from volatilization. Other forms of N can be side-dressed in areas where anhydrous ammonia cannot be used (Barker and Beuerlein, 2005).

Phosphorus (P) and potassium (K) fertilization is usually based on soil test results and in some areas applications may not be needed every year because of the soil type, the crop rotation schedule, and many years of previous application to agricultural soils (Barker and Beuerlein, 2005). Generally, all of the K and phosphate are row banded at planting. Phosphorus and K are not lost from the soil in the same way as N, thus application concerns are less restrictive. Phosphorus is most commonly lost as runoff. Utilization of conservation practices that minimize the risk of soil runoff to surface waters are good P management practices. Both P and K are often banded near the seed at planting. Band applications are generally 5 cm (2 in) to the side and 5 cm (2 in) below the seed to prevent fertilizer burn. The total amount of salts applied (N + K_2O) should not exceed 110 kg/ha (100 lb/acre) (Barker and Beuerlein, 2005).

Although field corn is increasing planted no-till, sweet corn acreage has been slower to convert because conservation tillage practices, particularly in cool season areas, are not conducive to early season emergence. In many areas, especially when a cover crop or heavy residue is present, sweet corn fields are plowed and worked with a disc harrow or rotovator until the seedbed is level and the soil texture is uniform. In the absence of residue, the ground can be disked or rotovated without plowing. Good drainage is important regardless of the soil type, so corn is often produced on raised beds especially on heavier soils or where furrow irrigation is used (Fig. 13.8).

Beds may be sloped to increase exposure to the sun for early season plantings.

Field establishment

Sweet corn is direct seeded into the field. Although occasionally used to decrease the time to first harvest, transplanting is not a common practice because corn does not transplant well (Welbaum et al., 2001). The importance of precision planting for high yields and uniform maturity cannot be overemphasized. Precision in-row spacing produces a desired stand, saves seed, and eliminates the need for thinning after emergence. Planting to a precise and optimum depth affects the uniformity of seedling emergence and maturity.

Spacing is dependent on cultivar with early cultivars that produce small plants requiring less space than later maturing large plants. The seeds are generally planted in rows 0.6–1.2 m (2–4 ft) apart with an in-row spacing of 15–46 cm (6–18 in).

Recommended plant populations range from around 37,064 plants/ha (15,000 plants/acre) for main season cultivars to around 61,774 plants/ha (25,000 plants/acre) for first-early cultivars. Plant spacing can be increased if irrigation is available. Seeding rates range from 9 to 16 kg/ha (8 to 14 lb/acre).

The appropriate planting depth varies with soil and weather conditions. For normal conditions corn is planted 1.5–2 in (3.8–5 cm) deep to provide frost protection and allow for adequate root development. Shallower planting often results in poor root development. In spring, when the soil is usually moist and there is little evaporation, seed can be planted shallower and no deeper than 3.8 cm (1.5 in). As the season progresses and soil evaporative losses increase, deeper planting ensures sufficient moisture for seed germination. Under dry conditions seed may be placed

Fig. 13.8. Sweet corn production on raised beds in Taiwan.

up to 5 cm (2 in) deep in non-crusting soils. Seed press wheels can help ensure good seed–soil contact, which is especially important as temperatures increase to 21°C–27°C (70–80°F) and air pockets form in the soil. Sweet corn is often planted with belt or vacuum seeders for precise placement. Planting of individual cultivars is usually made on at least eight-row blocks in order to ensure successful pollination (Barker and Beuerlein, 2005).

Plant populations depend on the time of planting, cultivar, cultural practices, and availability of irrigation. Wider in-row plant spacing increases ear size, although total marketable yield may be reduced per unit area (Fig. 13.9). Overcrowding may reduce ear length and tip-fill.

Clear plastic, spunbound polyester, low tunnels or other covering materials are sometimes used in northern areas to enhance germination and seedling growth of early-season plantings of sweet corn (Westgate and Hazzard, 2005). These coverings are applied at planting to raise soil temperatures and conserve moisture to maximize germination of earlier season plantings. Total yields may not be affected but maturity dates can be 10–12 days earlier when soil coverings are used. The most common procedure is to plant the seeds in double rows and lay the plastic after planting. Perforated clear plastic is used to increase air circulation and to avoid leaf burn. The plastic may be left over the row if slitted types are used. Tunnels, solid row covers, and clear plastic mulch must be removed after emergence to allow proper development and prevent overheating.

Irrigation

Sweet corn has a shallow fibrous root system with most roots in the top 0.6 m (2 ft) of the soil. Consequently, it is important to keep adequate moisture in the root profile, especially during stand establishment and early growth. A sweet corn crop typically uses 38–50 cm (15–20 in) of water during a growing season for best growth without water stress. The soil must provide a corn crop with enough water to offset the amounts lost through transpiration. If these needs are not met, the plant will wilt. Tightly rolled leaves early in the morning indicate significant drought stress (Barker and Beuerlein, 2005). Water requirements of corn vary according to the stage of development, cultivar, and climate (Table 13.4).

Through the vegetative growth phase, corn is fairly tolerant of dry soils and drought stress. Mild drought stress during the vegetative phase may stimulate deep root development that will benefit the crop later in development. From 2 weeks before to the 2 weeks following pollination, corn is very sensitive to drought, however, and dry soils during this period may cause serious yield losses. Drought stress will delay maturity and cause uneven harvest. Corn reaches its peak water use during pollination when plants are silking (Table 13.4). Drought stress during

Fig. 13.9. Sweet corn production at wider spacing to increase ear size and facilitate hand harvest.

Table 13.4. Water use required for optimum growth of corn at different growth stages (Thomison et al., 2012).

Growth stage	Water use (in/day)
Prior to 12-leaf stage	<0.20
12-leaf	0.24
Early tassel	0.28
Silking	0.30
Blister kernel	0.26
Milk	0.24

tasseling will reduce pollination causing kernels near the tip of the ear to abort and not develop (poor tip-fill), thus reducing both yield and the appearance of the ears. Most of these losses result from pollination failure, and the most common cause is the failure of silks to emerge from the end of the ear. When this happens, the silks do not receive pollen and the kernels are not fertilized and will not develop (Thomison et al., 2012).

Supplemental irrigation or rainfall totaling at least 1 in/week is a good rule of thumb to promote steady growth and tasseling. Crops grown on light sandy soils would benefit from additional water. Overhead sprinkler and furrow irrigation are used for sweet corn. Furrow irrigation is limited to areas where the land is level.

Weed Management Strategies

Between-row weeds can be controlled by shallow cultivation. There are many effective herbicides for controlling broad leaf weeds since corn is a monocot. Local herbicide recommendations should be consulted before selecting herbicides. Atrazine (2-chloro-4-(ethylamino)-6-(isopropylamino)-s-triazine) controls pre- and post-emergence broadleaf and grassy weeds in corn. This chemical is effective and inexpensive, and thus has been widely used for weed control in corn. Atrazine degrades in soil primarily by the action of microbes, but the half-life of atrazine ranges from 13 to 261 days depending on environmental conditions (EPA, 2003). Therefore, if this chemical is used on sweet corn, carry-over effects may inadvertently affect subsequent crops planted in the same field. Transgenic herbicide-resistant corn has been developed to allow post-emergence weed control (Boerboom, 2006).

Growth and Development Phases of Corn

A growing plant of corn passes through the following distinct growth and development phases (Barker and Beuerlein, 2005). The later stages correspond to changes in the condition of the endosperm as the kernels mature.

Seedling stage

This is the sprouting stage that comes about 1 week after sowing as the seed germinates, the plant emerges from the soil and develops to the two- to four-leaf stage. Cold temperatures and moisture stress may delay germination and early seedling growth.

Grand growth stage

This is knee height stage of the plant that arrives about 35–45 days after sowing depending on environmental conditions. This is a critical stage of corn growth. Boosting the growth of plants at this stage by proper irrigation, fertilization, and weeding can significantly improve yield.

Tasseling stage

This is also called the flowering initiation stage when the tassels or male flowers open and pollen is released. Final top dressing should be done at this stage.

Silking stage

This is also known as cob initiation stage when the female flowers or cobs are formed. The silks (styles) emerge from the husks during this stage and fertilization begins after pollen grains land on the silks, germinate, and the pollen tubes grows through the style to fertilize each individual ovule attached to the cob.

Blister stage

The blister stage is when the endosperm in the small developing kernels is a clear liquid. During this stage, the pericarp is thin and easily ruptured with a thumbnail, releasing the watery endosperm. This is an immature preharvest stage.

Milk stage

This is the harvest stage for sweet corn. The milk stage corresponds to a change in endosperm from clear to a thick white liquid. During this stage, the pericarp can be ruptured with a thumbnail releasing the milky endosperm. Sweet corn should be harvested while the endosperm is milky for best quality before it hardens to a doughy consistency.

Soft dough stage

As the milky endosperm is absorbed it becomes a soft dough. During this stage the kernels reach maximum size and the pericarp is gummy when chewed. The silks protruding from the ends of the ears begin to turn dry and brown. The husks and flag leaves are dark green and fully developed but beginning to brown at the margins. The rows of kernels can be prominently felt when the husks are held firmly. Roasting ears are often harvested early in this stage of development

Hard dough stage

This is physiological maturity when the plants die and turn brown. The kernels at this stage are too hard for use as sweet corn but Indian corn, popcorn, and grain are harvested at this stage of development. Corn for propagation can be saved at this stage if the cultivar is open-pollinated and not an F-1 hybrid.

Harvesting and Marketing

Harvest stage

Sweet corn is seldom harvested or sold based on brix readings or percent moisture. For processing or for fresh market use, more subjective means such as visual appearance or the way the ear feels when grasped are often used to assess maturity. Fresh-market corn is normally harvested when less mature than corn for processing. The percent moisture for *su* or *se* sweet corn cultivars for processing range from 72% to 74% while fresh-market corn usually ranges from 75% to 77%. Sh_2 corn can be harvested as high as 78% moisture content for fresh market (Suslow and Cantwell, 2012).

Sweet corn, particularly for processing, is often harvested using the growing-degree-days (GDD) system. This is a maturity evaluation system based on the exposure of the crop to units of heat above a threshold temperature (Barker and Beuerlein, 2005). It is more accurate in determining sweet corn development than the days-to-maturity because corn growth and development is directly related to the accumulation of heat over time rather than the number of calendar days from planting. The GDD system has several advantages over the days-to-maturity system. The GDD system provides information that allows the grower to follow the progress of the crop through the growing season, and aids in planning harvest schedules (Barker and Beuerlein, 2005).

The GDD 86/50 cutoff method is commonly used to track corn maturity in North America. GDD are calculated as the average daily temperature minus 50.

$$GDD = (T_{max} + T_{min} \div 2) - 50$$

If the maximum daily temperature (T_{max}) is greater than 86°F, 86 is used to determine the daily average. Similarly, if the minimum daily temperature (T_{min}) is less than 10°C (50°F), 50 is used to determine the daily average. The high cutoff temperature (30°C/86°F) is used because growth rates of corn do not increase above 30°C (86°F). Growth at the low temperature cutoff (10°C/50°F) is already near zero, so it does not decline at lower temperatures. GDD are calculated daily and summed to define thermal time for a given period (Barker and Beuerlein, 2005). An early-maturing cultivar may require 1,400 GDD to reach harvest maturity while a late maturing cultivar may require 2,000 GDD or more.

As with any system, the GDD system has several shortcomings. Growing-degree-day ratings of cultivars with similar days-to-maturity do not always agree, especially for cultivars produced by different seed companies. This is usually because some seed companies start counting GDDs from the day of planting, while others begin from the day of emergence. Variation of 100–150 GDD may occur when values are calculated from planting rather than emergence. Although most seed companies use the 86/50 cutoff method, others use different methods to calculate GDDs. Also if planting is significantly delayed until later in the season or under extreme environmental stress, GDD requirements for maturity may be reduced significantly (Barker and Beuerlein, 2005).

Harvest methods

Corn grown for processing is often machine harvested, while fresh-market corn is harvested both by hand and by machine. Hand harvest offers the additional advantage of allowing multiple trips through the field to select ears most ready for harvest. Pick-your-own operations are popular in some areas and allow consumers to select their own ears in the field, which greatly reduces production costs, but may lead to greater waste.

Mechanical harvesters are expensive and are only warranted for very large operations (Fig. 13.10). This harvester pulls down on the ear so the shank is retained, giving each ear the appearance of hand-harvested sweet corn. Fast uniform germination and emergence is essential to obtain the uniform maturity needed for mechanical harvesting so that all ears can be harvested at the same maturity. "Good" yields per acre are 12,000–14,000 ears for fresh market and 11–13 metric tonnes/ha (5–6 short tons/acre) for processing.

Postharvest handling

After harvest, sugar in sweet corn with the *su* gene is very rapidly converted from sugar to starch. To slow this conversion, the ears must be immediately cooled to near freezing if high quality is maintained. Cooling is usually accomplished commercially by hydrocooling or icing. For most commercial operations, ears are drenched with ice-cold water to remove field heat immediately after harvest and maintained under refrigeration until the time of sale (Brecht, 2004). There is more flexibility in the postharvest handling of *se* and sh_2 cultivars but they must also be cooled to maximize shelf life and quality after harvest. Corn is a warm-season, frost-intolerant crop, but is not chilling sensitive when exposed to temperatures less than 13°C (55°F) like many other warm-season vegetable crops.

Marketing trends

Fresh sweet corn is traditionally sold in the husk in many markets. At harvest, the silks are brown but

Fig. 13.10. Once-over, destructive mechanical sweet corn harvester.

the husks should be bright green. Some markets prefer that the flag leaves and even the shank remain attached during marketing.

Consumer preferences have changed in recent years in many markets. Today, many consumers prefer the convenience of husked and cleaned ears sold for fresh market. Cleaned ears have the husks and silks removed so that the ears can be rinsed and immediately cooked with minimal preparation. Husked corn is appealing because the quality of the ears can be better evaluated before purchase without removing the husks. Precleaning also eliminates the need to remove and dispose of the husks prior to cooking. Marketing husked and cleaned ears relieves the grower of the burden of keeping the husks looking fresh, attractive, and free of insect damage. Shrink-wrapped packages of cleaned ears are frequently sold with the tips clipped to remove waste and to hide poor tip-fill or insect damage, which most commonly occurs at the end of the ear. Shrink-wrapping may also generate a modified atmosphere to extend shelf life while protecting the ear from physical damage during marketing.

Diseases

Foliar fungal diseases may reduce yield by reducing photosynthetic capacity of the leaves. Fungal disease may also reduce eating quality and marketability because of damage to kernels and their unattractive effects on the appearance of the husks, respectively. Fungal diseases of corn include anthracnose (*Colletotrichum graminicola*), common rust (*Puccinia sorghi*), downy mildews (*Peronosclerospora* spp. and *Sclerophthora* spp.), northern corn leaf blight (*Exserohilum turcicum*), southern corn leaf blight (*Bipolaris maydis*), southern rust (*Puccinia polysora*), tropical rust (*Physopella zeae*), and yellow leaf blight (*Mycosphaerella zea-maydis*) (Sherf and MacNab, 1986). Other fungal diseases include common smut (*Ustilago maydis*) and head smut (*Sphacelotheca reiliana*) (Sherf and MacNab, 1986). Corn smut develops into grotesque-looking fungal masses that protrude from infected ears, leaving them unmarketable. In Mexico and other parts of the world, smut is a delicacy that is removed from the ear and eaten like a mushroom before the spores are released (Fig. 13.11).

Corn stalk rot is caused by *Diplodia maydis* or *Fusarium* spp. (Rubatzky and Yamaguchi, 1997). Soil-borne fungi such as *Pythium* and *Fusarium* spp. cause seed death and damping off prior to emergence (Rubatzky and Yamaguchi, 1997).

Fig. 13.11. Corn smut is a fungal disease that infects ears but is also consumed as a delicacy similar to mushrooms.

Stewart's bacterial wilt (*Erwinia stewartii*) is a bacterial disease transmitted by corn flea beetles that can significantly reduce yield. Cultivars resistant to Stewart's wilt are available. Sweet corn is susceptible to a number of viral diseases including cucumber mosaic virus, maize chlorotic dwarf virus, maize dwarf mosaic virus, maize mosaic virus, maize rough dwarf virus, maize streak virus, and sugarcane mosaic virus. Genetically engineered resistance to maize streak virus has been developed (Shepherd *et al.*, 2007). Aphids are common vectors of viral disease and virus-resistant cultivars have been developed. Other diseases that appear similar to virus diseases include corn bush stunt, caused by a mycoplasma, corn stunt, caused by a spiroplasma, and maize wallaby ear, a disease induced by feeding of the leafhopper *Cicadulina bipunctata* (Sherf and MacNab, 1986).

Insect Pests

Caterpillars are an important pest of sweet corn. Lepidopteran larvae feed on ears in many areas of the world (Rubatzky and Yamaguchi, 1997; Flood *et al.*, 2005). The European corn borer is one of the most aggressive caterpillars that feed on sweet corn because they can bore directly through the husk and may damage any part of the ear. Once inside the ear and protected by the husk, they are difficult to control.

Bacillus thuringiensis (Bt) is a naturally occurring, insect-specific (entomopathogenic) soil bacterium. Bt produces a crystal protein that is toxic to specific groups of insects including lepidopteran larvae, in particular the European corn borer (Tanada and Kaya, 1993). The protein is selective, generally not harming insects in other orders (such as beetles, flies, bees, and wasps). Bt, applied as a spray using conventional equipment, has been available as a commercial microbial insecticide since the 1960s and is sold under various trade names. Since the bacteria must be eaten by the larvae to be effective, good spray coverage is essential for control.

Bt sweet corn cultivars have been genetically engineered to express the bacterial Bt toxin by inserting a gene from the microorganism Bt into the sweet corn genome. This gene codes for the protein toxin, which causes the formation of pores in the lepidopteran larval digestive tract. These pores allow naturally occurring enteric bacteria, such as *E. coli* and *Enterobacter*, to enter the hemocoel, where they multiply and cause sepsis, killing the caterpillar (Broderick *et al.*, 2006). To kill a susceptible insect, a part of the plant that contains the Bt protein (not all parts of the plant necessarily contain the protein in equal concentrations) must be ingested.

Corn growers in the USA who plant Bt corn must by law plant non-Bt corn nearby. These non-Bt fields are often called refuges to harbor pests in an effort to slow the evolution of the pests' resistance to the Bt toxin. Resistance to Bt is likely to occur because insects will evolve a recessive allele by natural genetic mutation that will give resistance to Bt toxin. If a resistant pest is feeding on non-Bt corn in the nearby refuge, the resistance is neutral and offers no advantage to the pest over a nonresistant pest. Having a mixture of breeding pests nearby that are not resistant increases the chances of the resistant pests choosing a mate that is nonresistant to Bt. Since the gene is recessive, all offspring will have only one copy of the resistance gene and will not be resistant to Bt. Using this method, scientists and farmers hope to keep the number of resistant genes very low to slow the spread of resistance genes (Jaffe, 2009). However, transgenic sweet corn expressing the bacterial Bt toxin is not widely grown in the USA despite government approval, because of consumer concerns over genetically engineered vegetables.

Besides the European corn borer, other worms eat the silks and enter the open end of the husk, eating the kernels near the tip first and continue feeding toward the base of the ear. Worms that attack sweet corn include armyworms (*Spodoptera frugiperda* and *Pseudaletia unipuncta*), corn earworm (*Heliothis zea*), Asian corn borer (*Ostrinia furnacalis*) (He *et al.*, 2002), western bean cutworm (*Richia albicosta*), and the pink stem borer (*Sesamia nonagrioides*) (Velasco *et al.*, 2002). The seed corn maggot (*Hylemya platura*) can attack seeds or young seedlings. Root-feeding insects include wireworms (*Melanotus* spp.), white grubs (*Phyllophaga* spp.), and rootworms (*Diabrotica* spp.). Cutworms (*Agrotis* and *Feltia* spp.) and variegated cutworm (*Peridroma saucia*) can significantly reduce stands by feeding on young seedlings.

Corn flea beetles (*Chaetocnema pulicaria*) feed on foliage, silks, and exposed ends of ears, but more seriously, transmit the bacterium *E. stewartii* which causes Stewart's bacterial wilt. Other insect pests include the corn leaf aphid (*Rhopalosiphum maidis*), grasshoppers (*Melanoplus* spp.), spider mites (*Tetranychus urticae*), wheat curl mite (*Aceria tosichella*), stalk borer (*Papaipeema nebris*), tarnished plant bug (*Lygus lineolaris*), sap beetles (*Carpophilus* spp. and *Glischrochilus quadrisignatus*), and thrips (*Anaphothrips obscurus*) (Flood *et al.*, 2005).

Vertebrate Pests

Rodents and birds may feed on planted corn seed before or after germination, significantly reducing the sweet corn population in some areas. The developing ears are attractive food for birds, skunks, squirrels, raccoons, and deer and their feeding may result in significant yield losses particularly in small isolated fields (Barclay, 1996). High fences, electric fences, and noise-making devices may provide some degree of control.

References

Barclay, J.S. (1996) Animal pests of sweet corn. In: Adams, R.G. and Clark, J.C. (eds) *Northeast Sweet Corn Production and Integrated Pest Management Manual.* University of Connecticut Cooperative Extension System, Storrs, Connecticut, pp. 93–101.

Barker, D. and Beuerlein, J. (2005) Corn Production. In: *Ohio Agronomy Journal*, 14th edn. Ohio Cooperative Extension, Columbus, Ohio, pp. 31–56.

Bar-Zur, A. and Saadi, H. (1990) Prolific maize hybrids for baby corn. *Journal of the American Society for Horticultural Science* 65, 97–100.

Beadle, G.W. (1981) Origin of corn: Pollen evidence. *Science* 213, 890–892.

Boerboom, C. (2006) Pest Resistant Sweet Corn and Other Herbicide Developments. *Proceedings of the 2006 Wisconsin Fertilizer, Aglime & Pest Management Conference*, Volume 45, 238–243.

Brecht, J.K. (2004) Sweetcorn. In: Gross, K.C., Wang, C.Y. and Saltveit, M. (eds) *The Commercial Storage of*

Fruits, Vegetables, and Florist and Nursery Stocks. Beltsville, Maryland.

Broderick, N., Raffa, K. and Handelsman, J. (2006) Midgut bacteria required for *Bacillus thuringiensis* insecticidal activity. *Proceedings of the National Academy of Sciences* 103, 15196–15199.

Chutkaew, C. and Paroda, R.S. (1994) *Baby corn production in Thailand – a success story*. FAO Regional Office for Asia & the Pacific, Asia Pacific Association of Agricultural Research Institutions, APAARI Publication.

Darrah, L.L., McMullen, M.D. and Zuber, M.S. (2003) Breeding, genetics, and seed corn production. In: Ramstad, P.E. and White, P. (eds) *Corn: Chemistry and Technology*. American Association of Cereal Chemists, Minneapolis, Minnesota, pp. 35–68.

EPA (2003) *Interim Reregistration Eligibility Decision for Atrazine Case No. 0062 Report*. Office of Prevention, Pesticides and Toxic Substances, United States Environmental Protection Agency, Washington, DC.

FAO (2011) Online Database of Crop Production Statistics 2011. Food and Agriculture Organization (FAO), Rome. Available at: http://aostat.fao.org/site/567/DesktopDefault.aspx?PageID=567#ancor (accessed 27 December 2011).

Flood, B.R., Foster, R., Hutchison, W.D. and Pataky, S. (2005) Sweet corn. In: Foster, R. and Flood, B.R. (eds) *Vegetable Insect Management*. Meistermedia Worldwide, Willoughby, Ohio, pp. 39–63.

Fonseca, A.E., Westgate, M.E., Grass, L. and Dornbos, D.L., Jr. (2003) Tassel morphology as an indicator of potential pollen production in maize. Available at: www.plantmanagementnetwork.org/pub/cm/research/2003/tassel (accessed 15 December 2013).

Galinat, W.C. (1992) Evolution of corn. *Advances in Agronomy* 47, 203–231.

Gardner, F.P., Pearce, R.B. and Mitchell, R.L. (1985) *Physiology of Crop Plants*. Iowa State University Press, Ames, Iowa.

Goodman, M.M. (1988) The history and evolution of maize. *CRC Critical Reviews in Plant Science* 7, 197–220.

Hallauer, A.R. (2001) *Specialty Corns*, 2nd edn. CRC Press, Boca Raton, Florida.

He, K., Zhou, D., Wang, Z., Wen, L. and Bai, S. (2002) On the damage and control tactics of Asian corn borer *Ostrinia furnacalis* Guenee in sweet corn field. *Acta Phytophylacica Sinica* 29, 199–204.

Jaffe, G. (2009) *Complacency on the Farm: Significant Noncompliance with EPA's Refuge Requirements Threatens the Future Effectiveness of Genetically Engineered Pest-protected Corn*. Center for Science in the Public Interest, Washington, DC.

Lusas, E.W. and Rooney, L.W. (2001) *Snack Foods Processing*. CRC Press, Boca Raton, Florida.

Marshall, R.O. and Kooi, E.R. (1957) Enzymatic conversion of D-glucose to D-fructose. *Science* 125, 648–649.

Matsuoka, Y., Vigouroux, Y., Goodman, M.M., Sanchez G.J., Buckler, E. and Doebley, J. (2002) A single domestication for maize shown by multilocus microsatellite genotyping. *Proceedings of the National Academy of Sciences* 99, 6080.

Miles, C.A. and Zenz, L. (2000) Baby Corn. Available at: http://cru.cahe.wsu.edu/CEPublications/pnw0532/pnw0532.pdf (accessed 20 August 2012).

Ranere, A.J., Piperno, D.R., Holst, I., Dickau, R. and Iriarte, J. (2009) The cultural and chronological context of early Holocene maize and squash domestication in the Central Balsas River Valley, Mexico. Available at: www.pnas.org/content/106/13/5014.full (accessed 15 December 2013).

Rubatzky, V.E. and Yamaguchi, M. (1997) *World Vegetables*. Chapman & Hall, New York.

Sears, F.B. (1982) Fossil maize pollen in Mexico. *Science* 216, 932–934.

Shepherd, D.N., Mangwende, T., Martin, D.P., Bezuidenhout, M., Rybicki, E.P. and Thomson, J.A. (2007) Maize streak virus-resistant transgenic maize: a first for Africa. *Plant Biotechnology Journal* 88, 325–336.

Sherf, A.F. and MacNab A.A. (1986) *Vegetable Diseases and Their Control*. John Wiley & Sons, New York.

Suslow, T.V. and Cantwell, M. (2012) Corn, Sweet: Recommendations for Maintaining Postharvest Quality. Available at: http://postharvest.ucdavis.edu/pfvegetable/CornSweet (accessed 25 September 2012).

Tanada, Y. and Kaya, H.K. (1993) *Insect Pathology*. Academic Press, San Diego, California.

Thomison, P., Lipps, P., Hammond, R., Mullen, R. and Eisley, B. (2012) Corn Production. In: Ohio Agronomy Guide, 14th edn, Bulletin 472-05, The Ohio State University Cooperative Extension Service, Columbus Ohio. Available at: http://ohioline.osu.edu/b472/0005.html (accessed 20 August 2012).

Tracy, W.F. (2000) Sweet corn. In: Hallauer, A.R. (ed.) *Specialty Corns*, 2nd edn. CRC Press, Boca Raton, Florida, pp. 155–199.

Tracy, W.F., Whitt, S.R. and Buckler, E.S. (2006) Sugary1 and the origin of sweet maize. *Crop Science* 461, 49–54.

USDA (2012) Sugar and sweeteners outlook SSS-M-289. Available at: http://usda.mannlib.cornell.edu/MannUsda/viewDocumentInfo.do?documentID=1386 (accessed 25 September 2012).

USDA Nutrient Database (2011) The USDA Nutritional Database for Standard Reference. United States Department of Agriculture, Washington, DC. Available at: http://ndb.nal.usda.gov (accessed 11 July 2011).

Velasco, P., Revilla, P., Butrón, A., Ordás, B., Ordás, A. and Malvar, R.A. (2002) Ear damage of sweet corn inbreds and their hybrids under multiple corn borer infestation. *Crop Science* 42, 724–729.

Welbaum, G.E., Frantz, J.M., Gunatilaka, M.K. and Shen, Z.X. (2001) A comparison of the growth, establishment, and maturity of direct-seeded and transplanted sh_2 sweet corn. *HortScience* 39, 261–265.

Westgate, P. and Hazzard, R. (2005) New England Sweet Corn Crop Profile. Available at: http://ipmcenters.org/cropprofiles/docs/NewEnglandSweetcorn.pdf. (accessed 22 August 2012).

14 Family Amaryllidaceae, Subfamily Allioideae

Origin and History

Onion originated in Middle Asia and was domesticated in what are today Afghanistan, Iran, and Pakistan. Onion is a very ancient crop and has been under widespread cultivation dating back to as early as 600 BC. Onions were a popular food of the Greeks and Romans as early as 400–300 BC and were introduced into northern Europe about AD 500 at the start of the Middle Ages (Zohary and Hopf, 2000). Production occurs worldwide but the greatest concentration is in the northern hemisphere. In the tropics and much of Southeast Asia unfavorable climate and handling conditions limit onion production so shallots are preferred. Shallots are believed to be native to Asia, explaining their popularity in this region.

Garlic is believed to be of middle Asian origin with a history of human use of over 7,000 years (Ensminger, 1994). The culture of garlic parallels that of onion. Greek author Homer mentioned garlic in the ninth century BC (Zohary and Hopf, 2000). Garlic was worshipped by the ancient Egyptians who used it for both culinary and medicinal purposes, chewed by Greek Olympic athletes, and thought to be essential for keeping vampires at bay during the Middle Ages. The Spanish, Portuguese, and French introduced onions and garlic to the New World.

Botany and Life Cycle

Allioideae is the relatively new botanical name of a monocot subfamily in the family Amaryllidaceae, order Asparagales. Allioideae was formerly a separate family named Alliaceae. The subfamily name is derived from the generic name of the genus *Allium*, which includes several important vegetables: *A. cepa* (onion, shallot, top-set onion, multiplier onion), *A. sativum* (garlic), *A. ampeloprasum* (elephant garlic or great head leek), *A. schoenoprasum* (chive), *A. fistulosum* (Welsh onion, Japanese bunching onion), *A. chinense* (rakkyo), *A. tuberosum* (Chinese chive), and *A. cepa* × *A. fistulosum* (Beltsville bunching onion).

ONIONS AND GARLIC

Onion

Onion is an herbaceous biennial monocot grown commercially as an annual or by means of its bulb, a perennial. Onion is a cool-season crop that will tolerate light frosts but not prolonged periods far below freezing. High temperatures negatively affect bulb shape and quality. Onion morphology and development is novel in comparison to other vegetables. The leaf blades are tubular, slightly flattened on the upper side, hollow, and have a closed tip (Fig. 14.1).

Each leaf consists of a blade and sheath. The sheath forms a tube that encloses and protects younger leaves and the shoot apical meristem. The leaf blade and sheath join to form a pore for the succeeding leaf blade to emerge through. Leaves are initiated alternately, opposite one another so that all leaves lie in a single plane, and arise from a short, compressed stem (Figs 14.1, 14.2). Together, the composite leaf sheaths form the pseudostem between the leaf blades and stem.

Each leaf is larger than the preceding one until bulbing begins, and newly formed leaves become progressively shorter giving rise to bladeless storage leaves. Thus, the onion bulb consists of a vegetative stem, axis, and the bases of the concentric storage and vegetative leaves (Fig. 14.2a). Onion skins are the dry paper-like outermost leaf scales that lose their fleshiness during bulbing.

Bulb shapes range from spherical to nearly cylindrical and include flat and cone-like bulbs (Fig. 14.2b). Size variation is considerable as is skin color, which may be white, yellow, brown, red, or purple.

Fig. 14.1. Young onion plants in a production field. Onion leaves are hollow, upright, and form opposite to each other. New leaves emerge from the center of the plant.

Other features such as pungency and dry matter are important characteristics. Each of these traits is genetically determined but can be altered by environmental conditions (Rabinowitch and Currah, 2002).

Onion roots are shallow and only extend 15–20 cm (5.9–7.9 in) below the surface and less than 50 cm (20 in) horizontally. After germination, seedlings initially produce a primary root, but otherwise all roots are adventitious. Roots are initiated from the short disk-like stem at the base of the leaves. The roots grow downward, emerging through the stem disk. Roots rarely branch, have root hairs, or increase in diameter. Onion roots are short lived, but are continuously produced at a rate of three to four per week as old roots die. During early growth, the number of active roots increases, but at bulb maturity, roots

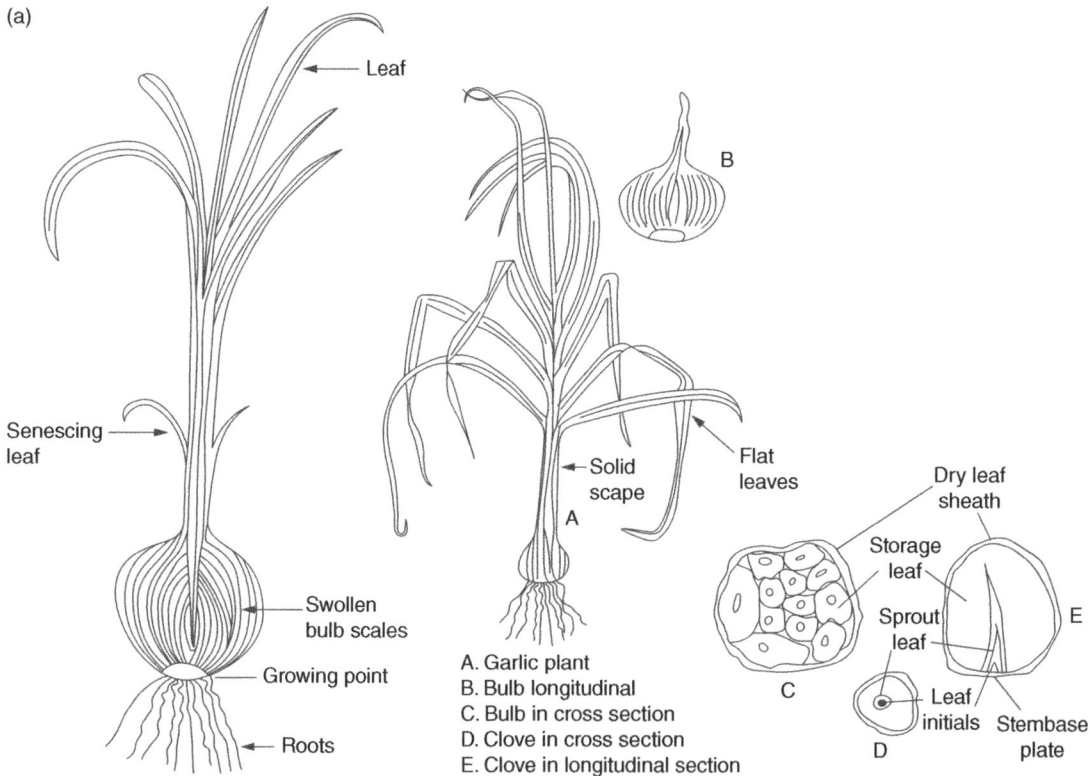

Fig. 14.2. (a) The drawing on the left shows an onion plant in cross section. The right diagram shows a garlic plant and its various component tissues. (b) The shape and size of onion bulbs varies widely among cultivars.

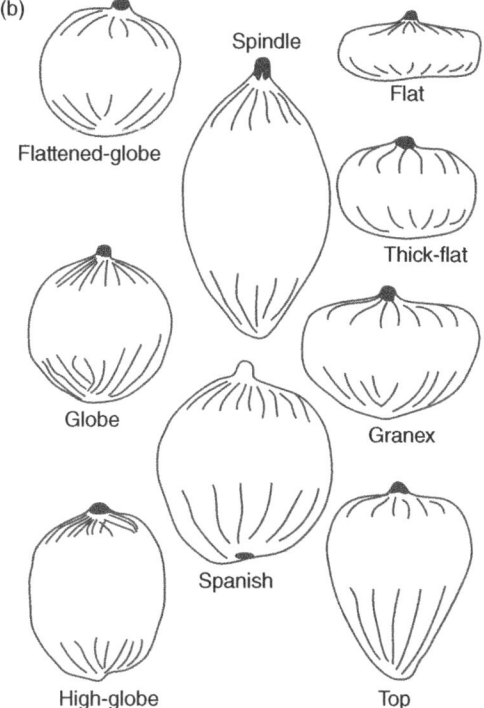

Fig. 14.2. Continued.

Fig. 14.3. Bolting onion plants in a seed production field. These plants show little bulb enlargement because flowering inhibits bulb production.

die more rapidly than they are formed (Rubatzky and Yamaguchi, 1997).

The terminal inflorescence develops from the apical meristem after vernalization. Flower stalks or scapes, one to several, generally elongate well above the leaves and range in height from 30–100 cm (12–39 in) or more (Fig. 14.3). The scape is the stem internode between the base of the inflorescence sheath and the last foliage leaf. Initially, the scape is solid but becomes thin-walled and hollow. The onion scape has a characteristic bulge at the lower third of its length. The number of scapes that develop depends on the number of sprouted lateral buds. A spherical flower structure called an umbel, ranging from 2–15 cm (0.8–6.0 in) in diameter, is borne atop each scape (Fig. 14.4). During early development, the inflorescence is initially enclosed within a spathe or sheath, which opens during development to expose florets. The umbel is an aggregate of many florets at different stages of development.

An umbel typically contains 200–600 small florets, but the number can range from 50 to 1,000 (Rubatzky and Yamaguchi, 1997). The flowering period may last 4 or more weeks, and individual florets are fertile for 1 week. Infrequently, bulbils are produced at the top of the inflorescence. Flowers are perfect with six white petals, six stamens, and a three-carpel pistil. Protandry (pollen shedding before the stigma is receptive) promotes out-crossing and a reliance on insect pollinators (George, 2009).

Garlic

Garlic (*A. sativum*) is a bulbous plant that grows to 0.6 m (2 ft) in height. Its USDA hardiness zone is 8, which means it can tolerate at least brief exposure to temperatures as low as −7°C and −12°C (19°F and 10°F). Garlic is an adaptable species, with some cultivars that grow well in cold climates, although much of the commercial production occurs in mild climates. Garlic cultivars are often referred to as clones, since they are vegetatively propagated and are genetically the same. Garlic requires a particular temperature and photoperiod for proper clove and bulb formation.

Family Alliaceae

Fig. 14.4. Onion seed field with umbels maturing on the scape at the top of each plant. This field will be combined in a few weeks for seed harvest.

There are two subspecies of garlic, *A. sativum* subsp. *ophioscorodon* and *A. sativum* subsp. *sativum*. Subspecies *ophioscorodon* is referred to as "hardneck" garlic, and is not significant for commercial production in the USA (Rosen *et al.*, 2008). Subspecies *sativum* is commonly called "softneck" garlic, which is the predominant commercial type grown in North America. While both types begin growth with leaf production, the hardneck types have a greater tendency to bolt and produce scapes. Some hardneck clones bolt reliably while others rarely if ever do (Rosen *et al.*, 2008).

The two most popular cultivars grown in the USA are 'California Early' and 'California Late'. Both are softneck types. 'California Early', which takes about 1 month less to mature than 'California Late', is grown primarily for dehydration. 'California Late' is usually grown for fresh market (WIPMC, 2004).

Allium sativum also grows wild and has become naturalized. The wild, crow, and field garlic of Britain are members of the species *A. ursinum*, *A. vineale*, and *A. oleraceum*, respectively. In North America, *A. vineale* and *A. canadense* (meadow or wild garlic or wild onion) are common weeds (McGee, 2004). Elephant garlic is a type of leek (*A. ampeloprasum*) and not true garlic.

Although resembling onion in growth and appearance, garlic differs from onion in several important characteristics. Like onion, garlic's disk-like stem is very short, but the adventitious root system is more extensive than onion's. Garlic has a solid, flattened, V-shaped, folded leaf blade. Another difference is that only the bladeless storage leaf of the clove stores reserves and not the leaf bases of garlic foliage (Fig. 14.2a; Rubatzky and Yamaguchi, 1997).

Garlic scapes are straight and solid, but vary in height because of differences among clones and growing conditions. The garlic umbel inflorescence develops at the top of the scape, is subspherical, and usually contains only bulbils or a combination of bulbils and flowers, which rarely if ever set seed in commercial clones. The infrequently formed flowers are lavender and usually wither and abort. However, bulbils can also form on the scape depending on the cultivar (Rosen *et al.*, 2008). With some bolting hardneck garlics, an inflorescence may not be evident, and bulbils are produced within the pseudostem just above the bulb. Over the past 25 years, researchers in Germany, Japan, and the USA have studied the genetics of garlic and reliably obtained viable seeds from certain clones

(Rabinowitch and Currah, 2002; Volk *et al.*, 2004). Plants most often producing true seed belong to subsp. *ophioscorodon* (Pooler and Simon, 1994).

The garlic bulb or head is more complex than onion and consists of a grouping of sessile lateral bulbs called cloves (Fig. 14.5). The number of cloves formed is quite variable and ranges from 1 to more than 25.

Most growers consider an average of eight to ten cloves ideal but many more may be produced (Fig. 14.5). Cloves are usually ovoid to ellipse-oblong and each one consists of two mature leaves. One is a paper-thin protective sheath that encloses a single second thickened storage leaf that contains a single central vegetative bud. Cloves are initiated preferentially from these vegetative buds located in the leaf axils. Each bud primordia forms from two to six growing points. Any of the primordia can develop into a lateral bud and subsequently a clove. When conditions are favorable for growth, and after completion of the rest period, the vegetative bud will sprout and begin developing into a clove. Irregularly shaped bulbs are caused by cloves formed in leaf axils further from the center, often as a result of an excessive period of low-temperature exposure (Rubatzky and Yamaguchi, 1997). Insufficient exposure to low temperatures results in no bulb formation and this often occurs when garlic is grown at constant high temperatures. The storage leaf, which accounts for most of the clove size, is fleshy and bladeless. The bulb or head consists of several cloves, each originating from a separate lateral bud and enclosed by the sheaths of the foliage leaves from the mother clove. Younger leaves emerge inside older ones of the pseudostem formed by the sheathing bases of successive leaves. The plant's outermost older leaves develop into the smooth parchment-like protective sheath surrounding the entire developing bulb or head. The uppermost apical meristem either forms a flower stalk (scape) or the final leaf (Rubatzky and Yamaguchi, 1997).

Types and Cultivars

Onion cultivars

Onion production can be divided into two categories: immature green and bulb. Separate cultivars are grown for each application. The leaf sheaths are eaten whole or after chopping and are used to flavor other fresh and processed foods. Green onions can also be chopped and preserved by refrigeration, drying, or freezing. The leaves of immature plants can be eaten as well but become fibrous with maturity (Smith *et al.*, 2011a).

Bulb onions take longer to mature. The bladeless leaves and leaf sheaths at the base of the plant enlarge in response to a critical daylength and the response is cultivar dependent. Bulbs are used fresh, dried, or processed.

There are many horticultural characteristics that distinguish plants of various onion cultivars, including uniformity among plants of the same cultivar, disease resistance, amount of "bloom" (a waxy layer on top of the cuticle that gives the leaves a whitish appearance), foliage color (varies from green to bluish green), leaf size and height (small and compact or large), yield, ease of bolting, and maturity. Maturity is complex and is determined by daylength as well as planting time. The bulbing response is often generally classified but ultimately is determined by the daylength requirement of each cultivar (Table 14.1). The classified seasonal maturity, e.g. early, midseason, etc., depends on specific minimum number of hours of daylight required for bulbing to begin.

Some important characteristics that differentiate cultivars and define bulb quality include skin color (ranges from tan to yellow, white, brown, purple, or red), size (small, large, or jumbo), shape (Fig. 14.1), uniformity of size, interior color (green, white, purple, red, or yellow), pungency, storage quality, skin thickness, solidity (cultivars vary in bulb density, and both soluble and insoluble solids), neck size (thinner allows bulbs to seal upon drying so

Fig. 14.5. Heads of dried garlic containing several cloves each.

Table 14.1. Maturity classes of onions are based on day length.

Maturity class	Day length (h)	Production location
Early	About 12 (short day)	Generally grown at lower latitudes
Midseason	13.5–14 (intermediate)	Mid-latitudes
Late	15	Mid-latitudes
Very late	Over 15 (long day)	Higher latitudes

pathogens cannot enter), durability and retention of skin (the skin protects the inner bulb so excessive sloughing is undesirable), inner leaf thickness (thin or thick), and freedom from doubles (doubles do not make suitable onion rings).

Pungency

Onion cultivars vary in their degree of pungency. Environmental factors also affect the degree of pungency (Yoo et al., 2006). Cultivars are often rated as sweet, mild, pungent, or strong. Onions and garlic contain no starch but accumulate sugars instead, which contribute to sweetness.

Methyl and propyl disulfides, and thiopropanyl sulfoxide in particular, are responsible for pungency (Rabinowitch and Currah, 2002). When a pungent onion bulb is sliced, chopped, or damaged, volatile sulfur compounds are released by enzymes called alliinases that break down amino acid sulfoxides and generate sulfenic acids. Sulfenic acid is broken down by a second enzyme, lachrymatory factor synthase, producing syn-propanethial-S-oxide, a volatile gas widely known as the onion lachrymatory factor (LF) (Block, 2009). When LF diffuses through the air and contacts the human eye, sensory neurons are activated, creating a stinging sensation and activating tear glands in an attempt to dilute and flush away the irritant. Cutting onions submerged under water prevents the release of volatile compounds responsible for eye irritation. The enzymatic production of LF is slower in cold onion so refrigeration before use also may reduce tears. By measuring pyruvic acid content, the pungency of onion can be chemically assessed (Yoo et al., 2006).

There is no genetic relationship between color and pungency in onions. Bud scales of bulb onions may be yellow or brown, white or purple and cultivars with any scale color can be either mild or pungent. Pungent types usually have longer shelf life than sweet cultivars.

F-1 hybrids

Many popular modern onion cultivars are F-1 hybrids developed through traditional plant breeding. In the middle of the 20th century, onion breeder Henry Jones pioneered development of low-cost hybrids using male sterility to cross inbreed lines without costly and time-consuming hand pollination. F-1 hybrid cultivars are more uniform and higher yielding than open-pollinated onions. Transgenic onions or garlic have not had an impact on world production (Rabinowitch and Currah, 2002).

Garlic cultivars

Currently all true garlic (A. sativum) is clonally propagated from cloves for commercial production. Private companies have begun garlic breeding using plants found in Asia that produce true seed. Seed propagation of garlic would have a significant impact on the management of clove storage, planting, and virus and nematode control. Because plant breeding has not been used for the genetic improvement of garlic until recently (Rabinowitch and Currah, 2002), the choice of cultivars is limited compared to other vegetables, because new cultivars are created by mutation and selection. Garlic cultivars vary with respect to plant size (compact to large), maturity, yield, disease resistance, and scape development. Bulb characteristics vary with respect to color (red, white, or tan), smoothness, shape, clove size, clove color and clove number, storage life, and pungency. The following section describes some popular garlic clones (Brewster, 2008).

'Artichoke' is a productive softneck type for cold climates that may partially bolt following cold winters. After cold treatment, bulbils may form just above the bulb, reducing market value. In a mild winter only 1–2% of plants will bolt, but following a cold winter without snow cover, 70–100% will bolt. Bulb color varies from all white to a purple blush, while bulbils when present are usually purple. Bulbs typically contain 12–20 cloves and 0.45 kg (1 lb) of bulbs will produce about 80 cloves. Cloves are difficult to peel. Bulbs can be stored for 6–9 months. Related cultivars selected from 'Artichoke' include 'Inchellium Red', 'California Early', 'Susanville', 'California Late', 'Early Red Italian', 'Machashi', and 'Red Toch' (Rosen et al., 2008).

'**California Late**' (CL) is one of the most widely grown garlic clones produced in California. It produces numerous pink or pink-brown cloves, in smooth, medium-sized white bulbs. Plants have small stems with narrow upright leaves that mature in August following establishment in October of the previous year. 'California Late' is a medium yielding cultivar with high solids and long storage life and is the leading US commercial cultivar (WIPMC, 2004).

'**California Early**' matures 2–3 weeks ahead of CL. 'California Early' plants are large, more vigorous than CL and better yielding. However, the bulbs are large and rough, containing tan-colored cloves with lower solids that do not store as well as CL (WIPMC, 2004).

'**Creole**' matures bulbs 1 month ahead of 'California Early'. Cloves are small with a deep purple skin. Keeping quality and solids content are intermediate between the California cultivars described above. 'Creole' is commonly grown in Mexico, the Imperial Valley of California, and Louisiana. Selections include 'Ajo Rojo', 'Burgundy', and 'Creole Red' (Rosen et al., 2008).

'**Elephant**' (*A. ampeloprasum*) produces heads and cloves that are larger than true garlic. 'Elephant' is large, mild, early maturing, and easier to grow than many other *A. sativum* clones, particularly in the eastern USA. Elephant garlic grows to about 0.9 m (3 ft), has a distinctive milder flavor, and does not store as well as many of the *A. sativum* clones (Rosen et al., 2008).

'**Rocambole**' produces a moderate-sized plant 0.9–1.2 m (3–4 ft) tall with the scape unfurled. The scape coils two to three times before becoming straight. 'Rocambole' performs well in cold climates. Bulbs are off-white with purple streaks and are prone to double cloves. Bulbils are numerous and generally purple. Clove skins are brownish and easy to peel. Bulbs have a storage life of about 4–5 months. Named selections include 'German Red', 'German Brown', 'Spanish Roja', 'Russian Red', 'Killarney Red', and 'Montana Giant' (Rosen et al., 2008).

'**Purple Stripe**' is a moderately sized plant 0.9–1.5 m (3–5 ft) tall with the scape uncurled, which is well adapted to cold climates. The scape forms a characteristic ¾ coil or a downwards "U" shape before growing straight. Bulbs have purple streaks, while bulbils are numerous and generally purple. Clove skins are brownish and relatively difficult to peel. The cloves are rarely double. Bulbs have a storage life of about 5–7 months. A typical bulb contains 8–12 cloves and 0.45 kg (1 lb) of garlic will produce about 60 cloves. Selections from 'Purple Stripe' include 'Chesnok Red' and 'Persian Star' (Rosen et al., 2008).

'**Porcelain**' is an older cultivar well adapted to cold climates. It reaches a height of 1.2–1.8 m (4–6 ft) with the scape unfurled. The scape has loose and somewhat random coils. Bulbs are large, typically containing four to six cloves. The large clove size is great for cooks, but growers need to save more for propagation. Clove skins are smooth and white and tend to be more difficult to peel than 'Rocambole'. Double cloves are rarely produced. Bulbils are numerous, small and generally white. Bulbs may be stored for 5–7 months. A weight of 0.45 kg (1 lb) of garlic will produce about 35 cloves. There are many selections of 'Porcelain' including 'Romanian Red', 'Georgian Crystal', 'Music', 'Polish Hardneck', 'Zemo', 'Georgian Fire', 'Northern White', 'German White', and 'Krasnodar White' (Rosen et al., 2008).

'**Asiatic**' plants grow to only about 0.9 m (3 ft) tall when the scape is fully-grown. 'Asiatic' is a hardneck type that performs well in cold climates. Cold exposure causes the formation of a scape with a long characteristic bulbil capsule. The dark purple bulbils are much larger than those produced on other clones. There are usually four to eight large single and double cloves per bulb. Cloves are brown and bulb color varies from white to pink to purple striped. Clove skins are somewhat tight and difficult to peel and are prone to splitting through the bulb skins if harvest is delayed. Bulbs typically can be stored for 5–7 months. Named selections include 'Asian Tempest', 'Japanese', 'Wonha', 'Sakura', and 'Pyong Vang' (Rosen et al., 2008).

'**Silverskin**' is a true softneck type with only a few flower stalks forming after a cold winter. The lack of a flower stalk makes this clone popular for braiding. Clove number per bulb ranges from 8 to 40, and 0.45 kg (1 lb) of bulbs will produce about 90 cloves. 'Silverskin' is best suited for warm climates with mild winters. Bulb size is small (usually less than 5 cm (2 in)) when grown under cold conditions but bulbs may be significantly larger following a mild winter. Bulb size can sometimes be improved by planting early in the spring as soon as the ground can be prepared. Because of their weak necks, the plants lodge (fall over) before harvest. Clove skins are somewhat tight and difficult to peel. Bulbs typically can be stored for up to 1 year. Cultivars selected from 'Silverskin' include 'Silver White', 'Nookota Rose', 'Mild French', 'S&H Silver', and 'Idaho Silver' (Rosen et al., 2008).

'Turban' is considered to be a softneck type, but will form a flower stalk under cold conditions. Scapes are weak and tend to turn downwards. There are typically 7–11 cloves per bulb and double cloves are rarely produced. The bulb color is usually dark purple striped and the cloves are brown. The purple bulbils are numerous and small. Clove skins are loose and easy to peel. Storage is poor, lasting only 3–5 months. 'Turban' matures about 1–3 weeks earlier than most other clones. Selections include 'Red Janice', 'Blossom', 'Xian', 'Tzan', and 'Chinese Stripe' (Rosen et al., 2008).

Other important garlic cultivars include 'Texas White', 'Chilean' (an important cultivar in Mexico), 'Mexican' ('Elephant Paw' or 'Tahiti'), and some locally important clones (Rubatzky and Yamaguchi, 1997).

Economic Importance and Production Statistics

Onion

Onion is a major world food crop consumed in most countries. The FAO estimates that world per capita consumption of onions is just less than 10 kg/person/year (22 lb/person/year) (FAOSTAT, 2011). Onion production statistics are divided into green onions and dried bulb onions. World green onion or shallot production totals over 237,000 ha (585,390 acres). The average yield for green onion production is 204,733 t/ha (9.13 ton/acre). The leading producers of green onions include China 27,429 ha (67,750 acres), Japan 24,200 ha (59,774 acres), Turkey 20,366 ha (50,304 acres), Republic of Korea 19,666 ha (48,575 acres), Iraq 19,195 ha (47,412 acres), Ecuador 15,720 ha (38,828 acres), Thailand 15,511 ha (38,312 acres), and Nigeria 14,000 ha (34,580 acres).

The world average yield of bulb onions is about 15 t/ha (7 ton/acre), but with excellent growth and management, yields of 45–60 t/ha (20–27 ton/acre) are possible. Dried bulb onions are produced on a larger area of almost 4.3 million ha (10.6 million acres) (FAOSTAT, 2011). The world average yield for dried onions is 198,980 hg/ha (8.9 ton/acre). India is the leading world producer of dried onions with just over 1.1 million ha (2.7 million acres) followed closely by China with just over 1.0 million ha (2.5 million acres). Other important producing countries of dried onions include Nigeria 192,000 ha (474,442 acres), Pakistan 147,600 ha (364,572 acres), Bangladesh 127,940 ha (316,012 acres), Russia 95,500 ha (235,885 acres), Uganda 74,581 ha (184,215 acres), Myanmar 72,400 ha (178,828 acres), Iran 69,752 ha (172,287 acres), and Ukraine 66,600 ha (164,502 acres) (FAOSTAT, 2011).

Onion was the second highest valued vegetable produced in the USA in 2010, with a fresh market farm value of US$1.2 billion (ERS, 2011). In 2010, the average American consumed 9.7 kg (21.4 lb) of onions (Table 14.2) (ERS, 2011). Demand for onions rapidly increased during the 1970s as the popularity of onion rings increased. In the 1980s, the rising popularity of restaurant salad bars increased onion consumption (Burden et al., 2012). Lightly processed ready-to-eat chopped fresh onion products have also stimulated demand.

In the USA, approximately 60,595 ha (149,670 acres) of onions were harvested, with average 2010 yields of 54.9 metric tonnes/ha (24.5 short tons/acre). Total 2010 US onion production was 3.3 million t. California led the USA in onion production in 2010, supplying 854,545 metric tonnes (941,975 short tons) in 2010, followed by Oregon (645,454 metric tonnes (711,491 short tons)) and Washington (640,909 metric tonnes (706,481 short tons)) (Burden et al., 2012).

Garlic

Total world garlic production in 2011 was over 1.4 million ha (3.5 million acres) with an average yield

Table 14.2. Changes in the per capita consumption of onion and garlic in the USA since 1930 (ERS, 2011).

Year	Consumption kg (lb) per person per year	
	Onions	Garlic
1930	5.9 (13.0)	0.1 (0.2)
1940	5.3 (11.7)	0.05 (0.1)
1950	5.4 (11.8)	0.1 (0.2)
1960	5.6 (12.3)	0.2 (0.4)
1970	5.6 (12.4)	0.2 (0.5)
1980	6.5 (14.3)[a]	0.4 (0.9)
1990	8.7 (19.1)[a]	0.6 (1.3)
2000	8.4 (18.6)[a]	1.1 (2.4)
2011	9.7 (21.4)[a]	1.0 (2.3)

[a]Includes processed onions.

of 167,167 metric tonnes/ha (7.5 tons/acre) (FAOSTAT, 2011). The People's Republic of China is the world's leading producer with over 833,000 ha (2,057,510 acres) followed by India 200,600 ha (495,482 acres), Bangladesh 41,997 ha (103,733 acres), and Myanmar 29,129 ha (71,949 acres). Leading European producers include Russia 26,800 ha (66,196 acres), Ukraine 21,000 ha (51,870 acres), and Spain 15,661 ha (38,683 acres) (FAOSTAT, 2011). In the Americas, Argentina produces 17,739 ha (43,815 acres), Brazil 12,928 ha (31,932 acres), and the USA 10,180 ha (25,145 acres) (FAOSTAT, 2011).

In the USA, California is the primary garlic production area. Much of the increased recent consumption of garlic in the USA and other western countries can be attributed to increased availability through world trade, an increased appreciation for its varied uses in food preparation, and documented health benefits (see Table 14.3). In 2010, the USA exported 8,431 metric tonnes (9,294 short tons) of fresh garlic, primarily to Canada and Mexico (ERS, 2011). For all uses, the USA exported 12.6 metric tonnes (13.9 short tons) of garlic. The USA is a net importer of garlic and the world's largest import market for fresh garlic. For all uses, the USA imported 4,119 metric tonnes (4,540 short tons) of garlic, mainly from China, Argentina, and Mexico (Boriss, 2011).

Nutrition

Dried onions are used primarily as a flavoring and contribute mainly carbohydrates and minerals to the human diet (Table 14.3). Onions are low in calories and are a source of dietary fiber. Green onions are higher in vitamins and minerals than onion bulbs (Table 14.3). Fresh onions also contain vitamin C, with one medium onion providing 15% to 20% of the US daily requirement (Ensminger, 1994).

Overall, garlic is better nutritionally than either green or dried onions. Garlic is high in dry matter, carbohydrate, and minerals (Table 14.3). However, garlic is eaten in much smaller quantities than other vegetables since it is used primarily as a seasoning to flavor bread or other foods (Ensminger, 1994).

Production and Culture

General

Onions are a cool-season crop, tolerant of frost but not prolonged temperatures below −6.7°C (20°F). Optimum temperatures are 12.8–23.9°C (55–75°F).

Table 14.3. Nutritional composition of onions and garlic (USDA, 2012).

Nutrient	Amount per 100 g (3.5 oz) edible portion		
	Onion (dry)	Onion (fresh)	Garlic (dry)
Water (%)	91	92	59
Energy (kcal)	34	25	149
Protein (g)	1.2	1.7	6.4
Fat (g)	0.3	0.1	0.5
Carbohydrate (g)	7.3	5.6	33.1
Fiber (g)	0.4	0.8	1.5
Ca (mg)	25	60	161
P (mg)	29	33	153
Fe (mg)	0.4	1.9	1.7
Na (mg)	2	4	17
K (mg)	155	257	401
Ascorbic acid (mg)	0	5,000	0
Vitamin A (mg)	8.4	45	31.2

The range for optimum seedling growth is somewhat higher, 18.9–25.0°C (66–77°F), but is still relatively narrow. Onions for dried bulb production are one of the most difficult crops to produce because cultivar selection is critical and environmental factors dramatically impact growth and development. Green onion production is much less challenging because the crop matures in about 45–50 days and environmental factors are less important.

Site selection and nutrient management

Onions, garlic, and related alliums grow well in fertile, well-drained, non-crusting mineral soil, or muck soils high in organic matter. Since good drainage is essential, alliums are often grown on raised beds (Fig. 14.6).

Onion should be grown in rotation with unrelated crops to help prevent the buildup of soil-borne disease. The soil must be finely tilled using rotovators or disks and freed of clods, stones, and other impediments, especially for direct seeding, to ensure uniform emergence and good stand establishment (Fig. 14.6). Pelleted onion seeds are often planted with precision seeders to improve singulation and spacing.

Generally speaking, alliums have a high requirement for micronutrients and moderate need for nitrogen (N), phosphorus (P), and potassium (K).

Fig. 14.6. A newly emergent winter onion field established by direct seeding for dried bulb production in south Texas.

The nutritional requirements for onion and garlic are similar. Garlic has a moderate to high fertilizer requirement, with row banding or fertigation the preferred application methods. Alliums have a shallow fibrous root system restricted to the top 30 cm (12 in) of the soil, so precision placement of fertilizer is essential to prevent waste. Fertilization needs vary depending on many factors, including soil type, so fertilizer inputs should be made in accordance with soil test results to ensure that costly mineral inputs are not wasted and do not become pollutants.

Nitrogen requirements vary depending on N-supplying capacity of the soil, irrigation efficiency, and the amount of N loss due to rainfall and other environmental factors. If the irrigation system used is efficient, a total of 280 kg/ha N (250 lb/acre) should be sufficient to maximize yield on most mineral soils, with less needed on muck soils or other soil types with high residual N contents. Higher amounts may be justified in fields receiving significant rainfall or if irrigation is less efficient. Nitrogen fertilizer should be delivered in multiple applications throughout the season, with not more than 25% of the seasonal total applied in one application (Smith *et al.*, 2011b). For garlic, approximately 20–30% of N is row banded at planting to support early growth. The remainder of the N can be split-applied in the spring through fertigation or as a top- or side-dress after shoots emerge and then again 2–3 weeks later. Later applications of N may delay maturity and reduce bulb quality and storage life. Excess N promotes unwanted secondary growth (Rosen *et al.*, 2008).

Soils with bicarbonate extractable P greater than 30 ppm require a preplant application of no more than 56 kg/ha of P_2O_5 (50 lb/acre), while soils testing at 10 ppm P or less may require as much as 224 kg/ha of P_2O_5 (200 lb/acre). With adequate preplant application, in-season P additions are rarely needed (Smith *et al.*, 2011b).

Soils exceeding 150 ppm ammonia-acetate-exchangeable K are unlikely to respond to additional inputs. However, if soils test at less than 100 ppm, K additions of up to 168 kg/ha (150 lb/acre) may be required to ensure adequate fertility for onion production. Muck soils generally require less N and P but more K than mineral soils (Smith *et al.*, 2011b).

Some growers row-band all P or P and K fertilizer in the bed before planting. Nitrogen, and K if needed, is applied in various combinations such as a preplant band application at seeding, a top dressing approximately 4 weeks into the season as bulbing begins, or weekly if required by fertigation in conjunction with petiole analysis results (Smith *et al.*, 2011b). Alliums are generally sensitive to ammonia and salts but tolerant of low pH in the range of 5.3–6.5. Above pH 6.5, deficiencies of copper, manganese, and zinc occur (Smith *et al.*, 2011b).

Propagation methods

Onions are propagated by direct seeding, transplants, or sets. The use of direct seeding has increased over the years as technologies such as priming, pelleting, and precision sizing have improved seed quality so that reliable and uniform stands may be obtained with precision belt or air seeders. Transplants and sets are used in short-season areas and to mature a crop early in the season when prices are highest. Transplants may produce higher yields especially under conditions where stand establishment is difficult.

Onion transplants are produced in fields, tunnels, or greenhouses. The most favorable growth is at alternating 17°C (63°F) day and 10°C (50°F) night temperatures. Seeding densities for transplant beds are 75–100 kg/ha (67–89 lb/acre). Under favorable conditions, seedlings are large enough for transplanting after 8–12 weeks of growth and/or when stem diameters are 3–4 mm (0.12–0.16 in). Onions are relatively easy to transplant and are handled bare-root, although the use of multiple-cell transplant trays is increasing to reduce transplant shock and produce uniform seedlings. If transplants are too large they are sensitive to vernalization when planted during cold weather below 10°C (50°F) and will bolt prematurely.

When transplants are used for bulb production, field populations are usually considerably less than those used for direct seeding. Green onion production is usually by direct seeding because the higher costs of transplants are not justified for a rapidly maturing crop with high plant populations.

Onions can also be established using "sets". Sets are small bulbs, with growth intentionally stopped for the purpose of resuming at a later time. The benefits of sets are earliness or accommodation of short growing seasons. Seeds of a short-day cultivar are sown in late spring and usually grown during long-day summer conditions so that bulbing will occur during set propagation. Plantings are scheduled so the appropriate daylength to induce early bulbing is reached when plants are still small. High densities of 1,000–1,300 plants/m^2 (90–117 plants/ft^2) produce maximum yield, limit bulb size, and suppress weeds. Adequate fertilization is needed to support early growth during set production.

Excessive N can cause succulent foliage and thick necks that are susceptible to disease and difficult to dry. Prior to harvesting, water is withheld and sets are ready to harvest when the leaf tops have dried. The dried leaves are mowed or left intact if small and dry. The plants are undercut and lifted above the soil, windrowed, and further dried in the field for several days. During field drying the sets must be protected from moisture and direct sunlight. Set yields in excess of 20 metric tonnes/ha (9 short tons/acre) are common. Sets' diameters should be 15–20 mm (0.59–0.79 in), each weighing 2–3 g (0.07–0.11 oz). Sets greater than 25 mm (1 in) are sensitive to vernalization and may bolt later in development if induced.

Minisets are less than 10 mm (0.4 in) in diameter and can be handled by some seed drills similar to seed. Minisets take less time to produce compared to full-size sets. However, minisets tend to produce smaller bulbs that are less uniform because of improper orientation of sets during planting.

Quality sets can be stored for 6–8 months at 0–5°C (32–41°F) and 60% relative humidity (RH) or at 20–30°C (68–86°F) and 60% RH in ventilated boxes or mesh bags. Temperatures above 20°C (68°F) reduce bolting tendencies, but also reduce storage life and weight. Storage at 28–30°C (82–86°F) also results in earlier bulbing than storage at 20°C (68°F). Between 5–20°C (41–68°F) and RH >75%, sprouting, root development, and/or decay are likely. Therefore, sets must be stored below 5°C (41°F) or above 20°C (68°F).

Planting

Alliums can be grown on many different bed widths and row configurations. For bulb production, seed, sets, or transplants are usually planted on 80 cm–1 m (32–40 in) wide raised beds in double rows, spaced 30–46 cm (12–18 in) apart (Fig. 14.7).

Onions are also planted using four rows on a 150 cm (60 in) wide bed (Fig. 14.6). Direct-seeded, precision-planted onions are generally not thinned, so high quality seed and uniform spacing is required to produce a high percentage of large bulbs to maximize yields. Seed is normally planted 0.64–2.45 cm (0.25–1.0 in) deep, with deeper planting on lighter, drier soils. Optimal germination occurs at 24°C (75°F). Seed size ranges from 220,000 to 286,000 seeds/kg (100,000 to 130,000 seeds/lb) (Smith *et al.*, 2011b).

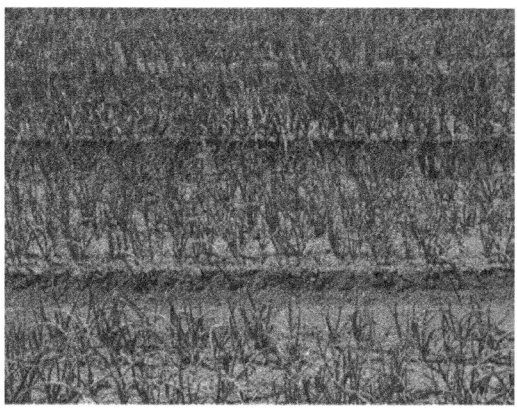

Fig. 14.7. A summer onion field with sandy soils and raised beds established by direct seeding for dried bulb production in the Annapolis Valley of Nova Scotia, Canada.

Since green onions are harvested at a very immature stage, the crop is densely seeded nine or ten rows wide on 1 m (40 in) or 18–20 rows wide on 2 m (80 in) beds. The seed is planted approximately 1.3 cm (0.5 in) deep in well-tilled soil and kept moist until emergence occurs. Plantings of green onions require 13–20 kg/ha seed (12–18 lb/acre) or 3.0–5.7 million/ha seed (1.2–2.3 million/acre) (Smith et al., 2011a).

Garlic is normally planted in the fall or early winter and bulbs are harvested during summer in temperate regions. In areas of very low winter temperatures, crops are planted in spring and harvested in late summer or fall. For planting, the bulb is separated into individual cloves, but the small central cloves are usually not planted because they produce undersized, less productive plants, with fewer cloves. Plant densities of 25–40 cloves/m^2 (2–4 cloves/ft^2) are recommended for garlic but for higher yields, densities of 60–70 cloves/m^2 (6–7 cloves/ft^2) are used. Approximately 1 ton of cloves will plant 1 ha (810 lb/acre). Individual clove size greatly influences yield potential with large cloves consistently out-yielding small ones (Rubatzky and Yamaguchi, 1997). Cloves are planted about 3–5 cm (1.2–2.0 in) deep, with the base side down so the emerging shoot will grow upward (Rubatzky and Yamaguchi, 1997).

Mosaic virus and stem and bulb nematode (*Ditylenchus dipsaci*) infestations of propagative materials are major problems for garlic producers. Garlic planting stock may be contaminated with mosaic virus, so propagules derived from tissue culture that are virus- and nematode-free should be planted when possible.

Irrigation

For direct-seeded crops, overhead sprinkler irrigation is often preferred to prevent onion seeds from drying out during germination and to keep the soil surface from crusting. Typically seedling emergence continues for a 10–20 day period depending on seed quality (Smith et al., 2011b). After establishment, sprinkler irrigation may be continued or a switch may be made to furrow or drip irrigation. Drip tape may be placed on the bed surface when plants are 15–20 cm (6–8 in) tall. Subsurface drip tape is increasing in popularity and may be placed 2.5 cm (1 in) beneath the surface prior to planting so no sprinkler irrigation is needed for emergence (Smith et al., 2011b).

Onions require frequent irrigation throughout the season. Moisture is needed to stimulate new root growth because most onion roots are non-branching, originate from the basal plate and grow within 30 cm (12 in) of the soil surface. Little water is extracted from deeper soil depths. Periods of water stress reduce bulb size, may increase doubles, encourage bulb splitting, and produce onions with greater pungency (Smith et al., 2011b). The amount and frequency of irrigation depends on factors including method, soil type, environmental conditions, and stage of crop development. Immediately after emergence, water needs are minimal because the crop is small and irrigations are scheduled less frequently or not at all if sufficient periodic rainfall occurs. As bulbing begins, water requirements increase and more frequent irrigations may be needed to maintain uniform moisture in the beds if rainfall is inadequate (Smith et al., 2011b). In dry climates, drip irrigation is necessary every 2–4 days or furrow irrigation every 5–6 days, throughout the season (Smith et al., 2011b). As the bulbs mature, irrigation is ended. Depending on the environment, a minimum of 1 month without water is needed for the crop to dry in the field prior to harvest. An onion bulb crop typically requires at least 50–75 cm (20–30 in) of water to meet minimum needs. If water is plentiful, 75–125 cm (30–49 in) of water may be applied to maximize yield (Smith et al., 2011b).

A combination of soil monitoring and weather-based irrigation scheduling can determine the water needs of onion and garlic. For optimum plant growth in most types of soil, the soil water tension should be maintained at less than 30 kPa (30 centibars) in the upper 30 cm (12 in) of soil. Keeping soil moisture too high (over 10 kPa (10 centibars)) wastes water and reduces the storage life of onion bulbs (Smith *et al.*, 2011b). Water use is highest when the crop reaches full canopy cover. Water use can be estimated using evapotranspiration (ET) data adjusted with a crop coefficient that relates water loss to leaf canopy cover. At a maximum leaf canopy cover of about 85% of the field the crop coefficient would be close to 1.0, meaning that the full amount of evapotranspiration should be applied back to the crop during irrigation. Since evaporation causes the majority of water loss during the earlier stages of growth, a crop coefficient of 0.3 to 0.5 better predicts water needs early in development when the leaf canopy covers less than 30% of the field. Maximum yields are achieved for onion when 100–150% of crop ET is applied (Smith *et al.*, 2011b). Evapotranspiration can be estimated by the pan evaporation method. This method requires placing a container of water with straight sides in a field at canopy height and periodically monitoring its level.

Crop Growth and Development

During onion seedling establishment, new foliage and roots are continually produced, along with a minor elongation and widening of the compressed stem. Initially, successive leaves tend to be longer and have wider bases but at the onset of bulbing, leaves tend to be shorter and smaller, change shape, and become bladeless. Bolting alters foliage growth as seed stalk formation takes precedence over leaf development.

In areas with extended periods of subfreezing temperatures, garlic plants are mulched to protect against the cold and sudden drops in temperature during the fall and winter (Rosen *et al.*, 2008). The mulch also helps control weeds and conserves moisture. After planting, rows are often covered with a 7.5–12.5 cm (3–5 in) layer of weed-seed-free organic mulch. Mulch may be removed in the spring after the threat of subfreezing temperatures is over. Some growers remove the mulch completely in the spring to allow the soil to warm faster, while others may leave the mulch in place to minimize weed pressure and conserve moisture (Rosen *et al.*, 2008).

Bulbing

Bulbing is a change in onion leaf morphology initiated when sufficient exposure to a critical daylength is exceeded, although temperature has an influence as well (Rubatzky and Yamaguchi, 1997). Each cultivar has a critical daylength for the induction of bulbing. During bulb formation food reserves are accumulated in the leaf bases, resulting in an enlargement that forms the storage structure called the bulb. Photosynthate partitioning differs with various growth phases. During early seedling growth when leaf development predominates, leaf blade growth has priority over the accumulation of storage reserves by the leaf sheaths. After the induction of bulbing, leaf sheath growth accelerates compared to leaf blade growth. As bulbing continues, inner scale or bladeless leaf growth dominates plant growth and development.

Onion cultivars are identified as short-, intermediate- or long-day (Table 14.1). All cultivars are long-day plants with respect to bulbing, because they bulb in response to increasing rather than decreasing daylength. The duration of light exposure is most important, and the exposure process is cumulative. A brief exposure to the appropriate daylength stimulus is not sufficient to initiate bulbing. When cultivars reach their critical daylength early in the seedling stage before adequate vegetative growth is made, developing bulbs will be small. Cultivars that require long days to bulb will not bulb when grown during short days. Short-day plants are usually grown for bulb production at less than 30° latitude, intermediate-day plants are grown between latitudes 30° and 38°, and those grown at latitudes greater than 38° are long-day types.

The influence of temperature on the bulbing process is less predictable than daylength. When scheduling planting dates and selecting cultivars, the daylength for a production area is the most important consideration. Once bulbing is initiated, temperature becomes very important and is the main factor influencing foliage growth and bulb enlargement (Rubatzky and Yamaguchi, 1997). Bulbing and maturation occur earlier and faster under the combination of long days and higher temperatures. The bulbing response at the critical daylength will be shortened at high temperatures but the shape of round-bulb cultivars becomes more oblong with increasing temperatures above the optimum during development (Fig. 14.8; Rubatzky and Yamaguchi, 1997).

Low temperatures may delay but will not prevent bulbing. Certain cultivars may not bulb at low temperatures even when the appropriate daylength is experienced, but will bulb at warm temperatures.

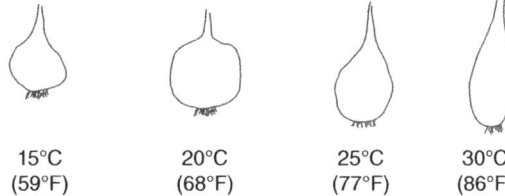

15°C (59°F) 20°C (68°F) 25°C (77°F) 30°C (86°F)

Effect of temperature on Spanish onion (*Allium cepa*) bulb shape

Fig. 14.8. High temperatures cause round onion bulbs to be oblong.

Bulb yield is highly dependent on the plant size and specifically the leaf area developed prior to bulb initiation. The rate of bulb growth and maturation is also influenced by cultivar, plant nutrition, moisture supply, plant competition, and both light intensity and quality. For a high yield, the plant should partition 70–90% of the accumulated dry weight into the bulb after development has been initiated (Rubatzky and Yamaguchi, 1997). Large and old plants are more responsive to bulbing than small and young plants when daylength requirements are achieved. However, given a strong photoperiodic stimulus, even a tiny one-leaf seedling can bulb. Nitrogen deficiency occurring near the critical daylength tends to accelerate the initiation of bulbing, while excessive N tends to delay the onset of bulbing. Water stress and competition with either plants or weeds may also accelerate bulbing.

Onions growing at daylengths shorter than those required to initiate bulbing will continue new leaf development and will not bulb. Bulbing is undesirable for green onion production, so usually long-day, white-skinned cultivars are preferred. Short-day cultivars grown during long-day conditions bulb early and produce small bulbs because of inadequate plant growth and photosynthetic capacity prior to bulbing. Short-day cultivars are intentionally grown during long-day conditions to produce sets for propagation or tiny bulbs for pickling or "cocktail" onions for processing.

Bolting

Bolting is the formation of a seed stalk and its associated inflorescence. Bolting greatly diminishes bulb quality and yield, so genotypes and conditions that induce bolting should be avoided for bulb production. However, conditions that cause bolting should be maximized for onion seed production. Seed stalk formation occurs after plants have been vernalized at low temperature, after developing past a juvenile phase. Plants with less than four or five leaves or "neck" diameters less than a pencil (6 mm, 0.24 in), including set diameters, are usually considered to be juvenile and are not vernalized. Large plants are generally more responsive to vernalization than small ones. Cultivars differ greatly in their response to low temperature and the duration of exposure needed to induce bolting. Exposure to 5–10°C (41–50°F) for 1–2 months is adequate to vernalize many cultivars. For some cultivars, temperatures between 10–15°C (50–59°F) are adequate to initiate bolting. Exposure to high temperature after cold treatment can reverse vernalization. Rapid bulb formation can suppress seed stalk emergence even if already initiated. It is possible to have bulbs and seed stalks developing at the same time. Because priority is given to the formation of a seed stalk and associated inflorescence, bulb size is greatly reduced during simultaneous development. Development of a seed stalk also reduces bulb quality because the hard core in the center is inedible and prevents the bulb from sealing for long-term storage.

Onion harvesting

Green bunching onions can be harvested from the time they are pencil-sized, 6.5–12.5 mm (0.25–0.5 in), in diameter. Green onions are undercut, pulled, and tied into bunches after size sorting and removal of decayed, aged, or damaged leaves. Roots are cleaned of attached soil, trimmed, and washed before packing. Green bunching onions are perishable but can be held for 3–4 weeks at 0–3°C (32–37°F) and 90–100% RH (Smith *et al.*, 2011a).

A bulb onion crop usually requires about 80–170 days from seeding to maturity but times may vary widely since bulbing is strongly influenced by both daylength and temperature. As onion bulbs reach their maximum size the foliage begins to senesce. Harvest of bulb onions generally occurs when 50–80% of the tops have lodged or fallen. To take advantage of favorable market prices, the tops can be broken with a roller to hasten maturity after about 10% have fallen naturally. Sometimes harvest occurs before tops have fallen. If onions are harvested while tops are erect and fleshy before natural senescence occurs, bulb yield is reduced and the potential for postharvest and storage problems increases. These immature onion bulbs have high moisture content and a relatively short postharvest life. As onions naturally senesce and mature, the foliage dries, the pseudostem (neck) shrinks and collapses. As the neck collapses it seals, reducing further desiccation from the bulb, decreasing the incidence of disease and improving storage potential. However, onions harvested at full leaf senescence tend to have reduced storage life as well. Therefore, the optimum time for harvest is a compromise between increasing bulb weight and a possible decrease of postharvest quality and storage life. During harvest, plants are mechanically undercut and placed on the soil surface in shallow windrows to dry. Sometimes undercutting precedes harvest to accelerate senescence. In many regions, bulbs, tops, and roots are harvested and trimmed by hand. Bulbs are protected from sunburn by windrowing or covering with shade material and dried in the field in arid regions. In humid areas, the onions are collected and force-air dried. Onions for dehydration or processing are likely to be machine harvested because their per unit value is less than fresh-market bulb crops and some bulb damage is acceptable.

Garlic harvesting

The harvesting, handling, and storage of garlic heads are similar to dried onion bulbs. Garlic maturation occurs in mid- to late summer, as indicated by drying leaves and falling tops. Garlic is usually harvested after tops have fallen and heads are well dried, so irrigation should be stopped at least 3 weeks prior to the anticipated harvest date. Bulbs are loosened by undercutting and are mechanically or hand-harvested and placed in shallow rows for drying in climates with low humidity (Fig. 14.9).

Bulbs are covered with dry leaves or other breathable shade-providing materials to prevent sun burning or exposure to moisture. Any direct contact with moisture should be avoided after harvest. Providing good ventilation is important to facilitate drying. Bulbs often are placed into slatted containers or netted sacks that allow air movement. Usually 7–10 days is sufficient to dry bulbs before trimming away roots and/or leaves in preparation for marketing or storage. In areas with frequent rainfall or high humidity, indoor drying is preferred to prevent disease.

Fig. 14.9. Garlic drying in the San Joaquin Valley of California.

Onion bulb curing

Following harvest, bulbs are cured to heal wounds. Curing improves postharvest handling characteristics and limits entry of organisms through the pseudostems or injured tissues. Curing also helps the formation of attractive, well-colored, intact outer skins. When weather permits, bulbs are cured in the field for several days, depending on environmental conditions, until the neck has sealed, the outer scales are papery and wounds have healed. For field or indoor curing at ambient temperature, bulbs are windrowed to air-dry or placed into well-ventilated boxes or bags, respectively. Any loss or damage to the outer skin reduces appearance, market value, and protection against mechanical injury and desiccation. Onions can also be cured with forced circulation of warm (30°C, 86°F), low-humidity air through bins or piles of onions placed on slatted floors for 12–24 h. During curing, onion can lose as much as 5% of its initial harvest weight.

Onion storage

Green onions should be stored at 0°C (32°F) and 95% RH for 10–20 days. At 5°C (41°F), storage life may be limited to 1 week (Smith *et al.*, 2011a). Best storage occurs after bulbs have been cured to seal bulbs and heal damage as described above. Onion bulbs do not heal cuts and surface wounds as successfully as other crops such as Irish potato, so mechanical damage during harvest and handling should be minimized prior to storage. Decay, sprouting, and rooting cause losses during onion bulb storage. These losses can be minimized by storage at or near 0°C (32°F) and 65–70% RH. Storage at between

25°C (77°F) and 35°C (95°F) is also satisfactory. At the higher temperatures, bulbs of some cultivars can be stored for 3–6 months without sprouting, but once removed from storage, sprouting may occur rapidly.

Dry matter and moisture are lost during storage. Most bulbs shrink in size due to respiratory losses and carbohydrate translocation from the outer to inner scales or sprouting roots and shoots. This reallocation of resources causes the succulent outermost scales to gradually desiccate, becoming a dry protective layer, further reducing water loss from the inner scales. Bulb diameter decreases because of the net loss of succulent scales over time. Onion respiration rates are temperature dependent, decreasing with refrigerated storage. Respiratory heat must be removed by ventilation or refrigeration. Relative humidity influences storage life by reducing shrinkage from water loss. Modified atmosphere storage with elevated CO_2 and reduced O_2 can extend onion storage life.

As a general rule, long-day types with high solids have longer storage potential than short-day types with low solids. Pungent cultivars generally have longer storage potential than sweet. A fully cured pungent bulb may be stored under optimal conditions for up to 1 year. In contrast, a sweet cultivar may only last 2–3 months under similar conditions.

Sales of pre-chopped onions and garlic have increased dramatically in the last 20 years because consumers and food service companies like the convenience of a ready-to-use product. Onion bulbs are chopped in a processing facility, placed in sealed containers to reduce moisture loss and oxidation and stored at 0°C (32°F) (Fig. 14.10).

Garlic storage

Cloves become dormant at maturity. The length of dormancy varies with clones and environmental conditions during maturation. Dormancy gradually declines during storage, gradually disappearing after several weeks, although it may persist for as long as 2 months or more in certain cases. For immature bulbs, dormancy is broken by treatment at 35°C (95°F), whereas with fully mature bulbs dormancy is rapidly depleted during storage at 5–10°C (41–50°F)(Rubatzky and Yamaguchi, 1997). Bulbs intended for food use are held at ambient temperatures and can remain in good condition for

Fig. 14.10. Hand peeling of Spanish-type bulb onions prior to chopping in a fresh-cut, minimal processing facility.

several months. However, for extended storage they should be held at 0°C (32°F) and at 60% RH. Well-dried mature bulbs stored at −2°C (28°F) can be maintained in good condition for as long as 8 months. When no longer dormant, cloves readily sprout at temperatures between 5°C (41°F) and 10°C (50°F). However, like onion, storage at 25°C (77°F) can prevent sprouting, but results in more rapid clove shrinkage. Garlic should not be stored at greater than 70% RH. Treatments with maleic hydrazide or gamma irradiation can extend storage life. Sprouting inhibition treatments should not be applied to cloves intended for propagation.

Cloves for planting are usually conditioned at 5–10°C (41–50°F) for a few days before planting. This is critical, because without conditioning, cloves held at less than 5°C (41°F) may mature prematurely producing small and/or rough bulbs. Storage above 18°C (64°F) may delay sprouting of cloves.

Bulb rest and dormancy

Onion bulb dormancy is complex with different phases. Onion bulbs first enter a state of rest at harvest maturity that lasts a period of 6–8 weeks. After the rest period, natural dormancy begins and bulbs will not sprout or grow because inhibitors synthesized in green leaves earlier in development remain. The inhibitors are gradually destroyed with time. Therefore, it is important to have normal senescence of onion leaves to ensure translocation of inhibitors to the bulb to improve storage life and reduce early sprouting. The change from rest to dormancy is gradual and highly dependent on cultivar. Dormant bulbs will not sprout at optimal storage temperatures. Once dormancy has passed, root emergence occurs, followed by leaf shoots at favorable temperatures and moisture conditions (Rubatzky and Yamaguchi, 1997).

Bulb sprouting and sprout inhibition

Sprouting is optimum at 10–15°C (50–59°F). Sprouting is inhibited at both low and high temperatures as explained above. For long-term storage, a spray inhibitor such as maleic hydrazide (MH) is sometimes used. MH at a concentration of 2,500 ppm is applied at 500 l water/ha (53 gal/acre). Spray inhibitors are generally applied in the field 1–2 weeks prior to harvest. If applied too soon, foliage injury occurs and when applied too late, foliar absorption is insufficient to be effective. To improve absorption, applications are made when most of the foliage is still green and when dew is not present. Onions treated with MH and held at a temperature between −2°C (28°F) and 0°C (32°F) and 65–70% RH can be stored for as long as 6–7 months without sprouting. Bulb rooting and sprouting are inhibited by gamma irradiation and controlled-atmosphere storage at reduced O_2, and continuous removal of newly formed roots tends to delay sprout emergence.

Diseases

A number of fungal diseases affect onion crops. A few of the more serious diseases are described here but many more exist (Schwartz and Mohan, 2008; CABI, 2013). "Botrytis leaf blight" (BLB) is a fungal disease that occurs in many of the onion-growing areas of the world. The organism, *Botrytis squamosa*, causes spots (lesions) and water-soaked areas on leaf tissue resulting in death and dieback. The lesions are whitish in color ranging from 1–5 mm (0.04–0.20 in) in length, and most are surrounded by greenish-white halos that appear water-soaked when first formed (Schwartz and Mohan, 2008). To control BLB, infected plant materials should be rogued from fields and nearby wild populations and destroyed during and after the season. Fungicides applied on 7–10 day intervals, starting in midseason before the disease is evident can provide effective control.

Purple blotch (*Alternaria porri*) is a common disease of onion worldwide. The fungus overwinters as mycelium in onion leaf debris. Periods of leaf wetness or 90% or higher RH for 12 h or more cause sporulation and the disease spreads by wind. Infection is greatest at 25°C (77°F) (Schwartz and Mohan, 2008). Purple blotch symptoms are first observed as small, elliptical, tan lesions that often turn purplish-brown. Concentric rings can be seen in lesions as they enlarge. A yellow halo surrounds lesions and extends above and below. Lesions usually girdle leaves, causing them to fall over (Schwartz and Mohan, 2008). Older plants are more susceptible to infection by purple blotch.

The soil-borne fungus (*Phoma terrestris*) causes pink root. Pink root is present in many soils and is easily recognized because the roots turn pink or maroon when infected. The main effect of the disease is reduced bulb size. In severe cases the roots may die and the plants become weakened or stunted, especially in drier areas. Unless the crop is suffering from heat or drought stress, yield losses are not likely to occur in good soils (Chaput, 2011).

Downy mildew is sometimes confused with purple blotch because both diseases cause purple lesions. Downy mildew (*Peronospora destructor*) is a common foliar disease of onion that seems to be worse on direct-seeded crops. The disease is favored by cool (<22°C, 72°F), humid or wet weather. Early symptoms include a purple-gray, velvety growth that usually begins in an isolated area and then spreads rapidly through the field (Schwartz and Mohan, 2008). Diseased leaves turn pale-green, yellow, and then collapse. Downy mildew management combines the use of cultural practices with fungicide applications where registered. Crop rotation for 3 years will reduce the overwintering populations in soil. Controlling alternative weed hosts in and around the field is also important to eliminate overwintering inoculum (Chaput, 2011).

The soil-borne fungus white rot (*Sclerotium cepivorum*) begins in both garlic and onion fields and can infect bulbs in storage (Schwartz and Mohan, 2008). The first above-ground symptoms are a yellowing and dieback of the leaf tips, followed by a collapse of the affected leaves. The bulbs and roots develop a white, fluffy mold and soft rot. Infected bulbs can rot in storage and stain other bulbs. White rot typically develops in patches in the field and is less serious in dry warm soil (>24°C, 75°F) (Schwartz and Mohan 2008). White rot is difficult to control. Damage may be severe one year and nonexistent the next. A rotation of at least 4–5 years with no other alliums can control white rot. Treating soil with fungicide prior to planting can provide control (Crete *et al.*, 1981).

The soil-borne fungus *Fusarium oxysporum* f. sp. *cepae* is the cause of basal rot and affects onions grown in warm soils (optimum 29°C, 84°F) (Schwartz and Mohan, 2008). The fungus survives indefinitely in soil. Infection occurs through wounds or in the vicinity of old root scars at the base of the bulb. The early field symptoms include yellowing of leaves and tip dieback. As the disease progresses, the whole plant may collapse, and if pulled from the soil, there are few or no roots attached to the bulb. The basal plate of the onion becomes pinkish-brown and secondary bacterial rots may develop. If infection occurs late in the season, the symptoms may not develop until the onions are in storage (Schwartz and Mohan, 2008). To control Fusarium basal rot, avoid growing onions in fields with a history of the disease and rotate for 3–4 years with non-alliums. Soil insects and foliage diseases that might create wounds that allow infection should be controlled if possible. Onions should be properly cured before storage. Bulbs should be stored at cool temperatures since fungal growth is favored by warm conditions (CABI, 2013).

Neck rot is a common storage disease caused by various *Botrytis* spp., including *B. aclada*, *B. byssoidea*, and *B. squamosa* (Schwartz and Mohan, 2008). The symptoms usually appear in storage, however some necks may become soft and rotten just before harvest. There is usually a separation between healthy and diseased scales within the onion. As the disease progresses, a gray mold may also develop (Schwartz and Mohan, 2008). Black sclerotia eventually appear on the affected tissue. The decay symptoms can easily be confused with bacterial decay and eventually the whole bulb will break down. Sometimes both types of diseases are present (Chaput, 2011).

Insect Pests

A broad range of insects and nematodes attack onions and other alliums. A few of the more serious insects are described below but many more exist. Please consult local sources for more information about insect pests of alliums and their control in your area (CABI, 2013).

Cutworms

The black cutworm (*Agrotis ipsilon*) and dark-sided cutworm (*Euxoa messoria*) are two of the most serious insect pests that affect bulb crops (OMFRA, 2009). There are several species of cutworm that can attack onion. Cutworm damage in onions and garlic occurs early in the season. Plants at the seedling stage are most vulnerable to cutworms. Affected plants fall over and appear to have been cut in two with scissors, at or just below the soil surface. The larvae are soft and fat and they roll into a circle when disturbed. A single larva may destroy several seedlings. Adult cutworms are a gray, night-flying moth about 2.5 cm (1 in) long. Adult populations can be monitored with black light traps and/or sex pheromones; however research has shown that light traps may underestimate early season densities (OMFRA, 2009). Migrating female moths are attracted to weedy fields in the early spring for egg-laying. Early-season cutworm control is most effective on small (less than 2.5 cm or 1 in) larvae. Larger larvae are difficult to control with insecticides. At more mature stages, when less than 2.5 cm (1 in) in length, they cease feeding as they prepare to pupate, and control becomes unnecessary. Insecticides, if used, should be applied in the early evening, as the cutworms come to the surface to feed. Insecticides are more effective on moist soils (OMFRA, 2009).

Onion maggot

The onion maggot (*Delia antiqua*) is a small, less than 5 mm (0.20 in) translucent white larva of a fly. The maggots are legless and oblong-shaped, while the adult is a small 5 mm (0.20 in) gray-black fly that emerges from overwintering sites in late spring when onion plants are most susceptible. There are multiple generations each year. First-generation onion maggot thresholds are determined by a combination of trapping and visual scouting. To monitor populations, sticky cards can be placed centrally and along the sides of the field. The percent damage can be assessed using plots containing 100 onions repeated four times within the field. Degree-day threshold models predict arrival of different generations (OMFRA, 2009). Maggots do not begin development unless the temperature is at least 4°C (39°F). Infestations often result in wilted transplants and poor stand establishment (OMFRA, 2009). Crop losses may occur if control measures are not used. Control measures include good field sanitation by removing onion residue from the field and cull pile management to reduce feeding and overwintering locations for future generations. Some granular insecticides may provide control if the correct rate and placement are used.

Thrips

Thrips (*Thrips tabaci*) are small, less than 3 mm (0.12 in), soft-bodied insects that can be a serious pest of alliums. Thrips are cream to light brown in color. They cause tissue damage by sucking on leaves, leaving silvery spots. Thrips overwinter on grasses, especially winter wheat and alfalfa. They migrate into fields of alliums in late spring and early summer. Thrips prefer hot, dry weather. Early detection of thrips is key to successful control. The threshold for dry onions, leeks, and garlic is one thrip per leaf. Insecticide should not be used until the infestation threshold is reached. High water volumes when spraying help insecticides penetrate into the leaf axils, improving control (OMFRA, 2009).

Leafminers

Adult leafminers (*Liriomyza* spp.) are small, 2–3 mm (0.08–0.12 in), shiny black and yellow flies. Eggs are laid in the leaves, leaving small bronzed puncture marks. Females pierce the leaves to feed on plant sap. Depending on the species, mines can be serpentine or straight. The larvae of leafminers are small maggots, which are pale yellow. The optimal temperature for leafminer feeding and egg-laying ranges from 21–32°C (70–90°F). Egg laying is reduced at cool temperatures below 10°C (50°F) (OMFRA, 2009).

Plant debris should be removed and disposed of immediately following harvest to aid in leafminer control. Debris should be totally covered, buried, or burned to reduce the dispersal of any emerging leafminer flies. The degree of infestation may vary with the type of crop, neighboring crops, weeds, temperatures, and leafminer species. Many wild plants and weeds are hosts for leafminers. Maintaining weed-free conditions is recommended. Crop rotation is an effective pest management tool. Alternating leafminer-susceptible crops with leafminer-resistant crops reduces the population (OMFRA, 2009).

A few pesticides are labeled for control of leafminers. Leafminers are known to develop resistance to insecticides very quickly. When available, always rotate between insecticide groups. The most effective control is obtained with systemic or translaminar products, which target the larvae. Thorough coverage of the crop is essential for effective leafminer control. Insecticide applications should be timed to have the most impact on susceptible stages. Spraying should only be based upon regular and consistent scouting information.

Wireworms

Wireworms (*Coleoptera* spp.) are copper-colored and cylindrical with hard bodies and three pairs of tiny legs near the head. Wireworms vary in size from a few millimeters to 2 cm (0.75 in) in length. Infested plants are stunted and seedlings lack vigor or fail to emerge (OMFRA, 2009). Damage is often scattered randomly across a field. Wireworms are present all season. Young plants are most susceptible, therefore early season control is critical. Wireworms have a life cycle of several years. They are likely to be present in fields with a history of insect problems or in fields that have recently grown sod. Good early growth will help reduce the losses to this pest where populations are low to moderate. Bait stations can be used to identify and avoid fields with high wireworm populations (OMFRA, 2009). In some areas granular insecticides can be placed in the seed furrow at planting.

Weed Management Strategies

Weed control is a critical part of onion production. The lengthy growing season required for garlic and bulb onion production allows for successive flushes of weeds at different times of the year. Common weeds that affect onion and garlic include yellow nutsedge, field morning-glory, cheeseweed, chickweed, henbit, lambsquarters, marestail, pigweed, purslane, shepherdspurse, annual bluegrass, barnyardgrass, and foxtail. Onion transplants and seedlings do not compete well with weeds, so hand cultivation or herbicides are necessary if weeds are a problem. Since onions are a monocot, there are a number of good herbicides available especially to control broadleaf weeds. Herbicides used in garlic are typically grouped as: (i) those used prior to planting; (ii) those used pre-emergence; and (iii) those used post-emergence. No-till production of bulb onions and garlic is not widely practiced at this time due to establishment and weed control issues.

LEEKS, SHALLOTS, AND CHIVES

Leek

Leek (*A. ampeloprasum* L., Leek Group) is a cool-season, frost-tolerant biennial that is cultivated around the world but is more popular in Europe than North America or Asia (McGee, 2004). Leek is a self-compatible tetraploid that outcrosses with other Groups of *A. ampeloprasum* such as great-head garlic. It is an ancient crop as documented in biblical writing. Leek is broadly adapted to soil and environmental conditions and has a profuse relatively shallow root system that is proficient in extracting nutrients and water from soil. Leeks are established from direct seeding or transplanting. Transplanting is more common in short-season areas because plants grow slowly and do not compete well with weeds. Leek spacing is dependent on the stem size desired at harvest. For thickest stems, seedlings should be spaced 15–20 cm (6–8 in) apart. For thinner leeks, plants should be spaced 7–10 cm (3–4 in) apart. Once established, leeks are winter hardy and some cultivars can be left in the ground during the winter. They hold well in the field and can be harvested over an extended period at a broad range of sizes. The plants do not form a bulb and are large and upright, reaching heights varying from 40 to 75 cm (16–30 in). Leek produces a long cylinder of bundled leaf sheaths that are generally blanched by mounding soil around the base of the plant to obtain milder flavor (McGee, 2004). Cultivars vary widely in maturity. Best temperatures for vegetative growth are between 20°C (68°F) and 25°C (77°F). Leeks are vernalized by low temperature and will bolt when nonjuvenile plants are exposed to temperatures under 15°C (59°F) for prolonged periods (Rubatzky and Yamaguchi, 1997). The temperature and duration of exposure required to cause bolting varies widely among cultivars. Leeks can be classified as summer, fall, or winter maturing with maturity from planting ranging from 70 to 140 days. Harvest and handling is similar to green onions. Leeks can be bunched and harvested early when they are about the size of a green onion or they can be thinned and develop to a much larger mature size. Both F-1 and open-pollinated cultivars are available.

Leeks have a mild onion-like flavor. The edible portions of the leek are the white leaf bases. The dark green portion is usually discarded because it has a tough texture and stronger flavor, but leaves can be sautéed or added to flavor broth, soup, or stew. The white leaf-base tissue of leek may be boiled, which turns it soft and mild tasting, to flavor soups stews or broth. Care should be taken to chop leek, or else the intact fibers that run the length of the leaf bases will tangle into a ball while chewing. It may also be fried, which makes the texture crunchier and preserves flavor, or eaten raw in salads.

Shallots

Shallots (*A. cepa* var. *aggregatum*) are a cool-season, frost-tolerant biennial that grows on a variety of well-drained soils including muck and loam (Brewster, 2008). Leaves and bulbs of shallot are similar to but smaller than onion. Well-developed bulbs are about 5 cm (2 in) in diameter (Rubatzky and Yamaguchi, 1997). Shallots have pear-shaped, reddish-brown skinned bulbs that form clusters at the base of the plant. Shallots are a popular vegetable in Europe, particularly France, and grow best in a pH range of 5.5–7.0. Because plants grown from seed are variable and unlike their parents due to heterozygosity, shallots have historically been propagated vegetatively from bulbs (Rubatzky and Yamaguchi, 1997). Bulbs should be planted 2.5–3.8 cm (1–1.5 in) deep and 7.6–10.0 cm (3–4 in) apart. Rows are spaced 30–64 cm (12–24 in) apart. New cultivars have been developed that grow true-to-type from seed (Rabinowitch and Currah, 2002). Shallots grown from seed are responsive to daylength and must be planted in early spring to respond to the lengthening days of summer for bulb formation. In mild climates, shallots can survive most winters and can be planted in fall. In cold climates where the ground freezes during winter, to obtain best yields shallots should be planted early in the season as soon as the soil can be tilled (Rubatzky and Yamaguchi, 1997). Seeds are planted 1.3 cm (0.5 in) deep and 0.6–1.9 cm (0.25–0.75 in) apart in a 5.0–10.0 cm (2–4 in) wide band across a bed at a density of 40–50 seed per 30 cm (12 in) of row. This spacing will produce a high percentage of single-bulb plants, while wider spacing may result in clusters of bulbs. Clusters may contain only a few bulbs or as many as 15. Because shallots have a rest period similar to onions, bulbs are often stored before planting. Harvest occurs after the leaves begin to senesce and wilt. Depending on growing conditions, maturity is 60–100 days after planting. Bulbs are pulled and dried before cleaning and bunching. Shallot bulbs are known for their subtle delicate flavor and their usage is similar to onion (McGee, 2004).

Chives

Chives (*A. schoenoprasum* L.) are native to Asia and eastern Europe. They are hardy, draught-tolerant perennials, which grow 20–51 cm (8–20 in) tall, in dense clumps from underground bulbs (Fig. 14.11). The leaves are round and hollow, similar to onions, but less fibrous and smaller in diameter. In June or July, chives produce large round heads of attractive purple to pink flowers (Davis, 1997).

Chives grow best in full sun in rich, moist soil that is high in organic matter and has a pH of 6–8. Chives will tolerate partial shade and a range of soil types. Specific nutritional recommendations are not available for chives, but a general recommendation is to incorporate 56–84 kg/ha (50–75 lb/acre) each of N, phosphate, and potash per acre at planting (Davis, 1997). Side-dressing with an additional 11–17 kg/ha (10–15 lb/acre) of N twice during the growing season will stimulate new growth after harvest. Chives should be kept well watered and weeded.

Chive seed germinates slowly. For transplant production, seed should be sown about 1.3 cm (0.5 in) deep in trays containing a peat-based soilless mix maintained at 16–21°C (60–70°F). Transplants are established in the field when 4–6 weeks old. Chives can be established by direct seeding when the soil is warm if weed control is maintained. Leaves should not be harvested the first year from direct-seed crops. Older clumps of chive plants can be divided for propagation and replanting at the same spacing as described above. Plantings should be divided every 2–3 years to prevent overcrowding. Space divided plants 10–38 cm (4–15 in) apart in rows at least 51 cm (20 in) apart. Few disease or insect pests bother chives compared to other vegetable crops (Davis, 1997).

Chive leaves are harvested and consumed as a garnish or flavoring. Leaves of chive plants can be harvested after established plants reach at least 15 cm (6 in) tall. Chives are harvested by cutting leaves 5 cm (2 in) above the ground. All leaves or a portion of the leaves can be harvested at one time. New growth from the same clump can be cut repeatedly throughout the growing season. Depending on growing conditions, new growth is ready for harvest several weeks after the previous harvest. Plants should be cut regularly to encourage succulent new growth, new bulblet development, and to prevent flower formation.

Freshly harvested leaves may be trimmed and sold in bunches tied together, in small plastic bags or in plastic "clamshell" containers. Storage is best just above freezing and with high humidity. Whole plants are sold in 5.0–7.6 cm (2–3 in) pots. Chives can be freeze-dried or frozen. When forced-air dried, they tend to discolor and lose flavor.

Chives are finely chopped and used fresh and are a common addition to soups, sour cream dips, and egg dishes. Scientific evidence suggests that chives improve digestion and reduce high blood pressure. The oil has antibacterial properties (Davis, 1997).

Fig. 14.11. Chive plants growing near Blacksburg, Virginia.

References

Block, E. (2009) *Garlic and Other Alliums: The Lore and the Science*. Royal Society of Chemistry, Cambridge, UK.

Boriss, H. (2011) Garlic Profile. Available at: www.agmrc.org/commodities_products/vegetables/onion-profile (accessed 11 June 2013).

Brewster, J.L. (2008) *Onions and Other Alliums*, 2nd edn. CAB International, Wallingford, UK.

Burden, D., Huntrods, D. and Morgan, K.L. (2012) Onion Profile. Available at: www.agmrc.org/commodities_products/vegetables/garlic-profile (accessed 11 June 2013).

CABI (2013) Datasheets: *Allium cepa* (onion) production and trade. Available at: www.cabi.org/cpc (accessed 28 August 2013).

Chaput, J. (2011) Identification of diseases and disorders of onions. Available at: www.omafra.gov.on.ca/english/crops/facts/95-063.htm (accessed 14 June 2013).

Crete, R., Tartier, L. and Devaux, A. (1981) *Diseases of Onions in Canada*. Agriculture Canada/Ministry of Supply and Services, Quebec, Canada.

Davis, J.M. (1997) Chives. Available at: www.ces.ncsu.edu/hil/hil-124.html (accessed 13 June 2013).

Ensminger, A.H. (1994) *Foods and Nutrition Encyclopedia*, Vol. 1. CRC Press, Boca Raton, Florida.

ERS (2011) Vegetables and melons yearbook. Available at: http://usda.mannlib.cornell.edu/MannUsda/viewDocumentInfo.do?documentID=1212 (accessed 11 June 2013).

FAOSTAT (2011) Onion and garlic production statistics. Available at: http://fao.org/site/567/DesktopDefault.aspx?PageID=567 (accessed 3 June 2013).

George, R.A. (2009) *Vegetable Seed Production*. CAB International, Wallingford, UK.

McGee, H. (2004) *On Food and Cooking*, revised edn. Scribner Publishing, London.

OMFRA (2009) Ontario Crop IPM: Onions. Available at: www.omafra.gov.on.ca/IPM/english/onions/index.html (accessed 14 June 2013).

Pooler, M.R. and Simon, P.W. (1994) True seed production in garlic. *Sexual Plant Reproduction* 7, 282–286.

Rabinowitch, H.D. and Currah, L. (2002) *Allium Crop Science: Recent Advances*. CAB International. Wallingford, UK.

Rosen, C., Becker, R., Fritz, V., Hutchison, B., Percich, J., Tong, C. and Wright, J. (2008) Vegetable management series. Growing garlic in Minnesota. Available at: www.extension.umn.edu/distribution/cropsystems/dc7317.html (accessed 15 June 2013).

Rubatzky, V.E. and Yamaguchi, M. (1997) *World Vegetables: Principles, Production and Nutritive Values*, 2nd edn. Chapman & Hall, New York.

Schwartz, H.F. and Mohan, S.K. (2008) *Compendium of Onion and Garlic Diseases*, 2nd edn. APS Press, St. Paul, Minnesota.

Smith, R., Cahn, M., Cantwell, M., Koike, S., Natwick, E. and Takele, E. (2011a) Green onion production in California. Available at: http://anrcatalog.ucdavis.edu (accessed 4 June 2013).

Smith, R., Biscaro, A., Cahn, M., Daugovish O., Natwick, E., Nunez, J., Takele, E. and Turini, T. (2011b) Fresh-market bulb onion production in California. Publication 7242. Available at: http://anrcatalog.ucdavis.edu (accessed 4 June 2013).

USDA (2012) USDA national nutrient database for standard reference, release 25. Available at: www.ars.usda.gov/nutrientdata (accessed 30 June 2013).

Volk, G.M., Henk, A.D. and Richards, C.M. (2004) Genetic diversity among U.S. garlic clones as detected using AFLP methods. *Journal of the American Society for Horticultural Science* 129, 559–569.

WIPMC (Western IPM Center) (2004) Crop profile for garlic in California. Available at: www.ipmcenters.org/cropprofiles/docs/CAgarlic.pdf (accessed 7 July 2013).

Yoo, K.S., Pike, L., Crosby, K., Jones, R. and Leskovar, D. (2006) Differences in onion pungency due to cultivars, growth environment, and bulb sizes. *Scientia Horticulturae* 110, 144–149.

Zohary, D. and Hopf, M. (2000) *Domestication of Plants in the Old World*, 3rd edn. Oxford University Press, Oxford, UK.

15 Family Convolvulaceae

SWEETPOTATO

Origin and History

The sweetpotato is an ancient crop of the New World. The sweetpotato was an important food crop of the Mayan and Inca cultures long before the arrival of the Europeans in South and Central America. In Central America, sweetpotatoes were domesticated at least 5,000 years ago while in South America, Peruvian sweetpotato remnants date back as far as 8000 BC (Austin, 1988).

The center of origin of the sweetpotato may have been somewhere between the Yucatán Peninsula of Mexico and the mouth of the Orinoco River in Venezuela (Austin, 1988; Zhang et al., 1998). Molecular genetic comparison studies suggest that Peru–Ecuador was a secondary center of origin for sweetpotato (Zhang et al., 1998).

Sweetpotatoes were introduced to Spain by European explorers about 1600, to western Africa by Portuguese traders and later into India, the East Indies, China, and Japan (Woolfe, 1992). The sweetpotato has been cultivated in Virginia since at least 1648 (O'Brien, 1972). The introduction of the sweetpotato in North America is unclear. Native American people in the southeastern USA had words for sweetpotato in their native language and were cultivating the crop when European colonists arrived. However, this may not be an indication of historical production related to Mayan domestication. It is possible that Native Americans acquired the sweetpotato from the Spanish explorers from Hispaniola and not as a natural migration through indigenous cultures from its center of origin in Central America (Austin, 1988).

A major mystery is the arrival of the sweetpotato in the islands of the South Pacific 300–400 years earlier. The sweetpotato was also grown in Polynesia before western explorers could have introduced it. The Polynesians, who had traveled to South America and back, may have introduced sweetpotato to central Polynesia around AD 700 (Bassett et al., 2004). The sweetpotato apparently reached New Zealand as early as AD 1300.

The name "potato" is derived from the Mexican word "batata" used for the sweetpotato. The sweetpotato has become far more important in subtropical and tropical areas than has the Irish potato, because it thrives in hot, moist climates whereas the Irish potato requires a cool climate.

Botany

Sweetpotato (*Ipomoea batatas* (L) Lam.) is a member of the morning glory family Convolvulaceae. Of the approximately 50 genera and more than 1,000 species of Convolvulaceae, *I. batatas* is the only crop plant of major importance. Some cultivars of *I. batatas* are grown as ornamental plants for their color foliage and flowers. The genus *Ipomoea* also includes species of popular garden flowers called morning glories. *Ipomoea aquatica* is also known as water spinach and is grown for its edible foliage in Southeast Asia (Fig. 15.1). Water spinach cannot be grown in the southern USA because it is an invasive noxious weed prohibited by the government.

The sweetpotato plant is an herbaceous vine, bearing alternate heart-shaped or palmately lobed leaves and medium-sized sympetalous flowers (Fig. 15.2). The leaf veins may be green or purple. Sweetpotato is genetically a hexaploid (6n = 90), so it does not grow true-to-type from seed and is vegetatively propagated from rooted sprouts that develop from the storage roots. The edible tuberous root is fusiform, with a smooth skin whose color ranges among yellow, orange, red, copper, brown, beige, and purple. The flesh of sweetpotato roots ranges from beige through white, red, pink, violet, yellow, orange, and purple. Sweetpotato roots with white or pale yellow flesh tend to accumulate more starch and have a drier texture than cultivars with red, pink, or orange flesh.

Yams and sweetpotatoes are often confused, particularly in parts of North America. The sweetpotato is botanically very distinct from a true yam, which is native to both Africa and Asia and belongs to the monocot family Dioscoreaceae. The domestication of the yams in Asia, Africa, and tropical America took place separately, so cultivars from each region are distinct. Plants from the Dioscoreaceae family show some primitive characters of both monocotyledons and dicotyledons, suggesting that they are among the earliest angiosperms. Not all species of yam are economically important. Some of the most important yams include: *Discorea alata*, *D. esculenta*, *D. opposita*, and *D. rotundata* (Fig. 15.3).

Yams produce a long and cylindrical true tuber rather than a fusiform storage root like the sweetpotato. True yams produce one to five tubers per plant. The yam tuber is dry and starchy and matures when the plant senesces. The white flesh of yam tubers is dry, starchy, and very low in β-carotene. Table 15.1 summarizes the difference between true yams and sweetpotatoes.

Life Cycle

The sweetpotato is a perennial dicot grown as an annual in areas where freezing temperatures occur. The sweetpotato does not tolerate frost and does poorly at cool temperatures because it is chilling sensitive and tissues are damaged by exposure to temperatures <13°C (55°F). The optimum average temperature for sweetpotato is 25°C (77°F) with

Fig. 15.1. Chinese water spinach is an important aquatic vegetable in parts of Asia. The flower strongly resembles its relative morning glory, which is a weed and/or ornamental in parts of the world.

Fig. 15.2. Sweetpotato leaves and flower. Flowering occurs under short day conditions, so is rare in summer production fields in temperate areas.

Fig. 15.3. True yam plant (*D. opposita*) and tuber.

Table 15.1. Comparison of sweetpotato and yam characteristics.

Convolvulaceae	Dioscoreaceae
Dicotyledon	Monocotyledon
2n=90 (hexaploid)	2n=20 (polyploid)
Monoecious	Dioecious
Origin: tropical America	Origin: West Africa, Asia
Grown in the USA	Imported from the Caribbean
Storage root	Tuber (hypocotyl)
Fusiform	Long or cylindrical
4–10 roots per plant	1–5 tubers/plant
Moist, sweet, no definite maturity	Dry, starchy, mature at senescence
High in β-carotene (orange fleshed cvs)	Very low in β-carotene
Propagated by transplants, vine cuttings	Propagated by tuber pieces
Season: 90–150 days	Season: 180–360 days
Cured at 27–29°C (80–85°F)	N/A
Stored at 13–16°C (55–60°F)	Stored at 12–16°C (54–61°F)

full sunlight and warm nights. Depending on the cultivar and conditions, tuberous roots mature in 3–9 months. The plant is a trailing vine that grows on top of the ground without supports. Sweetpotato is a short-day plant and rarely flowers when the daylight is longer than 11 h. Flowers are approximately 2.5–3.8 cm (1–1.5 in) in diameter and strongly resemble morning glory (Fig. 15.2). True seeds are used for breeding only.

Root System

The edible portion of the sweetpotato is an enlarged fleshy storage root with adventitious buds. Roots tend to extend deep into the soil and are moderately fibrous and extensive in lateral movement; hence plants are relatively drought tolerant.

The sweetpotato storage root is fusiform in shape and shows proximal (stem end) dominance. The edible organs are storage roots formed from thickened secondary roots. Only about 15% of the roots will thicken into storage roots. Typically four to ten storage roots form per plant. Much of the root growth occurs during the first 2 months (Fig. 15.4). The root diameter will continue to increase as long as the leaves are attached, so there is no definite maturity.

Types and Cultivars

Taste and texture preferences vary for sweetpotatoes grown in different regions of the world. In much of the world, high-starch, low-sugar, dry-textured sweetpotatoes are preferred. Typical of

Fig. 15.4. The storage roots can be easily differentiated from fibrous ones on this sweetpotato plant with only five fully expanded leaves.

this type are the white-fleshed sweetpotatoes with firm-dry flesh, higher starch content and mealy texture that are popular in the Caribbean, Africa, and parts of Asia. However, in the USA consumers prefer sweetpotato cultivars that are moist when cooked, have deep orange flesh, and dark red or copper skins. Such cultivars are marketed as "yams", particularly in the eastern USA. In the US market, sweetpotato cultivars are dry when cooked, have lighter-colored flesh, light-colored skins, and are often called "Jersey types". This confusing choice of names resulted in an attempt many years ago to differentiate the newly developed moist, orange-flesh "Porto Rico type" cultivars from the older dry, yellow-flesh fusiform "Jersey type" cultivars that were the standard of the time. "Porto Rico" types have reddish skin to identify their bright orange moist flesh inside and include the cultivars 'Covington', 'Centennial', 'Beauregard', 'Eureka', 'Jewel', and 'Porto Rico'. Examples of the tan skin yellow dry-flesh Jersey types include 'Nugget', 'Nemagold', and 'Jersey'. The moist flesh Porto Rico type cultivars are the most popular in the USA today (Fig. 15.5).

White flesh starchy sweetpotatoes are increasing in popularity in the USA and are now grown in south Florida and California where ethnic groups are familiar with these cultivars. These cultivars are also better suited for processing because their starch content does not caramelize at high temperatures as do sugars. Sweetpotato French fries and chips are products that are processed from cultivars with lower sugar content. Some starchy sweetpotato cultivars were developed for animal feed or industrial processing of starch and alcohol.

Ornamental cultivars have received considerable attention from horticulturists in recent years. Some ornamental cultivars have pale-green foliage while others are deep purple. Ornamental cultivars provide vivid color in the landscape in sunny areas with poor soil. The roots of many are also edible and have white flesh. The young leaves and shoots of sweetpotatoes can be eaten as greens and are popular in Southeast Asia (Loebenstein and Thottappilly, 2009).

Uses

The sweetpotato is an important staple in many countries, particularly in the tropics and subtropics. In developing countries, sweetpotatoes are grown mainly as a substitute for rice and corn and rank as the fifth most important food crop on a fresh-weight basis, after rice, wheat, corn, and cassava (FAOSTAT, 2010). The foliage is almost always used as animal feed in developing countries. However, sweetpotato use has diversified into food products like noodles, flour, starch and pectin, and desserts. In Africa and Asia, dried sweetpotato roots are eaten by humans and also fed to animals as a substitute for corn. In China, it is estimated that 30–50 million metric tonnes (33–35 million short tons) of sweetpotato roots and vine are fed annually to pigs and other livestock. In the northern USA, the sweetpotato is used only as human food, but in the south it is extensively used in regional cuisines and as livestock feed.

Economic Importance and Production Statistics

China is the largest producer with approximately 75% of the total world production (Table 15.2). The most tropical production occurs in Africa and production has steadily increased there. The world sweetpotato cultivation area peaked at 15 million ha (37 million acres) with a total production of over 130 million metric tonnes (143 million short tons) in the early 1970s. There has been a significant reduction in the world production area, to 9.7 million ha (24 million acres) with 105 million metric tonnes (116 million short tons) output in 2009.

Sweetpotato production and consumption in the USA declined rather dramatically from the 1930s until the early 1990s then began to increase. Between 1930 and 1989 per capita sweetpotato consumption in the USA fell from 8.3 kg (18.4 lb) of sweetpotato per person per year to 2 kg (4.5 lb) (Table 15.3). The reasons responsible for such a sharp decline are due

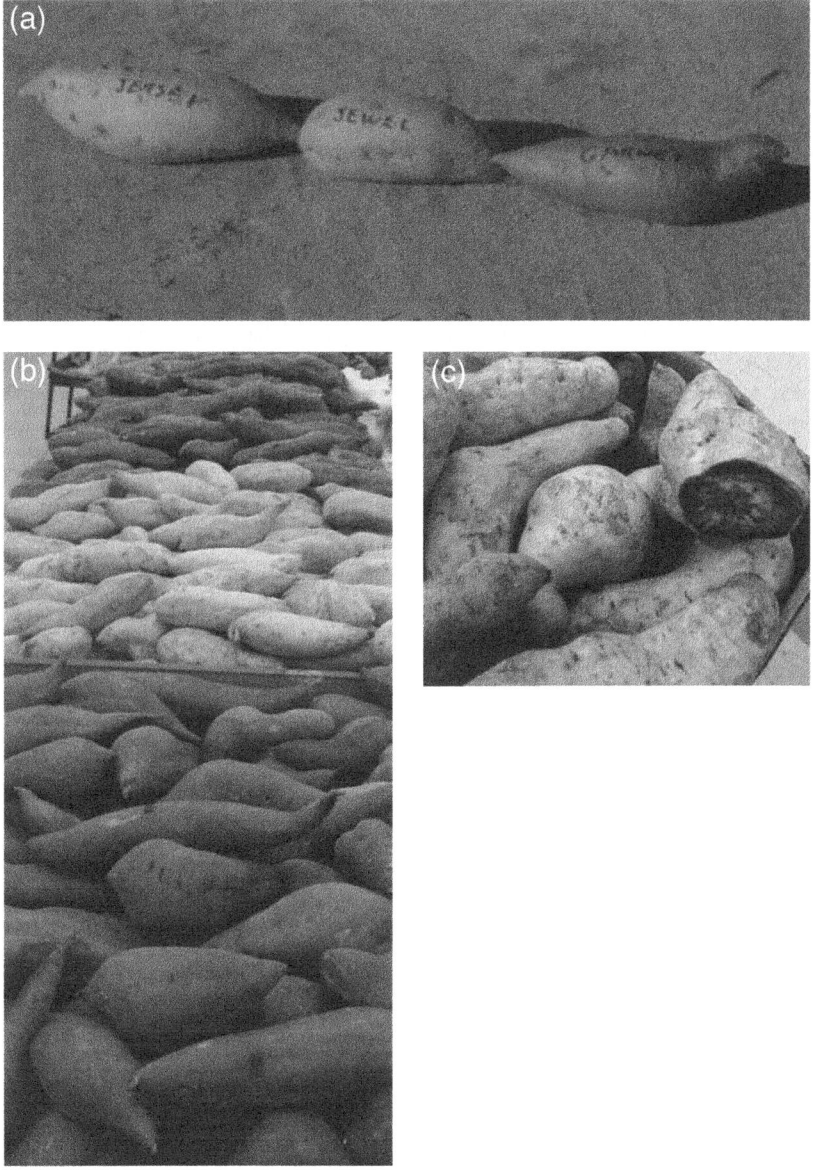

Fig. 15.5. Sweetpotato cultivars vary in both their interior and exterior color. (a) The root on the left is 'Jersey' and has tan skin and dry yellow flesh. 'Jewel' and 'Garnet' are moist-flesh types with copper and deep red colored skins, respectively, and orange moist flesh. (b) Sweetpotatoes in a grocery store with red, tan, and copper skins. The roots with tan skins have purple interiors (c).

to two factors. Unlike the Irish potato, the sweetpotato is not as well suited for processing because of its higher sugar content and textural differences. So processed products made from Irish potatoes increased in popularity during this period at the expense of the sweetpotato. The second reason is an image problem caused by the perception of the sweetpotato as a poor man's staple crop. From a dietary standpoint the decline in the popularity of the sweetpotato is unfortunate because sweetpotatoes are very nutritious and grow well on marginal lands and conditions unfavorable for many other crops.

In 2011, the per capita consumption had risen to 2.9 kg (6.3 lb) of sweetpotato per person per year. The rise from 20 years earlier is due to the development of new cultivars that are better adapted to processing into chips and fries. New cultivars developed for fresh market and an extensive advertising campaign to promote the sweetpotato industry have stimulated the US market as well.

Fresh and frozen exports of sweetpotatoes from the USA jumped to 90,859 metric tonnes (100,155 short tons) in 2010. Canada and the UK remained the two top foreign markets for US sweetpotatoes (USDA, 2010). US fresh and frozen sweetpotato imports continued to decline in 2010, falling to 9,555 metric tonnes (10,533 short tons).

Nutritional Values

Sweetpotatoes provide many nutrients that are important to human health. They are rich in complex carbohydrates, dietary fiber, β-carotene (provitamin A), vitamin C, and vitamin B6 (Table 15.4). Sweetpotato cultivars with dark orange flesh have more β-carotene than those with light-colored or white flesh. Sweetpotatoes are also high in dry matter and a good source niacin, thiamine, riboflavin, and certain minerals.

Production and Culture

There is a major difference between subtropical (temperate) and tropical sweetpotato culture. In the former, the plant is handled as an annual with winter storage of roots for food and as "seed" for subsequent vegetative reproduction. In the tropics, the plant is grown as a perennial with stem cuttings taken and rooted from the standing crop in a continuous planting cycle.

Site selection and field preparation

Sweetpotatoes can be grown on a wide range of soil types, but a well-drained, sandy or silt loam with a clayey subsoil is preferred. Heavier soils can

Table 15.2. World sweetpotato production 2009.[a]

Country	Production (million metric tonnes (short tons))
China	76.8 (84.7)
Uganda	2.8 (3.1)
Nigeria	2.8 (3.1)
Indonesia	2.1 (2.3)
Tanzania	1.4 (1.5)
Vietnam	1.2 (1.3)
India	1.1 (1.2)
Japan	1.0 (1.1)
World total	102.7 (113.2)

[a]For current information on sweetpotato production, see FAO Statistics.

Table 15.3. Sweetpotato production in the USA.

State	Area harvested, acres × 1,000 (ha × 1,000)	Yield, cwt/acre, (cwt/ha)
North Carolina	38.0 (15.4)	130 (321)
Louisiana	23.0 (9.3)	105 (259)
California	9.0 (3.6)	180 (445)
Texas	7.3 (3.0)	110 (272)
Georgia	6.4 (2.6)	140 (346)
Alabama	5.9 (2.4)	115 (284)
South Carolina	5.0 (2.0)	100 (247)
Mississippi	4.8 (1.9)	105 (259)
New Jersey	2.4 (1.0)	110 (272)
Total	103.5 (41.9)	125 (309)

Table 15.4. Nutritional value of raw uncooked sweetpotato (USDA, 2012).

Nutritional value per 100 g (3.5 oz)	
Energy	86 kcal
Water content	69.4%
Carbohydrates	20.1 g
– starch	12.7 g
– sugars	4.2 g
– dietary fiber	3.0 g
– fat	0.1 g
Protein	1.6 g
Vitamin A equivalents	709 µg (89%*)
– β-carotene	8509 µg (79%)
Thiamine (vitamin B1)	0.1 mg (9%)
Riboflavin (vitamin B2)	0.1 mg (8%)
Niacin (vitamin B3)	0.61 mg (4%)
Pantothenic acid (B5)	0.8 mg (16%)
Vitamin B6	0.2 mg (15%)
Folate (vitamin B9)	11 µg (3%)
Vitamin C	2.4 mg (3%)
Vitamin E	0.26 mg (2%)
Ca	30.0 mg (3%)
Fe	0.6 mg (5%)
Mg	25.0 mg (7%)
P	47.0 mg (7%)
K	337 mg (7%)
Na	55 mg (4%)
Zn	0.3 mg (3%)

*Percentages are relative to USDA daily recommendations for adults.

be used for sweetpotato culture but these may cause roots to be rough rather than smooth. The best bulk density of the soil is 1.3–1.5 g/ml (0.23–0.26 oz/tsp). Higher bulk densities tend to reduce storage root formation, resulting in reduced yields or poorly shaped storage roots. The optimum soil pH for sweetpotato is 5.6–6.8, but it grows well in soils with a pH as low as 4.2. Sweetpotato is sensitive to alkaline and saline soils. The maximum soil salinity threshold without a loss in yield is approximately 1.5 dS/m (0.46 dS/ft). As with most root crops, good aeration is essential and this may be a problem on heavier soils. The root systems may develop to a depth of 1.8 m (6 ft), so deep soils are advantageous for good root development. It is recommended to rotate sweetpotato plantings with other crops on a 3- to 4-year cycle.

Growth and development

In tropical countries, where sweetpotato is grown in a continuous cycle, new plants are rooted from vine cuttings. In countries where sweetpotatoes are grown as annuals, sweetpotato transplants are grown from "seed" roots and are called "slips". Sweetpotato transplants are usually produced in special beds during early spring. The seed roots used for propagation are either saved from the previous crop or purchased as seed roots for propagation. Storage roots have no natural dormancy and develop adventitious sprouts (slips) from their vascular tissues. The slips will form adventitious roots on opposite sides of each node along portions of the stems that contact soil. Approximately 250 kg (550 lb) of sweetpotato seed roots will produce 13,000 slips, enough to plant 0.4 ha (1 acre) of sweetpotatoes (Coolong et al., 2012).

Prior to planting, seed stock should be stored at 24–29°C (75–85°F) with 90–95% RH for about 3–4 weeks. There should be no condensation during this period and the room should be ventilated. Presprouting reduces transplant production time and increases transplant yields by two to three times. The seed stock commonly consists of small to medium-sized 2–3.3 cm (0.75–1.5 in) diameter roots that are planted longitudinally. The bed site for transplant production should be a well-drained loam or sandy loam soil. The transplant bed should not be in the same location used for sweetpotato production in the last 3 years and should be located near water for irrigation. The bed should have adequate fertility to produce transplants that are not nutrient stressed. Often a complete analysis fertilizer is applied prior to planting or accordance with soil test results (Coolong et al., 2012).

For rapid even sprouting, the soil temperature at a depth of 10 cm (4 in) should be 18°C (65°F) for several consecutive days prior to bedding. The seed stock should be handled with care to minimize skinning and bruising. Before planting, any mutated, off-colored or diseased roots should be removed. If certified roots are used, this inspection step was already conducted prior to purchase. Roots exhibit apical dominance. Shoots emerge from the proximal end of the root that developed closest to the mother plant more rapidly than at the distal or bottom of the root. To encourage even shoot growth, seed roots are sometime cut in half longitudinally before planting or aged, which also breaks root apical dominance (Rubatzky and Yamaguchi, 1997). The seed roots are often treated with fungicide to reduce infections from *Rhizopus* sp. and other soil rot-causing organisms before planting. Seedbeds are sometimes fumigated with steam or chemicals to reduce rot prior to planting (Coolong et al., 2012).

The transplant bed is cover with loose soil to a uniform depth of about 5 cm (2 in), (Fig. 15.6). Burying the seed roots too deeply can increase rotting due to suffocation, delay transplant production, and cause uneven emergence. The bed should be immediately watered to settle the soil and encourage sprouting. Depending on the bed history, a weed control strategy may be needed. Stale-bed tillage, bed fumigation, hand weeding, or herbicide treatment are examples of weed control strategies that can be employed. If soil and air temperatures are below optimum, the beds can be covered with clear plastic film for solar heating to increase temperatures to no greater than 29°C (85°F) in order to speed sprout emergence. In sunny weather, ventilation may be necessary to prevent overheating. Each parent root may produce up to 15 slips, although not all develop at the same time.

The slips are ready for transplanting when 25–30 cm (10–12 in) tall with approximately four to six leaves (Fig. 15.7). Slips can either be cut about 2.5 cm (1 in) above the sand line and re-rooted or rooted while attached to the mother root. Supporters of re-rooting slips say that this practice slows the transmission of root/soil-borne diseases. Shoots should be cut with an upward motion. The cut should always be made upward without letting

Fig. 15.6. Roots are covered in this transplant bed to encourage sprouting for the production of slips.

Fig. 15.7. This sweetpotato transplant bed is covered at night to protect against low temperatures.

the knife touch the soil. Sterilize the knife frequently by dipping in a 1:1 solution of chlorine bleach and water (Jett, 2006).

Rooted slips may be disease free if certified roots are used for propagation and the soil sterilized or fumigated prior to planting. By grasping a slip near its base and lifting and simultaneously twisting the slip with roots intact, it can be separated from the parent root and lifted free of the soil. Rooted slips will establish more rapidly than unrooted cuttings. Only clean disease-free boxes should be used to store and transport transplants. Transplants account for about 20% of the total production costs.

Used seed stock, if in good condition after transplant production, should not be sold for fresh market or processing. The seed roots may contain biological contaminates or chemicals that are dangerous for human health.

Certified seed

In many countries, certified sweetpotatoes are available for propagation. The advantage of certified sweetpotatoes is that they have been inspected in the field and storage to assure genetic purity, a minimum of genetic mutations, and freedom from serious diseases. When available, many growers prefer certified sweetpotato seed stock to help control disease and maintain genetic purity. Certified seed is the second crop produced after foundation seed stock. In the USA, several states maintain sweetpotato seed certification programs to ensure growers have a reliable supply of high quality roots for propagation (Jett, 2006).

Field establishment

In preparation for planting, the soil must be finely tilled using rotovators or disks and free of clods, stones, and other impediments that would result in misshapen roots. Conservation tillage techniques are not widely employed for commercial production in temperate regions. An ideal field for sweetpotato production can be shaped into high ridges to ensure adequate drainage from waterlogged soils (Fig. 15.8).

Poor aeration or an oxygen concentration of less than 10% in the soil in the initial phase increases the degree of lignification of stele cells and suppresses the primary cambium activity, resulting in young roots with excessive fiber. Flooding near harvest may cause roots to rot in the soil or in storage.

Soil should be washed from transplanters and other equipment between use in commercial and seed production fields to minimize possible contamination. Sweetpotato transplants are usually spaced in rows 0.81–1.21 m (32–48 in) apart with an in-row spacing of 20–46 cm (8–18 in) depending on the cultivar and how the roots will be used (Coolong *et al.*, 2012). To encourage root formation cuttings are sometimes planted at an angle to increase soil–stem contact. The soil should be damp to encourage rooting, so regular sprinkler irrigation will keep cuttings from drying out and keep the soil moist to encourage rooting. Rooted transplants are easier to establish but may need irrigation after transplanting to reduce shock and hasten establishment.

Fig. 15.8. A sweetpotato field successfully established on ridged rows in eastern North Carolina.

Irrigation

Sweetpotato is considered to be a moderately drought-tolerant crop because of its extensive root systems. However, it cannot withstand long periods of drought. Yield is considerably reduced if drought occurs at the time of storage-root initiation. Established sweetpotatoes can withstand severe drought but moisture stress at transplanting, especially when establishing cuttings with few roots, will reduce stands and ultimately yields. Irrigation at the time of establishment is a common practice in dry areas especially when planting unrooted cuttings, to decrease shock and improve establishment.

The water requirement is for about 2.5 cm (1 in)/week for optimum growth. Annual rainfalls of 750–1,000 mm (30–39 in) are considered most suitable, with a minimum of 500 mm (20 in) during the growing season. Sandy soils generally require more moisture compared to clay loams.

Dry weather favors the formation and development of storage roots. Soil moisture at 60–70% of field capacity is favorable for the initial phase, 70–80% for the intermediate phase, and 60% for the final phase (CABI, 2012). Overwatering can result in cracking and irregular growth. Irrigation should be withheld 3–4 weeks before harvest to avoid cracking and oversized roots.

In areas with periodic rainfall throughout the season, irrigation is not regularly required. Where rainfall occurs outside the growing season, irrigation is required. Drip, sprinklers, or sometimes carefully managed furrow irrigation is used with drip being the most water efficient but also the most expensive to install.

Crop production and development

Long days with high temperatures are required for optimum yields of sweetpotato. Temperatures of 24–29°C (75–85°F) during early summer favor vine and foliage development, which is essential to support root development. When temperatures fall below 12°C (54°F) or exceed 35°C (95°F), growth is retarded. Dry-matter production increases with soil temperatures from 20–30°C (68–86°F), but declines above 30°C (86°F). In late summer and early fall, relatively short days and moderate temperatures 21–34°C (70–75°F) favor an extended and efficient storage-root development period. It takes approximately 4–5 months to mature a sweetpotato crop with marketable-sized roots even though root enlargement begins shortly after transplanting (Woolfe, 1992).

Sweetpotato is a sun-loving crop; however, it can tolerate a 30–50% reduction of full solar radiation. Light saturation of single-leaf photosynthesis occurs at around 800 $\mu E/m^2/s$. The light intensity required for saturation in the canopy increases with leaf area index. Optimum leaf area index in the field is 3–4 at solar radiation of 380 $gcal/cm^2/day$ (CABI, 2012). The photosynthetic rate of the canopy in the field is highest between 10 a.m. and 2 p.m.

Plastic mulch and trickle irrigation is an effective technique for sweetpotato production, particularly in northern areas where soil warming is needed. Black plastic mulch provides weed control and increases soil temperature (Jett, 2006). For best performance the soil surface must be smooth so that the plastic contacts the soil surface. A plastic-laying machine is the most effective way to install mulch properly. Clear plastic mulch is very effective for increasing soil temperature but does not control weeds. IRT (infrared-transmitting) films provide soil warming and weed control. Early and total yields are increased and more than compensate for the increased materials and labor costs (Seavert, 2003).

Non-woven or spunbonded polyester and polypropylene, and perforated polyethylene field covers, increase heat unit accumulation by two to three times over ambient and may stimulate early-season growth in short-season areas. Soil temperatures and root growth are also increased under row covers as are early yields, and in some cases total yields. Two to four degrees of frost protection may also be provided (Seavert, 2003).

Fertilizer and nutrition

Sweetpotatoes are considered moderate feeders relative to other vegetable crops. Both the leaves and roots sequester significant amounts of mineral nutrients from the soil. The vines contain more nitrogen (N) and calcium (Ca) than the roots. To maintain soil fertility the nutrients removed by the roots must be replaced (Table 15.5). If the vines are cut and left in the field to decompose, their nutrients need not be replaced.

Sweetpotato does well on relatively poor soils and has a low requirement of phosphorus (P). Relative to other the crops, the N required is relatively low while the potassium (K) requirement is moderately high. Potassium deficiency can cause excessively slender fusiform roots. Sweetpotatoes have relatively low requirements for micronutrients.

Table 15.5. Nutrient removal by sweetpotato roots and foliage[a] (adapted from Scott and Bouwkamp, 1974).

Nutrient	Sweetpotato tissue		
	Roots kg/ha (lb/acre)	Leaves and vines kg/ha (lb/acre)	Total kg/ha (lb/acre)
N	47 (42)	52 (46)	99 (88)
P	19 (17)	8 (7)	27 (24)
K	179 (160)	101 (90)	280 (250)
Ca	11 (10)	46 (41)	57 (51)
Mg	9 (8)	9 (8)	18 (16)
Total	265 (237)	216 (192)	481 (429)

[a]Data are averaged from four cultivars.

Band placement is the most efficient application technique. Some growers band all fertilizer in the bed before, during, or just after transplanting, while others prefer a split application with one of the applications coming near planting and the other later in the season. Starter solution (15N-30P-15K) is used in the transplant water at a rate of 7.2 g/l (3 lb/50 gal). Side dressing of all N 3 or 4 weeks after transplanting is recommended for some cultivars. Excess N can overstimulate vine growth and retard development of storage roots (Motes and Criswell, 2003).

Harvesting and marketing

Early harvest minimizes the risk of losses due to cold and wet conditions later in the season. The timing of harvest depends on the market and cultivar. Sweetpotatoes hold very well in the field because they do not ripen and senesce like many other crops. However, roots should be harvested before the soil temperatures fall below 13°C (55°F) to avoid chilling injury. In advance of a frost, vines should be detached from the roots to prevent vine decay from spreading to the roots. Vines are usually cut from the roots as part of the digging operation or as a separate operation in advance of digging. Roots are dug either using a modified moldboard plow or chain-type potato harvester that separates the storage roots from the soil. In some areas, sweetpotatoes are dried in rows. After drying, the roots are collected and taken to a packinghouse for sorting and grading. Sweetpotatoes have thin skins that are easily damaged by mechanical harvesting, so sweetpotatoes are sometimes gathered by hand in the field after digging, even though automated harvesting is widely used.

Average yield of storage roots throughout the world in 1994 was 13 metric tonnes/ha (5.8 short tons/acre). The average yield in Asia was 15 metric tonnes/ha (6.7 short tons/acre), varying from 2 to 22 metric tonnes/ha (0.9–9.8 short tons/acre). China has a yield of 17 metric tonnes/ha (7.6 short tons/acre), which lifted the average yield of Asia and the world. Southeast Asia has a yield of less than 8 metric tonnes/ha (3.6 short tons/acre) and Africa, 6 metric tonnes/ha (2.7 short tons/acre) (FAOSTAT, 2010). The yield potential of sweetpotato is high. However, various abiotic and biotic stresses in the tropics prevent the full expression of this potential.

Curing and storage

Postharvest handling procedures differ between temperate and tropical regions. In the tropics sweetpotato roots are harvested as needed, often by hand and the vines are rooted for new production (Rubatzky and Yamaguchi, 1997). In temperate regions, sweetpotatoes are harvested and handled mechanically, often damaging the storage roots in the process. Harvested roots are cured at 29°C (85°F) and 90%+ RH for 7–10 days to heal cuts in the periderm. Curing should begin as soon after harvest as possible. The curing treatment promotes the formation of wound cork and the production of phenolics on damaged surfaces, thereby preventing excessive water loss and pathogenic infection. Curing also converts starch to sugar, which improves eating quality. Cured roots can be stored for 6–12 months if stored at 13–15°C (55–59°F) and 85–90% RH (Coolong et al., 2012).

Roots are susceptible to chilling injury and should not be exposed to low nonfreezing temperatures. Chilling injury is a serious physiological disorder that reduces storage life and quality of tropical plants when stored below 13°C (54°F). Cell membranes become leaky after chilling injury, leading to root shriveling, surface pitting, abnormal wound periderm formation, fungal decay, and "hardcore" when the interior of the root fails to soften after baking. Non-cured roots are more susceptible to chilling injury than cured roots (Rubatzky and Yamaguchi, 1997).

Weed Management Strategies

Weed management programs may combine timely mechanical cultivation with herbicide applications. However, the number of herbicides approved for use on sweetpotato, at least in North America, is small. Herbicides increase production costs, may drift and adversely affect adjacent crops, and may have harvest waiting intervals from 30 to 55 days after use. Chemical control is not an option for sustainable/organic producers and may not be the best option for some conventional growers as well. Since the sweetpotato vine growth is aggressive, the leaf canopy alone may shade out many weeds after establishment if there is effective control at the time of transplanting. Because sweetpotatoes are not transplanted until early summer, there is sufficient time to employ the stale-bed technique in fields prior to planting for effective early-season weed control without chemicals. With the stale-bed technique, weed seeds are encouraged to germinate then are killed by shallow cultivation prior to planting. Mechanical cultivation may be an effective weed control option if the field is not contaminated with perennial weeds or seed from previous seasons. Typically, three cultivations are required to provide effective weed control. After the stale-bed is cultivated and the crop planted, additional cultivations should begin within 10 days after planting and continue until vines begin to "run", covering most of the field. The final cultivation may cut some of the tips of vines, which usually is not a problem. On fields with very high weed seed banks mechanical cultivation may not be sufficient to reduce all weeds within rows (Coolong et al., 2012).

Another option for controlling weeds in sweetpotato fields is the use of mulch. Organic mulches can be spread after transplanting to smother weeds until the vine canopy can cover the soil. Some hand weeding may be necessary if weeds penetrate the mulch or the plant canopy.

A plasticulture system to warm soil, control weeds, and conserve moisture can be used effectively for sweetpotato culture as well. (Seavert, 2003).

Diseases

Fusarium wilt (*Fusarium oxysporum*) is controlled by using resistant cultivars and only nitrate forms of N. High soil pH will improve control of Fusarium wilt but may favor other soil-borne pathogens like pox. Rotation for 3 years away from sweetpotatoes is also helpful against most fungal pathogens. Sweetpotatoes and tobacco are believed to be susceptible to the same strains of *Fusarium*, so they should not be grown in rotation. Disease-free certified transplants are also important tools for controlling sweetpotato diseases.

Scab (*Sphaceloma batatas*) is one of the most prevalent diseases of sweetpotato. Small brown lesions form on the veins of the leaves. As the disease progresses, these lesions become corky, which causes the leaves to curl. The stem lesions are slightly raised with rusty-brown spots. Scab-like structures form on the stems as the spots coalesce. To control sweetpotato scab, use disease-free roots for propagation, rotate sweetpotato with unrelated crops, and practice proper sanitation. Some sweetpotato cultivars are resistant to the disease.

Sweetpotato witches' broom or little leaf disease is caused by a phytoplasma and causes severe stunting of affected plants (Gibb et al., 1995). Symptoms include vein clearing, rounded, small, chlorotic leaves with curled leaf margins, stunted plants, more erect growth habit, proliferation of axillary shoots, and a greatly reduced root system, resulting in weak plants with a compact bushy appearance. The number and quality of storage roots are reduced. Leafhoppers (*Orosius lotophagorum*) can vector the phytoplasm between infected and healthy plants of *Ipomoea* spp. As the pathogen has a very long latent period, up to 283 days by graft transmission, infected planting material can appear healthy (Jackson and Zettler, 1983). In sweetpotato certification programs, no phytoplasm infections are tolerated during production, so using certified roots can prevent the disease.

Black rot (*Ceratocystis fimbriata*) disease causes small, circular, slightly sunken, dark brown spots on sweetpotato roots. With further development, the spots enlarge and appear greenish black to black when wet and grayish black when dry. The rot usually remains firm and shallow. However, if secondary fungi or bacteria invade the tissue, which frequently occurs, the flesh beneath the spot turns black to the center of the root. Tissue near the discolored area may taste bitter. Eventually, the entire root rots in severe cases. Roots that appear healthy at harvest can rot in storage, during transit or at the market (Coolong et al., 2012).

The black rot fungus survives in the soil in crop debris, such as infected storage roots that escape detection at harvest or bedding. The fungus either colonizes the young shoots or infects the stem. Transplants, and subsequently, field plants are thereby infected as the disease is easily spread via soil, on equipment or through transplants.

Scurf (*Monilochaetes infuscans*) is a soil-borne fungus disease that causes black to brown surface discoloration of the roots. Making cuttings of shoots in seedbeds above the soil line and rerooting transplants can help successfully control scurf and Southern blight (*Sclerotinium rolfsii*). Using crop rotations of 3–4 years away from sweetpotatoes is also effective. Careful handling of roots during harvest to avoid bruising along with proper curing reduces the incidence of these diseases after harvest (Coolong et al., 2012).

To prevent buildup of pox soil rot (*Streptomyces ipomoea*) crop rotation should be practiced, disease-free roots and transplants used, and soil pH maintained below 5.5 in sweetpotato production fields. Soil fumigation is used to correct serious soil infection (Coolong et al., 2012).

The feathery mottle virus causes internal cork and russet cracking. Internal cork virus is another common disease of sweetpotato. The use of disease-free planting material and crop rotation are the most reliable means of controlling these diseases.

Harvest and handling conditions greatly influence susceptibility to postharvest decays. A common disease of storage is Rhizopus soft rot caused by the fungus *Rhizopus stolonifer*. This disease appears as a "hairy" fungal rot and typically invades sweetpotatoes through wounds that occur during harvest or the curing process. Following proper curing protocols can ensure adequate wound healing to prevent infection. Storing only blemish-free roots and discarding damaged or rotted roots can help control this disease (Coolong et al., 2012).

Insect Pests

Among the 300 insect and mite species that feed on sweetpotato in the tropics and subtropics, only sweetpotato weevil (*Cylas formicarius*) and vine borer (*Omphisa anastomosalis*) cause damage and yield loss over wide areas. The sweet-potato weevil is the most destructive insect pest in the tropics and subtropics (CABI, 2012). No resistant cultivars are available, so integrated pest management (IPM) is recommended consisting of crop rotation, eradication of *Ipomoea* spp. weeds, use of clean planting material, deep planting, regular hilling to fill soil cracks around plants, and the use of sex pheromone to trap male weevils.

Soil-borne insect pests are the most important insect pests of sweetpotatoes in temperate regions. Wireworms (*Melanotus communis*) and white grubs (*Phyllophaga smithi*) can be challenging to control because they can attack roots over an extended period. With sweetpotatoes, economic damage by wireworms can be common in some areas and may result in more than 40% damaged roots. Wireworms can be found in soils following any type of rotation, but they are usually more severe when crops follow established sod or the second year following sod. Wireworms feed upon the small roots of sweetpotatoes throughout the season. Most mature wireworm larvae are hard, brown, smooth, varying from 1.3–3.7 cm (0.5–1.5 in) in length. Most wireworms have lifecycles that last 2 or more years, so recent history of wireworms in a field often indicates increased risk (Coolong et al., 2012).

White grubs are soil-borne insects that attack developing roots. White grubs are the larval stage of May and June beetles. White grub feeding damage may extend 1.25–2.54 cm (½–1 in) in diameter. White grub numbers in the soil are influenced by past rotations, proximity to wooded areas and levels of organic matter in the soil. Fields previously in sod or with high organic matter are more likely to have high grub numbers. Soil-applied insecticides are sometimes used to control grubs and wireworms on sweetpotatoes before or during planting (Coolong et al., 2012).

A few insects feed in the foliage of sweetpotatoes in temperate regions, but economic infestations, while they do occur, are not common. Adult flea beetles (*Chaetocnema confinis*) feed on foliage leaving narrow channels or grooves in the upper surfaces of leaves, which may turn brown and die. Larvae live underground and feed on roots. Shallow winding tunnels etched into root surfaces indicate an infestation of flea beetle larvae. These tunnels eventually darken and split open leaving shallow scars. Cultural practices can effectively prevent flea beetle infestations. Controlling weeds along fencerows and plowing under crop debris destroy overwintering and egg-laying sites. The use of resistant cultivars such as 'Jewel' or 'Centennial' is the most effective means of preventing sweetpotato flea beetle injury (Sorensen and Baker, 2012).

Silverleaf whitefly (*Bemisia argentifolii*) and sweetpotato whitefly (*Bemisia tabaci*) suck sap from the foliage of plants. This feeding causes weakening and early wilting and reduces the plant growth rate and yield (Berlinger, 1986). It may also cause leaf chlorosis, leaf withering, premature dropping of leaves and, in severe cases, plant death. Indirect damage results from the accumulation of honeydew produced by the whiteflies. This honeydew serves as a substrate for the growth of black sooty mold on leaves (Berlinger, 1986).

There is no easy way of controlling the sweetpotato whitefly (Mau and Kessing, 2007). Egg mortality is usually minimal. Weather and predation may cause high mortality rates during the crawler and first nymphal stages, but has only moderate effects on the later nymphal stages. Sweetpotato whitefly has become resistant to chemical insecticides quite rapidly in some parts of the world. Regular insecticide applications can result in resurgence of other pests. A combination of cultural practices and chemical application provides the best chance for controlling whiteflies on sweetpotato (Mau and Kessing, 2007). The use of sound cultural practices that may avoid, delay, or lessen the severity of the sweetpotato whitefly infestation is a key to control. Careful selection and timely application of insecticides can help regulate whitefly populations as part of an IPM strategy.

Tortoise beetles (*Aspidomorpha furcata*) chew round holes in the leaves and may be found on the undersides of the leaves. Several natural enemies including egg and larval parasites (*Tetrastichus* sp., Eulophidae, Chalcidae) and predators (*Stalilia* sp., Mantidae) have been used effectively in Southeast Asia. Spotted (*Diabrotica undecimpunctata*) and banded (*D. balteata*) cucumber beetles may also feed on vines and leaves. These insects are controlled with foliar insecticides and trapping, and natural predators may provide control (Coolong *et al.*, 2012).

The root-knot nematode (*Meloidogyne* spp.) may also attack sweetpotato. Preplant fumigant materials are an option, although these are expensive and require specialized training and equipment for proper application. Most growers use crop rotation to manage nematodes. Rotating sweetpotato for 2 or more years with tall fescue (*Festuca arundinacea*), Bahia grass (*Paspalum notatum* Flugge), rapeseed (*Brassica napus*), or certain legumes may reduce populations without the use of nematicides (Rodriguez-Kabana and Canullo, 1992; Coolong *et al.*, 2012).

References

Austin, D.F. (1988) The taxonomy, evolution and genetic diversity of sweetpotatoes and related wild species. In: Gregory, P. (ed.) *Exploration, Maintenance, and Utilization of Sweetpotato Genetic Resources.* CIP, Lima, Peru, pp. 27–60.

Bassett, K.N., Gordon, H.W., Nobes, D.C. and Jacomb, C. (2004) Gardening at the edge: documenting the limits of tropical Polynesian kumara horticulture in southern New Zealand. *Geoarchaeology* 19, 185–218.

Berlinger, M.J. (1986) Host Plant Resistance to *Bemisia tabaci*. *Agricultural Ecosystems Environment* 17, 69–82.

CABI (2012) Sweetpotato. Crop protection compendium. Datasheet: *Ipomoea batatas* (sweetpotato). Available at: www.cabi.org/cpc/?compid=1&dsid=28783&loadmodule=datasheet&page=868&site=161 (accessed 22 July 2012).

Coolong, T., Seebold, K., Bessin, R., Woods, T. and Fannin, S. (2012) Kentucky Sweetpotato Production for Kentucky. Available at: www.ca.uky.edu/agc/pubs/id/id195/id195.pdf (accessed 25 July 2012).

FAOSTAT (2010) Food and Agriculture Organization of the United Nations. Available at: http://faostat.fao.org/site/567/DesktopDefault.aspx?PageID=567 (accessed 22 July 2012).

Gibb, K.S., Padovan, A.C. and Mogen, B.D. (1995) Studies on sweat potato little-leaf Phytoplasma detected in sweetpotato and other plant species growing in northern Australia. *Phytopathology* 85(2), 169–174.

Jackson, G.V.H. and Zettler, F.W. (1983) Sweetpotato witches' broom and legume little-leaf diseases in the Solomon Islands. *Plant Disease* 67(9), 1141–1144.

Jett, L.W. (2006) Growing Sweetpotatoes in Missouri. G-6368. Available at: http://extension.missouri.edu (accessed 22 July 2012).

Loebenstein, G. and Thottappilly, G. (2009) *The Sweetpotato.* Springer Verlag, New York.

Mau, R.F.L. and Kessing, J.L. (2007) Crop Knowledge Master. Sweetpotato Whitefly *Bemisia tabaci* (Gennadius). Available at: www.extento.hawaii.edu/kbase/crop/Type/b_tabaci.htm (accessed 26 July 2012).

Motes, J.E. and Criswell, J.T. (2003) Sweetpotato Production. Available at: http://osufacts.okstate.edu (accessed 22 July 2012).

O'Brien, P.J. (1972) The sweetpotato: Its origin and dispersal. *American Anthropologist* 74, 343–365.

Rodriguez-Kabana, R. and Canullo, G.H. (1992) Cropping systems for the management of phytonematodes. *Phytoparasitica* 20, 211–224.

Rubatzky, V.E. and Yamaguchi, M. (1997) *World Vegetables: Principles, Production, and Nutritive Values*, 2nd edn. Chapman and Hall, New York.

Scott, L.E. and Bouwkamp, J.C. (1974) Seasonal mineral accumulation by the sweetpotato. *HortScience* 9, 233–235.

Seavert, C. (2003) Sweetpotato (*Ipomoea batatas*) Commercial Production Vegetable Guides. Available at: http://nwrec.hort.oregonstate.edu/swpotato.html (accessed 22 July 2012).

Sorensen, K.A. and Baker, J.R. (2012) Insect and related pests of vegetables: Some important, common and potential pests in southeastern United States. Publication Ag-295. Available at: http://ipm.ncsu.edu/ag295/html/index.htm (accessed 22 July 2012).

USDA (2010) US sweetpotato exports to selected countries. Available at: http://usda.mannlib.cornell.edu/MannUsda/viewDocumentInfo.do?documentID=1492 (accessed 27 July 2012).

USDA (2012) Data from national nutrient database for standard reference, release 24. Available at: http://ndb.nal.usda.gov/ndb/foods/show/3273 (accessed 28 July 2012).

Woolfe, J.A. (1992) *Sweetpotato: An Untapped Food Resource*. Cambridge University Press and the International Potato Center (CIP), Cambridge, UK.

Zhang, D.P., Ghislain, M., Huamán, Z., Cervantes, J.C. and Carey, E.E. (1998) AFLP assessment of sweetpotato genetic diversity in four tropical American regions. *CIP Program Report 1997–1998*, pp. 303–310.

16 Family Brassicaceae

Botany

Brassicaceae is a very important family with over 1,800 species from more than 100 genera worldwide including many important vegetable, field, and oil crops (Table 16.1). Members of this family are also sometimes referred to by their archaic family name Cruciferae or are called crucifers for short (Nieuwhof, 1969; Rubatzky and Yamaguchi, 1997).

Plant and Flower Characteristics

The word Cruciferae means cross in Latin. The family was so named originally because of the characteristic cross-shaped flowers shared by all members of this family. Close examination reveals that each floret has four opposed flower petals that form a square cross (Fig. 16.1). Flower petals vary widely in color among species and may be white, cream, pink, or purple (Nieuwhof, 1969).

The flowers are bisexual with one pistil, and four long and two short stamens on each flower for a total of six. A superior ovary develops into a long fruit pod called a silique, 4.5–10 cm (2–4 in) in length, with a thin, translucent inner membrane, the replum, that separates the two chambers of the pod, and to which the seeds are attached (Fig. 16.2; Nieuwhof, 1969).

Flowers are usually cross-pollinated by insects, most often by bees. Although the flowers are perfect, having both male and female flower parts, they are not self-pollinated because of pollen incompatibility. Pollen can only successfully germinate and grow on a stigma of a flower from a different plant because it is genetically incompatible with flowers from the same plant where it was produced (Myers, 2006). The branched inflorescence may be 0.5 m (1.6 ft) across in some cases. When siliques dry they dehisce and separate into two halves releasing from 10 to 30 dark purple seeds (Musil, 1948). A well-pollinated *Brassica oleracea* plant may produce 0.23 kg (0.5 lb) of seed. There are approximately 320 seeds in a gram (9,000 seeds/oz; Lorenz and Maynard, 1988).

The immature siliques are edible but rarely are eaten. Immature radish siliques are sometimes used to flavor salads and have pungency very similar to radish roots. Many but not all plants in this family, e.g. mustard, horseradish roots, cress, and watercress foliage, have a strong pungency associated with roots, leaves, or seeds. The pungency of vegetables in this family is caused by allyl isothiocyanate, an organosulfur compound. Allyl isothiocyanate helps the plant defend against herbivory. Plants store allyl isothiocyanates in the harmless form as glucosinolates separate from the myrosinase enzyme (Fenwick *et al.*, 1982). When an animal or insect chews the plant tissue, the glucosinolates and myrosinase are combined, releasing bitter tasting and goitrogenic substances such as isothiocyanates, thiocyanates, nitriles, and goitrin that repel predators (Fenwick *et al.*, 1982).

GENUS *BRASSICA*

Brassica is one of the major genera, and contains probably more than 40 species of mostly annual, biennial, or sometimes perennial herbaceous plants or small shrubs of Old World origin. Cabbage, broccoli, cauliflower, Brussels sprouts, kale, collards, and kohlrabi are familiar crops in the genus *Brassica* (Nieuwhof, 1969). There is confusion about the correct botanical name for the various crops in this genus. This is due to molecular genetic studies that have shown that major morphological differences between different *B. oleracea* may be caused by single or few genes and therefore do not warrant classification into separate botanical varieties (Bancroft *et al.*, 2006). Over the past 20 years, there has been a growing consensus among plant biologists that the *B. oleracea* crops such as cabbage, broccoli, and cauliflower, should be not be

Table 16.1. A list of Latin binomial and common names of some of the more economically important crops from the family Brassicaceae (Griffiths, 1994).

Brassicaceae (Cruciferae outdated name) – mustard family

Armoracia rusticana	Horseradish
Sinapis alba	White mustard
Brassica juncea	Leaf mustard
Brassica napus (Napobrassica group)	Rutabaga
B. napus (Pabularia group)	Siberian kale
Brassica nigra	Black mustard
Brassica oleracea (Acephala group)	Kale, collard
B. oleracea (Albogiabra group)	Chinese kale
B. oleracea (Botrytis group)	Cauliflower
B. oleracea (Capitata group)	Cabbage
B. oleracea (Gemmifera group)	Brussels sprout
B. oleracea (Gongylodes group)	Kohlrabi
B. oleracea (Italica group)	Broccoli
B. oleracea (Costata group)	Tronchuda cabbage
Brassica rapa (Chinensis group)	Chinese cabbage (nonheading), pak-choi
B. rapa (Pekinensis group)	Chinese cabbage (heading), pe-tsai
B. rapa (Perviridis group)	Spinach mustard
B. rapa (Rapifera group)	Turnip
B. rapa (Ruvo group)	Broccoli raab, rapini
Lepidium sativum	Garden cress
Crambe maritime	Sea kale
Nasturtium officinale	Watercress
Raphanus sativus	Radish
Wasabia japonica	Japanese horseradish

Fig. 16.1. Brassica flower with four opposed petals.

Fig. 16.2. An immature Brassicaceae silique. When mature, the silique will turn brown and dehisce releasing the small round seeds.

subdivided into distinct botanical varieties. For example, many now classify cabbage as *B. oleracea* Group Capitata rather than botanical variety *capitata*, which is more common in the older literature. The term Group (gp.) is used for convenience by horticulturists to show groups of horticultural significance within a species that were previously classified as separate varieties (Griffiths, 1994).

In this first section, we will examine some common features and requirements for growing brassicas. In later sections, specific details about the production of cabbage, cauliflower, broccoli, Brussels sprouts, and Chinese cabbage will be discussed.

Species and cultivar selection

Selection of *Brassica* species and cultivars should be made with regard to physiological and environmental limitations and with knowledge of potential disease and insect problems. Prolonged exposure of some species to low temperatures (between 1.7–10°C; 35–50°F) can cause premature bolting, particularly in early-maturing cultivars (Nieuwhof, 1969). As flower stalk formation is genetically controlled, selection of cold-tolerant cultivars becomes important in regions with fluctuating temperatures. State or regional recommendations should be followed for cultivars best adapted to specific local conditions. Cultivar selection is also determined by market demand as related to morphological characteristics. Examples include round versus conical-headed cabbage or curly versus flat-leaf kale, purple versus green cabbage, or physical size of some vegetables (e.g. dwarf versus standard size).

Soil preparation

Deep tillage and open-soil cultivation are commonly used to produce brassicas. However, most brassicas are also well adapted to both plasticulture and conservation tillage methods depending on local conditions. Brassica vegetable crops are often planted in shaped, raised beds, about 15 cm (6 in) high, with a flat, uniform bed surface (Swaider et al., 1992). When wet saturated soil conditions are expected, bed height can be increased and tile drains used to compensate, because most brassicas prefer well-drained soils. Use of cover crops and resulting residues conserves moisture and reduces erosion, making conservation tillage practices desirable on highly erodible soils and in areas with limited rainfall. This method has proven effective for summer production in the more mountainous areas. Using strip-tillage, transplants can also be successfully established in a conservation tillage system (Hoyt et al., 1994).

Fertilization

Brassicas are generally cool-season crops that prefer deep, fertile, friable, sandy or silt loam soils (Nieuwhof, 1969). Optimum soil pH is 6.0–6.5 for most species, but some are more tolerant of soil acidity than others. Fertility requirements of a given crop are site specific and dependent upon many environmental factors, including soil type and structure, native soil fertility, previous crops in the rotation, and length of the growing season (Swaider et al., 1992). Therefore, application of fertilizers should rely on local production recommendations for specific crop species based upon current soil test results.

Nutrients can be applied preplant. For both transplanted and precision-seeded crops such as cabbage, broccoli, cauliflower, Brussels sprout, and some Asian vegetables, banding a part of the fertilizer 5–10 cm (2–4 in) from the young roots aids in establishment and early growth. For longer season cultivars with later maturity, from 30% to 50% of the total nitrogen (N) requirement is supplemented post-plant during the rapid vegetative phase of growth. For example, supplemental N for crops that form a head is often applied at the onset of head formation. Excess N in broccoli and cauliflower can predispose plants to hollow stems and delay head or curd formation. Most brassica species have a significant boron requirement, ranging from 1 to 4 kg/ha (0.9–3.6 lb/acre). Additional boron (B), sulfur (S), and other micronutrients may be needed in sandy soils or soils naturally low in these elements. Many brassica species, particularly those grown as vegetables, are sensitive to molybdenum (Mo) deficiency. The management of soil pH can be critical for these crops since Mo availability, like that of phosphorus (P), is reduced below soil pH 5.5. Application of excess micronutrients such as B, can cause plant injury and reduce yields. When drip irrigation is available, fertigation can be an effective means of providing all or most of the crop nutritional needs (Peirce, 1987; Swaider et al., 1992).

Direct seeding

The planting method used depends upon the crop to be grown, the growing region and target market. Vegetable brassicas such as cabbage, broccoli, cauliflower, greens, and many of the Asian vegetables are direct-seeded using precision seeders to optimize plant spacing and reduce the need for costly hand-thinning and reduce seed costs. Adequate soil preparation is essential for precision seeding. To promote best stand establishment, seeding should occur when soils are moist and soil temperatures optimum to promote rapid germination. Planting certified seed helps to eliminate seed-borne disease problems, enhances germination and early plant vigor, which is critical for uniform maturity and efficient harvest. Often, brassica vegetable seed is treated with fungicides to help prevent damping-off disease that occurs in cold wet soils. Seed "priming" is a controlled hydration treatment followed by drying that is sometimes used to improve the rate of seed germination. However, priming may reduce the storage life of seeds, particularly under adverse conditions (Welbaum et al., 1998).

Transplanting

For early production of brassicas, production in short-season areas and where temperature and moisture stress are common, crops are often established by transplanting. Growers may produce their own transplants or purchase them from growers or companies that specialize in transplant production. Plants can be produced in specially prepared seedbeds in the field and transplanted bare-rooted or grown in cell trays in a greenhouse or an outdoor covered bed and transplanted with a root ball into the field (Swaider et al., 1992).

Seeding depth should be 1.5–3.0 cm (0.6–1.2 in), depending upon soil temperature and available moisture.

Seeding should be planted deeper in dry conditions and in soils that are prone to drying or the seedbeds should be irrigated to ensure adequate growth for germination and seedling establishment. Approximately 28 g (1 oz) of seed should produce about 3,000 transplants. Germination and plant growth can be adversely affected by either excessively high or low temperature. Once seeds germinate, plants in the juvenile phase will tolerate short-term exposure to temperature as low as 0–5°C (32–41°F) with little adverse effect. However, if plants have more than four to five true leaves and the stem diameter is larger than 8 mm (0.3 in, roughly pencil size), they have reached the mature phase and are subject to vernalization and premature bolting. Germination is poor at temperatures exceeding 32°C (90°F), the ceiling temperature for many brassica species (Jett *et al.*, 1996). The order of tolerance to high temperature for the major vegetable members of *B. oleracea* is cabbage (most tolerant), broccoli, cauliflower, and Brussels sprouts (least tolerant) (Fig. 16.3).

Conditions that favor rapid growth, i.e. high levels of N and/or excess moisture, result in transplants with a large leaf area but few roots and should be avoided. Field-grown plants are ready to transplant 5–8 weeks after sowing, depending upon soil temperature and weather conditions. Ideally, bare-root transplants should be planted within 24–48 h after pulling from the seedbed, but can be stored up to 9 days at 0°C (32°F) or 5 days at 19°C (66°F). Pulled transplants should not be left in the sun or allowed to dry out (Swaider *et al.*, 1992).

Alternatively, transplants may be produced in trays or flats in which seeds are planted in individual cells filled with a soilless media. Although the initial cost of the tray-grown plants is higher, the survival of transplants in the field is improved because the root ball remains intact, decreasing transplant shock and the establishment period. The volume of transplant tray cells is important because plants with a severely restricted root system may be harder to mechanically transplant and slower to resume growth, thus delaying crop maturity. Temperature, nutrition, and moisture control is critical in producing quality containerized transplants. At 16–18°C (61–64°F), tray-grown plants should be ready for transplanting in 4–7 weeks. Optimum transplant

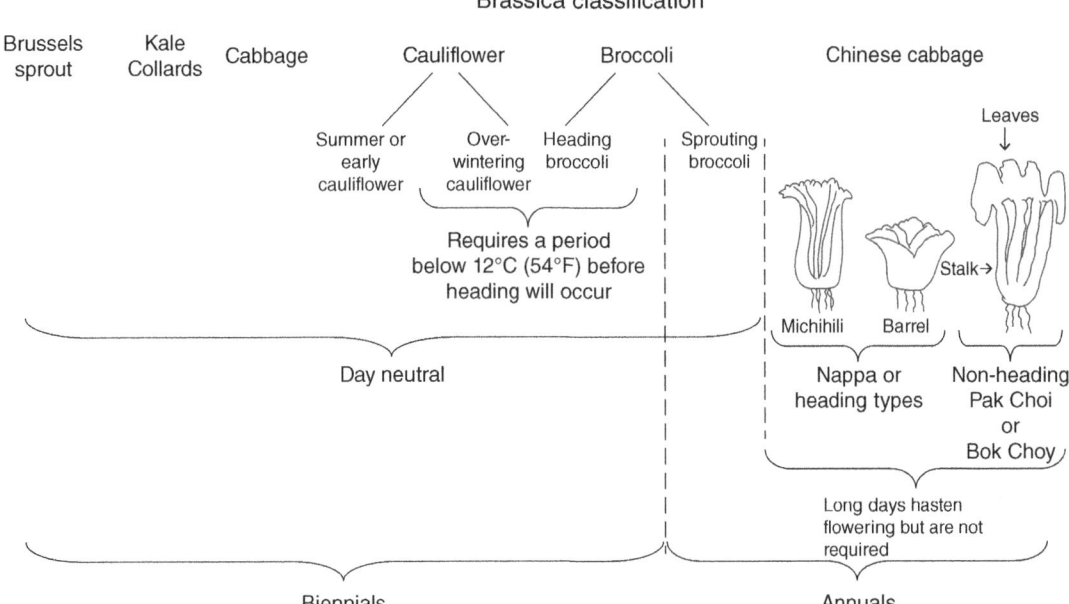

Fig. 16.3. Life cycle and hardiness relationships between members of *Brassica*.

height is 15–20 cm (6–8 in), and oversized plants can be more difficult to mechanically transplant and establish than younger, smaller plants. Transplants should not be pruned or trimmed as this may remove the apical meristem and stimulate shoot growth at the expense of root growth (Swaider et al., 1992).

Float-bed systems may also be used to produce containerized transplants. Polystyrene flats with individual cells are filled with a soilless media, seeded, and floated in water or nutrient solution. This system provides uniform watering and eliminates wetting of the leaves, which may cause foliar diseases, but may result in overly succulent plants (Frantz and Welbaum, 1998).

Transplants are hardened before field planting to reduce shock. Hardening is generally accomplished by withholding water, reducing N and/or by exposing plants to low temperature and direct sunlight. This results in plants with short, sturdy stems, well-developed root systems, and greater dry weight. Transplanting is best conducted under cool, moist conditions that limit water loss. A liquid fertilizer solution is usually applied at transplanting (Swaider et al., 1992).

Pest management

Control of plant pathogens, insect pests, and weeds should be an integrated component of production management. Development of an integrated pest management (IPM) system should rely upon regional recommendations and be in accordance with labeling restrictions for control measures (Flint, 1987).

Brassica crops should be grown in a 3–5 year rotation with crops from other families to reduce the risk of soil-borne diseases. Longer rotations may be needed when soil-borne diseases such as black leg (*Phoma lingam*), black rot (*Xanthomonas campestris*), or clubroot (*Plasmodiophora brassicae*) are present. Control of weeds is essential to avoid yield reduction from competition, and to increase air circulation around plants to reduce diseases. Introduction of plant pathogens, root-dwelling insects, nematodes, and weed seeds from soil-laden farm equipment, transplants, and in irrigation water should be avoided. The insect pest management program should keep pest populations below damaging levels, i.e. the economic threshold, while protecting natural enemies and competitors (Cornell, 2004; Rimmer et al., 2007).

Irrigation

Irrigation systems are often needed to ensure uninterrupted plant growth and to promote uniform maturity of *Brassica* species. The rule of thumb of applying at least 1 in (2.5 cm) of water per week applies to the brassicas, more in sandy soils (Saunders, 1993). Damage by nematodes, root diseases, insects, and weed competition is also more apparent on moisture-stressed crops. In many areas, the final irrigation is scheduled a few days to 1 week before harvest to ensure plants are fully hydrated. Several types of irrigation systems such as sprinkler, furrow, sub-surface, and trickle or drip are used to produce brassica crops. Crop moisture requirements generally increase with brassica crop development. Excessive moisture or poor drainage may contribute to the development of black rot (*X. campestris*) and Phytophthora stem rot (Rimmer et al., 2007).

Harvest and storage

For brassicas grown for fresh market, optimizing crop yield and maximizing quality are high priorities. Harvesting should be completed in a timely manner when the crop has reached commercial or physiological maturity, and if well planned, to coincide with optimal market prices. Precooling to remove field heat and storage at optimum temperature and relative humidity will prolong shelf life and reduce postharvest losses. Several precooling techniques may be used for brassica crops such as room cooling, forced-air cooling, hydro-cooling, vacuum cooling, and slurry icing. Early morning harvest before field heat accumulates can be an effective, low cost means for increasing quality and shelf life (Kader, 1992).

Cabbage

Botany

Brassica oleracea Group Capitata has nine chromosomes and can cross with other Groups or botanical varieties in this species. A cabbage head is a single large terminal bud. The head consists of tightly overlapping leaves and surrounding wrapper leaves enclosing the bud on the end of a short unbranched stem (Fig. 16.4).

Cabbage is a cool-season, frost-tolerant crop that is widely adapted and is commonly grown throughout temperate and subtropical regions of

Fig. 16.4. Mature heads of the green cabbage cultivar 'President'.

both the northern and southern hemispheres. The center of origin of cabbage is most often considered to be coastal areas of the Mediterranean, the British Isles, and western Europe. Today, cabbage is a popular crop in northern Europe, North America, central and northern Asia, Australia, southern and highland areas of Africa, and South America. Production in tropical and subtropical areas is limited to higher elevations where temperatures are cooler.

Cabbage is a biennial plant that grows best under full sunlight (Fig. 16.3). The optimum mean temperature for growth and quality head development is 15–18°C (60–65°F) with a minimum temperature of 4°C (40°F) and a maximum 24°C (75°F). Generally, young plants are more tolerant of heat and cold than plants nearing maturity. Well-hardened transplants will withstand −7°C (20°F) for short periods. Hardening is a process characterized by accumulation of dry matter by withholding water or subjecting plants to brief periods of non-lethal cold temperatures.

As with other biennials, vegetative growth continues while plants pass through a juvenile phase when they cannot be vernalized. Only plants that have developed beyond the juvenile phase, with stem diameters roughly greater than 7 mm (0.3 in), can be vernalized by prolonged exposure to temperatures between 1.7–10°C (35–50°F). Flowering begins with bolting or stem elongation, that occurs after the heads split open allowing rapid elongation of the flower stalk. After bolting has been initiated warmer temperatures speed the rate of seed stalk elongation. The flower stalk may elongate to up to 1 m (39.3 in) or more in length before flowers open. Sensitivity to bolting is largely dependent on the cultivar and some require longer chilling exposure than others before flowering. Obviously bolting is only advantageous for seed production and should be avoided when producing cabbages for human consumption.

There is a great diversity of cabbage cultivars that vary in color, size, leaf texture, disease resistance, and cuticular wax or bloom. Colors include white, green, blue, red, or purple (Fig. 16.4). Cultivars with red and/or savoy (crinkled- or blistered-leaf) leaves are grown for many markets to add variety even though the flavor is similar to other cultivars. Many of the current commercial cultivars are F-1 hybrids produced by crossing inbred lines, often with resistance to one or more pests. F-1 hybrids are popular because they exhibit heterosis and are more productive. Seeds developing on F-1 hybrid plants do not grow true-to-type when saved and planted the next year, so new hybrid seeds should be purchased each year. Cabbage cultivars are often classified into different types based on similarity of head characteristics. There are also several distinct types of cabbage heads, which vary in both size and shape.

Cultivar selection should be based upon several factors, which include adaptability, but in particular the preferences and timing of the target market should be considered. Examples of some of the major cultivar types with various characteristics are listed below (Rubatzky and Yamaguchi, 1997):

- 'Wakefield': small, pyramid-shaped head early maturing. Cold tolerant and resistant to bolting.
- 'Copenhagen': round head with few wrapper leaves and small core. Early maturing. Susceptible to bolting.
- 'Flat Dutch': large, flat and solid head with numerous wrapper leaves. Maturity varies from early to mid-season.
- 'Danish': moderate plant and head size, solid heads with few wrapper leaves, covered with fairly heavy bloom. Late maturing for fall production. Good keeping qualities for fresh market, processing, and storage.
- 'Alpha': small head. Early maturing (of little commercial importance).
- 'Volga': large head with thick, steel blue colored leaves that are solid on top, but open below. Late maturing (not widely grown).

Crack resistance is a very desirable cultivar trait. Wet rainy weather and market conditions may cause harvest to be delayed. Growing a cultivar that can hold in the field without splitting or bursting is very advantageous and can prevent losses. Cultivars vary greatly in their ability to hold in the field without cracking or splitting. Slower growing late cultivars tend to be more crack resistant than rapidly growing early cultivars.

Fertilization

Cabbage grows well in both mineral and organic soils with adequate moisture and fertility. Early planting can be achieved on sandy soils that warm more quickly, while heavier silt loam and organic soils generally are planted later for fall and early winter harvest. Soil pH in the range 6.0–6.5 is preferred, but cabbage will tolerate a soil pH range of 5.5–6.8 (Kemble *et al.*, 1999). Fertility varies with region and soil type, so soil should be tested for nutrient content before a fertilization strategy is developed for a crop. Cabbage has a high requirement for N and sufficient amounts are needed to maximize yields. Little additional N is needed for cabbage grown on organic soils but applications are needed on mineral soils. Depressed yields, delayed maturity, reduced keeping quality, toughness, and strong or objectionable flavors are indicative of N deficiency. Conversely, excessive N can promote reduced head density, secondary growth and split heads. A high level of N will also decrease storage life of cabbage. Combined with high temperatures, excess N can promote such rapid growth that plants show symptoms of tip burn in susceptible cultivars. Like most brassica crops, cabbage has a high requirement for B and Mo. Boron deficiency causes yellowing or chlorosis of youngest leaves and stems, which often starts from the base to the tip (Kemble *et al.*, 1999). Rosetting or even death of terminal shoots or buds occurs in extreme cases. The common symptoms of Mo deficiency in cabbage include a general yellowing, marginal and interveinal chlorosis, marginal necrosis rolling, scorching, and downward curling of margins usually on older leaves.

Production

There are many different ways to produce cabbage and we will describe but a few of the production practices. Cabbage may be produced using so called "conventional tillage" using a moldboard plow to prepare the soil for either transplanting or direct seeding of cabbage (Lal *et al.*, 2007). The plow cuts and inverts soil layers, so loose soil is brought to the surface and the top layer of residue is buried. After plowing, soils for conventional production are generally disked or tilled into small aggregates for planting. However, with conventional tillage the bare soil produced is prone to erosion, particularly after fall plowing or on soils that are not level. Also, weed seeds are brought to the surface where they can germinate. Weed control under conventional tillage systems has traditionally been obtained by cultivation, mulch, or the use of herbicides. Early season weed control can also be obtained using the "stale bed" technique, which requires the bed be prepared and then lay fallow several weeks before the intended planting date (CABI, 2012). This allows weed seeds to germinate before the crop is planted. Herbicides or shallow cultivation can kill weeds prior to planting. Care must be taken not to bring new weed seeds to the surface during cultivation. The stale-bed technique may be difficult to employ in some areas, because wet early-season soil conditions often prevent soil preparation in advance of planting. Cabbage is often planted on raised beds that are shaped from bare soil after plowing and tilling. This technique is popular on level soils where furrow irrigation is also used, to improve soil drainage for plasticulture production (Kemble et al., 1999).

The trend has been toward increased conservation tillage, particularly on steeply sloping soils prone to erosion (Hoyt et al., 1994). Conservation tillage reduces the number of field passes by farm equipment thus reducing labor, compaction and equipment wear, and conserving fuel. Conservation tillage systems cause minimum disturbance to the soil after the previous crop has been harvested. Crop residues are left in the field to reduce soil erosion, conserve moisture, inhibit weed growth, and act as green manure. There are several types of conservation tillage used for cabbage production as well as combinations of conservation and conventional tillage (Hoyt et al., 1994). A few of the more popular techniques include:

- **No-till:** the soil is left undisturbed from the time the previous cash or cover crop is harvested until the new crop is planted. Nutrients are applied to the root zone rather than broadcast. Transplanting or seeding is in a narrow seedbed or slot cut into the residue with special planting equipment to deal with greater residue using coulters, row cleaners, disk openers, in-row chisels, or roto-tillers. Weed control is accomplished through herbicides, rolling, or mowing. Cultivation is only used for weed control in an emergency.
- **Ridge-till or strip till:** the soil is left undisturbed after the previous crop has been harvested until the new crop is planted. Nutrients are injected into the soil rather than broadcast. Planting is on ridges or narrow lanes cleared in the residue using row cleaners, sweeps, disk openers, or coulters. Residue is left in the row middles. Weed control is accomplished using herbicides or cultivation.
- **Mulch-till:** the soil is disturbed prior to planting using chisel plows, spike-tooth harrows, field cultivators, disks, sweeps, etc. Much of the residue remains on top of the soil as mulch. Weed control is accomplished using herbicides or cultivation.

Disadvantages of conservation tillage include: lower soil temperatures, slower germination and emergence when direct seeding is used, slower early growth, delayed competition with weeds, delayed spring planting, more root disease, heavier crop residue, planting may be more difficult, weed spectrum changes, and increased insect pests (Hoyt and Walgenbach, 1995).

Cabbage can be established by either direct seeding or transplanting. Cabbage is easy to transplant because plants root easily and resist transplant shock (Swaider et al., 1992). In northern latitudes, early cabbage crops are grown from transplants. Transplants are ready for field planting 4–6 weeks after seeding. Cabbage transplants can be produced in plug trays in a protected structure or grown in outdoor beds then pulled and transplanted with little root mass into a production field. Storage of pulled, field-grown cabbage transplants should not exceed 9 days at 0°C (32°F) or 5 days at 19°C (66°F) prior to transplanting (Kemble et al., 1999).

In conservation tillage systems, transplants often improve establishment in the residue although direct seeding may be successfully used in some systems as well. When establishing a cabbage crop by conservation tillage, certain adjustments may be needed for planting in cooler soils. For example, when direct seeding is used, seed should be planted less deep (2.5 cm, 1 in or less), the planter operated more slowly (8 km/h, 5 mph) and the highest possible quality seed should be used. Planting seeds under drier conditions, treating seeds with fungicide/biological controls, and the use of disease-resistant cultivars may help to reduce root rot. Controlling weeds in conservation tillage systems may require killing weeds, especially perennial weeds preplant (Hoyt et al., 1994).

Where growing seasons are long and soil is finely worked, cabbage can direct-seeded using precision planters. However, seed should be sown 15–20 days in advance of the normal transplanting for the same maturity date. Early cultivars require 85–90 days from seeding to harvest and main-season crops require 110–115 days. Cabbage can be planted using equidistant spacing but is commonly planted in rows to facilitate mechanization.

Successful stand establishment of cabbage results in uniform plant populations that reduce differences in head maturity, making a single-pass harvest by either hand or machine possible. In many areas, cabbage is grown on raised beds 1.8 m (6 ft) from center-to-center to improve drainage or facilitate irrigation practices although planting on flat ground is still common. Plant spacing is determined by the target market and cultivar. Smaller heads are preferred for fresh market, while larger heads are used for processing. Therefore, fresh-market cultivars are planted closer than those for processing.

Within-row spacings of 25–40 cm (10–18 in) are common for early maturing cultivars that frequently have fewer wrapper leaves, while 40–70 cm (18–29 in) within-row spacing may be needed for late cultivars (Fig. 16.5).

Closer spacing reduces internal tip burn and heads are less likely to crack, but crop maturity may be delayed. Auxiliary heads, small heads that resemble Brussels sprouts, may be a problem with certain cabbage cultivars that are direct-seeded, but rarely with a transplanted crop. Incidence of auxiliary heads seems to be increased with wide spacing and high fertilization. Increased plant density, reduced fertilization, and/or a change in cultivar should reduce or eliminate the occurrence of auxiliary heads and reduce variation in head maturity.

Because of the dense, shallow root systems of cabbage, weed competition can be very detrimental to crop development and head quality, uniformity of maturity, and crop yield. Shallow cultivation may be used when the crop is small, but deep cultivation should be avoided to minimize root damage. In many countries, hand hoeing is no longer economical for commercial production. An integrated control program for weeds is often employed that includes a combination of cultural practices with or without the use of herbicide.

However, organic cabbage is gaining popularity in many countries through diverse cultural practices to manage pests. For example, to control troublesome weeds and to create a more diverse agro-ecosystem, brassicas can be grown using management practices called overseeding (also called underseeding in some literature) or interseeding (Stivers *et al.*, 1999). With overseeding and interseeding, a low-growing beneficial plant that is less competitive than a weed, such as a legume like white clover, is established in row middles to act as "living mulch" (Akobundu, 1980). This technique can be an effective alternative to conventional cultivation, reduce herbicide use, reduce erosion, build soil organic matter, and fix atmospheric N (Infante and Morse, 1996). The overseeded plant acts as a "managed weed" that serves as a "mulch" to out-compete more noxious weeds. The overseeded crop may compete with the cash crop for water, so this technique may not be effective in dry areas or when irrigation is not available. There is little evidence to suggest that legumes benefit the nutritional status of the established crop during the first season, however, the overseeded crop will accumulate nutrients and organic matter that will eventually be returned to the soil. Overseeding may also be used with conservation tillage (Infante and Morse, 1996).

Fig. 16.5. Fresh-market cabbage grown in Texas on raised beds with two rows per bed.

Cabbage can also be grown in a plasticulture system (Lamont, 2004). Thin plastic film is spread across raised beds and the edges of plastic covered with soil to create a tight covering of plastic mulch on top of the bed that will drain rather than collect water. Drip irrigation installed under the plastic provides water and nutrients since rain and overhead irrigation is excluded unless perforated plastic is used. Soil fumigants may be injected into the soil as the plastic is laid in order to kill soil-borne pests. Transplants are placed through holes cut or burned in the plastic after fumigant, if used, has dissipated. Plastic mulch provides several advantages: modified soil temperature and conservation of soil moisture, reduced soil compaction, prevention of crusting, reduced losses of fertilizer by leaching, prevents drowning of crops, and reduced competition from weeds depending on the type of plastic used. In plasticulture production, heads tend to develop more uniformly and cabbage wrapper leaves are cleaner at harvest, requiring less washing. Although mulch increases production costs, the earlier and larger yields of high-quality heads provide greater revenue. In order to spread the cost of plastic film over two seasons, growers may sometimes plant another crop (double cropping) into plastic mulch before or after the cabbage.

Irrigation

Generally, cabbage requires from 2.5–3.8 cm (1–1.5 in) of water per week for uninterrupted growth. More water may be required when cabbage is grown on sandy soils or when evapotranspiration is high. Yield reduction may occur when soil moisture remains below 50% of field capacity for extended periods, particularly on sandy soils (Saunders, 1993). Drought during the juvenile stage (three to four true leaves, with stem diameter <7 mm (0.3 in) may adversely affect plant development and uniformity of maturity. On the other hand, excessive moisture may delay maturity and reduce yield, especially in early cultivars. Too much moisture close to harvest can cause heads to crack open or burst, making them unsalable.

Harvest

Cabbage is ready for harvest when heads are firm to the touch but before cracking begins (Kemble et al., 1999). Generally, most fresh markets prefer heads that average 0.9–2.3 kg (2–5 lb). For processing into slaw or kraut or for long-term storage, larger-headed cultivars are used. When harvesting, stems are cut close to the ground near the base of the head. Typically, outer wrapper leaves are removed. Mechanical harvesters are used in areas where manual labor is limited or for cabbage destined for processing where head uniformity is not as critical. Historically, with hand-harvested open-pollinated cultivars, a given field was harvested two to four times to obtain heads of uniform size and maturity. Only one or two harvests of hybrid cultivars are usually required because of their greater uniformly. Use of uniform transplants and consistent growing conditions also helps reduce the number of harvests. Field packing of fresh-market cabbage is common because it is more efficient. Cabbage is field graded, packed in crates, mesh bags or cartons of various sizes often ranging from 23–27 kg (50–60 lb) each. However, if heads lack uniformity a grading and packing facility may be necessary. Mechanically harvested cabbage is usually conveyed into a truck or wagon driven alongside the harvester and taken to a processing facility. Yields will vary with the season of production, cultivar and production system. With proper management, cabbage can produce 22 metric tonnes/ha (10–12 short tons/acre) (Swaider et al., 1992).

Postharvest handling

Once cut, cabbage should be cooled to remove field heat and placed in an environment with high humidity to avoid moisture loss and wilting. Prompt removal of field heat will increase shelf life and reduce the development of storage rot diseases (Kader, 1992). With cabbage, hydro- and forced-air cooling are often used. Night and early morning harvests can reduced cooling costs and are used in some areas. The ideal storage temperature is 0°C (32°F). Cabbage should not be stored with fruits that produce appreciable amounts of ethylene, since exposure to ethylene causes yellowing of the wrapper leaves (Kader, 1992). Storage life varies dramatically, based upon the type of cabbage and the growing conditions. Most fresh-market cultivars can be stored from 30 to 60 days (0°C (32°F), RH >90%). Some cultivars of cabbage have been developed specifically for long-term storage and these can be kept at 0–1°C (32–34°F) with 92–98% RH for 5–6 months (Kader, 1992).

Uses

Cabbage is a very versatile commodity with many different uses. Cabbage is sold fresh, as lightly processed fresh-cut products or processed by canning. Processed products include those that are pickled in vinegar or fermented such as kraut or kimchi. Some fresh-cut or lightly processed products are marinated in dressing and sold as slaw. Many ready-to-eat salad mixes contain shredded cabbage. Both fresh cabbage and processed cabbage are traded internationally. Processed cabbage is often traded as canned product or in refrigerated bags. Production of late cabbage for storage was once popular in North America and parts of Europe. Today, consumers generally prefer fresh cabbage, when available, to stored cabbage. Some cabbage is stored for processing to provide product for canneries over a longer period.

World production and trade

Cabbage is grown in almost all countries located in temperate and subtropical regions of both the northern and southern hemispheres. Cabbage is also grown at high elevations under cooler conditions in tropical regions. According to FAO statistics, China and India are the world's largest cabbage producers, with China alone accounting for over one-third of total world production (FAOSTAT, 2012). Fresh cabbage and processed products are traded among countries in eastern Asia in many of the same areas where Chinese cabbage is popular. Cabbage is a popular crop in northern Europe and is widely traded among countries of the European Union and eastern Europe. Fresh cabbage is grown during the winter in southern Europe, northern Africa and the Middle East and exported to northern Europe.

In the USA, New York, California and Texas are the largest producers of fresh-market cabbage combining for almost 50% of all fresh cabbage. The USA imports approximately US$13 million of cabbage accounting for almost 90% of all cabbage imports, mostly from Canada, with Mexico being a distant second (USDA, ERS, 2010). Sauerkraut from Germany, Canada, and Poland is the most important form of processed cabbage. The USA is a net exporter of cabbage, but trade is not a major factor in the US cabbage industry. The USA exported only 3.5% of domestic production, mostly as fresh cabbage to Canada, Japan, or Mexico (USDA, ERS, 2010).

Pest management

Insects

Diamondback moth (DBM) caterpillars have become a serious worldwide pest of the brassicas because it has developed resistance to insecticides (Cornell, 2004). The caterpillar's name comes from the diamond-shaped markings on the adult moth. DBM caterpillars are most active in spring and early summer, producing as many as ten generations during the growing season. DBM larvae are smaller than the larvae of other caterpillar pests of brassica crops. Mature larvae are green, 10–15 mm (0.4–0.6 in) long and tapered at both ends. A larva wriggles violently when disturbed and often dangles from a leaf by a silken thread.

Because DBM populations can increase rapidly at temperatures above 27°C (80°F), at least twice-weekly scouting is recommended. Heavy rains may reduce populations dramatically. *Bacillus thuringiensis* (Bt) insecticides are usually effective and should be applied at the first sign of worm activity. Sprays may need to be applied at 5-day intervals when populations are high. A combination of Bt and pyrethroid insecticides may be used as well. DBM pupae suffer natural mortality from native parasitoids, and parasitized pupae have a broad white stripe around the pupa. Removing or destroy crop debris will prevent buildup of DBM and migration to adjacent fields.

Cabbage loopers (CL) are most active in early summer and fall. Mature larvae grow as large as 38 mm (1.5 in) in length and have three pairs of prolegs in the rear. The larvae loop their bodies when they move and grasp plants firmly when handled. Their green color blends well with the foliage and their dark frass, or excrement, is often seen before the larvae. CL larvae chew complete holes in the leaves, and more mature, larger larvae consume great amounts of plant material. Because large larvae are less susceptible to insecticides than young, sprays should be applied when larvae are still in the early growth stages. Bt insecticides are moderately to highly effective and may be used separately or in combination with pyrethroid insecticides if needed (Kemble *et al.*, 1999).

Imported cabbageworm (ICW) moths are white to yellowish-white butterflies with black spots on their wings. Unlike nocturnal CL moths, ICW moths fly during the day. The green larvae have a velvety appearance because of their dense coating of fine hairs. The mature larva has a faint orange stripe down its back and chews irregular holes in the leaves, bore into heads, and contaminate leaves and heads with frass. Damage is similar to that caused by the CL (Kemble et al., 1999).

Cabbage webworm larvae are 19 mm (0.75 in) long when mature and gray in color with five dark stripes. The head capsule is black with a distinct white V-shaped mark. Cabbage webworm is an occasional pest of the brassicas, feeding primarily around buds. The caterpillars produce webbing that can protect them against insecticide contact (Kemble et al., 1999).

Cross-striped cabbageworm has black and white transverse stripes on its back. It is controlled with Bt and other insecticides recommended for caterpillar control (Kemble et al., 1999).

Beet armyworm may infest fall plantings of cabbage and related crops. Armyworms migrate to cabbage in large numbers as other crops die or are harvested, so it is important to monitor late summer/fall plantings carefully and to apply insecticides when infestations are first detected. The moths lay egg masses on the undersides of leaves and cover the eggs with white, fuzzy scales. The larvae are light green to dark olive green and may have longitudinal stripes on the back or sides of the body. Control measures should be applied if egg masses or larvae are found on 2–3% of the plants (Kemble et al., 1999).

Cutworms are caterpillars that rest beneath the soil during the day and feed at night, causing damage to stems and foliage. They are dark gray-brown in color with a greasy appearance, and they often curl into a C shape at rest or when disturbed. Cutworm larvae overwinter in fields and therefore may be present at the time of planting, particularly in fields with high organic matter from previous crop residue. Cutworms can be detected when land is prepared for planting. Before planting remove weeds and plant debris to starve developing larvae. Beneficial nematodes will attack and destroy cutworms in the soil. Release *Trichogramma* wasps weekly for 3 consecutive weeks to parasitize cutworm eggs. Diatomaceous earth sprinkled around the base of plants is very effective. Scatter bran or corn meal mixed with Dipel Dust (Bt-kurstaki) and molasses on the soil surface to kill caterpillars. After harvest remove organic debris; plowing the field will disturb overwintering larvae. A preplant or at-plant soil insecticide can also provide effective control. Insecticide sprays should be directed to the base of the plants if cutworm damage is observed after planting (Kemble et al., 1999).

Seed and root maggot larvae, which develop into flies, are attracted to decaying organic matter and feed on young roots and stems, severely reducing cabbage stands. Damage usually occurs during cool, wet conditions when plants grow slowly. Delaying planting until cool and wet conditions are over may reduce the risk of maggot damage (Kemble et al., 1999).

Several aphid species, including cabbage aphids, may infest brassica crops. Cool, dry weather is most favorable for aphid development. Aphids do not usually infest transplants but may build up after field establishment. Large numbers of aphids may kill small plants, and their feeding can cause leaf curl on older plants. Often, the most serious problem associated with aphids is contamination. Aphids that colonize inside heads of cabbage are almost impossible to remove before marketing. Natural enemies can provide control, but may not be effective in controlling heavy aphid populations below damaging levels. Use of broad-spectrum insecticides like pyrethroids can increase aphid numbers by eliminating natural enemies; therefore, these insecticides should be used sparingly early in the season and only when necessary to control other insect pests (Kemble et al., 1999). Insecticidal soaps provide biological control.

Sweetpotato or silverleaf whitefly adults are tiny, winged insects about 2 mm (0.06 in) long with white bodies and yellow heads. Although small, they are easy to detect on the undersides of leaves or as they fly from plants when disturbed. The adults lay eggs on the undersides of leaves where the immobile nymphs develop. The nymphs are scale-like in appearance. Like aphids, whitefly adults and nymphs suck plant sap with piercing-sucking mouthparts, and they produce sticky honeydew on leaves where they are feeding. Black sooty mold often grows on the honeydew. Insecticides are not highly effective against whiteflies, so prevention is the best approach. Transplants should be free of whiteflies, and fields planted as far as possible away from other whitefly-infested crops. Destroying weeds and previous crop residues that may harbor whitefly populations are also important pest management techniques (Kemble et al., 1999). Whiteflies tend to be a pest in long-season areas with mild winters.

The harlequin bug insect is closely related to the common stinkbug. Harlequin bug is a brightly colored, shield-shaped bug with red and black markings and piercing-sucking mouthparts that feeds on the veins of cabbage leaves. Harlequin bug eggs are barrel-shaped and laid in clusters on leaves. The eggs are white, and each has two black bands around the circumference (Kemble et al., 1999).

Flea beetles are most common in the spring and in fields that are weedy or surrounded by weeds. The adults are small, dark-colored beetles with enlarged hind legs that enable them to jump great distances. They feed on the undersides of leaves, causing numerous small, round or irregular holes. Flea beetles are of greatest concern during the seedling and immature stages of cabbage development and usually cause little damage to mature plants (Kemble et al., 1999).

The vegetable weevil adult is 6–10 mm (0.24–0.39 in) long, brownish gray with two light-colored marks on its wing covers, and has a long snout. Larvae are legless grubs. Adults and larvae feed on the foliage and stems of plants and cause the most damage to seedlings, causing significant reductions in stands of young plantings, particularly in early fall and spring plantings. Treatment is recommended when more than 5% of the stand is damaged (Kemble et al., 1999).

Cabbage diseases

Black rot (*X. campestris*) is a bacterium that affects cabbage and related crops worldwide and is one of the most serious cabbage diseases in warm climates. Infected seedlings are yellow, stunted, have blackened margins on their cotyledons and may eventually wilt and die. Seedling infection can be difficult to diagnose since only a few plants in a lot may be infected. On older plants, the disease is easily recognized by the presence of yellow V-shaped or U-shaped areas extending inward from the margins of the leaves. As the disease progresses, the yellow lesions turn brown and the tissue dies. The veins of affected leaves darken and the midrib of leaves turns black. This vein discoloration progresses toward the base of the leaf. Eventually, the bacteria spread into the main stem. When infected stems are cut in cross-section, a black vascular ring may be evident where bacteria have moved into the water-conducting vessels. This vascular discoloration extends from the stem to the upper leaves and down into the roots. In later stages of infection, all central tissues of the main stem turn black. This symptom can be confused with Fusarium yellows disease; however, discoloration caused by Fusarium is dark brown rather than black. Cabbage heads infected with black rot often do not reach full size and lose their lower leaves. Frequently, symptoms are more severe on only one side of the head. Diseased heads may rot quickly before or after harvest because of secondary infection from bacterial soft-rot (Kemble et al., 1999).

The black rot bacterium can overwinter on infected cabbage seeds, in weeds from the family Brassicaceae including black mustard, field mustard, wild turnip, wild radish, shepherd's purse, and pepperweed or in infected plant material in the soil. The bacterium can persist in plant residue for 1–2 years or as long as the plant debris remains intact. Black rot is spread on seed and seedlings, by movement of contaminated plant material, in irrigation water or splashing rain, by insects, by cultivation equipment, and by field workers (Kemble et al., 1999).

In the spring when seedlings emerge, bacteria are typically carried from diseased plant refuse to leaf edges by splashing rain. Bacteria invade young leaves through natural openings or wounds. From the infected leaves, the bacteria move through the water-conducting vessels to the main stem, down into the roots and up into the leaves. Plant-to-plant infection in the field occurs through the hydathodes at leaf margins. Bacteria can also enter through insect feeding wounds. "Clipping" of oversized transplants with rotary or flail mowers also promotes spread of the disease if the bacterium is present. Root infections can occur through wounds and are most common when infested soil is saturated with water. Black rot is favored by warm, (26.6–30°C, 80–86°F) wet conditions. Free moisture in the form of rain, dew, or fog is required for infection to occur and for the disease to develop (Rimmer et al., 2007).

To control this devastating disease, disease-free seeds and transplants should be used. Establish crops in black rot free soils that have not grown crops from the family Brassicaceae for at least 3 years. Black rot is usually most severe in wet poorly drained soils. Black rot can also be seen moving down rows as the bacterium is spread during cultivation or through free-standing water. Growing cabbage on raised beds helps eliminate conditions that support black rot. When possible, remove, burn, or cleanly plow-under all crop debris immediately after harvest to prevent overwintering. Soil fumigation can be an effective control measure. Copper-containing pesticide may reduce damage from black rot and chemical treatments should begin when weather conditions favor disease development (Kemble *et al.*, 1999).

Black leg (*Ph. lingam*) symptoms appear as oval, sunken, light brown cankers, often with black or purple margins near the base of the stem. The fungal canker enlarges until the stem is girdled and the plant wilts and dies. Pale, irregular-shaped leaf spots gradually enlarge, becoming circular with gray centers. Small, black, pepper-like spots (pycnidia) can be seen within the leaf spots and these are the spore-producing structures of the fungus. Severely infected plants may have blue coloration, are stunted, wilt, and then suddenly collapse due to stem deterioration (Kemble *et al.*, 1999).

The fungus can survive in plant residue for 2–3 years and may be seed transmitted. Plants are often infected in the seedbed, where splashing water spreads the spores of the fungus among plants. Secondary spread of the disease may occur during transplanting as the fungus is spread by splashing rain, workers, and equipment during wet weather.

Transplants free of black leg should be planted. Crop rotation is an effective control and cabbage or related crops should not be grown in the same field for at least 3 years. Remove plant debris at the end of the season to prevent overwintering and spread to adjacent fields. Fungicide seed treatments, soil fumigation, or solarization can effectively control black leg (Kemble *et al.*, 1999).

Clubroot (*Pl. brassicae*) is another common fungal disease of brassicas (Cornell, 2004). The disease can be well established before aboveground symptoms become evident. Infected roots are unable to absorb water and nutrients, top growth is stunted and lower leaves may yellow and drop off. Affected plants may wilt during the day and recover at night. Secondary organisms may kill plants by invading root galls, causing decay. Control measures include maintaining a high pH by regular applications of lime. Foliar applications of B may be necessary if deficiency symptoms occur because of high soil pH. Long rotations between brassica crops can effectively control club root as well as improved soil drainage, preventing contamination from infected soils, and using irrigation water that is not contaminated with disease spores (Rimmer *et al.*, 2007).

Alternaria leaf spot (*Alternaria brassicae*) causes small, circular, dark spots on older leaf surfaces. As spots enlarge, concentric rings develop within lesions surrounded by a yellow halo. The tan-colored centers of lesions may eventually fall out, producing a hole, or under wet conditions may become covered with masses of black spores. In storage, spots enlarge and soft-rot bacteria may infect the lesions. The disease can overwinter on crop debris. Weeds from the family Brassicaceae may also harbor the fungus. Spores of *Alternaria* can be spread by wind and water. The disease is most damaging under wet, warm conditions 20–30.5°C (68–81°F). Cabbage should never be grown in fields where other brassica crops have been grown in the past 3 years. Infected plant debris should be removed or destroyed after the season and disease-free transplants should always be used. A fungicide spray program at 7–10-day intervals after the first sign of the disease may provide control (Kemble *et al.*, 1999).

Wirestem (*Rhizoctonia solani*) is a fungal disease that affects transplant beds and production fields. Affected plants show reddish-brown discoloration and constriction of the stem near the soil line causing distorted and twisted stems from which its common name is derived. Surviving plants produce weak undersized heads. The fungus may be present in all soils but is most common in soil where infected plant debris has not decomposed. Prolonged, overly moist soil favors the disease. Wirestem is greatly influenced by recent cropping history. Crop rotation is an effective control and cabbage or its relatives should not be grown in the same field for 3 years in a row. Eliminating plant debris following disease outbreak also prevent the disease from spreading. Soil fumigation, use of raised beds, and careful watering may reduce the incidence of this disease. Fungicide treatments applied at transplanting may provide effective control (Kemble *et al.*, 1999).

Downy mildew (*Peronospora parasitica*) is a fungus spread by wind-blown spores from infected plants. In moist weather, a white, fluffy, fungal growth develops mostly on the undersides of infected leaves. With time, yellow to tan spots develop on the corresponding upper leaf surface. Infected leaves on young plants may drop and the plants eventually die. On older leaves, the infected areas turn tan with a papery texture and show gray spore masses on the undersides of infected leaves, which usually remain attached. Downy mildew can predispose infected plants to bacterial soft rot. The fungus overwinters on seed, in weeds from the family Brassicaceae, and possibly in soil. Heavy fogs, light rains, prolonged dews, overhead irrigation, and night temperatures between 8–16°C

(46–61°F) with day temperatures below 23.8°C (75°F) favor this pathogen. Spores of the fungus drift long distances in cool, moist air (Kemble et al., 1999).

Fusarium wilt (*Fusarium oxysporum conglutinans*), also called "yellows", can attack all members of the cabbage family, including cabbage, cauliflower, broccoli, Brussels sprouts, kale, kohlrabi, collards, and radish and is closely related to *Fusarium* diseases in other vegetables. Infected plants will develop a yellowish-green color 2–4 weeks after emergence or transplanting. Symptoms are more noticeable on one side of the plant followed by curling of the stem and leaves. The lower leaves turn yellow first, and then symptoms move to the upper leaves. With time, the yellow tissue turns brown and becomes dry and brittle. The vascular tissues in the stems and leaves turn dark brown, expressing symptoms similar to black rot. However, Fusarium wilt develops internally and appears in the lower portions of the plant first. Fusarium grows in the soil and on plant debris. It also has two types of spores, one of which is short-lived and the other, heavy-walled and capable of withstanding long periods of low temperatures and drought. *Fusarium* can remain alive in the soil for many years and even increase in soil free of cabbage-family plants. The *Fusarium* fungus grows poorly at temperatures below 16°C (60°F) and reaches its maximum growth rate between 27–32°C (80–90°F). At 35°C (95°F) and above growth is inhibited. Since *Fusarium* can live free in the soil without a host, conventional controls such as rotation, fungicide treatments, and destruction of crop refuse are not effective. The use of resistant cultivars is one of the best control measures (Cornell, 2004).

Turnip mosaic virus (TuMV) is one of the most important and widespread viruses infecting crucifers. The host range for this virus is not limited to crucifers. This virus infects radish, turnip, lettuce, endive, spinach, and other brassicas. TuMV is efficiently transmitted in a nonpersistent manner by several aphid species, most notably the green peach aphid (*Myzus persicae*) and the cabbage aphid (*Brevicoryne brassicae*). TuMV causes mosaic and black necrotic ring spots in cabbage, cauliflower, and Brussels sprouts. Necrotic spots may not be evident on cabbage heads at harvest, but may appear after 2–5 months in storage. These spots are the result of infections that occurred during the growing season. Infections do not spread among heads while in storage. Spotting may be found several layers deep within the head and appears on the midribs, the side veins, and in the interveinal areas where the spots may coalesce. TuMV also causes mosaic with leaf distortion and necrosis on lower leaves of turnip, radish, and mustard (Cornell, 2004).

Cauliflower mosaic virus (CaMV) is another virus that infects crucifers, and its symptoms have often been confused with TuMV infections. Like TuMV, CaMV is transmitted by the same aphid species in a nonpersistent manner. CaMV has a host range limited to crucifers and is distributed mainly in the temperate regions of the world. The virus induces mosaic and a striking veinal chlorosis in most hosts. A masking of symptoms may occur in chronically infected plants, particularly at high temperatures. Cabbage heads displaying "pepper spotting" and "vein streaking necrosis" in storage were previously thought to be infected by CaMV but now these are believe to be distinct disorders (Cornell, 2004).

Broccoli

History

Sprouting broccoli was domesticated in the Mediterranean region, possibly in Cyprus or Crete, although the exact location is difficult to determine conclusively. Broccoli was derived from cultivated leafy cole crops in the northern Mediterranean region in about the 6th century (Maggioni et al., 2010). Broccoli's name comes from the Italian "brocco", meaning "sprout" or "shoot", which comes from the Latin "brachium", meaning "arm" or "branch" (Berg, 2012). Sprouting broccoli only recently became popular beyond its Mediterranean origin. Sprouting broccoli (calabrese) was introduced from Italy to the USA in the 1800s. However, it did not become popular in America until the 1920s following the successful marketing of the D'Arrigo Brothers Company who grew and promoted broccoli (Berg, 2012). By the 1940s it had become a common vegetable in America and over the course of several decades, it rapidly became an important vegetable overtaking cauliflower in terms of popularity (Thompson and Kelly, 1957). Its popularity in the USA spread to other parts of the world and its reintroduction and appreciation in Europe has been comparatively recent. Today, broccoli is widely grown as a significant economic crop in Europe, North America, parts of Central and South America, eastern Asia, and Australia (FAOSTAT, 2012). DNA sequence analysis shows that Chinese broccoli (*B. oleracea* Group Alboglabra) is closely related to kale and cabbage native to Portugal and may have been introduced to Asia by Portuguese traders (Bancroft *et al.*, 2006).

Botany

Sprouting broccoli (*B. oleracea* Group Italica) is an herbaceous cool-season, frost-tolerant annual or biennial dicot, which is widely adapted and grown throughout temperate and subtropical regions of both the northern and southern hemispheres (Fig. 16.3). Sprouting broccoli is more heat tolerant than some other brassicas, so annual cultivars can be grown in subtropical climates. Vegetative growth can occur over a wide range of temperatures but high-quality head development occurs over a temperature range from about 12–20°C (54–68°F). Above 25°C (77°F), compact heads usually do not form because of excess stem elongation and rapid floral development. Low temperatures during early plant development may cause premature heading. Growth is slow at temperatures below 5°C (41°F). The inflorescence, if properly acclimated, can tolerate temperatures as low as −7°C (20°F) for short periods without damage. Mature vegetative tissues are even more freezing tolerant, while young seedlings are less (Nieuwhof, 1969).

Broccoli has nine chromosomes and can cross with other groups or botanical varieties of *B. oleracea*. Sprouting broccoli is difficult to distinguish from related *B. oleracea* crops in the seedling stage. Leaves are thick, somewhat leathery, smooth oblong, simple, alternate, pinnate, and petiolate. Leaf colors range from grayish blue to green. Root systems are moderately shallow with a prominent taproot that branches readily, giving rise to many fibrous roots concentrated within the top 30–40 cm (12–16 in) of soil (Nieuwhof, 1969). Like cabbage, the inflorescence produces bisexual cross-shaped flowers with four yellow or white petals, one pistil, and six stamens. The branched inflorescence begins as a compact, slightly dome-shaped head reaching approximately 40 cm (16 in) across at harvest maturity before losing its compactness as the flower stalk elongates and flowers open (Fig. 16.6).

At full flower, the bushy plant will be roughly 1.2 m (47 in) high and approximately 0.5 m (20 in) across (Nieuwhof, 1969). Even though flowers are perfect they are usually cross-pollinated by insects,

Fig. 16.6. A broccoli head at edible maturity (a) and (b) a bushy flowering plant approximately 2 weeks after edible maturity.

most often by bees because of pollen incompatibility. The silique size and structure are very similar to cabbage with pods 4.5–10 cm (2–4 in) in length and 3–6 mm (0.12–0.24 in) wide. When siliques dry they dehisce and separate into two halves, releasing 20–40 seeds ranging in color from dark brown to brownish purple (Musil, 1948). Seeds usually mature 50–90 days after fertilization. A well-pollinated broccoli plant may produce 0.23 kg (0.5 lb) of seed. Like cabbage, there are approximately 300 seeds/g (8,490 oz).

Types and cultivars

Most commercial types of sprouting broccoli grown in the Americas, parts of Asia, and southern Europe tend to be annuals that do not require vernalization for head formation or flowering. Annual cultivars form an inflorescence of many closely spaced immature flower buds clustered in a head 50–70 days after emergence without environmental cues. Biennial types that require vernalization for head formation and flowering are more common in northern Europe (Rubatzky and Yamaguchi, 1997). For biennial cultivars, the stem elongates after plants have advanced beyond the juvenile phase and received sufficient chilling below 13°C (50°F) to initiate floral development. Cultivars of sprouting broccoli may be classified as early-, medium-, or late-maturing. Late-maturing and overwintering cultivars are biennial and require vernalization (Fig. 16.3). Important types of broccoli include purple sprouting (an overwintering branching biennial), purple cape (an overwintering single-headed biennial), purple Sicilian (pale purple single-heading annual sometimes incorrectly called "purple cauliflower" because its head consists of developed flower buds and not undifferentiated flower primordial), white sprouting (overwintering branching biennial), and calabrese (green sprouting, with both annual and biennial forms) (Rubatzky and Yamaguchi, 1997).

Calabrese types are the most widely grown commercial type and hybrids have generally replaced open-pollinated cultivars. Important characteristics that differentiate cultivars include head compactness and shape, extent of branching, size and color of individual flower buds, stem length, number and length of internodes, and auxiliary floral development. Many of the current commercial cultivars are F-1 hybrids produced by crossing inbred lines, often with resistance to one or more pests. F-1 hybrids are popular because they exhibit heterosis and are more productive and uniform (Swaider *et al.*, 1992). Seeds saved from F-1 hybrid plants do not grow true-to-type when planted and new seeds should be purchased each year.

Heads of true sprouting broccoli are composed of green or purple fully differentiated immature flower buds that with further maturation open into flowers. Using this convention, cultivars with heads composed of immature flower buds but having cauliflower plant architecture with wrapper leaves extending above the sides are considered broccoli based on their flower structure. In contrast, a cauliflower head is composed of dense clusters of usually white flower primordials, with a texture similar to callus that will only give rise to flower stalks after plants are vernalized or chilled. Cauliflower heads are denser than broccoli. Heads of broccoli are fully exposed without leaf covering during development (Rubatzky and Yamaguchi, 1997).

For broccoli, the primary inflorescence forms at the terminus of the elongated unbranched stem, although the inflorescence is highly branched (Fig. 16.6). The edible head grown for commerce is the inflorescence and tender portions of the upper stem. Stems are taller than cabbage or cauliflower ranging from about 40–100 cm (16–39 in) and have longer internodes as well. The green or purple bud clusters may be surrounded but are usually not covered by subtended leaves. With continued growth, the branches of the inflorescence tend to grow apart, resulting in loss of head compactness and shape. Following growth of the primary inflorescence, small inflorescences form in the axils of the lower leaves. Development of secondary inflorescences is influenced by apical dominance of the terminal inflorescence and the degree of suppression is cultivar dependent (Rubatzky and Yamaguchi, 1997).

In newer literature there is a tendency to refer to cultivars with several stalks that form multiple smaller heads with thin stalks grown for bunching as sprouting broccoli and cultivars with a single large head and thick stalk like calabrese as heading broccoli. However, using traditional notation both these types are sprouting broccoli separate and distinct from the overwintering types that only form a head after vernalization. For example, in some areas, late-maturing cultivars and overwintering types of cauliflower have been traditionally referred to in some circles as "heading broccoli" (Rubatzky and Yamaguchi, 1997).

In addition there are other distinct vegetables called "broccoli" that have a similar appearance to sprouting broccoli. Chinese broccoli (kan lan or jie lan) *B. oleracea* Group Alboglabra or botanical variety *alboglabra* has also been classified by some as *Brassica alboglabra*. Also known as Chinese kale and white-flowered broccoli, Chinese broccoli resembles both sprouting broccoli and kale, growing between 40–50 cm (16–20 in) tall. The edible portions of Chinese broccoli include the stem, accompanying leaves and an early developing inflorescence that does not require vernalization (CABI, 2012).

Broccoli raab (*Brassica rapa* L. rapa (DC.) Metzg. or Group Ruvo Bailey) is a popular leafy vegetable in parts of southern Europe. Broccoli raab is an annual that resembles sprouting broccoli but develops a much smaller head with a less compact inflorescence. Broccoli raab is also called Italian turnip, cima de rapa, or rapini. The remainder of this chapter describes sprouting broccoli unless stated otherwise (CABI, 2012).

Production

Crop rotation with crops outside the family Brassicaceae is highly recommended as a strategy to control soil-borne pests. Broccoli may be grown by direct seeding or from transplants. Bunching broccoli is often direct-seeded using precision planters on raised beds to promote drainage and accommodate furrow irrigation. For uniform emergence, a fine seedbed is needed to achieve good seed-to-soil contact for rapid uniform emergence. For bunched broccoli, in which two to four heads are banded together with a rubber band, high plant populations with in-row spacing as close as 10 cm (4 in) create the optimum head size for bunching. To ensure a good stand, broccoli is sometimes overseeded and thinned when plants are in the seedling stage (Swaider et al., 1992). For markets preferring large-head, single-stalk broccoli, lower plant populations are used with in-row spacing ranging from 20 to 60 cm (8–24 in) with between-row spacing of 50–90 cm (20–36 in) depending on the cultivar.

Broccoli for large head production or broccoli grown in short-season areas is sometimes transplanted (Swaider et al., 1992). Transplants are produced in densely seeded plant beds in fumigated or steam-treated field soil or in plug trays in a protected structure. Broccoli is easy to transplant because plants root easily with little root mass. Transplants are ready for field planting 4–5 weeks after seeding. Storage of pulled, field-grown broccoli transplants should not exceed 10 days at 0°C (32°F) or 5 days at 19°C (66°F) prior to transplanting.

Broccoli grows well in a wide range of mineral and organic soils with adequate moisture and fertility. Soil fertility varies with region and type and should be tested for nutrient content before planting broccoli. Soil pH in the range of 6.0–6.5 is preferred, but broccoli will tolerate a pH range of 5.5–6.8. Broccoli has a high N requirement and requires that fertilizer be placed near the plant since it has a relatively small root zone. Depressed yields, delayed maturity, reduced keeping quality, and strong or objectionable flavors are indicative of N deficiency. Often N is applied in a split application at planting and again at heading when plants respond to fertilization. Little additional N may be needed for broccoli grown on organic soils. Excessive N can promote rapid growth and poor quality heads that are open and have poor shelf life. Combined with high temperatures, excess N can promote such rapid growth that plants show symptoms of tip burn in susceptible cultivars. Like most brassica crops, broccoli has a high requirement for B and Mo. Boron deficiency causes yellowing or chlorosis of youngest leaves and stems, which often starts from the base to the tip. Rosetting or even death of terminal shoots or buds occurs in extreme cases. The common symptoms of Mo deficiency in broccoli include a general yellowing, marginal and interveinal chlorosis, marginal necrosis rolling, scorching, and downward curling of margins usually on older leaves (CABI, 2012).

Broccoli can be produced using the same conservation tillage techniques described above for cabbage including no-till, ridge-till, strip till, or mulch till (Hoyt et al., 1994). Conservation tillage systems cause minimum disturbance to the soil after the previous crop has been harvested, reducing soil erosion, conserving moisture, inhibiting weed growth, and building organic matter in the soil. The advantages of conservation tillage for broccoli production are generally the same as those listed for cabbage including less machinery needed to produce the crop, less labor, fuel saving, and reduced soil erosion and compaction (Hoyt et al., 1994). Disadvantages of conservation tillage include: lower soil temperatures, slower germination and emergence when direct seeding is used, slower early growth, delayed competition with weeds, delayed spring planting, more root disease, heavier crop residue, planter operation may be more difficult, weed spectrum changes, increased insect pests, and later maturity (Hoyt and Walgenbach, 1995). Growing broccoli using conservation tillage practices requires many of the same adjustments described above for cabbage to account for cooler and often wetter soil conditions (Infante and Morse, 1996).

Like cabbage, broccoli can be grown using a plasticulture system (Lamont, 2004). Sheets of thin plastic film are spread across raised beds and the edges of plastic covered with soil to create a tight covering of plastic mulch on top of the bed that will drain rather than collect water. Drip irrigation is usually used with this system since rain and overhead irrigation is excluded unless perforated plastic is used. Soil fumigants may be injected into the soil as the plastic is laid to kill soil-borne pests, and drip tape for irrigation or fertigation is laid beneath the film as well. Transplants are placed through holes in the plastic. When a fumigant is used, sufficient time, usually about 7 days, must pass before transplanting occurs. Polyethylene (plastic) mulch helps by modifying soil temperature and conserving moisture. Plastic mulch also reduces soil compaction, crusting, leaching of fertilizer, drowning of crops, and competition from weeds, particularly if the soil is fumigated or light is not transmitted through the plastic (Lamont, 2004). With plasticulture, plants are cleaner at harvest, requiring less washing and enabling efficient field packing. Although using mulch increases production costs, those costs can be offset by earlier and larger yields of high-quality heads. In order to spread increased costs over two seasons, growers can plant another crop (double cropping) into plastic mulch before or after the broccoli.

Generally, broccoli requires 2.5–3.8 cm (1–1.5 in) of water per week for uninterrupted growth. Yield reduction may occur when soil moisture remains below 50% field capacity, particularly on sandy soils, for extended periods (Saunders, 1993). Drought during the juvenile stage (three to four true leaves) may adversely affect subsequent plant development and uniformity of maturity. Drip irrigation is efficient and can be used with or without plastic mulch. Overhead irrigation uses water less efficiently and may contribute to foliar disease. In flat areas with loam soils, furrow irrigation can also be used. However, excessive moisture may delay maturity and reduce yield, especially in early cultivars. Shallow tillage, herbicides, mulching, or a combination of methods are used for weed control (Swanton and Weise, 1991).

Harvesting

Most broccoli is hand harvested by crews that simultaneously pass across several rows assisted by a harvesting aid that moves boxes and facilitates trimming and field packing (Fig. 16.7). For bunching broccoli production, heads are hand-cut in the field, while the buds are tightly closed, and bunched in groups of three or four small heads approximately 15–20 cm (6–8 in) diameter held together by a rubber band. The stalks are trimmed to approximately 20 cm (8 in) length. For large head single-stalk or crown-cut broccoli, much larger single heads up to 40 cm (16 in) diameter are cut and packed without bunching (Jett *et al.*, 1995). The goal is to harvest broccoli in a single pass through a field because when workers have to search for broccoli to cut, harvesting is less efficient. Use of precision seeding, irrigation, and F-1

Fig. 16.7. Field packing of bunched broccoli in California.

hybrid cultivars provides sufficiently uniform maturity to make once-over harvesting possible after ideal growing conditions. Under non-uniform conditions, multiple harvests may be necessary at 2–4 day intervals. Field packing in cardboard boxes is popular because it is efficient and reduces handling and bruising. Broccoli degrades quickly and must be rapidly transported to a postharvest handling facility immediately after cutting for cooling to remove field heat (Kader, 1992).

Postharvest handling

Because of its high respiration rate, sprouting broccoli must be cooled immediately after harvest to maintain head quality and color. Broccoli should be stored at 0°C (32°F) and 95–100% RH (Kader, 1992). Broccoli should be cooled to 0°C (32°F) by ice, forced-air, or other means immediately after harvest to remove field heat and preserve quality. Slurry ice is often added directly to field-packed boxes to remove field heat and preserve freshness (Kader, 1992). Although the trend has been toward more field packing to reduce costs, harvested broccoli may sometimes be transported to a grading facility where it is trimmed, bunched, cooled, and stored until shipment. Because of concerns about the disposal and slow decomposition of the expended waxed boxes used to ship fresh broccoli, alternatives to top icing are now used in some areas. Shrink film wrapping, after cooling to remove field heat, followed by refrigerated storage can be an effective technique to preserve freshness during long distance shipment and marketing (Jett *et al.*, 1995). The film wrapping creates a microclimate around the head with a modified atmosphere that helps preserve freshness. The shelf life of broccoli is relatively short (7–10 days) and may be significantly shorter if delays in cooling or fluctuations in storage temperature occur. Broccoli should not be stored with fruits that produce appreciable amounts of ethylene, since exposure to ethylene causes yellowing (Kader, 1992). For processing, broccoli is hand harvested and collected in bulk, then transported to a processing facility where it is washed, cut, and flash frozen.

Uses

Large single heads or bunches of smaller heads are sold fresh and boiled or steamed for eating. Fresh uncooked broccoli can be eaten in salads without cooking. Sometimes broccoli is sold as "crowns" or "crown cut" with only a minimal amount of stem attached. The market demand for "lightly processed" or "fresh cut" broccoli is increasing. Broccoli heads are washed, trimmed, and sold in plastic containers or bags as florets or as a salad mix with other vegetables.

Broccoli is processed by freezing and drying. Dried broccoli florets are convenience packaged with rice or pasta for a ready-to-prepare meal. Broccoli florets, diced broccoli (cut pieces of stems and heads), or diced stems, or chopped crowns are sold frozen. Diced broccoli is sometimes mixed with other ingredients such as pasta and sold frozen in ready-to-cook bags.

World production and trade

The USA is the world's largest producer of broccoli with almost 60,000 ha (150,000 acres) (FAOSTAT, 2012). Most of the broccoli harvested in the USA is sold fresh but some is frozen. California is the leading broccoli-producing state accounting for about 90% of the US cropland. Fresh-market broccoli is the fourth most exported vegetable in the USA. About 50% of US broccoli exports go to Canada while Japan received 44% (USDA, ERS, 2010). The USA imports broccoli primarily from Mexico. Broccoli also is grown on a large scale in China, Italy, Ecuador, Guatemala, Europe, and the Far East. China has become a major exporter of broccoli in Southeast Asia and has surpassed the USA as the largest exporter to Japan.

Pest management

Insects

Broccoli shares many of the same insect pests with cabbage. Please refer to the cabbage section of this chapter for a detailed description of the following pests and their control measures. Diamondback moth caterpillar, cabbage looper, imported cabbageworm, cabbage webworm, cross-striped cabbageworm, beet armyworm, cutworms, seed and root maggots, aphids, sweetpotato or silverleaf whitefly, harlequin bug, flea beetle, and vegetable weevil are all insects that attack both broccoli and cabbage (Kemble *et al.*, 1999). Additional insect pests that may attack broccoli include: armyworm (*Pseudaletia unipuncta*), corn earworm (*Heliothis obsoleta*), cabbage white fly (*Aleurodes proletella*), cabbage seed pod weevil (*Ceutorrhynchus assimilis*), cabbage stem weevil (*Ceutorrhynchus napi*), cabbage moth (*Barathra brassicae*), leaf webber (*Crocidolomia binotalis*), oblique-banded caterpillar (*Cacoecia costana*), salt marsh caterpillar (*Estigmene acrea*), thrips (*Thrips tabaci*), and leafminers (*Liriomyza* spp.) (CABI, 2012). Local recommendations should be consulted for the best control measures available to combat these pests of broccoli.

Disease

There are many diseases that affect broccoli and many of these also affect both cabbage and cauliflower. In addition to reducing yield, infection may also cause cosmetic damage that reduces marketability. For a detailed discussion of each of the following diseases, please refer to the section on cabbage diseases above: black rot (*X. campestris*), blackleg (*Leptosphaeria maculans*), Alternaria leaf spot (*Al. brassicae*), wirestem (*Rh. solani*), and downy mildew (*Pe. parasitica*) (Kemble *et al.*, 1999). Some other important diseases that may infect broccoli include: bacterial leaf spot (*Pseudomonas syringae* pv. *maculicola*), leaf spot (*X. campestris* pv. *armoraciae*), soft rot (*Erwinia carotovora* or *Pseudomonas* spp.), Cercospora leaf spot (*Cercospora brassicicola*), Cercosporella leaf spot (*Cercosporella brassicae*), clubroot (*Pl. brassicae*), damping off (*Pythium*, *Fusarium* or *Rhizoctonia* spp.), Sclerotinia rot (*Sclerotinia sclerotiorum*), Verticillium wilt (*Verticillium alba-atrum*), and Fusarium wilt (*F. oxysporum*) (Cornell, 2004; Rimmer *et al.*, 2007).

Cauliflower

History

Cauliflower (*Brassica oleracea* Group Botrytis) is a cool-season, frost-tolerant crop that is widely adapted and grown throughout temperate and subtropical regions of both the northern and southern hemispheres (Griffiths, 1994). The exact location where cauliflower was domesticated is in doubt but believed to be somewhere in the eastern Mediterranean region. It is likely that cauliflower arose from a genetic mutation that occurred in cabbage that resulted in a head consisting of callus-like flower primordials instead of a large vegetative bud. One of the first written references to cauliflower is found in the writings of the Arab scientists Ibn al-'Awwam and Ibn al-Baitar, between the 12th and 13th centuries (Fenwick *et al.*, 1982). Cauliflower gained popularity in France in the mid-16th century and was subsequently cultivated in northern Europe and the British Isles. Heat-tolerant tropical forms of cauliflower have been developed in India during the past 200 years by selecting for heat tolerance from European genotypes. Today, cauliflower is widely grown in Europe, North America, parts of Central and South America, Asia, and Australia. The USA, France, Italy, India, and China are leading producing countries (FAOSTAT, 2012).

Botany

Cauliflower is an herbaceous dicot that is grown as an annual or biennial depending on the cultivar. Cauliflower can be sexually crossed with other Groups or botanical varieties of *B. oleracea* because they are in the same species and genetically very similar. As young seedlings, cauliflower is difficult to distinguish from related crops such as cabbage, broccoli, and Brussels sprouts. Cauliflower has a moderately shallow root system with a branched taproot and many fibrous roots in the top 40 cm (16 in) of soil. Leaves are oblong, pinnate, thick, somewhat leathery, glaucous, smooth, and usually alternate (Nieuwhof, 1969).

For biennial cultivars, the stem elongates after plants have advanced beyond the juvenile phase and received sufficient chilling to induce flowering (Fig. 16.3). At full flower, the bushy plant will be roughly 1 m (39 in) high and approximately 0.5 m (20 in) across (Nieuwhof, 1969). The inflorescence is a raceme that elongates rapidly with many small flowers forming at the terminus. The inflorescence produces bisexual, cross-shaped flowers with four yellow petals, one pistil, six stamens, and a two-celled ovary like broccoli and cabbage (Nieuwhof, 1969). The flowers are insect pollinated, usually by bees, which are attracted to the many flowers and nectar. The perfect flowers are not self-pollinated because of pollen incompatibility. This means that pollen can only successfully germinate and grow on a stigma of a flower from a different plant (Rubatzky and Yamaguchi, 1997). After pollination the ovary develops into fruit called a silique containing from 20 to 40 seeds as in cabbage and broccoli.

A cauliflower head or curd is composed of undifferentiated flower primordia reminiscent of plant calli, a mass of unorganized parenchyma cells, derived from plant tissue (Fig. 16.8). The cauliflower curd consists of tightly clustered undifferentiated, usually white, shoot apices formed upon thick, hypertrophied, repeatedly branched fleshy terminal portions of the short thick stem (Rubatzky and Yamaguchi, 1997). Slow elongation associated with rapid thickening of lateral branches results in short, thick and compressed dome-shaped curd in most cultivars while some cultivars produce curds with pointed or pyramidal shaped heads. Continued growth of the curd beyond edible maturity results in elongation of its many branches, causing spreading with loss of compactness and shape. In general, curd initiation occurs in the post-juvenile plant stage. For early-maturing cultivars, post-juvenile curd formation occurs after 15–20 leaves have formed. For late-maturing cultivars, initiation occurs after 25–30 or more leaves have formed. Tropical and summer cultivars of cauliflower produce heads at relatively high temperature and flower with minimal vernalization. Following vernalization, curds of strongly biennial, late-maturing cultivars produce shoot apices with differentiated flower primordia. Flower primordia become evident in most cases after the curd is well past edible maturity. Flower buds are formed in the axils of the elongated branches of the curd (Rubatzky and Yamaguchi, 1997).

Plant heights in the vegetative stage are variable, but most are 40–80 cm (16–32 in) tall. Leaves are usually upright and oblong and longer and narrower than those of cabbage (Nieuwhof, 1969). Foliage colors range from grayish to blue green with waxy bloom and smooth or curly leaf margins. Small inner leaves initially envelop and protect the curd from discoloration due to sunlight. As curd size increases and heads mature, the innermost leaves are unable to overlap and protect the head. The extent of inner and outer leaf cover varies greatly among cultivars. Late-maturing cultivars produce more and larger leaves and therefore provide better covering of the head. Large leaf size is generally correlated with large curd size. Summer and tropical cultivars have less leaf growth and often require leaf tying to protect heads from light-induced discoloration (Rubatzky and Yamaguchi, 1997).

Types and cultivars

Both open-pollinated and hybrid cultivars of cauliflower are available. Cauliflower cultivars grown in temperate regions can be broadly grouped into three major maturity types: early (summer or fall harvest), intermediate (late fall and early winter), and late (winter and spring). One important group of cauliflowers (Cavolofiore di Jesi) include cultivars from Italy with varied curd form including a cream to yellow colored type with pyramidal curd and two genotypes with green curd, one that is smooth and the other pyramidal (cvs. Romanesco, Flora Blanca). Northern European cultivars grown as annuals during the summer and autumn include (cvs. Alpha and Snowball). North-western European cultivars are grown as biennials for late winter and spring production such as cvs. Roscoff and St. Malo. Australian cultivars, developed from European germplasm and grown as annuals for late, early winter, or spring harvest include cv. Barrier Reef. Asian cultivars are grown as early annuals and adapted for high temperature regions (cv. Panta) (Rubatzky and Yamaguchi, 1997).

Some cultivars of cauliflower will not form a head until they have undergone a cold treatment. All cauliflowers require some vernalization in order to flower and the required chilling period varies among cultivars (Rubatzky and Yamaguchi, 1997). Late-maturing cultivars require longer exposure and lower temperatures than early cultivars. Important characteristics that define cultivars are head size, shape, compactness, surface texture, and color. White cultivars are preferred in most markets although cultivars producing, cream, purple, green,

Fig. 16.8. A cauliflower head at edible maturity. The head is composed of flower primordia that give rise to flower stalks with further development for annual cultivars or after vernalization for biennials.

and orange curds are available (Rubatzky and Yamaguchi, 1997). For the white-curd fresh market, cauliflower heads should be uniform in shape, with clean, un-bruised, pure white to light cream curds. For the processing market there is greater acceptance of off-color curds that are whitened during processing. Curd tissues typically lack chlorophyll, but one cultivar of cauliflower has green curds and is marketed as "broccoflower" or "broccoliflower". Broccoflower is not a hybrid between broccoli and cauliflower, but simply a cultivar of cauliflower that produces chlorophyll in its curd.

In northern Europe, some later maturing overwintering cauliflower cultivars are sometimes referred to as broccoli or heading broccoli (Rubatzky and Yamaguchi, 1997). These designations should not be confused with sprouting broccoli, which is an annual and has an upright plant architecture. Heads of true sprouting broccoli are composed of green to purple fully differentiated immature flower buds that with further maturation open into flowers. On the other hand, a cauliflower head is composed of dense clusters of usually white undifferentiated flower primordia that only give rise shoot apices after the plants receive a chilling treatment.

Fertilization

Deep, loamy, well-drained soils are needed for the production of cauliflower. Soil fertility varies with region and soil type and soil should be tested for nutrient content prior to planting. The soil pH and fertility should be carefully maintained as described for cabbage. Soil pH in the range from 6.0 to 6.5 is preferred, but cauliflower can tolerate pH over a broader range from 5.5 to 6.8 (Swaider *et al.*, 1992).

Cauliflower has a high N requirement and fertilizer should be placed near the plant since it has a relatively small root system. Depressed yields, delayed maturity, and reduced storage are associated with N deficiency (Peirce, 1987). Often N is applied at planting and again at heading when plants are growing rapidly and have a high nutrient requirement. Less N is needed for cauliflower produced on organic soils. Long-season cultivars of cauliflower should receive supplemental N at the time of heading or when dictated by the results of petiole analysis (Swaider *et al.*, 1992). For processed cauliflower, higher than normal N fertilization, maximize head size, curd density and yield, since reduced shelf life caused by high N is less important. Excessive N can promote rapid growth and poor quality "open" heads with poor shelf life. Combined with high temperatures, excess N can promote tip burn in susceptible cultivars.

Like most brassica crops, cauliflower has a high requirement for B and Mo (Swaider *et al.*, 1992). Boron deficiency causes yellowing or chlorosis of youngest leaves and stems, which often starts from the base to the tip. Rosetting or even death of terminal shoots or buds occurs in extreme cases. Boron stress can cause stem cracking and browning of cauliflower curd. Water-soaked areas on the stems and branches of the curd may alo develop. Cauliflower is very sensitive to Mo deficiency, particularly on well-fertilized light sandy soils with pH levels close to or below pH 6. Whiptail can be eliminated in many cases by raising soil pH above 6.5 by liming or by adding a few ounces per acre of sodium or potassium molybdate to increase available Mo. The common symptoms of Mo deficiency in cauliflower include a constriction and yellowing of the leaf lamina, called whiptail, marginal and interveinal chlorosis, and scorching and downward curling of margins usually on older leaves (Swaider *et al.*, 1992). In severe cases, Mo deficiency can abort head development. Nitrogen deficiency, particularly when combined with low temperature or other environmental stresses, can cause premature heading, which is also called "buttoning" (Rubatzky and Yamaguchi, 1997). Cauliflower is sensitive to Mg and Mn deficiencies, which are expressed as interveinal chlorosis on the older and younger leaves, respectively.

Production

Crop rotation outside the family Brassicaceae is highly recommended as a strategy to control soilborne pests. Cauliflower is relatively easy to transplant because plants root easily and little root mass is required for successful establishment (Swaider *et al.*, 1992). However, excessive stress during transplanting and establishment can result in premature heading or barren plants as pointed out above (Rubatzky and Yamaguchi, 1997). To minimize stress, cauliflower transplants are produced in plug trays in protected structures rather than pulled bare root from outdoor transplant beds as some cabbage seedlings are produced. Transplants are ready for field planting 4–6 weeks after seeding. Cauliflower may be established by direct seeding but are more often transplanted to avoid environmental stress and its negative consequences. Cauliflower is often grown on raised beds to promote drainage and accommodate furrow irrigation. Depending on the cultivar, typical in-row spacing ranges from 30–60 cm (12–24 in) with between-row spacing of 60–90 cm (24–36 in) (Swaider et al., 1992).

Warm temperatures tend to inhibit or delay head formation and flowering but promote vegetative growth of cauliflower. On the other hand, cool or cold temperatures favor curd formation. It is desirable for curd formation to be delayed until plants have sufficient leaf area and photosynthetic capacity to support head development in order for the curd to attain adequate size. Therefore, the goal of many producers is to grow a large vegetative plant with large leaf area before curd formation is initiated (Rubatzky and Yamaguchi, 1997). Premature curd formation is a common problem for some annual and tropical cultivars. Winter cultivars will not produce curd unless low temperature exposure is adequate. If winter cultivars are grown in the tropics, they remain vegetative. Information about temperature and cultivar interaction is needed in order to determine planting schedules for optimum productivity.

Once initiated, reproductive growth is not easily reversed. Curd initiation diminishes leaf development and causes lateral buds to elongate into shoot apices that comprise the convex surface of the domed head. For many cultivars, the best curd quality develops at 16–18°C (61–64°F) and quality begins to decline at temperatures above 20°C (68°F). Active curd formation in certain winter cultivars occurs at temperatures as low as 10°C (50°F), while curd development in some tropical cultivars may continue at temperatures up to 30°C (86°F) (Rubatzky and Yamaguchi, 1997). Once initiated, high temperature tends to accelerate the rate of curd development but also tends to reduce compactness.

Curd disorders related to temperature include buttoning, riceyness, bracting, and misshapen curds. Buttoning is caused by premature head development before adequate vegetative development occurs, resulting in small curd development. A velvet-like curd surface is called riceyness and is caused by precocious flower bud formation (Rubatzky and Yamaguchi, 1997). It is attributed to cold inductive temperatures followed by warm temperatures that promote rapid growth of stem apices. Variation in the appearance of riceyness is possibly due to cultivars maturing at different times as well as differing exposures to cold. Bracting results at high temperatures from rapid vegetative growth of small leaves that are normally suppressed between lateral branches of the curd. Rapid and/or extended growth will similarly elongate lateral branches of the curd resulting in misshaped heads as well as bracting. A condition referred to as "blindness" is likely caused by low temperature or other damage to the apical meristem. An apical meristem is lacking in "blind" plants resulting in reduced leaf numbers and no heads (Rubatzky and Yamaguchi, 1997).

Some markets require that curds be white and not discolored by sun exposure. For many cultivars, foliage development is adequate to protect heads when small, but with head enlargement, reflexing of the expanding subtended leaves exposes the curd (Rubatzky and Yamaguchi, 1997). Self-blanching cultivars were developed specifically for long upright leaves that shade the curd without tying. High-density plantings and high N fertilization also help produce large erect leaves that shield curds. However, for summer cauliflower production in California where sunlight is intense, workers must tie wrapper leaves with rubber bands above the heads to prevent discoloration by sunlight. Rubber bands are color-coded so workers can determine the likely harvest date based on when the leaves were tied. Winter types generally produce vigorous leaf growth and may not need to be tied. Because winter types require a long growing period, their production has declined in many areas except northern France, some areas of England, Denmark, and the Netherlands. Effects of sun discoloration are of less concern for tropical cultivars. The open growth habit of tropical cultivars makes it difficult to blanch heads with foliage cover.

Cauliflower can be produced using conservation tillage techniques as discussed previously for cabbage and broccoli. Conservation tillage systems minimally disturb the soil after the previous crop has been harvested.

Conservation tillage for cauliflower production requires less machinery, labor, and fuel, as well as reduces soil erosion and compaction. Disadvantages of conservation tillage are the same as discussed earlier for cabbage and broccoli.

Cauliflower can be grown using a plasticulture production system as described previously for cabbage and broccoli. Please refer to the cabbage section for a detailed description of plasticulture production.

Generally, cauliflower requires at least 2.5–3.8 cm (1–1.5 in) of water per week for uninterrupted growth. Yield reduction may occur when soil moisture remains below 50% field capacity, particularly on sandy soils, for extended periods (Saunders, 1993). Drought during the juvenile stage (three to four true leaves) may reduce canopy size, cause blindness, buttoning and decrease crop uniformity at maturity. Drip irrigation is most efficient and can be used with or without plastic mulch. Overhead irrigation uses water less efficiently and may contribute to disease. In flat areas with loam soils, furrow irrigation may be an option if water is plentiful. Excessive moisture may delay maturity and reduce yield, especially in early cultivars. Shallow tillage, herbicides, mulching, or a combination of methods generally provides weed control in cauliflower.

Harvesting

Fresh-market cauliflower is harvested by hand when the heads are compact before they expand. Accompanying the loss of compactness is a change in shape. The usually convex-shaped curd becomes flattened or even concave when heads expand beyond saleable size. Broccoli destined for distant markets is often field packed, but may be transported to a grading facility where it is sorted and graded, field heat removed, and packed for shipment. A once-over complete hand harvest is the goal of growers but multiple harvests are often needed because of plant-to-plant variation and non-uniform curd development. Selective mechanical harvesting machinery has been developed but most is harvested by hand using harvesting aids that move slowly across fields to move packed boxes to trucks for transport out of the field. When harvested, the stem is cut well below the base of the curd. Some basal leaves surrounding the curd are retained but trimmed to the level of the curd to allow viewing while providing protection against physical damage. Head weights at harvest maturity range from 0.5–2 kg (1.1–4.4 lb) with diameters from 15–30 cm (6–12 in). Field packing of individually plastic-film-wrapped trimmed heads is practiced to reduce desiccation and decrease handling. Individually wrapped heads are placed in cardboard boxes in the field to reduce handling and bruising, which shortens shelf life. Cauliflower heads marketed close to production areas may not be wrapped. After field packing cauliflower is transported to a cooling facility where forced-air or vacuum-cooling rapidly removes field heat. The cooled boxes are maintained in refrigerated storage until shipment. For processing, bulk bins are loaded and taken to the processing facility immediately after harvest without cooling. The market demand for fresh lightly processed cauliflower is increasing, so the heads are often trimmed, washed, divided, and marketed in bags as ready-to-eat pieces.

Postharvest handling

Cauliflower has a high respiration rate, is prone to desiccation and should be cooled immediately after harvest to maintain quality and color. Cauliflower can be forced-air, vacuum-, or hydro-cooled prior to storage or marketing (Kader, 1992). Vacuum-cooling works well for field-packed heads and is more effective if cauliflower is damp at the time of cooling. Commercial storage at 0°C (32°F) with 95–98% RH can maintain good quality for up to 3 weeks. However, freeze damage can occur at just −1°C (30°F) resulting in discoloration and softening of the curd (Kader, 1992). Maintaining 98–100% RH can further reduce weight loss and maintain turgidity of heads, but free water accumulation on the curd must be avoided. Storage-life is reduced to approximately about 7–10 days at 5°C (41°F), 5 days at 10°C (50°F), and 3 days at 15°C (59°F) depending on initial quality. Loss of quality during prolonged storage includes wilting, browning and spreading of curds, yellowing of leaves, and decay. The benefits from controlled atmosphere and modified atmosphere are modest. Low O_2 (<2%) in combination with 3–5% CO_2 may delay leaf yellowing and the onset of curd browning. However, injury may occur at <2% O_2 and/or >5% CO_2 (Kader, 1992). Injury may not be apparent until curds are cooked, when they become soft, grayish, and develop off-flavors. Storage in >10% CO_2 can induce injury within 2 days. Unlike broccoli, fresh cauliflower does not produce strong off-odors when held in low O_2 environments. Cauliflower has a very low ethylene production rate of <1 μl/kg (0.2 tsp/short ton), but it is extremely sensitive to ethylene, with the most prevalent symptoms being curd discoloration and leaf yellowing and abscission (Kader, 1992).

Growing conditions can strongly influence quality of fresh cauliflower. Heads must be protected from the sun to prevent yellowing and strong flavor development in the curd. Only high quality heads should be stored or shipped long distances. Heads must be handled gently to avoid bruising that results in rapid browning and decay. In some cases, harvested cauliflower may also be transported to a grading facility where it is trimmed, bunched, cooled, and stored until shipment. For processing, cauliflower is harvested by hand or machine collected in bulk, transported to a processing facility, washed, cut, and flash frozen.

Uses

Single heads are sold fresh and boiled, baked, or steamed for eating. Fresh cauliflower can be eaten in salads without cooking. Cauliflower is well adapted and widely used for fresh cut "lightly processed" products. Cauliflower heads are washed, trimmed, or added to fresh salad mixes with other vegetables.

Cauliflower is processed by freezing or drying. Dried cauliflower are convenience packaged with dried rice or pasta. Diced cauliflower (cut pieces of stems and heads), or diced stems or chopped curds are sold frozen. Diced cauliflower is sometimes mixed with other ingredients such as pasta and sold frozen in ready-to-cook bags.

World production and trade

China and India are the largest producers of cauliflower and approximately 70% of world cauliflower production occurs in Asia (FAOSTAT, 2012). Europe produces approximately 20% of the world's cauliflower and Italy and the UK are leading European producers. Significant cauliflower production also occurs in North America, where the USA is the leading producer (FAOSTAT, 2012).

Pests

Insects

Cauliflower shares most of the same insect pests with cabbage and broccoli. Please refer to the cabbage section of this chapter for a detailed description of the following pests and their control measures: diamondback moth caterpillar, cabbage looper, imported cabbageworm, cabbage webworm, cross-striped cabbageworm, beet armyworm, cutworms, seed and root maggots, aphids, sweetpotato or silverleaf whitefly, harlequin bug, flea beetle, and vegetable weevil are all insects that attack both broccoli and cabbage. Additional insect pests that may attack broccoli include: armyworm (*Ps. unipuncta*), corn earworm (*Heliothis obsoleta*), cabbage white fly (*Al. proletella*), cabbage seed pod weevil (*Ce. assimilis*), cabbage stem weevil (*Ce. napi*), cabbage moth (*Ba. brassicae*), leaf webber (*Cr. binotalis*), oblique-banded caterpillar (*Ca. costana*), salt marsh caterpillar (*Es. acrea*), thrips (*Th. tabaci*), and leafminers (*Liriomyza* spp.). Local recommendations should be consulted for the best control measures available to combat these pests.

Diseases

Many diseases that affect cauliflower also affect both cabbage and broccoli. In addition to reducing yield, infection may also cause cosmetic damage that reduces marketability. For a detailed discussion of each disease, please refer to the section on cabbage diseases above: black rot (*X. campestris*), blackleg (*Le. maculans*), Alternaria leaf spot (*Al. brassicae*), wirestem (*Rh. solani*), downy mildew (*Pe. parasitica*) (Kemble *et al.*, 1999). Some other important diseases include: bacterial leaf spot (*Ps. syringae* pv. *maculicola*), leaf spot (*X. campestris* pv. *armoraciae*), soft rot (*Er. carotovora* or *Pseudomonas* spp.), Cercospora leaf spot (*C. brassicicola*), Cercosporella leaf spot (*Ce. brassicea*), clubroot (*Pl. brassicae*), damping off (*Pythium*, *Fusarium* or *Rhizoctonia* spp.), Sclerotinia rot (*Sc. sclerotiorum*), Verticillium wilt (*V. alba-atrum*), Fusarium wilt (*F. oxysporum*), turnip mosaic virus (TuMV), and cauliflower mosaic virus (CaMV) (Cornell, 2004; Rimmer *et al.*, 2007).

Blackspeck is a physiological disorder typified by black specks on cauliflower curds. Blackspeck is common in the popular 'Snowball' cultivar series and is apparently due to an imbalance in Ca nutrition. Blackspeck seems to be related to the disorders tip burn and internal browning that are common in cabbage and Chinese cabbage and are caused by inadequate transport of Ca to rapidly growing tissues. Foliar spray, with calcium nitrate, may control the problem or alternatively control may be obtained by reducing N fertilizer applications (Cornell, 2004).

The major causes of postharvest decay are bacterial soft rot caused by *Erwinia* and *Pseudomonas* spp. and brown rot caused by *Alternaria* spp. Storing only good quality, disease-free heads and maintaining good temperature control can effectively control these organisms (Cornell, 2004).

Brussels Sprout

History

The origin of the Brussels sprout is believed by many to be northern Europe (Field, 2001). The development of Brussels sprouts probably occurred in northern Europe possibly as early as early as the 5th century (Field, 2001). Brussels sprouts likely arose from natural mutation(s), possibly in the Middle Ages, in a cabbage plant introduced to northern Europe by the Romans. One source claims that the plant was cultivated near Brussels in the 13th century. The first recorded description was in 1587, so Brussels sprouts are of comparatively recent origin (Rubatzky and Yamaguchi, 1997). The mutation(s) resulted in axillary bud development along an elongated vegetative stem. The stable mutation(s) were collected and recurrent selection for longer stems and axillary bud development lead to the Brussels sprouts we know today. Brussels sprouts first reached North America with French settlers, who grew them in Louisiana. The first plantings in California's Central Coast began in the 1920s, but significant production did not begin until the 1940s as Brussels sprouts gained popularity in North America. The highlands of Guatemala became a major producer of Brussels sprouts for export starting in the 1980s, because of lower labor costs and favorable climate. The crop has had limited distribution and/or production beyond northern Europe where it is a major crop. Brussels sprouts are also grown commercially in North America, Australia, Guatemala, and New Zealand (FAOSTAT, 2012).

Botany

Brussels sprouts (*B. oleracea* Group Gemmifera) are a cool-season, frost-tolerant crop, and is an herbaceous dicot biennial (Griffiths, 1994).

Brussels sprouts can intercross with cabbage, cauliflower, broccoli, and other members of *B. oleracea* because they all are in the same species. In the seedling stage, Brussels sprouts are difficult to distinguish from broccoli, cabbage, and cauliflower. Leaves are thick, somewhat leathery, smooth oblong, simple, alternate, pinnate, and petiolate. Leaf colors range from purple to green. Root systems are moderately shallow with a prominent taproot that branches readily, giving rise to many fibrous roots concentrated within the top 30–40 cm (12–16 in) of soil (Nieuwhof, 1969). Brussels sprouts plants are readily identified by the numerous vegetative buds (sprouts) that develop in leaf axils of an elongated unbranched stem approximately 5–8 cm (2–3 in) in diameter (Nieuwhof, 1969). Brussels sprouts plants exhibit strong apical dominance. The tall (50–90 cm, 20–35 in) stem terminates with a large dominant apical bud that strongly inhibits axillary bud development near the top of the plant (Nieuwhof, 1969). Axillary buds continue to be produced from the bottom of the plant toward the top until growth ceases (Fig. 16.9).

The production of more than 100 buds per plant is common (Rubatzky and Yamaguchi, 1997). Early-maturing cultivars tend to be shorter and produce fewer sprouts than tall-growing late cultivars. However, tall cultivars are more prone to lodging. Bud compactness, size, shape, color

Fig. 16.9. A harvest-mature Brussels sprout plant in early winter. The lower leaves senesce as axillary buds or sprouts form from the base of the plant toward the top.

intensity, harvest period, and productivity are important attributes. Cool temperatures or subfreezing temperatures will increase the compactness and flavor of sprouts.

The inflorescence is a raceme that elongates rapidly from the terminal bud after plants have been vernalized (Fig. 16.3). The inflorescence is similar to other brassicas with many small flowers. The inflorescence produces bisexual cross-shaped flowers with four yellow petals and one pistil and six stamens (Nieuwhof, 1969). At full bloom, the bushy plant will be roughly 1.2 m (47 in) high and up to 0.5 m (20 in) across. Even though flowers are perfect they are cross-pollinated, most often by bees. Flowers are not self-pollinated because of pollen incompatibility. This means that pollen can only successfully germinate and grow on a stigma of a flower from a different plant (Rubatzky and Yamaguchi, 1997). Incompatibility genes are exploited by plant breeders to insure cross-pollination of the inbred lines that must be crossed to produce F-1 hybrid seeds. A superior ovary with one or two cells develops into a silique, very similar in size and shape to those described above for cabbage, broccoli, and cauliflower (Nieuwhof, 1969). Seeds usually mature 50–80 days after fertilization. A well-pollinated Brussels sprout plant may produce 175 g (6 oz) of seed. There are approximately 350 seeds/g (9,900 seeds/oz).

Production

Crop rotation with crops outside the family Brassicaceae is highly recommended to control soil-borne pests. Brussels sprouts should not be grown in rotation with related crops for at least 3 years before production. Brussels sprouts grow well in a wide range of mineral and organic well-drained soils with adequate moisture and fertility. Soil fertility varies with region and soil type and soil should be tested for nutrient content before planting Brussels sprouts. Soil pH in the range of 6.0–6.5 is preferred, but Brussels sprouts will tolerate a pH range of 5.5–6.8. Brussels sprouts have a moderately high N requirement and fertilizer placement should be near the plant since it has a relatively small root zone. Depressed yields, delayed maturity, reduced keeping quality and strong or objectionable flavors are indicative of N deficiency. Often N is applied at planting and again during sprout formation. Little additional N may be needed for Brussels sprouts grown on organic soils. Excessive N can promote rapid growth and large sprouts that are not solid and have poor shelf life. Like other brassicas, Brussels sprouts have a high requirement for B and Mo. Boron deficiency causes yellowing or chlorosis of youngest leaves and stems, which often starts from the base to the tip. Rosetting or even death of terminal shoots or buds occurs in extreme cases. The common symptoms of Mo deficiency in Brussels sprouts include a general yellowing, marginal and interveinal chlorosis, marginal necrosis rolling, scorching, and downward curling of margins usually on older leaves.

Vegetative growth occurs over a wide range of temperatures but high quality sprout development occurs at cool temperatures ranging from 5–18°C (41–64°F). Above 25° (77°F), sprouts are less compact and have a strong flavor. If temperatures exceed 30°C (86°F), sprout development may be inhibited. Growth is slow at temperatures below 5°C (41°F) even though the plants tolerate prolonged exposure to cold temperatures. Sprouts that develop when temperatures are cool have a milder flavor than those that develop at high temperatures. Some people feel that best Brussels sprout flavor is obtained after the plants have been exposed to freezing temperatures. F-1 hybrid Brussels sprout cultivars are popular because they exhibit heterosis (hybrid vigor) and have greater uniformity compared to open-pollinated cultivars. Seeds saved from F-1 hybrid plants do not grow true-to-type when planted and new seeds should be purchased each year. Cultivars of Brussels sprout are often grouped as early, main season, or late according to maturity, which ranges from 85 to 125 days from transplanting. Cultivars with smaller sprouts are often grown for freezing. Early cultivars require 85–90 days from field planting to harvest, and main-season crops require 110–120 days.

Brussels sprouts are most often grown from transplants with in-row spacing ranging from 25 to 60 cm (10–24 in) with between-row spacing of 50–90 cm (20–36 in) depending on the cultivar. Brussels sprouts are usually transplanted to shorten the growing season and improve uniformity at harvest. Transplants are produced in densely seeded plant beds or in plug trays filled with growing media in a protected structure. Brussels sprouts can be established with little root mass and, like cabbage, are considered relatively easy to transplant. Transplants are ready for field planting 4–6 weeks after seeding. Storage of pulled, bed-grown transplants should not exceed 10 days at 0°C (32°F) or 5 days at 19°C (66°F) prior to transplanting.

Brussels sprouts can be produced using conservation tillage techniques although in sloping plantings along California's central coast traditional tillage systems are used. The techniques described for conservation tillage of broccoli, cauliflower, and cabbage apply to Brussels sprout as well. Transplanting is the most common method for establishing Brussels sprouts in a conservation tillage seedbed. Since Brussels sprouts do well under cool soil conditions, are transplanted, and are often planted under warm conditions for fall or spring harvest, the disadvantages of conservation tillage often listed for warm-season vegetables such as lower soil temperatures, slower germination and emergence when direct seeding is used, and slower early growth are not negatives for Brussels sprout.

Brussels sprouts are well adapted to regions of cool temperatures and moderately severe winters such as those of northwestern Europe, where this crop is sometimes overwintered for spring harvesting. Vegetative growth can continue, although slowly, at temperatures as low as 5°C (41°F). Plants have frost tolerance and, when acclimated, survive freezing temperatures as low as −5 to −10°C (23–14°F). High temperatures can suppress stem elongation, and although axillary bud growth may be advanced, compactness decreases.

Generally, Brussels sprouts require 2.5–3.8 cm (1–1.5 in) of water per week for uninterrupted growth. Yield reduction may occur when soil moisture remains below 50% field capacity, particularly on sandy soils, for extended periods. Drought during the juvenile stage (three to four true leaves) may adversely affect subsequent plant development and uniformity of maturity. Drip irrigation is efficient. Overhead irrigation uses water less efficiently and may contribute to foliar disease. In flat areas with loam soils, furrow irrigation can also be used. Shallow tillage, herbicides, mulching or a combination of methods can be used to obtain weed control.

Harvesting

Brussels sprouts are grown as an annual for their small axillary buds (sprouts) that can be removed individually during the season or at one time in a single destructive harvest. Harvest commences when Brussels sprouts are firm and well developed, approximately 2.5–5 cm (1–2 in) in diameter (Swaider *et al.*, 1992). Several successive hand harvests may be made starting at the base of the plant for fresh-market production. For hand harvest, individual axillary buds (sprouts) are removed by snapping them from the stem. As the lower sprouts are removed, those further up the stalk continue to develop. Hand harvest can occur over a period of several weeks and as long as 4 months. For multiple hand-harvesting, the apical bud is not removed and the stem can continue to elongate and produce additional axillary buds. Individual sprouts weigh from 20 g (0.71 oz) to more than 50 g (1.76 oz), with a diameter range from 20–60 mm (0.8–2.4 in), the preferred diameter for freezer processing being between 15–25 mm (0.6–1 in). Yields range from 6–20 metric tonnes/ha (2.5–8.9 short tons/acre) and yields for multiple harvested fresh-market crops are generally higher than those obtained for processing. However, as height increases, plants may lodge, which interferes with cultural and harvest operations (CABI, 2012).

While Brussels sprouts can be hand harvested, the practice of a single destructive harvest using labor aids or partial mechanization has increased, especially for freezer processing. In this procedure, plants are stripped of leaves, the stems cut, and then sprouts are mechanically sheared from the stem by a sprout stripper. For once-over mechanical harvest, application of a tip inhibitor or physically removing the growing point at the top of the plant when the lower sprouts are about 1.3 cm (0.5 in) in diameter will break apical dominance, stimulating uniform sprout development along the entire stalk. The growth retardant SADH (succinic acid 2,2-dimethyl hydrazine) may substitute for apical bud removal to arrest apical dominance (Rubatzky and Yamaguchi, 1997). Although further stem elongation is prevented, total sprout growth is not hindered, and axillary buds near the stem apex enlarge at a faster rate than those near the base. Most of the bud growth is achieved by redistribution of N and carbohydrates from the leaves. Accordingly, apical bud removal should not occur before plants have achieved appropriate growth.

Postharvest handling

Brussels sprouts should be cooled immediately after harvest to 0°C (32°F) by hydrocooling, forced-air or other means to remove field heat and preserve quality (Kader, 1992). The shelf life of fresh Brussels sprouts is up to 30 days if stored at 0–5°C (32–40°F) and 95–100% RH and may be significantly shorter if delays in cooling or fluctuations in storage temperature occur. Brussels sprouts should not be stored with fruits that produce appreciable amounts of ethylene, since exposure to ethylene causes yellowing (Kader, 1992). For processing, Brussels sprouts are stripped from the stalk usually with a harvesting aid, collected in bulk, and then transported to a processing facility where they are washed, cut, and flash frozen.

Pest management

Insect pests

Since Brussels sprouts are closely related to cabbage, broccoli, and cauliflower they share many of the same pests. Please refer to the cabbage section of this chapter for a detailed description of the following pests of Brussels sprouts and their control measures: diamondback moth caterpillar, cabbage looper, imported cabbageworm, cabbage webworm, cross-striped cabbageworm, beet armyworm, cutworms, seed and root maggots, aphids, sweetpotato or silverleaf whitefly, harlequin bug, flea beetle and vegetable weevil (Kemble *et al.*, 1999). Additional insect pests that may attack Brussels sprouts include: armyworm (*Ps. unipuncta*), corn earworm (*Heliothis obsoleta*), cabbage white fly (*Al. proletella*), cabbage seed pod weevil (*Ce. assimilis*), cabbage stem weevil (*Ce. napi*), cabbage moth (*Ba. brassicae*), gramma moth (*Phytometra gamma*), leaf webber (*Cr. binotalis*), oblique-banded caterpillar (*Ca. costana*), salt marsh caterpillar (*Es. acrea*), thrips (*Th. tabaci*), and leafminers (*Liriomyza* spp.) (Cornell, 2004). Local recommendations should be consulted for the best control measures available to combat these pests of Brussels sprouts.

Diseases

Many diseases that affect cauliflower, cabbage, and broccoli also affect Brussels sprouts. In addition to reducing yield, infection may also cause cosmetic damage that reduces marketability. For a detailed discussion of each disease, please refer to the section on cabbage diseases above. The diseases include: black rot (*X. campestris*), blackleg (*Le. maculans*), Alternaria leaf spot (*Al. brassicae*), wirestem (*Rh. solani*), and downy mildew (*Pe. parasitica*) (Kemble *et al.*, 1999). Some other important diseases include: bacterial leaf spot (*Ps. syringae* pv. *maculicola*), leaf spot (*X. campestris* pv. *armoraciae*), soft rot (*Er. carotovora* or *Pseudomonas* spp.), Cercospora leaf spot (*C. brassicicola*), Cercosporella leaf spot (*Ce. brassicae*), clubroot (*Pl. brassicae*), damping off (*Pythium*, *Fusarium* or *Rhizoctonia* spp.), Sclerotinia rot (*Sc. sclerotiorum*), Verticillium wilt (*V. albaatrum*), Fusarium wilt (*F. oxysporum*), turnip mosaic virus (TuMV) and cauliflower mosaic virus (CaMV) (Cornell, 2004; CABI, 2012).

Uses

Brussels sprouts are primarily consumed fresh or frozen. Individual sprouts are sold in bulk or pre-packaged in bags or plastic containers. Brussels sprouts are generally cooked before eating. Brussels sprouts are well adapted to freezing. Freezing helps preserve quality and flavor. Brussels sprouts are sometimes preserved in vinegar as well.

Trade of Brussels sprouts

Brussels sprouts are most widely produced in northern Europe. In Continental Europe, the largest producers are the Netherlands, at 82,000 metric tonnes (90,390 short tons), and Germany, at 10,000 metric tonnes (11,023 short tons). The UK has production comparable to that of the Netherlands, but the crop is consumed domestically and not generally exported. Brussels sprouts are also a significant crop in Australia and New Zealand. Brussels sprouts are a minor crop in North America with approximately 1,000 ha (2,470 acres) in California, which produces approximately 98% of all commercially grown Brussels sprouts in the USA. A substantial majority of California Brussels sprouts, 80–85%, are processed frozen and the rest are sold fresh (NASS, 1999). Since Brussels sprout production is relatively labor-intensive, production has shifted to countries where labor is less expensive and plentiful. This trend has caused an increase in the international trade of both fresh and frozen Brussels sprouts. Guatemala is major exporter of both fresh and frozen Brussels sprouts to the USA and Canada. Brussels sprouts are grown in Africa for export to the European Union.

Kohlrabi

Kohlrabi is a cool-season biennial plant that is grown as an annual vegetable for its enlarged stem (Rubatzky and Yamaguchi, 1997; Fig. 16.10).

Kohlrabi is more popular in Europe than in North America. Cultivar selection in the USA is limited, with 'White Vienna' being the most common. Green and purple cultivars are popular in some markets. Kohlrabi is similar to turnip in terms of its taste and usage. In the USA, kohlrabi is only grown outdoors. In Europe, kohlrabi is grown both outdoors and in greenhouses.

Rich, loamy soil or generous applications of organic matter aid production of tender stems. Fertilization and cultivation are similar to that needed for cauliflower. Excess N and water can cause stem cracking in some cultivars. The best quality stems are obtained by encouraging rapid and continuous growth. The optimum temperature for kohlrabi growth and development is 16–21°C (61–70°F). Exposure to temperatures below 7°C (45°F) can result in vernalization and bolting, while high temperatures retard growth and promote tough, stringy tissue.

Although kohlrabi can be transplanted, it is more typically direct-seeded and thinned to 10–16 cm (4–6 in) between plants or it may be precision-seeded to a final stand. Successive plantings every 2–3 weeks ensure a continuous supply throughout the season. Kohlrabi should be harvested when the swollen stem is 5–8 cm (2–3 in) in diameter, the size of a small apple, and before it is tough and woody, normally 55–65 days after seeding (Swaider et al., 1992). The root is cut off, and stems may be sold individually, bunched by the top or bagged prior to sale. The recommended storage temperature is 0°C (32°F), but quality can only be maintained a few days when the tops are left on. Removal of the tops and storage at 0°C (32°F) can extend storage life up to 4 weeks (Kader, 1992).

Greens

Greens are a diverse group of non-heading brassica species, including mustard, turnip, kale, and collards. Greens are grown for their leaves that are either used fresh or cooked. In parts of Canada, young rutabaga leaves are also used as greens. The greens are cool-weather crops that are planted in summer, or in the fall in the southern USA. Certain cultivars of turnips have been developed for use only as greens and do not form enlarged roots. The Siberian smooth-leaf and the Scotch, curly-leaf types of kale are widely grown. Currently, most kale grown in the USA is of the curly type, of which cultivars with either blue or green leaves are available (Fig. 16.11; Rubatzky and Yamaguchi, 1997).

Production management is the same for all greens and similar to that of cabbage. All of these are cool-weather crops that are planted in early spring or late summer or fall in the southern latitudes and harvested during the winter and spring. Although some kale and collards have been transplanted in the past, most are direct-seeded and thinned only if necessary. The growing season may be as short as 40–50 days from seeding for kale and mustard. The growing season for collards is usually 75–90 days from seeding (Swaider et al., 1992). Greens are best grown on a well-drained sandy loam soil. In some areas, double or triple seeded, cell-grown transplants are planted in plastic mulch for the production of bunch-harvested collards. Although some kale and collards are transplanted, most are direct seeded and thinned only if necessary. Rapid, uniform growth is needed to maintain tenderness and to facilitate harvest and field packing. Adequate fertility is needed to maintain vegetative growth, but excessive N may reduce cold tolerance and storage quality. Collards, mustard, and kale in particular, tolerate temperatures as low as −9°C (16°F), if not preceded by warm weather (Rubatzky and Yamaguchi, 1997). These crops can be harvested after frost until they are eventually killed by

Fig. 16.10. A box of kohlrabi for sale at a vegetable auction in the Netherlands.

Fig. 16.11. A field of green curled kale at harvest maturity in northern Germany.

sub-freezing weather in northern areas. Leaves of collards and kale are harvested sequentially as they reach full size but are still tender. Collards may also be harvested by cutting the rosette at ground level when plants are 15–30 cm (6–12 in) tall. Two or three individual plants are often bunched together for retail sales. Trimmed leaves may be marketed either loose or bunched and packed in cardboard boxes for market. An increasing percentage of greens are sold pre-washed, chopped, and packed in polyethylene bags. Leafy greens are highly perishable, and field heat should be removed quickly by icing, hydro-cooling, or vacuum-cooling. The optimum storage temperature is 0°C (32°F) and 95–100% RH, but storage life is usually only 1–2 weeks (Kader, 1992).

ASIAN BRASSICAS

The genus *Brassica* includes many vegetables that are widely grown in Southeast Asia and to a lesser extent in other parts of the world. A discussion of these vegetables is complicated by the many common names derived from different languages. However, Asian vegetables may be grouped as Chinese cabbage (*B. rapa*), Asian mustard greens (*B. juncea*), and Chinese broccoli or Chinese kale (*B. oleracea*) (Myers, 1991).

Chinese Cabbage

Botany

There is confusion about the correct botanical name for Chinese cabbage. The debate has been raging for decades and this explains the fact that several different scientific names are sometimes used. Recently, molecular genetic studies have shown that the major morphological differences among *B. rapa* may be caused by fewer genes than diverse phenotypes suggest and, therefore, classification of subspecies or botanical varieties under *B. rapa* may not be warranted. The term Group (gp.) is gaining popularity among horticulturists and botanists to show groups of horticultural significance within a species that were previously classified as separate botanical varieties or subspecies (Griffiths, 1994).

Chinese cabbage can be classified as heading and nonheading types (Figs 16.3, 16.12). These distinct types are frequently classified as Pekinensis and Chinesis, Groups of *B. rapa* L., respectively. A few of the common names used to describe heading types of Chinese cabbage (Pekinensis Group) include: nappa or napa, hakusai, pai-tsai, won bok, pechay, and tsina. Common names for nonheading types (Chinesis Group) include: bok choy, pak choy, celery cabbage, celery mustard, chongee, petsay, and pei tsai. Heading types may be further classified according to head shape. Heading types are classified as open, erect, or

Fig. 16.12. Heading (Pekinensis) (a) and nonheading (Chinesis) (b) types of Chinese cabbage.

cylindrical, and ovoid or barrel (Barlow, 2007; CABI, 2012).

The stem of Chinese cabbage is compressed and unbranched until flowering occurs. Leaf blades of sessile, mostly ovate and slightly wrinkled leaves extend to the bottom of the broad flat and lighter colored midrib. Many cultivars of heading Chinese cabbage resemble romaine lettuce in appearance. A cabbage head is a single large terminal bud comprised of tightly overlapping leaves attached to and enclosing much of the short unbranched stem. After vernalization the bud develops an inflorescence with bisexual flowers with one pistil, six stamens, and a superior ovary with one or two cells that develops into a long pod called a silique that resembles a thin green bean. Flowers are usually cross-pollinated by insects, most often by bees. Flower petals are either yellow or sometimes white. Although the flowers are perfect, they are not self-pollinated because of pollen incompatibility. Pollen can only successfully germinate and grow on a stigma of a flower from a different plant because it is genetically incompatible with flowers from the same plant where it was produced (Rubatzky and Yamaguchi, 1997).

Chinese cabbage is the most widely grown Asian brassica and the most important vegetable in *B. rapa*. Chinese cabbage is grown as a cool-season annual. A productive, relatively short maturity and broad adaptation along with many cultivar choices makes Chinese cabbage a popular world crop (Wittwer, 1987). Many cultivars are biennial, although some exhibit annual flowering behavior. The height of most cultivars ranges from 20–60 cm (8–24 in). Chinese cabbage has a finely branched extensive root system with the majority of roots developing in the top 30 cm (12 in) of soil. Moderate day and cool night temperatures result in high productivity and quality. Chinese cabbage does best when grown under daily average temperatures between 13–21°C (55–70°F) (Rubatzky and Yamaguchi, 1997).

For biennial types the stem elongates after plants have advanced beyond the juvenile phase and have received sufficient chilling. Both cold temperatures 4–10°C (40–50°F) and long days of 15–16 h or more for 4–5 weeks will induce flowering in cultivars susceptible to bolting. This limits the production of the Chinese cabbage in lower latitudes to the fall or winter so that the crop can mature under short days and cooler temperatures (Fu *et al.*, 1991). The optimum mean temperature for growth and head quality is 15–18°C (60–65°F) with a minimum temperature of 4°C (40°F) and a maximum of 24°C (75°F). Temperatures greater than 25°C (77°F) tend to delay heading and reduce quality. Generally, young plants are more tolerant of heat and cold than plants nearing maturity. However, well-hardened transplants will withstand −7°C (20°F) for short periods. Some believe that exposure to subfreezing temperatures improves flavor (Rubatzky and Yamaguchi, 1997).

Types and cultivars

Cultivars with erect cylindrical heads are called chichili, while round and compact heading types are called chefoo (Fig. 16.12a; Rubatzky and Yamaguchi, 1997). Having been grown in China since the 5th century, natural crossing has occurred resulting in distinctive and intermediate types in addition to the two described above. Some cultivars have greater heat tolerance and these develop narrow leaves having a high ratio of midrib to leaf blade tissues (Fig. 16.12b). There are many local distinctions in Asia for the various forms that exist (Barlow, 2007). Many of the current commercial cultivars are hybrids, often with resistance to one or more pest problems. Maturity characteristics are used to select cultivars and to schedule planting and harvest periods. State or regional recommendations should be consulted for locally adapted cultivars. Early cultivars require 55–70 days from direct seeding to harvest, and main-season crops require 80–110 days. Days to harvest are strongly influenced by environmental conditions (Rubatzky and Yamaguchi, 1997).

Chinese breeding work has created a vast diversity of cultivars over its centuries of cultivation. F-1 hybrid Chinese cabbage cultivars are available and widely grown throughout the world. F-1 hybrids are generally recognized to be more uniform and productive than open-pollinated cultivars. Significant progress has been made in breeding Chinese cabbage cultivars with greater heat tolerance and bolt resistance (Fu *et al.*, 1991). Cultivars with improved disease resistance are also available. *Agrobacterium*-mediated transformation of Chinese cabbage has been successful and this will enable the introduction of foreign genes from outside *B. rapa* to be introduced to improve Chinese cabbage in the future. The commercial use of transgenic Chinese cabbage is limited at this time.

Production

Chinese cabbage requires a rich, well-drained soil with a high water-holding capacity. Soil pH should range from 5.5 to 7.0. The fertility program for Chinese cabbage should follow state or regional recommendations based on soil testing and foliar analysis during the growing season (Kalb and Chang, 2005). Usually about half the N should be applied at seeding and the other half shortly after thinning. If the crop is transplanted, a sidedressing of N can be applied just prior to the onset of head formation. During cool and wet weather, P deficiencies may occur, particularly in heavier soils. All the P and K can be applied before or at the time of planting (Kalb and Chang, 2005). Alternatively, fertigation can be used to apply nutrients throughout the season based on soil and plant tissue analysis and local recommendations based on soil and climate. Excess N, particularly in combination with high temperatures, results in rapid succulent growth, loose heads, fewer marketable heads, and tip burn, which gives raise to bacterial soft rot (Cornell, 2004). Excessive N increases susceptibility to rot during shipping and storage. Additional B may be needed in some soils (Stephens, 1994).

There are many different ways to produce Chinese cabbage. Conservation tillage techniques can be used to reduce labor, maintain moisture and reduce soil compaction. Using conservational tillage, transplants are set into residue of the previous crop with minimal soil disturbance. In some cases, cover crops are used to protect and build the soil between cash crops (Kalb and Chang, 2005).

Uniform plant populations are needed to optimize crop uniformity and reduce differences in head maturity. Chinese cabbage may be grown on either flat or raised beds, depending on the soil type and climate. Raised beds are recommended to improve drainage, particularly on heavy soils. In some areas, Chinese cabbage is grown on raised beds 1.8 m (6 ft) from center-to-center (Kalb and Chang, 2005). In northern latitudes or for early-season production, early Chinese cabbage crops are grown from transplants.

Plant spacing is determined by the target market and cultivar. Transplants are generally used for spring crops, while fall plantings are most often direct-seeded 1.3 cm (0.5 in) deep. Transplants should be started 4–5 weeks before field planting, depending on environmental conditions, for the same maturity as direct-seeded crops (Stephens, 1994). Chinese cabbage does not form adventitious roots as easily as *B. oleracea* so bare-root transplants are not recommended. Plug tray transplants with the root system intact have the greatest chance for success (Swaider *et al.*, 1992). Direct seeding is usually in rows, and the plants are later thinned to the desired stand. Because of the small seed-size of Chinese cabbage, a well-tilled soil with small particle size should be used for direct seeding to ensure good seed to soil contact. Precision seeders direct seed to a final stand.

Spacing depends on mature head size. Chinese cabbage can be planted using equidistant spacing but is commonly planted in rows to facilitate mechanization. Row spacing of 50–60 cm (20–24 in) with an in-row spacing of 50 cm (20 in) would be suitable for medium-sized, main-season cultivars. Within-row spacings of 25–40 cm (10–16 in) are common for early-maturing smaller cultivars with fewer wrapper leaves, while 40–70 cm (16–28 in) within-row spacing may be needed for late cultivars (Stephens, 1994).

Because of its dense shallow root system, weed competition can be very detrimental to Chinese cabbage development and head quality, uniformity of maturity, and crop yield. Shallow cultivation may be used when the crop is small, but deep cultivation should be avoided to minimize root damage. In many areas, hand hoeing is no longer economical. An integrated control program for weeds includes a combination of chemical control and cultural practices to reduce competition from weeds.

Generally, Chinese cabbage requires 2.5–3.8 cm (1–1.5 in) of water per week for uninterrupted growth depending on soil type using sprinkler, drip, or furrow irrigation. Greater amounts are needed on sandy soils and lesser on heavier soils. Temperature and RH also play a role in determining irrigation requirements. Yield reduction may occur when soil moisture remains below 50% field capacity, particularly on sandy soils, for extended periods. Drought during the juvenile stage (three to four true leaves, with stem diameter less than that of a pencil) may adversely affect subsequent plant development and uniformity of maturity. However, excessive moisture may delay maturity and reduce yield, especially in early cultivars. Drought stress particularly in conjunction with high temperatures can cause tip burn and head rot (Stephens, 1994).

Mulch

Growing Chinese cabbage in mulch offers several advantages. In warmer areas straw or other types of organic mulch can help lower soil temperatures while decomposing organic matter can act as slow-release fertilizer based on its C to N ratio. Mulch also conserves soil moisture, reduces compaction, crusting of soil and competition from weeds (Kalb and Chang, 2005). At harvest, wrapper leaves are cleaner, requiring less washing, which greatly facilitates field packing.

Plastic mulches can also be used for Chinese cabbage production, especially when crops are established from transplants. Plastic mulches enable soil fumigation, prevent drowning of crops and leaching of fertilizer during periods of heavy rainfall and offer even greater modification of soil temperatures. Although using mulch, particularly plastic mulch, increases production costs, those costs can be offset by the benefits listed above.

Harvesting

Chinese cabbage is ready to harvest when the heads are fully developed after 55–110 days from direct seeding depending on the cultivar and environmental conditions. Chinese cabbage is most frequently hand-harvested by cutting the entire head at ground level near the base of the head. Typically, loose outer wrapper leaves are removed. Optimum harvest is when heads are closed and firm but before they burst or seed stalk development begins (Wittwer, 1987).

Historically with hand harvesting, a given field was harvested three to five times to obtain uniform sizing and maturity. Only one to three harvests of hybrid cultivars are required because they mature more uniformly. Mechanical harvesters have been developed and are used in areas where manual labor is limited and for processing. Yields will vary with the season of production, cultivar, and production system. With proper management, cabbage can produce 20–27 metric tonnes/ha (9–12 short tons/acre). Chinese cabbage is often packaged and graded in the field and placed in crates or mesh bags. For processing, cabbage may be harvested and transported in bulk. Generally, most fresh markets prefer heads that average 0.9–3.0 kg (2.0–6.6 lb). For processing and long-term storage, larger headed cultivars are preferred (Wittwer, 1987).

Postharvest handling

Chinese cabbage may be stored for 2–6 months, depending on cultivar, at temperatures just above freezing 1–2°C (approx. 34–36°F) and <90% RH. Air circulation is essential for long-term storage. Storage life is severely shortened by higher temperatures. There are several postharvest disorders that occur in Chinese cabbage. Black leaf speck is a physiological disorder that appears as lesions on the midrib and leaf veins during storage at 0°C (32°F) for several weeks (Kader, 1992). Exposure to ethylene at concentrations greater than 100 ppm causes leaf abscission and leaf discoloration (Kader, 1992). Chinese cabbage should not be stored with ethylene-generating crops like apples. Storage at near 1–2°C (approx. 34°F) with high RH and a properly controlled or modified atmosphere can decrease the likelihood of this disorder. The optimum postharvest head quality can be achieved in controlled atmospheres of 1–2% O_2 and 2–5% CO_2. Storing Chinese cabbage at oxygen levels below 1% is likely to cause off odors and flavor because of anaerobic respiration (Kader, 1992). The use of film wraps and other packaging materials may prevent water loss during storage.

Trade of Chinese cabbage

Chinese cabbage is of major importance in China, Korea, and Japan. Other Asian countries also have significant production. It is grown to a lesser extent in cooler climates of North America, Europe, Australia, and South America, where it is a minor vegetable. The major producers of Chinese cabbage are China, South Korea, and Japan with over 300,000, 50,000, and 35,000 ha, respectively. Because of the perishable nature of Chinese cabbage, international trade is somewhat limited. Chinese cabbage production is typically consumed near production areas. Use of refrigerated and controlled atmosphere storage has increased storage life, making greater international trade possible. International trade is anticipated to increase as these technologies become more widely available. Processed Chinese cabbage products, such as kimchee, a traditional fermented Korean dish often made with Chinese cabbage, are more widely traded than fresh (CABI, 2012).

Pests of Chinese cabbage

Since Chinese cabbage is closely related to cabbage, broccoli, and cauliflower they share many of the same diseases. Please refer to the cabbage section of this chapter for a detailed description of the following pests of Chinese cabbage and their control measures: diamondback moth caterpillar, cabbage looper, imported cabbageworm, cabbage webworm, cross-striped cabbageworm, beet armyworm, cutworms, seed and root maggots, aphids, sweet-potato or silverleaf whitefly, harlequin bug, flea beetle, and vegetable weevil are all insects that attack both broccoli and cabbage (Kemble et al., 1999). Additional insect pests that may attack Chinese cabbage include: armyworm (*Ps. unipuncta*), corn earworm (*Heliothis obsoleta*), cabbage white fly (*Al. proletella*), cabbage seed pod weevil (*Ce. assimilis*), cabbage stem weevil (*Ce. napi*), cabbage moth (*Ba. brassicae*), leaf webber (*Cr. binotalis*), oblique-banded caterpillar (*Ca. costana*), salt marsh caterpillar (*Es. acrea*), thrips (*Th. tabaci*), and leafminers (*Liriomyza* spp.) (Cornell, 2004). Local recommendations should be consulted for the best control measures available to combat pests of Brussels sprouts.

Cotton leaf worm (*Spodoptera littoralis*) or rice cutworm is basically a leaf eater and affects a very wide range of brassicas including Chinese cabbage. Heavy infestations of cotton leaf worms can defoliate plants. The larvae prefer young leaves to old leaves, always feeding on the underside of the leaf (Rimmer et al., 2007).

Chinese cabbage diseases

Many of the same diseases that affect other brassicas also affect Chinese cabbage. Please consult the list in the cabbage section for diseases that may also affect Chinese cabbage. Turnip mosaic virus is a Potyvirus that infects most brassica plants, but is particularly damaging in Chinese cabbage. The most common symptom in these crops is a distinct mosaic of light and dark green colors in the leaves. Necrotic streaks, flecks, or ring spots may also occur. The virus is transmitted by several aphid species from infected crops or weeds, and conditions that favor migrating aphid populations will lead to a high incidence of infection (Parker et al., 1995). Resistant cultivars of Chinese cabbage are available, but none are resistant to all five strains known to occur in this crop.

Chinese cabbage is similar to lettuce in that high temperatures may cause brown and black necrotic areas on the leaf margins of external and internal foliage. This disorder is attributed to rapid growth and insufficient translocation of Ca to rapidly expanding tissues. The problem is compounded by high N, water stress, and insufficient Ca and B in the plant. Some Chinese cabbage cultivars possess more tolerance to tip-burn than others (Fu *et al.*, 1991). Spraying with calcium citrate may be effective in reducing tip burn and associated rotting. Calcium citrate at a concentration of 25 g/100 l (3.3 oz/100 gal) water sprayed twice weekly from 4 weeks after transplanting is recommended before tip burn occurs. Where soil Ca content and pH are low, liming is generally recommended to adjust the soil pH and meet the large Ca needs of the crop (Stephens, 1994).

When Chinese cabbage is grown at excessive high temperatures, and particularly when Ca is limiting in the soil or plants are under drought stress, soft rot often occurs in conjunction with tip burn. Bacterial soft rot, caused by *E. carotovora*, attacks wounds such as leaf scars, insect injury, mechanical injury, lesions caused by other pathogens and tip burn etc., and these are primary avenues for soft rot bacterial invasion (Cornell, 2004). Infected plant tissues first develop water-soaked lesions that enlarge rapidly in diameter and depth. The affected area becomes soft and collapses, turning dark in advanced stages of disease development. Soft rot-infected plants have an offensive odor. Disease losses from soft rot may occur in the field, transit, or storage.

Soft rot bacteria are transmitted through infected plant debris, by plant roots, through infected soils, and by several insect species. Rainfall and high temperatures enhance infection in the field. Transit and storage infection may develop from bacterial contamination that occurred in the field or postharvest from handling equipment and storage containers. Soft rot bacteria can grow over a temperature range of 5–37°C (41–99°F) with an optimum temperature of approximately 25°C (77°F) (Cornell, 2004).

Disease management is based primarily on sanitation and cultural practices. Sufficient time should be allowed for crop residues to decompose before planting a second crop. Vegetable crops should be rotated with cereals or other non-susceptible crops. Fields should be well drained to reduce soil surface moisture and plants should be spaced sufficiently to allow ventilation for rapid drying of foliage. Mulch and irrigation practices that prevent soil splash and foliage wetting should also reduce soft rot.

Black leaf speck is a physiological disorder that appears as moderate sized discolored lesions on the midrib and veins of the leaves. The symptoms can occur after low temperatures in the field, but usually are associated with transit and storage conditions, and often appear within 1 month of storage at temperatures around 0°C (32°F). Controlled atmosphere storage reduces the incidence of black speck in Chinese cabbage. Ethylene may accelerate or promote its development.

When Chinese cabbage is grown under long days, it will "bolt" producing a seed stalk instead of a solid head. When plants bolt they do not produce a mature head and therefore cannot be sold. Cold temperatures during seedling growth also induce bolting. Factors that cause a check in plant growth such as nutrient deficiencies or water stress may also induce bolting. Certain cultivars may bolt more readily than others. When planting in the spring, "slow-bolting" cultivars should be used.

Asian Mustard Greens

Asian mustard greens (*B. juncea*) are a large and diverse group that include Chinese green mustard or gai choi, mizuna, and various potherb mustards (Barlow, 2007). Mustard leaves may be eaten raw in mixed salads, and the gai choi cultivars can be prepared like spinach.

Asian mustard greens are grown as annuals and should be produced at the same time of year as other cool-season greens. The requirements for climate, soil fertility, irrigation, and postharvest handling are similar to those described for nonheading types of Chinese cabbage (Rubatzky and Yamaguchi, 1997). Most Asian mustard greens are direct-seeded into flat or raised beds at a depth of 1.3 cm (0.6 in), in rows 30–40 cm (12–16 in) apart and thinned to 15–30 cm (6–12 in) after emergence. Leaves may be harvested when plants are at least 3 weeks old. If Asian mustard greens are to be harvested when 10–15 cm (4–6 in) tall, the seed can be broadcast and incorporated to a soil depth less than 1.3 cm (0.6 in). Some cultivars of Chinese green mustard have particularly small seeds that should be planted 0.6 cm (0.2 in) deep, 2 cm (0.8 in) apart, in rows spaced 30 cm (12 in) apart and later thinned to 10 cm (0.4 in). Larger seeded types may be planted 1.3 cm (0.6 in) deep, 5 cm (2 in) apart and later thinned to 25 cm (10 in). Larger plants have greater pungency and may be harvested 45–70 days after planting, when they are 15–20 cm (6–8 in) tall (Myers, 1991).

Mizuna is widely grown in Japan and is gaining popularity in North America and other parts of the world because it withstands light frost, is relatively heat tolerant, and resists bolting better than Chinese cabbage. Plants are 30–46 cm (12–18 in) tall with yellow-green, smooth, and slightly pubescent leaves that have a deeply notched, narrow, feathery shape.

Chinese Broccoli

Chinese broccoli (*B. oleracea* var. *alboglabra*) is also known as gai-lon (or gai-lohn), kailan, or Chinese kale. The Chinese broccoli plant resembles sprouting broccoli, except that the leaves are a bit broader, the stems longer, and the head is much smaller. Flowers form first in diminutive heads and then elongate rapidly into stalks bearing yellow or white flowers similar to broccoli raab. Chinese broccoli is grown as an annual like other types of broccoli. Phosphorus and K requirements of Chinese broccoli are similar to those of Chinese cabbage, but the N requirement is higher. Seeds are planted 1.3 cm (0.6 in) deep, 2.5–5 cm (1–2 in) apart and in rows spaced 46 cm (18 in) apart. Plants are generally thinned to 15 cm (6 in) when established. The crop needs at least 2.5 cm (1 in) of water per week and matures about 55–60 days after seeding. Heads, including about 20 cm (8 in) of the stalk and a few leaves, are harvested just before the flowers open. The timing of harvest is critical since the plants grow rapidly and can only be marketed successfully at the correct stage of development. However, each plant can be harvested several times. Harvested heads should be stored at 0°C (32°F) and 95–100% RH. Under optimal conditions, storage life is 10–14 days (Rubatzky and Yamaguchi, 1997).

ROOT CROPS

Annual root crops include radish (*Raphanus sativus*), turnip (*B. rapa* subsp. *rapa*), and rutabaga (*B. napus* subsp. *rapifera*). Horseradish (*Armoracia rusticana*) is an herbaceous perennial. The annual root crops prefer cool temperatures for growth and development. Long days and warm temperatures promote seed-stalk elongation. An adequate supply of N, P, and K is required to promote rapid, continuous growth of annual root crops. On some soils B may be required or brown heart may develop in the root tissue. For all direct-seeded root crops, a fine seedbed, free of stones and clods, is needed to facilitate precision seeding and uniform root development. Stones may increase the incidence of branched and crooked roots, especially in radishes. Stones in the soil also increase bruising of root crops during harvest.

Radish

Wild species are found in Europe eastward to the Volga River in Russia, and in the Mediterranean basin eastward to the Caspian Sea. Some taxonomists suggest the origin is further eastward, i.e. China, but most agree it is somewhere east of the Mediterranean. Radishes were known to the Egyptians in at least 2000 BC, in China 500 BC, Japan AD 700, Germany 1400, England 1548, and Mexico 1500 (Zohary and Hopf, 2000).

Radishes are gown for their enlarged hypocotyl and root tissues. Radish (*R. sativus*) types can be grouped by root maturity (Fig. 16.13).

The early spring or garden types mature in 25–40 days and have round or long roots that are usually white, red, or bicolored. Spring and summer cultivars are sold in the grocery store as bunched or bagged. They are never transplanted. All are annuals that bolt in response to long days and high temperatures. Common root colors of summer radish cultivars include red, white, or bicolor. Popular cultivars include: 'Red', 'Early Scarlet Globe', 'Scarlet Knight', 'Cherry Bell', 'Champion', 'Cavalrondo', and 'Comet' (Swaider *et al.*, 1992).

The late maturing winter types include 'Spanish', 'Daikon', and 'Chinese' (Fig. 16.14). These take longer to develop, have larger roots weighing up to 20 kg (44 lb), a wide range of root colors and much longer storage life. Round-rooted spring-type cultivars are more likely to bolt than long-rooted cultivars that are more resistant to high temperatures (Rubatzky and Yamaguchi, 1997).

Radish is more tolerant of low soil pH of 5.5–6.0 than broccoli, cabbage, cauliflower, or Chinese cabbage, but grows better at pH 6.0–6.8. Light, friable soil is considered best, although radish can be grown on most soil types. Sandy or sandy-loam soils are preferred for an early crop, while a cool, moist soil is better for production of winter radish. Radish are produced on muck soils in some areas. Spring radish is usually planted at quite high populations on either flat or raised beds, while winter radish requires more space and are thinned to 5–10 cm (2–4 in) within rows. Successive seedings of spring radish are made, since the crop matures in only 3–5 weeks. Winter radish requires twice as long to mature. Spring radish matures rapidly and requires high levels of N for sustained growth and optimum root quality. Harvest of spring radish begins as soon as roots are of marketable size. High temperatures cause roots to become pithy, increase pungency, and have poor quality. Roots split during excessively rapid growth, e.g. high

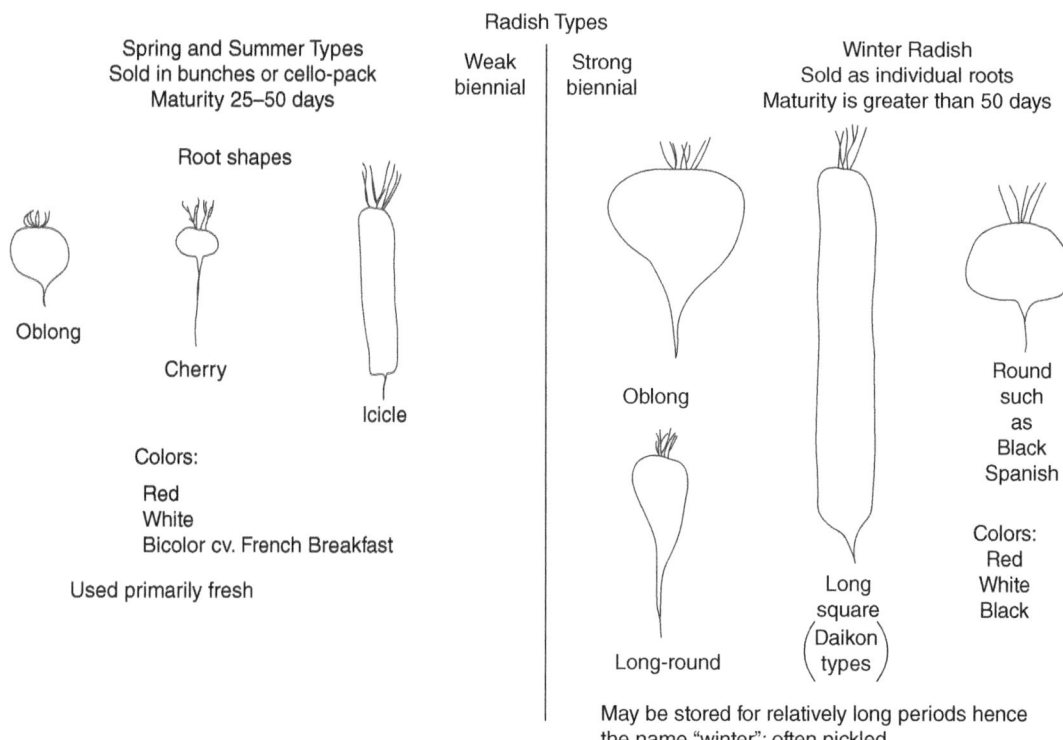

Fig. 16.13. Radish types are often classified as summer or winter types based on their life cycle and maturity range.

water, high N, and high temperatures. Optimum temperatures are 13–15.5°C (55–60°F).

At one time, radishes were bunched with the tops left on, but now most are harvested mechanically, hydro-cooled, sorted, and packed in polyethylene bags with trimmed tops. Spring radish cultivars can be stored at 0°C (32°F) and 95–100% RH for 3–4 weeks. Winter radishes are handled carefully to avoid bruising, hydro-cooled and packed in polyethylene bags, or pieces of long-rooted cultivars are sold individually. At 0°C (32°F) storage temperature winter radish cultivars should keep for 3–4 months.

Turnip and Rutabaga

Culture of turnips is ancient. Turnips were known to the Romans prior to the Christian Era and grown in France about 2,000 years ago. Cartier brought turnips to North America in 1541. The Mediterranean region is thought to be a primary center of origin along with eastern Afghanistan and western Pakistan. Turnips are important cool-season crops used as forage (both roots and tops) for livestock in addition to their use as human food throughout the world (Swaider et al., 1992).

It is uncertain whether or not rutabaga exists in a truly wild form. The domestication of rutabaga is recent with first reports occurring in England about 1600, earlier on the European continent, and in the USA about 1800.

Turnips are a cool-season biennial. They can be overwintered in some areas for an early spring harvest. However, with sufficient cold treatment below 13°C (55°F) they will produce a flower stalk (Rubatzky and Yamaguchi, 1997). For commercial root production, turnip is often grown as an annual. Turnips may develop several different root shapes including round, top shaped, flattened, and cylindrical roots. Turnips have a distinctive taproot from which most secondary roots originate. Usually they have little or no neck (Fig. 16.15).

Turnip cultivars are often classified as root types, foliage types, grown only for the leaves, which are used as "greens", or dual purpose when both tops

Fig. 16.14. Winter radishes come in many shapes sizes and colours. 'Tokinashi' is a long-rooted Daikon type (a), while 'Spanish' is round with a black exterior and white interior (b).

Fig. 16.15. 'Tokyo Cross' is a popular white-rooted slightly cylindrical F-1 hybrid turnip that matures in about 35 days from direct seeding. Both the roots and leaves are edible.

and roots can be eaten. Turnips occur in a wide range of root colors (purple, green, white, pale yellow, or bronze) and roots may be either solid or bicolor. Bicolor cultivars generally have different top and bottom colored roots. The leaves and petioles are pubescent and yellowish green. Culture is similar to that of radish. Rows are seeded about 46 cm (18 in) apart with in-row spacing of about 5 cm (2 in). Spacing depends on the cultivar. Turnip roots achieve their best quality when 5–10 cm (2–4 in) in diameter. Turnips are grown from seed and mature in approximately 50–55 days. Optimum growth temperatures are 15.5–18°C (60–65°F). A good yield for roots is 22 metric tonnes/ha (10 short tons/acre). The initiation of flower stalk development terminates root enlargement and vegetative leaf development, and causes roots to become hard and pithy, ruining their eating quality. Roots become pithy and strong when they are overly mature as well. Root splitting occurs when growth is too fast and is caused by excessive water, high N, and warm temperatures.

Rutabaga is a cool-season frost-tolerant crop that prefers temperatures <22°C (72°F). It is a biennial that is grown as an annual crop because flower stalk development reduces yields and root quality. The rutabaga roots are larger than turnip with a rounded and/or elongated root that can become very large, weighing 2.3 kg (5 lb) or more each (Fig. 16.16; Rubatzky and Yamaguchi, 1997).

Secondary roots arise from the underside of the enlarged root as well as from the taproot. The neck or crown is prominent. Leaves are smooth, bluish white, with thick petioles and covered with bloom. Cultivars of rutabaga, also known as swede, and turnip are available with white or yellow root color (Rubatzky and Yamaguchi, 1997). Roots of most rutabaga cultivars are yellow and larger in comparison than the white-fleshed turnip cultivars. The color on the root shoulder of both crops may be bronze, green or purple. Turnip and rutabaga are also grown as animal forage crops or winter cover crops in some areas.

Turnip and rutabaga can be grown on all types of soil, though yields are better on deep, rich, well-drained loams. Both vegetables are grown as row crops but rutabaga requires more space than turnip. Turnip and rutabaga are sensitive to B deficiency.

Rutabagas are a relatively insignificant crop in the USA and are produced predominantly in high latitude production areas like northern Europe and Canada. A major portion of the US supply is provided by Canada. Rutabaga is a relatively important crop in Canada, England, and northern Europe. Rutabagas are direct seeded and mature in about 90 days. Row spacing is generally 56 cm (24 in) between rows with in-row spacing of 10–15 cm (4–6 in), depending on the cultivar. Some people describe the taste of rutabagas as bland. However, flavor is improved by exposure to low temperatures or light frosts (Rubatzky and Yamaguchi, 1997).

Turnip and rutabaga may either be machine harvested or they may be pulled by hand. For commercial production, the roots are mechanically dug and topped similar to carrots or beets. When marketed directly after harvest, rutabaga roots are sometimes waxed to reduce shriveling and weight loss. Both crops require a cool temperature near freezing and high RH during storage and will keep for 2–6 months. Turnip tops are very nutritious and are high in vitamins (A and C in particular) and minerals. The roots are less nutritious. Club root and black rot are troublesome diseases. Maggots, flea beetles, and aphids are serious insect pests (Rubatzky and Yamaguchi, 1997).

Horseradish

Horseradish (*A. rusticana*) is an herbaceous perennial grown for its pungent roots, which are grated and eaten fresh or added as an ingredient to other food products such as seafood sauce. It is grown primarily as an annual but may also be grown in a perennial production system. It is a long-season crop, requiring warm temperatures during the vegetative phase, followed by cooler temperatures in late summer and fall that enhance root development. In the annual system the crop is planted in the previous fall or early spring, and harvested after the first killing frost. In the perennial system, upright, thickened shoots arising from a "mother" plant are harvested every other year, with the original plant left in the field for regeneration. Perennial fields may stay in production for 10–20 years. Horseradish has wide climatic adaptation, with the exception of the warmer zones of the far southern USA (Bratsch, 2009).

Fig. 16.16. Round rutabaga roots with purple shoulders in a Canadian market.

Horseradish is propagated vegetatively by root cuttings (sets), 1.25–3.5 cm (0.5–1.4 in) in diameter and 20–40 cm (8–16 in) in length. In the annual system, this set enlarges and becomes the primary marketable product at season's end, along with the many secondary roots forming along the medial and distal regions (Bratsch, 2009). At harvest, cuttings are taken from these branched roots arising from the original, trimmed to an appropriate length, tertiary rootlets are removed, and then stored for propagation the next season. Because horseradish roots exhibit distinct polarity, and since the diameter is similar for the length of the root cutting, sets are trimmed straight across at the top (head) and oblique at the lower end (tail) to mark polarity (Bratsch, 2009). Sets are uniformly packed by head/tail polarity in wooden or plastic storage crates lined with heavy plastic. The plastic is loosely sealed to cover the sets, and they are kept in cold storage at 0–1°C (32–34°F) until spring. Sometimes underground storage pits or caves are used in lieu of modern coolers. There appears to be no vernalization requirement, and sets can be planted any time after separation from the main plant (McClung and Schales, 1982).

Horseradish grows best on rich, moist, deep friable loam, or sandy loam soils rich in organic matter. Good drainage is needed for the production of quality roots. Sandy soils and soils with hardpan subsoil frequently produce highly branched roots that are of poor quality. Following deep plowing and cultivation, rows and spacing in the field are marked with a paddle wheel device, which creates a "pillowed" indentation for set placement. Generally, rows are spaced 0.9–1.2 m (35.4–47.2 in) apart, with 45–60 cm (18–24 in) between plants (Bratsch, 2009). Sets are manually planted in the field at a slight angle, with the head placed "up" on the pillowed side. Proximal direction is maintained for multiple rows, depending on cultivation equipment width. Cuttings are set in the field at approximately a 45° angle with the square end up and approximately 7.5–10 cm (3–4 in) below the soil surface. Approximately 2,940–3,538 root cuttings/ha or 450–560 kg/ha (400–500 lb/acre) of roots are needed to establish a new crop (McClung and Schales, 1982).

Horseradish has a high fertilizer requirement. Horseradish has a high K requirement, moderate P needs, and low to moderate N requirements. Fertilizer application should supply at least 111 kg/ha (100 lb/acre) of N, P, and K for most soils. However, fertilizer applications may be reduced when concentrations of extractable soil P and K are high. Irrigation during dry periods can improve marketable yield. Good drainage is needed for the production of quality roots. Irrigation during dry periods, particularly in late summer to fall, can improve marketable yield. Roots are hardy and can be overwintered in the ground. However, roots are usually dug and stored for easier access. Roots are plowed out of the ground and then the tops and side roots removed by hand cutting (McClung and Schales, 1982). Large, straight roots are considered the best quality. It is important that all roots are removed during harvest because horseradish can become a serious weed problem in subsequent years.

Wasabi

Wasabia japonica is Japanese horseradish, frequently called wasabi, grown for its rhizomes (Collins, 2003). Wasabi is a much different crop than horseradish. Wasabi is cultivated in semi-aquatic conditions and is also grown wild near flowing steams. Water temperature is critical and should be 10–13°C (50–55°F). The green paste sold with sushi in some US restaurants and groceries is made from horseradish rather than wasabi as a cost-cutting measure (Bratsch, 2009).

OTHER CROPS
Culinary Mustards

Yellow mustard (*Sinapis alba*), brown, and oriental mustard (*B. juncea*) are grown for their seed, which is processed into paste, culinary mustard, or powder for use as condiments. Other members of the species *B. juncea* are grown as an oilseed crop, greens, and stem vegetables (Rubatzky and Yamaguchi, 1997). Yellow mustard comprises about 90% of the culinary mustard crop in North America, while brown and oriental mustards are grown on a small scale. *Brassica juncea*, however, is more heat and drought tolerant compared to *S. alba*. In Europe, yellow mustard is also known as white mustard (Oplinger *et al.*, 1989; Pouzet, 1995).

Mustard is an annual herb that is planted in the spring. It emerges rapidly but grows slowly. It can be grown on variable soil types with good drainage but is best adapted to fertile, well-drained, loamy soils. Soil crusting on heavier soils may inhibit stand establishment. Growth will be stunted on waterlogged soils. Drought-prone sand and sandy loam soils should also be avoided. Since mustard seeds are very small, it is important to have good seed to soil contact for successful establishment. The seedbed should be well worked into small beds, firm, fairly level, and free of weeds and crop residue. Shallow tillage, sufficiently deep to kill weeds, is preferable for maintaining soil moisture close to the surface while preparing a seedbed for planting. Minimum tillage systems have also been successfully used for mustard production. Soil is firm enough for seeding when only a shallow depression is made by a shoe heal when walking across a field. If necessary, the seedbed should be packed before planting (Oplinger et al., 1989).

Fertilization requirements are similar to oilseed rape and other brassicas and are applied pre-plant based on soil test results. Mustard often responds well to N additions and to S on some soils. On soils deficient in B (less than 0.5 ppm), 0.6–1.1 kg/ha (0.5–1.9 lb/acre) should be applied. Soils with a pH near 7.0 are desirable, nevertheless, mustards will tolerate an alkaline pH and slightly saline soils (Oplinger et al., 1989).

Yellow mustard is seeded at 9–16 kg/ha (8–14 lbs/acre), although higher rates are appropriate on heavy, fertile soils or when emergence is difficult. Brown and oriental mustards have smaller seeds and should be planted at 6–8 kg/ha (5.4–7.1 lb/acre) and no deeper than 1.3–2.5 cm (0.5–1 in). Under dry soil conditions, seeding depth should be increased to 3.5 cm (1.4 in).

Seedlings are usually somewhat tolerant to mild frosts after emergence, but severe frosts can destroy the crop. The taproots will grow 1.5 m (59 in) into the soil under dry conditions, allowing for efficient use of stored soil moisture. Plants cover the ground in 4–5 weeks with favorable moisture and temperature conditions. Flower buds are visible about 5 weeks after emergence. Yellow flowers begin to appear 7–10 days later and continue blooming for a long period if water supply is adequate. If moisture stress occurs, yields are reduced.

Cultivars of yellow mustard usually mature in 80–85 days, whereas brown and oriental types require 90–95 days. Plant height at maturity varies from 76 to 114 cm (30–45 in) depending on type, cultivar and environmental conditions. Yellow mustard does not shatter readily, and if it is free from green weeds, the crop can be combined directly at 12–13% moisture followed by artificially drying of the seed. If the crop is weedy or uneven in maturity, it should be windrowed when 60–70% of the seed has turned yellow. Plants are cut just beneath the lowest seedpods, so the top will settle into the stubble and reduce the effect of high winds. Brown and oriental cultivars shatter more readily when ripe and should be windrowed when the overall field color has changed from green to yellow-brown, when pods from the middle of the racemes have 75% yellow or brown seeds. The remaining green seeds will mature in the windrow before combining. The harvested seed should be handled carefully since it will crack easily. Air temperatures for drying should not exceed 50°C (122°F), and seed temperature should stay below 35°C (95°F) during the drying process. Mustard seed is optimally stored when it reaches moisture content of 10% (Pouzet, 1995).

Watercress

Watercress (*Nasturtium officinale*) production requires large quantities of clean, continuously flowing water. A supply of 3.78×10^6 l/ha/day (1.0×10^6 gal/acre/day) of water is needed for optimal growth, provided water temperature and nutrition are adequate. Water temperatures above 25.5°C (78°F) will cause slow or poor growth (McHugh et al., 1987).

Bright, sunny days with little or no cloud cover throughout the growing cycle are also critical. Watercress is grown in shallow ponds or beds with 0.6–5 cm (0.25–2.0 in) of water flowing continuously at a rate of 0.6–1.2 m (2–4 ft)/s. The water movement throughout a bed should be uniform to obtain even growth. The crop will not produce luxuriant growth in areas where the water circulation is poor (McHugh et al., 1987).

Optimum daytime air temperatures for watercress growth are 21–29°C (70–85°F). An intermittent overhead sprinkler system (4 min/30 min) during the day can moderate excessively high daytime temperatures by evaporative cooling for production during warm periods (McHugh et al., 1987).

Watercress is propagated by seed, stem, or terminal shoot cuttings. Bunches of four to six cuttings are placed on the surface of the bed at 30 cm (1 ft) intervals. The cuttings should be 30 cm (1 ft) long and placed lengthwise with the rooted basal ends pointing downstream in the water flow. In 10–14 days the root system will be suitably developed to anchor each plant. When the plants are properly anchored additional water can be directed into the beds as needed.

Watercress obtains most of the nutrients necessary for its growth from the water. Nitrogen, in nitrate form, is the most important nutrient and optimal growth occurs when the nitrate content of the water is 5 ppm. High chloride or high pH of the water can also inhibit watercress production. In general, water with chloride content greater than 1,000 ppm or pH above 7.5 will inhibit growth. Iron deficiency also causes chlorosis. Chelated iron sulfate can be applied as a foliar spray to prevent this deficiency (McHugh et al., 1987).

The crop requires about 45 days from planting to harvesting. As the crop grows, the water depth should be increased slowly to 5 cm (2 in) and then reduced after harvest. Subsequent crops are produced from the stems left in the bed after harvesting as well as an additional planting of vegetative tip cuttings on top of the residue (McHugh et al., 1987).

Watercress is harvested when the plants reach a height of 30–35 cm (12–14 in) above the water level. The harvesting is done by grasping the stems in one hand, cutting them with a sickle knife, removing any yellow or spotted lower leaves, tying cut stems in 30–35 cm (12–14 in) bunches, and washing in clean water, with 30 bunches to a bundle. After packing, the watercress is cooled and stored at 1°C (34°F). Refrigerated watercress may be stored for up to 1 week (McHugh et al., 1987).

References

Akobundu, I.O. (1980) Live mulch: A new approach to weed control and crop production in the tropics. In: *Proceedings 1980 British Crop Protection Conference – Weeds.* British Crop Protection Council, Thornton Heath, UK, pp. 2:377–382.

Bancroft, I., Lydiate, D., Osborn, T., Renard, M., Friedt, W., Lim, Y-P., Sadowski, J., Meng, J., Edwards, D. and King, G. (2006) Brassica Genome Gateway. The Multinational Brassica Genome Project. Available at: http://brassica.bbsrc.ac.uk/welcome.htm (accessed 5 April 2006).

Barlow, Snow (2007) Multilingual Multiscript Plant Name - Brassica Names. Available at: www.plantnames.unimelb.edu.au/Sorting/Brassica_rapa.html#chinensis (accessed 31 July 2007).

Berg, L. (2012) History of broccoli. Available at: http://weightlossninja.org/history-of-broccoli (accessed 3 July 2012).

Bratsch, A. (2009) *Specialty Crop Profile: Horseradish Virginia Cooperative Extension, publication 438-104.* College of Agriculture and Life Sciences, Virginia Polytechnic Institute and State University, Blacksburg, Virginia.

CABI (2012) Crop protection compendium. Available at: www.cabi.org/cpc (accessed 14 October 2013).

Collins, R. (2003) Growing wasabi in western North Carolina. Available at: www.cals.ncsu.edu/specialty_crops/publications/reports/collins2.html (accessed 27 June 2012).

Cornell (2004) Integrated crop and pest management guidelines for commercial vegetable production. Available at: www.nysaes.cornell.edu/recommends (accessed 14 October 2013).

FAOSTAT (2012) Cabbage and other brassicas 2010, Online database of crop production statistics. Available at: http://faostat.fao.org/site/567/DesktopDefault.aspx?PageID=567 (accessed 3 July 2012).

Fenwick, G., Heaney, R.K., Mullin, W., VanEtten, J. and Cecil, H. (1982) Glucosinolates and their breakdown products in food and food plants. *CRC Critical Reviews in Food Science and Nutrition* 18, 123–201.

Field, R.C. (2001) Cruciferous and Green Leafy Vegetables. In: Kiple, K.F. and Ornelas, K.C. (eds) *Cambridge World History of Food*, Vol. 2. Cambridge University Press, Cambridge, UK, pp. 1738–1739.

Flint, M.L. (ed.) (1987) *Integrated Pest Management for Cole Crops and Lettuce.* University of California Statewide Integrated Pest Management Project. Division of Agriculture and Natural Resources, 1987, Pub. 3307, 112 pp.

Frantz, J.M. and Welbaum, G.E. (1998) Horticultural crop production using hydroponic transplant production systems. *HortTechnology* 8, 392–395.

Fu, I., Shennan, C. and Welbaum, G.E. (1991) Evaluating Chinese cabbage cultivars for high temperature tolerance. In: Janick, J. and Simon, J.E. (eds) *Proceedings of the Second National Symposium on NEW CROPS, Exploration, Research, Commercialization.* Timber Press, Portland, Oregon, pp. 570–573.

Griffiths, M. (1994) *Index of Garden Plants.* Timber Press, Portland, Oregon, 1234 pp.

Hoyt, G.D. and Walgenbach, J.F. (1995) Pest evaluation in sustainable cabbage production systems. *HortScience* 30, 1046–1048.

Hoyt, G.D., Monks, D.W. and Monaco, T.J. (1994) Conservation tillage for vegetable production. *HortTechnology* 4, 129–135.

Infante, M.L. and Morse, R.D. (1996) Integration of no tillage and overseeded legume living mulches for transplanted broccoli production. *HortScience* 31, 376–380.

Jett, L.W., Morse, R.D. and O'Dell, C.R. (1995) Plant density effects on single-head broccoli production. *HortScience* 30, 50–52.

Jett, L.W., Welbaum, G.E. and Morse, R.D. (1996) Effects of matric and osmotic priming treatments on broccoli seed germination. *Journal of the American Society for Horticultural Science* 121, 423–429.

Kader, A.A. (1992) *Postharvest Technology of Horticultural Crops, Publication 3311.* Division of Agriculture and Natural Resources, University of California, Oakland, California.

Kalb, T. and Chang, L.C. (2005) *Suggested Cultural Practices for Heading Chinese Cabbage International Cooperators' Guide, Publication 05-642*. Asian Vegetable Research and Development Center, Shanhua, Tainan, Taiwan.

Kemble, J.M., Zehnder, G.W., Sikora, E.J. and Patterson, M.G. (1999) *Guide to Commercial Cabbage Production, Publication ANR-1135*. Alabama Cooperative Extension System, Alabama A&M University and Auburn University, Auburn, Alabama.

Lal, R., Reicosky, D.C. and Hanson, J.D. (2007) Evolution of the plow over 10,000 years and the rationale for no-till farming. *Soil and Tillage Research* 93, 1–12.

Lamont, W.J. (ed.) (2004) *Production of Vegetables, Strawberries and Cut Flowers Using Plasticulture, Publication 133*. National Resource, Agriculture, and Engineering Service Cooperative Extension Service, Ithaca, New York.

Lorenz, A.O. and Maynard, D.N. (1988). *Knott's Handbook for Vegetable Growers*, 3rd edn. Wiley New York, p. 76.

Maggioni, L., von Bothmer, R., Poulesen, G. and Branca, F. (2010) Origin and Domestication of Cole Crops (*Brassica oleracea* L.): Linguistic and Literary Considerations. *Economic Botany* 64, 109–123.

McClung, C.A. and Schales, F.D. (1982) *Commercial Production of Horseradish, Publication HE 127-82*. University of Maryland Extension, College Park, Maryland.

McHugh, J.J., Fukuda, S.K. and Takeda, K.Y. (1987) *Hawaii Watercress Production, Publication 088*. University of Hawaii Extension, Hawaii.

Musil, A.F. (1948) *Distinguishing the species of Brassica by their seed, Circulation 857*. US Department of Agriculture, Washington, DC.

Myers, C. (1991) *Specialty and Minor Crops Handbook, Publication 3346*. University of California, Small Farm Centre, Davis, California.

Myers, J.R. (2006) *Outcrossing Potential for Brassica Species and Implications for Vegetable Crucifer Seed Crops of Growing Oilseed Brassicas in the Willamette Valley, Report 1064*. Oregon State University Extension Service, Corvallis, Oregon.

NASS (1999) National Agricultural Statistics Service Vegetables 1999 Summary. Available at: www.ipm-centers.org/cropprofiles/docs/cabrusselssprouts.html (accessed 4 July 2012).

Nieuwhof, M. (1969) *Cole Crops (World Crops Books Series)*. Leonard Hill, London.

Oplinger, E.S., Hartman, L.L., Gritton, E.T., Doll, J.D. and Kelling, K.A. (1989) Canola (Rapeseed). In: *Alternative Field Crops Manual*. University of Wisconsin Cooperative Extension, Madison, Wisconsin.

Parker, B.L., Talekar, N.S. and Skinner, M. (1995) *Field guide: Insect pests of selected vegetables in tropical and subtropical Asia*. AVRDC, Shanhua, Tainan, Taiwan.

Peirce, L.C. (1987) *Vegetables: Characteristics, Production, and Marketing*. Wiley, New York.

Pouzet, A. (1995) Agronomy. In: Kimber, D. and McGregor, D.I. (eds) *Brassica Oilseeds - Production and Utilization*. CAB International, Wallingford, UK, pp. 65–92.

Rimmer, S.R., Shattuck, V.I. and Buchwaldt, L. (eds) (2007) *Compendium of Brassica Disease*. American Phytopathological Society, St. Paul, Minnesota.

Rubatzky, V.E. and Yamaguchi, M. (1997) *World Vegetables – Principles, Production, and Nutritive Values,* 2nd edn. Chapman Hall, New York.

Saunders, D.C. (1993) Vegetable crop irrigation. Available at: www.ces.ncsu.edu/hil/hil-33-e.html (accessed 7 March 2012).

Stephens, J.M. (1994) *Cabbage, Chinese – Brassica campestris L. (Pekinensis group), Brassica campestris L. (Chinensis group)*. University of Florida, Gainesville, Florida.

Stivers, L.J., Brainard, D.C., Abawi, G.S. and Wolfe, D.W. (1999) Cover crops for vegetable production in the northeast. Available at: http://ecommons.library.cornell.edu/handle/1813/3303 (accessed 3 July 2012).

Swaider, J.M., George, W. and McCollum, J.P. (1992) *Producing Vegetable Crops,* 4th edn. Interstate Printers and Publishers, Danville, Illinois.

Swanton, C.J. and Weise, S.F. (1991) Integrated weed management: The rationale and approach. *Weed Technology* 5, 657–663.

Thompson, H.C. and Kelly, W.C. (1957) *Vegetable Crops*. McGraw-Hill, New York.

USDA, ERS (2010) Cabbage statistics. Available at: http://usda.mannlib.cornell.edu/MannUsda/viewDocumentInfo.do?documentID=1397 (accessed 3 July 2012).

Welbaum, G.E., Shen, Z.-X., Oluoch, M.O. and Jett, L.W. (1998) The evolution and effects of priming vegetable seeds. *Seed Technology* 20, 209–235.

Wittwer, S. (1987) Chinese Cabbage-Year-Round. In: Wittwer, S., Yu, Y., Sun, H. and Wan L. (eds) *Feeding a Billion: Frontiers of Chinese Agriculture*. Michigan State University Press, Lansing, Michigan, pp. 271–277.

Zohary, D. and Hopf, M. (2000) *Domestication of Plants in the Old World,* 3rd edn. Oxford University Press, Oxford, UK.

17 Family Amaranthaceae, Subfamily Chenopodiaceae

BEETS AND CHARD

Origin and History

Beet, also known as beetroot, is a very ancient crop with a long history of cultivation dating back to the second millennium BC. Beet was likely domesticated somewhere in the Mediterranean region, taken to Babylonia around the 8th century BC and introduced to China by approximately AD 850 (Zohary and Hopf, 2000). The writings of Aristotle and Theophrastus suggest that leafy beets were grown extensively during the crop's early history. However, once spinach was available, the use of beet as a leafy vegetable diminished significantly. The Romans ate beets in the second and third centuries because they were believed to be an important food for promoting good health. Chard was apparently of later origin and was likely not eaten as a vegetable until the 13th century. Colonists introduced beets and chard to the western hemisphere from Europe. George Washington conducted experiments with them at his Mount Vernon home. Chard, and red, white, or yellow beetroot was widely grown in US gardens in the early 1800s.

Botany and Life Cycle

Beet (*Beta vulgaris* L.) was classified as a member of the family Chenopodiaceae for many years. However, many plant biologists now include the chenopods as a subfamily in the family Amaranthaceae (USDA Plants, 2010). This important family contains many widely distributed dicot, annual, biennial, and perennial plants, all with small inconspicuous flowers. Beet plants are bisexual, unisexual, or dioecious. In some cases two or more flowers grow together in a dense cluster and form a fruit with multiple embryos called a utricle. The beet calyx continues to grow after flowering, becoming corky and completely covering the seeds inside the fruit.

The vast majority of chenopods are weeds, and many are salt and drought tolerant. In addition to garden beets, related crops in the subfamily Chenopodiaceae include fodder beet, sugarbeet, spinach, chard (also called Swiss chard or spinach chard), quinoa (*Chenopodium quinoa*), and orach (*Atriplex hortensis* L.). Lamb's quarters or goosefoot (*Chenopodium album*) is a fast growing weedy annual that is also cultivated as a vegetable in some regions but is more widely considered to be a weed (Board, 2004; USDA, ARS, 2012a).

Beta vulgaris is sometimes divided into three subspecies: *B. vulgaris* subsp. *vulgaris*, *B. vulgaris* subsp. *maritima*, and *B. vulgaris* subsp. *adanensis*. *Beta vulgaris* subsp. *vulgaris* includes many common crops such as chard, sugarbeet, fodder beet, and garden beet. The other two subspecies are wild relatives that have been used for genetic improvement. *Beta vulgaris* subsp. *maritima*, often called sea beet, is a wild ancestor of cultivated beets, and is found throughout the Mediterranean, the Atlantic coast of Europe, the Near East, and India. *Beta vulgaris* subsp. *adanensis* is another wild subspecies that is distributed from Greece to Syria (USDA, ARS, 2012b).

Beta vulgaris subsp. *vulgaris* has historically been subdivided by some into distinct botanical varieties: crassa and cicla (Hanelt *et al.*, 2001). However, the former botanical variety cicla is now considered to be the Leaf Beet Group within *B. vulgaris* subsp. *vulgaris*. The group designation denotes horticultural similarities between Swiss chard and spinach beet cultivars consumed primarily for their foliage when significant genetic differences do not warrant separate classification into botanical varieties. Similarly, the former botanical variety crassa is now considered to be a Fodder Beet Group that denotes horticultural similarities among garden beets, fodder beets, and sugarbeet types consumed primarily for their roots (USDA, ARS, 2012b).

Beets are a cool-season, frost-tolerant herbaceous biennial. After germination, the seedling forms a crown, consisting of an un-elongated stem from which a rosette of leaves forms, attached to a fleshy storage taproot that penetrates deep into the soil during the first year (Fig. 17.1).

The color of leaves varies from light green to dark red. Petioles (leaf stems) vary in length and tend to be thin. Leaf shapes vary from oblong to triangular, leaf margins may be curved or straight, and leaf-blade surfaces smooth or savoy (wrinkled). The beet storage root is an enlarged hypocotyl-root axis formed near the soil surface along with the narrowing tapered true root portion that extends deeper into the soil, anchoring the plant. The mature beetroot grown for commerce is a spherical or elongated fleshy axis of varying shape, composed morphologically of the crown, the neck, and the root. The neck is exposed extending from below the crown to the soil line. The broadest portion of the beet lies below the neck, has no lateral roots, and is completely smooth. Below the hypocotyl-root axis is the root. The root tapers to a slender taproot, which may be quite long. Lateral or secondary roots arise all along the primary root. Secondary rooting tends to develop in the top 40–60 cm (16–24 in) of soil beneath the surface even though taproots may penetrate deeper. Wild beetroots were white, long, thin, and well-branched, so human selection and plant breeding has produced the more familiar modern beet plant that is distinct from its wild relatives.

The enlarged beet hypocotyl-root axis is the tissue most commonly consumed and for convenience often referred to as "root". The root interior consists of alternating zones of vascular and storage tissues that appear as rings. Increases in thickness result from cambial growth accompanied by division and enlargement of parenchyma tissues.

Garden beets occur in a wide range of colors ranging from deep red, to purple to white to deep yellow. Purple is the most popular color of commercial garden beets sold in North America and much of Europe. Interior color differences between xylem and phloem tissues, commonly called zoning, are due to different levels of pigmentation. The red and purple colors of beets are caused by betacyanin, a pigment with properties similar to anthocyanin, which is the source of purple color in many vegetables. Beets also contain betaxanthin, a yellow pigment. The ratio of betacyanin and betaxanthin varies among cultivars and can change because of environmental conditions. White cultivars lack both pigments while intense red color means that betaxanthin is present and yellow means that little betacyanin is present.

Shapes of beets are variable and may be globed, cylindrical, top-shaped, or flattened. The size of storage roots is quite variable, ranging from 2–20 cm (0.8–8 in) in diameter.

Fodder beets

There are three distinct types of *B. vulgaris* grown for distinct uses: fodder beets, sugarbeets, and garden (or table) beets. All three emerged from a common ancestor through human selection and plant breeding. Fodder beets are an agronomic crop grown for livestock feed in cool-temperate climates. Average fodder beet maturity is 200 or more days depending on location and climate. The fodder beet or "mangel-wurzel" is sown in autumn and can be grown as a winter crop in warm-temperate to subtropical climates or as a summer crop in North America, northern Europe, and other short-season areas. After development from smaller fodder beets in Germany and Holland, the large-rooted beets, known as the mangel-wurzel, were used as livestock feed in Europe and England in the 1770s. An unfortunate English mistranslation of the German mangold-wurzel ("beet-root") as mangel-wurzel ("scarcity root") resulted in a widely held belief that this plant would be excellent food for the poor during periods of famine. However, the mangel-wurzel turned out to be better suited for livestock feed. Contemporary use is primarily for

Fig. 17.1. A developing redbeet (garden beet) plant, 'Detroit Dark Red' type, from a commercial field in Nova Scotia, Canada.

cattle, pig, and other livestock feed, although humans sometimes eat immature plants as a vegetable. Both leaves and roots may be eaten. Leaves can be lightly steamed for salads or lightly boiled (Wright, 2001).

Sugarbeets

Sugarbeets are large plants similar in size and maturity to fodder beets (Fig. 17.2). The sugarbeet is a comparatively new crop created through plant breeding and natural selection. During the mid-1700s, the German chemist Andreas Margraff discovered that both the white and the red beetroot contained sucrose, which was indistinguishable from that produced from sugarcane (Hill and Langer, 1991). Prior to this time, tropical sugarcane was the source of all sugar and was prohibitively expensive for most Europeans. After the supply of sugarcane was cut off by the English blockade of continental Europe during the Napoleonic Wars, the demand for sugar grew throughout Europe.

One of Margraff's students, Franz Karl Achard, worked to establish the beet as an economic source of sucrose in Europe. Achard is considered by many to be the father of the sugarbeet industry. He built the first sugar factory in what is today Poland, and developed processing methods using white fodder beets available at that time that had relatively high sugar content. Napoleon encouraged French research on sugarbeets, and between 1810 and 1815 over 31,995 ha (79,000 acres) went into production with more than 300 small beet-processing plants. After Napoleon's fall, tropical sugar became readily available and prices collapsed due to excess supplies. The majority of the sugarbeet factories were closed and new development proceeded slowly. However, after the decline of slavery in the West Indies, the European industry became more competitive with the tropical sources of sucrose, and by the 1850s the sugarbeet industry was well established (Hill and Langer, 1991). Over the years, plant breeding has increased the sugar content to between 15% and 21% of the root's total weight, comparable to sugarcane.

Garden beets

The most commonly consumed beet vegetable is simply referred to as red, garden, or table beet. Vegetable beets are much smaller plants in comparison to fodder and sugarbeets. Garden beets also develop rapidly in 50–80 days from direct seeding. Beets are a very versatile vegetable that can be cooked and preserved in a number of different ways. The dark-green leaves and stems of young garden beets are eaten raw in salads, steamed, or boiled or sometimes stir-fried. The hypocotyl axis or "root" of garden beet can be peeled, boiled, and eaten with butter as a side dish. Garden beets are often pickled in vinegar and then eaten cold as a side dish or condiment. Beets may also be peeled, shredded raw, and then eaten as a salad. In parts of Europe, beet soup, such as cold borsch, is a popular dish.

Beet juice, obtained from cutting or crushing the hypocotyl axis (root), contains betanins that are

Fig. 17.2. Sugarbeet is closely related to garden beet and is the main source of sugar in the northern latitudes where sugarcane cannot be grown. Sugarbeet was created from fodder beets or mangel-wurzels by plant breeders who selected roots with high sugar content with many weighing 4.5–6.8 kg (10–15 lb).

used industrially as natural red food colorants to intensify the color of sauces, desserts, jams and jellies, ice cream, sweets, and cereals.

There is a great deal of variation among garden beet cultivars with respect to colors, shapes, and horticultural characteristics. Both open-pollinated and F-1 hybrid cultivars of beets are available. A few of the most popular types of cultivars are summarized below.

Red garden beet cultivars are often divided into two distinct groups based on their root shapes and horticultural characteristics. 'Crosby Egyptian' and related cultivars tend to produce an early maturing, flattened-round attractive root that is smooth and uniformly shaped with a small taproot. The inside of the 'Crosby Egyptian' has prominent zoning with rings of alternating red and white corresponding to regions of xylem and phloem tissue. Egyptian types are well suited for early market, whole use, pickling, and sale in bunches but have less attractive internal root coloration.

In contrast, 'Detroit Dark Red' and related cultivars are later maturing, round to oval with a rougher appearance and prominent taproot. Interiors are solid red with indistinct zoning when viewed in cross section. In summary, 'Detroit Dark Red' cultivars have less attractive exteriors but uniformly and intensely colored interiors and are better suited for slicing. Since 'Crosby Egyptian' and 'Detroit Dark Red' cultivars can be crossed, some modern cultivars combine the best characteristics of each, although the original cultivars and their relatives remain popular.

While most markets prefer beets that are dark purple, other colors, such as white, red, and yellow, exist and are grown as novelties (Fig. 17.3). Golden beets are a popular novelty cultivar grown primarily for local markets and home consumption because they provide a colorful alternative to purple beets (Grubben and Denton, 2004). Leaf venation may also be yellow with green blades. Golden beets may be less sweet than red cultivars and have what some describe as a mellow, less earthy flavor.

Chioggia beets are a popular heirloom cultivar in some markets because of their contrasting zoned interiors of alternating light/dark coloration of xylem and phloem tissues when viewed in cross section (Burge, 1991). The interior characteristics are similar to 'Crosby Egyptian' cultivars described above, only more pronounced. The color may vary among cultivars and growing conditions, some are a subtle yellow-and-orange combination while others alternate red-and-cream colors. Chioggia cultivars are visually interesting, although the striping often fades when cooked.

Fig. 17.3. Garden beets (*Beta vulgaris*) come in many colors and shapes. The roots on the left could be called red, in the center golden, and on the right, dark red or purple.

Consumers are often attracted to vegetables that are harvested early at an immature state of development because they are perceived to be more tender, have higher nutrient content, require less cooking, and are more likely to free of pesticides since they are often harvested early before pests can attack. Beet seedlings often with only three to five true leaves are harvested and sold as microgreens or mixed with other immature vegetables and herbs to create microgreen mixes. 'Yellow Beet', 'Bull's Blood', and 'Early Wonder Tall Top' are examples of cultivars used for microgreen production. Another immature product is "baby" beets. "Baby" is not a specific cultivar but rather beets that are harvested prematurely when very small and often sold at a premium price (Schrader and Mayberry, 2002). Baby beets may be planted at close in-row spacing to reduce size.

Some important characteristics that differentiate beet cultivars include plant growth rate, root and leaf color, and foliage size (often referred to as "top size"), with small, medium, or large cultivars available. Leaf midrib color varies from white to red to yellow. Leaf shape, petiole thickness, and petiole length vary and are important when bunched plants are sold or the tops are consumed. For whole-beet sales, less distinct and smaller shoulders and collars are considered more attractive. Strong leaf attachment is an important trait when selecting cultivars for bunched beet sales. The neck and crown size should be small and is important for determining the strength of the leaf attachment and the number of beets that can be bunched together.

Root uniformity is important for presentation of fresh beets but less important for canned, frozen, or minimally processed beets. High quality roots are free of fiber and cracks. Fiber usually develops as beets age and is not strictly related to size. Rapidly growing large young roots may be free of fiber while smaller older beets may be more fibrous. Minimal branching and side roots are preferred for both fresh and processed products.

Production and Culture – Beets

Growth and development

Beets are a cool-season biennial that are grown as an annual. Beetroots require 50–80 days to reach harvest maturity for use as a vegetable from direct seeding based on environmental conditions and cultivar. Beets are hardy and can tolerate subfreezing temperatures as low as $-9.4°C$ ($15°F$) for brief periods once established, although extended exposures are lethal. Low temperatures tend to increase leaf thickness (Rubatzky and Yamaguchi, 1997). Yield and quality are best in regions where cool night temperatures slow the respiration rate, enhancing retention of stored carbohydrates. Optimum production temperatures are 12–24°C (55–75°F). In contrast to spinach, beets are more tolerant of higher temperatures. The crop will grow at warmer temperatures particularly during seedling growth, but roots develop too rapidly and become lignified and fibrous at above optimum growth temperatures. High root yields depend on a prolonged vegetative growth phase, so beets should be grown at temperatures above 15°C (59°F) to prevent vernalization (exposure to cold temperatures) and the initiation of reproductive growth.

Most beet cultivars grown in northern latitudes are biennials. Post juvenile-phase beets form a flower stalk after vernalization. After vernalization, usually during the second year of growth, the stem elongates and develops into a flower stalk. Bolting, a term for rapid seed-stalk elongation, is initiated after plants develop beyond the juvenile size, which equates to roughly pencil diameter of the axis, and are exposed to temperatures below 10°C (50°F) for a period of 21 days or more. Little research has been done on the vernalization requirements for garden beet and much of the available information is based on sugarbeet. The optimal temperature for vernalization of sugarbeet is about 8°C (46°F). At the threshold temperature of 15°C (59°F), vernalization requires much longer exposure times (Smit, 1983).

The dominant B gene (B allele) controls bolting in beets (Abegg, 1936). All cultivated biennial beets grown commercially in northern Europe and North America are genotype bb and require extensive vernalization to induce flowering under long days. Most beet cultivars must be vernalized, prior to 14 h long day exposure, a process called photothermal flower induction (Owen *et al.*, 1940; Owen, 1954). The vernalization requirement can be quantified by the length of exposure to low temperature. For many cultivars, the amount of chilling required for vernalization is much greater than some other biennial vegetables, so early-seeded plantings are usually not at risk of premature seed-stalk development (Goldman and Navazio, 2007). Many cultivars need at least 2 months below 15°C (59°F) to be vernalized although other cultivars may bolt after exposure to as little as 3 weeks of temperatures below 15°C (59°F).

Flower stalk formation ruins crop quality and reduces yields because most of the storage reserves go into flower stalk development rather than root growth. Beetroots also become hard and fibrous once the flower stalk develops.

Site selection and field preparation

Soils must be porous (well drained) and friable to achieve high yields. Thus, sandy loams, silt loams, and well-drained organic soils are well suited for beet production (Fig. 17.4).

The soil must be finely tilled using rotovators or disks and free of clods, stones, and other impediments to ensure uniform emergence and good stand establishment. Deep plowing (moldboard or chisel), cultivation to eliminate clods, removal of rocks, and clearing of heavy debris all help to achieve a rooting environment that promotes development of uniform straight roots that are free from defects.

Field establishment

Beet utricles are established by direct seeding. Utricles may be polygerm, containing multiple embryos, or monogerm, containing a single embryo. Monogerm types are widely used for precision seeding of commercial beet fields to eliminate the need for thinning and to promote uniform maturity for once-over mechanical harvest. Beets germinate at soil temperatures above 4°C (40°F), but optimum germination occurs at 29°C (85°F).

Polygerm embryos may germinate all at once or at different times, thus requiring thinning. While variable emergence is not advantageous for mechanical harvesting, it does provide home gardeners and small producers with a prolonged harvest period as small plants develop after larger neighboring plants are hand-harvested.

Beets are direct seeded in single or multiple rows 1.3–2.6 cm (0.5–1 in) deep. Beets may be scatter-planted within a narrow band as well. Row spacing depends on the application. For fresh market, beets are often spaced in rows 38–51 cm (15–20 in) apart with final in-row spacing of 4 cm (1.5 in) (see Fig. 17.4). For processing, beets are often crowded to produce small whole beets. Large beets, with diameters greater than 15 cm (6 in), are susceptible to cracking, excessive fiber and poor coloration (Rubatzky and Yamaguchi, 1997).

Fig. 17.4. Red beet production field on sandy soil in Nova Scotia, Canada.

Fertilizer and nutrition

Beets are tolerant of salinity like other members of the subfamily Chenopodiaceae but are sensitive to soil acidity. An old weed-control practice was to apply salt to beet fields (Rubatzky and Yamaguchi, 1997). Beets grow best at pH 6.4–6.8. Nitrogen (N) is the most limiting nutrient for beet production. Proper N management is critical to maximize growth. Nitrogen fertilizer applications should be based on anticipated yields, minus the contribution for residual soil N in the soil root zone, as well as N mineralized from soil organic matter decomposition. Soil testing can help determine residual N and organic matter present. Nitrogen in soil organic matter becomes available to plants through mineralization that depends on both moisture and temperature, so general recommendations on mineralization rates are difficult. Research results show that approximately 33.6 kg/ha of N (30 lb/acre) will become available during the growing season for each 1.0% organic matter in the soil surface layer (Westfall and Davis, 2009). If uncomposted animal manure is used, much of the N in manure may be released as NO_3-N during the latter part of the season and may not benefit early growth (Westfall and Davis, 2009). Composted manure tends to release N faster. Overall, beets require moderate amounts of N, phosphorus (P), and potassium (K). A beet crop yielding 28.4 t/ha (11.5 t/acre) removes via harvested roots 74, 9, 90, 8, and 13 kg/ha (66, 8, 80, 7, and 12 lb/acre) of N, P_2O_5, K_2O, CaO, and MgO, respectively, while the tops remove 96, 0, 60, 108, 102 kg/ha (86, 0, 54, 96, 91 lb/acre) of N, P_2O_5, K_2O, CaO, and MgO, respectively (Knott, 1962).

Phosphate potassium, sulfur (S), and micronutrient fertilizers should be applied based on soil test results because levels vary greatly with soil type. Beets require high levels of most trace elements, especially boron (B) and zinc (Zn), for normal growth. Cavity spot is a physiological disorder caused by B deficiency, in which black necrotic lesions form on the roots. Boron deficiency also inhibits meristem and shoot growth, often leading to heart rot. Beets are one of the most B-intensive crops, possibly because their wild progenitors were constantly exposed to sea spray in their native habitat in coastal regions. On commercial farms, a 60 t/ha (26.8 ton/acre) beet harvest requires 600 g/ha of B (8.6 oz/acre) for best growth.

Irrigation

Beets have a shallow root system and must receive a consistent supply of water to keep the roots actively growing so they do not become woody. Spring plants may be grown without irrigation in areas that receive seasonal rainfall because the crop matures quickly and field water loss from evapotranspiration is relatively low. However, when beets are grown during the summer months, evapotranspiration is higher, so overhead irrigation is often used although other types of irrigation are possible. Beets should receive at least 2.5 cm (1 in) of water per week as a rule, to ensure continuous growth and approximately 300 mm (12 in) is required to produce a crop (Rubatzky and Yamaguchi, 1997).

Harvesting and marketing

Good yields of garden beets approximate 49 metric tonnes/ha (22 short tons/acre). Yields could be higher except that a premium is often paid for small sizes that are not fibrous, especially for whole processing. Bunched beets are hand-pulled or mechanically dug and bunched, often in the field, to keep the foliage looking fresh and in good condition. Bunched beets are hydro-cooled and shipped in slurry ice or temperatures just above freezing to retain freshness, particularly of the tops. Beet greens and bunched beets can be maintained in marketable condition for 10–15 days if stored at 0°C (32°F) and 95% RH. Beet leaves respire at a higher rate than roots and require rapid postharvest cooling and continuous low temperatures and high humidity during transit and storage to maintain quality. Beetroots sold in bags without tops can be stored for 4–6 months under the same conditions, although long-term storage may decrease tenderness and increase sweetness as complex carbohydrates are converted to simple sugars. Beets for processing are mechanically topped, dug, and bulk handled much like other root crops.

Diseases

Although there are pests that attack beets, pest problems are not as serious as those for many other vegetables. Powdery mildew (*Erysiphe polygoni*) is often a serious fungal pest of sugarbeets and not a serious pest of garden beets because harvest occurs before the disease becomes too severe. Powdery mildew is more serious in areas with mild winters where the disease can overwinter on related species close to a beet field. Small, white, powdery spots appear first, usually on the under surface of older beet leaves. Under suitable conditions, the fungus spreads rapidly over the entire surface of the leaf, and eventually to all leaves on affected plants. Ideal conditions for powdery mildew development are warm, dry weather, with optimum temperatures between 15–30°C (60–86°F). Cultivars with moderate resistance are available and chemical fungicides are effective. Downy mildew (*Peronospora parasitica*) is transmitted by seed and may cause red-rimmed spots on the leaves of adult plants (Kaffka *et al.*, 2010).

Another fungal foliar disease is caused by *Cercospora beticola*. Symptoms of Cercospora leaf spot first appear as individual, circular spots that are tan to light brown with reddish purple borders. As the disease progresses, individual spots coalesce. Heavily infected leaves first become yellow and eventually turn brown and necrotic (Kaffka *et al.*, 2010). This disease was primarily a problem on garden beets and sugarbeets in areas where warm nights combine with high humidity (Kaffka *et al.*, 2010; CABI, 2013). Optimum daytime temperatures for *Cercospora* development are 25–35°C (77–95°F) with night temperatures above 16°C (61°F) and 90–95% RH. The primary source of inoculum is residue from a previously infected crop, but the fungus can be carried on seed and hosted by numerous weeds. Control of Cercospora leaf spot can be achieved by removing inoculum from a field, plowing to incorporate crop residues, and using 3-year rotations with nonhost crops. Cultivars vary considerably in resistance.

Pleospora betae is a fungal pathogen causing seed-transmitted disease that can cause damping-off with symptoms similar to *Pythium* and *Rhizoctonia* (CABI, 2013). Black spot or cavity spot is a nonpathogenic physiological disorder caused by B deficiency that appears as unsightly necrotic black spots or pits on the root surface.

There is a series of virus diseases that affects beets. Some of the most common include beet curly top virus, beet mild yellowing virus, beet mosaic virus, beet pseudoyellows virus, and beet soil-borne mosaic virus. Symptoms for many of these virus diseases include leaves that are dwarfed, crinkled, and rolled upward and inward. Veins on the lower side of infected leaves are irregularly swollen with bumps. In cross-section, dark rings of vascular tissue can be seen in large roots. Young roots of infected plants are dwarfed and become twisted, distorted, and are often killed at more advanced stages. Death of rootlets is followed by production of new ones, leading to a "hairy root" symptom (Kaffka *et al.*, 2010).

Beet curly top virus is vectored by the beet leafhopper *Circulifer tenellus*, which has an extensive host range, a high reproductive capacity, and can migrate long distances from its breeding grounds. The leafhopper overwinters on a wide range of annual and perennial weeds and readily acquires the virus when it feeds on infected plants. Once acquired, a vector can usually transmit the virus for the rest of its life. The severity of the curly top virus on beet depends on climatic factors that influence the prevalence of weed hosts and the reproductive capacity and migration of the leafhopper vector. Beet curly top virus can also cause significant losses in tomatoes, beans, peppers, and occasionally cucurbits. Cultivars resistant to certain viral diseases are available (Kaffka *et al.*, 2010).

Insect Pests

Beet armyworm (*Spodoptera exigua*) and the yellow striped armyworm (*Spodoptera praefica*) are small, mottled gray- or dusky-winged moths that deposit pale greenish or pinkish striated eggs on leaves in small or large masses covered with white cottony material. Eggs hatch in a few days and tiny caterpillars begin feeding. The olive green caterpillars have a yellow stripe on each side of the body. They reach full size in approximately 2–3 weeks, about 3.2 cm (1.3 in) long. Beet armyworms may cause severe injury in summer and fall (Natwick *et al.*, 2010). Armyworms skeletonize leaves, leaving the veins largely intact. In severe infestations, as food becomes scarce they will consume the veins and petioles and will even feed on the exposed portions of the beetroot. Because of their ability to reach high numbers and cause severe defoliation, armyworms should be closely monitored later in the season. Controls include beneficial insects or periodic insecticide applications.

Larvae of beet webworms (*Hymenia perspectalis*, *Loxostege sticticalis*, *Spoladea recurvalis*) produce webs and feed on beet foliage, stunting growth, and reducing yields in severe cases. Sometimes aphids, flea beetles (*Chaetocnema confinis*), and beet leafminers (*Pegomya hyoscyami*) are a problem on garden beets. Meloidogyne root knot and other nematodes may adversely affect beetroots, causing distorted development and reduced yield (Natwick *et al.*, 2010). Other notable insect pests of beet include vegetable weevil (*Listroderes costirostris*), beet bug (*Piesma quadratum*), and green looper caterpillar (*Chrysodeixis eriosoma*) (CABI, 2013).

Weed Management Strategies

Beet does not compete well with weeds. For commercial plantings an effective weed-control strategy is necessary, especially when crops are planted to a final stand without thinning. Uncontrolled weeds can reduce beet yield significantly. Dense weeds make hoeing, the use of electronic thinners, cultivation, and harvest difficult.

Problem weeds vary with location but in general annual weeds such as mustard species and annual bluegrass can be problematic. Winter annual weeds die out in summer, but summer annuals begin germinating in March and continue throughout the summer growing season. Troublesome summer annual weeds include barnyardgrass, cocklebur, pigweed, velvetleaf, and knotweed (Hembree and Norris, 2010).

Selection of an effective weed management strategy is dependent on a number of factors: geographic location, which influences planting date, weed spectrum, and irrigation or rainfall; planting date, which influences weed spectrum and irrigation or rainfall; weed species present, which determines choice of weed control method and choice of herbicides; the availability and cost of labor for weeding, which determines if hand weeding can be considered within the program; availability of equipment, which determines how well cultivation can be conducted and if herbicides can be applied accurately and properly incorporated into soil (Hembree and Norris, 2010).

Economically acceptable weed control can only be achieved with a management program that integrates several methods, as no currently available weed control practice provides complete weed control in a beet crop. Band applications of an herbicide in-row and/or between row cultivation are two of the most common practices. These combined practices reduce herbicide usage and minimize labor, which lowers production costs and herbicide usage (Hembree and Norris, 2010).

Swiss Chard

Swiss chard is a close relative of the garden beet but does not form an enlarged root. Chard is essentially a foliage beet developed for its large fleshy leafstalks and broad, crisp leaf blades (Schrader and Mayberry, 2002). Swiss chard withstands hot weather better than many other greens. The leaves are prepared and eaten like spinach. Its culture is similar to beets or spinach. Because Swiss chard can be serially harvested it is widely grown in home gardens. There are relatively few cultivars. Swiss chard leaves can be flat or savoy. Flat leaf blades have no undulation while savoy leaf cultivars are cupped between the veins with a very convoluted appearance. Some cabbage and spinach cultivars also have the savoy leaf-blade characteristic. 'Lucullus' is probably the most popular, having large savoy, dark-green leaves with a broad, white midrib (petiole) (Fig. 17.5).

Other cultivars include 'White Silver', 'Large Ribbed', 'Fordhook Giant', and 'Bright Lights' (Schrader and Mayberry, 2002). Swiss chard cultivars can have red, yellow, or white leaf veins. 'Ruby Red' is a cultivar with red petioles often used for microgreen production. Red-veined cultivars are often called "Rhubarb" chard (Fig. 17.5). This is confusing because rhubarb is also an unrelated vegetable (discussed in chapters in this book). True rhubarb leaves contain high concentrations of oxalic acid and should not be eaten.

Economic Importance and Production Statistics for Garden Beets

Although garden beets are grown in many regions of the world including North and South America,

Fig. 17.5. Swiss chard is a foliage vegetable closely related to beets with a larger leaf blade and thicker petiole and no root enlargement. Leaf blades vary in color and may be light to dark green or reddish and flat, savoy, or semi-savoy depending on the cultivar. Leaf veins may be white, red, or yellow.

Australia, New Zealand, Europe, and Asia, much of the world's garden beet production is consumed locally and not traded internationally. Therefore, UN FAO and other groups do not track international garden beet production since sugarbeet is economically a much more important crop than garden beets.

In the USA, beet production, particularly for processing, is located in northern states where temperatures are cooler and growth and fiber development proceeds at a slower pace allowing for greater flexibility in harvest scheduling. New York, Oregon, and Wisconsin are major producers of beets for processing in terms of hectares. Some fresh-market production of garden beets occurs in the southern USA during the winter months when northern production is not possible.

Per capita consumption of fresh garden beets has declined steadily over the past 70 years in the USA from 10.8 kg (7 lb)/person/year in 1940 to less than 0.1 kg (0.2 lb)/person/year today. Processed beets, mostly canned, have remained consistently near 0.5 kg (1.0 lb)/person/year over this same period (USDA, NASS, 1999).

US consumption of fresh beets has likely declined because of the effort required to top, peel, and cook them as well as the increased availability of more exciting fruits and vegetables in the marketplace for much of the year. Consumers who enjoy beets prefer the convenience of canned beets that are easier to prepare and have more consistent quality compared to fresh beets.

Nutritional Values

The nutritional value of beet greens, beetroots, and Swiss chard greens are compared in Table 17.1. Beetroots are a source of K, P, and carbohydrate but are relatively low in many other nutrients. Beet leaves in many regards are much more nutritious than the roots. Many people incorrectly believe that beetroots are a good source of Fe, possibly because many popular cultivars are red. The leaves have over three times more Fe than beetroots. The red color is caused by betacyanin, a water-soluble plant pigment related to anthocyanin, and not the presence of iron (Fe). Swiss chard contains high levels of oxalic acid, which binds calcium (Ca) when ingested. Therefore, chard is not considered to be as good nutritionally as some of other green leafy vegetables.

Table 17.1. Nutritional composition (100 g (3.5 oz) sample uncooked) (USDA Nutrient Database, 2012).

Nutrient	Crop		
	Beet greens	Beet roots	Chard greens
Vitamin A (IU)	6,100	20	3,300
Thiamine (mg)	0.1	0.05	0.04
Riboflavin (mg)	0.22	0.02	0.09
Niacin (mg)	0.4	0.4	0.4
Ascorbic acid (mg)	30	11	30
Vitamin B6 (mg)	0.11	0.05	0
Water (%)	92	87	93
Energy (kcal)	19	44	19
Protein (g)	1.8	1.5	1.8
Fat (g)	0.1	0.1	0.2
Carbohydrate (g)	4.0	10	3.7
Fiber (g)	1.3	0.8	0.8
Ca (mg)	119	16	51
P (mg)	40	48	46
Fe (mg)	3.3	0.9	1.8
Na (mg)	201	72	213
K (mg)	547	324	379

SPINACH

Origin and History

Spinach is native to southwestern Asia, Iran, and nearby areas of the Middle East. It was unknown to the ancient Greeks and Romans, but was probably transported to other areas of Asia Minor about 2,000 years ago. Spinach was also mentioned in Chinese literature around the year 650. Spinach reached Spain via the Moors about 1100 and was brought to America by early colonists in the 1600s (Zohary and Hopf, 2000).

Botany and Life Cycle

Spinach (*Spinacia oleracea* L.) is a leafy annual, also in the subfamily Chenopodiaceae of the family Amaranthaceae (USDA Plants, 2010). Spinach is an annual plant grown exclusively for its leaves and is sometimes called a winter annual because it can overwinter if not exposed to prolonged temperatures below −12°C (10°F).

Spinach has a distinctive vegetative phase, when the plant forms a short compressed stem and a rosette of leaves, before flowering in response to environmental stimuli. It grows to a height of up to 30 cm

(12 in). The leaves are alternate, simple, and vary in length from about 2–30 cm (0.8–12 in) and 1–15 cm (0.4–6 in) across, with larger leaves at the base and small leaves higher on the flowering stem.

Both the leaf petiole and blade are harvested during the vegetative rosette stage (Fig. 17.6).

After the seedling stage, the leaves form a rosette with fleshy leaves attached to a short stem. Spacing and environmental conditions help to determine both leaf size and number. Leaf blades range from ovate to elongated to almost triangular and narrow arrowhead shapes in some older cultivars (Rubatzky and Yamaguchi, 1997). Leaf growth habit varies from prostrate to upright with plant spacing having an effect. Leaf margins are smooth or curved, while leaf blade topography varies from smooth to semi-savoy to heavily savoyed. The blistered appearance of the savoyed leaf blade occurs because of differential growth of parenchyma tissues between leaf veins (Rubatzky and Yamaguchi, 1997). Spinach petioles are usually the same length as the leaf blade and hollow at maturity.

At the induction of the reproductive state, the primary stem elongates (bolts) with a disruption of the rosette to form a long flower stalk. Bolting renders the crop unmarketable because vegetative growth is severely restricted and any new leaves that do form tend to be small and narrow. After bolting occurs, the crop is no longer suitable for harvest for commercial sale as a vegetable. The flower induction process in spinach is very similar to leaf lettuce as discussed in the chapter on the family Asteraceae. Seed-stalk formation is induced by long days, while temperatures above 20°C (68°F) accelerate the process. The critical photoperiod that initiates flowering ranges between 12.5 and 15 h (Rubatzky and Yamaguchi, 1997). As the central stem elongates, lateral branches form, containing clusters of up to 20 small inconspicuous yellow-green 3–4 mm (0.13–0.16 in) diameter wind-pollinated flowers that mature into a small, hard, dry, lumpy fruit cluster 5–10 mm (0.2–0.39 in) across, containing several seeds (Fig. 17.7).

The spinach root system consists of many shallow fibrous laterals that develop from an enlarged taproot with little branching. Some cultivars resist flowering better than others, so slow-bolting cultivars should be selected if plantings must be scheduled to mature during long days and warm temperatures.

The sex expression of spinach plants is difficult to determine based solely on vegetative growth, although male plants tend to be smaller and produce fewer leaves. Spinach sex expression is generally dioecious, although other types also occur. The most common plant types are male, female, or monoecious. Occasionally, hermaphroditic flowers occur

Fig. 17.6. Winter production of semi-savoy spinach in south Texas with overhead irrigation (background). Rows on the left have been cut for the first of multiple mechanical harvests.

Fig. 17.7. Spinach flower clusters.

as well. In addition to these general characterizations of sex expression, other conditions exist as well. "Extreme" males occur that produce only staminate flowers, minimal foliage, flower early, and often die shortly after flowering. Vegetative males produce only staminate flowers but have more foliage and flower later than the "extreme" males. Female plants produce only pistillate flowers, have well-developed foliage and flower later. Monoecious plants produce staminate and pistillate flowers on the same plant, have well-developed foliage and are slow to flower (Rubatzky and Yamaguchi, 1997). The apetalous staminate flowers are clustered on spikes. Plant breeders rogue extreme males and poor leaf producers from seed fields so modern F-1 hybrid spinach cultivars tend to have homogeneous sex expression, while older open-pollinated heirloom cultivars tend to be more variable (Rubatzky and Yamaguchi, 1997).

Pistillate flowers lack petals and are attached to the base of the calyx that encloses the ovary. Following wind pollination, the ovary develops into a single-seeded fruit called a utricle that is often referred to as a "seed". Spinach seeds may be either smooth and round or irregularly shaped with pointed spines. Seeds with spines are considered to be winter types, while smooth types are often referred to as a summer types. Spiny-seeded cultivars are often heirloom cultivars. The weight of 100 spinach seeds is about 1 g (0.04 oz) (Rubatzky and Yamaguchi, 1997).

Types and Cultivars

Spinach is an important green leafy vegetable in temperate climates. In Asia spinach is used primarily as a fresh vegetable, most often consumed after light cooking, but in western Europe and North America spinach is increasingly used fresh or as a lightly processed, bagged, ready-to-eat product. Spinach is also canned or frozen.

As a general rule, older or heirloom cultivars tend to bolt early under long-day conditions. Heirloom cultivars tend to have narrower leaves, stronger flavor, and more bitterness. Newer cultivars often used for commercial production tend to be F-1 hybrids that grow more rapidly but have improved resistance to bolting, broader leaves, and milder flavor. Spinach cultivars may be divided into three major categories based on leaf characteristics:

- Savoy has dark green, crinkly, and curly leaves. Because of its showy appearance, this type was preferred for fresh bunched spinach. An example of savoy spinach is the heirloom cultivar 'Bloomsdale', which has been replaced commercially by higher yielding predominately female F-1 hybrid cultivars that are sometimes collectively referred to as Bloomsdale types. 'Long Standing Bloomsdale' was a popular cultivar that resisted bolting better than the original 'Bloomsdale'.
- Flat- or smooth-leaf spinach has broad, smooth leaves that are easier to clean than savoy but the flat leaves tend to stick together when packaged. This type is often grown for canning, ready-to-eat bagged spinach, frozen spinach, as well as soups, baby foods, etc.
- Semi-savoy has leaf blades that are intermediate between flat and savoy. Semi-savoy cultivars have a texture similar to savoy, are not as difficult to clean, and leaves resist compression and clumping during packaging and transport.

Baby spinach is popular in some markets because the immature plants are tender with less bitterness. Although cultivars with smaller leaves and plant architecture are sometimes used, conventional cultivars are often seeded at higher densities and plants harvested with only five to seven leaves to produce baby spinach. High-density plantings have the extra benefit of crowding out weeds and early harvest often occurs before insect or disease attacks, making pesticide applications unnecessary. Baby spinach may be flat, savoy, or semi-savoy (Fig. 17.8).

In addition to leaf-blade variations described above, there is also a wide variety of leaf shapes, ranging from heart shaped to oval. Cultivars are available with genetic resistant to both downy mildew and cucumber mosaic virus, two major diseases that affect spinach. Characteristics of commercial importance

Fig. 17.8. Organic savoy baby spinach production along the coast of central California. Three rows per bed are densely seeded to choke out weeds. This baby cultivar is much smaller than regular spinach and will be sold in a lightly processed, bagged, ready-to-eat, salad mix.

that vary among cultivars in addition to those mentioned above include plant uniformity, yield, size, high-temperature resistance, cold resistance, and leaf variation in size, color, shape, and petiole length. Commercially desirable, high-quality, cultivars would have the following characteristics: short petioles, medium to dark green leaves, "sweet" flavor free of bitterness, tender thick leaves, and disease and bolting resistance (Rubatzky and Yamaguchi, 1997).

Economic Importance and Production Statistics

Spinach is an important crop that is produced in all regions of the world. In 2010, there were 825,207 ha (2,039,130 acres) of spinach harvested in the world. China was the leading producing country with 640,370 ha (1,582,388 acres), Indonesia was second, producing 48,844 ha (120,696 acres), Turkey was third with 23,000 ha (56,834 acres), Japan was fourth with 21,000 ha (51,892 acres), and the USA was fifth with 17,520 ha (38,444 acres) (FAOSTAT, 2010).

Total world production was over 20 million metric tonnes (22.1 million short tons) in 2010. China alone produced 18.1 million metric tonnes (20 million short tons). The USA was second with almost 400,000 metric tonnes (440,925 short tons), followed by Japan with 269,000 metric tonnes (296,522 short tons), and Turkey with 218,291 metric tonnes (240,625 short tons). Leaders among European countries include Belgium and France with 93,150 metric tonnes (102,680 short tons) and 80,101 metric tonnes (88,296 short tons), respectively (FAOSTAT, 2010).

The average world spinach yield was calculated to be 24,364 kg/ha (21,737 lb/acre). The gross production value of Chinese spinach was estimated to be US$5.1 billion in 2010, compared to US$367 million for the US crop (FAOSTAT, 2010).

In the USA, approximately 25% was grown for processing and the rest for fresh market. California was the leading producing state in 2010 with about 10,886 ha (26,900 acres), including 7,931 ha (19,600 acres) fresh and 2,954 ha (7,300 acres) processed as predominately frozen spinach. Other important production states include Arizona, Texas, and New Jersey. California also recorded the greatest yields per hectare, partially because repeated harvests are possible due to the long season and cool temperatures in the coastal valleys where spinach is produced (USDA, NASS, 2012).

Nutritional Values – Spinach

The nutritional value of fresh spinach leaves is shown in Table 17.2. Spinach is a good source of

Table 17.2. Nutritional value of fresh uncooked spinach per 100 g (3.5 oz) (USDA Nutrient Database, 2012).

Energy	97 kJ (23 kcal)
Carbohydrates	3.6 g
– Sugars	0.4 g
– Dietary fiber	2.2 g
Fat	0.4 g
Protein	2.9 g
Water	91.4 g
Vitamin A equiv.	469 µg (59%)
Vitamin A	9,377 IU
– β-carotene	5.6 mg (52%)
– Lutein and zeaxanthin	12.2 mg
Thiamine	0.078 mg (7%)
Riboflavin	0.189 mg (16%)
Niacin	0.724 mg (5%)
Vitamin B6	0.189 mg (15%)
Folate (vitamin B9)	194 µg (49%)
Vitamin C	28 mg (34%)
Vitamin E	2 mg (13%)
Vitamin K	483 µg (460%)
Ca	99 mg (10%)
Fe	2.71 mg (21%)
Mg	79 mg (22%)
Mn	0.897 mg (43%)
P	49 mg (7%)
K	558 mg (12%)
Na	79 mg (5%)
Zn	0.53 mg (6%)

Percentages are relative to USDA daily recommendations for adults.

vitamin A, and is especially high in lutein, vitamin C, vitamin E, vitamin K, Mg, Mn, folate, betaine, Fe, vitamin B2, Ca, K, vitamin B6, folic acid, Cu, protein, P, Zn, niacin, selenium, omega-3 fatty acids, and antioxidants. Spinach also contains relatively high levels of oxalic acid, which can bind to the Fe and Ca in the body decreasing absorption and use.

Production and Culture – Spinach

Growth and development

Spinach requires approximately 40–70 days from seeding for spring plantings, depending on cultivar and environment. Overwinter production delays maturity due to cooler temperatures and reduced light intensity during winter months, although spinach is more tolerant of low light conditions than many other vegetables. In addition to the influence of temperature on development, earliness is also related to growth rate, and early-maturing cultivars tend to grow faster than later ones (Rubatzky and Yamaguchi, 1997). Much like other vegetables, cultivar selection is very important for success. The goal of many spinach producers is to match a cultivar to growing conditions, so rapid development and high yield are obtained but bolting is avoided.

Temperature requirements

Spinach is a cool-season, annual crop that is direct-seeded. The optimum temperature for spinach growth is 15–20°C (59–68°F) with a minimum growth temperature of approximately 5°C (41°F) and a maximum of 32°C (90°F). Spinach is fairly winter hardy and can survive temperatures as low as –12°C (10°F). Spinach is planted in early spring for late spring or early summer production or overwintered with fall planting and spring harvests in areas where winters are not severe. Temperature affects leaf qualities, with low temperature increasing leaf thickness but decreasing size and smoothness (Rubatzky and Yamaguchi, 1997). High temperatures reduce foliage color and savoying, and increase petiole length.

Site selection and water needs

Spinach can be grown on a wide range of soil types but poorly drained soils prone to becoming waterlogged should be avoided. Soils best suited to spinach production are well drained but should have adequate moisture-holding capacity. Raised beds are often used to improve drainage on heavier soils. Since spinach is a cool-season crop grown during times of the year when transpiration is generally low, water requirements are lower compared to many other crops. In areas with consistent rainfall, spring and overwinter crops can be produced without irrigation. About 250 mm (9.8 in) of water is required to produce a crop.

Because of its shallow root system, spinach plants are susceptible to drought stress during dry periods. An obvious effect of drought stress is reduced leaf size, which directly affects yields and leaf tenderness. In dry areas, such as California and Texas, overhead irrigation is sometimes used to establish the crop. Overhead furrow or drip irrigation is used later in the season as the crop matures (Fig. 17.6).

Approximately 2.5 cm (1 in) of water is needed each week to maintain active growth of the crop. Spinach can be grown hydroponically although management may be more challenging than lettuce or tomato culture. Spinach performs best hydroponically in well-aerated recirculating systems.

Establishment and culture

Spinach germinates at soil temperatures above 1.7°C (35°F) but optimum germination occurs at approximately 21°C (86°F). However, thermoquiescence occurs in excess of 30°C (35°F), reducing emergence (Harrington and Minges, 1954). High temperature germination performance can be improved by seed priming. Acid treatment or presoaking can stimulate germination by physiological enhancement, softening pericarp tissues, and removing inhibitors (Atherton and Farooque, 1983a, b). Finely tilled soils, freed of clods using rotovators or disks, help to ensure uniform emergence and good stand establishment although conservation tillage is also an option (Hoyt et al., 1994). Spinach can be direct-seeded at depths from 1–3 cm (0.4–1.2 in) in multirows on a standard 102 cm (40 in) raised bed using precision seeding equipment to achieve near optimal plant spacing without thinning, although other row configurations are possible. Spinach may also be scatter-planted in narrow 10 cm (4 in) bands. Seed spacing varies with cultivar and the intended use. Plant densities often vary from 60 to 120 plants/m^2 (/10.1 ft^2). High plant populations promote upright leaf growth, which is desirable for mechanical harvesting, but plant density that is too high produces weak petioles and accelerates seed stalk development (Fig. 17.9).

Weed control is important, especially for nonselective mechanical harvests, because weeds contaminate the product. Some weeds are similar to spinach and are difficult to remove. Frequent cultivation, dense plantings, the stale-bed technique, and selective herbicides can all be used to effectively control weeds.

Fertilizer and nutrition

Spinach plants are sensitive to acidity but tolerant of salinity. Like beets, spinach is tolerant of saline

Fig. 17.9. Spinach can be mechanically harvested multiple times by cutting the leaves but leaving the crowns intact. (a) Pull-type spinach harvester near Salinas, California. (b) The beds on the left have been harvested while the beds on the right are ready for harvesting.

Family Chenopodiaceae

soil conditions and is considered to be a halophyte. The pH optimum would be considered to be within a range of 6.5 to 8.0. Spinach is not an especially heavy feeder. On fertile alluvial soils, usually only modest fertilization is needed in conjunction with soil test results. On lighter sandy soils, high quantities of N-K-P may be needed depending on cropping history. Nitrogen inputs often increase production of winter-grown spinach because little nitrification occurs at low soil temperatures. Most growers prefer to keep spinach crops well fertilized to increase leafiness and to meet the demands of very rapid growth that occur during the short period between emergence and harvest. About 70% of crop biomass is produced during the final third of preharvest growth. Because of its rapid production cycle, most often fertilizer is applied preplant or at planting and supplemental fertilization is only used in extreme cases.

Harvesting and marketing

Spinach is harvested when plants have reached the desired size, which can occur anywhere from 30 to 150 days in the case of overwinter production. Most plants have five to nine fully developed leaves at harvest.

Spinach may still be sold in some markets in hand-wrapped bunches that are manually cut at the soil line in the field. However, this increases the cost and bunched spinach is more difficult to keep fresh and undamaged during shipment. Bunched spinach is being replaced or has been replaced in many markets by fresh, bagged, minimally processed spinach. Bagged spinach is mechanically cut, bulk handled, and washed. Packaging increases shelf life, makes handling easier and provides consumers with a convenience product that is ready to eat with minimal preparation. Spinach for canning, freezing and fresh, bagged, spinach is mechanically harvested (Fig. 17.10).

Harvesters are adjusted to cut 10–15 cm (4–6 in) above the growing point in the center of the rosette to reduce petiole length to enable grow-back and repeated harvests in 3–4 weeks under favorable conditions. Processing crop yields range from 5–24 metric tonnes/ha (2–11 short tons/acre) with the latter value reflecting multiple harvests. Multiple harvests are possible in areas with cool temperatures and shorter daylengths such as winter production in Texas as well as spring production in the Salinas Valley of California where up to three or four harvests per season are possible before bolting begins (Fig. 17.9). Fresh-market yields are often toward the lower range.

Smooth or semi-savoyed cultivars are often preferred for canned, frozen, and bagged spinach because they grow faster, produce higher yields, and are easier to wash free of soil. Savoy and semi-savoy cultivars are often preferred for fresh-market

Fig. 17.10. A self-propelled harvester for processed spinach near Uvalde, Texas.

production, as the leaves resist compression during packaging and thus improve aeration, cooling, and storage life.

Commercial greenhouse production of spinach occurs in northern Europe during winter and early spring where winter conditions prevent overwinter production. Rapidly growing cultivars have been specially developed to grow well in this unique environment of low light intensities, cold temperatures and short days (Rubatzky and Yamaguchi, 1997).

Spinach is highly perishable because of its relatively high respiration rate and delicate leaf structure. Storage at room temperature or above quickly destroys quality, so rapid cooling, and refrigeration is required to prevent wilting and weight loss. Hydro-cooling and vacuum cooling are effective technologies for spinach. Hydro-vac cooling, where moisture is added during vacuum cooling, may reduce wilting. Bagged lightly processed spinach is often packed in a modified atmosphere to reduce respiration and extend storage life. In 0.8% O_2 atmosphere, spinach O_2 uptake was reduced by an average of 53%, and CO_2 production was reduced by 35% relative to those stored in air, while leaf deterioration was reduced by 30–54%. Weight loss and chlorophyll content were not affected by 0.8% O_2 atmosphere (Ko et al., 1996).

However, low oxygen levels can produce off-odors in bagged spinach. Baby spinach stored to 1% O_2 with CO_2 had the lowest quality at the end of storage due to development of off-odors, while storage at 10% O_2 and CO_2 had fewer off-odors (Tudela et al., 2012). Spinach can be stored at 0°C (32°F), 95% RH, in modified atmosphere packaging for 14 days. Bagging may provide a favorable environment for microbial pathogens that cause human disease, particularly at higher storage temperatures and extended storage periods (Lopez-Velasco et al., 2010).

Diseases

Downy mildew (*Peronospora farinosa*) is one of the most serious diseases of spinach. Downy mildew is difficult to control using fungicides particularly after it appears on the plant. It thrives under cool wet conditions. Initial symptoms of downy mildew are dull to bright yellow spots that may form on cotyledons and leaves at any stage of development. With time, these spots may enlarge and become tan and dry. Purple growth of the fungus occurs on the underside of the leaf. Host resistance to all four physiological races is available in modern F-1 hybrid cultivars (LeStrange and Koike, 2012).

Cucumber mosaic cucumovirus (CMV) is another serious disease that affects spinach. Symptoms include slight chlorosis and narrow or "puckered" young leaves. Leaves may have margins that roll inward. In advanced stages of infection, the plants may appear stunted and the crown may be completely blighted, killing the growing point. Control of the vector aphids (*Aphis fabae* and *Myzus persicae*) reduces incidence of CMV (LeStrange and Koike, 2012).

Other important viral diseases of spinach include beet western yellows virus. This infection appears as interveinal and leaf margin chlorosis on older leaves. Secondary fungi often invade older infected leaves, leading to necrosis that supports dark green to black fungal growth. Beet curly top virus infections stunt leaf development and cause chlorosis. Younger leaves in the center of the rosette are often very chlorotic, extremely curled, and rigid. Plants usually die a few weeks after symptoms appear. Impatiens necrotic spot virus and tomato spotted wilt virus each may cause ringspots, circular leaf spots, and necrotic spotting on foliage. Tobacco rattle virus causes yellow and necrotic spotting, mottling, and leaf crinkling on spinach leaves (LeStrange and Koike, 2012).

Weeds may serve as virus reservoirs and should be removed from areas in and around production fields. However, weed removal will not necessarily prevent virus infections from occurring. No virus-resistant spinach cultivars are available. Insect vector populations should be managed if possible (LeStrange and Koike, 2012).

Damping-off affects spinach production throughout the world. Cultivar, soil texture, soil moisture, and pathogen populations influence the severity of damping-off. Severe damping-off is associated with clay or poorly draining soils used repeatedly for spinach production. While all stages of spinach can be infected by root rot organisms, newly emerging plants, and young seedlings are most susceptible.

A complex of pathogenic soil-borne fungi cause damping off, including one or more of the following: *Fusarium oxysporum*, *Pythium* (several species), and *Rhizoctonia solani*. Aboveground symptoms are similar to symptoms of root rot. Seedling damping-off caused by *Pythium* spp. and *Rhizoctonia* spp. can be prevented by reducing soil moisture levels, seed-dressing with fungicides and using fungicide-treated seeds (LeStrange and Koike, 2012).

Insects

A partial list of spinach insect and nematode pests includes: *Ditylenchus dipsaci* (nematode), *Clavigralla tomentosicollis* (African pod bug), *Helicotylenchus dihystera* (common spiral nematode), *Heterodera schachtii* (beet cyst eelworm), *Liriomyza sativae* (vegetable leaf miner), *Liriomyza trifolii* (American serpentine leafminer), *Lis. costirostris* (vegetable weevil), *My. persicae* (green peach aphid), *Nacobbus aberrans* (false root-knot nematode), *Tetranychus cinnabarinus* (carmine spider mite), *Trichoplusia ni* (cabbage looper), *Helicoverpa zea* (corn earworm), and *Sp. exigua* (beet armyworm) (LeStrange *et al.*, 2012).

Caterpillars (loopers, earworms, and armyworms) are some of the most problematic pests because they directly feed on the leaves and can quickly destroy a crop. These caterpillars damage the crown of spinach plants, severely stunting or killing them. The potential for damage and contamination continues until harvest. Cutworms (*Peridroma saucia* and *Agrotis subterranea*) feed at or below ground level. Beet armyworms lay their eggs in distinctive cottony masses on leaf surfaces as described earlier in this chapter. The western yellow-striped armyworm can also develop high populations, rapidly damaging spinach fields.

Several management techniques are effective for controlling spinach caterpillars. Biological control using natural enemies that systematically attack spinach caterpillars can be very effective. Among the most common parasites are the wasps *Hyposoter exiguae* and *Chelonus insularis*, and the tachinid fly *Lespesia archippivora*. Naturally occurring viral diseases may also kill significant numbers of caterpillars in a healthy balanced ecosystem. Cultural control methods may also be effective. For example, disking fields immediately following harvest to kill larvae and pupae and destroying weeds along field borders are effective cultural control techniques (LeStrange *et al.*, 2012).

There are organic methods that can provide acceptable control of caterpillars as well. The cultural and biological controls mentioned above along with sprays of *Bacillus thuringiensis* or the Entrust® formulation of spinosad are approved for organic production in the USA and certain other countries (LeStrange *et al.*, 2012). Spinosad is a fast-acting, somewhat broad-spectrum material that acts on the insect primarily through ingestion or by direct contact, causing loss of muscle control. Continuous activation of motor neurons causes insects to die of exhaustion within 1–2 days after ingestion. Foliar applications of spinosad are not highly systemic although some movement into leaf tissue has been demonstrated. The addition of a penetrating surfactant increases absorption by tissues and activity on pests that mine leaves (Larson, 1997).

Effective control of insect pests on spinach requires careful monitoring before seedlings emerge by checking for eggs and young larvae in surrounding weeds. If populations are high on weeds, larvae may also be present on seedlings. Once seedlings emerge, they should be checked twice weekly for egg masses and young larvae. Insecticides, if used, are more effective against young larvae than eggs, so applications should be delayed until a majority of eggs have hatched before treatment begins. For beet armyworm control, the best time to apply insecticide is at dawn during twilight hours when worms are active (LeStrange *et al.*, 2012).

NEW ZEALAND SPINACH AND ORACH

New Zealand spinach (*Tetrogonia expansa*) is a member of the family Tetrogoniaceae, and is not true spinach. The leaves resemble spinach leaves and are often used in cooking in the same way. The plants are well branched and will spread up to 0.91–1.21 m (35.8–47.6 in) and grow to a height of 0.30–0.61 m (11.8–24 in). The leaves are thicker than spinach, smaller, darker green, and somewhat triangular in shape. The stems are larger, more fleshy, and firm. The stem tips with leaves are harvested several times during the growing season. The plant is sensitive to frost but will perform well in hot weather, making it an excellent substitute for spinach. The plant is a perennial but is usually replanted after 2–3 years.

Orach is believed to have originated in the Indian subcontinent with medicinal and food uses (Rubatzky and Yamaguchi, 1997). Orach (*Atriplex hortensis*), also called mountain spinach or French spinach, is resistant to hot weather and drought. Orach is not a significant crop in the USA although it is included in some bagged salad mixes and microgreens. Orach is an important salad vegetable in France and parts of Great Britain. It is a hardy monoecious annual that is tolerant of drought, salinity, and a broad range of temperatures. Orach is often grown as a substitute for spinach in warm climates because flowering is slow and plants tolerate higher temperatures even though stems may rapidly elongate.

Seed are sown about 1 cm (0.39 in) deep in rows approximately 50–75 cm (19.7–29.5 in) with in-row spacing of about 3–5 cm (1.2–2 in). The plant initially forms a rosette of leaves followed by a seed stalk 2–3 m (79–118 in) high. Whole plants are harvested when 10–15 cm (3.9–5.9 in) tall and cooked before eating (Fig. 17.11).

Tender shoots and leaves may also be harvested individually. Orach remains edible after spinach has bolted, so it can be grown throughout the summer months. Orach has smooth, heart- to shield-shaped leaves. Three main cultivar types are pale green (white), red, and dark green (Rubatzky and Yamaguchi, 1997). 'Fire Red' is a cultivar with red leaves and petioles sometimes used in microgreen and leafy green mixes.

References

Abegg, F.A. (1936) A genetic factor for the annual habit in beets and linkage relationships. *Journal of Agricultural Research* 53, 493–511.

Atherton, J.G. and Farooque, A.M. (1983a) High temperature and germination in spinach. I. The role of the pericarp. *Scientia Horticulturae* 19, 25–32.

Atherton, J.G. and Farooque, A.M. (1983b) High temperature and germination in spinach. II. Effects of osmotic priming. *Scientia Horticulturae* 19, 221–227.

Board, N. (2004) *Handbook On Herbs Cultivation And Processing.* Asia Pacific Business Press, Delhi.

Burge, W. (1991) *Grow the Best Root Crops*, Vol. 117. Storey Publishing, North Adams, Massachusetts.

CABI (2013) *Beta vulgaris*, list of pests. Crop protection compendium. Available at: www.cabi.org/cpc/?compid=1&dsid=8778&loadmodule=datasheet&page=868&site=161 (accessed 31 January 2013).

FAOSTAT (2010) Spinach production statistics. Available at: http://faostat.fao.org/site/567/default.aspx#ancor (accessed 8 January 2013).

Goldman, I.L. and Navazio, J.P. (2007) Table Beet. In: Prohens-Tomás, J. and Nuez, F. (eds) *Handbook of Plant Breeding*, Vol. 1, *Vegetables: Asteraceae, Brassicaceae, Chenopodiaceae, and Cucurbitaceae.* Springer, New York, pp. 219–238.

Grubben, G.J.H. and Denton, O.A. (2004) *Plant Resources of Tropical Africa. 2. Vegetables.* PROTA Foundation, Wageningen, the Netherlands.

Hanelt, P., Büttner, R., Mansfeld, R. and Kilian, R. (2001) *Mansfeld's Encyclopedia of Agricultural and Horticultural Crops.* Springer, New York.

Harrington, J.F. and Minges, P.A. (1954) *Vegetable seed germination.* Mimeo leaflet. University of California Cooperative Extension, Davis, California.

Hembree, K.J. and Norris, R.F. (2010) UC IPM Pest Management Guidelines: Sugarbeet weeds. UC ANR Publication 3469. Available at: www.ipm.ucdavis.edu/PMG/r735700111.html (accessed 22 January 2013).

Hill, G. and Langer, R.H.M. (1991) *Agricultural Plants.* Cambridge University Press, Cambridge, UK.

Hoyt, G.D., Monks, D.W. and Monaco, T.J. (1994) Conservation tillage for vegetable production. *HortTechnology* 4, 129–136.

Fig. 17.11. Orach (*Atriplex hortensis*) is a vegetable in the subfamily Chenopodiaceae. Orach has leaves that are eaten like spinach. However, unlike spinach, orach leaves remain tender under warm temperatures and long days even after the plant flowers.

Kaffka, S., Turini, T.A., Wintermantel, W.M., Lewellen, R.T. and Frate, C.A. (2010) Management guidelines: Sugarbeet diseases. UC ANR Publication 3469, Statewide IPM Program. Available at: www.ipm.ucdavis.edu/PMG/r735100611.html (accessed 31 January 2013).

Knott, J.E. (1962) *Knott's Handbook for Vegetable Growers*. Wiley and Sons, New York, 245 pp.

Ko, N.P., Watada, A.E., Schlimme, D.V. and Bouwkamp, J.C. (1996) Storage of spinach under low oxygen atmosphere above the extinction point. *Journal of Food Science* 61, 398–401.

Larson, L.L. (1997) Effects of adjuvants on the activity of Tracer™ 480SC on cotton in the laboratory. *Arthropod Management Tests* 22, 415–416.

LeStrange, M. and Koike, S.T. (2012) UC IPM Pest Management Guidelines: Spinach. UC ANR Publication 3467, Diseases, UC Cooperative Extension, Statewide IPM Program. Available at: www.ipm.ucdavis.edu/PMG/r732100111.html#SYMPTOMS (accessed 7 January 2013).

LeStrange, M., Koike, S.T. and Chaney, W.E. (2012) UC IPM Pest Management Guidelines: Spinach, UC ANR Publication 3467, Insects and Mites, UC Cooperative Extension, Statewide IPM Program, Agriculture and Natural Resources. Available at: www.ipm.ucdavis.edu/PMG/r732300411.html (accessed 7 January 2013).

Lopez-Velasco, G., Davis, M., Boyer, R.R., Williams, R.C. and Ponder, M.A. (2010) Alterations of the phylloepiphytic bacterial community associated with interactions of *Escherichia coli* O157:H7 during storage of packaged spinach at refrigeration temperatures. *Food Microbiology* 27, 476–486.

Natwick, E.T., Summers, C.G., Haviland, D.R. and Godfrey, L.D. (2010) Sugarbeets: insects and mites. UC IPM Pest Management Guidelines: Sugarbeet, UC ANR Publication 3469. Statewide IPM Program, Agriculture and Natural Resources, University of California. Available at: www.ipm.ucdavis.edu/PMG/r735300811.html (accessed 22 January 2013).

Owen, F.V. (1954) The significance of single gene reactions in sugar beets. *Proceedings of the American Society of Sugar Beet Technology* 8, 392–398.

Owen, F.V., Carsner, E. and Stout, M. (1940) Photothermal induction of flowering in sugar beet. *Journal of Agricultural Research* 61, 101–124.

Rubatzky, V.E. and Yamaguchi, M. (1997) *World Vegetables: Principles, Production and Nutritive Values*, 2nd edn. Chapman & Hall, New York.

Schrader, W. and Mayberry, K. (2002) *Beet and Swiss Chard Production in California. ANR Publication 8096*. University of California, Division of Agriculture and Natural Resources, Oakland, California.

Smit, A.L. (1983) Influence of external factors on growth and development of sugar beet (*Beta vulgaris* L.) Ph.D. Dissertation. Landbouwhogeschool te Wageningen, Agricultural Research Reports 914. Centre for Agricultural Pub. and Documentation, Pudoc, Wageningen, the Netherlands.

Tudela, J.A., Marín, A., Garrido, Y., Cantwell, M., Medina-Martínez, M.S. and Gil, M.I. (2012) Off-odour development in modified atmosphere packaged baby spinach is an unresolved problem. *Postharvest Biology and Technology* 75, 75–85.

USDA, ARS (2012a) *Chenopodium album* L. National genetic resources program. Germplasm resources information network. Available at: www.ars-grin.gov/cgi-bin/npgs/html/taxon.pl?10178 (accessed 28 December 2012).

USDA, ARS (2012b) *Beta vulgaris* L. subsp. vulgaris. National genetic resources program. Germplasm resources information network. Available at: www.ars-grin.gov/cgi-bin/npgs/html/taxon.pl?7057 (accessed 28 December 2012).

USDA, NASS (1999) *Vegetable and Specialties Yearbook. ERS-VGS-278*. Economic Research Service, US Department of Agriculture, Washington, DC.

USDA, NASS (2012) Vegetables 2011 summary. Available at: www.nass.usda.gov (accessed 7 January 2013).

USDA Nutrient Database (2012) Available at: http://ndb.nal.usda.gov/ndb/foods/show/3151?fg=&man=&lfacet=&format=&count=&max=25&offset=&sort=&qlookup=spinach (accessed 7 January 2013).

USDA Plants (2010) PLANTS profile for *Beta vulgaris* (common beet). Available at: http://plants.usda.gov (accessed 21 November 2010).

Westfall, J.G. and Davis, D.G. (2009) Fertilizing sugar beets. Available at: www.ext.colostate.edu/pubs/crops/00542.html#top (accessed 24 January 2013).

Wright, C.A. (2001) *Mediterranean Vegetables: A Cook's ABC of Vegetables and their Preparation in Spain, France, Italy, Greece, Turkey, the Middle East, and North Africa with More than 200 Authentic Recipes for the Home Cook*. Harvard Common Press, Boston, Massachusetts.

Zohary, D. and Hopf, M. (2000) *Domestication of Plants in the Old World*. Oxford University Press, Oxford, UK.

18 Family Asparagaceae

ASPARAGUS

Origin and History

Asparagus is a very ancient crop native to the eastern Mediterranean region, Asia Minor, and possibly as far east as the Caucasus mountains (Rubatzky and Yamaguchi, 1997). The Ancient Greeks (200 BC) and Romans considered asparagus a delicacy that also had medicinal qualities for relieving toothaches. Asparagus was gathered from the wild until the Romans began cultivation. After the Roman Empire ended, there was little mention of asparagus during medieval times (Vaughan and Geissler, 2009).

There are historic references to asparagus production in England in 1538 and Germany in 1567, and by the end of the 16th century asparagus was produced in France. Louis XIV enjoyed asparagus so much that he constructed hothouse beds for out-of-season production (Ilott, 1901). Asparagus was introduced to North America from Europe by 1672, and President Thomas Jefferson grew asparagus at his home in Virginia in the late 1700s. Asparagus was first planted in California in the 1860s in the San Joaquin delta where it is still an important crop. Today asparagus has spread around the world and is an important vegetable in over 60 countries, including many parts of Europe, Asia, Australia, and New Zealand (FAOSTAT, 2010).

Botany

Asparagus (*Asparagus officinalis*) is a perennial monocot in the family Asparagaceae, subfamily Asparagoideae (Chase *et al.*, 2009). Asparagus plants grow to 100–150 cm (39–59 in) tall, with rigid stems and highly branched feathery foliage that develops from the tender immature shoots or spears that are eaten as a vegetable (Rubatzky and Yamaguchi, 1997). There are almost 300 species of plants in this family and most are evergreen, long-lived perennial plants, including *Asparagus aethiopicus*, the Sprenger's asparagus, a common ornamental plant (WCSP, 2011). *Asparagus officinalis* is one of several edible species in the genus.

Asparagus produces new above-ground stems each year from a fleshy underground "crown" (Fig. 18.1). These stems are harvested as tender immature spears and eaten before lignification occurs. The crown consists of unelongated basal internodes of old stems (Blasberg, 1932). The gradual increase in crown size that occurs with age is due to the development of both apical and lateral buds. New roots develop from the bases of the actively growing buds. These roots enlarge and develop into storage roots that store food reserves for spear production. The roots make up the bulk of a mature crown because they are very thick with few root hairs, have a heavily suberized epidermis, and are fleshy because of a loosely fitting cortex (Rubatzky and Yamaguchi, 1997).

Many buds are located on the crown and each is potentially capable of producing the spears for commerce. There is a regional dominance in the crown in the vicinity of the growing spear similar to the more familiar apical dominance exhibited by many plants. The regional crown dominance prevents the adjacent sprouts from growing until the dominant, actively growing spear is removed (Tiedjens, 1926). Each bud is developed in the axil of the one that preceded it. A spear is composed of an apical meristem with many tightly closed buds covered with bud scales. With continued growth, lateral branches elongate from the nodes under each bud scale along the spear, producing the fern-like cladophylls that appear to be leaves but are actually modified stems. The small bracts at each node of the spear are the true leaves but are not photosynthetically active. The cladophyll is the photosynthetically active tissue of asparagus. Spear size is positively correlated with bud size so that large buds produce larger spears. Bud size tends to be more variable than bud number. Most of the cell differentiation

Fig. 18.1. An asparagus crown and extensive root system.

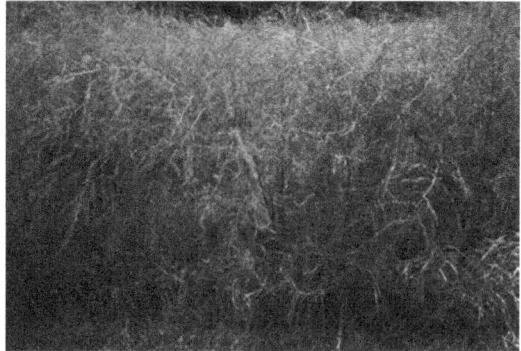

Fig. 18.2. Asparagus spears elongate rapidly giving rise to large bushy plants that are mostly fern. Another name for fern is cladophyll, the photosynthetically active tissues of the plant. Asparagus is dioecious, so plants producing bright orange berries are female and barren plants are male.

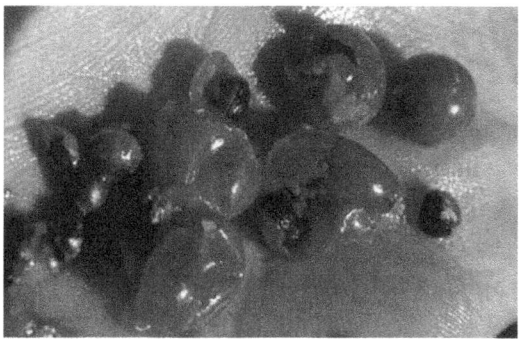

Fig. 18.3. Asparagus berries have tough skins, are bright orange at maturity, and contain black seeds.

occurs prior to elongation so the buds can be thought of as miniature spears in an arrested state until growth begins. In other words, spear growth and emergence is caused by cell elongation rather than cell division (Rubatzky and Yamaguchi, 1997).

Asparagus plants are dioecious, which means there are separate male and female plants (Fig. 18.2). Occasionally, perfect flowers may appear on male plants. The flowers are bell-shaped, greenish-white to yellowish, 4.5–6.5 mm (0.18–0.26 in) long, and form singly or in clusters of two or three in the junctions of the branches. Pollination for seed production is by bees. The fruit are initially green when immature, turning bright orange when mature. The fruit is a small red berry 6–10 mm (0.25–0.31 in) in diameter. The berries contain glucosides and should not be eaten. The juice of the berries may cause minor skin irritation (Anonymous, 2012). Each berry contains dark black seeds (Fig. 18.3). The birds feed on berries and disperse seeds, which germinate in the wild to become weeds.

Climactic Requirements

Mature plants, especially when in the fern stage, are also relatively tolerant of a broad range of environmental stresses, including heat, drought, and salinity. Mature asparagus crowns can tolerate subfreezing cold when dormant, but the newly emerging "spears" and fern are damaged by frost (Swaider et al., 1992). Average day temperatures of 25–30°C (77–86°F) and 15–20°C (59–68°F) at night are optimum for spear and fern growth as well as the accumulation of photosynthetic reserves in the roots (Rubatzky and Yamaguchi, 1997). Soil temperatures above 10°C (50°F) are sufficient to initiate bud sprouting in the spring. Cooler temperatures slow the rate of development, resulting in higher quality spears. Therefore asparagus is often classified as a winter-hardy, cool-season crop. There is considerable variation among cultivars, particularly with regard to cold tolerance. Some cultivars are adapted to tropical or semitropical climates and these do not perform well in short-season areas or vice versa.

Production and longevity are enhanced when plants have a dormant period, although dormancy is not required for commercial production of asparagus. When the fern is killed by frost and soil temperatures fall below 10°C (50°F), plants enter a dormant period that is characterized by low respiration. Dormancy tends to preserve the carbohydrates stored in the roots so they are available for subsequent spear production (Robb, 1984). Withholding water can partially substitute for dormancy in warmer climates. In tropical or subtropical regions with mild winters, foliage growth may be continuous. Since the plants are continuously growing and never go dormant, few storage reserves are accumulated, so alternative management practices must be used.

Cultivars

A single gene controls the sex of asparagus with maleness being dominant, although other genes influence the development of reproductive structures (Bracale et al., 1991). Plants that are homozygous for maleness are called supermales because they producer large spears and tend to be more vigorous. Male plants are preferred for commercial asparagus production and many popular cultivars are predominantly male hybrids. Supermale plants have been developed by anther culture of male plants to yield haploids that can be subsequently doubled using colchicine or by selfing the rarely occurring perfect flower on a male plant. Supermales are used as parents for hybrid crosses with female lines (Rubatzky and Yamaguchi, 1997). Clonal propagation in tissue culture is used to clone elite male and female plants for commercial-scale hybrid asparagus seed production (Murashige et al., 1972). Male plants typically have a longer life, exhibit greater disease resistance and are higher yielding. One reason why male plants are more productive is because they do not produce berries so they can store more reserves in their crowns (Robb, 1984).

Rutgers University in New Jersey, USA has produced several high-yielding predominantly male asparagus hybrid cultivars that are widely adapted to temperate regions of North America, such as 'Jersey Gem', 'Jersey Giant', 'Jersey King', 'Jersey Knight', and 'Jersey Supreme' (Cantaluppi and Precheur, 1993). Open-pollinated older cultivars containing both male and female plants that are still popular include 'Precoce D'Argenteuil' and 'Mary Washington'. 'Marte' is a predominately male hybrid adapted to white asparagus production. Cultivars adapted to the eastern USA and Canada do not perform well on the west coast of North America where winters are mild. Hybrid cultivars grown in the western USA include 'Atlas', 'Grande', 'Apollo', 'DePaoli', and 'UC 157' (Cantaluppi and Precheur, 1993).

While most asparagus cultivars are genetically diploid (2n number of chromosomes), 'Purple Passion' is a selection of the Italian cultivar 'Violeta d'Albinga'. 'Purple Passion' is an open-pollinated tetraploid (4n number of chromosomes) that is purple when harvested but turns green upon cooking due to the leaching of the water-soluble plant pigment anthocyanin. 'Purple Passion' produces larger diameter but fewer numbers of male and female spears with average sugar content 20% higher than green asparagus cultivars. Some traditional open-pollinated cultivars of asparagus grown in southern Europe include 'Altedo', 'Eros', 'Dariana', 'Larac', and 'Minerva'.

Some of the more important spear traits that define cultivars include spear diameter, tip shape, tip compactness, tip color (green, pink, or purple), uniformity, yield, sugar content, flavor, and fiber content. Important plant characteristics include disease resistance, vigor, resistance to lodging in the fern stage, season (early, midseason, or late), and harvest pattern (e.g. concentrated spear production or extended production) (Cantaluppi and Precheur, 1993).

Economic Importance and Production Statistics

Asparagus is cultivated in 60 countries around the world in climates ranging from cool temperate to tropical. According to FAO statistics (2010), an estimated 1.29 million ha (3.19 million acres) of asparagus were harvested worldwide, totaling some 7.8 million metric tonnes (8.6 million short tons). China is the largest asparagus producer at 6.9 million metric tonnes (7.6 million short tons), followed by Peru (335,000 metric tonnes (369,274 short tons)), Germany (92,000 metric tonnes (101,413 short tons)), Mexico (75,000 metric tonnes (82,673 short tons)), Thailand (63,000 metric tonnes (69,446 short tons)), Spain (50,000 metric tonnes (55,116 short tons)), Italy (44,000 metric tonnes (48,502 short tons)), and the USA (36,000 metric tonnes (39,683 short tons)) (FAOSTAT, 2010).

Peru is one of the few countries where high-quality asparagus is harvested year-round, taking advantage of different climatic zones within the country. Peru produces asparagus for two different markets: green asparagus for the USA and white asparagus for the European market. Green asparagus, which is about 45% of total production, is sent fresh to the USA packed in 5 kg (11 lb) boxes, while white asparagus is processed and exported to Europe.

Two types of asparagus are produced, white and green. White asparagus is almost exclusively consumed in the Netherlands, Spain, France, Poland, Belgium, Germany, and Switzerland. Also called "blanched asparagus", "white gold", or "edible ivory", white asparagus is preferred in some markets because it has a less bitter, delicate, mild flavor.

In the USA, where the majority of asparagus production is green, the leading production states are California, Washington, and Michigan, producing in excess of 12,000, 9,000, and 7,000 ha (30,000, 23,000, and 17,000 acres), respectively (USDA, 2012b). In recent years, production in the eastern USA has decreased while production in the western states has increased. Part of this has been caused by the greater availability of seasonal farm workers in the west since asparagus is a very labor-intensive crop.

Asparagus used to be a seasonal vegetable available primarily in the late winter, spring, and early summer. Exports from other countries such as Chile make it possible to buy fresh asparagus during the winter months in the northern hemisphere as well. Processed white asparagus is grown in countries where labor is cheaper, like China, Thailand, and Peru and exported to other countries (USDA, 2010).

Asparagus is a common and popular garden vegetable in many parts of the world because it is easy to grow and a perennial. Once established in the garden, it can produce spears for many years with little or no maintenance. Spears from escaped plants are sometimes harvested from the wild.

Nutritional Value

Asparagus is a high-value food crop, low in sodium and calories (25 kcal/100g (25 kcal/3.5 oz); Table 18.1). A serving of four spears containing only

Table 18.1. Nutritional composition of green asparagus (Rubatzky and Yamaguchi, 1997; USDA, 2012a).

Amount	100g (3.5 oz) edible portion
Water (%)	91.7
Protein (g)	2.5
Fat (g)	0.2
Carbohydrate (g)	5.0
Dietary fiber (g)	2.1
Sugars (g)	1.8
Ca (mg)	22
P (mg)	62
Fe (mg)	1
Na (mg)	2
K (mg)	278
Vitamin A (IU)	900
Thiamine (mg)	0.18
Riboflavin (mg)	0.20
Niacin (mg)	1.5
Ascorbic acid (mg)	33

10 calories. Asparagus is a good source of vitamin B6, calcium (Ca), magnesium (Mg), and zinc (Zn), and a source of dietary fiber, protein, vitamin A, vitamin C, vitamin E, vitamin K, thiamin, riboflavin, rutin, niacin, folic acid, iron (Fe), phosphorus (P), potassium (K), copper (Cu), manganese (Mn), and selenium (Se) (USDA, 2012a). Asparagus takes its name from the amino acid asparagine, which accumulates in very high levels. Asparagus is also a diuretic.

Certain compounds in asparagus spears give urine a distinctive smell when metabolized due to ammonia and various sulfur-containing degradation products, including various thiols and thioesters (White, 1975). These sulfur-containing compounds originate in the asparagus as asparagusic acid and its derivatives. The onset of the asparagus urine smell is remarkably rapid. It has been estimated to start 15–30 min after ingestion.

Green and white asparagus have different nutritional compositions. In a comparison of both types, white had a lower phenolic (bitter components) content, less vitamin C, and lower protein content, but was higher in simple sugars. Fiber content is similar between green and white asparagus spears (Makus and Gonzalez, 1991).

Makus (1994) compared mineral nutrients between green and white 'Jersey Giant' asparagus spears and found that green spears generally contained higher concentrations of mineral nutrients than white spears. Most of the measured mineral nutrients decreased from the spear tip to the base, regardless of color.

Production and Culture

Site selection

Loams, sandy loams, and sands are generally preferred for asparagus production. The ideal pH range for asparagus is between 6.7 and 7.0. Asparagus does not tolerate acid soils and will not grow well at a pH of less than 6.0. Since the disease Fusarium wilt survives better at a low pH, keeping the pH near 7.0 helps provide control (Cantaluppi and Precheur, 1993). Seedlings are sensitive to salinity but mature plants are quite tolerant. Asparagus roots may extend 1.2 m (48 in) below the surface, so asparagus plantings should be in deep, well-drained, and friable soils (Lorenz and Maynard, 1988). The field should not have been planted in asparagus for at least 4 years previously, especially if there is a history of Fusarium wilt. Well-drained soils and use of raised beds to improve drainage help control Fusarium.

Asparagus plantings are established from crowns, transplants, or direct seeding. Some commercial nurseries specialize in crown or transplant production. Direct seeding of asparagus into production fields is a less-common commercial establishment technique in many areas because seedling growth is slow and emerging seedlings do not compete well with weeds. For the production of crowns in propagation nurseries, seeds are usually sowed in sterilized or fumigated seedbeds with a linear spacing about 0.3 m (12 in) apart. The crowns are dug from the seedbed and sold to commercial growers when they are 2 years old, although 1-year-old crowns are sometimes sold as well (Cantaluppi and Precheur, 1993).

Prior to planting, 50–60 cm (20–24 in) deep trenches are made in production fields to accommodate the placement of crowns or transplants. In areas with shorter growing seasons, 2-year-old crowns are planted in furrows at least 15–25 cm (6–10 in) deep (Fig. 18.4; Cantaluppi and Precheur, 1993).

In areas with longer growing seasons, like California or the Mediterranean region, 10–15-week-old transplants are planted instead of crowns. Transplants are grown in greenhouses in plug trays and set into the field in early spring. The transplants are easier to produce and establish successfully if the growing season is sufficiently long (Takatori *et al.*, 1980).

Spacings of both crowns and transplants range from 30–91 cm (12–36 in) apart in rows, and 107–152 cm (42–60 in) between rows depending on cultivar vigor and soil conditions. After the plants or crowns are placed early in the spring, they are covered with a shallow layer of soil. The shallow trench or furrow is gradually filled in as the spears develop during the season, until the field is level again at season's end (Fig. 18.5; Takatori *et al.*, 1980).

Fig. 18.4. Planting crowns in the Rhine River valley of Germany.

Fig. 18.5. These transplants were set in a trench that will be gradually filled during the season.

No spears are harvested during the first year of planting. The spears are allowed to grow into fern to facilitate the accumulation of photosynthetic reserves in the roots for future production. The second year after planting, the field may be harvested for a short period, 3 or 4 weeks for example. The third year and subsequent years, the plot can be harvested for a full 6–8-week period. The most productive years generally occur from 4 to 6. Production generally declines as diseases accumulate in the plants or the crowns become too large and crowded. Plantings are renewed when productivity has declined to the point where the field is no longer profitable to harvest (Cantaluppi and Precheur, 1993).

The spring harvesting period in temperate regions leads to saturated markets in the spring and early summer and lower market prices for farmers. Management techniques have been developed to produce asparagus out of season when market prices are more favorable for growers.

For early-season production, low tunnels of plastic film are placed over the rows in early spring to warm the soil and stimulate or "force" spear production before the main spring harvest season occurs. These temporary tunnels are opened so that early emerging spears can be harvested and covered afterwards to raise temperatures. The tunnels are removed completely later in the season to allow fern growth after spear harvest has concluded.

Summer production is possible during the final year before the crop is renewed. Since there is no need for the crowns to accumulate reserves to produce future crops, the harvest period may be extended until the spears are no longer of marketable size.

Another technique is to use spring growth to build reserves for a fall harvest. For this system, the first spears in the spring are allowed to immediately go to fern. In late summer, irrigation water is withheld to prevent vegetative growth and substitute for winter dormancy. The dried fern is removed and irrigation is restored to stimulate fall spear production before the onset of freezing conditions.

The "mother-stalk" production system was developed for tropical and semitropical regions to overcome the lack of a winter-imposed dormancy period. This system was designed to provide continuous photosynthetic accumulation to support continuous spear production throughout the year. Four or five spears develop into fern (mother stalks) to produce photosynthates before any harvesting is done. Newly emerging spears are harvested as they emerge around the mother stalks. New spears develop into fern to take the place of senescing mother-stalks. Using this system, a longer period of spear production and higher yields can be realized, although the production costs are higher and plant longevity reduced (Rubatzky and Yamaguchi, 1997).

Fertilizer and nutrition

Fertilizer is applied in the furrow at the time of establishment in accordance with soil test results. Organic matter in the form of animal manure or composted waste can also be incorporated into the row prior to establishing crowns or transplants to improve fertility and soil health. An adequate initial application of P and K is needed for good plant establishment because later additions of particularly P are difficult to supply without injuring the roots (Cantaluppi and Precheur, 1993). A small quantity of nitrogen (N) is often applied as a liquid starter solution for both crowns and transplants to simulate early-season growth for spring planting. Later nutrient applications during the season should be applied based on tissue analysis of the fern. Nitrogen should be applied during fern development when nutrients can be better utilized for growth rather than before spear harvest when less nutrient absorption occurs (Rubatzky and Yamaguchi, 1997). During the late summer or early fall each year before the fern dies back, asparagus fields should be top-dressed with fertilizer in conjunction with soil-test results to replace nutrients removed during spear harvest and fern production.

Irrigation

Because of its extensive deep root system, asparagus is often not irrigated in areas that receive adequate spring rainfall except during establishment or extreme drought. In areas where irrigation is needed, an annual supply of 1.2 times evapotranspiration is recommended (Rubatzky and Yamaguchi, 1997). In areas with seasonal rainfall like California or the Mediterranean Region, asparagus crops are periodically irrigated with drip, furrow, or overhead irrigation during the dry season to keep plants photosynthetically active. Surface irrigation methods are generally preferred to help control foliar diseases such as rust and Botrytis and apply water more efficiently (Cornell, 2003). Withholding water is used in subtropical and tropical regions as a substitute for the dormant period that normally occurs during the winter months (Rubatzky and Yamaguchi, 1997).

Weed control and field maintenance

Since asparagus is a perennial crop, weeds tend to accumulate in fields, competing with the crop and complicating harvest if a control strategy is not implemented. Annual weeds are killed each fall by frost, but the seeds they deposit may germinate and cause problems the following season. Therefore, it is important to prevent weeds from shedding their seeds in asparagus fields. To make it easier for farm workers to spot emergent spears most commercial fields are bare at the time of harvest. Weeds are generally controlled in production fields by shallow tillage before the first spears emerge in the spring or after the ferns die in the fall when the field is cleared (Swaider et al., 1992). Care must be taken to avoid damaging crowns and subterranean buds with deep tillage. Effective pre- and post-emergence herbicides have been developed for asparagus. Asparagus seeds from female plants can germinate and become "weeds" in subsequent years. Planting cultivars of predominantly male plants can eliminate this problem (Rubatzky and Yamaguchi, 1997). If cultivars with both male and female plants are grown, removing fern with mature berries from the field before they drop may prevent volunteer asparagus plants. Generally, the fern should remain standing as long as it is green. Green fern is photosynthetically active and exports reserve materials to the crown. Once the fern has turned brown, it is no longer accumulating reserves and should be removed by chopping, disking, or burning. If rust is a problem, fern should be removed and not returned to the soil.

Harvesting and marketing

Asparagus for fresh market is largely hand-harvested and very labor intensive. The newly emerging spears are harvested by hand for a 6–8-week period each spring or when spear numbers and spear diameter are rapidly decreasing. A longer harvest period may be possible in areas with a long growing season that will enable reserve materials to be accumulated over a longer period or before a field is scheduled to be removed from production (Robb, 1984).

Asparagus spears grow very rapidly. This is because spear growth is mainly due to cell expansion and not cell division. The spear cells are all formed in the below-ground buds so growth of the emerging spear is due to water uptake until the tips begin to open and produce lateral shoots. Spears often do not grow straight and tend to bend toward the wind (Fig. 18.6). This phenomenon occurs because water is lost from the side facing the wind. This causes the side away from the wind to grow faster, causing the spear to bend toward the wind direction (Rubatzky and Yamaguchi, 1997). In areas with strong prevailing winds, windbreaks can be planted to reduce spear curvature.

When temperatures are above 10°C (50°F), spears must be harvested daily or they will become excessively tall. For hand harvesting, spears are selected when 20–30 cm (8–12 in) tall, while the tips are still very tight, with a special long-handled knife (Fig. 18.7).

Care must be taken not to damage adjacent spears that have not emerged through the soil. Harvesting aids have been developed that allow workers to sit on a moving platform and cut marketable spears by hand as the unit passes over the row. The harvested spears are placed in a container on the platform that is unloaded at the ends of the field. Selective harvesters scan the bed, identify emergent spears and mechanically cut and collect them. Nonselective mechanical harvesters cut spears of varying lengths for freezing and canning.

Fig. 18.6. Asparagus spears typically bend in the direction of the wind, which can complicate packing.

An excellent yield for hand-harvested asparagus is 0.45 kg (1 lb) of spears per plant per year. In many cases 0.34 kg (0.75 lb) is a more typical yield. Mechanically harvested asparagus yields are usually lower. In terms of area, asparagus yields range from 1.7 to 3.4 metric tonnes/ha (0.75 to 1.5 short tons/acre) (Swaider *et al.*, 1992).

White Asparagus Production

White asparagus spears do not develop chlorophyll because soil is mounded over the row to exclude light. The spears elongate underneath the soil and are harvested as they reach marketable size before emerging. Harvesters carefully dig soil around the spear as the tip emerges, cutting the base by hand with a long-handled knife (Rubatzky and Yamaguchi, 1997). Black plastic film is sometimes placed over the mounded row to exclude light from emerging tips, to keep soil from drying out, and keeping heavy rainfall from eroding the mounds of soil. The plastic film is removed long enough to harvest the spears and then is replaced (Fig. 18.8).

Rigid plastic covers can also be placed over the row to exclude light for the production of white asparagus. The covers allow spears to develop in darkness without soil being mounded over the row. Workers remove the covers to cut the spears similar

Fig. 18.7. Hand harvest is difficult work that requires bending to cut and collect spears. To ease the process, long handled knives are used to cut the spear below the soil surface.

to the harvesting of green asparagus and no digging is required (Cantaluppi and Precheur, 1993). The covers are replaced immediately after harvest. The labor costs for producing white asparagus are high. Therefore, white asparagus that is canned or frozen is often grown in countries with lower labor costs.

Postharvest Handling

Asparagus grown for fresh markets is removed from the field and taken to a packinghouse. In the packinghouse, field heat is removed and the spears are sometimes recut, cleaned, and packed in boxes for shipment. Spears have a high respiration rate so warm temperatures dramatically shorten shelf life. At high postharvest temperatures, spears rapidly elongate even after harvest, sugars are lost, and spears senesce. Exposure to ethylene hastens senescence as well. High-quality spears have high sugar content (4–5% soluble solids) and low fiber. To slow fiber formation and the loss of sugars, asparagus should be hydrocooled immediately after harvest to 2°C (36°F) (Fig. 18.9; Lipton, 1990).

When maintained at 2°C (36°F) and 95% RH asparagus can be stored for up to 10–14 days in good condition. Prolonged temperatures below 2°C (36°F) can damage spears and reduce shelf life (Lipton, 1990). Increased fiber development is directly correlated to spear length, age, and high temperatures. The portion of the spear that snaps easily often corresponds with the beginning of the most fibrous region.

Fresh-market asparagus is often marketed in bundles of uniformly sized and trimmed spears. Before shipment, spears are graded by length and diameter, bundled, and packed upright with the

Fig. 18.8. White asparagus is grown by excluding light by mounding soil over rows and covering with black plastic (a). Soil is carefully removed around the spear so that it can be cut as close to the base as possible (b).

Fig. 18.9. Asparagus is removed from the field in plastic baskets (a) and immediately hydrocooled to remove field head and, in the case of white asparagus, remove residual dirt that may adhere to the spears (b).

basal end down (Fig. 18.10). If placed horizontally, spears will exhibit negative geotropism and turn upward. The basal ends of the bundles are placed on a moist pad to keep the spears fresh. The spears will continue to elongate in transit after harvest, so headspace is provided in packaging to avoid tip damage (Lipton, 1990).

Uses

Both white and green asparagus spears are cooked in a number of different ways and typically served as a vegetable side dish or sometimes as an appetizer. Asparagus is sometimes eaten raw in salads. Fresh green asparagus consumption continues to increase relative to the production of white asparagus. It is common to peel the white spears before cooking, whereas the green spears are normally eaten unpeeled. Both white and green asparagus may be canned, pickled, frozen, or dehydrated for preservation. Whole spears, both peeled and unpeeled, are a premium product in contrast to canned cut pieces. The production of frozen asparagus has been increasing in many markets because of its superior quality. Asparagus seeds have been used as a substitute for coffee (Rubatzky and Yamaguchi, 1997).

Diseases

One of the major asparagus diseases in much of the world is caused by two species of fungi called Fusarium. *Fusarium moniliforme* causes decay of storage roots, stems, and crowns. This fungus also infects corn, grasses, and other monocotyledonous plants as well as asparagus. *Fusarium oxysporum* f. sp. *asparagi* causes root rot and seedling blight, and may also plug the xylem vessels (water-conducting tissue), causing the spears and fern to wilt (Howard *et al.*, 1994).

Both species are prolific, long-lived soil fungi capable of living on dead or decomposing organic matter in soil and feeding on decaying asparagus residues. *Fusarium* colonizes old roots and crowns, invading directly through root tips or through wounds from implements, cutting tools, or insect feeding. Asparagus plants under stress are more susceptible to infection than those that are growing vigorously (Howard *et al.*, 1994).

Affected spears may shrivel and rot in the spring before or after emergence, while infected crowns have hollow, rotted roots. The inside of the infected crown has a reddish-brown discoloration. The fern is stunted, yellow to brown in color, and has fewer stalks per crown. Plants infected with *Fusarium* decline and eventually die (Cornell, 2003).

Fusarium species are present in most agricultural soils and are difficult to avoid. Minimizing infection early in the life of crowns through careful management is one of the most important control measures. Suggested management practices to reduce *Fusarium* include the following. Practice crop rotation so asparagus is planted in fields where it has not been grown for at least 4 years. *Fusarium* builds up to extremely high levels in soil during the long periods of perennial asparagus culture and survives for many years even after the crop is removed. Soil fumigation with a suitable

Fig. 18.10. After cooling to remove field heat, asparagus is trimmed, bundled, and packed in boxes for shipment.

material will reduce infection in crown nurseries or production fields. Seeds disinfested with 1 part sodium hypochlorite bleach in 5 parts water for 2 h, then rinsed in clean water and dried before planting can prevent seed transmission. Only vigorous disease-free crowns should be planted. Minimize stress in young and established plantings. Use *Fusarium*-resistant cultivars if available (Howard et al., 1994).

Rust (*Puccinia asparagi*) is another serious widespread disease of asparagus. Rust has a complicated life cycle consisting of several stages that all occur on asparagus. There are no alternate hosts as is common with other types of rust diseases, like wheat and oats. Rust overwinters on asparagus debris. Spores germinate in the spring and are blown on to emerging spears, which causes the infection. Warm weather with heavy dew, fog, or light rainfall enhances rust development. Late in summer, black teliospores form to complete the yearly life cycle (Howard et al., 1994).

Rust develops in the tissue of asparagus fern and competes for essential nutrients. The foliage then senesces prematurely, further reducing the production of photosynthates and their storage in the crown. Successive years of rust infection will weaken asparagus crowns.

Breaking the continuity of the rust life cycle is one of the best control measures. Inoculum may overwinter in fern residue and spread from wild or neighboring cultivated asparagus. It is important to destroy wild and volunteer asparagus, clean-cut all fields during harvest, and isolate asparagus propagation nurseries and young plantings from established fields if possible (Cornell, 2003). Burning or burying may also be effective ways to destroy infected plant tissues. Use resistant asparagus cultivars such as 'Viking', which has a degree of tolerance to rust, but will become infected under favorable environment conditions and abundant inoculum. The fungicide zineb (zinc ethane-1,2-diylbis(dithiocarbamate)), if used properly, may provide limited control (Howard et al., 1994).

Botrytis (*Botrytis cinerea*) is a fungal disease that occurs during summer, causing browning of the lower fern canopy. Botrytis progresses most rapidly during hot, moist weather when the fern remains wet. This fungus attacks many other crops, such as small fruit, vegetables, and many ornamentals (Howard et al., 1994).

Botrytis begins on senescing flowers and injured fern. The spores are spread by wind and rain within the dense canopy. Individual lesions are tan with dark brown borders, often surrounded by a yellow halo. During persistent wet weather, newly emerged spears may turn brown or black and become covered with a gray, spore-bearing fungal mass. Botrytis is caused by excessively wet conditions so any management practice to improve drainage and limit wetting of the foliage will provide some control. The fungicide Zineb may also give effective control (Howard et al., 1994).

There are a several viruses that may affect asparagus. These include arabis mosaic virus (hop bare-bine), asparagus virus 1, asparagus virus 2, strawberry latent ringspot virus (latent ring spot of strawberry), and tobacco streak virus (stunt of asparagus) (Cornell, 2003).

Other diseases that may affect asparagus include: *Penicillium aurantiogriseum* (crown rot of asparagus), *Cercospora asparagi* (leaf spot: *Asparagus* spp.), *Phomopsis asparagi* (stemblight: *Asparagus* spp.), *Phytophthora cryptogea* (tomato foot rot), *Phytophthora megasperma* (root rot), *Rhizobium radiobacter* (crown gall), *Rhizobium rhizogenes* (gall), *Sclerotinia sclerotiorum* (cottony soft rot), *Pleospora tarda*, and *Stemphylium vesicarium* (onion leaf blight) (Howard et al., 1994; Rubatzky and Yamaguchi, 1997; Cornell, 2003).

Insect and Nematode Pests

There are many insect pests that attack asparagus, but two of the worst pests that are specific to asparagus are the common asparagus beetle, *Crioceris asparagi*, and the spotted asparagus beetle, *Crioceris duodecimpunctata*. Distinguishing between the two species is important because the common asparagus beetle is the more damaging of the two (Delahaut, 2000).

Sheltered locations such as loose tree bark or in the hollow stems of old asparagus are locations were common asparagus beetle adults overwinter. The adults appear in fields just as the asparagus spears are emerging in the spring. The beetles may lay dark brown, oval-shaped eggs in rows on the spears. The eggs hatch within 1 week. The light gray, slug-like larvae have black heads and legs and migrate to the fern where they begin feeding. The larvae feed for about 2 weeks and then fall to the ground to pupate in the soil. About 1 week later, adults emerge to start another generation. The common asparagus beetle adult is about 6 mm (0.25 in) long, bluish-black colored with prominent cream-colored spots on its back (Howard et al., 1994).

The spotted asparagus beetle has a similar life cycle but usually appears in fields later in the season than the common asparagus beetle. The spotted asparagus beetle adult is reddish-orange with six black spots on each wing and is 6 mm (0.25 in) long. The spotted asparagus beetle lays greenish eggs on the ferns. The orange larvae typically feed on the asparagus berries (Howard *et al.*, 1994).

Adult asparagus-beetle feeding can cause browning and scarring on the spears, which often makes them unmarketable. Eggs of the common asparagus beetle make the spears unappealing to the consumer. The common asparagus beetle larvae and adults can also devour large quantities of fern later in the season, weakening the plant and making it more susceptible to invasion by *Fusarium*. Serious defoliation can also impair the accumulation of storage reserves for the following season. In contrast, when the spotted asparagus beetle larvae feeds on berries, it does not do much long-term damage to the plant (Howard *et al.*, 1994).

The best cultural control is to destroy crop residues to eliminate overwintering sites. Biological controls include release of the metallic green wasp *Tetrastichus asparagi*, which parasitizes 70% of asparagus beetle eggs. Lady beetle larvae and other predators will consume both eggs and larvae. Most insecticides that may control the asparagus beetle will also kill beneficial predators, parasites, and possibly bees if used during the morning hours when bees are most active (Cornell, 2003).

Other insect pests that attack asparagus include: *Adelphocoris lineolatus* (lucerne bug), *Agrotis ipsilon* (black cutworm), *Agrotis segetum* (turnip moth), *Frankliniella intonsa* (thrips, flower), *Longidorus* (longidorids), *Loxostege sticticalis* (beet webworm), *Myzus persicae* (green peach aphid), *Parasaissetia nigra* (pomegranate scale), *Pinnaspis strachani* (lesser snow scale), *Popillia japonica* (Japanese beetle), *Saissetia oleae* (olive scale), *Scirtothrips dorsalis* (chili thrips), and *Xiphinema* (dagger nematode) (Howard *et al.*, 1994; Rubatzky and Yamaguchi, 1997).

References

Anonymous (2012) Poisonous asparagus. Available at: www.asparagus-friends.com/asparagus-knowledge/poisonous-asparagus (accessed 18 July 2012).

Blasberg, C.H. (1932) Phases of the anatomy of *Asparagus officinalis*. *Botanical Gazette* 94, 204–214.

Bracale, M., Caporalia, E., Gallia, M.G., Longoa, C., Marziani-Longoa, G., Rossia, G., Spadaa, A., Soavea, C., Falavignab, A., Raffaldib, F., Maestric, E., Restivoc, F.M. and Tassic, F. (1991) Sex determination and differentiation in *Asparagus officinalis* L. *Plant Science* 80, 67–77.

Cantaluppi, C. and Precheur, R. (1993) *Asparagus Production, Management, and Marketing*. Publication 826. The Ohio State University Cooperative Extension Service, Columbus, Ohio.

Chase, M.W., Reveal, J.L. and Fay, M.F. (2009) A subfamilial classification for the expanded Asparagalean families Amaryllidaceae, Asparagaceae and Xanthorrhoeaceae. *Botanical Journal of the Linnean Society* 161(2), 132–136.

Cornell University (2003) Vegetable disease ID and management. Available at: http://vegetablemdonline.ppath.cornell.edu (accessed 19 July 2012).

Delahaut, K. (2000) University of Wisconsin extension. University of Wisconsin garden facts asparagus beetle. Available at: http://learningstore.uwex.edu/Assets/pdfs/A3760-E.pdf (accessed 19 July 2012).

FAOSTAT (2010) Food and agriculture organization of the United Nations. Available at: faostat.fao.org/site/567/DesktopDefault.aspx?PageID=567#ancor (accessed 20 July 2012).

Howard, R.J., Garland, J.A. and Seaman, W.L. (1994) *Diseases and Pests of Vegetable Crops in Canada: An Illustrated Compendium*. Entomological Society of Canada and Canadian Phytopathological Society, Ottawa, Canada.

Ilott, C. (1901) *The Book of Asparagus*, 1st edn. John Land, London.

Lipton, W.J. (1990) Postharvest biology of fresh asparagus. *Horticultural Reviews* 12, 69–155.

Lorenz, O.A. and Maynard, D.N. (1988) *Knott's Handbook for Vegetable Growers*, 3rd edn. Wiley-Interscience, New York.

Makus, D.J. (1994) Mineral nutrient composition of green and white asparagus spears. *HortScience* 29, 1468–1469.

Makus, D.J. and Gonzalez, A.R. (1991) Production and quality of white asparagus grown under opaque rowcovers. *HortScience* 26, 374–377.

Murashige, T., Shabde, M.N., Hasegawa, P.M., Takatori, F.H. and Jones, J.B. (1972) Propagation of asparagus through shoot apex culture. I. Nutrient medium for formation of plantlets. *Proceedings of the American Society for Horticultural Science* 97, 158–161.

Robb, A.R. (1984) Physiology of asparagus (*Asparagus officinalis*) as related to the production of the crop. *New Zealand Journal of Experimental Agriculture* 12, 251–260.

Rubatzky, V.E. and Yamaguchi, M. (1997) *World Vegetables: Principles, Production, and Nutritive Values*. Chapman & Hall, New York.

Swaider, J.M., Ware, G.W. and McCollum, J.P. (1992) *Producing Vegetable Crops*. Interstate Publishers, Illinois, 626 pp.

Takatori, F.H., Souther, F.D., Sims, W.L. and Benson, B. (1980) *Establishing the Commercial Asparagus Plantation*. University of California, Division of Agriculture and Natural Resources UC Cooperative Extension Service, Berkeley, California.

Tiedjens, V.A. (1926) Some observations on root and crown bud formation in *Asparagus officinalis*. *Proceedings of the American Society for Horticultural Science* 23, 189–196.

USDA (2010) World asparagus: Import value, 1961–2007. Economics, statistics and market information system. Available at: http://usda.mannlib.cornell.edu/MannUsda/viewDocumentInfo.do?documentID=1771 (accessed 18 July 2012).

USDA (2012a) Nutrient data library. Asparagus raw. National nutrient database for standard reference, release 24. Available at: http://ndb.nal.usda.gov/ndb/foods/show/2892 (accessed 18 July 2012).

USDA (2012b) National agricultural statistics service, vegetables 2011 summary. Available at: www.usda01.library.cornell.edu/01-26-012.pdf (accessed 18 July 2012).

Vaughan, J.G. and Geissler, C. (2009) *Stem, inflorescence, and bulb vegetables*. In: *The New Oxford Book of Food Plants*, 2nd edn. Oxford University Press, Oxford, UK.

WCSP (2011) World checklist of selected plant families, '*Asparagus aethiopicus*.' Royal Botanic Gardens, Kew, UK. Available at: www.kew.org (accessed 12 July 2012).

White, R.H. (1975) Occurrence of S-methyl thioesters in urines of humans after they have eaten asparagus. *Science* 189, 810–811.

19 Family Polygonaceae

RHUBARB

Origin and History

Rhubarb has an ancient history. Its roots were used medicinally as a laxative in cool regions of Asia by the Chinese 4,500 years ago (Grubben, 2004). Traders introduced rhubarb to Europe through Italy from the east in about 1608 (Thompson and Kelly, 1957). However, rhubarb did not become an important food crop until the 18th century in Great Britain (Grubben, 2004). The use of rhubarb as food is a relatively recent innovation and coincided with the availability of affordable sugar to common people. In 1815 it was accidently discovered that rhubarb could be "forced" to produce petioles during the winter when warm soil was placed over a quiescent plant at a construction site during the winter months. Forced rhubarb became very popular as a fruit substitute because it provided a colorful, fruity-tasting vegetable in the winter when fresh fruits and vegetables were otherwise not available. The appreciation of rhubarb as a spring and summer garden vegetable also grew during the 1800s. Rhubarb was introduced to the USA, most likely from Italy in the late 1700s, and by 1806 it was widely grown in New England (Thompson and Kelly, 1957). Rhubarb remained a popular vegetable in the USA until World War II (Foust and Marshall, 1991).

Production of staple crops was a priority over rhubarb during the war, causing a decline in consumption because of shortages of labor, energy, and sugar. After the war, the US rhubarb industry recovered for a few years, but went into a long steady decline because of the greater availability of better-tasting fruits and vegetables that were grown in southern climates during the winter and shipped by truck to northern cities on the newly completed Interstate Highway System in the 1950s and 1960s (Foust and Marshall, 1991).

Rhubarb serves as an example of how consumer preferences for vegetables may change with time. Today in the USA, rhubarb is a minor commercial vegetable that is unknown to a large percentage of the population, particularly younger Americans. Much of the fresh rhubarb that is produced is sold to retirees in Florida and other southern states who nostalgically remember it from the 1950s and 1960s (Foust and Marshall, 1991). Rhubarb is still grown as a home garden vegetable in the northern USA because of its extreme winter hardiness, ease of production, and early maturity in short-season areas.

Rhubarb remains a popular and important commercial vegetable crop in other parts of the world. Although production has declined somewhat, rhubarb is still an important commercial crop in Canada and parts of northern Europe (Foust and Marshall, 1991).

Botany

Rhubarb is an herbaceous, perennial dicot from the Polygonaceae or buckwheat family. Rhubarb is grown for its large, thick leafstalks or petioles that are consumed as food (Marshall, 1988) (Fig. 19.1).

New growth arises from the thickened semi-woody stem or crown that consists of fleshy rhizomes, buds, and storage roots. In addition to the thickened storage roots, an extensive fibrous root system also develops.

There is some debate over the correct species name for cultivated rhubarb. The genus *Rheum* consists of approximately 50 species, both wild and domesticated. Some "wild" species are occasionally cultivated in some parts of the world. *Rheum rhabarbarum* L. is often used to describe cultivated rhubarb, while *R. undulatum* and *R. rhaponticum* appear in some older literature as well. Commercial cultivars are increasingly designated as *Rheum* × *hybridum* Murray to reflect that they were derived from interspecific hybridization (Grubben, 2004). Some other species include common garden rhubarb, *R.* × *cultorum*, ornamental rhubarb, *R. acuminatum* and *R. alexandrae*, and wild rhubarb, *R. alpinum* (Marshall, 1988).

Fig. 19.1. The enlarged petioles develop from a central crown of the rhubarb plant. Only the petioles are used for food.

The overwintering structure in mature plants is a crown that develops several inches below the ground and consists of thickened roots. Rhubarb leaves are large (20–50 cm × 15–50 cm) (8–20 in × 6–20 in), dark green, and either ovate or cordate. The apex is obtusely rounded and palmately three- to seven-veined with pubescence on the lower veins with long, fleshy petioles. Cauline leaves are gradually shorter and narrower (CABI, 2008).

Rhubarb produces small flowers grouped in large compound leafy greenish-white to rose-red panicles that rise up to 1.5 m (58 in) above the crown. The flowers are bisexual, greenish-white with tepals in two whorls of three each, nine stamens, and three styles. The fruit is an ovoid achene usually more than 1 cm (0.4 in) long (CABI, 2008).

Freezing temperatures kill rhubarb leaves and cause the plants to go dormant during the winter. However, the rhubarb crown is extremely winter-hardy and can tolerate temperatures as low as −40°C (−40°F) and very dry periods during portions of the year (Thompson and Kelly, 1957).

Types and Cultivars

Petiole color is associated with rhubarb quality and in most markets the order of preference is red, pink, and green. The red-crimson color results from the presence of anthocyanin pigment in the petioles or stalks and varies according to both the rhubarb cultivar and production technique. The green-stalked rhubarb cultivars contain chlorophyll and are higher yielding but less popular (Marshall, 1988). High-quality rhubarb has a stalk diameter of at least 2.5 cm (1 in) at the base and at least 36 cm (14 in) in length. Fresh uncooked stalks have a crisp texture similar to celery and a sharp, tart taste. Although all rhubarb tends to be tart, there is variation among cultivars with less tartness being desirable. Productive cultivars should produce at least six to ten marketable stalks per season. Absence of seed-stalk production is another characteristic that defines rhubarb cultivars. Little or no seed-stalk production is a desirable trait because seed stalks interfere with harvesting and inhibit production of leaves and petioles (Zandstra and Marshall, 1982).

Some important cultivars include: 'Canada Red', 'Crimson Red', 'MacDonald', 'Strawberry', 'German Wine', 'Valentine', 'Sutton', 'Victoria', 'Red Cherry', 'Riverside Giant' (green), and 'Giant Cherry', the leading cultivar in California (Zandstra and Marshall, 1982; Schrader, 2000). 'Victoria' is versatile and suitable for forcing or annual production and propagation from seed. Most of these cultivars do not grow true-to-type from seed. 'Hawke's Champagne', 'German Wine', 'Sutton', and 'Timperley' are popular cultivars suitable for forcing (Anonymous, 2004; CABI, 2008). 'Victoria', 'McDonald', and 'Strawberry' produce more seed stalks than other cultivars (Zandstra and Marshall, 1982).

Production Statistics

Rhubarb is considered a minor crop in the USA and the world, so the USDA or the FAO do not record production statistics. The cultivation of vegetable rhubarb has mostly spread in the northern hemisphere, especially in west and central Europe, the USA, Scandinavian countries, Canada, Russia, Japan, and Zimbabwe (Grubben, 2004; CABI, 2008).

In Southeast Asia, rhubarb is cultivated as a vegetable in the cool mountainous regions of Java in Indonesia, Cameron Highlands in Malaysia, and around Baguio in the Philippines. It is also grown to a limited extent in the mountains of central and East Africa, India, and the West Indies (CABI, 2008).

The following data illustrate the decline in US commercial production. In 1940, there were an estimated 2,900 rhubarb growers in the USA who produced over 5,000 acres of rhubarb (Foust and Marshall, 1991). In 2010, there were an estimated 60 commercial growers producing less than 405 ha (1,000 acres). Leading rhubarb-producing states in the USA include Washington, Oregon, Michigan, and California. In Canada, British Columbia, Ontario, and Nova Scotia have significant production as well.

Forced rhubarb production in the UK is concentrated around the noted "Rhubarb Triangle" of Wakefield, Leeds, and Morley (Wakefield, 2012). However, production in this region has also diminished since its peak in the 1930s when 200 growers provided 90% of the world's forced rhubarb. In 2010 it is estimated that only a dozen or so growers remain in the Rhubarb Triangle, which now covers a much-reduced production area. An estimated 80 ha (175 acres) of rhubarb are forced commercially in Washington, Oregon, and Michigan, USA (Schrader, 2000).

Nutritional Value

Rhubarb is not a highly nutritious vegetable; it consists of 93% water and significant amounts of potassium (K) and dietary fiber. Rhubarb contains very little protein, fat, carbohydrates, and less provitamin A, vitamin B, or vitamin C than many other vegetables (Table 19.1; Rubatzky and Yamaguchi, 1997).

Rhubarb leaves contain oxalic acid and should not be eaten. While the oxalic acid content of rhubarb leaves can vary, a typical value is about 0.5% (Smolinske et al., 2007). This means that 5 kg (11 lbs) of the extremely sour leaves would have to be consumed to cause death (LD_{50}). The leaves are believed to contain another toxin, anthraquinone glycoside (also known as senna glycosides) (Cooper and Johnson, 1984).

The petiole juice has a sharp acidic flavor. The oxalic acid in the petioles is much lower, only about 2–2.5% of the total acidity, which is dominated by malic acid (McGee, 2004). This means the stalks are not hazardous to eat, but have a very tart taste. The pH of rhubarb juice has been measured to be 3.2, so large quantities of sugar are often used to prepare rhubarb dishes.

Production and Culture

Perennial rhubarb

Rhubarb can be grown on a wide range of soil types, but does best on deep, well-drained fertile loams, sandy loams, and silt loams. Rhubarb will tolerate pH as low as 5.0 but generally the best growth is obtained between pH 6.0 and 6.8. Rhubarb should be planted as early in the spring as the soil permits. The plants can be propagated from crown divisions, buds, or by seed (Fig. 19.2). Buds at the outer portion of the central stem form new shoots. Production from seed is not usually true-to-type, although some cultivars are reportedly better than others.

Table 19.1. Nutritional composition of rhubarb petioles.

Nutrient	Amount/100 g (3.5 oz) edible portion
Water (%)	93.3
Energy (Kcal)	18
Protein (g)	0.74
Fat (g)	0.13
Carbohydrate (g)	3.8
Fiber (g)	0.75
Ca (mg)	130
P (mg)	21
Fe (mg)	0.9
Na (mg)	6
K (mg)	360
Vitamin A (IU)	100
Thiamine (mg)	0.03
Riboflavin (mg)	0.04
Niacin (mg)	0.3
Ascorbic acid (mg)	10

Fig. 19.2. Most cultivars of rhubarb do not produce offspring like their parents from seed and are propagated vegetatively by planting crowns.

Crowns or transplants are covered with at least 5 cm (2 in) of soil at a spacing of 1.2 × 1.2 m (4 ft × 4 ft) or 0.6 × 1.8 m (2 ft × 6 ft) in trenches 6 in (15 cm) deep (8,892 plants/ha; 3,600 plants/acre).

For mechanical harvesting, plants are spaced 46 cm (18 in) apart in rows 1.2 m (4 ft) apart (17,784 plants/ha; 7,200 plants/acre). After 2–3 years, rhubarb plants can become up to 1.2 m (4 ft) in diameter and up to 0.9 m (3 ft) tall, so wide spacing is recommended to reduce crowding (Schrader, 2000).

After planting, no harvest should occur during the first year to allow crown growth and development. The first shoots that develop the second season can be harvested. However, a full harvest should be delayed until year three, particularly in short-season areas. Restricting the harvest period allows reserves to be stored in the crowns for subsequent years. Harvesting in late summer and fall is only recommended if the field is to be renewed at the end of the season. Fields can remain in production for 10 years or more if properly maintained (Thompson and Kelly, 1957). Eventually, plants will become crowded because of crown growth. When productivity declines and/or stalks become hollow and small the planting may need to be replaced. Seed stalks use crown reserves, decreasing productivity, and should be removed as soon as they are visible. The plants can be dug, crowns divided, trimmed and replanted, but regenerating commercial plots with newly propagated disease-free plants in a new field to practice crop rotation is suggested to control disease. Tissue culture is used to mass-propagate elite disease-free cultivars in some areas.

Fertilizer is applied in the furrow at the time of establishment in accordance with soil test results. Rhubarb responds well to organic matter in the form of animal manure or compost that can be incorporated into the row prior to establishing crowns. Rhubarb is a heavy feeder and requires adequate fertilization to maximize productivity. An annual fertilization program often includes a split side-dress application both before and after harvest in conjunction with soil test results. Typically 45–68 kg (100–150 lb) of N-P-K per year is needed to maintain productivity in a mineral soil (Thompson and Kelly, 1957). Rhubarb is sometimes mulched with straw or other clean organic residue to keep petioles clean and conserve moisture, and in the winter months to protect the crowns.

Rhubarb is generally one of the first vegetables to produce a crop in spring. Fresh rhubarb grown outdoors is mostly available in late winter through spring depending on the climate (Marshall, 1988). In temperate climates, plants enter a dormant period with the onset of cold weather in the fall as the leaves die back to the ground. Sprouting will not occur in the late winter or spring until a chilling requirement has been met. There is conflicting information concerning the chilling requirement necessary to induce dormant crowns to produce new stalks. Some reports state that exposure to temperatures below 4.4°C (40°F) is required to break dormancy (Helsel *et al.*, 1981; Anonymous, 2004). Other accounts say that cold units are accumulated between 9°C and –2°C (49°F and 28°F) and the cultivars 'Victoria' and 'German Wine' require the accumulation of 470–500 cold units before petiole growth occurs in dormant crowns (Anonymous, 2004). The amount of chilling needed to break dormancy varies among cultivars.

Under tropical conditions, dormancy does not usually occur (Grubben, 2004). Dormancy is induced by very short days (<10 h) rather than by low temperatures. Increasing daylength, combined with higher temperatures and irrigation after drought, will break the dormancy. Under daylengths greater >10 h and favorable conditions, plants normally grow fast without dormancy under tropical conditions. Plants remain vegetative and continuously produce new leaves. Rhubarb grown in the tropics can be harvested for many years, but needs a period of rest after each production season (Grubben, 2004).

Rhubarb can be grown under a wide range of conditions and is extremely cold and drought tolerant. Temperatures that average 23.9°C (75°F) during the day with minimum temperatures above 10–12.8°C (50–55°F) are optimum for stalk development. Clear plastic covers are sometimes placed atop the rows to raise temperatures and stimulate early-season growth. Sufficient room should be allowed between the soil and plastic to accommodate stalk enlargement. Plastic covers are removed prior to harvest.

Weeds may cause significant yield loss and should be removed, preferably before planting. Shallow cultivation can remove weeds when rhubarb plants are dormant. During the growing season, weeds can be removed by cultivation or herbicides. Maintenance of a rhubarb field is a multi-faceted and labor-intensive process.

Annual rhubarb

Rhubarb may be grown as an annual in subtropical and tropical climates that lack sufficient chilling for perennial production as long as there is a period of cool weather. Most cultivars are sensitive to high temperatures and this results in production of spindly, weak, sometimes hollow petioles with poor color. At higher temperatures >32°C (90°F) plants will cease growth, the foliage may die and the crowns become quiescent until cooler conditions return. An alternative for warmer climates is to grow rhubarb as an annual from seed. Rhubarb seeds can be planted in transplant trays 0.5 cm (0.25 in) deep and protected from intense direct sunlight during early growth.

Transplants can be established in the field when 8–10 cm (3–4 in) tall after acclimation to outdoor temperatures and light intensities. Most rhubarb seeds do not grow true-to-type, with the exception of 'Victoria' and a few others, so seedlings may be variable with different stalk colors. Plants rarely flower at high temperatures in combination with short days, but flower heads that do develop should be removed. The developing seed stalk competes with the crown for resources, reducing petiole size and yield. Annual rhubarb produced at higher temperatures tends to develop smaller stalks but with intense red color. The quality of annual rhubarb is generally inferior to perennial crops. The stalks are harvested as they mature. Florida growers report good results planting seed in August and harvesting stalks in March through May. 'Victoria' is reportedly well adapted to annual production (Maynard, 1990).

Forcing rhubarb

The rhubarb crowns are grown without harvest in a field for 2 years prior to forcing to build up storage reserves in the crown. In the autumn after the plants have gone dormant and had their chilling requirement satisfied so they can produce new stalks, the crowns are dug from the field, briefly washed to clean off soil, and closely packed on the surface of soil in special forcing buildings with no windows in darkness at a temperature of 55°F (13°C) (Fig. 19.3). Alternatively, crowns can be stored in a cold room after digging if their chilling requirement has not been satisfied, otherwise it is more efficient to leave the plants in the field. Gibberellic acid can substitute for part of the cold treatment needed to break crown dormancy by injection directly into the crown (Anonymous, 2004). The crowns are covered with a thin layer of organic mulch and watered periodically to prevent desiccation. No light is needed and some believe that the quality is superior when the stalks are forced in total darkness. After 3–4 weeks the petioles will

Fig. 19.3. Rhubarb crowns can be forced to produce petioles indoors during off-season periods when outdoor production is not possible.

begin to grow from the crowns (Fig. 19.3). The nutrients stored in the crowns provide the resources for petiole growth and development.

The forced stalks are hand-harvested using artificial light carried by the workers. The exhausted rootstock is discarded at the end of the harvest, which takes place between late December and March. The best-quality forced rhubarb is 40–50 cm (16–20 in) long and about 2 cm (1 in) thick. The skin of the forced stalks ranges from salmon pink to blood red depending on the cultivar (Fig. 19.3). The inside of the stalk is almost white. The long production cycle and higher costs require a premium price be charged for forced rhubarb. In modern times the greater availability of a wide range of fresh fruits and vegetables supplied from other areas has reduced consumer interest in forced rhubarb and led to a decline in the industry. 'Victoria' and 'German Wine' are well suited to forcing (Anonymous, 2004).

Irrigation

Rhubarb is well adapted to high rainfall as long as drainage is good because crowns cannot tolerate waterlogged soils. Rhubarb should receive about 2.5 cm (1 in)/week of water throughout the production season to prevent moisture stress that could decrease yield and quality. Less water can be tolerated at other times of the year. Rhubarb has an extensive root system and since much of its active growth is in the early spring when conditions are generally wet, rhubarb is often not irrigated in the eastern USA and Europe. In California, furrow irrigation or other systems can be withheld after the spring crop during the dry months to cause the plants to become quiescent. Irrigation can be reapplied during late summer and fall to stimulate crown growth for a second late-season fall crop (Schrader, 2000).

Harvesting

Stalks are harvested as they reach mature length. By firmly pulling up on the base of the stalks by hand, clean separation from the rhizome occurs without breaking the petiole or using a knife. Plants should not be over-harvested. At the end of the harvest period, which may last 8–10 weeks, new shoots emerge but are not harvested (Schrader, 2000). Harvesting should last for only 4–6 weeks in short-season areas. The new leaves that develop will produce photosynthates that are translocated to the roots and stored in the crown to produce the next year's crop. In most operations, all developing stalks are picked in early spring until the end of the harvest period when all emergent leaves are allowed to fully develop to replenish the reserves used by the crown. However, an alternative is to let one or two "mother" stalks develop to provide photosynthates to help replenish reserves while the other emerging stalks are harvested. Leaving some of the early stalks to develop as mature leaves allows harvest to continue over a longer period. Harvesting stalks around developed leaves slows the picking process and can be tedious for workers. For processing, both ends of the petiole are trimmed so that no leaf tissue remains. For fresh market, 6 mm (0.25 in) of leaf tissue is sometimes left attached to the petiole and the basal end is not trimmed. In long season areas, a second crop may be harvested in the fall if sufficient reserves can be stored after the winter/spring harvest. Weak crowns with insufficient storage reserves often produce pithy stalks. Stalks that are frozen or limp should not be sold. Freezing increases oxalic acid levels in the stalks due to migration from the leaf blades (Schrader, 2000). Rhubarb grown for processing may be mechanically harvested in some areas although this is not yet a standard practice. For mechanical harvesting, developed petioles are all cut simultaneously as the harvester passes over the field (Marshall, 1986).

Yield

Very good commercial yields for rhubarb are approximately 1.5–3 kg (3.3–6.6 lb) of petioles per plant per year and between 34 and 40 metric tonnes/ha/year (15–18 short tons/acre/year). Average commercial yields are 22 metric tonnes/ha (10 short tons/acre) (Zandstra and Marshall, 1982). Red cultivars are the most commercially important but yield as much as 50% less than green ones. Perennial fields may last as long as 15 years, but yields usually peak between year 4 and 6, then gradually decline due to crowding and pest damage (Zandstra and Marshall, 1982).

Postharvest Handling

Rhubarb is commonly packaged in 9 kg (20 lb) cartons or in 0.5 kg (1 lb) plastic bags in cartons of ten bags each. Stalks must be rapidly hydrocooled or forced-air cooled to remove field heat and prevent wilting. The temperature of the stalks should reach 0°C or 0.6°C (32°F or 33°F) within 1 day of harvest. Rhubarb stalks can be stored for 2–4 weeks at 0°C (32°F) and 95–100% RH (McGregor, 1987).

Uses

The fleshy petioles or stalks of rhubarb are used as a vegetable. The stalks can be stewed, made into pies, tarts, sauces, wine, or jams and jellies. Rhubarb is primarily served as a side dish or dessert in contrast to most other vegetables. Rhubarb has a tart flavor, so quantities of sugar must be added to remove the sourness. In the USA, fresh rhubarb stalks are increasingly difficult to find in grocery stores because of its declining popularity as a vegetable and seasonal nature. However, many groceries still offer frozen rhubarb that has been processed by being chopped into pieces about 2.5 cm (1 in) in length and sold in plastic bags (Fig. 19.4).

The medicinal use of rhubarb pre-dates its use as a food crop. Rhubarb roots have long been used in traditional Chinese medicine. The rhizomes are rich in anthraquinones, such as emodin and rhein, which are laxatives. Rhubarb rhizomes have been used as a dieting aid because of its cathartic properties. The rhizomes contain stilbenoid compounds, including rhaponticin, which lower blood glucose levels in diabetic laboratory mice (Chen et al., 2009).

Pests

Rhubarb is relatively disease and insect free, so insecticides and fungicides are not as frequently used as in many other crops. However, there are a large number of pests and diseases that may affect rhubarb

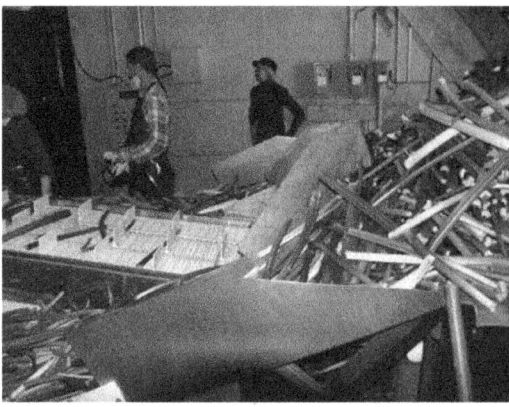

Fig. 19.4. Fresh rhubarb petioles are processed by washing, chopping and flash freezing.

(Howard et al., 1994; CABI, 2008). The following is a partial listing of pests that attack rhubarb.

Rhubarb curculio (*Lixus concavus*) is a large, rusty-colored snout beetle approximately 1.9 cm (0.75 in) long that causes only minor damage by puncturing the stalk. The rhubarb curculio lays its eggs in the stems of wild dock plants. Elimination of these weeds near the rhubarb in midsummer, after the eggs are laid, will help control curculios. The rhubarb stalk borer (*Papaipema nebris*) overwinters in the egg stage on grassy weeds, so eliminating these weeds near rhubarb will control stalk borer. The potato stem borer (*Hydraecia micacea*) may become a rhubarb pest during midsummer. Stem borer caterpillars are about 8.9 cm (3.5 in) long, pinkish-white in color and bore into the rhubarb stalks. The adult moth lays its eggs on the stems of grasses in August that hatch the following spring. Controlling couch grass and other weeds in and around rhubarb fields will help control stem borers. The adult tarnished plant bug (*Lygus lineolaris*) is light reddish-brown and about 5 mm (0.2 in) long. It primarily attacks new rhubarb plantings by feeding on young leaves. Keep areas directly adjacent to the field free of weeds and avoid planting rhubarb beside legumes to control this pest. *Korscheltellus lupulina* (swift moth), *Pantomorus cervinus* (Fuller's rose beetle), *Popillia japonica* (Japanese beetle), *Pratylenchus penetrans* (nematode, northern root lesion), *Pseudococcus calceolariae* (scarlet mealybug), *Chrysodeixis eriosoma* (green looper caterpillar), *Hydraecia micacea* (potato skin borer), *Heterodera schachtii* (beet cyst eelworm), and slugs (*Arion rufus*) have all been reported as pests on rhubarb as well (Howard et al., 1994).

A number of diseases may cause postharvest losses of rhubarb. Anthracnose (*Colletotrichum erumpens*) causes oval, soft, watery lesions on petioles. Bacterial soft rot, commonly caused by species of gram-negative bacteria such as *Erwinia*, *Pectobacterium*, and *Pseudomonas*, produces a soft, slimy decay of stocks in storage. Gray mold (*Botrytis cinerea*) causes soft, brown lesions on petioles. Postharvest decay is usually traced to poor sanitation of hydrocooling water, so proper sanitation with recommended storage temperature is essential to avoid infection (Snowdon, 1992). Rhubarb may also be susceptible to the following diseases: *Peronospora destructor* (downy mildew), *Cercospora* spp. (Cercospora leaf-spot), *Erwinia rhapontici* (rhubarb crown rot), *Phytophthora cactorum* (apple collar rot), *Pseudomonas marginalis* pv. *marginalis* (lettuce marginal leaf blight), *Rhizobium radiobacter* (crown gall), *Rhizobium rhizogenes* (gall), *Sclerotinia sclerotiorum* (cottony soft rot), Arabis mosaic virus (hop bare-bine), strawberry latent ringspot virus, turnip mosaic virus (TuMV), and cabbage A virus mosaic (CABI, 2008).

References

Anonymous (2004) Rhubarb, *Rheum rhabarbarum*, in commercial vegetable production guides. Oregon State University College of Agricultural Sciences. Available at: http://nwrec.hort.oregonstate.edu/rhubarb.html (accessed 8 July 2012).

CABI (2008) Crop protection compendium. Datasheets: *Rheum hybridum* (rhubarb). Available at: www.cabi.org/cpc/?compid=1&dsid=47109&loadmodule=datasheet&page=868&site=161 (accessed 12 July 2012).

Chen, J., Ma, M., Lu, Y., Wang, L., Wu, C. and Duan, H. (2009) Rhaponticin from rhubarb rhizomes alleviates liver steatosis and improves blood glucose and lipid profiles in KK/Ay diabetic mice. *Planta Medica* 75, 472–477.

Cooper, M.R. and Johnson, A.W. (1984) *Poisonous Plants in Britain and their Effects on Animals and Man*. Her Majesty's Stationery Office, London.

Foust, C.M. and Marshall, D.E. (1991) Culinary rhubarb production in North America: History and recent statistics. *HortScience* 26, 1360–1363.

Grubben, G.J.H. (2004) *Rheum xhybridum* Murray. Available at: http://database.prota.org/search.htm (accessed 7 July 2012).

Helsel, D., Marshall, D. and Zandstra, B. (1981) *Rhubarb, cultural practices for Michigan*. Extension Bulletin E-1577. Michigan State Cooperative Extension Service, Michigan State University, East Lansing, Michigan.

Howard, R.J., Garland, J.A. and Seaman, W.L. (1994) *Diseases and Pests of Vegetable Crops in Canada:*

An Illustrated Compendium. Entomological Society of Canada and Canadian Phytopathological Society, Ottawa, Canada.

Marshall, D.E. (1986) Design and performance of a mechanical harvester for field grown rhubarb. *Transactions of the American Society of Agricultural Engineers* 29, 652–655.

Marshall, D.E. (1988) *A Bibliography of Rhubarb and Rheum species*. United States Department of Agriculture, National Agricultural Library and Agricultural Research Service Bibliographies and Literature of Agriculture No 62, Beltsville, Maryland.

Maynard, D.N. (1990) Annual rhubarb production in Florida. *Proceedings of the Florida State Horticultural Society* 103, 343–346.

McGee, H. (2004) *On Food and Cooking: The Science and Lore of the Kitchen*. Scribner, New York.

McGregor, B.M. (1987) *Tropical products handbook*. USDA Agric. Handbook No. 668, 158 pp.

Rubatzky, V.E. and Yamaguchi, M. (1997) *World Vegetables: Principles, Production and Nutritive Values*, 2nd edn. Chapman & Hall, New York.

Schrader, W.L. (2000) *Rhubarb Production in California, Agricultural and Natural Resources Publication 8020*. University of California, Division of Agriculture and Natural Resource, Communication Services, Oakland, California.

Smolinske, S.C., Daubert, G.P. and Spoerke, D.G. (2007) Poisonous plants. In: Shannon, M.W., Borron, S.W. and Burns, M.J. (eds) *Haddad and Winchester's Clinical Management of Poisoning and Drug Overdose*, 4th edn. Saunders Elsevier, Philadelphia, Pennsylvania.

Snowdon, A.L. (1992) *Color Atlas of Postharvest Diseases and Disorders of Fruits and Vegetables*, Vol. 2: *Vegetables*. CRC Press, Boca Raton, Florida.

Thompson, H. and Kelly, W. (1957) *Vegetable Crops*, 5th edn. McGraw Hill Book Co., New York, 611 pp.

Wakefield Council (2012) Rhubarb - welcome to the rhubarb triangle. Available at: www.wakefield.gov.uk/CultureAndLeisure/HistoricWakefield/Rhubarb/default.htm (accessed 7 July 2012).

Zandstra, B.H. and Marshall, D.E. (1982) A grower's guide to rhubarb production. *American Vegetable Grower* December 1982, 6–10.

20 Family Fabaceae

BEANS AND PEAS

Origin and History

Fabaceae, also known as the legume, pea, or bean family, is large and economically important. Fabaceae is the third-largest plant family, behind only the Orchidaceae and Asteraceae, with 730 genera and over 19,400 species (Stevens, 2012). The family was known as Leguminosae for many years but was redesignated Fabaceae in the 1980s. Members are often simply referred to as legumes.

The common bean dates back approximately 7,000 years ago based on radiocarbon dating. Singh *et al.* (1991) identified two distinct gene pools of common bean, one of Andean origin and the other in Central America and Mexico. The primary center of origin for bean is southern Mexico and warm regions of Guatemala, while the second center is in Peru, Ecuador, and Bolivia. In the wild, the common bean is found in both low and high elevations as well as dry and humid locations. European explorers spread the New World bean (*Phaseolus* sp.), especially *P. vulgaris*, to other regions where they were quickly adapted and rapidly accepted (Zohary and Hopf, 2000).

During domestication, branching decreased while flower number, pod, and seed size increased. Although seed size increased, seed number per pod decreased. Overall, pod dehiscence and pod fiber development were reduced, and pod fleshiness increased in snap bean types. A shift from short-day photoperiod response to daylength neutrality occurred with many biotypes (Rubatzky and Yamaguchi, 1997).

Soybeans

Soybean (*Glycine max* L. Merr.) originated in north China where both vegetable (immature seeds) and agronomic (mature dry seeds) types are major crops that have been grown for centuries (Hymowitz, 1970). The first domestication of soybean was likely in the 11th century BC. Soybean production was localized in China and Korea for centuries before spreading to Japan in the first century. From Japan, soybean spread throughout Southeast Asia. Soybean production in the western hemisphere is more recent. The exact time that Europeans began growing soybeans is unclear but was likely in the 1700s when missionaries introduced seed from China. American colonists planted soybean seeds from China as early as the mid-1700s. Soybean seeds were introduced to American farmers in Illinois during the 1850s but it took almost 100 years before they became a major agronomic crop in the USA. Today, soybeans are a leading agronomic crop in the USA, second only to corn.

Vegetable soybeans, eaten in the mature green stage, are growing in importance in the USA and other parts of the world (Fig. 20.1). However, agronomic soybeans dominate world production. Agronomic soybeans are harvested as dried grain and used to make many different products, including animal feed, cooking and industrial oils, tofu (bean curd), soy sauce, soymilk, and many other food and industrial products. Soybeans are an important source of protein for various cultures in Southeast Asia.

Peas

Peas (*Pisum sativum*) are among the oldest cultivated crops. The earliest archaeological finds of peas date from the Neolithic era and in what is today Syria, Turkey, and Jordan. The pea was also present in what is today the Republic of Georgia in the 5th millennium BC. In Egypt, early finds date from ca. 4800–4400 BC in the Nile delta region and from ca. 3800–3600 BC in other parts of northern Egypt. The propagation of pea apparently moved to further east.

Peas were present in Afghanistan ca. 2000 BC, in Harappa, Pakistan, and in northwest India in 2250–1750 BC (Zohary and Hopf, 2000). In the first century, Roman legionaries gathered wild peas in Palestine, to supplement their diets, and at this time peas first appeared in China (Makasheva, 1983). Pea was grown in Europe during the Middle Ages (Davies *et al.*, 1985). Pea was introduced into the Americas soon after Columbus. Segregation of seed, fodder, and vegetable type peas occurred after domestication. Edible-podded peas are of more recent origin, likely the result of spontaneous mutations that were collected and preserved.

Lima bean (*Phaseolus lunatus* L.)

Remnants of large-seeded lima beans found in archeological sites in Peru date back several thousand years. Evidence suggests small-seeded lima beans were grown in Central America 2,500 years ago. Lima bean was a novelty for the Spanish explorers, who introduced them to Europe and cultivation spread to Asia and Africa. Since lima beans can tolerate humid tropical weather, they have become an important widespread crop with significant production in northern Brazil as well as parts of Indonesia and Southeast Asia.

Botany and Life Cycle

Most of the vegetable legumes are herbaceous dicot annuals or perennials. In temperate areas, most types of beans are grown as annuals but in the tropics some, such as the wing bean (*Psophocarpus tetragonolobus*), are grown as perennials (Fig. 20.2).

Legumes display a broad range of temperature tolerance. Many are warm-season, frost-intolerant crops (snap beans and soybeans) while others are cool-season, frost-tolerant (pea or fava bean; Table 20.1).

Plant architecture ranges from vining or indeterminate to bush or determinate. Leaves are generally alternate and mostly compound, and pinnate, trifoliate or shaped like an open hand. Flowers are perfect and characteristically butterfly shaped, consisting of an upright dorsal petal, two lateral petals or wings, and two lower petals often more or less united in a keel-like structure. Enclosed within the petals are the stamens and pistil (Rubatzky and Yamaguchi, 1997). Legumes are predominantly self-pollinated although some are cross-pollinated by bees.

Legumes produce simple dry dehiscent fruit consisting of a mature ovary wall or pericarp containing from few to many seeds. The pod is a single-carpel fruit that splits along two sutures at maturity releasing the seeds (Rubatzky and Yamaguchi, 1997). The pod itself may be green, purple or yellow or solid color with speckles of a second color. Cultivars of snap bean pods have thicker pod walls and develop seed at a slower rate than agronomic beans harvested as grain. In some cultivars, often designated as string beans, the vascular bundle lignifies with maturity and becomes inedible.

Legume seed consists of two large cotyledons surrounding a tiny but well-developed embryonic plant. The cotyledons contain carbohydrates, fats,

Fig. 20.1. Soybean is a very important world crop that can be harvested as a dry grain or as a vegetable in the immature green stage as pictured here.

Fig. 20.2. Wing beans have distinctive lobed fruit that grow on long vines. The perfect self-pollinating flowers are pale blue.

Table 20.1. A list of some vegetable legumes (Rubatzky and Yamaguchi, 1997).

Common name	Genus and species
Adzuki bean	*Vigna angularis*
African yam bean	*Sphenostylis stenocarpa*
Ahipa	*Pachyrhizus ahipa*
Apois	*Apois americana*
Australian pea	*Dolichos lignose*
Bambara groundnut	*Vigna subterranea*
Broad bean	*Vicia faba*
Catjang cowpea	*Vigna unguiculata* Group cylindrica
Chickpea	*Cicer arietinum*
Cluster bean	*Cyamopsis tetragonolobus*
Cowpea	*Vigna unguiculata* Group unguiculata
Fenugreek	*Trigonella foenum-graecum*
Garden and field pea	*Pisum sativum*
Grass pea	*Lathyrus sativus*
Hyacinth bean	*Lablab purpureus*
Jack bean	*Canavalia ensiformis*
Kudzu	*Pueraria lobata*
Lentil	*Lens culinaris*
Lima bean	*Phaseolus lunatus*
Lupines	*Lupinus* spp.
Marama bean	*Tylosema esculentum*
Mat bean	*Vigna aconitifolia*
Mung bean	*Vigna radiata*
Peanut	*Arachis hypogaea*
Pigeon pea	*Cajanus cajan*
Potato bean	*Pachyrhizus tuberosus*
Rice bean	*Vigna umbellata*
Scarlet runner bean	*Phaseolus coccineus*
Common and snap bean	*Phaseolus vulgaris*
Soybean	*Glycine max*
Sword bean	*Canavalia gladiata*
Tapary bean	*Phaseolus acutifolius*
Urd bean	*Vigna mungo*
Winged bean	*Psophocarpus tetragonolobus*
Yam bean	*Pachyrhizus erosus*
Yardlong bean	*Vigna unguiculata* Group sesquipedalis

and protein in varying proportions depending on the species. Around the whole seed is a tough seed coat or integument. Some legumes have "hard" seed coats that resist rapid water uptake. The seeds vary widely in their size, shape, and color. Some legumes such as pea have hypogeal germination where the cotyledons and seed coat remain beneath the soil during germination because of limited hypocotyl elongation. The epicotyl is fully differentiated prior to germination and pushes through the soil. Epigeal germination occurs in the common bean where the cotyledons emerge above the soil surface because of rapid hypocotyl elongation. Peanut, *Arachis hypogaea*, exhibits a third type of legume emergence with the cotyledons remaining at the surface rather than below or above the soil. Further growth of peanut seedlings is due to above-ground epicotyl elongation and the length of the hypocotyl depends on the sowing depth.

Bean botany

The root system of common bean tends to be shallow and not particularly large. However, the prominent but often short taproot can extend to as much as 1 m (3.3 ft) in deep friable soils under optimal conditions. Nodules develop on lateral roots in the presence of symbiotic bacteria (*Rhizobium* spp.). Bean plant architecture is classified as indeterminate or climbing and nonclimbing or bush plant types. Half-runner beans have an intermediate growth habit between bush and indeterminate vining types. Modern common bean determinate bush cultivars differ from earlier indeterminate climbing types because they have less apical dominance and little or no short-day photoperiod response. The stem length of indeterminate climbing types can grow to as long as 3 m (10 ft) with more than 25 flowering nodes. However, these forms lodge severely and are grown on poles or a trellis for support. Determinate bush forms are short, approximately 60 cm (2 ft) tall, with fewer nodes compared to indeterminate types, and the inflorescences are terminal (Rubatzky and Yamaguchi, 1997).

Bean flowers are large and showy and may be white, pink, or purple with ten stamens, nine of which are united into a tube enclosing the long ovary while one upper stamen is free of the rest. Flowers are self-pollinating and there is generally little outcrossing. The common bean foliage is pinnately trifoliate. Many modern bush cultivars have been developed with small leaves to improve light penetration into the canopy for high-density plantings used for once-over, destructive, mechanical harvesting. Smaller leaves tends to increase yield, but is also genetically linked to smaller pod size (Rubatzky and Yamaguchi, 1997).

Common bean pods are longer than wide with lengths ranging from 8–20 cm (3–8 in) or more, with widths of less than 1 cm (0.4 in) to several centimeters (Fig. 20.3). Depending on cultivar,

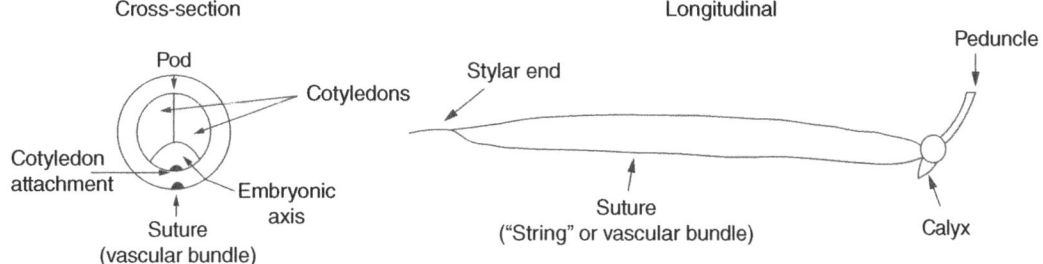

Fig. 20.3. Drawing of common bean fruit, often called a pod or simply bean.

pod tips may be pointed or blunt. Cross-sectional pod shapes may be round, elongated flat or heart shaped. Pods of most modern cultivars are relatively straight, although some are naturally curved.

The fruits of most cultivars are light to dark bluish-green but novelty cultivars may have yellow, purple, or multicolored pods. Cultivars with yellow pods are often called wax beans. The amount and rate of pod fiber development also varies. The word "string" describes the long thread-like fibers associated with the vascular bundles that run along the dorsal and ventral pod sutures. The dorsal fiber or string is the strongest and objectionable if eaten. Through selective plant breeding, the pod fiber has been greatly reduced over time. The stringless character was introduced more than 100 years ago (Rubatzky and Yamaguchi, 1997). Calvin Keeney, a seed grower from LeRoy, New York, is credited with introducing the first stringless cultivar in about 1800 (Rubatzky and Yamaguchi, 1997). Stringless is a recessive genetic trait and is incorporated in most modern cultivars so the string does not have to be removed before consumption. Stringless-type cultivars also contain less fruit wall fiber. Despite the development of the stringless character, "string bean" remains a generic term to describe all cultivars of snap or green beans in some circles. Presently, only heirloom and other old cultivars possess the strong string-like suture fiber. The term "snap" was adapted to describe stringless cultivars, possibly from the sound resulting when fresh pods are broken.

The number of seeds per pod varies among bean types (Fig. 20.4). Most snap bean cultivars contain three to five seeds, while dry beans tend to have more.

Mature seed sizes exhibit variation in size and weight, ranging between 5–20 mm (0.2–0.8 in) in length and individual seed weights vary from 0.15g to >0.80 g (0.005–0.03 oz). Seed shapes are variable and maybe round, orbicular, ovoid, oblong,

Fig. 20.4. Drying adzuki beans contain eight to ten seeds per pod. Adzuki bean has vivid yellow perfect flowers.

or kidney-shaped. Seed coat colors vary and are cultivar specific. In different countries of Latin America, certain seed coat colors are preferred: black seed coat in Brazil, El Salvador, Mexico, and Venezuela, red in Colombia and Honduras, yellow in Peru, and white in Chile (Rubatzky and Yamaguchi, 1997). Cultivars grown for snap bean processing usually have white or light-colored seeds so the liquid in canned bean products is not dark colored from the leaching of water-soluble pigments.

Pea botany

Peas are frost-tolerant, cool-season, herbaceous, annual dicots, having alternate leaves with the tip of the compound leaf modified as a tendril. Growth habits range from indeterminate vine to determinate bush or dwarf forms. The foliage of pea cultivars varies from those having extensive leaflets to the so-called leafless types that form tendrils instead. Foliage surfaces have an obvious waxy cuticle layer and leaf colors ranging from yellowish green to deep blue green. Pea stems are singular, slender, angular, and hollow, but solid at the base. Some cultivars have an upright habit but are generally not self-supporting and branching is usually limited (Rubatzky and Yamaguchi, 1997).

Taproots may reach a depth of 80 cm (31 in) but root systems are generally not extensive. Pea flowers develop at leaf axils and are perfect, containing both male and female parts. Flowers are self-pollinating, often before fully open, and cross-pollination is rare. Flowers are usually white, but may be pink, purple, or a blend of colors. The number of nonflowering nodes is a cultivar characteristic. Very early-maturing cultivars produce flowers in as few as five or six nodes. Some late cultivars flower only after 15 or more vegetative nodes have developed (Rubatzky and Yamaguchi, 1997).

Pod number per node is a genetically controlled trait that is also influenced by the environment. For example, environmental stresses like high temperatures or drought may reduce pod numbers. Early-flowering cultivars average one or perhaps two pods per node, while late-flowering cultivars average more than two per node. Some cultivars develop four or more pods per node. Determinate cultivars tend to produce flowers that are grouped in a terminal inflorescence instead of among the nodes. Development of a terminal inflorescence stops further vegetative growth (Rubatzky and Yamaguchi, 1997).

The pea pod is a dehiscent fruit formed from the carpels of the flower. Seed are attached alternately to either side of the fused carpels. Pod walls of garden and field peas have a hard, lignified-parchment layer. The pod walls of snow and sugar snap edible pod peas have no lignified layer. Pod size and seed number vary among cultivars. The pea seed consists of two large cotyledons surrounding the embryo and enclosed by either a pigmented or colorless seed coat that may be either smooth or wrinkled. Mature smooth-seeded types have higher starch content, a mealy texture, better cold tolerance, and are often used for canning. Peas with wrinkled seed coats generally have higher sugar content and less starch and are processed by freezing.

Lima bean botany

The taxonomic classification of lima beans is controversial relative to the species designations of *Pisum lunatus* and *Pisum limensis*. The species name *lunatus* comes from the moon-shaped seed. Some classifications consider the thick-podded, large-seeded limas as *P. limensis*, and thin-podded, small-seeded forms *P. lunatus*. Classifying different types of limas into separate species is questionable and probably not justified because all types are interfertile. Many botanists agree that all wild and cultivated limas are *P. lunatus* (Rubatzky and Yamaguchi, 1997). Lima beans are often divided into three groups within *P. lunatus*. Small-seeded lima beans are Sieva types. Large-seeded limas are a second type sometimes referred to as Butterbean and Madagascar bean. The third type is "potato" lima, also known as Fordhook (Fofana et al., 1997).

Limas are considered perennials or long-lived annuals but are grown commercially as annuals. Both climbing and bush cultivars are grown. Climbing cultivars can reach heights of 3–4 m (118–157 in), whereas bush types grow to only 50–90 cm (20–35 in). Limas have a highly branched root system reaching depths of more than 1 m (39.3 in). Roots develop nodules containing *Rhizobium* like other legumes.

Flowering is indeterminate and flowers are self-fertile although some outcrossing may occur. The slightly curved oblong pods range from 5–15 cm (2–6 in) in length to 2–3 cm (0.8–1.2 in) wide. Most cultivars usually contain two to four seeds while some pods may contain as many as six. Pods of some cultivars are thick and others are relatively thin. The large, flat, oblong seeds of large-seeded limas are as long as 3.5 cm (1.4 in). Sieva seed types are also flat but more rounded and about 1 cm (0.4 in) long. Fordhook types are intermediate.

Lima commercial cultivars in North America tend to have a light green or white seed coat but in other areas may be red, purple, brown, or black. Two large cotyledons account for most of the seed volume. Seeds of wild limas found in South and Central America have a high cyanogenic glucosides content and must be leached before or during cooking. Modern cultivars, especially those with light-colored seeds, contain little or no glucosides. The consumption of raw lima beans is not recommended (Rubatzky and Yamaguchi, 1997).

Bacterial symbiosis

Rhizobia are soil bacteria that fix atmospheric nitrogen after developing a symbiotic relationship with plants from the family Fabaceae. Rhizobia require a host plant to fix nitrogen (N) and cannot do so by themselves as free-living organisms.

Rhizobia living in the soil can sense flavonoid compounds secreted by the roots of potential legume hosts. Flavonoids trigger the secretion of *nod* factors by the rhizobia, which in turn are recognized by the host plant and lead to root hair deformation and other cellular adaptations to accommodate the rhizobia. The best-known mechanism is called intracellular infection where the rhizobia enter legume roots through a deformed hair. A second mechanism called "crack entry" occurs when the bacteria enter between cells through cracks produced by lateral root emergence (Jones *et al.*, 2007). Eventually, the bacteria are internalized in cells through tubular cell wall invaginations called infection threads. The infection triggers cell division in the cortex of the root where a new organ, the nodule, develops. A nodule is a specially adapted round growth or knot that develops on the root of a legume containing the N-fixing bacteria. Infection threads grow to the nodule, releasing the rhizobia into the central tissue. While in the nodule, the rhizobia differentiate morphologically into bacteroids and fix gaseous atmospheric nitrogen (N_2) into ammonium ($NH_3 + H^+ \rightarrow NH_4^+$), a form that can be utilized by the plant, by the enzyme nitrogenase (Jones *et al.*, 2007). The reaction for all N-fixing bacteria is:

$$N_2 + 8\ H^+ + 8\ e^- \rightarrow 2\ NH_3 + H_2$$

The legume–rhizobium symbiosis is a classic example of mutualism because rhizobia supply ammonia or amino acids to the plant that in return provides the bacterium with organic acids, carbohydrates, and proteins. Leghemoglobins, plant proteins similar to the hemoglobins found in human blood, help maintain oxygen levels high enough for respiration while keeping the free concentrations sufficiently low so that nitrogenase activity is not inhibited (Nelson *et al.*, 2008).

Legumes normally get most of their N from the atmosphere through rhizobium fixation. Legumes typically need less N fertilizer but do respond to it in the seedling stage before the symbiosis with *Rhizobium* is established. Supplementing legume crops after establishment can be counterproductive, because N fixation declines rapidly when soil N is high. Additions of phosphorus (P) and/or potassium (K) may increase nodule numbers, fresh weight, and the amount of N fixed. An important micronutrient for N fixation is molybdenum (Mo). Soils should be tested for Mo prior to growing legumes.

A pH below 6.0 may reduce Mo availability. Molybdenum fertilizer may be applied as a seed treatment, since only small amounts are required. Some *Rhizobium* inoculants have Mo incorporated (Erker and Brick, 2006).

Legume inoculation is the process of introducing commercially prepared sources of rhizobia to promote N fixation. When there is doubt about the rhizobial population in a field, it is a good practice to apply inoculum, especially if the legume has never or not recently been grown (Erker and Brick, 2006). Commercial inoculants are rhizobia selected for their N-fixation potential. Inoculation is often applied directly to the seed or by metering the inoculum into the furrow during planting. If the legume crop was grown in the field previously, there is a good chance that the soil already contains the correct rhizobial species for nodulation, since rhizobia can survive in soil without a host. Rhizobia survive poorly in acid soils. Often native rhizobia have less potential to fix N. A nodule that is actively fixing N is pink to reddish inside (Erker and Brick, 2006).

Vegetable legume seeds are often pre-inoculated before purchase. Inoculates can also be purchased separately from seeds and added by the grower at the time of planting. There are three basic forms of commercial inoculation, solid, liquid, and freeze-dried (Erker and Brick, 2006). The most common is solid, peat-based inoculation that can be applied dry to the seed or as a slurry. The dry method has the disadvantage of uneven distribution and poor adhesion. The slurry is prepared just prior to planting by mixing the inoculum with water for better seed coverage (Erker and Brick, 2006). Liquid inoculants are sold in culture broth or as a frozen concentrate. Broth or frozen concentrates are mixed with water and sprayed into the seed furrow at planting. Because liquid inoculants must be frozen or refrigerated during shipment and storage, their availability through normal distribution channels is limited (Erker and Brick, 2006).

Types and Cultivars

Common beans can be classified into different categories based on plant type and fruit characteristics. Bean cultivars are usually classified as indeterminate, semi-determinate, or indeterminate (Table 20.2). Determinate bush beans are compact and self-supporting cultivars and need no external support. With a heavy fruit load bush types may

Table 20.2. Common bean plant architectures.

Name	Plant type	Description	Representative cultivars
Pole bean	Indeterminate	Climbing types can grow to as tall as 3 m (118 in)	Kentucky Wonder, Blue Lake Pole
Half-runner bean	Semi-determinate	Vine to a length of 1 m (39.3 in)	White Half Runner, Mountaineer Half Runner
Bush	Determinate	60 cm (23.6 in)	Provider, Top Crop. Green Pod

lodge after heavy rain, particularly on shallow soils when deep anchor roots fail to develop. Bush cultivars are better adapted to high-density plantings and their concentrated fruit set is well suited for once-over, destructive, mechanical harvest, which predominates for both fresh-market and processing crops. The concentrated set of bush cultivars also makes hand harvest easier and more efficient. Bush snap bean cultivars are well suited for pick-your-own operations because customers can focus on harvesting the concentrated set of uniform pods.

Pole beans have a vining habit and reach heights of 1.8 m (72 in) or more. They require more space than bush types and a substantial pole, trellis, or support system for most efficient growth and harvesting (Fig. 20.5).

Pole beans mature over a longer period and are not suited for mechanical harvest, which is why bush cultivars have come to predominate commercially. Pole beans are well suited for home gardening where a steady supply of pods is needed over a longer period of time and they are easier to hand-harvest because there is less need to bend over.

Half-runner beans have an intermediate growth habit between pole and bush types. Half-runners can be grown without support but need room to grow because they spread more than bush beans, covering as much as 1 m (39.3 in) across. Half-runner beans remain important in some markets such as the Appalachian region of the USA.

Bush cultivars tend to be stringless, half-runner beans may be string or stringless, and pole beans tend to have strings. Some consumers prefer the stronger bean taste of pole beans and half-runner cultivars, which develop seeds faster and have pods that are less fleshy compared to bush cultivars. Snap bean cultivars for processing usually have less pod fiber than ones developed for fresh market. Greater pod fiber helps reduce breakage and maintain

Fig. 20.5. Indeterminate beans are grown on a support to make harvest easier.

a rigid appearance in the marketplace even with moderate desiccation.

Most present-day snap bean cultivars are photoperiod insensitive. Nevertheless, a few cultivars flower only under short days. Beans grow best in full sun and yields are reduced when plants are shaded. Bean leaves face and follow the sun across the sky during the course of a day, which is an adaptation to improve photosynthetic efficiency.

However, during periods of excessive heat and low soil moisture, leaves will turn parallel to the sun's rays to decrease leaf temperature.

Common beans can also be grouped according to how they are used (Table 20.3). French beans, which are simply called green beans by many, have stringless fleshy edible pods that are consumed when immature and tender before significant seed development has occurred. Other names included snap bean and Italian green bean. Another edible pod bean is the string bean. Although similar to the French beans, the string or lignified vascular bundle is removed before cooking and eating. String beans have faster seed development and less fleshy pods. Haricot filet or vert is a third type of edible pod bean with a stronger bean flavor and extremely thin, small pods. Both string and stringless types exist and the pod is less fleshy than the French bean.

Horticultural are dual use beans. The entire pod can be eaten like a green bean early in development but later the seeds are shelled from the pod and consumed separately after cooking. Cultivars that are grown for harvesting only the fresh seeds inside a mature green pod are called shelly or shellie beans. Haricot beans are very similar to shelly beans because the fresh seed inside the pod is used. Haricot cultivars tend to have white seed but this is not an absolute requirement. Only the fresh seeds are eaten because the pods contain string and fiber and generally are not consumed unless when very immature. Immature fresh kidney beans harvested when the pods are still green are considered a haricot type. However, this popular bean may also be harvested when fully mature at low moisture content as a dry bean.

Many bean seeds are harvested when they are mature and dry (Table 20.3). Dry bean seeds are destructively harvested as a grain at full maturity by combining or hand-harvested after the plants have senesced and turned brown but before the pods dehisce. Because of this, dried beans are considered to be an agronomic crop in some circles. Dry bean seeds are less perishable than fresh seeds and are amenable to long-distance shipping and long-term storage if maintained at low moisture content.

The term pulse is also widely used to describe grain legumes. A pulse or grain legume is an annual leguminous crop yielding from 1 to 12 seeds of variable size, shape, and color within a pod (Table 20.4). The word pulse may refer to only the seed or the entire plant (Fig. 20.6).

Pulses are used as human and animal food and exclude vegetable, oil, and forage legumes. The word gram is also used to describe some legume seed-bearing plants. Gram bean is a general term for any one of several leguminous plants, such as *Phaseolus mungo* (black gram or urd) and *P. aureus* (green gram), used as food in India (Rubatzky and Yamaguchi, 1997).

Types and cultivars: garden and field peas

Peas can be divided into three major types: (i) plants producing well-developed, succulent but immature seed; (ii) plants producing immature, succulent, edible pods and seed; and (iii) plants producing fully developed, mature, dry seed. In parts of Southeast Asia, the tender shoots of pea plants are eaten as greens. English, garden, vining, or green peas are synonymous terms referring to the first type described. The name English pea may have come from the country of early plant breeders who improved the crop. Frozen pea cultivars are much different than canned. Frozen peas have attractive dark green color, wrinkled thick skins, and higher sugar content. In contrast, canned pea cultivars have smooth, thin pale-green skins and higher starch content.

With edible-pod peas, the whole fruit, both pods and seeds, is eaten. The pods of edible pod peas remain relatively succulent because the fibrous parchment layer of the interior pod wall fails to develop. Edible pod peas are often classified into one of two types. Snow, sugar, or China pea (*P. sativum* subsp. *saccharatum*) are harvested early in development and are primarily flat, wide, and thin-walled pods with very little seed development. Another edible pod type (*P. sativum* subsp. *macrocarpon*) is called the sugar snap pea. The sugar snap pea has a succulent, edible pod that somewhat resembles snap bean. The Gallatin Valley Seed Company of Twin Falls, Idaho, first released the sugar snap pea in 1979, so it is of recent origin. Field or dried peas are grown for their mature dry seed and are usually considered an agronomic crop.

Types and cultivars: lima bean

Many characteristics described for common bean cultivars also apply to *P. lunatus*. In the USA and Canada, small cultivars such as the sieva types are

Table 20.3. Common beans grouped by utilization.

Name	Characteristics	Description	Representative cultivar or type
Horticultural or shelly (shellie) bean	Dual purpose, green bean/seed bean easier to shell when immature than snap bean	Seeds are green or bicolor. Can be eaten as green beans when immature or shelled with beans eaten separately later in maturity. Similar to haricot	French Horticultural Taylors Horticultural
Green bean, French bean, snap bean, Italian green bean	Whole fruit is eaten when immature, slow seed development	Green, yellow, or purple fleshy pods with under-developed seeds are eaten intact. Pods do not have string or parchment layer, in cross-section pods are round, flat, or heart shaped	Provider Green Pod Greencrop Topcrop Contender
String bean	Pod fibrous string must be removed before eating, pod less fleshy than green beans, prominent seed development	Indeterminate, long pods, later maturing than bush snap beans, stronger flavor. Indeterminate growth habit common	Kentucky Wonder Blue Lake Pole
Haricot filet (vert)	Similar to snap bean with a stronger bean flavor, extremely thin, tender, small pods	Pods contain a string or are stringless, the fleshy immature pods with seeds are eaten. Pods are green, purple, or yellow	Maxibell Tavera Conca
Haricot	Cultivars tend to have white seed but this is not an absolute requirement	Fresh seed are eaten after removal from pods. Pods contain a string and fiber and generally are not consumed unless very immature similar to shelly and horticultural	Cannellini, certain cultivars of Navy or Kidney
Dry (field) bean	Plants are mature and dry at harvest. Harvest is often destructive mechanical before pods dehisce	Dry shelled seed is treated as grain. Pods are fibrous and not eaten	Pinto, Marrowfat, Kidney, Navy, Black

called baby limas. Lima beans often require a longer growing season and do not pollinate well at extremely high temperatures. Lima beans are grown for their maturing seeds harvested either dry or while the pods are still green. Lima bean cultivars are differentiated by seed size, shape, color, and growth habit: indeterminate, determinate, or semi-determinate.

Genetic improvement

Seeds of bean and pea cultivars are open-pollinated rather than F-1 hybrids. Beans and peas are self-pollinated and the flower emasculation and hand-pollination needed to cross-pollinate the inbred lines necessary to produce F-1 hybrid seed would be prohibitively expensive. Beans and peas will grow true-to-type if the seed is saved and replanted. While some cultivars of agronomic soybeans have been genetically engineered for herbicide and insect resistance, this technology has not been widely applied to vegetable soybeans.

Economic Importance and Statistics

In 2011, over 29 million ha (72 million acres) totaling over 23 million metric tonnes (25 million short tons) of dry beans were harvested in the world. Leading producers of dry beans include: India (>4.5 million metric tonnes (5 million short tons)), Myanmar (>3.7 million metric tonnes (4.1 million short tons)), Brazil (>3.4 million metric tonnes (3.7 million short tons)), China, the USA, Tanzania, Kenya, Mexico, Uganda, and Cameroon. The same year, over 1.5 million ha (3.7 million acres) of green beans totaling over 20 million metric tonnes (22.1 million short tons) were harvested. The leading producers of green beans were China (>15.7 million

Table 20.4. The 11 pulse crops recognized by the FAO.

Type or genus and species	Example
Dry beans (*Phaseolus* spp. including several species now classified as *Vigna* spp.)	Kidney bean, haricot bean, pinto bean, navy bean (*Phaseolus vulgaris*)
	Lima bean, butter bean (*Phaseolus lunatus*)
	Adzuki bean (*Vigna angularis*)
	Mung bean, golden gram, green gram (*Vigna radiata*)
	Black gram, urad (*Vigna mungo*)
	Scarlet runner bean (*Phaseolus coccineus*)
	Ricebean (*Vigna umbellata*)
	Moth bean (*Vigna aconitifolia*)
	Tepary bean (*Phaseolus acutifolius*)
Dry broad beans (*Vicia faba*)	Horse bean (*V. faba* subsp. *equina*)
	Broad bean (*V. faba*)
	Field bean (*V. faba*)
Dry peas (*Pisum* spp.)	Garden pea (*P. sativum* subsp. *sativum*)
	Protein pea (*P. sativum* subsp. *arvense*)
Cicer arietinum	Garbanzo, chickpea, and Bengal gram
Blackeye bean (*Vigna unguiculata*)	Dry cowpea, black-eyed pea
Gandules (*Cajanus cajan*)	Pigeon pea, arhar /toor, cajan pea, Congo bean
Lens culinaris	Lentil
Vigna subterranea	Bambara groundnut, earth pea
Common vetch (*Vicia sativa*)	Vetch
Lupinus spp.	Lupins
Minor pulses	Lablab, hyacinth bean (*Lablab purpureus*)
	Jack bean (*Canavalia ensiformis*), sword bean (*Canavalia gladiata*)
	Winged bean (*Psophocarpus teragonolobus*)
	Velvet bean, cowitch (*Mucuna pruriens* subsp. *utilis*)
	Yam bean (*Pachyrrizus erosus*)

Fig. 20.6. Chickpea, also called garbanzo or Bengal gram, is an important world legume crop that is often referred to as a pulse (Table 20.4). The seeds are processed into hummus or used in other preparations. The seeds can be harvested fresh without drying or as dried grain. Whole immature plants with fruit are sold at this market in Greece.

metric tonnes (17.3 million short tons)), Indonesia (>883,000 metric tonnes (973,340 short tons)), India (>617,000 metric tonnes (680,126 short tons)), Turkey, Egypt, Thailand, Morocco, Spain, Italy, and Bangladesh (FAOSTAT, 2011).

Over 6 million ha (14.8 million acres) totaling over 9.5 million metric tonnes (10.5 million short tons) of dry peas were harvested in the world in 2011. Leading producers of dry peas include: Canada (>2.1 million metric tonnes (2.3 million short tons)), Russia (>2.0 million metric tonnes (2.2 million short tons)), China (almost 1.2 million metric tonnes (1.3 million short tons)), India, France, Australia, Ukraine, Ethiopia, the USA, and Spain. The same year, over 2.2 million ha (5.4 million acres) of green peas, totaling almost 17 million metric tonnes (18.7 million short tons), were harvested. The leading producers of green peas were China (>10.2 million metric tonnes (11.2 million short tons)), India (>3.5 million metric tonnes (3.9 million short tons)), the UK (>424,000 metric tonnes (467,380 short tons)), the USA, France, Egypt, Algeria, Morocco, Kenya, and Turkey (FAOSTAT, 2011).

World production of pulse crops in 2011 totaled in excess of 4.9 million ha (12.1 million acres) with

Table 20.5. Green pea per capita consumption in the USA (kg/person/year (lb/person/year)) (ERS, 2011).

Year	FM[a]	Canned[b]	Frozen[b]	Total
1930	1.2 (2.6)	2.1 (4.6)	0.0 (0.0)	3.3 (7.2)
1940	1.0 (2.1)	2.5 (5.5)	0.1 (0.2)	3.5 (7.8)
1950	0.3 (0.7)	2.4 (5.4)	0.4 (0.9)	3.2 (7.0)
1960	0.1 (0.3)	2.0 (4.4)	0.8 (1.8)	2.9 (6.5)
1970	–	1.8 (3.9)	0.9 (1.9)	2.8 (6.1)
1980	–	1.5 (3.3)	0.7 (1.5)	2.3 (5.1)
1990	–	0.9 (2.0)	1.0 (2.2)	1.9 (4.2)
2000	–	0.7 (1.5)	1.0 (2.1)	1.6 (3.6)
2010	–	0.5 (1.2)	0.7 (1.6)	1.3 (2.8)

[a]Farm-weight basis
[b]Processed-weight basis

Table 20.6. Green bean per capita consumption in the USA (kg/person/year (lb/person/year)) (ERS, 2011).

Year	FM[a]	Canned[b]	Frozen[b]	Total
1930	2.0 (4.5)	0.9 (2.0)	0.0 (0.0)	2.9 (6.5)
1940	2.3 (5.0)	1.0 (2.3)	0.0 (0.0)	3.3 (7.3)
1950	1.8 (3.9)	1.5 (3.4)	0.2 (0.4)	3.5 (7.7)
1960	1.2 (2.6)	1.9 (4.2)	0.4 (0.8)	3.4 (7.6)
1970	0.8 (1.7)	2.6 (5.8)	0.5 (1.1)	3.9 (8.6)
1980	0.6 (1.4)	2.4 (6.3)	0.5 (1.0)	3.9 (8.7)
1990	0.5 (1.1)	1.7 (3.7)	0.9 (1.9)	3.0 (6.7)
2000	0.9 (2.0)	1.8 (4.0)	0.8 (1.8)	3.5 (7.8)
2010	0.9 (1.9)	1.7 (3.7)	0.9 (2.0)	3.4 (7.6)

[a]Fresh market (farm-weight basis)
[b]Processed (processed-weight basis)

a total production over 3.5 million metric tonnes (3.9 million short tons). The leading producers were India (700,000 metric tonnes (771,618 short tons)), the UK (>351,000 metric tonnes (386,911 short tons)), Mozambique (>229,084 metric tonnes (252,522 short tons)), Russia, Poland, Pakistan, Vietnam, China, Thailand, and Tanzania.

Some interesting trends are apparent when observing pea and snap bean consumption trends in the USA over the past 80 years (Table 20.5). Fresh green peas were popular in America many years ago but consumption slowly declined to nil during the 1960s. The drop in green pea production is likely linked to greater time to shell and prepare fresh peas compared to canned and frozen pea products, which offered similar quality. Recently, preshelled fresh peas have begun to reappear in the market as a convenience item. It will be interesting to see whether these consumer-friendly, ready-to-eat fresh peas products can increase consumption in the USA, reversing the long-standing decline.

Data show that Americans prefer frozen peas over canned. The consumption of canned green peas has also gradually declined over time while the consumption of frozen peas has slowly increased (Table 20.5). Overall consumption of green peas experienced a 42% drop over the past 80 years. The drop in overall green pea consumption suggests that consumers prefer other vegetables to pea, since per capita consumption of vegetables in the USA has increased over this period (Cook, 2011).

Compared to green peas, the consumption of green beans has been much more stable over the past 80 years (Table 20.6). Total US green bean consumption is actually greater in 2010 than it was in 1930. During this period, fresh bean consumption dropped dramatically, although it appears to have stabilized in recent years at 0.9 kg (2 lb) per person per year. Canned bean consumption increased to a peak in 1980 and has since declined. Frozen beans did not appear in the US market until the 1950s and consumption has gradually risen since that time. The consistent consumption of beans compared to the dramatic drop in peas since 1930 suggests that US consumers find green beans preferable to green peas although differences in price may also affect consumer choices. These data also highlight the fact that consumption patterns are dynamic and change over time. It is important for producers and retailers to be aware of such changes as well as the reasons for why they occur.

Nutritional Values

Dry beans and peas are a major source of protein for many cultures (Table 20.7). Dried beans and peas contain approximately 24% protein, which is twice that of wheat (12%) and three times that of rice (8%).

Production and Culture

Bean site selection and nutrient management

Well-drained, noncrusting, medium-textured loams are suited for bean production. Growth is strongly reduced by soil compaction and poor drainage. A favorable soil temperature range is 18–30°C (68–77°F). A mean air temperature of 20–25°C (68–77°F) is optimal for vegetative growth and yield of common bean. Temperature stress reduces pod set, although some cultivars are more tolerant than others (Rubatzky and Yamaguchi, 1997).

The majority of beans, especially bush cultivars, have relatively small root systems, so the absorption

Table 20.7. Nutritional composition of some vegetable legumes.

Nutrient	Amount/100 g (3.5 oz) edible portion				
	Lima (fresh green seeds) (*Phaseolus lunatus*)	Bush snap (*Phaseolus vulgaris*)	Vegetable soybean (*Glycine max*)	Green pea immature (*Pisum sativum*)	Edible pod pea (*Pisum sativum*)
Water (%)	68.3	90.9	67.5	74.6	87.6
Calories	105	33	141	87	38
Carbohydrate (g)	22.2	6.6	12.5	15.4	7.9
Protein (g)	7.9	2.0	13.7	6.5	2.6
Fat (g)	1.2	0.19	5.1	0.55	0.13
Fiber	1.6	1.2	1.8	2.2	1.8
Ash	1.75	0.76	1.7	0.86	0.58
Vitamin A (IU)	250	550	410	780	405
Vitamin C (mg)	29	19	25	33	52
Vitamin B1 (mg)	0.22	0.11	0.44	0.33	0.13
Vitamin B2 (mg)	0.11	0.11	0.18	0.15	0.09
Niacin (mg)	1.4	0.7	1.7	2.6	0.7
Ca (mg)	102	50	107	30	50
P (mg)	165	41	205	118	50
K (mg)	460	220	436	285	185
Na (mg)	3	8	6	4	4
Mg (mg)	5	34	64	35	24
Fe (mg)	2.8	0.9	2.8	1.8	1.3

of water and mineral nutrients occurs close to the plant. Mineral nutrients must be placed in the root zone for most efficient uptake and utilization. Beans are sensitive to salt damage, so fertilizer should not be placed in contact with seeds or seedlings. If fertilizer is applied at planting it should be placed a minimum of 5 cm (2 in) to the side and below the seed (Davis and Brick, 2013). The optimum pH range is between 6.0 and 6.5. Fertilizer damage to seedlings is greater on acid soils when the pH is below 5.5.

A snap bean crop yielding 11.2 metric tonnes/ha (5 short tons/acre) will remove approximately 134-11-62 kg/ha (120-10-55 lb/acre of N-P-K), respectively. Bean plants remove an additional 56-67.5-50 kg/ha (50-60-45 lb/acre) of N-P-K, respectively (Lorenz and Maynard, 1988). A crop management strategy for beans should provide these amounts as a minimum and more if high yields are anticipated. Many incorrectly believe that legumes do not require N fertilization because of their symbiotic relationship with rhizobia. This may only be true if beans follow crops that have been fertilized with large amounts of N or crops that leave large amounts of N-rich residue in the soil (Davis and Brick, 2013). Since symbiosis does not occur immediately after planting, particularly for determinate cultivars, early growth may be limited by insufficient N before rhizobium-fixed N is available (Rubatzky and Yamaguchi, 1997). Therefore, in many cases N fertilization is necessary for vigorous early development. Applications of 56–90 kg/ha (50–80 lb N/acre) are often needed when beans are grown in fields with a history of heavy fertilization and intensive culture. From 90–123 kg/ha (80 to 110 lb N/acre) are required if forage legumes or heavily fertilized vegetable crops were not grown the preceding year. Too much N causes excessive vining, delayed maturity, and reduced pod set. Nitrogen application depends on soil test results, the amount of organic matter present and soil texture. Split applications are used on sandy soils with approximately one-half of the N applied about 10 days after planting and the remainder 2 weeks later. On heavier soils where leaching potential is less, all N can be applied before planting or as a side-dress treatment later in the season.

Phosphorus is especially important during early bean growth. Beans respond to applied P on soils with low or medium levels of extractable P based on soil tests. Fertilizer is often banded in the root zone because P is not mobile in soil. Phosphate fertilizers may also be surface applied and tilled into the soil prior to planting so seeds and seedlings are not damaged by fertilizer salt in dry soil. Monoammonium phosphate (MAP, 11-52-0), diammonium phosphate (DAP, 18-46-0), and ammonium polyphosphate (10-34-0) are equally effective and the choice depends on availability, equipment and cost per unit of P (Davis and Brick, 2013).

Depending on their cropping history and soil testing results, some soils may not require K fertilization to produce a bean crop. However, sandy or highly eroded soils may be low in extractable K. No application is recommended when soil K concentrations exceed 60 ppm as tested by M-1 extraction. In Florida, snap bean yields responded to soil K concentrations through 73 and 78 ppm M-1 soil K (Hochmuth and Hanlon, 2010). Incorporating K into soil before planting or row-banding away from the seed are common application methods. On heavier soils, a single application of K just before or at planting is effective. A split-K application is recommended for unmulched production on sandy soils with overhead irrigation to decrease the danger of fertilizer burn and leaching. Preplant application of 25–50% of K fertilizer with the remainder applied as side-dressing after 2 or 3 weeks' growth, similar to N, is a standard recommendation on sandy soils (Hochmuth and Hanlon, 2010).

Combined N and potash (K_2O) applications should not exceed 101 kg/ha (90 lb/acre) when between-row spacings are 0.9 m (36 in), or 121 kg/ha (108 lb/acre) for 0.75 m (30 in) spacing because of the danger of seedling injury from the high concentration of salt. Beans are highly sensitive to excessive soil boron. Zinc may limit production of beans particularly at high pH.

Crusted soils can inhibit the epigeal emergence of beans and cause loss or partial loss of cotyledons that can slow seedling growth and reduce stands. Irrigating soils or tillage with a rotary hoe or other implement can prevent soil crusting. Plant breeders have selected against the hard seed-coat trait in legumes to improve germination. However, reducing seed hardness increases susceptibility to physical damage, such as seed-coat cracking, during seed harvest, handling, and planting. Rubberized harvesting and handling equipment softens impacts and reduces physical damage during bean seed handling.

Seed germination is optimum between 25°– 30°C (77–86°F). At temperatures <10°C (50°F) and >35°C (95°F) germination is slow and percentages are reduced. In warm moist soils, high quality seed emergence can be expected within 6–10 days of planting. Planting in cold soil slows germination and increases seed decay. Seed-coat cracking and leakage historically have been greater in white-seeded cultivars, reducing emergence in cold soils and increasing susceptibility to pathogenic attack, compared to black-seeded snap bean cultivars (Mohamed-Yasseen, et al., 1994). Poor germination is not characteristic of all white-seeded cultivars, and crack resistance and low temperature performance in white-seeded beans has been improved through plant breeding (Dickson, 1971). Fungicide treatments are standard for bean seed planted in cold soils to protect against fungal attack. Bean germination is often vigor-tested in cold soil to assess establishment potential under stressful field conditions.

Sowing depths range from about 2–8 cm (0.8–3.1 in) depending on seed size. Dry seeds are less dense than soil and tend to migrate to the surface after heavy rain or irrigation if planted too shallow. If seeds are close to the surface, they may be too dry to germinate or may be eaten by predators. Researchers at Oregon State University working with bush cultivars of snap bean bush showed 232 cm^2 (36 in^2)/plant (429,780 plants/ha, 174,000 plants/acre) is the optimum plant population density (Hemphill, 2010). With the same number of days from seeding to harvest, a higher grade (smaller sieve bean) was obtained, resulting in higher quality and dollar value than obtained at the conventional spacing of 290–387 cm^2 (45–60 in^2) per plant (Hemphill, 2010). Seeding rates of 26–40 seeds/m (8–12 seeds/ft) in rows 38–76 cm (15–30 in) apart may be necessary to accommodate cultivating, spraying, or harvesting equipment. Close spacing decreases air movement thus increasing diseases such as gray and white mold in humid climates or if overhead irrigation is used. However, in dry climates disease is less prevalent and can be avoided by increasing in-row spacing and/or an effective fungicide program (Hemphill, 2010).

Increasing the plant populations reduces the number of fruit per plant but increases pod yield. At close spacing, plants are more erect, pods form closer to the stem and higher in the canopy. A final stand of bush beans ranges from 296,400–432,250 plants/ha (120,000–175,000 plants/acre) depending on the spacing between rows. This usually equates to 84–123 kg seed/ha (75–110 lb seed/acre) depending upon seed size and germination percentage (Hemphill, 2010).

Wider spacing is used for pole, snap and runner beans that are repeatedly harvested during the season. Trellised pole snap beans are usually spaced about 10 cm (4 in) apart in row with between-row spacings ranging from 120–150 cm (47–59 in). Pole beans are sometimes planted in hills rather than rows. Pole-supported hill plantings are spaced equidistant, from 90–120 cm (35–47 in). Hill plantings are usually sown with four to six seed per hill, later thinned to about three plants.

Growth and development

Bush snap bean cultivars produce a crop within 55–75 days of direct seeding. Pole cultivars require about 10–20 days longer and half-runner beans are intermediate between pole and bush cultivars. Pods of most green bean cultivars are large enough for harvest about 7–15 days after anthesis. Dry beans such as kidney, navy, cranberry, etc., require from 100 to 120 days from seeding to field dry-seed harvest (Fig. 20.7).

Initiation and flower development is greatly delayed by suboptimum temperatures. If temperatures are <10°C (50°F), fertilization is poor and pods are small and misshapen. Blossom drop and ovule abortion can also be caused by high temperatures >35°C (95°F). Lateral branching is beneficial because it provides more flowering nodes and extends the flowering period and yield potential. Pole bean cultivars generally have more lateral branches, making them better adapted to stress. Determinate cultivars are more susceptible to stress that interferes with pod set, which can reduce yields for destructive single harvests. However, under favorable conditions, determinate plants enjoy the advantage of concentrated fruit set and highly uniform pod development characteristics that are well suited for mechanical harvesting.

To extend harvest maturity while maintaining edible quality, plant breeders have developed snap bean cultivars with slow seed development (Rubatzky and Yamaguchi, 1997). Production methods for dry bean cultivars are similar to those for snap beans. Dry beans are commonly grown on a larger scale than snap beans because they are less perishable and usually require less management and labor.

Irrigation

Moisture management in snap beans is extremely important for stand establishment, disease control, pod set, and pod quality. To avoid soil crusting and promote uniform emergence in areas with insufficient rain, irrigation should applied prior to planting (Hemphill, 2010). When irrigating, adequate moisture must be applied for pod growth while minimizing wetting the canopy to avoid disease. Sprinkler irrigation is sometimes used until emergence and then a switch is made to furrow if available. Drip irrigation is less common for snap bean production because of the rapid maturity. Soil should remain at a minimum of 50% field capacity during snap bean production (Hemphill, 2010). Beans are most sensitive to moisture stress during flowering and pod sizing. Water deficits during this

Fig. 20.7. Commercial dry bean production for mechanical harvest in the delta region of Sacramento Valley of California.

period will reduce both quality and yield. Peak water use for green bean production ranges from 0.5–0.4 cm (0.20 to 0.16 in)/day for summer crops. After full bloom, only irrigation early in the day is recommended to allow canopy drying as quickly as possible (Hemphill, 2010). In areas without frequent rain, weekly irrigation during the peak is adequate; however, with sandy and sandy loam soils, irrigation may be required as frequently as every 3–4 days. Ideally, 250–450 mm (10–18 in) of moisture evenly distributed throughout the growth season should be sufficient for bean production.

Harvesting and marketing

Harvesting of snap beans is determined by the stage of pod development. The harvest window is relatively large and depends somewhat on the market. For high yields, snap bean pods should be harvested at their maximum length while the pods are succulent and free of fiber before significant seed enlargement occurs (Fig. 20.8).

For destructive mechanical harvest, the ideal situation is to have all pods at the same stage of development. Degree-day measurements are frequently used for predicting and scheduling harvests of snap beans. To accurately predict the maturity date of a snap bean crop is important for farm managers and processors. The total heat unit model is most often used to predict bean development. Heat-units can be calculated as either growing-degree days (GDD) or growing-degree hours (Delahaut and Newenhouse, 1997). A degree day is a measure of the departure (°C or °F) of the mean daily temperature above a base temperature. A degree hour is the departure (°C or °F) of the hourly temperature above a base temperature. The base temperature used for snap beans is 10°C (50°F) (Delahaut and Newenhouse, 1997). If the mean daily temperature is lower than the base temperature then no GDD are accumulated. For example, a day with a high of 23°C (73°F) and a low of 12°C (54°F), and using a base of 10°C (50°F), would contribute 7.5 GDD^Cs (13.5 GDD^Fs) according to the following equation:

$$GDD = 7.5 = (23 + 12)/2 - 10$$

It takes from 1,000 to 1,200 GDD^F (556 to 667 GDD^C) for many snap bean cultivars to mature. GDDs may be calculated using either Celsius or Fahrenheit and can be converted using the relationship:

$$5\ GDD^C = 9\ GDD^F$$

In many developed countries most of the snap beans are grown from bush cultivars that are destructively harvested by a self-propelled or pull-type mechanical harvester that collects beans from several rows simultaneously. Beans are sequentially planted so harvests can be made continually throughout a season. A common harvester design strips the leaves and beans from each plant and then lighter fractions of leaves and other debris are blown back into the field while the beans are collected. A root system that strongly anchors the plant is an important feature for mechanical harvesting. Mechanically harvested snap beans for fresh market are often hand-sorted and graded before packaging. Snap bean yields for processing range from 13–17 metric tonnes/ha (6–8 short tons/acre) (Hemphill, 2010). Fresh market bean yields average approximately 5–11 metric tonnes/ha (2.5–5 short tons/acre).

Hand-harvest of bush, half runner, and pole snap beans is common in most of the world where automation is not feasible. Pole beans may be harvested over a longer period relative to bush types, so total yields may be higher. Besides higher yield potential, pole beans are better adapted to high-rainfall conditions, because the canopy and pods dry faster so there is a lower incidence of disease. Additionally, because pole bean pods are less likely to contact the soil they are clean and grow straight. Despite these advantages, bush bean production continues to increase relative to pole beans because of their lower production costs.

Haricot, shelly, or horticultural beans are harvested when seeds have achieved full size and are relatively

Fig. 20.8. Box of uniform snap beans grown in Florida.

firm. Seed moisture is much higher compared to dry bean seeds, which are generally combined or hand-harvested when the pods are brown and the seeds dried to a moisture content of <15%. Hand labor and/or a mechanical harvester collects pods that are then shelled to obtain the seeds. For most cultivars the pods are discarded but in some dual-purpose horticultural beans the pods may be eaten. Shelled haricot, shelly, or horticultural beans remain firm after cooking, much like dry beans. When seeds of snap beans are cooked they soften and lose their form.

Ideally, green beans are processed immediately after harvest without storage. Harvested beans should be protected from heat and sun and delivered to the processor as soon as possible. Fresh-market snap beans are highly perishable and should be cooled rapidly to 4–6°C (40–43°F) after harvest. Green beans can be vacuum cooled, forced-air cooled, or hydrocooled. The cold water of hydrocooling rapidly removes field heat and prevents wilting or shriveling. After hydrocooling beans should be thoroughly drained so that water is not trapped in packaging. Snap beans lose moisture quickly if not properly protected by packaging or if stored at a relative humidity (RH) of <95%. When the RH approaches saturation, as occurs in consumer packaging, temperatures above 7°C (45°F) must be avoided or decay is likely to occur within a few days. Green beans can be stored short-term at 4–7°C (40–45°F) and 95% RH (Hemphill, 2010). Green beans should be stored for only 7–10 days between 4–7°C (40°F and 45°F) because these temperatures may cause chilling injury. Snap beans are susceptible to chilling injury at temperatures of 3°C (38°F) or below. Chilling injury causes surface pitting and russeting from 24 to 48 h after transfer to warmer temperatures. Russeting is aggravated by free moisture and is especially noticeable where condensation occurs. 'Tendergreen' beans can be held for about 2 days at −0.6°C (31°F), 4 days at 1.6°C (35°F), or 12 days at 5.5°C (42°F) before chilling injury is induced (Hemphill, 2010). Cultivars differ significantly in their sensitivity to chilling. Beans can be held about 10 days at 4°C (40°F) if they are eaten or processed immediately after storage. Longer storage at temperatures above 7°C (45°F) will hasten yellowing and the development of fiber. Beans stored too long or at high storage temperatures are subject to various decays, including water soft rot (*Sclerotinia* spp.), cottony leak (*Pythium butleri*), gray mold (*Botrytis cinerea*), and Rhizopus rot (*Rhizopus* spp.) (Hemphill, 2010).

Storage containers of beans should be stacked to allow sufficient air circulation. If containers are packed too close, the temperature may rise from the heat of respiration and the beans will deteriorate rapidly. Beans can be stored in controlled atmospheres (2–3% oxygen and 5–10% carbon dioxide) and 4–7°C (40–45°F) to retard yellowing. The discoloration of broken ends of bean pods can be reduced by storage at 20% or 30% carbon dioxide for 24 h (Hemphill, 2010).

Diseases

Beans

White mold is caused by the fungus *Sclerotinia sclerotiorum* and develops as a cottony growth on the stem, stem branches, and pods of bean plants. The fungus also produces black, hard mats of mycelium called sclerotia near the cotton-like growth, which survive through the winter. The sclerotia germinate the next season when the leaf canopy completely covers the field and the soil surface is cool (15–18°C; 59–65°F) and wet for 10–14 days. White mold kills plants and infects pods, which severely reduces yield. High humidity and plant canopy temperatures from 15–18°C (68°F to 76°F) favor the spread of white mold (Meronuck et al., 1993).

Crop rotation helps prevent the buildup of inoculum. A rotation of 3–4 years between susceptible crops is necessary. Small grains and corn are recommended in a rotation with beans because they are not susceptible. Bean cultivars with an upright growth habit in wide rows and the use of recommended fertility and seedling rates will alleviate disease pressure. Careful irrigation management is also important because white mold is worse when there is excess moisture in the plant canopy. Timely applications of fungicides provide good control (Meronuck et al., 1993).

Bean rust is a fungal disease caused *Uromyces appendiculatus*. Symptoms first appear as small pale spots or lesions that turn yellow with small dark centers. Spots enlarge and produce brick-red rust spores that spread the disease. Lesions develop black overwintering spores later in the season. Ten hours or more of constant moisture and temperatures between 17–27°C (63–81°F) cause infection and symptoms appear within 10–15 days. Early infection increases the potential for yield reduction. Crop rotations with unrelated crops for 3–4 years provide control. Crop residue is the primary source of inoculum for the next season, so residue should be buried by plowing. If the disease is identified early, fungicide applications can provide control (Meronuck et al., 1993).

Common bean blight is caused by *Xanthomonas campestris* pv. *phaseoli*, halo blight is caused by *Pseudomonas syringae* pv. *phaseolicola* and brown spot is caused by *P. syringae* pv. *syringae*. Common bacterial blight first appears as small translucent water-soaked spots on the leaf. As these spots enlarge, this tissue dies leaving brown spots with a narrow yellow margin. Lesions are often large and irregular in shape. Water-soaked sunken pod lesions eventually turn brownish-red. Bacteria may infect the vascular system and kill the main stem and branches. Relatively high temperatures favor common blight (Meronuck et al., 1993).

Halo blight-infected plants exhibit symptoms similar to common blight. Halo blight lesions are surrounded by a large, pale-yellow circle 1.3 cm (0.5 in) in diameter. Temperatures between 16–20°C (60–68°F) favor halo blight. The systemic plant infection produces stunted plants with small chlorotic trifoliolate leaves. Red or brown water-soaked lesions appear on pods (Meronuck et al., 1993).

Brown spot lesions are similar to those of common blight with a narrow yellow border, but no halo, surrounding some of the lesions but without water-soaking of the leaf tissue. Brown spot infection is favored by relatively cool conditions. As lesions mature, the dead tissue in the center falls out, resulting in a shot-hole. Infected pods may be twisted or kinked at the infection point (Meronuck et al., 1993).

Infected seed-spread and damp conditions favor the development of all three diseases. A 3–4-year rotation, incorporation of old bean debris, the reduction of volunteer beans and the use of certified streptomycin-treated seed helps to manage these diseases. Certain copper-containing compounds may prevent the spread of bacteria (Meronuck et al., 1993).

Fusarium solani f. sp. *phaseoli*, *Rhizoctonia solani*, and various *Pythium* sp. commonly cause root rot and damping-off diseases in beans. These root-rotting fungi are present in the soil and live on decomposing vegetation and can attack plants whenever the population is large enough and the soil is cool, wet and compacted (Meronuck et al., 1993).

Rotation is very important in controlling root rots. Rotations of 3–4 years between bean crops are effective. Poor drainage and waterlogged conditions can contribute to damping-off and root disease. Other crops such as sunflower, potato, sugarbeet, and soybean that are susceptible or support the survival of these fungi, should not be used in rotation.

Bean common mosaic virus (BCMV) stunts plants and causes malformed and mottled leaves. Leaves infected with BCMV are irregular shaped and puckered with light yellow and green patches. Infected leaflets are often narrower and longer than normal, with a downward cupping. Bean plants infected early in the season are yellowish, dwarfed, and spindly. Dark necrotic lesions and spots are sometimes found on the roots, petioles, pods, and leaves. The virus is spread by direct contact from aphids and by seed. The only methods of control are resistant cultivars and noninfected seed. Removing infected plants from the field prevents secondary spread (Meronuck et al., 1993). Other viruses that infect beans include bean golden mosaic virus (BGMV), bean pod mottle virus (BPMV), bean southern mosaic virus (BSMV), bean yellow mosaic virus (BYMV), the bean strain of cucumber mosaic virus (CMV), the bean strain of sugar beet curly top virus (SBCTV), and peanut stunt virus (PSV) (Rubatzky and Yamaguchi, 1997).

Several *Alternaria* sp. fungi cause leaf spot diseases of bean. Spots have an irregular shape and are gray-brown colored. When the weather is cool and humid, lesions coalesce, forming large contiguous areas. Infected leaves may tear and have a ragged appearance. *Alternaria* survives in infested crops and weed residue and may infect leaves through wounds. The disease is favored by wet and damp conditions and is found most prevalent on mature or senescent leaves. Satisfactory control measures have not been developed, although crop rotation and wider row spacings help to limit damage. No fungicide control is available (Meronuck et al., 1993).

Angular leaf spot is a fungal disease caused by *Phaeoisariopsis griseola*. The disease affects the foliage and pods of beans and can be a problem when warm moist conditions accompany abundant inoculum from infected plant residues or contaminated seed. All above-ground tissues are susceptible. Leaf lesions are gray or brown with a chlorotic halo. After approximately 10 days lesions turn brown and necrotic and assume a distinct angular shape. In advanced stages, the lesions may drop out, leaving holes in the leaf blade. Primary inoculum comes from seed or infested residue.

Infection and disease occur between 16–28°C (61–82°F) with an optimum of 24°C (75°F). Planting disease-free certified seed, resistant cultivars, and rotating 2–3 years between bean crops provides control. Fungicides should be applied when the disease first appears and conditions are favorable for development (Meronuck et al., 1993).

Anthracnose (*Colletotrichum lindemuthianum*) can affect all above-ground tissues of a bean plant. Lesions first appear as water-soaked areas then become elongate, angular, and brick-red to purple before turning dark brown to black. Lesions appear on the petioles and on both the upper and lower surfaces of leaves and leaf veins. A spore-bearing structure ruptures the host cuticle. Pod lesions are tan- to rust colored and develop into sunken cankers outlined by a slightly raised black ring surrounded by a reddish-brown border. Tan- to salmon-colored spores may form in the lesions. The disease is favored by temperatures of 13–21°C (55–70°F) with an optimum at 17°C (63°F). The disease is most severe on susceptible plants under frequent rainfalls accompanied by wind and splashing rain. Free water is needed for all stages of development. The fungus survives in crop debris and can be spread on seed, in the air and water. Seed treatment, certain copper fungicides, crop rotations to reduce the buildup of disease inoculum and planting disease-resistant cultivars provide control (Meronuck et al., 1993).

Insect Pests

Bean

Many insects are pests of common bean (CABI, 2013a). Lepidopterous caterpillars from the genera *Ascotis*, *Spilosoma*, *Amsacta*, and *Euproctis* all attack bean crops. Caterpillars feed on leaves, buds, flowers, and pods. Many predators and parasites attack the eggs of caterpillars including several species of *Trichogramma*. Most parasitized eggs turn black, but there may be a lag period before they do so. Generalist predators such as lacewings, minute pirate bugs, and damsel bugs feed on corn earworm eggs and small larvae as well. Sprays of *Bacillus thuringiensis* are an effective biological control. Chemical insecticide sprays may also be used for conventional production (UC IPM, 2013).

Several species of leafhoppers are found in dry beans with *Empoasca fabae* and *E. solana* being two of the more common (CABI, 2013a). Empoasca leafhoppers are small (3 mm long, 0.13 in), bright-green, wedge-shaped insects. The small, wingless nymphs or immatures are also wedge shaped and green and move rapidly forward, backward, and from side to side. Both adults and nymphs live on the underside of leaves. Leafhoppers cause a symptom called "hopperburn", where the leaf margins turn yellow, particularly at the leaf tip, and these areas soon become necrotic. The entire leaf may become yellowed resembling virus symptoms. Insecticides are most effective when leafhoppers first appear (UC IPM, 2013).

There are several types of chrysomelid or leaf beetles that feed on beans. Leaf beetles belong to the family Chrysomelidae. This family includes the pests bean leaf beetle (*Ootheca bennigseni*) and redheaded flea beetles (*Systena frontalis*). They are oval to oval-elongate shaped and usually less than 13 mm (0.5 in) long. Their color and shape is so variable, however, that they may be difficult to identify consistently. Adults feed on flowers and foliage while the larval forms generally attack roots (CABI, 2013a).

Darkling beetles (*Blapstinus* spp.) are not a type of leaf beetle. They are from 3.5–6 mm (0.13–0.25 in) long and vary from black or bluish-black to rusty brown. They are often hidden by dust or a thin layer of soil and difficult to detect. Larvae are cylindrical, wireworm-like, soil-inhabiting worms that are light yellow to dark brown and range from 0.8–8 mm (0.03–0.33 in) in length (UC IPM, 2013). They are often referred to as false wireworms. Darkling beetle development from egg to adult may require 50 days during summer. Eggs hatch in 3–6 days and there can be five larval instars. Beetles are frequently numerous in spring and early summer and may run on the ground but are more frequently found under clods or organic debris during daylight hours. Damage is often caused when young plants are girdled or cut off at or below the soil surface. Older plants are generally not affected. Plants should be inspected for darkling beetle damage when the crop emerges. Treatment is applied only when darkling beetles reduce stands of the young plants (UC IPM, 2013). Chemical insecticides provide control.

Mexican bean beetle (MBB, *Epilachna varivestis*) adults are coppery brown with black spots. When adults arrive in a bean field, they lay yellow-orange egg masses on the underside of bean leaves. Eggs hatch into bright yellow, spiny, oval larvae that feed and molt several times. They pupate on the underside of leaves (UC IPM, 2013). Feeding damage is by both adults and larvae, primarily on leaves. Pod injury can occur if numbers are high. The adult MBB overwinters and seeks out beans for feeding and reproduction. Overwintering adults colonize beans in June. There are two or three generations per season, usually increasing in numbers with each generation. The complete life cycle takes 30–40 days during the summer. *Pediobius foveolatus* is a commercially available biological control agent for MBB control (UC IPM, 2013). This small (1–3 mm, 0.04–0.1 in), non-stinging parasitic wasp lays its eggs in MBB larvae. Wasp larvae feed inside the MBB larva, kill it, and pupate inside, forming a brownish case or "mummy". Chemical sprays are also available.

Some of the most serious pests of dried beans are weevils (cowpea weevil, *Callosobruchus maculatus*; broad bean weevil, *Bruchus rufimanus*; bean weevil, *Acanthoscelides obtectus*). Bean weevils and cowpea beetles are widespread, attacking filling and dried beans (CABI, 2013a). The adults are relatively small beetles, 3.5–5 mm (0.13–0.2 in) in length, somewhat teardrop or triangular in shape, and dull-colored with white, reddish or black markings (UC IPM, 2013). The eggs may be glued to the bean or the pod in the case of cowpea weevil, glued to green pods (broad bean weevil) or laid loosely among beans or through cracks in the pods (bean weevil). The larval and pupal stages are spent inside the bean thus destroying them for commercial use. Infestations begin in the field as adults are introduced through discarded beans left in containers, harvesters, planters, or feed areas. Broad bean weevil infestations also start in the field, but this pest does not create a problem in storage. As with the cowpea weevil, bean weevil will attack dried beans, and can be a serious pest in stored beans consuming nearly the entire bean contents (UC IPM, 2013). Pupation occurs in the beans and adults emerge through a round hole in the seed coat. Damage is a combination of the feeding and contamination. Sanitation offers the most practical means of control. Because field infestations originate from beans, potential sources of weevils in production areas should be eliminated (UC IPM, 2013). Pyrethrins are effective in controlling weevils.

Thrips (*Thrips tabaci*, *Heliothrips* spp.) feed on leaves and buds of young bean seedlings, making plants look ragged. A sign of serious thrip infestation is distorted leaves that turn brownish around the edges and cup upward. Usually the plants will outgrow the problem. Foliage-feeding thrips are effective natural predators on early season spider mite (*Tetranychus urticae*) infestations of beans. Both adult and immature thrips may be found in spider mite colonies feeding on spider mite eggs. Biological control and unfavorable weather generally reduce thrip populations before treatment is necessary. Minute pirate bugs (*Orius tristicolor*) play a major role in controlling thrip populations. Thrip populations tend to build up on weeds. Cultivating nearby weedy areas before beans emerge will reduce the potential of thrips migrating when the weeds begin to die. Cultivating weedy areas after bean emergence will increase thrip problems (UC IPM, 2013).

The bean fly (*Ophiomyia phaseoli*) and stem fly (*Ophiomyia centrosematis*) are serious pests of many types of beans in Africa and Asia (CABI, 2013a). The most serious damage by adults occurs when plants are at the single-leaf stage. Single leaves typically show a large number of feeding and oviposition punctures on the upper side with corresponding light yellow spots, especially on the basal portion of the leaf (UC IPM, 2013). Larvae begin feeding soon after hatching, producing numerous larval mines visible on the underside of the leaves just under the epidermis. The developing larvae in second and third instar mine downward into the cortex just underneath the epidermis. The third instar continues to feed downwards into the taproot and returns to pupate still inside the stem close to the soil surface (UC IPM, 2013). The feeding tunnels are clearly visible on the stems. If the bean fly larvae population is high, larval feeding destroys the cortex tissue at the root–shoot junction. Initial symptoms of bean fly larvae are yellow leaves, stunting, and plant mortality. Unfortunately there is no completely effective biological control, although research is continuing. Carbaryl sprays may provide some control (UC IPM, 2013).

Bean aphids include cowpea aphid (*Aphis craccivora*), bean aphid (*Aphis fabae*), pea aphid (*Acyrthosiphon pisum*), and green peach aphid (*Myzus persicae*). Aphids damage plants by: (i) sucking plant sap, which causes heavily infested leaves to curl and stunts plants; (ii) excreting honeydew, which causes sticky shiny leaves to ultimately turn black because of a sooty-mold fungus growth; and (iii) spreading plant diseases (a large number of viruses are vectored by aphids) (UC IPM, 2013). Infestations frequently are localized with heavily infested leaves curled downward. Often natural parasites and predators prevent aphid infestations throughout an entire field. Temperatures >29°C (85°F) inhibit buildup of pea and green peach aphids. Some common predators of aphids in beans include lady beetles (ladybugs), syrphid or hover flies, and lacewings. Parasitic wasps attack each of the common aphid species, turning them into hard, crusty mummies. Insecticide may also provide effective control, but prolonged use may lead to resistance (UC IPM, 2013).

The adult lygus bug (*Lygus hespersus*, *Chauliops fallax*) is about 6 mm (0.25 in) long and about half as wide. It is generally brownish but varies from green to straw-colored, tawny or light brown. The body is marked with a pattern of different shades of brown and occasionally yellow or red marks (UC IPM, 2013). A prominent V-shaped yellowish area is present near the center of the body at the base of the wings. Lygus eggs are laid within plant tissue so that only the oval-shaped cap is visible. These eggs are difficult to see even with magnification. Active green nymphs hatch from the eggs. Red coloration on the tips of the antennae helps to distinguish early instar *Lygus* from aphids. Lygus bugs may be present throughout the growing season and can be highly destructive. They have sucking mouthparts that pierce and consume plant tissue. The type of damage varies with plant age. During early bud and flowering stages, lygus bugs cause bud and flower loss, reducing yields. Lygus-bug feeding on young, developing seed pods causes distortion, pitting, blemishes, and reduced germination in seed beans (UC IPM, 2013).

Lygus bug eggs are often parasitized and killed by a small parasitic wasp, *Anaphes iole*. General predators, such as lacewings and damsel bugs, may prey on lygus bug nymphs. Avoiding the use of broad-spectrum insecticides will protect natural enemies. As a preferred host, alfalfa hay might be managed to suppress movement of lygus into dry bean fields by staggering cuttings to preserve habitat. Leaving a small, uncut strip of alfalfa will limit the movement of lygus bugs into neighboring bean fields (UC IPM, 2013).

The most common nematode pests of beans are species of root knot nematodes (*Meloidogyne* spp.). Lesion nematodes, *Pratylenchus* spp., are also common. Management of nematodes in beans requires integration of several cultural practices, including choice of cultivar, crop rotation, sanitation and fallow periods. Nematode-resistant bean cultivars have been developed including 'UC-301', 'White Ventura N', 'Maria', 'UC-90', 'UC-92', and 'Cariblanco N' (UC IPM, 2013).

Site Selection and Field Production: Pea

Peas can be grown in many types of soil from light sandy loams to heavy clays. The pH range is from 6.0 to 7.5. Poor soil drainage and compaction reduce productivity and increase susceptibility to root disease. Crop rotations help reduce root-rot diseases, and peas should be grown in the same soil only once every 4 years.

Peas grow best during prolong periods of cool weather. In temperate regions, peas are usually planted in the spring or in late fall and early winter in mild climates where prolonged periods below freezing do not occur. Peas can tolerate light frosts before flowering. In the tropics and subtropics, peas are grown at high elevations and during seasons when temperatures are relatively cool.

Planting

High-quality seed is very important, especially for planting in cold soil. Germination can occur over a wide range of soil temperatures but is best at 20°C (68°F). At temperatures >25°C (77°F), germination percentage decreases. Seed may be treated with fungicides to minimize diseases during emergence and early seedling growth.

Peas are often planted using grain drills, but vacuum and precision-belt seeders may also be used. Pea seed should be inoculated immediately before seeding to insure an adequate supply of N-fixing bacteria as described above for beans. A fresh, effective, live culture of the correct strain of rhizobia should be used. The need for inoculation is reduced in a field where peas are grown in rotation and yields have been satisfactory. Seeds are planted in moist, well-tilled soil from 3–5 cm (1.2–2 in) deep, with in-row spacing of 3–5 cm (1.2–2 in) and between-row spacing of 20–30 cm (8–12 in). Plant densities of 80–90 plants/m^2 (7–8 plants/ft^2) are common for mechanical harvest. Hand-harvested, fresh-market peas are grown in wider-spaced rows 75–90 cm (30–35 in) apart to allow access for picking (Fig. 20.9). At higher populations, the development of tillers or sucker shoots is minimized. Pods that develop on tiller stems mature too late, especially for machine harvest.

Fig. 20.9. Immature determinate green pea cultivar producing both flowers and fruit. The wide row spacing is to accommodate hand harvest. Commercial production is generally mechanically harvested.

Therefore, single-stem plants are preferred and achieved by cultivar selection or cultural practices.

Lodging complicates mechanical harvest and reduces yield. High plant populations reduce lodging by providing mutual support through intertwining growth. Determinate and short cultivars have less tendency to lodge. Except for home gardens and edible-podded pea production, indeterminate cultivars and trellising are not commonly used commercially. Seeding rates are variable because of differences in seed size among cultivars, so the amount of seed sown can vary from 70 to more than 200 kg/ha (62–178 lb/acre) (Rubatzky and Yamaguchi, 1997).

Growth and development

Peas are very temperature responsive, especially during vegetative development. Optimum mean temperatures for vegetative growth of peas are between 13–18°C (55–64°F) and growth is very slow above 29°C (84°F). Although peas are classified as a cool-season vegetable, subfreezing temperatures may damage plants. Young plants are more tolerant of low temperatures than mature ones. Blossoms and pods are more sensitive to subfreezing temperatures than leaves and stems.

Growing degree days, as explained previously for bean, are commonly used to predict harvest dates and to separate planting dates for processing peas (Delahaut and Newenhouse, 1997). For peas, a base temperature of 4°C (39°F) with an upper limit of 29°C (84°F) is usually the range for calculating GDD because growth is negligible outside this range and few heat units are accumulated. Cultivars are often identified by their heat unit requirements. Early-producing cultivars require as few as 1,000 GDD to achieve harvest maturity. Some late-maturing cultivars require more than 1,600 GDD (Delahaut and Newenhouse, 1997). Generally, early-flowering cultivars are day neutral but flowering is accelerated by long days for late-maturing cultivars. Moderate temperatures also decrease the time to flowering, but temperatures above 30°C (86°F) can cause flower or ovule abortion. Moderate diurnal temperature fluctuations improve plant growth (Rubatzky and Yamaguchi, 1997).

Harvesting and marketing: pea

The optimum stage for green pea harvest is when pods are well filled and the seeds are still soft and immature. As seeds mature, they increase in firmness, their coats thicken, and sugars are converted to starch. Peas mature rapidly during high temperatures and are at optimum quality for only 1–2 days, similar to sweet corn. Peas for processing are often grown in cooler climates to slow development and give greater flexibility in harvest times.

To extend production, cultivars with different maturity dates are grown or sequential plantings made. Although extremely labor intensive, hand-harvesting minimizes physical damage and helps to retain quality. Increasingly, fresh green peas are sold with pods removed so they can be cooked without hulling. Mechanically harvesting peas for processing, and increasingly for fresh market, reduces labor. These harvesters separate pods from vines and then peas from pods. The peas are collected while the pods and vines are discarded. In some cases, the pods are mechanically or hand harvested from the vine and shelled with other machinery that shells the peas from pods. In order for the harvest equipment to work effectively, cultivars should be determinate, have short stiff stems, anchoring root systems, and pods concentrated near the top of the plant. Harvesters function more effectively with cultivars that have sparse foliage, such as the leafless-types.

The use of single destructive mechanical harvest can result in a range of pea maturities. Peas of different maturity of the same cultivar are commonly separated by size. Because shelled peas of different maturity have different specific gravities, they can also be separated by floatation using different concentrations of brine.

Because pea seeds develop rapidly over a brief period, it is important to consider edible quality and not just total yield. Edible quality is usually objectively determined using a tenderometer or instrument that measures their physical characteristics. These instruments measure resistance of peas to crushing pressure; a greater resistance to crushing means tougher, over-mature peas. Crushing resistance readings correlate well with other edible quality indicators such as sweetness, tenderness, and starch content (Rubatzky and Yamaguchi, 1997).

Edible-podded or snow pea and snap pea cultivars are almost exclusively hand-harvested, sometimes daily, in order to avoid over-maturity and preserve the delicate pods from mechanical damage. The proper harvest stage for snow pea harvest is when the pods are about full size but before seed development is apparent. With snap peas, the seeds are more developed but still immature. Harvests of indeterminate, edible-podded pea plants can continue for many weeks, and frequent harvesting tends to favor additional flowering. Production occurs where there is an abundance of inexpensive labor. For example, much of the US production of edible pod peas has relocated to the highlands of Central America.

Field-pea harvests are delayed as long as possible to allow late-setting pods to mature. However, harvest delays can cause seed shatter. Foliage desiccants are sometimes used to facilitate harvesting. Early harvesting can reduce yield and require artificial drying. Field peas are often harvested with a moisture content of about 40%. In their fully dried state, moisture content should less than 15% to prevent molding (Rubatzky and Yamaguchi, 1997).

Postharvest handling and storage: pea

Similar to sweet corn, temperature is a fundamental factor in the conversion of sugar to starch in green pea seeds. To maintain sugar content, it is important that harvested garden peas are not exposed to high temperatures and are cooled as soon as possible to remove field heat. Harvested pods or shelled peas should be cooled to 0°C (32°F) to limit sugar conversion and fiber development. Hydrocooling is the most effective method. A significant loss in sugar content occurs in as little as 3 h at ambient temperature. Unshelled peas retain higher quality better because the pod reduces desiccation. The high CO_2 content within the pod provides a modified atmosphere that reduces seed respiration. Dried field peas present few postharvest or storage concerns when maintained at <15% moisture content (Rubatzky and Yamaguchi, 1997).

Weed Management Strategies

Pea plants compete poorly with weeds. Weed control early in the season is essential for producing a successful crop. To accommodate narrow row spacing without a dependency on mechanical or hand cultivation, several selective herbicides are widely used. Wider row spacing is needed for cultivation. It is important when selecting herbicide to consider the crop grown after peas. Crop rotations and stale-bed preparations to kill weeds are also important components of a weed-management program.

Diseases

Although a few pea diseases are common to beans many are unique. The following is a partial list of some of the most common diseases of pea: Aphanomyces root rot (*Aphanomyces euteiches*), anthracnose (*Colletotrichum pisi*), Ascochyta blight complex (*Ascochyta pisi, A. pinodella, Mycosphaerella pinodes*), bacterial blight (*Pseudomonas syringae* pv. *pisi*), black root rot (*Thielaviopsis basicola*), Botrytis (*Botrytis cinerea*), downy mildew (*Peronospora pisi, P. viciae*), Fusarium wilt and near wilt (*Fusarium oxysporum* f. *pisi*, other races of *F. oxysporum*), Fusarium root rot (*Fusarium solani* f. sp. *pisi*), leaf and pod spot (*Phoma medicaginis* var. *pino-della*), powdery mildew (*Erysiphe polygoni* and other *Erysiphe* spp.), Pythium root rot (*Pythium ultimum* and other *Pythium* spp.), rusts (*Uromyces pisi, U. fabae, U. trifolii*), Sclerotinia (*Sclerotinia sclerotiorum*), and Septoria blight (*Septoria pisi*).

Virus diseases are a major problem for pea producers. Most of the following list are vectored by aphids and other insects: bean yellow mosaic virus (BYMV), pea strain, pea enation mosaic virus, pea streak virus, pea stunt virus, pea seed-borne virus, pea early browning virus, pea top yellow virus, common pea mosaic virus, cucumber mosaic virus

(CMV), and bean leaf roll virus (BYMV) pea strain (CABI, 2013b).

Hollow heart, also called marsh spot, is a physiological disorder that occurs in crops deficient in manganese (Mn). High soil pH can reduce Mn availability.

Insect Pests, Mites, and Nematodes

There are many insects, mites, and nematodes that attack peas including: pea aphid (*Acyrthosiphon pisum*), lucerne bug (*Adelphocoris lineolatus*), black cutworm (*Agrotis ipsilon*), pea weevil (*Bruchus pisorum*), pea leaf miner (*Chromatomyia horticola*), green looper caterpillar (*Chrysodeixis eriosoma*), white leafhopper (*Cofana spectra*), devil grasshopper (*Diabolocatantops axillaris*), flower thrips (*Frankliniella intonsa*), western flower thrips (*Frankliniella occidentalis*), red-legged earth mite (*Halotydeus destructor*), brown stink bug (*Halyomorpha halys*), spiral nematode (*Helicotylenchus dihystera*), cotton bollworm (*Helicoverpa armigera*), native budworm (*Helicoverpa punctigera*), pea cyst eelworm (*Heterodera goettingiana*), pea blue butterfly (*Lampides boeticus*), serpentine leafminer (*Liriomyza huidobrensis*), vegetable leafminer (*Liriomyza sativae*), American serpentine leafminer (*Liriomyza trifolii*), cabbage moth (*Mamestra brassicae*), bean flower thrips (*Megalurothrips usitatus*), peanut root-knot nematode (*Meloidogyne arenaria*), false root-knot nematode (*Nacobbus aberrans*), stemfly (*Ophiomyia centrosematis*), scarlet mealybug (*Pseudococcus calceolariae*), crown gall (*Rhizobium radiobacter*), bean bug (*Riptortus clavatus*), pea leaf weevil (*Sitona lineatus*), Costa Rican armyworm (*Spodoptera albula*), beet armyworm (*Spodoptera exigua*), cotton leafworm (*Spodoptera littoralis*), field thrips (*Thrips angusticeps*), honeysuckle thrips (*Thrips flavus*), and red flour beetle (*Tribolium castaneum*) (CABI, 2013b).

Production and Culture: Lima Bean

Lima bean is more sensitive to the environment than snap bean and slightly warmer temperatures are needed for best growth. Favorable mean temperatures range between 15–25°C (59–77°F). Larger-seeded limas generally prefer lower temperatures and higher humidities than small-seeded cultivars, especially with regard to pollination. For large-seeded limas, temperatures above 30°C (86°F) and <60% RH may cause poor pollination and blossom drop. This sensitivity restricts lima bean production to specific climates. The growth regulator NAA may improve pod set (Rubatzky and Yamaguchi, 1997). Moisture stress during flowering may also reduce fruit set.

Seeds germinate best between 15°C–30°C (59–86°F). Planting depths vary depending on seed size and range from 3–6 cm (1.2–2.4 in), with large-seeded types planted deeper. Emergence is most rapid at about 25°C (77°F). Like beans, seed are very susceptible to mechanical injury. Seeds are inoculated with rhizobium before planting, especially if other legumes have not been grown recently (Rubatzky and Yamaguchi, 1997). For bush-type plantings, a usual spacing is 60–90 cm (24–35 in) between rows with in-row spacing of 10–15 cm (4–6 in). Pole-limas are spaced wider to accommodate hand-harvesting and are often grown on a support similar to pole beans (OSU, 2004).

Lima bean is well suited for production on light-textured, warm, well-drained soils. A slightly acid soil with pH between 6 and 7 is preferred. Other than temperature restrictions, lima bean culture is similar to snap beans. Limas require longer growth and consequently more moisture and nutrients than green beans. The crop is fertilized before planting based on soil test results. Supplemental fertilizer is often side-dressed or applied through fertigation during the long growing season. Although they are N fixers, lima beans show a greater response to N fertilization compared to snap bean. Depending on soil type, from 67–112 kg/ha (60 to 100 lb N/acre) is applied in split applications before planting, 40 days after planting and sometimes at full bloom to meet crop needs. Care should be taken not to over-fertilize. Excessive N early in the season will inhibit nodulation. Too much N during crop development will delay flowering and increase disease. A lima bean crop will remove about 22 kg/ha (20 lb/acre) P, 112 kg/ha (100 lb/acre) K, 11 kg/ha (10 lb/acre) S, and 11 kg/ha (10 lb/acre) Zn. Many of the same diseases and pests that affect snap and other bean species also affect limas.

Harvest and postharvest: lima bean

Depending on cultivar and growing temperatures, the period from seeding to harvest may vary from

70 to 110 days or longer. Compared to snap beans, limas take longer to produce the enlarged but immature seed that is consumed. As the seed matures, the coat changes color from green to cream or white. During seed maturation, the pod bulges as a result of seed enlargement. At the proper harvest stage for fresh use, pods are green and seed moisture is at 60–70%. Because of different flowering times, pods do not mature uniformly.

Lima beans are difficult to separate from the plant, so hand harvest is challenging. The trend for commercial production in developed countries has been to grow small bush cultivars that are mechanically harvested to reduce labor costs. Yields of shelled fresh green limas range from 2.5–4 metric tonnes/ha (1.1–1.8 short tons/acre) (OSU, 2004).

Machine-harvested, shelled lima beans have a short postharvest life and are sold fresh or processed by freezing or canning as soon as possible. Lima beans of different maturity are separated based on specific gravity using brine solutions. Freshly harvested lima beans in pods have longer shelf life and can be held in marketable condition for several weeks at 5–7°C (41–45°F) and 90% RH. Chilling injury occurs at lower temperatures. Elevated atmospheric CO_2 extends storage (OSU, 2004).

Significant quantities of limas are harvested as dry, fully mature seed. A dry lima crop is also machine harvested and handled much like dry beans. When dry, most limas are white or tan colored. Dry limas store well at <15% moisture content and are sold in bags or boxes and are rehydrated for cooking.

References

CABI (2013a) Datasheets: *Phaseolus vulgaris* (common bean). Crop production: Notes on Pests. Available at: www.cabi.org/cpc (accessed 17 July 2013).

CABI (2013b) Datasheets: *Pisum sativum* pea. Crop production: List of pests, major host of. Available at: www.cabi.org/cpc (accessed 17 July 2013).

Cook, R. (2011) Tracking demographics and USA fruit and vegetable consumption patterns. USDA Factbook Chapter 2. Available at: www.usda.gov/factbook/chapter2.pdf (accessed 27 July 2013).

Davies, D.R., Berry, G.J., Heath, M.C. and Dawkins, T.C.K. (1985) Pea (*Pisum sativum* L.). In: Summerfield, R.J. and Roberts, E.H. (eds) *Grain Legume Crops*. Williams Collins Sons and Co., London.

Davis, J.G. and Brick, M.A. (2013) Fertilizing dry beans. Publication 0.539. Colorado State University Extension. Available at: www.ext.colostate.edu/pubs/crops/00539.html (accessed 17 July 2013).

Delahaut, K.A. and Newenhouse, A.C. (1997) *Growing Beans and Peas in Wisconsin: A Guide for Fresh-market Growers. Publication A3685.* University of Wisconsin–Extension, Cooperative Extension. Madison, Wisconsin.

Dickson, M.H. (1971) Breeding beans, *Phaseolus vulgaris* L., for improved germination under unfavorable low temperature conditions. *Crop Science* 11, 848–850.

Erker, B. and Brick, M.A. (2006) Legume seed inoculants. Publication No. 0.305. Colorado State University Extension. Available at: www.ext.colostate.edu/pubs/crops/00305.html (accessed 4 July 2013).

ERS (2011) Vegetables and melons yearbook. Economic research service. Available at: http://usda.mannlib.cornell.edu/MannUsda/viewDocumentInfo.do?documentID=1212 (accessed 11 June 2013).

FAOSTAT (2011) Beans, peas and pulses production statistics. Available at: http://faostat.fao.org/site/567/DesktopDefault.aspx?PageID=567 (accessed 3 June 2013).

Fofana, B., Vekemans, X., Du Jardin, P. and Baudoin, J.P. (1997) Genetic diversity in Lima bean (*Phaseolus lunatus* L.) as revealed by RAPD markers. *Euphytica* 95, 157–165.

Hemphill, D. (2010) Oregon vegetables: Snap beans-green roma, yellow wax. Available at: http://horticulture.oregonstate.edu/content/beans-snap-green-romano-yellow-wax (accessed 12 July 2013).

Hochmuth, G.J. and Hanlon, E.A. (2010) A summary of N, P, and K research with snap bean in Florida. Publication# SL 313. Available at: http://edis.ifas.ufl.edu/cv234 (accessed 15 July 2013).

Hymowitz, T. (1970) On the domestication of the soybean. *Economic Botany* 24, 408–421.

Jones, K.M., Kobayashi, H., Davies, B.W., Taga, M.E. and Walker, G.C. (2007) How rhizobial symbionts invade plants: the Sinorhizobium–Medicago model. *Nature Reviews. Microbiology* 5, 619–633.

Lorenz, O.A. and Maynard, D.N. (1988) *Knott's Handbook for Vegetable Growers*. Wiley, Hoboken, New Jersey.

Makasheva, R.K. (1983) *The Pea*. Oxonian Press Private Ltd., New Delhi, India.

Meronuck, R.A., Hardman, L.L. and Lamey, H.A. (1993) Crop pest management series edible bean disease and disorder identification. North Central Regional Extension Publication 159. Available at: www.extension.umn.edu/distribution/horticulture/dg6144.html#Bacterial (accessed 17 July 2013).

Mohamed-Yasseen, Y., Barringer, S.A., Splittstoesser, W.E. and Costanza, S. (1994) The role of seed coats in seed viability. *The Botanical Review* 60, 426–439.

Nelson, D.L., Lehninger, A.L. and Cox, M.M. (2008) *Lehninger Principles of Biochemistry*. Macmillan Publishing, New York.

OSU (2004) Commercial vegetable production recommendations. Lima beans *Phaseolus lunatus*. Available at: http://nwrec.hort.oregonstate.edu/lima.html (accessed 12 July 2013).

Rubatzky, V.E. and Yamaguchi, M. (1997) *World Vegetables: Principles, Production and Nutritive Values*, 2nd edn. Chapman & Hall, New York.

Singh, S.P., Gepts, P. and Debouck, D.G. (1991) Races of common bean (*Phaseolus vulgaris*, Fabaceae). *Economic Botany* 45, 379–396.

Stevens, P.F. (2012) Angiosperm phylogeny website. Version 12. Available at: www.mobot.org/MOBOT/research/APweb (accessed 2 July 2013).

UC IPM (2013) How to manage pests: Dry beans. Available at: www.ipm.ucdavis.edu/PMG/selectnewpest.beans.html (accessed 18 July 2013).

Zohary, D. and Hopf, M. (2000) *Domestication of Plants in the Old World*, 3rd edn. Oxford University Press, Oxford, UK.

21 Family Apiaceae

Origin and History

This family, consisting mostly of herbs, is one of the largest families of flowering plants, consisting of approximately 434 genera and 3,700 species (Stevens, 2012). Many of the species can be found in temperate regions and in tropical highlands located throughout many regions of the world. The family is defined by its distinctive umbrella-like inflorescence, the umbel. The name Umbelliferae still persists in some older literature but is outdated and largely replaced (Rubatzky *et al.*, 1999). Members of this family have had both culinary and medicinal uses although some are also poisonous. Many of the species are biennial.

CARROT

Progressing from its initial use as a medicinal plant, carrot (*Daucus carota* L. subsp. *sativus* (Hoffm.) Arcang.) has become a major world vegetable (Ross, 2005; USDA, 2013). Carrot is one of the most widely grown and important vegetables in the family Apiaceae. Wild carrot species are common around the world and include *D. carota* L. subsp. *maritimus*, *D. pusillus* Michx., *D. carota* subsp. *hispanicus* (Gouan) Thell, *D. carota* subsp. *gummifer* (Syme) Hook, *D. carota* L. subsp. *drepanensis* (Arcang.) Heywood, and *D. carota* subsp. *major* (Vis.) Arcang. (Heywood, 1983; USDA-GRIN, 2013). *Daucus carota* subsp. *carota* is a wild carrot known as "Queen Anne's lace" in North America and is a weed in Europe and Asia (USDA, 2013). Queen Anne's lace foliage resembles domesticated carrot while its white roots are narrow and irregularly shaped with a strong carrot odor. Queen Anne's lace is a biennial and readily crosses with the cultivated carrot, which may contaminate carrot seed production. It is classified as a noxious weed in some areas (USDA, 2013). The modern carrot probably evolved from a wild plant resembling Queen Anne's lace.

Carrots originated in present-day Afghanistan about 3000 BC, probably with purple or yellow roots. By the year AD 1000, carrots had spread to Syria. By the year 1100, yellow and purple carrots were known in Spain, and by the 1200s they had spread to Italy and China. During the 1300s, white, yellow and red types were grown in France, Germany, and the Netherlands. By the 1500s, the first reports of orange carrots appeared, and carrots were known throughout Europe (Banga, 1957, 1963).

It is speculated that mutation and selection were more responsible for the development of the cultivated carrot than hybridization with wild germplasm (Banga, 1957, 1963). Plant breeding during the 17th century in the Netherlands improved root smoothness and orange color leading to the 'Long Orange' and 'Horn-type' carrots (Banga, 1957, 1963). The horticulturist Louis de Vilmorin, who laid the foundations for modern plant breeding, developed 'Nantes' and 'Chantenay' carrots in France during the 19th century. These types served as the basis of much of the germplasm used in the development of modern cultivars (Banga, 1957, 1963). Early European colonists brought orange carrots to America (Banga, 1957, 1963). Over the centuries, orange carrots came to dominate commercially in North America, Europe, and much of the western world. Purple, yellow, and white carrot cultivars are regionally popular.

Botany and Life Cycle

Plant characteristics

Carrot is a cool-season, biennial dicot, whose fleshy axis is the edible tissue consumed as food. The leaves or tops are inedible. Carrot plants form a rosette of leaves and a large fleshy storage taproot during the first year. The stem is very compressed like a plate during the first year of growth, with foliage height generally ranging between 25–60 cm (10–24 in). Leaves that arise from the stem have long petioles that flare at their basal attachment. Leaf blades are divided into many small, narrow, highly lobed sections. The ratio of foliage to root varies among cultivars. Cultivars developing large tops tend to produce larger roots but grow slowly, while cultivars with short tops tend to produce smaller roots but mature quickly (Rubatzky et al., 1999).

The taproot is comprised of hypocotyl and primary root tissues, with the proportion of each varying among cultivars. For many cultivars, the enlarged, edible carrot "root" is largely hypocotyl tissue, although the upper taproot may be enlarged as well. Primary and lateral roots develop along the lower regions of the hypocotyl with highly branched fibrous roots, reaching depths of more than 75 cm (30 in). It is often difficult to identify the transition from hypocotyl to root since much of the hypocotyl is root-like in appearance (Rubatzky et al., 1999). Rapid root elongation begins 12–24 days after emergence, when the maximum length may be determined. Although yield potential is correlated with root length, those longer than 30 cm (12 in) often break during harvest and handling.

Anatomically the root is comprised of an outer layer of periderm, with a layer of pericycle beneath, and the interior is primary xylem and phloem tissues joined by a ring of cambium (Fig. 21.1).

The cambium produces secondary xylem toward the interior and secondary phloem toward the outside during root enlargement. For good eating quality, the root ideally should have a minimum of core (xylem) relative to cortex (phloem) tissue. Typically xylem color is less intense than the phloem tissue, but ideally both should be uniformly colored (Rubatzky et al., 1999).

The pericycle is a cylinder of parenchyma or sclerenchyma cells that lies just beneath the periderm. Oil ducts in the intercellular spaces of the pericycle contain essential oils responsible for the characteristic carrot flavor and aroma. The taproot serves as a storage organ accumulating sucrose and other sugars (Rubatzky et al., 1999).

The fleshy axis and short-stemmed rosette with silvery foliage develops into an elongated flower stalk during the second year, following vernalization. During bolting, multiple floral stalks elongate, producing many rough branches covered with stiff hairs or bristles. Flower stalks typically range from 1–2 m (39.3–78.7 in ft) tall (Rubatzky et al., 1999).

The inflorescence is a terminal compound umbel consisting of many umbellets bearing small white flowers. "Umbel" is the botanical name for the distinctive compound flower cluster that forms at the end of the seed stalk. The large primary umbel of a flower stalk may contain 50 umbellets each with up to 50 flowers. Progressively smaller secondary, tertiary, and quaternary umbels develop later at the terminus of the lateral branched shoots (Oliva et al., 1988). The tertiary and quaternary umbels are less productive, and their inferior quality seed may be small or partially developed (Hawthorn et al., 1962). Flowers are usually bisexual but with protandrous and centripal floral development, which requires insect pollination (Peterson and Simon, 1986). At maturity, peduncles of the outer umbellets curve inward and the umbels are concave. Seeds are flat, ribbed and spiny, and vary greatly in size, with a range from 500 to 1,000 seeds/g (14,300–28,500/oz) (Oliva et al., 1988).

Types and Cultivars

There is more diversity of carrot root shapes and sizes then usually seen in commercial markets. Preferences for carrot root shape and color vary regionally. Carrot cultivars are sometimes grouped into two broad classes: eastern or Asian and western or European. European cultivars have strong biennial characteristics and are more tolerant of low temperatures that induce bolting. European cultivars are firm textured, sweet, highly flavored, yellow-orange to strongly orange in color, slow

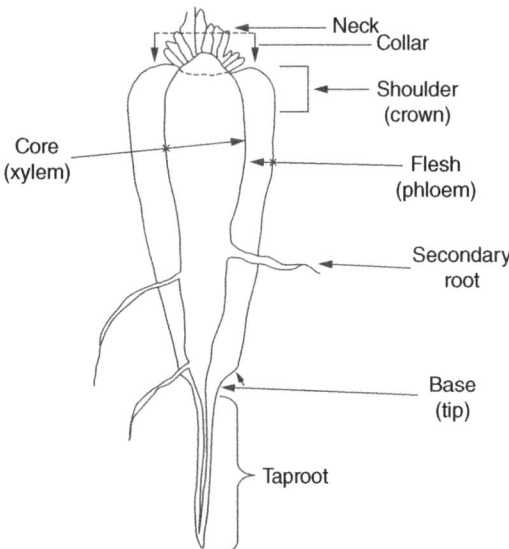

Fig. 21.1. Diagram of carrot root anatomy.

bolting, and acclimated to cool temperatures. Asian cultivars tend to have softer texture, are less sweet, less flavored, adapted to warm temperatures, bolt easily, and may be red or red-orange colored. Asian cultivars with an annual life cycle character bolt after minimal exposure to cold temperatures. Generally, temperate-zone European and North American cultivars are biennials, while tropical Asian cultivars exhibit an annual habit. Tropical types also appear to have a short-day preference for flowering (Rubatzky and Yamaguchi, 1997).

Carrots are vernalized below 5°C (41°F) after development beyond a juvenile growth phase that roughly correlates to plants that have fewer than eight leaves or a diameter smaller than pencil sized, 4–8 mm (0.16–0.31 in). The juvenile stage lasts longer for large-rooted cultivars. The exposure to vernalizing temperatures needed to induce flowering varies from several to as many as 12 weeks for some bolt-resistant cultivars. Some tropical Asian cultivars can be induced to bolt at temperatures less than 15°C (59°F). Following vernalization, seeds mature in 4–5 months (Rubatzky and Yamaguchi, 1997).

Cultivars are often grouped into types that reflect similar morphology or horticultural characteristics (Fig. 21.2). Western cultivars are often classified into three categories based on root shape and length. Short-rooted or baby cultivars, which mature quickly (50–60 days) for early season fresh-market production, include 'Oxheart', 'Early Mokum', 'Parisian Market', 'Early Nantes', and 'Amsterdam Forcing'. Medium-rooted cultivars are used for main-season production for both fresh market and processing. Cultivars mature in 60–75 days from seeding, and include 'Mokum', 'Flakkee', 'Autumn King', 'Danvers', and 'Royal Chantenay'. Long-rooted cultivars such as 'Imperator' are usually grown in thoroughly tilled deep loam or muck soils and mature 60–75 days from seeding.

Some of the major commercial carrot types or classes of carrot roots are described as follows:

- 'Nantes' is nearly cylindrical in shape, blunt, and rounded at both the top and tip. Its length is from 18 to 23 cm (7–9 in). The cultivar was developed in the late 1800s in the vicinity of Nantes, France. Shorter cultivars 'Nantes Half Long' and 'Early Nantes' followed. 'Nantes' cultivars are known for their sweet and pleasing carrot flavour. 'Nantes' cultivars have greater susceptibility to mechanical damage when compared to 'Imperator' because they are more likely to crack when dropped on a hard surface.
- 'Chantenay' have broad shoulders and taper towards a blunt, rounded tip. 'Chantenay' carrots are shorter than other cultivars with average length of 15–20 cm (6–8 in), but have greater girth, sometimes growing up to 8 cm (3 in) in diameter. They are most commonly processed

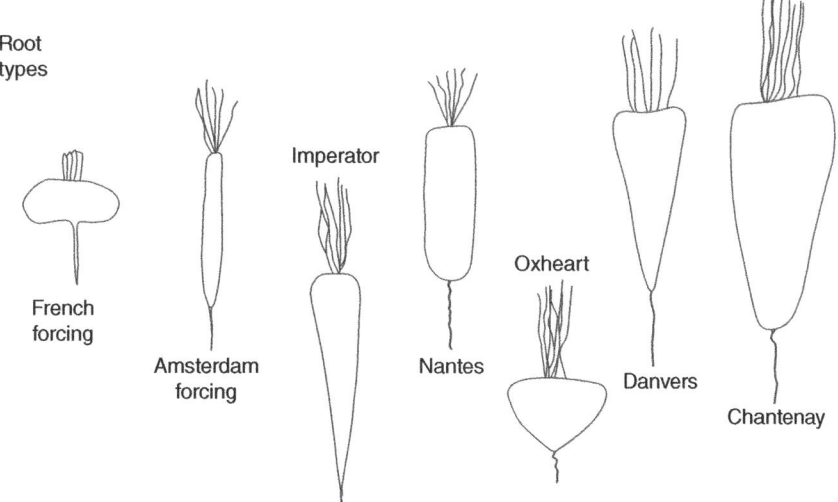

Fig. 21.2. Diagram of some common carrot cultivar root shapes.

into diced and other prepared foods. They grow better on heavy soils than many other cultivars.

- **'Danvers'** was developed in the late 1800s in a town in Massachusetts, USA of the same name. Danvers carrots have a conical shape, having well-defined shoulders and tapering to a point at the tip similar to 'Chantenay' but not as broad. The average 'Danvers' root length is 15–20 cm (6–8 in) and up to 5 cm (2 in) in diameter. 'Danvers' cultivars are somewhat tolerant of heavy soil and are often processed as baby food or other canned and frozen products.
- **'Imperator'** is a result of a cross between 'Nantes' and 'Chantenay' and was introduced in the 1920s. 'Imperator' roots are 25–30 cm (10–12 in), longer than other cultivars and taper to a point. They are very durable and withstand mechanical damage well, so they are popular for long-distance shipping. However, they have sometimes been criticized for lacking the flavor and eating quality of 'Nantes'. 'Imperator' are most commonly sold whole in the USA or are sliced and peeled for sale as "Baby Cut" carrots, a lightly processed product that is sold fresh in bags as snack food and a cheaper alternative to true baby or small-rooted cultivars.
- **'Oxheart'** or **'Guerande'** is an introduction from France. This nearly round cultivar grows well on shallow heavier soils. All cultivars can be used for fresh market, but some are better suited for processing. For processing, larger-rooted cultivars like 'Danvers' or 'Chantenay' are preferred.

Different cultures prefer specific root colors and shapes, so additional types are also grown. In Japan, where carrots are seldom eaten raw, long reddish orange carrots, like 'Kurodo', or even red or purple cultivars with a thick cylindrical shape are preferred (Rubatzky and Yamaguchi, 1997). In contrast, there is a preference for relatively short, slender, yellow-orange 'Nantes' and 'Nantes'-like cultivars in Europe, while in North America the long, deep orange 'Imperator' and related cultivars predominate.

Researchers have developed new cultivars of carrots with novel traits through traditional breeding methods. Transgenic carrots produced through genetic engineering have not had an impact on world markets as of 2013. Novel yellow, white and red cultivars are gaining popularity in some markets because they add variety and variation to salads and other foods containing carrots.

As a carrot flowers, anthers mature and fall before pistils become receptive. This characteristic is called protandry, and is a strategy to increase cross-pollination by insects, primarily bees. Pollen must come from different plants or from later maturing umbels on the same plant for successful pollination (Fig. 21.3). Therefore, carrots seeds are a mixture of self- and cross-pollination, which increases variability.

There has been a gradual conversion to F-1 carrot cultivars since the 1970s in the USA and Europe because of their greater uniformity and productivity. There is little variation among individual plants in a high quality carrot cultivar. Cytoplasmic male sterility (CMS) is used to ensure that inbred lines are cross-pollinated to produce seeds of hybrid cultivars. Open-pollinated cultivars remain popular for certain applications where the extra cost of F-1 seeds is not worth the investment, such as for processing into cut products.

Quality characteristics

In addition to uniformity, other major plant-breeding objectives for carrot cultivar improvement include increased pest resistance, growth rate and yield, root surface smoothness, cracking resistance, improved flavor, consistent texture, high-temperature tolerance, and bolting resistance. Root color is less intense at elevated temperatures, and some cultivars are naturally higher in carotenoids than others to compensate. Except for seed production, floral seed stalk formation is undesirable because it inhibits root enlargement. Bolting causes roots to

Fig. 21.3. Carrot flowers are umbels containing many small florets. Male and female flower parts develop at different times on the same flower to ensure cross-pollination.

become woody and secondary root development. Bolting can be avoided or limited by delayed planting schedules that minimize plant exposure to low temperature. Slow-bolting cultivars have greatly improved resistance to low temperatures.

Crack resistance is an important attribute, particularly for long-distance shipping and retail sale. The carrot root surface should be smooth, relatively free of branches and free of root hairs for best visual appearance and ease of preparation. A small uniformly colored core is desirable. Edible quality characteristics include freedom from bitterness, sugar content, texture, and vitamin content. High sugar content increases root rotting during storage and seed production.

Alpha and beta-carotene, responsible for yellow and orange color, respectively, are the major carotenoid pigments in carrot. Beta-carotene usually accounts for 50% or more of the total carotenoid content. Carotene is related to vitamin A, which occurs in two forms. The most common form in humans is vitamin A1 or retinol. Enzymatic reactions in the intestine and liver convert plant α- and β-carotene into vitamin A (Lietz et al., 2013).

The carotenoids are not uniformly distributed in the root. Phloem tissue usually contains about 30% more pigment than the xylem (Fig. 21.1). Differences in carotene content are also influenced by temperature, plant maturity, and cultivar. The carotene content of the most widely grown carrot cultivars ranges from 41 to >78 mg/g (41–78 mg/0.04 oz) fresh weight (Rubatzky et al., 1999). Breeders have greatly increased the carotene content in some new cultivars compared to heirlooms. A USDA carrot breeding program has developed lines with over ten times the carotene content of 'Chantenay' that are used for new cultivar development (Simon et al., 1990). One cultivar that is high in tocopherol (vitamin E) has white roots (Goldman and Breitbach, 2002).

Other pigments in carrot roots include lycopene and anthocyanin. Lycopene also functions as an antioxidant and is relatively low in most carrots except in some red-fleshed cultivars like 'Kurodo'. Anthocyanin is responsible for the reddish-purple color in some cultivars.

Economic Importance and Production Statistics

Total world production of carrots and turnips was 35.7 million metric tonnes (15.9 million short tons) for 2011 (Table 21.1). China was by far the largest producer accounting for 45.4% of the world total. The production area was 1.2 million ha (2.7 million acres) and the average yield was 30,109.7 kg/ha (26,797.6 lb/acre) (FAOSTAT, 2011).

In the USA, total production for fresh market, including lightly processed, is just less than 31,250 ha (70,000 acres) with approximately 5,580 ha (12,500 acres) for processing. California is the leading fresh-market production state and Washington, Wisconsin, and California are important producers of processed carrots. In 2009, the value of US carrot crops was over US$557 (USDA-NASS, 2009).

Per capita consumption

Trends in carrot consumption reflect changes in product availability and improved information about the nutritional benefits from eating carrot. After remaining constant for many years, US carrot consumption increased dramatically to a peak in 1990 (Table 21.2). This increase was likely caused by scientific information showing that carrots are an excellent source of dietary fiber and carotenoids, two factors known to be important for human nutrition. The available of convenience-packaged carrot items such as lightly processed baby-peeled carrots and shredded carrots in salad mixes also helped to increase consumption during the period. Since 1990, total carrot consumption has declined, with frozen carrot consumption declining sharply. Fresh per capita consumption plateaued at about 2.5 kg (5.5 lb/year) (Table 21.2) (USDA-ERS, 2011).

Table 21.1. World carrot production 2011 (statistics include turnip; FAOSTAT, 2011).

Rank	Country	Million metric tonnes (million short tons)
1	China	16.2 (17.86)
2	Russian Federation	1.74 (1.92)
3	USA	1.31 (1.44)
4	Poland	0.89 (0.98)
5	Ukraine	0.86 (0.95)
6	UK	0.69 (0.76)
7	Italy	0.54 (0.60)
8	Japan	0.60 (0.66)
9	Germany	0.55 (0.61)
10	The Netherlands	0.48 (0.53)

Nutritional Values

Carrot roots are recognized nutritionally as a good source of carbohydrates, dietary fiber, and minerals. They are also high in α- and β-carotene, precursors of vitamin A (Holland et al., 1991; Table 21.3).

Table 21.2. US per capita carrot consumption (kg (lb) per person per year).

Year	FM[a]	Canned[b]	Frozen[b]	Total
1970	2.7 (6.0)	1.0 (2.1)	0.6 (1.4)	4.3 (9.5)
1980	2.8 (6.2)	0.8 (1.7)	0.8 (1.7)	4.4 (9.6)
1990	3.8 (8.3)	0.5 (1.2)	1.0 (2.3)	5.4 (11.8)
2000	2.5 (5.5)	1.2 (2.7)	0.5 (1.0)	4.2 (9.2)
2010	2.5 (5.6)	0.6 (1.4)	0.4 (0.9)	3.6 (7.9)

[a]Fresh market (farm-weight basis).
[b]Processed (processed-weight basis).

Table 21.3. Raw carrot nutritional value per 100 g (3.5 oz). Percentages are relative to USDA daily nutritional recommendations for adults (USDA National Nutrient Database, 2013).

Component	Amount
Energy	173 kJ (41 kcal)
Carbohydrates	9.6 g
Sugars	4.7 g
Dietary fiber	2.8 g
Fat	0.24 g
Protein	0.93 g
Vitamin A equivalent	835 µg (104%)
β-carotene	8285 µg (77%)
Lutein and zeaxanthin	256 µg
Thiamine (vitamin B1)	0.066 mg (6%)
Riboflavin (vitamin B2)	0.058 mg (5%)
Niacin (vitamin B3)	0.983 mg (7%)
Pantothenic acid (vitamin B5)	0.273 mg (5%)
Vitamin B6	0.138 mg (11%)
Folate (vitamin B9)	19 µg (5%)
Vitamin C	5.9 mg (7%)
Vitamin E	0.66 mg (4%)
Ca	33 mg (3%)
Fe	0.3 mg (2%)
Mg	12 mg (3%)
Mn	0.143 mg (7%)
P	35 mg (5%)
K	320 mg (7%)
Na	69 mg (5%)
Zn	0.24 mg (3%)
Fluoride	3.2 µg

Production and Culture

Growth and development

Carrot is classified as a half-hardy vegetable, meaning that it is resistant to frost and light freezes when acclimated (Lorenz and Maynard, 1988). Young seedlings tolerate low temperatures and frosts better than older plants. Carrot seedlings can survive frosts and temperatures as low as −6.5°C (20°F) if properly acclimated. The top growth is slow at temperatures below 4°C (39.2°F), and consecutive severe frosts will kill leaves. Soil tends to buffer roots from fluctuating low air temperatures that damage leaves, so the crown can continue to produce new growth. Once the soil is frozen, crown damage also occurs as small hairline and horizontal splits develop on the surface of storage roots. Crowns are killed by exposure to prolonged subfreezing temperatures (Rosenfeld et al., 1997).

Both root and foliage growth is optimum between 16–21°C (61–70°F). Growth slows at temperatures below 10°C (50°F) (Lorenz and Maynard, 1988). Temperatures above 21°C (70°F) tend to make roots short and stubby, while temperatures below 16°C (61°F) tend to favor long slender roots. Foliage growth is less affected by temperature than root growth. The foliage is more tolerant of high temperatures, but above 30°C (86°F) foliage growth is inhibited and roots develop strong flavors. Large diurnal temperature fluctuations are conducive to rapid growth, and if night temperatures are sufficiently cool, carrots can be grown in the tropics (Rubatzky and Yamaguchi, 1997).

The optimum temperature range for carotene synthesis is from 16°C (61°F) to 25°C (77°F) and poor below 16°C (61°F) or above 25°C (77°F). Pigment synthesis lags behind root growth, a reason why young roots are pale. With continuing growth carotene accumulates, attaining a maximum after about 90–120 days and then usually remains constant or may slowly decline (Rubatzky and Yamaguchi, 1997).

Site selection and field preparation

Ideal soils for carrot production are deep, friable, fertile, well-drained sandy loam or peat (Fig. 21.4). Carrots, especially long-rooted cultivars, are markedly shortened by shallow or compacted soils. Carrots are tolerant of a wide range of soil pH but generally 5.5–6.5 is preferred on organic soils and

Fig. 21.4. Carrot field on sandy soil in Nova Scotia, Canada.

a pH of 6.0–6.8 on mineral soils. Fertilization rates vary dramatically with soil type and should be based on soil test results and foliar analysis throughout the season so inputs are not wasted and do not become pollutants. An average carrot crop can be expected to remove from the soil 162 kg/ha (145 lb/acre) N, 28 kg/ha (25 lb/acre) P, and 386 kg/ha (345 lb/acre) K, so carrots have a relatively high uptake of K. Excessive N fertilization tends to promote foliage growth over root enlargement; however, a side dressing of N is commonly applied 4–6 weeks after seeding to maximize production on mineral soils (Lorenz and Maynard, 1988).

Field establishment

Carrot seed quality is variable because seeds form on separate umbels that mature at different times. The mixture of seed of different physiological maturities often leads to variability in germination and emergence. Carrot seed germination is typically slow, so uniform emergence and stand establishment are constant concerns. Seed thermodormancy is another factor that may reduce germination at high temperature (Nascimento et al., 2008).

Carrots are precision planted to a final stand in single or multilinked rows using belt and vacuum seeding equipment and uniformly sized and/or coated seed. Three or four parallel seed lines may be sown about 5.1 cm (2 in) apart. In some cases, carrots are established by planting a narrow band of randomly scattered seeds (Fig. 21.5).

To accommodate mechanical harvesting equipment, the widths of seed lines or bands are usually less than 10–12 cm (3.9–4.7 in). Seeding rates vary from 33–197 seeds/m (10–60 seeds/ft). Higher densities are used for band seeding, where soil conditions are unfavorable for emerging, or when small-rooted cultivars are grown. Lower densities are used for large processing cultivars or where high quality pelleted carrot seeds are precision planted to a final stand. Carrot seed are sown between 5 and 20 mm (0.2–0.8 in) deep (Rubatzky and Yamaguchi, 1997). Pregerminated seeds embedded in a water-based gel carrier medium or in water to protect delicate emerging radicles from damage may be planted to improve field emergence. Regardless of establishment technique, ideal spacing should provide uniform distances between roots.

Seeding densities ranging from 1 to 3 million seeds/ha (405,000–1.2 million seeds/acre) are determined based on germination percentage, seed vigor, and the anticipated effect of soil and environmental

Family Apiaceae

conditions on emergence. Selecting the optimum seed density is important because post-emergence thinning adds to production costs and the proper plant density is essential for maximizing yield and determining root size for specific uses.

Fresh-market carrot plant stand densities generally range from 80 to 100 roots/m^2 (7–9 roots/ft^2). Field populations for small-rooted, baby cultivars range from 100 to 200/m^2 (9–18 roots/ft^2) and 5 million seed/ha (2 million seeds/acre) may be sown to achieve these target populations. For processing, target populations for large-rooted cultivars range from 40 to 70 roots/m^2 (4–6.5 roots/ft^2). In general, for a given cultivar, roots will be larger at wider spacing and smaller at close spacing.

Poor seeding performance is one of the biggest problems in carrot production. Seeds are very small and slow to germinate. Carrots germinate at soil temperatures above 4.4°C (40°F), but optimum germination occurs at 23.9°C (75°F). In addition to inferior quality seed, poor stands may occur because seedlings cannot grow properly due to soil crusting, adverse temperatures, weed competition, and dry conditions (Fig. 21.6).

Adequate moisture for uniform emergence and subsequent early seedling growth is critical. To optimize germination and improve the uniformity of emergence, seed priming, precision seeding, and sprinkler irrigation are often used. Hybrid cultivars are available in many areas and offer improved uniformity of germination, seedling growth, and vigor.

Early seedling growth is often slow because emergence can vary from 7 to more than 20 days and the first true leaves may not develop for 3–4 weeks after planting. Late-emerging seedlings often do not produce marketable roots because they compete poorly with the first plants that emerge. Slow growth makes carrot a poor competitor with weeds, which can overwhelm a planting before the crop is established. Selective herbicides are effectively used for weed management in some cases.

Fig. 21.5. Carrot row established in a narrow band.

Fig. 21.6. Fresh market carrot field near Salinas, California grown on raised beds with furrow irrigation. Carrot seedlings compete poorly with weeds. Several weed control strategies can be employed. This field was treated with post-emergence herbicide to kill broadleaf weeds. Hand weeding and cultivation can also be effective.

Irrigation

Uniform soil moisture is essential because water stress will slow growth and result in thickened, woody cells, reduced sugar content, and bitter flavor. Carrot crops require approximately 30–50 mm (1.2–2.0 in) of water per week or from 450–600 mm (17.7–23.6 in) for a cropping season depending on soil type and evapotranspiration (Rubatzky and Yamaguchi, 1997). Water should be applied whenever about 40% of available moisture has been depleted from the root zone. Crops are often established with overhead irrigation before switching to furrow or drip irrigation after emergence. Overhead irrigation may be used the entire season. Excessive soil moisture can cause root splitting or cracking and may inhibit proper color development.

Harvesting and marketing

Carrot harvesting is not determined by a clearly defined maturity stage. Depending on cultivar and growing conditions, the period from planting to harvest varies from less than 70 to more than 150 days. Earliness is achieved by small size and/or rapid growth rate. To take advantage of high prices or processing demand, carrots are sometimes harvested before the potential full root size or maximum yield is obtained. Yields of carrot roots exceeding 53 metric tonnes/ha (20 short tons/acre) are often obtained, particularly with processing types, and yields for small-rooted cultivars are significantly less.

Processing carrots are grown longer to increase weight, color, sweetness, or dry matter. Maximum root-color development occurs from 15.5–21.1°C (60–70°F). High dry matter content is associated with better storage and handling. Under certain circumstances, roots are "stored" in the field and harvested as needed. However, long delays in harvesting may be accompanied by increases in fiber and strong flavor. Cultivars with conical roots are stronger and resist splitting during growth, breakage during harvest and damage associated with postharvest handling better than cylindrical roots.

Hand harvesting is a difficult task and is performed primarily for small-scale production. Mechanical harvesting is routinely used for large-scale commercial production (Fig. 21.7).

One type of mechanical harvester cuts the foliage and discards it in the field before the roots are dug, similar to a potato harvester. Another harvester design undercuts roots as belts simultaneously grasp

Fig. 21.7. A tractor-pulled carrot harvester leaves foliage in the field and drops roots in a wagon pulled alongside (a). Close-up of a digging mechanism that lifts roots from the soil by the foliage (b).

the foliage, lifting the entire plant from the soil. Strong and healthy foliage is important for efficient mechanical harvesting using this system. In some cases, the harvester may remove the tops after digging and discard them in the field, while the roots are collected in containers, trucks, or wagons.

Bunched carrots with intact tops are still marketed, although much less often than in the past. A bunch is formed by tying the foliage of several intact plants together around the leaf bases just above the root (Fig. 21.8). Following bunching, carrots are washed to clean the roots and hydrocooled or iced to reduce temperature and respiration. Carrots should be cooled to 1–2°C (33.8–35.6°F)

Fig. 21.8. In many markets, bunched carrots, like the 'Imperator' cultivar shown here, have been replaced by more efficient bagging with tops removed. Most consumers discard carrot foliage after purchase although special effort had to be made to keep the tops looking fresh during transport and marketing.

as soon as possible to retain quality and reduce wilting, which is especially important with bunched carrots. The primary reason for selling carrots with tops is to show freshness of product, since the leaves are discarded and not consumed. Bunched carrots are more expensive because extra effort is required to keep the tops looking fresh without wilting. Bunched carrots have been largely been replaced by roots sold in plastic bags.

Carrots without tops are handled in bulk until grading. Fresh carrot roots are often washed after harvest and marketed in bulk or prepackaged in small plastic bags. Bagged carrots have become very popular because the package maintains fresh clean roots that are ready to eat after purchase.

Because of variation in emergence or competition, many roots are undersized at harvest and fail to contribute to marketable yield. For many years undersized carrots were culled and sold for animal feed, juicing, or other less valuable uses. To add value to unmarketable carrots, the "cut and peel" or "baby cut" carrot products were developed from undersized elongated roots. During this fresh-cut processing, long carrots are cut into shorter sections and skinned mechanically using abrasive rollers to yield carrot pieces that resemble "baby" carrots. "Baby cut" carrots are a successful value-added, washed, and peeled convenience product that is ready to eat at purchase. Shredded carrots are a common ingredient in many premade salad mixes because they add color, nutritional content, and texture.

Miniature cultivars are also being sold as "baby" carrots without cutting. While any carrot can be harvested at an immature tender stage, fast-maturing cultivars that produce smaller roots have been intentionally bred to produce true baby carrots. These small cultivars are also more tolerant of heavy or stony soil than long-rooted cultivars such as 'Nantes' or 'Imperator'. True baby carrots are produced for a limited specialty market and are usually more expensive than baby-cut carrots because yields are significantly lower.

Commercial production of organic carrots has increased in some markets. In North America, organic production is centered in the semi-arid western USA where disease pressure is low. As in conventional production, the main criteria for organic cultivars are good eating quality, disease resistance, and economic yield.

Carrot juice is extracted from fresh carrot roots, usually of 'Chantenay' and other large-root cultivars grown for processing. Roots for juice processing are washed, mechanically peeled, blanched in boiling water to inactivate enzymes, and extracted in a paddle pulper finisher to obtain the juice (Wu and Shen, 2011). Carrot juice is a good source of β-carotene and is commonly mixed with other vegetable and fruit juices to provide nutrients, color or flavor.

Storage

For fresh-market use, carrots are usually mechanically dug, washed, and stored at 0°C (32°F) and 95% RH during shipment and marketing. At low temperatures (0°C; 32°F), film-packaged carrots can maintain high quality for 6–7 weeks. Carrot quality deteriorates during long-term storage due to a loss of sugars to respiration. Root respiration rates are relatively low compared with other vegetables and decline further with refrigeration. Soluble sugars increase slightly during cold storage. However, bunched carrots store poorly and roots lose firmness rapidly because moisture is lost through the tops, significantly reducing shelf life to as little as 7 days. Bunched carrots are generally shipped packed in crushed ice to keep the tops looking fresh, which increases shipping costs. Postharvest storage of "minimally processed", film-packaged, cut, and peeled small carrot sections is usually limited to approximately 20 days.

Carrots are stored for longer periods in areas where production is possible for only a few months per year. Carrot roots are generally not washed before placement into long-term storage at 0°C (32°F) and high relative humidity. Controlled atmosphere storage (CA) increases long-term carrot storage. Controlled atmosphere storage of 1°C (34°F) with 2–6% O_2 and 3–4% CO_2 reduces respiration rates and the loss of sugar in roots compared to conventional storage. However, deviations from these O_2 and CO_2 ranges may reduce carrot quality. Well-managed refrigerated storage is simpler to manage and may produce results similar to CA (Rubatzky *et al.*, 1999).

Bitter-tasting compounds form in roots upon exposure to ethylene gas. Therefore, carrots should not be stored with ethylene-producing commodities, such as apples and melons. Carotenoid content declines during long-term storage.

Diseases

Fungal pathogens responsible for carrot root diseases include Phytophthora root rot (*Phytophtora medicaginis*), violet root rot (*Helicobasidium brebissonii*), white rot (*Sclerotinia sclerotiorum*), gray mold (*Botrytis cinerea*), yellows (*Fusarium* spp.), purple root rot (*Helicobasidium brebissonii*), and Sclerotinia rot (*Sclerotinia sclerotiorum*). *Pythium* spp. are associated with many diseases of carrot, including damping off, die back, forking, brown root, lateral root die back, rubbery slate rot, rusty root, cavity spot (*Pythium violae*), and Pythium brown rot (*P. sulcatum*). Root diseases are more severe in heavy soils with poor structure. Control measures for root rot include rotations for at least 5 years between carrot and related crops, liming to prevent acid conditions, and well-drained soil (Strandberg, 2000; CABI, 2008a; Minnis *et al.*, 2013).

Fungal diseases affecting primarily carrot foliage include powdery mildew (*Erysiphe betae* and *E. heraclei*), white rust (*Aecidium foeniculi* or *Uromyces graminis*), Alternaria leaf blight (*Alternaria dauci*), and Cercospora leaf spot (*Cercospora carotae*) (Strandberg, 2000; Takaichi and Oeda, 2000; CABI, 2008a).

Bacterial diseases include soft rot (*Erwinia chrysanthemi, E. carotovora* subsp. *carotovora*), bacterial leaf blight (*Xanthomonas campestris* pv. *carotae*), carrot bacteriosis (*Xanthomonas campestris*), crown gall (*Agrobacterium tumefaciens*), hairy root (*Agrobacterium rhizogenes*), milky disease (*Bacillus popilliae*), and scab (*Streptomyces scabiei*) (Strandberg, 2000; CABI, 2008a; Minnis *et al.*, 2013).

There are many virus and viroid diseases of carrot, including alfalfa mosaic genus *Alfamovirus* (AMV), carrot latent genus *Nucleorhabdovirus* (CtLtV), carrot mottle genus *Umbravirus* (CMoV), carrot red leaf genus *Luteovirus* (CaRLV), carrot thin leaf genus *Potyvirus* (CTLV), carrot yellow leaf (CYLV), celery mosaic genus *Potyvirus* (CeMV), cucumber mosaic genus *Cucumovirus* (CMV), beet curly top genus *Hybrigeminivirus* (BCTV), motley dwarf genus *Luteovirus*, carrot red leaf virus (CaRLV) and genus *Umbravirus*, carrot mottle virus (CMoV) (Strandberg, 2000).

The following nonpathogenic disorders (and their causes) also may occur in carrot: crown rot disorder (no known pathogen), heat canker (high soil surface temperature), hollow black heart (boron deficiency), ozone injury (ozone pollution), root scab (physiological disorder), and speckled carrot (genetic disorder) (Strandberg, 2000).

Insect and Other Pests

The carrot aphid *Semiaphis dauci* (Aphididae), the willow-carrot aphid (*Cavariella aegopodii*), the carrot sucker (*Trioza apicalis*), and other aphids feed on foliage and vector virus diseases. Thrips (*Frankliniella tritici* and *F. occidentalis*), flea beetles (*Systena blanda*), and in tropical areas whiteflies (*Bemisia* and *Trialeurodes* spp.) also feed on the foliage. The leafhopper (*Macrosteles quadrilineatus*) is a vector of aster yellows. Two of the most serious pests of carrot in temperate areas are the carrot weevil (*Listronotus oregonensis*) and the carrot root or rust flies (*Psila rosae*). Larvae of both insects create feeding tunnels in the root but the affected areas differ. The carrot weevil larvae tunnels in the upper one-third of the root, while the rust fly maggots tunnel mainly in the lower two-thirds of the root. The tunnels created by the rust fly maggot are narrower and more winding than those of the carrot weevil. The turnip moth also (*Agrotis segetum*) attacks carrot roots. Several cutworms (*Agrotis* spp.) are chronic pests of carrot. Another pest that causes serious crop losses is armyworm (*Spodoptera exigua*), whose larvae feed on leaves and roots. The lygus bug (*Lygus hesperus* and *L. elisus*) feeds on seed crops, reducing seed yield and quality. Affected plants may exhibit a virus-like foliage disorder or premature bolting (Rubatzky *et al*., 1999; CABI, 2008a).

Another pest of carrot roots is nematodes. Some nematodes that affect carrot include the lance nematode (*Hoplolaimus uniformis*), lesion nematode (*Pratylenchus penetrans* and *Pratylenchus* spp.), sting nematode (*Belonolaimus longicaudatus*), cyst nematode (*Heterodera carotae*), and the root knot nematode (*Meloidogyne hapla*). Crop losses by root-knot nematodes can be controlled by crop rotation and the application of manure (Strandberg, 2000).

Snails (*Helix aspera*) and field slugs (*Deroceras reticulatum*) damage crops and their presence in the marketplace is undesirable. Spider mites (*Tetranychus telarius*) and crown mites (*T. dimidiatus*) are pests whose feeding decreases foliage color and may cause desiccation and defoliation in extreme cases (Rubatzky *et al*., 1999).

CELERY

Origin and History

The origin of celery and related crops is unclear. Celery is believed to be native to marshy lowlands of temperate Europe and western Asia with a habitat and wild forms that extend from Sweden southward to Algeria, Egypt and Abyssinia, and to the mountains of India. The eastern Mediterranean region appears to be the center of domestication, but the broad distribution of wild types raises some doubt (Rubatzky and Yamaguchi, 1997). Wild celery was apparently used for medicinal purposes for many years before it was used as food. The Egyptians first cultivated celery around 400 BC as a medicinal plant. Early literature mentions the use of seeds and seed oil for medicinal purposes. Iranians used seed vapor to cure headaches, for example (Rubatzky *et al*., 1999).

The earliest plant form was likely "smallage", a leaf type of celery. When cultivation of celery began in Europe in the 16th century it was still a primitive plant. Cultivation of the plant for food was first recorded in France in 1623. By the early part of the 18th century, there had been genetic improvement of celery in Europe (Rubatzky and Yamaguchi, 1997). In the 1700s, soups and stews were being flavored with celery in England. European colonists introduced celery to North America. In the USA, the first cultivation of stalk celery occurred in Michigan about 1874 and in Florida in 1897 (Rubatzky *et al*., 1999).

Botany and Life Cycle

Celery (*Apium graveolens* subsp. *dulce* (Mill.) Schübl. & G. Martens) is a cool-season biennial that is grown as an annual. The plant is easily vernalized by exposure to below 10°C (50°F) for several weeks after developing past the juvenile phase, which roughly equates to plant diameters smaller than a pencil. The amount of chilling required varies among cultivars. Some cultivars may be vernalized after exposure to as little as 10 days to 2 weeks of temperatures below 10°C (50°F). Long days during vernalization inhibit flower initiation and bolting. Long days after vernalization accelerate flower stalk development. If vernalized, the plant can complete its life cycle within 1 year. A vernalized plant quickly produces a flower stalk (bolts), which inhibits further stalk development. Unless the crop is being grown for seed, bolting must be avoided because it reduces vegetative growth and stalk quality, diminishing economic value. Hardened celery plants can briefly withstand temperatures a few degrees below freezing.

Celery stalks are comprised of several enlarged, long, thick, fleshy and solid edible petioles in a rosette pattern. Younger petioles develop in the apical meristem and are hidden by the larger outer ones (Fig. 21.9). Leaf petioles are broad and erect with sheathing or shingling bases. A mature celery plant has 7–15 clearly distinguishable petioles. Interior leaves become progressively smaller and self-blanched because of light exclusion from the larger surrounding leaves and petioles. This grouping of small, tender, inner leaves is referred to as the celery "heart". Generally, the petiole is the most important edible portion and celery leaves are not consumed because of their strong flavor but are sometimes cooked or used as a garnish. Celery petioles are crescentic in trans-section, with prominent ribbing on the abaxial (outer) surface and a smooth adaxial

Fig. 21.9. A celery plant in a production field before trimming. Much of the foliage is trimmed away, usually in the field, to produce the more familiar celery stalk sold commercially.

Fig. 21.10. Field-grown celeriac in Germany.

(inner) surface. Ribbing is a result of separate collenchyma bundles that form on the outside of each petiole. Collenchyma tissue is strong, four times stronger than vascular tissue (Esau, 1936). Vascular bundles provide strength to the petiole and also contribute to its fibrous texture. Vascular bundles and collenchyma strands collectively make celery petioles stringy. However, the bulk of the petiole is composed of succulent, watery parenchyma tissue.

Celery tends to sucker, forming axillary stems around the base of the mother plant. Because suckers form late in development, they contribute little to the mother stalk and are usually trimmed off during harvest. The amount of tillering is greatly reduced in modern cultivars because plant breeders have selected against this trait.

Celeriac (*A. graveolens* subsp. *rapaceum*), a celery relative, develops a white, fleshy, globe-shaped, enlarged storage organ 10–15 cm (3.9–5.9 in) in diameter composed of hypocotyl and upper root tissues with the top portion exposed above the soil (Fig. 21.10).

Celeriac is popular in Europe and produces fewer smaller leaves and petioles compared to celery. Smallage (*A. graveolens* subsp. *secalinum*) is a primitive form of celery that is a strongly scented erect biennial used for flavoring. Smallage, also called foliage celery, forms a rosette of leaves with long, thin petioles. Celery plants may grow to a height of 60–90 cm (24–35 in), celeriac about 50–60 cm (20–24 in) and smallage >50 cm (20 in) (Rubatzky and Yamaguchi, 1997).

The root systems of celery develop a large mass of fine, well-branched roots within a relatively short radius of the taproot. With the exception of the taproot, lateral roots are relatively shallow, developing within 30 cm (12 in) of the surface (Rubatzky and Yamaguchi, 1997). During transplanting, the taproot is often damaged or destroyed, causing a proliferation of adventitious lateral roots at the base of the plant. The stem of the vegetative celery plant is short, highly compressed and fleshy. When the plant becomes reproductive the stem rapidly elongates or "bolts", and becomes fibrous.

The color of celery foliage varies with cultivar from yellowish green to dark green. Celeriac foliage is generally a darker green than most celery cultivars, while smallage leaves are also dark green and cultivars with reddish leaves exist.

Flowers of celery, celeriac and smallage are small, greenish white and borne in compound umbels. Flowers are mostly insect pollinated. Although individual florets are "perfect", containing both male and female parts, flower development limits natural selfing because male and female reproductive tissues do not develop in the same floret simultaneously (protandry). The fruit is a very small 1 mm (0.04 in) schizocarp that splits when mature into two single-seeded mericarps. Celery plants are prolific seed producers. The oval seed are very small and about 2,500 seeds weigh 1 g (0.04 oz) (Rubatzky and Yamaguchi, 1997). Celery seed is an important but lesser used flavoring with stronger aroma than celery stalks. Celery and celeriac seed can be ground and mixed with salt (Geisler, 2012).

Types and Cultivars

In the early part of the 20th century blanched celery was commonly grown for many markets because existing cultivars of green celery had an objectionable strong flavor. Blanching was a cultural production practice to exclude light from petioles to reduce the harsh flavor and fiber while increasing the succulence of the celery stalk. The blanching was performed several weeks before harvest by placing soil, boards, paper, or other opaque materials against petioles without covering the leaves. To facilitate blanching with soil, seedlings were often transplanted into shallow furrows, which were filled in during the season. Peat soils were preferred because of the ease of moving soil against celery petioles. Soil blanching also provided some frost protection. The application of ethylene gas, a natural plant hormone, following celery harvest was an alternative to field blanching that was occasionally used.

Major reasons for the decline of blanching were labor and material costs, and the presence of soil within the harvested plant that required careful washing. Later in the 20th century, blanched celery gave way to self-blanching cultivars that were naturally pale yellow without shading the stalks. These cultivars, such as 'Golden Self Blanching', resembled blanched celery because they had light yellow-green leaves and lighter petiole color. In the 1940s and 1950s the 'Pascal' cultivars were widely grown and were the first true green types sold in the USA. However, the 'Pascal's had poorer quality because the petioles were not as long and straight as the 'Utah' types popular in many markets today. 'Tall Utah' and its various selections became the predominant types grown in North America because of their exceptionally long, straight green petioles and mild flavor. There are cultivars with pink or red coloration that are used for flavoring and decoration but not eaten fresh (Rubatzky and Yamaguchi, 1997).

Traits that differentiate cultivars include growth rate, color, number of petioles produced per head, degree of suckering, plant diameter, petiole shape and depth of curvature, heart development, stalk compactness, susceptibility to bolting, keeping ability both in the field and during shipment, petiole length, and susceptibility to disease and physiological disorders.

Economic Importance and Production Statistics

Celery is used fresh and smaller quantities are dried, canned, and frozen, usually in combination with other vegetables and in soups. Essential oils are also extracted from celery plants for various uses. Approximately 36 metric tonnes (39.7 short tons) of celery oil are used annually in the flavor and fragrance industry. India produces half of the celery oil in the world. Other producer countries are China, France, the Netherlands, Hungary, and the USA (Falzari and Menary, 2005). Celery foliage and seeds are sometimes used as flavoring herbs. The stalk is served in raw vegetable dishes or stuffed. Celery is also sliced and cooked as an important ingredient in many soup- and stew-like dishes. The leaves may be chopped and used as a garnish similar to parsley.

In the USA and the UK celery cultivars with fleshy stems are popular while on the European continent, celeriac cultivars (also known as root celery or turnip-rooted celery) are common. In much of Asia, and particularly China, smallage is frequently used as salad, flavoring or as a cooked vegetable (Rubatzky et al., 1999).

The per capita consumption of celery in the USA is approximately 2.7 kg (6 lb) of fresh celery per year. In the USA celery consumption has remained fairly constant for decades. The majority of the US celery crop is for the fresh market but some is also processed into canned, frozen, or dehydrated products. According to the USDA, the 2011 crop of celery totaled 783,477 metric tonnes (863,636 short tons). The total value of celery in the USA was nearly US$382 million. Production acreage in the USA totaled 10,898 ha (26,930 acres) (Geisler, 2012). California produces about 75% of US celery and the remaining 25% is grown in Michigan, Texas, and Florida. California's industry is located in the cool coastal valleys. Celery requires a long, cool growing season and early spring plantings in Michigan, New York, and Ohio may sometimes be vernalized by cold weather or damaged by high temperatures in midsummer. This is not a problem in California where celery is produced year-round, although the heaviest volume occurs in the fall and early winter. In 2011, California harvested 10,684 ha (26,400 acres) of celery, producing 750,489 metric tonnes (827,273 short tons) valued at nearly US$369 million. That same year, Michigan harvested 728 ha (1,800 acres) of celery, producing 36,370 metric tonnes (40,091 short tons) worth US$13 million. In 2011, the USA exported 100,014 metric tonnes (110,247 short tons) of fresh celery valued at nearly US$71 million. Canada buys 80% of the fresh US celery exports (Geisler, 2012). The USA imports fresh celery with over half of the imports arriving between August and mid-April mainly from Mexico. The USA imported 41,915 metric tonnes (46,203 short tons) of celery in 2011, valued at US$18.3 million (Geisler, 2012). Italy is the largest producer of celery in the European Community, followed by Spain, Germany, France, the Netherlands, Greece, the UK, and Belgium (CABI, 2008b).

Nutritional Values

Fresh celery is mostly water and contains significant quantities of vitamins A and C, folic acid, and nutritional minerals (Table 21.4). It is also a source of dietary fiber. However, it is low in protein, lipids, and carbohydrates so it is low in calories. Celery is eaten primarily for its unique texture and crispness and is an excellent diet food. Some dieticians suggest the process of eating and digesting celery uses more calories than it provides (Nestle and Nesheim, 2012).

Celery is a very aromatic plant with a characteristic smell that comes primarily from 3-butylphthalid. As a medicinal plant, celery has a wide range of uses as an aphrodisiac, anthelmintic, antispasmodic, carminative, diuretic, emmenagogue, laxative, sedative, stimulant, and tonic (Simon et al., 1984). Celery is also used to treat asthma, bronchitis, and rheumatism. Celery oil is sold as a dietary supplement to promote and regulate healthy blood pressure, joint health (anti-arthritic, antirheumatic properties) and uric acid levels (improve kidney function and treating gout), combating bladder infections, and preventing cancer (Geisler, 2012). Celery contains the flavones apigenin and luteolin (Harnly et al., 2006). Celery has also shown hypoglycemic activity.

Production and Culture

Field establishment

Celery seed germination and emergence are slow even under favorable conditions. Germination at 10°C (50°F) often requires more than 15 days and 30 days or more at 5°C (41°F) depending on seed-lot vigor. At optimum temperatures between 15–20°C (59–68°F) germination requires 7–12 days. Some seed lots of celery exhibit thermodormancy and do not germinate at temperatures above 30°C (86°F) (Rubatzky et al., 1999). The presence of light, periods of low temperature alternation, and growth regulators help alleviate thermodormancy. There are significant cultivar differences in germination performance.

In addition to temperature effects at planting, poor germination may result from variable seed quality and postharvest seed dormancy. As with carrots, celery seed are of variable maturity because they develop sequentially on umbels over an extended period. Newly harvested seed exhibit dormancy due to germination inhibitors, which is removable by seed soaking to leach inhibitors after ripening, osmotic priming or treatment with plant growth regulators (Rubatzky et al., 1999).

Seed priming treatments are often used to improve the germination performance of celery seeds. Priming is a controlled hydration process followed by redrying, which allows imbibition and initial germination processes to occur in seeds but prevents full hydration and radicle emergence. Following priming and

Table 21.4. Nutritional value of celery. Percentages are relative to USDA daily nutrition recommendations for adults (USDA National Nutrient Database, 2013).

Celery, raw	Nutritional value per 100 g (3.5 oz)
Energy	16 kcal (67 kJ)
Carbohydrates	3 g
Sugars	1.4 g
Dietary fiber	1.6 g
Fat	0.2 g
Protein	0.7 g
Water	95 g
Vitamin A equivalent	22 μg (3%)
Thiamine (vitamin B1)	0.021 mg (2%)
Riboflavin (vitamin B2)	0.057 mg (5%)
Niacin (vitamin B3)	0.323 mg (2%)
Vitamin B6	0.076 mg (6%)
Folate (vitamin B9)	36 μg (9%)
Vitamin C	3 mg (4%)
Vitamin E	0.27 mg (2%)
Vitamin K	29.3 μg (28%)
Ca	40 mg (4%)
Fe	0.2 mg (2%)
Mg	11 mg (3%)
P	24 mg (3%)
K	260 mg (6%)
Na	80 mg (5%)
Zn	0.13 mg (1%)

drying to the original moisture content, celery seeds germinate and emerge more uniformly and rapidly than when unprimed. Osmotic seed priming or soaking seeds at 10°C (50°F) in solutions containing growth regulators such as GA4/7 (gibberellic acid) and ethephon at 1,000 ppm can overcome thermo-dormancy (Rubatzky *et al.*, 1999).

Celery seed is often pelleted to increase size to facilitate precision seed placement for both transplant production and direct-field seeding. Sowing of pregerminated seed in gel or liquid carriers has been used to accelerate seedling emergence (Rubatzky *et al.*, 1999). Because of the difficulty in establishing stands and the longer production times involved, direct seeding is rarely used to establish a crop. The majority of celery plantings are established using transplants (Fig. 21.11). Direct-seeded crops mature in 160–180 days and transplanted fields are usually harvested in 90–125 days (Rubatzky *et al.*, 1999).

Good quality transplants are critical for successful celery production. Transplant production has become a highly specialized operation. Because container-grown or plastic plug-tray transplants rapidly establish in the field and produce uniform growth, bare-rooted transplants are rarely used. Although more costly to produce, container-grown transplants are of high quality and better adapted to mechanical transplanting equipment, particularly fully automated transplanters. Mechanical transplanting is faster, requires less labor, and results in more uniform planting depth and spacing than hand transplanting. Celery has a small root system so it is important to have a well-tilled root zone to ensure uniform and rapid establishment of transplants. No- or minimum-till systems can be successfully used only if there is good soil-to-plug transplant contact at the time of establishment. Spacing arrangements generally range from 12–20 cm (5–8 in) in rows and from 50–75 cm (20–30 in) between rows. Typical celery field populations are from 50,000–80,000 plants/ha (20,000–32,000 plants/acre) (Rubatzky *et al.*, 1999).

A major disadvantage of celery transplanting is its higher cost. A growth period of 4–6 weeks is required to produce suitable transplants. However, this is partially compensated for because transplanting reduces the growth period in the field. Transplanting provides full stands and uniform spacing while reducing some cultural inputs, such as fertilization, irrigation, and pest and weed management (Rubatzky *et al.*, 1999).

Celeriac is mostly direct-seeded because transplanting causes excessive lateral rooting and roughness of the enlarged globe-like root. Celeriac spacing ranges from 30–60 cm (12–24 in), with field populations of 40,000–50,000 plants/ha (16,000–20,000 plants/acre). Smallage may be

Fig. 21.11. Establishing celery from transplants in California.

transplanted or direct seeded with field populations of about 100,000 plants/ha (40,000 plants/acre) (Rubatzky et al., 1999).

Celery site selection, field preparation, fertilization, and nutrition

Celery is one of the most challenging cool-season crops to grow in terms of managing fertility inputs. Celery requires high levels of nutrients, especially N, so fertile organic or muck soils, fine-textured loam or silt loam soils are preferred. Celery prefers a soil pH of about 6.5 on mineral soils and about 5.8 on organic soils. Poorly drained or waterlogged soils should be avoided (Rubatzky et al., 1999).

Celery planted in close rotation with other crops may benefit from considerable amounts of residual soil N depending on soil and environmental conditions. Soil nitrate testing can assess N fertility prior to planting, and nitrate levels greater than 20 ppm in the top 30 cm (12 in) of soil are adequate for celery growth (Daugovish et al., 2008). Celery crops may remove in the range of 225–280 kg/ha (200–250 lb/acre) N from soil (Daugovish et al., 2008). Little biomass is produced during the first month after transplanting. Small quantities of N, 22–34 kg/ha (20–30 lb/acre), are applied pretransplant to adequately supply plants during field establishment. The N crop requirements increase with maturity. The greatest nutrient need occurs the last 4–6 weeks before harvest. During the most rapid growth phase, 17–22 kg/ha (15–20 lb/acre) of N per week are required and may be added through drip tape during irrigation or alternatively by side-dress applications if fertigation is not possible (Daugovish et al., 2008). Under most field conditions, a seasonal fertigation total of 168–152 kg/ha (150–225 lb/acre) or 224–308 kg/ha (200–275 lb/acre) of total N is typically applied on most soils except for organic soils, where totals may be less (Daugovish et al., 2008).

Phosphate fertilization should be based on soil testing levels of extractable P. Phosphate uptake by a celery crop ranges from 45–50 kg/ha (40–45 lb/acre). Levels above 60 ppm are adequate for growth. For lower soil test results, 45–90 kg/ha (40–80 lb/acre) of P_2O_5 (phosphorus oxide) are recommended, particularly prior to planting in cold soils (Daugovish et al., 2008).

Potassium uptake by a high-yielding celery crop ranges from 390–505 kg/ha (350–450 lb/acre). Potassium fertilization needs should be determined by soil testing before inputs are applied. Soils with greater than 150 ppm of ammonium-acetate-exchangeable K are sufficient for crop production. Fertilization to replace the K removed by crop production is appropriate to maintain soil fertility although split application with a side dressing of K is commonly applied 6–8 weeks before harvest, particularly if fertigation is not used. (Daugovish et al., 2008)

Celery has a high requirement for other nutrients besides N-P-K that can negatively affect the crop growth and cause visual symptoms that reduce crop quality. Calcium is important for cell wall formation and Ca deficiency can cause a physiological disorder called "black heart", particularly in conjunction with drought stress, which results in the death of the apical meristem in the center of the plant (Daugovish et al., 2008). Celery is sensitive to boron (B) deficiency, which may result in lateral cracking across the petioles (Rubatzky et al., 1999). Fertilizer supplemented with B is commonly used in the Great Lakes region of the USA and other areas where soil levels are chronically low. Zinc fertilization is recommended if the extractable soil level is less than 1.5 ppm. Zinc (Zn) fertilization is commonly practiced in soils where soil P levels are high, reducing Zn uptake by celery plants (Daugovish et al., 2008). Insufficient magnesium (Mg) causes leaf chlorosis. Although soil applications made in advance of planting are preferred and more effective, topical sprays are sometimes used to treat these symptoms.

Celery growth and development

Celery is one of the more demanding vegetables to grow because of its slow growth and very specific environmental requirements. Celery is a labor-intensive crop, which requires fertile soils and a temperature range between 16–21°C (61–70°F) for optimum production (Rubatzky et al., 1999). Some modern cultivars developed through plant breeding have extended upper temperature limits, allowing celery to be produced in some subtropical regions. Celery is sensitive to freezing temperatures, while celeriac and smallage are slightly less sensitive. When acclimated, celery can tolerate a light frost of −2°C (28°F) for a short time with little or no damage. Leaf celery is more heat tolerant than either celeriac or stalk celery (Rubatzky et al., 1999).

The growth rate steadily increases so that most of the biomass accumulation occurs in the last 3–4 weeks before harvest. Stalk celery requires a long growing season maturing from about 90–125 days after transplanting depending on the cultivar and growing conditions. Spray applications of gibberellic acid (GA) accelerate celery growth and earliness but are not a standard production practice because treatment occasionally results in long, thin, and light-colored petioles. Crispness and tenderness are major quality attributes of celery petioles. Wilting, pithiness, or excessive petiole fiber greatly reduces value. Pithiness is a physiological disorder characterized by softening and degradation of parenchyma tissues in the petiole resulting in a loss of density (Rubatzky and Yamaguchi, 1997). When growth is excessively rapid, the maturing petioles become susceptible to pithiness, particularly if plants are stressed. Moisture or freezing stresses help induce pithiness. If severe, petioles develop hollow areas and the stem plate may also be affected. The causes of hollow celeriac roots are similar to pithiness in stalk celery.

When grown continuously at 15°C (59°F) or higher, celery plants remain vegetative. Seedlings with less than four or five true leaves are considered to be in the juvenile stage and insensitive to vernalization, but once beyond the juvenile stage, plants are susceptible. Bolting is initiated when sensitive plants are exposed to low temperatures, usually less than 10°C (50°F). Cultivars vary as to their sensitivity threshold for vernalization with regard to temperature and duration. Some tolerate low temperatures and long exposure better, while others are sensitive to vernalization at temperatures as high as 13–14°C (55–57°F). Plant stress can also increase the predisposition to vernalization (Rubatzky et al., 1999).

To avoid bolting, field plantings should be scheduled to minimize low-temperature exposure and transplants should not be exposed to temperatures of less than 15°C (59°F) during production. Acclimating transplants at between 25–30°C (77–86°F) for 10–20 days prior to transplanting reduces transplant sensitivity to low field temperatures (Rubatzky and Yamaguchi, 1997).

Celery irrigation

Celery is a shallow-rooted crop that requires a continuous supply of water to maximize yields and quality, especially during rapid growth. Celery has its highest moisture needs during the final month before harvest. If rainfall is insufficient, overhead sprinklers, drip, furrow, or combinations of these methods are used to ensure the crop receives the approximately 50 mm (2 in)/week or 750–900 mm (30–35 in) per season needed to optimize production depending on soil characteristics and environmental factors (Fig. 21.12; Rubatzky and Yamaguchi, 1997).

The moisture requirements for celeriac and smallage are less than celery. Soil moisture monitoring and weather-based irrigation scheduling can be used in combination to determine the water needs of a celery crop throughout the season. Maintaining soil moisture tension less than 0.03 MPa (30 centibars) prevents water stress and maximizes yield (Daugovish et al., 2008). Crop water use can be estimated using reference evapotranspiration data and a crop coefficient that relates water loss from the canopy. Since crop coefficients vary based on environmental conditions and crop development and are determined experimentally, local references should be consulted to determine the correct values to use when calculating how much irrigation to apply to a celery crop when soil tension reaches 0.03 MPa (30 centibars) (Daugovish et al., 2008).

Celery is often sprinkler-irrigated after transplanting until the first fertilizer side-dressing. Sprinkler irrigation permits more frequent and lighter irrigations that are helpful in activating side-dressed fertilizer or herbicides that may be applied early in establishment. After successful crop establishment, a switch to furrow irrigation is often made when available. Furrow irrigation may provide better uniformity, particularly in windy areas or when crop height exceeds sprinkler riser height (Daugovish et al., 2008).

The use of surface drip irrigation has increased in many areas in recent years to conserve water and more precisely provide nutrients through fertigation. Subsurface drip is an evolving technology that requires different management practices because the sub-irrigation lines are installed before transplanting and usually remain in the soil for more than one season.

Drip irrigation distributes water more uniformly than furrow or sprinkler irrigation. If properly managed, celery can be grown from transplants exclusively using drip irrigation although drip lines are often installed following planting and fertilizer side-dressing. Drip lines help attain uniform growth throughout fields with variable soil textures because it is easier to maintain high soil moisture levels, and

Fig. 21.12. Furrow irrigation of a celery field near Oxnard, California a few weeks prior to harvest. Drip irrigation is increasingly used because of its greater efficiency.

through fertigation, better nutrient management (Daugovish *et al.*, 2008). Drip irrigation helps minimize the leaching of N by frequently applying low rates of fertilizer and less water than may be applied by furrow or sprinkler irrigation. Some growers may supplement drip irrigation with sprinkler or furrow applications to saturate the entire bed in the last few weeks prior to harvest (Daugovish *et al.*, 2008). Soils that become too dry can cause Ca deficiency, which is induced by water stress. Well-watered soils also help prevent pithiness, a physiological disorder caused by plant stress. Overwatering should be avoided because it increases the risk of fungal diseases.

Celery harvesting and marketing

Celery stalks are normally harvested when the majority of plants in a field reach marketable size, although some size variation is inevitable. A delay in harvest results in pithy petioles, but early harvest results in fewer large stalks. Celery is primarily hand harvested. Rather than selectively harvesting large plants and letting smaller plants develop further, all plants are destructively harvested in a single operation and then graded according to size (Fig. 21.13). Important celery quality characteristics include plant weight, width, height, and the thickness and number of petioles.

Trimming and packing are performed directly in the field or on mobile field units whenever possible, as field packing is more efficient than trimming and packing celery stalks in a stationary packing shed at a remote location. At harvest, celery plants are undercut to sever roots, lifted, and trimmed of side roots and suckers. Further trimming removes much of the leaf blades and upper petioles to a length that meets packaging requirements. Untrimmed leaf blades are prone to wilting. Leaf celery has a shorter growing season and multiple harvests are possible if leaves are cut to encourage regrowth of new foliage.

A limited amount of machine harvesting of celery occurs, mostly for processing. Celeriac is mechanically dug for hand trimming and washing of the root globes. Celery crops frequently yield about 60–75 metric tonnes/ha (27–33 short tons/acre) and celeriac yields are about 40 metric tonnes/ha (18 short tons/acre).

Harvested celery is often hydrocooled following packing, although vacuum (hydrovac) and forced-air cooling are also used. Hydrovac equipment introduces water during the vacuum process to limit moisture lost from celery stalks, as even a slight amount of water loss results in visible wilting.

Most US celery is sold in the fresh market, but a portion is processed for use in prepared foods such as soups, juices, and convenience dinners. Celery is also lightly processed into ready-to-eat

Fig. 21.13. Hand harvesting, trimming, and packing of celery in a Californian field.

convenience-packaged products such as celery sticks and sliced celery. Celery is an expensive crop to produce but the return per hectare or acre can be among the highest of any vegetable crop. Celery is marketed in plastic bags or in bulk. Celery hearts are a product sold in the USA derived from short petioles less than 35 cm (14 in) in length that do not meet the US grade standards for stalk celery.

For the fresh vegetable market, celery must be harvested within a few days after reaching a marketable size and cooled to remove field heat before its quality deteriorates. Celery stalks must be carefully handled and should not be stacked more than four high during storage or in shipping crates to avoid bruising. Usually fresh celery stalks are packed in an upright position in their crates and require constant temperature and moisture control to ensure marketable quality. Optimum storage conditions for celery and celeriac are 0°C (32°F) and 95% RH. Under these conditions both types of celery can be stored for at least 1 month. Although now rare, some off-season field storage in root cellars or in plastic-covered, straw-insulated trenches occurs in some parts of the world. Controlled atmospheric storage can be used to maintain marketable quality for relatively long periods. Such storage requires 0°C (32°F) and high RH in an atmosphere of 1–2% O_2, 4% CO_2 and with ethylene removal (Rubatzky et al., 1999).

Celery Diseases

Celery transplants, particularly those grown in greenhouses, may become infected with late blight (*Septoria apiicola*) and early blight (*Cercospora apii*). Late blight can be successfully managed using disease-free seed, fungicide treatments and by using irrigation systems that keep the foliage dry, e.g. furrow, drip, or subsurface capillary mats for transplant production (Daugovish et al., 2008).

Fusarium (*Fusarium oxysporum* f. sp. *apii*) is a devastating soil-borne disease that can destroy entire crops. Several races exist and the pathogen can persist for years without a host, so short crop rotations are ineffective. Soils are sometimes fumigated to control *Fusarium* and other celery soil-borne pests, especially in production areas where crop rotation options are limited. Pink rot (*Sclerotinia sclerotiorum*) and crater rot (*Rhizoctonia solani*) are soil-borne fungal diseases that affect the lower petiole when conditions are wet (Daugovish et al., 2008). Fungicides sometimes provide effective control to both diseases. Other fungal diseases that attack celery include leaf curl of celery (*Colletotrichum acutatum*), which is confined to Australia, downy mildew, powdery mildew, crown and root decay of celery, brown spot, crater spot, phoma crown and root rot, southern blight, and violet root rot (Davis and Raid, 2002). The main postharvest diseases of celery are gray mold (*Botrytis cinerea*) and licorice root (*Mycocentrospora acerina*) (Davis and Raid, 2002).

The most important bacterial diseases are bacterial leaf spot of celery (*Pseudomonas syringae*), bacterial blight, and soft rot (*Erwinia carotovora* and *Pseudomonas marginalis*). Choosing resistant cultivars, limiting foliage moisture, using disease-free seed, and crop rotation can control bacterial disease. A reduced plant density will increase air movement within the canopy and create a drier environment. Bacterial blight often does not persist in the field.

Celery is susceptible to aster yellows mycoplasma (Daugovish *et al.*, 2008). Celery is also affected by a host of viral diseases with widespread distribution, including celery mosaic, cucumber mosaic, calico mosaic, alfalfa mosaic, arabis mosaic, celery latent, celery yellow mosaic, tomato black ring, curly top, beet curly top, parsnip yellow fleck, southern mosaic of celery, strawberry latent ring spot, celery yellow net, strap leaf, and tomato spotted wilt. Control of insect vectors, such as leafhoppers and white flies, is essential for stopping the spread of virus disease (Davis and Raid, 2002).

Celery Insect Pests

Depending on the region and the time of year, the most damaging insects to celery are leaf miners, caterpillars, and aphids (Daugovish *et al.*, 2008). The aster leafhoppers (*Macrosteles uadrilineatus*), which are vectors for the disease "Aster yellows", can be particularly problematic as well (CABI, 2008b). The pea leafminer (*Liriomyza langei*) and the serpentine leafminer (*L. trifolii*) are serious pests in California (Daugovish *et al.*, 2008). The vegetable leafminer (*L. sativae*) is also a pest of celery. The primary damage from leaf miners is caused by their larvae, which feed on leaf mesophyll tissue, forming hollow spaces in the center of affected leaves. Leafminer larvae are parasitized by several species of wasp, which naturally suppress populations. Control strategies targeting larvae are generally more successful than the mobile adults, which are insecticide resistant in some areas (Daugovish *et al.*, 2008).

If insecticides are used, they should be rotated, using different modes of action to slow insect resistance.

Other pests include carrot rust fly (*Psila rosae*), carrot weevil (*Listronotus oregonensis*), tarnished plant bug (*Lygus lineolaris*), spider mites, variegated cutworm (*Peridroma saucia*), celery looper (*Anagrapha alcifera*), beet armyworm (*Spodoptera exigua*), celery leaf tier (*Udea rubigalis*), green peach aphid (*Myzus persicae*), black bean aphid (*Aphis fabae*), and nematodes. Nematodes are controlled by crop rotation, disinfection of the soil by solarization or fumigation (CABI, 2008b).

Celery Weed Management Strategies

Integrated weed-management strategies should be implemented prior to celery transplanting. A weed management program usually includes crop rotation, removing weeds before they produce viable seed, preplanting irrigation to stimulate germination that is followed by cultivation to kill the first flush of weeds, timing of planting to reduce weed impact, careful preparation and spacing of beds, and precision transplanting so that cultivation equipment can be accurately aligned to the row (Daugovish *et al.*, 2008). For celery, weed control is most critical until transplants are established and form a canopy, since small plants compete poorly with weeds. Preplant and post-plant herbicides have been developed for celery and if used properly can be very effective. Mechanical cultivation and hand weeding can also be effective but require extensive use of labor.

MINOR CROPS IN THE FAMILY APIACEAE

There are many minor vegetable crops and herbs used for food in the family Apiaceae. These diverse family members can be divided into many categories but one useful classification is by root crops, foliage crops, and herbs or condiments (Rubatzky *et al.*, 1999). However, some crops such as lovage may fit into multiple categories, with foliage consumed as a vegetable, edible roots and seeds used as a condiment. A brief description of a few of the other umbelliferous crops follows.

Condiments and Herbs

Anise (*Pimpinella anisum*)

Anise is an annual whose strongly flavored seed is used for flavoring breads, cookies or cakes. Leaves may also be used in salads or as a garnish. Anise has a distinctive flavor reminiscent of licorice (*Glycyrrhiza glabra* L.).

Asafoetida (*Ferula assa-foetida*)

Asafoetida refers to the dried latex (gum oleoresin) exuded from the living underground rhizome or taproot of several species of plants from the genus *Ferula*, which is a perennial herb 1–1.5 m (3.3–5 ft) tall. The species is native to the mountains of Afghanistan and is mainly cultivated in nearby India. Asafoetida has a pungent, strong smell when raw, but when cooked it produces a smooth flavor reminiscent of leek (Rubatzky *et al.*, 1999).

Caraway (*Carum carvi*)

Caraway is a biennial plant native to western Asia, parts of Europe, and northern Africa. The plant has finely divided, feathery leaves with thread-like divisions similar to carrot, reaching a height of 20–30 cm (7.9–11.8 in). After bolting, the flower stalk grows to a height of 40–60 cm (15.7–23.6 in), with small white or pink flowers. Caraway fruits, erroneously called seeds, are crescent-shaped achenes, around 2 mm (0.8 in) long. The fruits are often used whole and have a pungent, anise-like flavor and aroma that is caused by essential oils. The fruits are used to flavor breads (especially rye), desserts, liquors, curry, and other foods.

Coriander (*Coriandrum sativum*)

Coriander is an annual. The name coriander is used to describe the fruit and seeds, which are used as a condiment for baking, confections, pickling, and making curry. When the foliage of the same plant is used it is called cilantro.

Florence fennel (*Foeniculum vulgare* var. dulce)

Florence fennel is a perennial that is grown as an annual. It is a popular vegetable in Europe but used less in other parts of the world (Fig. 21.14).

Plants are erect, glaucous green, reaching a height of 2.5 m (8.25 ft), with hollow stems. The finely dissected, threadlike leaves grow to 40 cm (15.7 in) long and about 0.5 mm (0.02 in) wide.

Fig. 21.14. Florence fennel production in Germany.

The bulb-like base is widely used cooked or raw as a vegetable and in salads. The seeds are strongly flavored and used in soups and breads. Leaves are used as a garnish and for pickling. The flavor is similar to licorice.

Root Crops

Arracacha (*Arracacia xanthorrhiza*)

Arracacha is an important New World crop. In some South and Central American locations, arracacha is a fairly important source of calories in the diet. Arracacha roots are consumed after cooking in a multitude of different preparations including boiling, baking, and frying, for use as whole, mashed, or pureed vegetables added to soups and stews (Rubatzky et al., 1999).

Chervil (turnip rooted) (*Chaerophyllum bulbosum*)

Turnip-rooted chervil is a biennial that produces swollen roots resembling a small short carrot in shape but with a dark gray color and yellowish-white flesh. The plant develops quickly but root quality improves when left in the soil after leaves die back. The roots have a long storage life. Roots are usually cooked by boiling and have a sweet aromatic flavor.

Parsnip (*Pastinaca sativa*)

Parsnip is a slow growing biennial grown similarly to carrots. Parsnip is native to Europe and Asia and was popular with early Greeks and Romans. The edible portions are long 25–38 cm (10–15 in) roots with broad shoulders. Parsnips are very cold tolerant and will survive freezing in the field. Many believe that the flavor of white to whitish-yellow fleshed roots is sweeter after frost. Parsnips have a long storage life.

Skirret (*Sium sisarum*)

Skirret is of Chinese origin, but apparently arrived in Europe in ancient times. It is a perennial plant with white roots that resemble sweet potato, but longer (15–20 cm, 5.9–7.9 in). The plant grows to about 1 m (39.3 in) high and is resistant to cold, pests, and diseases. It is usually grown from seed, but may also be started from root divisions. Skirret roots are boiled, stewed, or roasted. The woody core is inedible (Rubatzky et al., 1999).

Foliage Crops

Chervil or salad chervil (*Anthriscus cerefolium*)

Chervil is a cool-season annual. Leaves are used in salads and soups and as a garnish. Chervil is cultured similar to parsley and both curled- and plain-leaf cultivars are available. Chervil matures in 50–65 days and does not tolerate heat but can overwinter if winters are mild. Serial harvests are made throughout the season.

Cilantro (*Coriandrum sativum*)

Cilantro comes from the same plant as coriander (*C. sativum*). The name coriander is used to describe the fruit and seeds, while the young foliage used for flavoring soups, salads, and other dishes is called cilantro.

Dill (*Anethum graveolens*)

Dill is grown as both an annual and a biennial. The young leaves (dill weed) are used in salads and the flavoring of other foods. Seeds have a strong flavor and are used for making pickles. Dried leaves and seeds are commonly used in place of fresh material (Fig. 21.15).

Mitsuba (*Cryptotaenia japonica*)

Mitsuba is a cool-season perennial that is usually grown outdoors and harvested during the spring. It can also be forced in a greenhouse for production during the winter. It is an important crop in Japan and other parts of Southeast Asia. Mitsuba has a distinctive aromatic taste and is used much like celery, either raw or cooked in salads and soups, and as a flavoring.

Parsley (*Petroselinum crispum*)

Parsley is a biennial or short-lived, cold-tolerant perennial. Native to Europe, parsley has been used as a vegetable for more than 2,000 years. Plain and curled leaf cultivars are available. Parsley is used fresh or dried in salads and as a garnish. Plain-leaf cultivars usually are used for flavoring while curled are more often used as a garnish (Fig. 21.16). Parsley is serially harvested throughout the season.

Fig. 21.15. Dill seed field near Salinas, California.

Fig. 21.16. A curled-leaf parsley plant whose foliage is typically used as a garnish.

Multiple Use Crops

Angelica (*Angelica archangelica*)

Angelica is a genus of about 60 species of tall biennial and perennial herbs native to temperate and subarctic regions of the northern hemisphere. Plants grow to 1–3 m (39.3–118 in) tall, with large bipinnate leaves and large compound umbels with white or greenish-white flowers. *Angelica archangelica* is often called garden angelica because its roots are peeled, boiled, and eaten as a vegetable. Stems and petioles are peeled and used in salads. Angelica oil extracted from roots and powdered roots are used as flavoring and for medicinal purposes. The seeds are also used medicinally and for flavoring.

Lovage (*Levisticum officinale* Koch)

Lovage is an erect, herbaceous, perennial plant growing 1.8–2.5 m (70.9–98.4 in) tall, with a basal rosette of leaves and stems. The flowers are produced in umbels at the top of flower stalks arising from the center of the rosette. Lovage has long been cultivated in Europe. The leaves are used as an herb, the roots as a vegetable and the seeds as a spice, especially in southern European cuisine.

References

Banga, O. (1957) Origin of the European cultivated carrot. *Euphytica* 6, 54–63.

Banga, O. (1963) *Main Types of Western Carotene Carrot and their Origin*. Tjeenk Willink, Zwolle, the Netherlands.

CABI (2008a) *Daucus carota* (carrot) carrot. Notes on pests. Available at: www.cabi.org/cpc (accessed 28 May 2013).

CABI (2008b) Datasheets: *Apium graveolens* celery production and trade. Available at: www.cabi.org/cpc (accessed 28 May 2013).

Daugovish, O., Smith, R., Cahn, M., Koike, S., Smith, H., Aguiar, J., Quiros, C., Cantwell, M. and Takele, E. (2008) Celery production in California. Vegetable production series Publication 7220. Available at: http://anrcatalog.ucdavis.edu (accessed 4 June 2013).

Davis, R.M. and Raid, R.N. (2002) *APS Compendium of Umbelliferous Crop Diseases*. American Phytopathological Society, St. Paul, Minnesota.

Esau, K. (1936) Ontogeny and structure of collenchyma and of vascular tissues in celery petioles. *Hilgardia* 10, 431–476.

Falzari, L. and Menary, R. (2005) *Development of a celery oil and extract industry*, Publication No. 05/133, Project No. UT-35A. Australian Government, Rural Industries Research and Development Corporation, Kingston, Australia.

FAOSTAT (2011) Carrot and turnip production statistics. Available at: http://faostat.fao.org/site/567/DesktopDefault.aspx?PageID=567#ancor (accessed 1 March 2013).

Geisler, M. (2012) Vegetables: Celery profile. Available at: www.agmrc.org/commodities_products/vegetables/celery-profile (accessed 31 May 2013).

Goldman, I.L. and Breitbach, D.N. (2002) Reduced pigment gene of carrot and its use. Patent number: 6437222. Available at: http://assignments.uspto.gov/assignments/?db=pat (accessed 28 February 2013).

Harnly, J.M., Doherty, R., Beecher, G.R., Holden, J.M., Haytowitz, D.B. and Bhagwat, S. (2006) Flavonoid content of USA fruits, vegetables, and nuts. *Journal of Agricultural and Food Chemistry* 45, 9966–9977.

Hawthorn, L.R., Toole, E.H. and Toole, V.K. (1962) Yield and variability of carrot seeds are affected by position of umbel and time of harvest. *Proceedings of the American Society of Horticultural Science* 80, 401–107.

Heywood, V.H. (1983) Relationships and evolution in the *Daucus carota* complex. *Israel Journal of Botany* 32, 51–65.

Holland, B., Unwin, I.D. and Buss, D.H. (1991) *Vegetables, Herbs and Spices. The fifth supplement to McCance & Widdowson's The Composition of Foods*, 4th edn. Royal Society of Chemistry, Cambridge, UK.

Lietz, G., Oxley, A. and Boesch-Saadatmandi, C. (2013) Consequences of common genetic variation on B-carotene cleavage for Vitamin A supply. In: Sommerburg, O., Siems, W. and Kraemer, K. (eds) *Carotenoids and Vitamin A in Translational Medicine*. CRC Press, Boca Raton, Florida, pp. 383–391.

Lorenz, O. and Maynard. D (1988) *Knott's Handbook for Vegetable Growers*, 3rd edn. Wiley Interscience Publications, New York.

Minnis, A.M., Farr, D.F. and Rossman, A.Y. (2013) Fungal nomenclature database, systematic mycology and microbiology laboratory, ARS, USDA. Available at: http://nt.ars-grin.gov/fungaldatabases/nomen/nomenclature.cfm (accessed 23 May 2013).

Nascimento, W.M., Vieira, J.V., Silva, G.O., Reitsma, K.R. and Cantliffe, D.J. (2008) Carrot seed germination at high temperature: Effect of genotype and association with ethylene production. *HortScience* 43, 1538–1543.

Nestle, M. and Nesheim, M. (2012) *Why Calories Count: From Science to Politics*. University of California Press, Berkeley, California.

Oliva, R.N., Tissaoui, T. and Bradford, K. (1988) Relationship of plant density and harvest index to seed yield and quality in carrot. *Journal of American Society of Horticulture Science* 113, 532–537.

Peterson, C.E. and Simon, P.W. (1986) Carrot breeding. In: Bassett, M.J. (ed.) *Breeding Vegetable Crops*. AVI Publishing Company, Westport, Connecticut, pp. 321–356.

Rosenfeld, H.J., Baardseth, P. and Skrede, G. (1997) Evaluation of carrot varieties for production of deep fried carrot chips. IV. The influence of growing environment on carrot raw material. *Food Research International* 30, 611–618.

Ross, I.A. (2005) *Medicinal Plants of the World: Chemical Constituents, Traditional and Modern Medicinal Uses*, Vol. 3. Humana Press, New York.

Rubatzky, V.E. and Yamaguchi, M. (1997) *World Vegetables: Principles, Production and Nutritive Values*, 2nd edn. Chapman & Hall, New York.

Rubatzky, V.E., Quiros, C.F. and Simon, P.W. (1999) *Carrots and Related Umbelliferae*. CAB International, Wallingford, UK.

Simon, J.E., Chadwick, A.F. and Craker, L.E. (1984) *Herbs: An Indexed Bibliography. 1971–1980. The Scientific Literature on Selected Herbs, and Aromatic and Medicinal Plants of the Temperate Zone*. Archon Books, Hamden, Connecticut.

Simon, P.W., Peterson, C.E. and Gabelman, W.H. (1990) B493 and B9304, carrot inbreds for use in breeding, genetics, and tissue culture. *HortScience* 25, 815.

Stevens, P.F. (2012) Angiosperm phylogeny website. Version 12. Available at: www.mobot.org/MOBOT/research/APweb (accessed 12 June 2013).

Strandberg, J.O. (2000) Common Names of Plant Diseases - Carrot. Available at: www.apsnet.org/publications/commonnames/Pages/Carrot.aspx (accessed 28 May 2013).

Takaichi, M. and Oeda, K. (2000) Transgenic carrots with enhanced resistance against two major pathogens, *Erysiphe heraclei* and *Alternaria dauci*. *Plant Science* 153, 135–144.

USDA (2013) Natural resources conservation service plant profile. *Daucus carota* L. Available at: http://plants.usda.gov/java/profile?symbol=daca6 (accessed 25 February 2013).

USDA-ERS (2011) US carrot statistics. Available at: http://usda.mannlib.cornell.edu/MannUsda/viewDocumentInfo.do?documentID=1577 (accessed 1 March 2013).

USDA-GRIN (2013) National genetic resources program. Germplasm resources information network (GRIN). Available at: www.ars-grin.gov/cgi-bin/npgs/acc/display.pl?1593961 (accessed 26 February 2013).

USDA-NASS (2009) USA carrots: Area, production, price & value, 1950–2009. Available at: http://usda.mannlib.cornell.edu/MannUsda/viewDocumentInfo.do?documentID=1577 (accessed 1 March 2013).

USDA National Nutrient Database (2013) Standard reference release 25. Nutrient data for 11124, carrots, raw, celery raw. Available at: http://ndb.nal.usda.gov/ndb/foods/show/2886?qlookup=11124 (accessed 18 May 2013).

Wu, J.S.B. and Shen, S.-C. (2011) Processing of vegetable juice and blends. In: Sinha, N.K., Hui, Y.H., Özgül Evranuz, E., Siddiq, M. and Ahmed, J. (eds) *Handbook of Vegetables and Vegetable Processing*. Wiley-Blackwell Publishing Ltd., Ames, Iowa, pp. 343–344.

22 Family Agaricaceae

MUSHROOMS

Introduction

Mushrooms are a very important world crop that fit the definition of a vegetable given in Chapter 1. However, unlike the other vegetables we have discussed thus far, the mushroom is not a plant but a fungus (Carluccio, 2003). So mushroom-production practices are unique compared to the traditional vegetables discussed in previous chapters.

Mycology

Mushrooms are heterotrophic organisms, which means they must find and absorb food from their environment. This is contrast to most other vegetables, which are autotrophic plants that can fix carbon dioxide (CO_2) from the atmosphere via photosynthesis. Mushrooms must acquire carbon from sources other than gaseous atmospheric CO_2, so organic matter (rather than soil) is required as substrate for mushrooms to acquire their nutrients and support mycelial growth (Chang and Hayes, 1978; Del Conte et al., 2008). The types of organic matter needed vary widely with the kinds of mushrooms grown (Carluccio, 2003).

Types of Cultivated Mushrooms

There are many diverse types of mushrooms, both wild and cultivated, that are consumed and used medicinally (Bon, 1987; Royse, 1997; Stamets, 2000). Only a few of the most economically important mushrooms will be discussed here. The following is a brief description of some economically important mushrooms that are cultivated as vegetables.

Button mushroom (*Agaricus bisporus*) from the family Agaricaceae is one of the most commonly and widely consumed mushrooms in the world. In the USA and Canada, button mushrooms are the most important commercial type. The growth and development of mushrooms is very different from higher plants. Under the proper environmental conditions, mushroom mycelia grow vigorously from spawn on organic substrates, and when conditions are suitable a small "button" emerges (Fig. 22.1; Beyer, 2003).

The "button" is actually the sexual structure of the fungus known as a fruiting body, and is the edible portion of the mushroom. The stem or stipe of the fruiting body is often thick and smooth and bears a thickened cap or pileus (Yamaguchi and Rubatzky, 1997). The underside of the pileus ruptures at maturity, exposing gills from which spores are released. The spores are collected, purified, multiplied in a laboratory, and prepared as an aggregate of mycelia and spores known as spawn to propagate the next generation (Stamets and Chilton, 1983). Mushroom spawn, which is a mixture of released spores and organic matter or grain, is used to "seed" the organic-matter substrate (Beyer, 2003). Under the right conditions the spawn gives rise to mycelia, which spread throughout the substrate.

Button mushrooms have many common names. Button or white mushroom describes fruiting bodies with a closed cap and either pale white or light brown flesh. Immature fruiting bodies of strains with darker flesh may be marketed as crimini, baby portobello, baby bella, mini bella, portabellini, Roman, Italian, or brown mushrooms (Fig. 22.2).

During the later stages of development, the cap begins to open and at maturity when the gills are exposed, the fruiting bodies are marketed as "portobello", "portabella", and "portabello", depending on location (Zeitlmayr, 1976; Carluccio, 2003).

The shiitake (*Lentinula edodes*) is an edible mushroom native to East Asia that has a golden or dark-brown to blackish cap and a tough stipe that may not be used for cooking. In Chinese it is called "fragrant mushroom", "winter mushroom", or "flower mushroom" because of the flower-like cracking pattern that is present on the mushroom's cap at lower

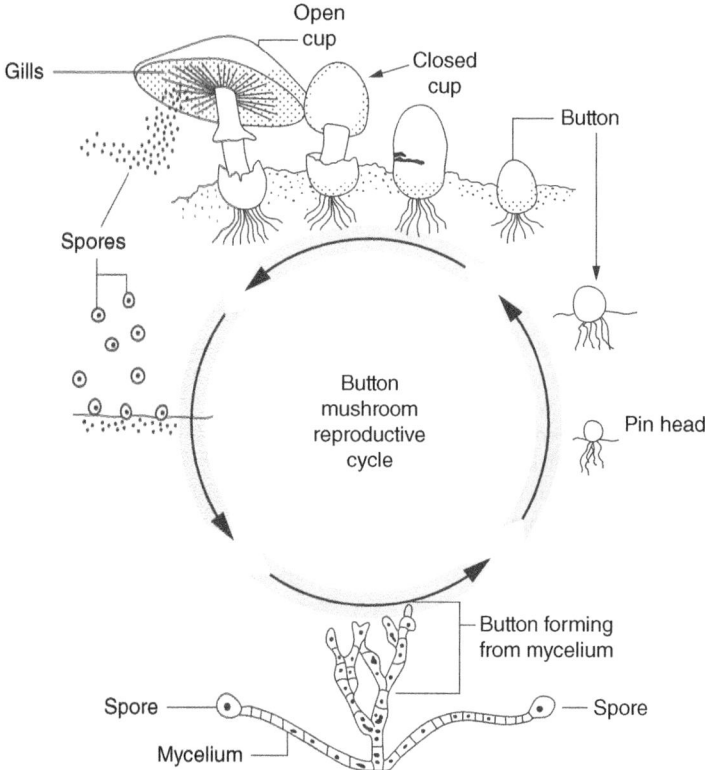

Fig. 22.1. The life cycle of a button mushroom.

Fig. 22.2. Brown button mushrooms (*Agaricus bisporus*).

temperatures. Common names in English include "Chinese black mushroom", "black forest mushroom", "black mushroom", "golden oak mushroom", or "oak-wood mushroom" (Stamets, 2000). Shiitake are cultivated and consumed fresh in Asia but are also dried and exported around the world. In Japanese, shiitake means "shii mushroom" from the Japanese name for the tree *Castanopsis cuspidate*, which is traditionally used for cultivation. Historically, the shii trees were cut and the logs placed by trees producing shiitake or containing shiitake spores. Today, shiitake are produced internationally by placing spawn in holes drilled in staked logs of several tree species (Leatham, 1982).

In addition to button and shiitake, other popular mushrooms include the following. Oyster (*Pleurotus ostreatus*), also called abalone or tree mushrooms, are some of the most commonly cultivated in both tropical and temperate climates throughout the world (Chang and Miles, 2004b). Most species of *Pleurotus* are white-rot fungi with a soft consistency that are grown on hardwood, although some also grow on decaying conifers (Fig. 22.3).

The caps may be attached directly to the substrate with no stem (stipe). If a stem is present, the lamellae or "gills" run along the stipe or stem (Chang and Miles, 2004b). The term pleurotoid is used for mushrooms having this general shape (Bon, 1987).

Another mushroom is enokitake (*Flammulina velutipes*), also called enoki or "golden needle mushroom" (Fig. 22.4). There are significant differences in the appearance of the enoki mushroom among wild and cultivated types. Cultivated enoki, often used in East Asian cuisine, are long, thin, and white because they are grown in a dark, CO_2-rich environment to encourage stem elongation (Royse, 1997). Wild forms are brown with a shorter, thicker stipe, a larger pileus or cap, and grow naturally on stumps of the Chinese hackberry tree ("enoki" in Japanese) as well as mulberry and persimmon trees. In most areas, only cultivated enoki are commercially available.

Enoki are usually cultivated in a plastic bottle or plastic bag for 30 days on a substrate of sawdust or corn cobs combined with other ingredients. The environmental conditions for cultivation are 15°C (59°F) and 70% humidity. Later, the enoki are transferred to a paper cone and a slightly cooler and more humid environment for an additional 30 days to encourage development of a long, thin stem (Yamanaka, 1997).

The straw mushroom (*Volvariella volvacea*), also called paddy straw mushroom, is another popular mushroom in East and Southeast Asia. Straw mushrooms are sold fresh in Asia, but are more frequently sold in cans or dried in other parts of the world for use in Asian cooking.

Straw mushrooms, as the name implies, are grown on rice-straw beds and picked when immature, during the button or egg phase and before the veil ruptures (Chang and Quimio, 1982). They take approximately 4–5 days to mature under optimal conditions, and are most successfully grown in subtropical humid climates with high annual rainfall.

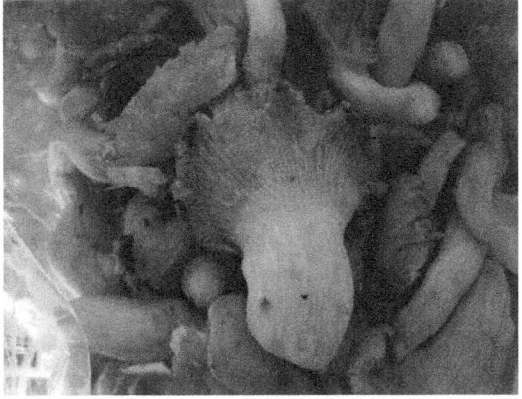

Fig. 22.3. Stem and fruiting body of an oyster mushroom.

Origin and History

Wild races of the common "button" mushroom (*A. bisporus*) occur throughout temperate areas of the northern hemisphere (Bon, 1987). It is likely that wild mushrooms have been collected and consumed for several hundred years, but mushroom cultivation is more recent, originating in England and France in the mid-18th century (Spencer, 1985). Some wild relatives of the cultivated button mushroom are poisonous, while others are edible (Zeitlmayr, 1976).

French botanist Joseph Pitton de Tournefort described commercial button mushroom cultivation in the early 1700s (Spencer, 1985). French agriculturist Olivier de Serres reported that transplanting mushroom mycelia, the vegetative part of a fungus consisting of branching, thread-like hyphae, increased productivity. Originally, cultivation was unreliable as mushroom growers would observe flushes of mushrooms in fields and transplant the mycelia to beds of composted manure or inoculate "bricks" of compressed litter, loam soil and manure (Genders, 1969). In 1893, sterilized or pure culture spawn was discovered and produced at the Pasteur Institute in Paris (Genders, 1969).

Fig. 22.4. Cultivated enoki packaged for retail sale.

Button mushrooms were brought to the USA from Europe in the late 1890s when farmers imported spawn from Europe. William Swayne, who some consider to be the father of the US mushroom industry, cultivated mushrooms in the space under the flower beds in his greenhouse by hanging flaps of burlap from the sides of raised benches to create an environment of stable temperature and humidity in the town of Kennett Square located in southeast Pennsylvania (Flammini, 1999).

These early efforts led to the building of the world's first mushroom house. Word of Swayne's success spread and other farmers in Kennett Square began to grow mushrooms as well. By the 1920s the mushroom industry began to grow rapidly aided by the development of pure culture spawn in the early 1900s by Edward H. Jacob, from spawn he had imported from England. The growth and development of the mushroom was due to several factors. The mushroom industry in Kennett Square flourished because of its proximity to the major markets in cities like Philadelphia and Baltimore, good infrastructure and an ample supply of horse manure that could be used as compost (Flammini, 1999).

Initially commercial button mushrooms were a light brown color. In 1926, a farmer in Pennsylvania, USA found a clump of common mushrooms with white caps in his production bed. The farmer collected and propagated these white mushrooms to preserve the mutation (Genders, 1969). Many of the white-, cream-, and some brown-colored commercial mushrooms available today were collected from naturally occurring genetic mutations that have been preserved over time, although some cultivars have been produced through mushroom breeding (Yamanaka, 1997). Today, button mushrooms are cultivated in at least 70 countries around the world (Cappelli, 1984).

Economic Importance and Production Statistics

In 2010, world production of mushrooms, including truffles (the fruiting body of a subterranean Ascomycete fungus not discussed in this chapter), covered 18,227 ha (45,040 acres) with a total estimated production of 5,987,144 metric tonnes (6,599,697 short tons) and a calculated yield of 328.5 metric tonnes/ha (146.7 short tons/acre) (FAOSTAT, 2010). China is the world leader in mushroom production with 4,182,079 metric tonnes (4,609,953 short tons), followed by the USA with approximately 400,000 metric tonnes (440,925 short tons). Other major producers include the Netherlands, Poland, Germany, Spain, Italy, Ireland, Japan, and France (FAO STAT, 2010).

Most of the US production is of button mushrooms. The US total presents total production of both fresh and processed mushrooms. About 75% of the US crop is fresh, while 25% is processed. A long-term trend in the USA is toward greater production of fresh mushrooms relative to processed. The total value of the mushroom crop in the USA is just less than US$1 billion, placing mushrooms among the top five vegetables in terms of production value (USDA, 2011b). About 65% of the mushrooms in the USA are produced in the northeast with southeastern and western Pennsylvania, New Jersey, and New York accounting for about 50% of the region's production. California is also an important production region (USDA, 2011b).

The per capita consumption of mushrooms in the USA is about 1.6 kg (3.5 lb)/year (USDA, 2011a). In many countries in western Europe and Asia, the per capita consumption is greater than 3.0 kg (6.6 lb)/year.

Nutritional Value

The button mushroom is an entirely no-waste vegetable that needs no peeling or special preparation other than washing. Button mushrooms are very versatile and can be eaten alone or as a complementary addition to a wide variety of dishes. They are a good source of B-complex vitamins and amino acids (Table 22.1; Rodwell, 1979). Mushrooms have a very high digestible protein value in comparison to other vegetables (Chang and Miles, 2004a). The amino-aci/d content of mushrooms ranks just below pork, chicken, beef, and milk. Button

Table 22.1. Nutritional composition of fresh button mushrooms (Rodwell, 1979; Chang and Miles, 2004).

Nutrient	Amount/100g (3.5 oz) edible portion
Water (%)	92
Energy (kcal)	25
Protein (g)	2.1
Fat (g)	0.4
Carbohydrate (g)	4.7
Fiber (g)	0.8
Ca (mg)	5
P (mg)	104
Fe (mg)	1.2
Na (mg)	4
K (mg)	370
Vitamin A (IU)	0
Thiamine (mg)	0.10
Riboflavin (mg)	0.45
Niacin (mg)	4.12
Ascorbic acid (mg)	3.5
Vitamin B6 (mg)	0.10

mushrooms are low in calories, with only 66 calories/ 450 g (66 calories/lb) (Rodwell, 1979). They are also low in carbohydrates and fat and do not contain cholesterol (Chang and Miles, 2004a). Many mushrooms, including shiitake, can produce significant quantities of vitamin D upon exposure to sunlight or UV light (Lee et al., 2009). Mushrooms are high in many nutrients; however, unlike many other vegetables they are not a good source of vitamin A or C (Chang and Miles, 2004a).

Despite the popularity of raw mushrooms in some cultures, many contain harmful compounds that can cause allergic reactions or more serious health effects after prolonged consumption. The common button and portobello mushrooms contain hydrazines, which are naturally occurring compounds found in uncooked mushrooms. Cooking mushrooms may deactivate hydrazines, such as the 4-(hydroxymethyl) benzenediazonium ion, which has been shown to cause cancer in animal feeding studies using *A. bisporus* (Toth and Erickson, 1986; Chang and Miles, 2004a). Eating raw or slightly cooked shiitake may cause an allergic response in some people. Lentinan, a natural compound isolated from shiitake, may be responsible for the allergic response. Cooking shiitake eliminates the allergic effects. Lentinan has also been used intravenously as an anticancer treatment because studies have demonstrated lentinan possesses antitumor properties. Clinical studies have associated lentinan with a higher survival rate, improved quality of life, and lower recurrence of cancer in patients after treatment (Nakano et al., 1999; Oba et al., 2009). Overall, there have been relatively few direct studies of mushroom consumption in humans. In human trials, mushrooms and their extracts have been generally well tolerated with few, if any, side-effects (Roupas et al., 2012). Despite some studies that have shown cancer risks from eating raw uncooked button mushrooms (Toth and Erickson, 1986), there are also studies that have pointed out many health benefits from eating mushrooms. The immunomodulating and antitumor effects of mushrooms and their extracts may hold potential health benefits. The most promising data are those indicating an inverse relationship between the consumption of certain types of mushrooms and breast cancer risk (Roupas et al., 2012).

Production and Culture

Button mushrooms

Producing button mushrooms is a science with acknowledged practices and principles, but it is also an "art". Although many practices have been standardized, methods still vary to a certain degree. Each crop is different and subtle practices that vary among farms and production cycles may affect production for reasons that are not always clear. Generally, cultivation is a continuous cycle, enabling picking or harvesting throughout the year (Wuest et al., 1985). In some countries, during either very hot or very cold months, growing is suspended because it is no longer economical. There are four basic cultivars of button mushrooms grown. These are differentiated primarily on the basis of color: white, off-white, cream, and brown.

Facilities

Early cultivated mushroom production often occurred in caves because of the cool, damp, and consistent temperature and high humidity they provide. Although some mushroom production still occurs in caves, the majority of production occurs in specially constructed mushroom houses. A typical mushroom house is constructed of wood or concrete blocks. Generally, mushroom houses have several 1.5–1.8 m (5–6 ft) wide beds stacked in multiple tiers. The beds are constructed of wood or metal (Beyer, 2003). Typically, each bed has three or four vertically stacked tiers with sufficient clearance to allow access. The total surface area provided by each bed is approximately 372 m^2 (4,000 ft^2).

Since mushrooms do not conduct photosynthesis, they do not need light, and consequently most growhouses are built without windows. Electric lights allow workers to see inside the houses. Ventilators are necessary for airflow and to control temperature and humidity and to prevent the buildup of certain gases.

Compost

Compost can be made from many materials. The most commonly used are animal manure or poultry litter. If these are not available, then a synthetic "formula" can be made by mixing inorganic fertilizers such as urea, ammonium sulfate, superphosphate, and calcium carbonate with other types of composted raw materials such as corn cobs, brewer's waste, and bran that are sources of nitrogen (N) and carbohydrates (Rodwell, 1979).

Other materials, such as cottonseed meal, soybean meal, castor-oil cake, sunflower seed, and cottonseed hulls, that are high in protein and decompose easily are sometimes used in small quantities to enrich mushroom compost. Whether natural or synthetic manure is used, the compost must be conditioned to encourage decomposition by microorganisms. To do this, compost is usually placed in piles 1.8 m (72 in) wide and 1.8 m (72 in) high on a concrete slab floor. It is then wetted, aerated, and mixed thoroughly (Beyer, 2003). Large operations use mechanical aerators to turn and mix the manure (Fig. 22.5). This process is repeated six or seven times at 2-day intervals. Often, it is necessary to spray the compost pile with materials to control flies. The conditioning process takes about 12–14 days (Beyer, 2003).

Disinfestation

Before spawning, the compost is disinfested to kill any nematodes or harmful pathogens (Wuest et al., 1985). After the last mixing, compost should have a moisture content of approximately 70% on a dry-weight basis for pasteurization to be most effective. The beds in the mushroom house are then filled to a depth of 15 cm (6 in) with compost, the house doors and ventilators closed, and the temperature in the beds is allowed to rise to 57–60°C (135–140°F) from the heat generated by bacteria in the manure (Beyer, 2003). It generally takes about 3 days to reach pasteurization temperatures. The 57–60°C (135–140°F) temperatures are maintained until the compost has no odor of ammonia (Rodwell, 1979).

The house is then well-aired and bed temperature is allowed to drop to 24–27°C (75–80°F). The entire process takes about 10–12 days (Wuest et al., 1985). The beds are now ready for planting.

Spawn

Spawn refers to mycelia usually grown from spores in specially prepared organic matter for planting in beds in a mushroom house. Spawn is obtained and multiplied in a laboratory. There are three steps involved in spawn production: (i) raising of pure culture; (ii) preparation of master culture/mother spawn; and (iii) multiplication of spawn (Stamets and Chilton, 1983). The spores are produced in the vale (gills) or underside of the cap, by mature mushrooms. Spawn is commercially available from a number of producers throughout the world and the mycelium is propagated mostly on grain, either millet or rye, although it may be propagated in manure or other organic matter (Stamets and Chilton, 1983). When spawn is propagated on grain it is sold in small bags containing about 2 kg (4.4 lb) that looks much like cereal. Grain-spawn is usually broadcast over the surface of the beds. 4.5 kg (1 lb) of spawn is sown across approximately 4.65 m^2 (50 ft^2) of bed. There are many strains and different types of spawn. It takes approximately 2 weeks for mycelium to populate the bed from the spawn (Beyer, 2003).

Fig. 22.5. Mixing animal manure during the production of button mushroom compost in California.

Casing

In addition to "seeding" the surface of the compost with spawn, mushroom production requires a thin layer of casing. The casing is needed to encourage the fungus to grow into the substrate below. Casing acts similar to the organic mulch applied to vegetable production fields because it holds water and prevents the substrate beneath from drying out. The material used as casing varies widely depending on availability. Clean friable soil, peat moss, or a mixture of materials, such as sisal waste, coconut fibers, peanut hulls, crushed bark, crushed limestone, and ash, are some of the popular materials commonly used (Rodwell, 1979). Good casing should remain friable, with a crumbly or open texture so it can be easily penetrated when the mushrooms emerge (Fig. 22.6; Rodwell, 1979). Therefore, materials such as sand, stone chips, and clay are not suitable.

The pH of the mixture can be adjusted by the incorporation of calcium carbonate, but should be slightly alkaline in a pH range of 7.2–7.6 (Rodwell, 1979). From the time a 36 mm (1.4 in) thick layer of casing is applied to the top of the compost, which is impregnated with mycelium, it takes approximately 16–21 days before the first mushrooms are ready to be picked.

Climate or environmental requirements for growth

Like most vegetables, a mushroom is mostly water and therefore requires moisture, both in the compost and in the casing, throughout its growth. The mushroom is reasonably tolerant within quite a wide range of environmental conditions and thrives best in climates that are slightly cooler than those that would suit most people, i.e. 14–18°C (58–60°F), with a reasonably high level of fresh air and a humidity of 85–90% (Beyer, 2003).

When the spawn gives rise to new mycelium that grows through the compost, more precise environmental control is required. The optimum temperature during this phase of production is 24°C (75°F) with a high humidity and little fresh air for a period of 10–21 days (Rodwell, 1979).

Modern spawns are more stress tolerant and can be grown in some tropical and semitropical parts of the world at higher temperatures, and higher and lower humidity, than traditionally believed (Beyer, 2003). Often high or low temperatures for short periods, such as 4–8 h, do not affect productivity.

Harvesting and postharvest handling

The industry in developed countries is moving towards mechanical harvesting since hand harvesting is monotonous, backbreaking work that increases production costs. Mushrooms may be mechanically harvested by machines that pass over the beds cutting at the surface or by hand with a sharp knife just before the cap expands to expose the gills (Rodwell, 1979). At harvest stage, button mushrooms may range from 2.5–7.5 cm (1–3 in) in diameter. Each crop will produce a series of flushes, or fruiting, at approximately weekly intervals. The flushes are heavier at the beginning until the nutrients required for mushroom production in the compost are gradually exhausted. Normally, a bed of compost gives a 6–8 week harvest period (Wuest *et al.*, 1985). The used compost is still rich in organic matter and is widely reused throughout the horticulture and nursery industries as fertilizer or for potting media (Fig. 22.7; Beyer, 2003).

Mushrooms are usually graded by size and sold in film-wrapped boxes to reduce water loss and shriveling. Fresh mushrooms can be stored for up to 3 weeks at 2–3°C (36–38°F) and high humidity (Fig. 22.8; Beyer, 2003).

Some specialty mushrooms are sold either fresh like button mushrooms or dried in plastic bags. Dried mushrooms behave much like a sponge and rehydrate in water (Royse, 1997).

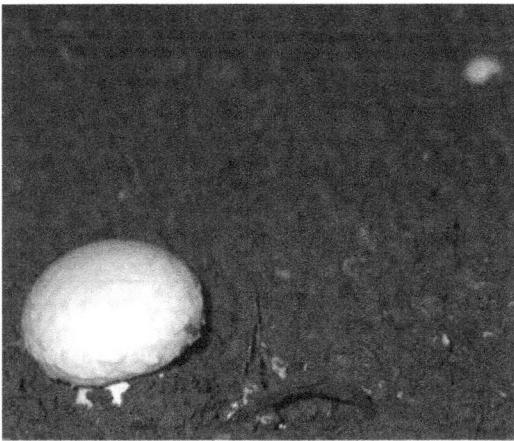

Fig. 22.6. A clean layer of soil is used as casing in this mushroom production house near Nanjing in China.

Fig. 22.7. A bag of mushroom compost for sale at a farmers' market in Nova Scotia, Canada. The compost continues to produce mushrooms after the production cycle has ended.

Fig. 22.8. Mushrooms are washed, graded, and kept under refrigerated storage until sale (a). Mushrooms are sold sliced or whole in containers covered with plastic film to reduce water loss, shrinkage, and discoloration (b).

Yields

It is important to have benchmarks for cropping yields and production performance. Mushroom crop yields are often measured for each growing cycle in pounds of picked mushrooms per square foot of growing area or kilograms per square meter. Yield values measured in this way often range from 29–39 kg/m^2 (6–8 lb/ft^2) (Beyer, 2003). Another frequently used measurement of yield is weight of mushrooms harvested per ton of compost. Yields range from 150–250 kg/metric tonnes (300–500 lb/short tons) of compost used. A valuable statistic is mushrooms picked per square foot or square meter of building area per year, and values range from 127–176 kg/m^2 (26–36 lb/ft^2).

If good cultural techniques are practiced and environmental conditions optimized, the production cycle can be reduced to 12 weeks, but under adverse conditions production may exceed 18 weeks. To maintain continuous weekly production, a minimum of 12 units is required so that at least six are producing one of the series of flushes at any time during the year. Using a weekly cycle, each growing room will produce approximately 4.3 crops/year (Table 22.2).

Shiitake cultivation

The shiitake (*Lentinula* (*Lentinus*) *edodes*) has been popular for centuries in Japan, where it is known as the forest mushroom and originally grew wild on the shii tree, a relative of the oak (Stamets, 2000). The Japanese were the first to commercialize shiitake production. A growing international appreciation of shiitake has increased demand and foreign production (Leatham, 1982). Japan remains a leading producer, although today they are grown commercially in many countries, including the USA. Shiitake are prized for their unique flavor and medicinal uses.

Table 22.2. Summary of the approximate cultural time cycle for production of button mushrooms (Rodwell, 1979; Yamaguchi and Rubatzky, 1997).

Step	Time (weeks)
Compost	3–4
Pasteurization	1
Spawn growth	2
Casing to first pick	3
Picking period	6–8
Total	15–18

The texture of shiitake is chewier, and the odor more aromatic with a stronger garlic-like flavor compared to the mild flavor of the button mushroom (Fig. 22.9; Royse, 1997; Stamets, 2000).

Of all the cultivated mushrooms, the shiitake has seen the greatest growth in the last decade in terms of both indoor and outdoor cultivation. Although a wide variety of media can be used to grow shiitake, including hay and rice, the most popular substrate is wood. A range of tree species can be used, but there is general agreement that oaks are among the best, particularly white oak (Leatham, 1982). Logs are cut from living, healthy trees that are dormant so that the wood contains maximum quantities of stored carbohydrates (Anderson and Marcouiller, 2012). Kits using alternative media can be purchased for small-scale production. The logs should be at least from 7.5–15 cm (3–6 in) in diameter and can be cut to many different lengths, but 91 cm (36 in) logs are one common size (Anderson and Marcouiller, 2012). The bark should remain intact. The wood for shiitake production can be cut from overgrown woodlots of immature trees that cannot be marketed for other uses.

The logs are inoculated with spawn to start the growth of the shiitake fungus. The logs should be inoculated within 2 weeks of harvest for best results. Spawn comes either as wooden plugs made from hardwood dowels or as sawdust (Anderson and Marcouiller, 2012). There are several inoculation techniques, but one of the most common is to drill holes 0.8 cm (0.3 in) in diameter and 2–2.5 cm (0.75–1 in) deep into the log.

Fig. 22.9. Fresh shiitake mushrooms. Shiitake are also commonly shipped and sold when dried, then rehydrated for cooking.

These holes are then filled with spawn and covered with wax or other material to retain moisture and protect against contamination. Holes should be staggered evenly around the log. Rows running the length of the log are spaced 3.8–6.3 cm (1.5–2.5 in) apart (Anderson and Marcouiller, 2012). The holes within a row should be spaced 15–25 cm (6–10 in) apart and alternating with the holes in the adjacent row. Heavier inoculation will accelerate the growth of the fungus within the log but also represents additional investment. Sawdust spawn is packed by hand or by a special injector into the drill holes.

The first "fruiting" will normally occur between 6 and 18 months after inoculation depending on the strain, the inoculation rate, the incubation conditions, and tree species (Royse, 1997). During the first 2 months, logs should be stacked closely to help retain at least 35–45% moisture content (Anderson and Marcouiller, 2012). Growth becomes poor when the moisture content falls below 35% or above 60%. When the moisture content becomes low, the log should be soaked or continuously watered for 48 h. After watering, air circulation is needed to keep the bark of the logs dry to prevent disease. Best growth occurs when the bark remains dry but the wood remains moist (Anderson and Marcouiller, 2012).

Shiitake spawn will grow at temperatures between 4–32°C (40–90°F), but the optimum is 22–25.5°C (72–78°F). Stacking logs in 60–70% shade helps to maintain moisture content while preventing overheating (Anderson and Marcouiller, 2012). If the logs dry out or overheat, the shiitake fungus may be killed. Common stacking methods include vertical X or horizontal crisscross patterns several logs high (Fig. 22.10; Anderson and Marcouiller, 2012).

On slopes, a lean-to pattern can also be employed effectively. Logs should be checked periodically and turned or restacked to keep the moisture content evenly distributed. Log-moisture content can be monitored periodically by weighing several logs of known dry weight (Anderson and Marcouiller, 2012).

Shiitake are cut from the logs by hand with a sharp knife. One 91 cm long (36 in) log should produce about 1.4 kg (3 lb) of mushrooms over a 4-year production cycle (Anderson and Marcouiller, 2012). Shiitake are often dried and sold in packages as preserved food. The stems are more fibrous than the caps so shiitake are sometimes sold without the stems. Shiitake mushrooms must be rehydrated in water before cooking (Leatham, 1982).

Fig. 22.10. Logs stacked for shiitake production in a wooded area of Virginia.

References

Anderson, S. and Marcouiller, D. (2012) *Growing Shiitake Mushrooms.* Oklahoma Cooperative Extension Service NREM-5029. Oklahoma Cooperative Extension Service, Oklahoma State University, Stillwater, Oklahoma.

Beyer, D.M. (2003) *Basic Procedures for Agaricus Mushroom Growing.* Catalog number UL210. The Pennsylvania State University College of Agricultural Sciences, Agricultural Research and Cooperative Extension, University Park, Pennsylvania.

Bon, M. (1987) *The Mushrooms and Toadstools of Britain and North-western Europe.* Hodder & Stoughton Ltd, London.

Cappelli, A. (1984) *Fungi Europaei: Agaricus.* Giovanna Biella Publishing, Saronno, Italy.

Carluccio, A. (2003) *The Complete Mushroom Book.* Quadrille Publishing Ltd, London.

Chang, S.T. and Hayes, W.A. (1978) *The Biology and Cultivation of Edible Mushrooms.* Academic Press, New York.

Chang, S.T. and Miles, P.G. (2004a) *Mushrooms - Cultivation, Nutritional Value, Medicinal Effect and Environmental Impact*, 2nd edn. CRC Press, Boca Raton, Florida.

Chang, S. and Miles, P.G. (2004b) *Pleurotus – A Mushroom of Broad Adaptability.* In: Chang, S.T. and Miles, P.G. (eds) *Mushrooms: cultivation, nutritional value, medicinal effect, and environmental impact*, 2nd edn. CRC Press, Boca Raton, Florida.

Chang, S.T. and Quimio, T.H. (1982) *Tropical Mushrooms: Biological Nature and Cultivation Methods.* The Chinese University Press, Shatin, N.T., Hong Kong.

Del Conte, A., Laessoe, T. and Campbell, S. (2008) *The Edible Mushroom Book.* DK Publishing, London.

FAO STAT (2010) FAOSTAT. Available at: http://faostat.fao.org/site/567/DesktopDefault.aspx?PageID=567 (accessed 20 June 2012).

Flammini, S.E. (1999) The Evolution of the Mushroom Industry in Kennett Square. Report on the history of the Mushroom Industry in Kennett Square. Available at: http://courses.wcupa.edu/jones/his480/reports/mushroom.htm (accessed 9 September 2013).

Genders, R. (1969) *Mushroom Growing for Everyone.* Faber Publishing, London.

Leatham, G.F. (1982) Cultivation of shiitake, the Japanese forest mushroom, on logs: a potential industry for the United States. *Forest Products Journal* 32, 29–35.

Lee, G.S., Byun, H.S., Yoon, K.H., Lee, J.S., Choi, K.C. and Jeung, E.B. (2009) Dietary calcium and vitamin D2 supplementation with enhanced *Lentinula edodes* improves osteoporosis-like symptoms and induces duodenal and renal active calcium transport gene expression in mice. *European Journal of Nutrition* 48, 75–83.

Nakano, H., Namatame, K., Nemoto, H., Motohashi, H., Nishiyama, K. and Kumada, K. (1999) A multi-institutional prospective study of lentinan in advanced gastric cancer patients with unresectable and recurrent diseases: effect on prolongation of survival and improvement of quality of life. Kanagawa Lentinan Research Group. *Hepato-Gastroenterology* 46, 2662–2668.

Oba, K., Kobayashi, M., Matsui, T., Kodera, Y. and Sakamoto, J. (2009) Individual patient based meta-analysis of lentinan for unresectable/recurrent gastric cancer. *Anticancer Research* 29, 2739–2745.

Rodwell, J.H.D. (1979) *A Brief Summary of Information on the Mushroom Industry: Cultural factors*. Del Norte Foods, Inc., Port Hueneme, California.

Roupas, P., Keogh, J., Noakes, M., Margetts, C. and Taylor, P. (2012) The role of edible mushrooms in health: Evaluation of the evidence. *Journal of Functional Foods* 4, 687–709.

Royse, D.J. (1997) Specialty mushrooms and their cultivation. *Horticultural Reviews 19*. Wiley, New York, pp. 59–97.

Spencer, D.M. (1985) The mushroom–its history and importance. In: Flegg, P.B., Spencer, D.M. and Wood, D.A. (eds) *The Biology and Technology of the Cultivated Mushroom*. Wiley, New York, pp. 1–8.

Stamets, P. (2000) *Growing Gourmet and Medicinal Mushrooms*. Ten Speed Press, Berkeley, California.

Stamets, P. and Chilton, J.S. (1983) *The Mushroom Cultivator: A Practical Guide to Growing Mushrooms at Home*, 1st edn. Agarikon Press, Seattle, Washington.

Toth, B. and Erickson, J. (1986) Cancer induction in mice by feeding of the uncooked cultivated mushroom of commerce *Agaricus bisporus*. *Cancer Research* 46, 4007–4011.

USDA (2011a) Food Consumption and Nutrient Intakes. Available at: www.ers.usda.gov/data/foodconsumption/spreadsheets/mushroom.xls (accessed 21 June 2012).

USDA (2011b) Mushrooms Annual Report. Available at: http://usda01.library.cornell.edu/usda/nass/Mush//2010s/2011/Mush-08-19-2011_revision.txt (accessed 22 June 2012).

Wuest, P.J., Duffy, M.D. and Royse, D.J. (1985) *Six Steps to Mushroom Growing*. Special Circular 268. The Pennsylvania State University Extension Bulletin, University Park, Pennsylvania.

Yamaguchi, M. and Rubatzky, V.E. (1997) *World Vegetables Principles, Production, and Nutritive Values*, 2nd edn. Chapman and Hall, New York.

Yamanaka, K. (1997) Mushrooms II. Breeding and Cultivation. I. Production of cultivated edible mushrooms. *Food Reviews International* 3, 327–333.

Zeitlmayr, L. (1976) *Wild Mushrooms: An Illustrated Handbook*. Garden City Press, Hertfordshire, UK.

Index

Note: page numbers in *italics* refer to figures

abiotic stress 31, 299
abscission 147–149, 153, 162, 171, 194, 198–199, 203–205
acclimation 177, 181
accumulation: photosynthetic 374
acidity 50, 363
    soil 10, 142, 151, 229, 306, 355, 373
aeration 39, 215, 231, 295, 365
    poor 297
aeroponics 102–103
Agaricaceae 441–451
agitation: mechanical 194
agribusiness 110
agrichemical systems 107
Agricultural Marketing Service (AMS) 111
agriculture 2, 31, 64, 70
    controlled-environment (CEA) 104–106
    conventional 114
    organic 114, 114–116, 121, 424
    settled 4
agroecosystem imbalance 112
agronomic crops 1, 47, 131, 248, 252–254, 257, 397–398
air
    circulation 95, 338–339, 405, 449
    pollution 108
algae 10, 39, 44
alley cropping 24
Allioideae 267–288
allium crops 286–287
Alternaria 144, 204, 330
    leaf spot 155, 164, 317
Amaranthaceae 349–368
Amaryllidaceae 267–288
ammonia: anhydrous 59, 259
ammonium injury 231
andromonoecious plants 137, 141, 149, 153, 157
angular leaf spot 155, 173, 406
anhydrous ammonia 59, 259
animal manure 116, 130, 155, 355, 374, 385, 445
    composted 103
animals 114, 130–132, 256
    feces 131
    feed 185, 253, 256–257, 292, 350–351, 390, 424
annual crops 116, 240–242, 385–386
anthocyanin 188, 205, 350, 358, 371, 383, 419
anthracnose (*Colletotrichum orbiculare*) 164, 173, 198, 204, 237, 388, 407

antioxidants 362
aphids 186, 200, 216, 246, 315, 339, 408
    green peach 164, 204, 237, 246, 318, 408, 409
    see also whitefly
Apiaceae 415–440
Armenian cucumber 136, 140–141
armyworms 165, 200, 218, 265, 323–324, 426
    beet 200, 238, 356, 366
artichoke 272
    globe 25–27, 239–246
    Jerusalem 239
Asian brassicas 335–341
Asian mustard greens 335, 340
Asparagaceae 369–381
asparagus 3, 36, 57, 369–380
    white 372, 376–378
Asteraceae 222–247, 359, 390
asthma 429
aubergine 201
autotrophic plants 441

Bacillus thuringiensis 118–119, 202, 238, 264, 314, 366, 407
back-flow regulators 62, 78
bacterial leaf spot 237
bacterial soft rot 186, 197–200, 217, 316, 388
bacterial spot 198, 217
bacterial wilt (*Erwinia tracheiphila*) 145–146, 155, 173, 180, 186, 217
bacterium: soil-borne 237
band placement 299
band seeding 421
banding
    root 61
    row 61–63, 193, 276, 402
Batavian lettuce 224–225
bean common mosaic virus (BCMV) 406
bean golden mosaic virus (BGMV) 406
bean yellow mosaic virus (BYMV) 406, 411–412
beans 4, 9, 200, 390–413
    bush 402
    common 390–392, 395–397, 400, 403, 407
    dry 413
    gram 397
    green 393, 400, 405
    haricot 397, 404–405

beans (*continued*)
    horticultural 397, 404–405
    Madagascar 394
    pole 396, 402–404, 412
    runner 402
    shelly 397, 404–405
    snap 391, 396–405, 412–413
    string 393, 397
bees 161, 194
    pollination 144, 153, 171, 205, 304, 336, 370
beet curly top hybrigeminivirus (BCTV) 156, 425
beetles
    common asparagus 379–380
    corn flea 264–265
    cowpea 408
    cucumber 145–146, 155–156, 164, 174, 302
    darkling 407
    flea 204, 301, 316, 356
    ground 44
    lady 238, 409
    Mexican bean (MBB) 407–408
    rove 44
    sap 265
    spotted asparagus 379–380
    tortoise 302
beets 57, 349–358, 357
    baby 353
    canned 358
    commercial 354
    fodder 349–351
    fresh 358
    garden 349, 349–356, 350–356
    wild 350
bell peppers 188–190, 193–200
benefits: nutritional 419
beta-carotene 290, 294, 419–420, 424
Bhopal disaster (1984) 109, 127
big vein disease 236–237
binding: soil 159
biodiversity 112
    increased 60, 112
    reduced 115
biofuels 166
biomass 113–116, 431–432
    crop 364
biopesticides 118, 123
biotechnology 33–35
biotic stress 31, 299
black leaf speck 339, 340
black leg 308, 317
black rot 173, 300, 308, 316, 324, 344
black spot 186–187, 356
blanching 428
    self- 428
blight 198–199
    Botrytis leaf 283
    common bean 406

    cucumber 164
    early 186, 204, 434
    gummy stem (*Didymella bryoniae*) 155, 164
    halo 406
    late 176–177, 186, 217, 434
blindness 327–328
blossom drop 403
blossom-end rot 58, 67, 98–100, 199–200
bolting 228–229, 271–272, 305, 334, 367, 416–419, 426–428
    premature 307, 426
    resistance 337, 418
    slow- 359, 419
boron deficiency 321, 326, 331, 355
Botanical Classification System 14
botany 117, 136–138, 146–147, 156–157, 166–169, 267, 304
    asparagus 369–370
    bean 391–395
    beets 349–353
    broccoli 319
    Brussels sprout 330–331
    cabbage 308–310
    carrot 415–416
    cauliflower 324–325
    celery 426–427
    chard 349–353
    chicory 223–224
    Chinese cabbage 335–337
    cucumber 136–138
    eggplant 201–202
    endive 223–224
    globe artichoke 239–240
    lettuce 223–224
    lima bean 394
    melons 146–147
    pea 391–395
    potato 177–178, 187–188
    pumpkins 166–169
    rhubarb 382–383
    spinach 358–360
    squash 166–169
    sweet corn 248–252
    sweetpotato 289–290
    tomato 205–206
    watermelon 156–157
Botrytis 245–246
Botrytis leaf blight (BLB) 283
bottom rot 232, 237
branching 249–250, 353, 359, 390
Brassica oleracea 304, 307, 319, 337
Brassicaceae 304–348
brassicas 38, 42, 57, 304–335
    Asian 335–341
breeding 122–123, 139, 190, 206, 272, 402, 431
    cross- 222
    selective 393

broadcasting 60, 159
broccoli 56, 104, 243, 304–307, 318–324, 327–333, 339–341
    Chinese 319–320, 341
    heading 320, 326
    sprouting 318–320, 323, 341
broccoli raab 320, 341
bronchitis 429
broomrape 216
brown mustard 345–346
brown root 425
brown spot 186–187, 406
bruising 204
Brussels sprouts 304–307, 312, 318, 324, 330–334
buffalo gourd 166
bugs
    squash 156, 174
    stink 218
    tarnished plant 218, 265, 388
bulbing 267, 276–280
bulbs
    characteristics 272
    formation 271, 279
bush cultivars 157, 163–173
butterhead lettuce 224–225, 228–229, 233–235
butternut squash 168
button mushrooms 441–449
buttoning 326–328

cabbage 56, 191, 199, 304–321, 326–338, 341
    Chinese 305, 335–341
    white fly 323–324
calcium 209
    deficiency 58
cancer 445
canning 197, 314, 358, 364, 400
Cantaloupensis 148–149
cantaloupes 53, 56, 104, 146–150, 154–157, 162–163, 212
capsaicin 190–191
Capsicum
    annuum 187–191
    baccatum 187–188
    chinense 187–190
    frutescens 187–192
    pubescens 187–188
carbon
    sequestration 52
    to nitrogen (C:N) ratio 22, 56, 60, 80–81
cardoon 239–241
carotene: beta- 290, 294, 419–420, 424
carotenoids 188, 196
carrot weevil 425–426
carrots 6, 299, 415–426, 429, 439
    baby 424
    Chantenay 415–419, 424
    imperator 418, 424

    juice 424
    Nantes 415–417, 424
    organic 424
    Oxheart 418
    shredded 424
cash crops 23–25, 60, 82, 113–116, 312, 337
catfacing 98, 216
cauliflower 304–307, 318–320, 324–334, 339–341
cauliflower mosaic virus (CaMV) 318, 333
cavity spot 356, 425
cayenne pepper 189
celeriac 427, 430–433
celery 56–57, 199, 426–435
    foliage 427–428
    leaf 431–433
    stalk 431–434
cells
    differentiation 369–370
    division 67
central-pivot system 72–73, 75
certification 121–123
    organic 218
    seed 297, 306, 407
Chantenay carrots 415–419, 424
chard 349–358
    Swiss 357–358
chemical contamination 121
chemical control 299
Chemical Examination of the Potato (Parmentier) 176
chemical fertilizer 110
chemical fumigation 97, 210
chemical insecticides 302
chemicals 3, 28, 42
    synthetic 122–123, 160
Chenopodiaceae 349–368
cherry peppers 189
chervil 437–439
chicory 222–239
    Witloof 226–227
chili peppers 187–190, 196–197
    pungent 198
chilling injury 193, 196, 204, 209, 212–214, 299, 405
Chinese broccoli 319–320, 341
Chinese cabbage 305, 335–341
Chinese kale 320, 335
chisel plowing 17, 116
chives 287
chlorophyll 57–58, 162, 184, 188, 196, 212, 376
chloropicrin 84, 87
chlorosis 55–57, 245, 310, 321, 326, 347, 365, 431
    interveinal 58, 331
    leaf 301
    veinal 318
chlorotic spotting 164
Chrysomelidae 407
circulation: water 346
civilization 5

cladophyll 369
classification 1–15
    botanical 10
    horticultural 188
    plant 13–14
    soil 50
    thermo- 8
clay loams 258–259, 297
climate 289, 340, 371, 447
    change 108
clipping 316
clubroot 308, 317, 344
cold frames 90–91
collards 304, 318, 334–335
collection: mechanical 183
Colletotrichum orbiculare (Anthracnose) 164, 173, 198, 204, 237, 388, 407
commercial cultivars 309
common asparagus beetle 379–380
common bean blight 406
common beans 390–392, 395–397, 400, 403, 407
compaction 116, 179, 321
    reduced 327, 338, 409
    soil 18–20, 82, 132, 229, 328, 337, 400
companion planting 24, 112
compatibility: sexual 10
competition 280
    less 37
    plant-to-plant 53
    weed 53, 154, 242
complete analysis fertilizer 59, 172, 295
compost 113–115, 445–446
compression injury 204
condiments 436–437
cone traps 119–120
conservation
    moisture 92, 151, 260, 279, 300, 311, 331
    soil 111
    tillage 159, 259, 297, 306, 311–312, 327–328, 332
    water 111
consumer
    acceptance 34, 110
    demand 33, 121–122, 127
consumption: human 1, 176, 185, 256
contamination 71, 109, 128, 315, 366, 408
    bacterial 340
    biological 70, 115, 123, 127–132, 155, 231, 297
    chemical 121
    cross- 183
    drinking water 158
    groundwater 232
    livestock 131
    microbial 115, 131
    prevention 317
    protection 449
    seed 27

contour cropping 19–20
controlled atmosphere storage (CA) 425
controlled-environment agriculture (CEA) 104–106
convection 90
convenience: horticultural 14
Convolvulaceae 289–303
cooling 265
    forced-air 204, 212, 234, 308, 328, 387, 405
    hydro-vac 365
    postharvest 155, 355
    rapid 204
    vacuum 234, 308, 328, 335, 365, 405
corky root 217, 237
corn 1, 123, 216
    baby 253–255
    dent 253–254
    field 252–255, 259
    flint 253–254
    fresh 256–257, 262–263
    frozen 256
    Indian 253–256, 262
corn earworm 265, 323–324
corn flea beetles 264–265
corn smut 256
corn stalk rot 264
cos lettuce 224–225
cotton 1, 218
cotton leaf worm 339
cover crops 23, 80–81, 111–116, 165, 170
    legume 81
    winter 182
cowpea beetles 408
crack resistance 310, 418–419
cracking 187
    fruit 67, 216
    radical 216
    root 423
    russet 301
    stem 334
crisis 5
    food 132
crisphead lettuce 5, 224–225, 228–234
crop rotation 155–158, 216–218, 235–239, 284–285, 301–302, 405–406, 434–435
    limited 176–177
    planned 113
Crop Water Stress Index (CWSI) 68
cropping 16–26
    double- 80, 88
    sequential 24, 112
    strip- 20, 132
    triple- 80, 88
crops 435–438
    agronomic 1, 47, 131, 248, 252–254, 257, 397–398
    allium 286–287
    annual 116, 240–242, 385–386
    bulb 9, 57

cash 23–25, 60, 82, 113–116, 312, 337
commercial 33, 382
cover 23, 80–81, 111–116, 165, 170, 182
development 69, 98
establishment 27–46, 179
foliage 435–438
fresh 332
perennial 5–6, 242–244, 386
root 9, 341–345, 435–437
shallow-rooted 116
transgenic 33–34, 190, 202, 207, 226, 261, 272
trap 23, 120
vegetable 8
cross-pollination 149, 169, 250–252, 255, 304, 391, 398
crown
    damage 420
    growth 385
Cruciferae 304
crusting 328
    soil 35–37, 229, 402–403
cryopreservation 30
cucumber 90, 99, 103, 136–148, 173, 200
    Armenian 136, 140–141
    pickling 142–145
    seedless 139
    Sikkim 136, 140
    slicing 139, *140*, 142–144
cucumber beetles 145–146, 155–156, 164, 174, 302
cucumber blight 164
cucumber mosaic virus (CMV) 145, 173, 217, 264, 360, 365, 406, 412
Cucumis melo (netted melon) 136, 146–156
Cucurbita
    maxima 165–168, 171
    moschata 165, 166
    pepo 165–169
Cucurbitaceae 136–175
cultivars 137, 138–141, 147–149, 157–158, 206–207
    asparagus 371
    beans 395–398
    broccoli 319–321
    bush 157, 163–173
    carrot 416–419
    cauliflower 325–326
    celery 427–428
    chicory 224–228
    Chinese cabbage 337
    commercial 309
    cucumber 138–141
    determinate 190, 205–206, 210–211, 394–395, 401–403
    development 31, 178
    endive 224–228
    fresh 211, 311
    garlic 271–274
    globe artichoke 240–241
    heirloom 31–32, *150*, 158, 206, 223, 256, 352
    indeterminate 190, 194, 205–206, 211, 395
    lettuce 224–228
    melons 147–150
    naked-seed 168–169
    new 34
    onion 271–274
    ornamental 190, 254–255, 292
    parthenocarpic 139–140, 161, 193, 202, 212
    peas 395–398
    peppers 188–191
    potato 178–179
    rhubarb 383
    selection 31–33, 275, 305, 309–310
    spinach 360–361
    sweet corn 253–257
    sweetpotato 291–292
    watermelon 157–158
    winter 327
cultivation 115–116, 227, 311–312, 334, 345, 385, 391
    between-row 357
    clean 121, 216
    conventional 25
    hand- 216, 286
    in-row 357
    indoor 449
    mechanical 300, 435
    open-soil 305
    outdoor 449
    protected 235
    shallow 261
    soil 17
culture 141–145, 149–155
    asparagus 373–376
    beans 400–405
    beets 353–355
    carrot 420–425
    celery 429–434
    cucumber 141–145
    eggplant 202–204
    garlic 275–279
    ground 211
    lettuce 228–236
    lima bean 412–413
    melons 149–155
    mushrooms 445–450
    onions 275–279
    peas 400–405
    pepper 191–197
    potato 179–184
    protected 90–106, 234–236
    pumpkins 169–173
    rhubarb 384–387
    spinach 362–365
    squash 169–173
    sweet corn 257–261

culture (*continued*)
    sweetpotato 294–299
    tomato 208
    watermelon 158–163
curing 281, 299
cuttings: shoot 346
cutworms 165, 174, 200, 246, 265, 284, 315, 366, 426

damage 366
    environmental 60, 87
    mechanical 154, 411
    root 38
damping-off 153–155, 160–162, 173, 210, 237, 366, 406
    preventing 306
darkling beetles 407
days-to-maturity 262–263
decay
    bacterial 284
    postharvest 388
    seed 402
decomposition 49, 81, 446
defoliation 173, 356, 380
degradation 432
    environmental 66
    soil 107
dehydration 110, 270
demand: water 70
density: plant 191, 362–363
dent corn 253–254
desiccation 182, 204, 216, 245, 256, 281–282, 426
    moderate 396
    preventing 386
    reduced 197, 328
    tolerance 30, 36
deterioration: postharvest 213
determinate cultivars 190, 205, 210–211, 394, 401–403
    semi- 206, 395
developed countries 123–124, 127, 181, 185, 206, 404, 412, 447
developing countries 127, 179–181, 184–185, 292
development 56, 86, 169–173, 298, 313, 410, 420
    beans 403
    beets 353–354
    celery 431–432
    crop 69, 98
    cultivar 31, 178
    delayed 258
    foliage 327
    fruit 142
    garlic 279–283
    onions 279–283
    peas 403
    peppers 193
    root 21, 298
    seedling 90
    shoot 242
    sustainable 108
    sweet corn 261–262
    sweetpotato 295–297
diamondback moth (DBM) 314
disease 22–23, 41–43, 144–146, 154–155, 177–179, 185–186, 208–210
    asparagus 378–379
    bacterial 435
    beans 405–407, 411–412
    beets 355–356
    Brussels sprouts 333
    cabbage 316–318
    carrot 425
    cauliflower 329–330
    celery 434–435
    chard 355–356
    Chinese cabbage 339–340
    control 16, 26, 160, 259
    crop 113
    cucumber 145–146
    eggplant 204
    fighting 187
    foliar 102, 153, 180–182, 209, 322, 332, 375
    foodborne 130
    fruit 153
    fungal 186
    garlic 283–284
    globe artichoke 245–246
    human 365
    leaf 53
    lettuce 236
    management 113–114
    melons 155–156
    minimizing 209
    mitigating 36
    onions 283–284
    peas 411–412
    pepper 198–200
    postharvest 194
    potato 185–186
    preventing 43
    pumpkins 173
    reducing 172, 183, 231
    resistance 109, 146, 166, 311, 371, 424
    root 328
    seedling 191–192
    soil-borne 113–114, 158–159, 217, 229, 258, 275, 295
    spinach 365–366
    spread of 101, 181
    spreading 408
    squash 173
    sweet corn 264
    sweetpotato 300–301
    tomato 217
    watermelon 164

disinfection 446
disposal 88
    water 70
disturbance
    root 191–192
    soil 337
ditching 21, 132
diversity 3
    genetic 5
    microbial 105
    microorganism 64
    plant 4, 6
    potato 177
domestication 3–6, 205, 239, 390
dominance 32
    proximal 291
dormancy 230, 282–283, 295, 371, 385
    crown 386
    postharvest 223
    seed 429
    winter 374
double-cropping 80, 88
downy mildew (*Peronospora cubensis*) 264, 317, 324, 356, 360, 365, 434–435
drainage 19–22, 36, 39, 182, 231, 297, 327
    good 259, 275, 345, 386
    improved 143, 159, 311, 317, 362, 373, 379
    internal 151
    natural 19
    poor 153, 400, 406, 409, 431
    subsurface 2, 21–22
    superior 82
    tile 21
drip fertigation 215, 452
drip irrigation 74–84, 151–153, 231–232, 321–322, 423, 431–433
drip tubing 83, 100–103
drought 144, 313, 332, 367, 370, 375, 385
    resistance 159
    stress 66–67, 228, 256, 260–261, 283, 340, 362
    tolerance 209, 287, 291, 297
dry matter 44, 268, 298, 309
drying 197, 271
    forced-air 287

early blight 186, 204, 434
earworm: corn 265, 323–324
ecology 108, 111–113
economic importance 141, 165, 169
    asparagus 371–372
    beans 398–400
    carrot 419
    celery 428–429
    chicory 228
    cucumber 141
    eggplant 202
    endive 228
    garden beets 357–358
    garlic 274–275
    globe artichoke 241
    lettuce 228
    mushroom 444
    onions 274–275
    peas 398–400
    peppers 197–198
    potato 184–185
    pumpkins 169
    spinach 361
    squash 169
    sweet corn 257
    sweetpotato 292–294
    tomato 207–208
    watermelon 165
edible maturity 203, 325
edible quality 206, 403, 411, 416, 419, 424
eggplant 103, 187, 190–192, 209
electrical conductivity (EC) 41, 42
elephant garlic 270, 273
elongation 309, 325, 370
    internode 58
    stem 332–333, 443
emergence 318, 422
    seedling 92, 259, 430
    shoot 182
    sprout 180–181
emigration 6
emulsifiable concentrates (EC) 84
endive 222–239, 318
endosperm 250–252, 258, 262
energy 3, 110, 121–122, 184, 382
    non-renewable 107
engineering: genetic 34, 110, 119, 123, 155, 207, 226
enoki mushrooms 443
enrichment: atmospheric 215
environmental degradation 66
enzyme-linked immunosorbent assay (ELISA) 236
erosion 18, 310, 311
    increased 115
    preventing 114
    reduced 19–22, 25, 116, 132, 306, 312
    soil 2, 17, 22, 121, 132, 321, 327
    water 132
    wind 132
Erwinia tracheiphila (bacterial wilt) 145–146, 155, 173, 180, 186, 217
Erysiphe cichoracearum (powdery mildew) 155, 164, 173, 245, 355–356, 425, 435
Escherichia coli 127–131, 155, 265
ethnobotany 4
ethylene 196–197, 204–207, 212–214, 234–236, 333, 339–340, 377

evaporation
    pan 70, 114, 232, 279
    reduced 75
    soil 92
evapotranspiration 67–69, 171, 232, 243, 279, 355, 375
    reduced 91
Extended Shelf Life (ESL) 207

Fabaceae 390–414
failure: crop 6, 177
famine 6, 176, 350
    Irish (1845–6) 176
farming 128, 372
    dryland 66
    rain-fed 66
farmscaping 24–25, 118
feces 130
    animal 131
feed: animal 185, 253, 256–257, 292, 350–351, 390, 424
fern 10, 380
fertigation 82–83, 151–152, 193, 306, 321, 337, 431–433
    drip 215
Fertile Crescent 4
fertility 41–42, 50, 159, 310, 321, 335
    adequate 276, 295
    crop 152
    increased 4, 97, 374
    managing 2, 63, 125, 142, 431
    see also soil fertility
fertilization 47–65, 262, 306, 310, 321, 326–327, 430–431
    additional 63
    adequate 277, 385
    conventional 113
    flower 211
    inorganic 59–60, 64
    modest 364
    nitrogen 142, 145, 171, 209, 231, 276, 355
    over- 42, 152, 412
    phosphorus 171, 193, 231, 259
    poor 403
    potassium 259, 431
    prescription 60–62
    supplemental 116–117, 203
    zinc 431
fertilizer 50–51, 58–59, 151–152, 242–243, 298–299, 363–364, 374–375
    chemical 110
    commercial 2
    concentrated 114
    granular 59, 83, *84*, 114, 159, 171
    mineral 64
    organic 58–60, 130
    solid 61
    synthetic 2, 59, 107
    see also liquid fertilizer
field corn 252–255, 259
field environments: modifying 91–95
field establishment 152–153, 159–160, 181–182, 259–260, 354, 421–422, 429–431
field harvesting 245
    mobile 233
field maintenance 375
field packing 323, 335
field preparation 16–18, 158–159, 208–209, 258–259, 294–295, 354, 420–421
Filial 1 (F-1) hybrids 32–33, 138–139, 149–150, 190–191, 206–207, 320–323, 418
filtering: gravitational 70
flat-leaf spinach 360
flea beetles 204, 301, 316, 356
flint corn 253–254
float-bed systems 38, 308
floating row covers 92, 123
flooding 20, 153
flowering 153–154, 160–161, 211–212, 245, 305, 320, 344
flowers
    bean 392
    bisexual 416
    characteristics 250–251, 304
    female 250–251
    hermaphroditic 149, 359–360
    male 250
fly
    fruit (*Drosophila melanogaster*) 130
    house (*Musca domestica*) 130
    hover 237, 409
    melon fruit (*Dacus* spp.) 146, 156, 164
    rust 425–426
    silverleaf whitefly 238, 301
fodder beets 349–351
foliage
    celery 427–428
    crops 435–438
    development 327
    growth 420–421
    injury 283
foliar analysis 62–63, 337, 421
foliar disease 102, 153, 180–182, 209, 322, 332, 375
foliar testing 171, 193
Food and Agriculture Organization (FAO) 132, 149, 165, 274, 314, 358, 371
food crisis 132
Food and Drug Administration (FDA) 109, 132
food poisoning 128
footprint: environmental 3
forced-air cooling 204, 212, 234, 308, 328, 387, 405
forced-air drying 287
fossil fuels 121

freeze tolerance 6
freezing 271, 364
    injury 204
fresh market 194, 208, 264, 308, 313, 364–365, 377
frost protection 92, 298
fruit 142, 153
    rot 151, 155, 164, 170, 173, 204
fruit fly
    Dacus spp. (melon) 146, 156, 164
    Drosophila melanogaster 130
fumigation 83–88, 151, 170, 216–218, 235–239, 321
    bed 295
    chemical 97, 210
    gaseous 87, 114
    soil 80–82, 301, 313, 317, 321, 328, 378–379
    steam 295
fungal disease 186
fungi 441
    soil-borne 28, 264, 283–284
    spores 317–318
fungicides 2, 145, 237, 402, 434
funnel traps 119–120
furrow irrigation 73–74, 151–153, 259–261, 327–328, 386–387, 432–433
Fusarium wilt (*Fusarium oxysporum* f. sp. *melonis*) 155, 164, 173, 237, 300, 318, 373

garden beets 349, 350–356
garden peas 397
garlic 267–286
    elephant 270, 273
    true 272
gene codes 265
gene flow 33
genetic engineering 34, 110, 119, 123, 155, 207, 226
genetic improvement 4, 141, 398
genetic mutation 5, 35, 122–123, 207, 297
    natural 265
genetics 270
    improved 31
    plant 36
    poor 206
    sweet corn 251–252
geographic information system (GIS) 63
geotropism 378
germination 35–39, 90–92, 209–210, 248, 258–259, 409, 421–422
    improved 402
    inhibiting 61
    optimum 160, 354, 422
    poor 252
    preventing 223, 429
    rapid 162, 230, 306
    uniform 30, 263
germplasm 5–6, 30
gibberrellic acid (GA) 242, 386, 432

gigantism 6
glassiness disorder 236
global positioning system (GPS) 3, 63
globalization 132
globe artichoke 25–27, 239–246
Good Agricultural Practice (GAP) 115, 123, 131–133
gourd: buffalo 166
grafting 44, *45*, 160, 210
    pepper 192
grain 262
gram bean 397
granular fertilizer 59, 83, *84*, 114, 159, 171
grasshoppers 265
gray leaf spot 199
gray mold 237, 388, 405
green beans 393, 400, 405
green manure 22, 26, 60, 64, 111, 114–116, 311
green peach aphids 164, 204, 237, 246, 318, 408, 409
green peas 399–400, 410
Green Revolution 107, 123
greenhouse gases 108
greenhouses 100–106, 139, 144, 194, 203, *214*, 215–217
    commercial 365
    managing 43
    permanent 91, 95–96, 234
    Quonset-style 95
    role of 105–106
greening 182, 187
greens: Asian mustard 335, 340
grocery chains 3
groundwater
    contamination 232
    declining 66
    irrigation 108
    subsurface 70
growing season 56, 88, 91–93, 334, 404, 428
    long 136
    short 143, 153
growing-degree-days (GDD) 262–263, 404, 410
growth 38, 56, 169–173, 280, 410, 420
    beans 403
    beets 353–354
    celery 431–432
    crown 385
    enhanced 159
    foliage 420–421
    habits 393
    leaf 359
    maize 261–262
    peas 403
    peppers 193
    poor-quality 113
    rate 142, 418
    reproductive 212, 353
    root 37, 278
    seedling 37–40, 43, 259–260, 409
    sweetpotato 295–297

growth (*continued*)
  transplant 92
  vegetative 47–65, 209, 212, 319, 359, 374, 394
  weed 17, 321
growth hormone production 58
grubs: white 265, 301
Guide to Minimize Microbial Food Safety Hazards for Fresh Fruits and Vegetables (1998) 132
gummy stem blight (*Didymella bryoniae*) 155, 164
gynoecious plants 137–139

habanero peppers 190–192, 196
halo blight 406
hand-harvesting 281, 313, 338, 375–376, 404–405, 409–411
hand-weeding 235, 295, 357
handling: postharvest *see* postharvest handling
hardening 44–45, 90, 143, 153, 309, 426
haricot beans 397, 398, 404–405
harlequin bug 315–316
Harrington's Rule 30
harvesting 144–145, 154–155, 160–163, 172–173, 332–337, 357–358, 375–377
  asparagus 375–376
  automated 299
  beans 404–405
  beet 355
  beets 355
  broccoli 322
  Brussels sprout 332–333
  carrot 423–424
  cauliflower 328
  celery 433–434
  Chinese cabbage 338
  cucumber 144–145
  destructive 1, 397, 403
  eggplant 203–204
  field 233, 245
  floodwater 74
  globe artichoke 243–244
  hand- 281, 313, 338, 375–376, 386, 404–405, 409–411
  immature 139
  lettuce 232–234
  lima bean 412–413
  melons 154–155
  mushrooms 447–448
  onion 280–281
  over- 387
  peas 404–405, 410–411
  peppers 194
  potato 182–183
  pumpkins 172–173
  rhubarb 387
  selective 375
  serial 437

  spinach 364–365
  squash 172–173
  sweet corn 262–264
  sweetpotato 299
  tomato 212–213
  watermelon 162–163
  see also mechanical harvesting
head lettuce 224, 232, 238
head rot 226, 338
heading broccoli 320, 326
health 2, 128–132, 445
  adverse 34–35
  human 109, 294
  maintaining 1
  public 70
  soil 63–64, 374
heat 180, 235, 370
  tolerance 337
heating
  soil 86, 91
  solar 91, 295
hedging 132
heirloom cultivars 31–32, *150*, 158, 206, 223, 256, 352
herbicides 115–116, 295, 299, 322, 332, 357, 375
  reducing 312
  selective 422
herbs 436–437
heterosis 32, 309, 320
heterotrophic plants 441
heterozygosity 286
high performance liquid chromatography (HPLC) 191
high-density polyethylene (HDPE) 87
high-fructose corn syrup (HFCS) 253–254, 257
history 1–15, 176–177, 222, 267, 415
  asparagus 369
  beets 349
  broccoli 318–319
  Brussels sprout 330
  celery 426
  chard 349
  chicory 222
  cucumber 136
  eggplant 201
  endive 222
  globe artichoke 239
  lettuce 222
  melons 146
  mushrooms 443–444
  peppers 187
  potato 176–177
  pumpkins 165–166
  rhubarb 382
  spinach 358
  squash 165–166
  sweet corn 248
  sweetpotato 289
  tomato 205

vegetable 3–4
watermelon 156
homozygous plants 371
horseradish 341, 344–345
horticultural beans 397, 404–405
horticultural characteristics 271
hotbeds 90–91
hotboxes 90
hotcaps 91–92
housefly (*Musca domestica*) 130
hoverfly 237, 409
human disease 365
humidity 105–106, 196–198, 204, 211, 287, 313, 355
    high 91–92, 177, 182, 186, 236, 445
    low 281
humus 47–49
hunter-gatherer society 4
hybridization 5, 415
hybrids 123, 152, 250, 313, 325, 338, 371
    Filial 1 (F-1) 32–33, 138–139, 149–150, 190–191, 206–207, 320–323, 418
    high-yielding 110
hydration 66–68, 102
    controlled 191, 429
hydro-vac cooling 365
hydrocooling 328, 333–335, 342, 355, 377, 387–388, 405
hydroponics 38, 44, 97–106, 142, 227, 234–236

iceberg lettuce 224, 228–229
imperator carrot 418, 424
imported cabbageworm (ICW) 314
indeterminate cultivars 190, 194, 205–206, 211, 395
Indian corn 253–256, 262
Indorous Group 147–149
industrialization 185
    rapid 108
infection 284
    intracellular 395
    parthogenic 299
    plant-to-plant 316
    secondary 200
    stem 199
infestation 285, 315, 408–409
infiltration 19
inflorescence 222–224, 270, 319–320, 324, 331, 336, 415
    branched 304
    terminal 269
infrared radiation 85
infrared transmitting (IRT) mulch 85–86, 170, 298
innoculation 395, 406–407, 449
innovation
    agrichemical 107
    technological 107, 110
inorganic fertilization 59–60, 64

insect pests 28, 113–115, 146, 164–165, 173–174, 333
    asparagus 379–380
    beans 407–409, 412
    beets 356–357
    carrot 425–426
    celery 435
    chard 356–357
    cucumber 146
    eggplant 204
    garlic 284–285
    globe artichoke 246
    lettuce 237–239
    managing 25
    melons 156
    onions 284–285
    peas 412
    pepper 200–201
    potato 186
    pumpkins 173–174
    reducing 26
    squash 173–174
    sweet corn 264–265
    sweetpotato 301–302
    tomato 217–218
    watermelon 164–165
insecticides 161, 238, 246, 301, 315, 366, 380, 409
    chemical 302
    periodic 356
insects 25, 43–44, 118, 314–316, 323–324, 329
    attacks 118, 172
    control 120, 259
    damage 53, 154, 264
    diseases 28
    outbreaks 127
    pollination 223, 304, 416, 428
instability: political 108
Integrated Pest Management (IPM) 119, 123, 218, 246, 301, 308
interbreeding 14
intercropping 24, 112, 170–171
International Code of Botanical Nomenclature 14
International Seed Testing Association (ISTA) 27
international trade 31, 339
interseeding 312
Irish famine (1845-6) 176
Irish potato 1, 6, 177, 281, 289, 293
iron
    deficiency 58, 347
    fertilization 58
irrigation 66–79, 143–144, 278–281, 311–313, 322–323, 403–404, 434–435
    celery 432
    drip 74–84, 151–153, 231–232, 321–322, 423, 431–433
    efficient 276
    furrow 73–74, 151–153, 259–261, 327–328, 386–387, 432–433

irrigation (*continued*)
    groundwater 108
    lettuce 232
    manual 74
    overhead 71–73, 243, 317–318, 322, 332, 346, 362
    post-plant 153
    potato 182
    rainfall 82
    rhubarb 386–387
    scheduling 278
    sub- 74, 432
    subsurface 209, 243
    supplemental 261
    surface 71, 108, 144, 209, 375, 432
    sweet corn 260–261
    sweetpotato 297–298
    trickle 2, 74–78, 131, 144, 209, 298
irrigation water 39–41, 374
    polluted 131
    tainted 128

jalapeño peppers 189, 196–197
Jerusalem artichoke 239
juvenile growth 417

kale 304, 318, 334–335
    Chinese 320, 335
kohlrabi 304, 318, 334

labor
    efficiency 110
    reduction 18, 337, 382
lacewings 238, 409
lachrymatory factor (LF) 272
lady beetles 238, 409
lance nematode 426
landrace 13
Larsen 100
late blight 176–177, 186, 217, 434
latent ringspot virus 379
latex: dried 222
leaching 131, 328, 402, 433
    fertilizer 159, 338
    nutrient 151
leaf
    celery 431–433
    disease 53
    growth 359
    sheaths 249, 267, 279
leaf blight: Botrytis 283
leaf speck: black 339, 340
leaf spot 198, 204
    Alternaria 155, 164, 317
    angular 155, 173, 406
    bacterial 237
    gray 199
leaf water potential (LWP) 68–69
leaf worm: cotton 339
leafhoppers 264, 300, 407, 425
leafminers 174, 238, 285, 356
legumes 30, 81, 113–114, 390–392, 395–397, 401, 412
lettuce 36–38, 99, 127–131, 199, 222–239, 243, 318
    Batavian 224–225
    butterhead 224–225, 228–229, 233–235
    cos 224–225
    crisphead 5, 224–225, 228–234
    head 224, 232, 238
    iceberg 224, 228–229
    loose-leaf 224–229
    romaine 224–225, 228–229, 233, 336
    stem 224–226
lettuce mosaic virus (LMV) 236
life cycle 146–147, 156–157, 166–169, 267
    beans 391–395
    beets 349–353
    carrot 415–416
    celery 426–427
    chard 349–353
    chicory 223–224
    eggplant 201–202
    endive 223–224
    lettuce 223–224
    melons 146–147
    peas 391–395
    potato 177–178, 187–188
    pumpkins 166–169
    spinach 358–360
    squash 166–169
    sweet corn 248–252
    sweetpotato 290–291
    tomato 205–206
    watermelon 156–157
lignification 369
lima bean (*Phaseolus lunatus* L.) 391, 397–398, 412–413
liquid fertilizer 59–62, 159, 171, 193, 308, 374
    organic 103
    starter 153
livestock *see* animals
loams (soil) 68, 142, 158–159, 242, 322, 328, 332–345, 373
    clay 258–259, 297
    fertile 384
    sandy 258, 354, 384, 404, 409
    silt 354, 384
lodging 193–194, 410
longevity 371
    plant 374
    seed 30
loose-leaf lettuce 224–229
low-density polyethylene (LDPE) 87
lutein 362

lycopene 158, 165, 205, 212, 419
lygus bug 204, 246, 409, 426

machine harvesting *see* mechanical harvesting
macronutrients 47, 50, 59–61
Madagascar beans 394
maggots: seed corn 165, 265
maize 4–5, 248, 261–262
maize mosaic virus 264
maleic hydrazide (MH) 283
management
    crop 63, 141–142, 401
    environmental 118
    nutrient 60, 64, 275–277, 433
    residue 22–23
    soil 113, 123
    temperature 234
    water 114
    see also weed management
manganese deficiency 58
manipulation: environmental 118
manure 90, 114–115, 444
    animal 103, 116, 130, 155, 355, 374, 385, 445
    composted 113, 131, 231, 355
    contaminated 128
    green 22, 26, 60, 64, 111, 114–116, 311
    natural 446
    synthetic 446
    uncomposted 231, 355
market
    commercial 416
    fresh 194, 208, 264, 308, 313, 364–365, 377
    visibility 299
marketing 144–145, 154–155, 162–163, 172–173
    asparagus 375–376
    beans 404–405
    carrot 423–424
    celery 433–434
    cucumber 144–145
    eggplant 203–204
    lettuce 232–234
    melons 154–155
    peas 404–405, 410–411
    peppers 194
    potato 182–183
    pumpkins 172–173
    spinach 364–365
    squash 172–173
    sweet corn 262–264
    sweetpotato 299
    tomato 212–213
    trends 263–264
    watermelon 162–163
marsh spot 58
matriconditioning 30

maturation 212, 279–282, 326
    crop 142, 228–229
    seed 413
maturity 86–88, 147–149, 153–154, 307–309, 331–332, 429–431
    advanced 187, 207
    bulb 269
    delayed 56, 193, 276, 307, 321, 328, 401
    edible 203, 325
    fruit 162
    harvest 263, 283, 319, 328, 353, 403, 410
    market 203
    over- 411
    physiological 141, 180, 194, 197, 308
    plant 419
    quick 157
    root 341
    seasonal 271
    seedling 259
    uniform 263
mechanical harvesting 143–144, *195*, 211–212, *213*, 338, 362–363, 375–376
    destructive 145, 194, 392, 404
mechanical injury 194, 281
mechanization 107, 127, 311
media 38–39
    soilless 103, 210, 307–308
    synthetic 102
medicine 387, 426
melon fruit fly (*Dacus* spp.) 146, 156, 164
melons 9, 53, 60, 93, 147–156
    mixed 146–156
    netted (*Cucumis melo*) 136, 146–156
    winter 146–147, *150*, 155
Mexican bean beetle (MBB) 407–408
microbes 63–64, 90
microclimate 90–91
microgreens 353, 367
micronutrients 47–50, 59–61, 100, 142, 298, 306
    deficient 57, 117
microorganisms 128
mildew
    downy (*Peronospora cubensis*) 264, 317, 324, 356, 360, 365, 434–435
    powdery (*Erysiphe cichoracearum*) 155, 164, 173, 245, 355–356, 425, 435
mineral deficiencies 54–55
mineral nutrition 47–65
mineralization 355
minerals: natural 117
mites 164, 216, 412
    spider 156, 204, 265, 408, 426
    two-spotted spider 246
moisture 68, 310, 362, 411
    conservation 92, 151, 260, 279, 300, 311, 321
    excess 307
    loss 313

moisture (continued)
    management 403
    retention 91, 449
    seed 30, 405
    stress 216, 243, 346, 386, 432
    sufficient 259
    see also soil moisture
mold: gray 237, 388, 405
moldboard plowing 16–17, 115, 299, 310
molybdenum
    deficiency 57, 306
    fertilization 395
monoculture 2, 25, 64, 110–114, 127
    intensive 23
monoecious plants 137–140, 149, 157, 360
morphology 1, 10, 14, 267
mosaic virus 186, 278
    bean common (BCMV) 406
    bean golden (BGMV) 406
    bean yellow (BYMV) 406, 411–412
    cauliflower (CaMV) 318, 333
    cucumber (CMV) 145, 173, 217, 264, 360, 365, 406, 412
    lettuce (LMV) 236
    maize (MMV) 264
    tobacco (TMV) 199, 217
    turnip (TuMV) 318, 333, 338–339
    watermelon (WMV) 164
    zucchini yellow (ZYMV) 173
moths 119, 284
    diamondback (DBM) 314
    turnip 426
mulch 20–22, 80–89, 116, 159, 162, 229, 338–340
    biodegradable 88
    infrared transmitting (IRT) 85–86, 170, 298
    living 312
    organic 216, 300, 338, 386
    reflective 120, 215
    see also plastic mulch
mulch-tillage 17–18, 311, 321, 327
multi-cropping 24
mushrooms 1, 441–450
    breeding 444
    button 441–449
    cultivated 441–443
    dried 447
    enoki 443
    fresh 447
    oyster 442
    portobello 441, 445
    shiitake 441–442, 445, 448–450
    specialty 447
    straw 443
    wild 443
muskmelon 56, 146–148, 153, 173
mustard 316, 334–335, 345–346
    brown 345–346
    oriental 345–346

mustard greens: Asian 335, 340
mutation 5, 178, 252, 272
    genetic 5, 34, 35, 122–123, 207, 265, 297
mutualism 395
mycelium 446, 447
mycology 441

naked-seed cultivars 168–169
Nantes carrot 415–417, 424
National Center for Genetic Resources
    Preservation (NCGRP) 5
National Organic Production (NOP) 111, 115–116, 121–123
natural selection 4–6
neck rot 284
necrosis 55–58
necrotic spot 57–58, 238, 365
needle nematodes 238–239
needs
    consumer 122
    plant 132
    water 362
nematodes 113–114, 159, 217, 239, 412, 446
    bulb 278
    needle 238–239
    pests 217–218, 379–380
    resistance 44
    root knot 156, 204, 218, 238–239, 302, 409, 426
New World 136, 146, 248, 267, 289, 390, 437
nickel deficiency 58
nightshade 176, 199
    deadly 205
nitrification 364
nitrogen 137, 152, 159, 346–347
    deficiency 280, 326
    fertilization 142, 145, 171, 209, 231, 276, 355
    recycling 113
    toxicity 56
no-tillage 80–81, 123, 182, 192, 286, 311, 321
nomenclature 1–15
novelty cultivars 141, 393
nutrient film system (NFS) 235
nutrient film technique (NFT) 215
nutrient uptake 102
    rapid 99
nutrients 38–39, 54–55, 97, 100, 103–106, 152, 203
    absorption 248, 374
    available 50
    deficiency 42–44, 53
    gaseous 51–54
    immobile 57–58
    management 60, 64, 275–277, 433
    managing 97–102
    mobile 55–57
    plant 51, 62

recirculating 101
recycling 114
soluble 113
nutrition 151–152, 193, 231, 242–243, 298–299, 363–364, 374–375
    garlic 275
    improved 207
    mineral 47–65
    natural 112–113
    onions 275
    poor 154
    soil 26
nutritional values 141, 156, 165, 169, 197
    asparagus 372
    beans 400
    beets 358
    carrot 420
    celery 429
    chard 358
    chicory 228
    cucumber 141
    eggplant 202
    endive 228
    globe artichoke 241
    lettuce 228
    melons 156
    mushrooms 444–445
    peas 400
    potato 184
    pumpkins 169
    rhubarb 384
    spinach 361–362
    squash 169
    sweet corn 257
    sweetpotato 294
    watermelon 165

Odda process 59
Old World 304
olericulture 1
onion rot 284–285
onions 267–286, 280, 281
    dried 275
open-pollination 32–33, 149, 207, 272, 286, 313, 325
orach 349, 366, 367
Organic Materials Review Institute (OMRI) 111
organic matter 47–50, 64, 80, 84, 275, 321, 441
organic production 81, 103–104, 107–126, 131
oriental mustard 345–346
origin 176–177, 222, 267, 415
    asparagus 369
    beets 349
    celery 426
    chard 349
    chicory 222
    cucumber 136

eggplant 201
endive 222
globe artichoke 239
lettuce 222
melons 146
mushrooms 443–444
peppers 187
potato 176–177
pumpkins 165–166
rhubarb 382
spinach 358
squash 165–166
sweet corn 248
sweetpotato 289
tomato 205
watermelon 156
ornamental cultivars 190, 254–255, 292
osmoconditioning 30
osmoregulation 56
overcrowding 260
overhead irrigation 71–73, 243, 317–318, 322, 332, 346, 362
overseeding 25, 312
overwatering 298, 433
overwintering 283–285, 316–317, 345, 362, 379–380, 383, 388
oxalic acid 384, 387
Oxheart carrots 418
oyster mushrooms 442

pan evaporation 70, 114, 232, 279
pan traps 119–120
paprika 188, 197
parasitic wasps 237–238, 408–409
parthenocarpic cultivars 139–140, 161, 193, 202, 212
pasteurization 446
pathogenic attack 35–37
    reducing 197
pathogens 170, 446
    foodborne 128
    fungal 425
    human 130, 131
    lettuce 236–237
    microbial 365
    soil-borne 114, 192, 300
peaches: green peach aphids 164, 204, 237, 246, 318, 408, 409
peas 390–413
    canned 400
    dry 399
    fresh 400
    frozen 400
    garden 397
    green 399–400, 410
    snap 411
    snow 397, 411

peas (*continued*)
    sugar 397
    sugar snap 394, 397
pelleting 28, 230, 275–277
peperin 190–191
pepper speck 200
pepper spotting 318
pepper spray 191
pepper weevils 200
peppers 30, 56, 90, 103, 187–201, 203, 209
    bell 188–190, 193–200
    chili 187–190, 196–198
    dried 198
    fresh 198
    habanero 190–192, 196
    jalapeño 189, 196–197
    nonpungent 188–190, 198
    pungent 189–190
    Scotch bonnet 189–190
    sweet 188–190
perennial crops 5–6, 242–244, 386
performance: seed 28
Peronospora cubensis (downy mildew) 264, 317, 324, 356, 360, 365, 434–435
pest control 113–115, 179, 235–236, 308, 314–318, 323–324, 333
    biological 112
    chemical 2
    effective 87
    nonchemical 110
pesticides 23, 108–110, 246
    biorational 117–120
    chemical 118
    residue 110, 113, 127
    synthetic 2, 110, 120
pests 108, 216, 229, 329–330, 339, 387–388, 425–426
    nematode 217–218, 379–380
    soil-borne 313, 321, 326–328, 331
    see also insect pests
pheromones 118–120, 246
    sex 118, 284
photodormancy 223, 230
photosynthesis 52–58, 66–67, 215, 280, 327, 369–370, 374–375
    efficient 396
    maximized 142–144
    reduced 154, 171
photosynthetic accumulation 374
physical traps 119–120
physiological disorders 98, 199
    lettuce 236
    pepper 198–200
    potato 186–187
    tomato 216
pick-your-own operations 263
pickling 137–139, 189–190
    cucumber 142–145

pigment 350, 420
pimento 188, 197
pink rot 186, 434
plant science 2, 33
planting 82–83, 277–278, 337, 402, 409–410
    companion 24, 112
    continuous 294
    field 45, 308, 332
plants
    andromonoecious 137, 141, 149, 153, 157
    autotrophic 441
    characteristics 149, 304, 415
    cool-season 6
    disease-free 385
    gynoecious 137–139
    heterotrophic 441
    homozygous 371
    monoecious 137–140, 149, 157, 360
    poorly nourished 55
    seed 10
plastic mulch 75–77, 80–83, 86–88, 91–92, 93–94, 151–153, 158–159
    light-transmitting 93
    types 85–87
plasticulture 80–83, 121, 142–143, 159, 300, 306, 311–313
plowing 16, 22, 115, 345, 354
    chisel 17, 116
    moldboard 16–17, 115, 299, 310
plug transplants 36–38, 162, 191–192, 203, 230, 326, 337–338
Poaceae 248–266
polarity 345
pole beans 396, 402–404, 412
pollen 160, 203, 211, 248–252, 258
pollination 153–154, 160–162, 199, 212, 255–258, 370
    bee 144, 153, 171, 205, 304, 336, 370
    cross- 149, 169, 250–252, 255, 304, 391, 398
    greenhouse 194
    hand- 272
    inadequate 216
    insect 223, 304, 416, 427
    manual 194
    open- 32–33, 149, 207, 272, 286, 313, 325
    self- 205, 215, 223–227, 324, 331, 336, 391–394
    successful 260, 418
    sweet corn 251
    wind 248, 251, 359–360
pollution
    air 108
    groundwater 105
    increased 104
    minimized 232
    water 113, 132
polyculture 24
Polygonaceae 382–389
polyvinyl chloride (PVC) 95

popcorn 254, 262
portobello mushrooms 441, 445
Portulaca oleracea 186
postharvest handling 180, 226, 234, 281, 299, 313, 340
    asparagus 377–378
    broccoli 323
    Brussels sprout 333
    cabbage 313
    cauliflower 328–329
    Chinese cabbage 338–339
    globe artichoke 244–245
    mushroom 447–448
    peas 411
    pepper 194–197
    potato 183–184
    rhubarb 387
    sweet corn 263
    tomato 213–214
    watermelon 163
potassium deficiency 298
potassium fertilization 259, 431
potato 27, 176–187, 199–200, 204, 289, 299, 406
    Irish 1, 6, 177, 281, 289, 293
    seed 179–180
potato leafroll luteovirus (PLRV) 186
potato X potesvirus (PVX) 186
potato Y potyvirus (PVY) 186
poverty 108
powdery mildew (*Erysiphe cichoracearum*) 155, 164, 173, 245, 355–356, 425, 435
precision seeding 143, 210, 230, 259, 277, 306, 321–323
precooling 308
preparation: soil 16, 305–306
prescription fertilization 60–62
preservation treatment 130
priming 30, 277, 429–430
    osmotic 429–430
    seed 30, 36, 191, 210, 422
    soil 60
production 111, 141–145, 149–155, 169–173, 310, 371
    asparagus 373–376
    beans 400–405
    beets 353–355
    broccoli 321–322
    Brussels sprout 331–332
    cabbage 310–313
    carrot 420–425
    celery 429–434
    Chinese cabbage 337–338
    commercial 344, 371, 413, 424
    conventional 107–108
    crop 27, 107, 298, 432
    cucumber 141–145
    eggplant 202–204
    garlic 275–279
    global 314, 323, 329
    globe artichoke 241–245
    lettuce 228–236
    lima bean 412–413
    melons 149–155
    mushrooms 445–450
    onions 275–279
    out-of-season 91, 369
    peas 400–405
    pepper 191–197
    potato 179–184, 182
    pumpkins 169–173
    rhubarb 384–387
    single-crop 88
    spinach 362–365
    squash 169–173
    sustainable 107–127, 160, 192
    sweet corn 257–261
    sweetpotato 294–299
    tomato 208
    transplant 39–41, 295, 430
    watermelon 158–163
production statistics 141, 165, 169
    asparagus 371–372
    beans 398–400
    carrot 419
    celery 428–429
    chicory 228
    cucumber 141
    eggplant 202
    endive 228
    garden beets 357–358
    garlic 274–275
    globe artichoke 241
    lettuce 228
    mushroom 444
    onions 274–275
    peas 398–400
    pepper 197–198
    potato 184–185
    pumpkins 169
    rhubarb 383–384
    spinach 361
    squash 169
    sweet corn 257
    sweetpotato 292–294
    tomato 207–208
    watermelon 165
productivity 63–64
    agricultural 64
    increased 2, 123, 232, 385
    reduced 109
    soil 26, 60, 108
profitability 3, 123
propagation 180–181, 262, 273, 277, 282, 295–297, 300
    asexual 177
    asparagus 379
    lettuce 229–231
    seed 272

propagation (*continued*)
    tomato 209–210
    vegetative 27, 179–180, 185
protection 214–215
    environmental 121
    frost 92, 298
protein inhibition 57
pruning: root 38, 102
pseudostem 270, 281
public attention 127
public health 70
pulses 9, 397–399
pumpkins 36, 56, 136, 165–174
    giant 171–172
pungency 187–191, 268, 272, 278, 304, 340–341, 436
purity: genetic 297

quality 185, 424
    bulb 276, 280
    edible 206, 403, 411, 416, 419, 424
    food 109, 132
    fresh 207
    fruit 144, 154, 171
    harvest 163
    improved 3
    maximized 153
    reduced 98
    root 341
    seed 27
    soil 107–108
    transplant 42
    vegetable 47, 53
    water 3, 41, 66, 70, 108, 131
quinoa 349

radiation 85–86
radicchio 226
radish 90, 304, 318, 341–342
rainfall 131, 160, 313, 321
    inadequate 66
    insufficient 403
    irrigation 82
raised beds 80–83, 143, 151, 159, 231, 237, 327–328
ready-to-eat produce 130, 233, 274, 360, 400, 433–434
recycling 88
    nitrogen 113
    nutrient 114
    water 70
reflective mulch 120, 215
refrigeration 271
relative humidity (RH) 234, 280–283, 333, 338–344, 355–356, 412–413, 424–425
reproduction 212, 353, 371
    asexual 27
    sexual 27, 32

residue 17–19, 115, 142, 315, 328, 346, 405–406
    contaminated 16
    management 22–23
    pesticide 110, 113, 127
    plant 317
    testing 121
resistance 218, 407
    bolting 337, 418
    crack 310, 418–419
    disease 109, 146, 166, 311, 371, 424
    drought 159
    genetic 190
    nematode 44
resources
    natural 132
    renewable 111
respiration 53–54, 58, 282, 423–424
    rate 282, 377
    seed 411
rheumatism 429
rhizobia 401, 409
Rhizobium 22, 28, 394–395, 412
Rhizoctonia 43, 164, 366
Rhizomonas suberifaciens 237
Rhizopus rot 217, 405
rhubarb 1, 382–389
    forced 382, 386
    perennial 384–385
Rhubarb Triangle 384
rice 1, 4
riceyness 327
ridge-tillage 17–18, 116, 311, 321, 327
rindworms 174
ring spot virus: tobacco (TRSV) 204
ripening 148, 205–207, 429
    after- 191
    blotchy 216
    delayed 155, 214
    disorders 213
    fruit 212
rockwool 100–102, 144
romaine lettuce 224–225, 228–229, 233, 336
Roman Empire 136, 369
root knot nematodes 156, 204, 218, 238–239, 302, 409, 426
root zone 75–78, 90, 153, 423
roots
    development 21, 298
    disease 328
    growth 37, 278
    rot 19, 311, 365
    storage 298–299
rootworms 265
rot
    black 173, 300, 308, 316, 324, 344
    blossom-end 58, 67, 98–100, 199–200
    bottom 232, 237
    corn stalk 264

fruit 151, 155, 164, 170, 173, 204
head 226, 338
neck 284
onion 284–285
pink 186, 434
Rhizopus 217, 405
sclerotinia stem 199
sour 217
wet 199
see also soft rot
rotary tine tillers (RTT) 16
rotation 142, 240, 243, 284
see also crop rotation
rove beetles 44
rows 19, 345
runner beans 402
runoff 20, 131–132, 259
russet spotting 234–236
rust 264, 379, 405
rust fly 425–426
rutabaga 334, 341–344

safety 2, 122, 123
food 115, 132
vegetable 127–135
worker 29
salad 9, 128
bagged 234, 235, 367
packaged 228, 424
ready-to-eat 127
salinity 192, 363, 367, 370, 373
soil 132, 151, 295
Salmonella 127–128, 155
salt
accumulation 162
concentration 61
tolerance 10
sanitation 115, 239, 246, 408
field 285
good 43, 237
improved 130
poor 128, 155, 388
sap beetles 265
saturation: soil 20, 217, 232
sauerkraut 314
scab 155, 173, 179, 300
common 186
scarcity
physical 66
water 66
sclerotia 199
sclerotinia stem rot 199
Scotch bonnet peppers 189–190
Scoville heat units (SHU) 191
security: food 107–108, 132
seed corn maggots 165, 265

seed potatoes 179–180
seed priming 30, 36, 191, 210, 422
seed science 27
seedbeds 142–143, 151, 295, 306
seeding
band 421
direct 159, 191–192, 227–232, 277–278, 286–287, 321, 334–338
precision 143, 210, 230, 259, 277, 306, 321–323
seedling
development 90
disease 191–192
seedling growth 37–39, 43, 259–260, 409
early 409
rapid 259
seeds 6, 27–46, 157, 158
planting 35–45
synthetic 27
selection 272, 415
cultivar 31–33, 275, 305, 309–310
human 4–5
self-pollination 205, 215, 223–227, 324, 331, 336, 391–394
semitropical climates 371, 447
senescence 173, 280–283
sewage: contaminated 128
sex expression 137, 140, 149, 153, 359
homogeneous 360
monoecious 157, 160, 171, 248, 251
sex pheromones 118, 284
shallots 286–287
shelf life 272, 331
improved 150
long 228
reduced 377
shelly beans 397, 404–405
shiitake mushrooms 441–442, 445, 448–450
shipping: long-distance 225, 228
shock 191–192, 297
reduction 308
transplant 37–38, 44, 143, 210, 277, 307, 311
shoot
cuttings 346
development 242
side-dressing 152, 171, 231, 259, 299, 337, 402
Sikkim cucumber 136, 140
silt loams 354, 384
silverleaf whitefly 238, 301
singulation 28, 35, 230, 275
site selection 158–159, 229, 258–259, 275–277, 294–295, 373–374, 420–421
bean 400–402
celery 431
peas 409–411
tomato 208–209
skin irritation 370
slash and burn 4

slavery 351
slicing cucumber 139, *140*, 142–144
slips 148, 154, 295
    rooted 296
smallage 427
smooth-leaf spinach 360
smut: corn 256
snails 426
snap beans 391, 397–405, 412–413
    bush 396, 403
snap peas 411
snow peas 397, 411
snowball 13, 330
soaps 117–118
soft rot 97, 100, 284, 317, 340, 405
    bacterial 186, 197–200, 217, 316, 388
softening 432
soil 43
    acidity 10, 142, 151, 229, 306, 355, 373
    bare- 16, 80, 310
    characteristics 68, 243
    compaction 18–20, 82, 132, 229, 328, 337, 400
    degradation 107
    fumigation 80–82, 301, 313, 317, 321, 328, 378–379
    loam 68, 142, 158–159, 242, 322, 328, 332–345
    mineral 421, 431
    non-crusting 260
    organic 310, 321, 354, 420–421, 431
    preferred 47–49
    requirements 142, 241–242
    sandy 341, 345, 402
    sterilized 296
    synthetic 37–38
    tension 232, 432
    testing 62–63, 171, 193, 209, 337, 374–375, 412
    texture 179, 401
    type 50, 131, 143, 306, 421–423
    warming 298–300, 374
soil erosion 17, 121, 132
    preventing 22
    reduced 2, 321, 327
soil fertility 61, 113, 179, 306, 326, 331, 340, 431
    improved 26
    long-term 59
soil moisture 84–85, 116, 279, 298, 313, 328, 346
    preserving 159, 338
    uniform 243, 423
soil structure 22, 406
    improved 26, 47–49, 59, 84, 113
soil temperature 84–85, 141–142, 152, 170, 193, 298, 370–371
    increased 88, 250
soil-borne disease 113–114, 158–159, 217, 229, 258, 275, 295
Solanaceae 176–221
solanine 184

solarization 87–88, 97, 216, 235, 317
    soil 113–114, 217, 435
sorting: hand- 404
sour rot 217
soybeans 1, 123, 390, 398, 406
spacing 143–144, 151, 160, 210–211, 275–277, 345, 359
    between-row 211, 261, 357, 402, 409
    common 203
    equidistant 311
    in-row 76–77, 211, 242, 259, 321, 343, 353–354
    linear 373
    plant 170, 191, 194, 337
    row 36, 143, 344, 354
    uniform 277
    wide 181, 406
    within-row 211, 312, 338
spaghetti squash 169
species 14, 25, 112, 305
Species Plantarum (Linnaeus) 10
spider mites 156, 204, 265, 408, 426
    two-spotted 246
spinach 127, 130, 228, 318, 349, 353, 357–366
    baby 360
    flat-leaf 360
    fresh 361
    frozen 361
    smooth-leaf 360
sporulation 119
spot
    brown 186–187, 406
    cavity 356, 425
    necrotic 57–58, 238, 365
    varnish 237
    see also leaf spot
spotted asparagus beetles 379–380
spotted wilt virus: tomato (TSWV) 199–201, 365
spotting
    chlorotic 164
    pepper 318
    russet 234–236
sprinkler irrigation 71–73, 243, 317–318, 322, 332, 346, 362
sprouting broccoli 318–320, 323, 341
squash 4, 56, 136, 165–174
    butternut 168
    summer 166, 169–173
    winter 169–173
squash bugs 156, 174
squash leaf curl bigeminivirus (SLCV) 173
stale beds 16, 310
stalk
    celery 431–434
    enlargement 385
stalk rot: corn 264
stand establishment 28, 153, 162, 229–231, 259–260, 354

starch 183–184, 250–253, 256–257, 272, 290–291, 299, 411
steam sterilization 97, 104, 235
stem: lettuce 224–226
stem blight: gummy (*Didymella bryoniae*) 155, 164
stem rot: sclerotinia 199
sterilization
    soil 105, 144
    steam 97, 104, 235
sticky traps 119
stink bugs 218
storage 180, 299, 308, 339
    ability 225
    carrot 424–425
    garlic 282
    life 207, 276, 313, 339, 365
    onion 281–282
    optimum 434
    peas 411
    seed 30
    tomato 213–214
storage roots 298–299
streak virus: tobacco 379
stress
    abiotic 31, 299
    biotic 31, 299
    drought 66–67, 228, 256, 260–261, 283, 340, 362
    environmental 37, 137, 326–327
    moisture 216, 243, 346, 386, 432
    nutrient 216
    plant 69, 154
    see also water stress
stress tolerance 44, 56, 153, 447
    environmental 207
string beans 393, 397
strip-cropping 20, 132
strip-tillage 17–18, 116, 311, 321, 327
structure: soil *see* soil structure
subsidence: ground 66
subsurface drip irrigation (SDI) 77, 78
subtropical climates 319, 375, 385–386
sugar 51–52, 178, 183–184, 254, 258, 272, 291
    accumulation 53–56, 152, 161
    concentration 162
    content 158, 251–252, 256, 351, 411, 419
    loss 382, 424
    retention 54
sugar peas 397
sugar snap peas 394, 397
sugarbeets 349–355, 358, 406
sugarcane 1, 351
sulfur 372
    deficiency 57
summer squash 166, 169–173
sunburn 182, 281
sunflower 239, 406
Sunken Colorado Pan 69

sunscald 216
supermarkets 3
surface delivery systems 73–78
susceptibility 186
sustainability 108, 111
    improved 3
    organic 121–125
    social 132
sustainable development 108
sustainable production 107–127, 160, 192
Svalbard Global Seed Vault 5
sweet corn 248–266
sweet peppers 188–190
sweetpotato 4, 27, 218, 289–302, 437
swiss chard 357–358
symbiosis 64, 401
    bacterial 392–395
symptoms: visual 54–55

Tabasco 189, 190
target spot 186
tarnished plant bugs 218, 265, 388
tasseling 258, 261–262
taste: improved 207
taxonomy 136
technology: advanced 2, 104
temperature 209, 280, 429
    extreme 192, 216
    fluctuations 96
    greenhouse 235
    increased 385
    management 234
    moderation 96
    optimum 212, 290–291, 353, 410
    reduced 423
    requirements 141–142, 362
    storage 283
    tolerance 193, 418
    see also soil temperature
tensiometers 68–70, 114
tension: soil 232, 432
testing: soil 62–63, 171, 193, 209, 337, 374–375, 412
texture: soil 179, 401
Thai chili peppers 190
thermo-classification 8
thermodormancy 30, 223, 429–430
    seed 421
thermoquiescence 362
thinning 35
    fruit 161
    post-emergence 422
thrips 156, 200–201, 216, 265, 285, 408, 425
tillage 16–26, 115–116, 131–132, 402
    conservation 159, 259, 297, 306, 311–312, 327–328, 332
    conventional 310

tillage (continued)
    deep 305
    fine 362
    minimum 346
    mulch- 17–18, 311, 321, 327
    no- 80–81, 123, 182, 192, 286, 311, 321
    ridge- 17–18, 116, 311, 321, 327
    shallow 322, 332, 375
    stale-bed 295
    strip- 17–18, 116, 311, 321, 327
    up-and-down-hill 20
Time-Domain Reflectometry (TDR) 68–69
tipburn 236, 338–340
tobacco 1
    wild 199
tobacco etch 200, 217
tobacco mosaic virus (TMV) 199, 217
tobacco ring spot virus (TRSV) 204
tobacco streak virus 379
tolerance
    drought 209, 287, 291, 297
    heat 337
    stress 44, 56, 153, 207, 447
    temperature 193, 418
tomatine 205
tomato 36–38, 56–57, 103, 163, 187, 190–194, 203–218
    bush 123
    fresh 206
tomato spotted wilt virus (TSWV) 199–201, 365
tortoise beetles 302
Totally Impermeable Film (TIF) 87
toxicity 370
    nitrogen 56
    nutrient 42, 50
trace availability 50
trade 314, 323, 329, 333–334
    agreements 104
    international 31, 339
traditional Chinese medicine (TCM) 387
traits 428
transgenic crops 33–34, 190, 202, 207, 226, 261, 272
translocation 52, 53, 283
transpiration 99
transplant shock 37, 38, 311
    reduced 44, 143, 210, 277, 307
transplants 151–153, 229–232, 296–300, 306–308, 311–313, 316–318, 331–334
    bare-root 37, 192, 203, 210, 307, 337
    beds 295, 296
    elongated 42
    establishment 232
    finishing 44–45
    growth 92
    hand- 430
    lettuce 237
    mechanical 430

peppers 191–192
plug 36–38, 162, 191–192, 203, 230, 326, 337–338
production 39–41, 295, 430
vegetable 36–45
transport 104, 178
    food 109
    long-distance 3, 245
    postharvest 204
trap crops 23, 120
trapping 285
trauma: root 37
treatment 70
    seed 28–30, 155, 317
trickle irrigation 2, 74–78, 131, 144, 209, 298
trimming 433
triple-cropping 80, 88
tropical climates 371, 375, 385–386, 447
true garlic 272
true potato seed (TPS) 180–181
truffles 444
tubers 176–187
tunnels
    high 93–95
    low 92–93, 93–94, 158
    plastic 151, 234
    temporary 374
turnip 334, 341–344, 419
turnip mosaic virus (TuMV) 318, 333, 338–339
turnip moth 426
two-spotted spider mite 246

uniformity 32, 37, 233, 259, 313, 338
    root 353
Union Carbide disaster (1984) 109, 127
United Nations (UN) 111, 132, 165, 358
United States Department of Agriculture (USDA) 132, 240, 254, 269, 383, 419, 428
urbanization 70, 104
uses 167–169, 314
    asparagus 378
    broccoli 323
    Brussels sprout 333
    cabbage 314
    cauliflower 329
    rhubarb 387
    sweetpotato 292

vacuum cooling 234, 308, 328, 335, 365, 405
variation: genetic 249
varnish spot 237
vegetable weevils 316, 357
vegetative growth 47–65, 209, 212, 319, 359, 374, 394
velocity differentiation domain (VDD) 68

ventilation 91–98, 235–236, 281, 295, 340, 445
    improved 92
vernalization 240, 307–309, 320, 325, 334–336, 342, 416–417
Verticillium wilt (*Verticillium dahlia*) 209, 237, 245
vigor 32
    cultivar 373
    plant 192
    seed 27–28, 160, 252, 258, 421–422
Virtually Impermeable Film (VIF) 87
virus 145, 164, 174, 217, 411
volatilization 259

warming: soil 298–300, 374
wasabi 345
wasps 366
    parasitic 237–238, 408–409
waste
    handling 3
    natural 80
    organic 231
water 70–71, 74, 105–106, 375
    erosion 132
    importance 66–67
    irrigation 39–41, 128, 131, 374
    pollution 113, 132
    recycled 70
    surface 20, 70–71
water quality 3, 41, 70, 108
    decreased 66, 131
water stress 68–69, 182, 193, 216, 243, 280, 432–433
    eliminating 172
water-holding capacity 100, 151
watercress 346–347
waterlogging 154, 192, 236, 297, 362, 386, 406
watermelon 1–3, 38, 136, 146, 156–165, 200
    seedless 159–162
watermelon mosaic virus (WMV) 164
waterways 19–22
webworms: beet 356
weed control 16–19, 91–92, 113–115, 286–287, 298–301, 310–311
    chemical 107, 116
    effective 231
    long-term 82
weed management strategies 25, 422, 430
    beans 411
    beets 357
    celery 434
    chard 357
    garlic 285–286
    lettuce 231
    onion 285–286
    peas 411

    potato 186
    sweet corn 261
    sweetpotato 299–300
    tomato 216–217
weeding 262
    hand- *235*, 295, 357
    thermal 115
weeds 127, 217, 375
    beneficial 24
    between-row 261
    competition 53, 154, 242
    growth 17, 321
    perennial 19
weevils
    carrot 425–426
    pepper 200
    vegetable 316, 357
welfare 132
    animal 132
West Indian gherkin 136–138
wet rot 199
whiptail 57, 326
white asparagus 372, 376–378
white grubs 265, 301
whitefly 156, 216
    silverleaf 238, 301
wilt
    bacterial (*Erwinia tracheiphila*) 145–146, 155, 173, 180, 186, 217
    Fusarium (*Fusarium oxysporum* f. sp. *melonis*) 155, 164, 173, 237, 300, 318, 373
    Verticillium (*Verticillium dahlia*) 209, 237, 245
wilting 144, 164, 199, 204, 245, 313
    marginal 57
    reduced 365, 424
wind
    erosion 132
    pollination 248, 251, 359–360
winter cultivars 327
winter melon 146–147, *150*, 155
winter squash 169–173
wirestem 317, 324
wireworms 204, 265, 285, 301, 407
Witloof chicory 226–227
worker safety 29
World Bank 107
World Health Organization (WHO) 34
World War II (1939–45) 107, 222, 382
worms
    cotton leaf 339
    cut 165, 174, 200, 246, 265, 284, 315, 366, 426
    damage 256
    imported cabbageworm (ICW) 314
    rindworms 174
    rootworms 265
    wireworms 204, 265, 285, 301, 407
    yellow-striped armyworm 200, 356, 366

xanthophyll 205

yams 290, 292
yellow-striped armyworm 200, 356, 366
yield 111, 173, 185, 338, 387, 448
    bulb 280
    increased 44
    maximized 144, 153

    potential 32, 67
    reduced 98

zinc
    deficiency 58
    fertilization 431
zoning 350, 352
zucchini yellow mosaic virus (ZYMV) 173

Printed and bound by CPI Group (UK) Ltd, Croydon, CR0 4YY
17/02/2026

14827796-0001